eleventh edition

Human Genetics

Concepts and Applications

WITHDRAWN

Ricki Lewis

Genetic Counselor
CareNet Medical Group
Schenectady, New York

Adjunct Assistant Professor of
Medical Education
Alden March Bioethics Institute
Albany Medical College

Writer, Medscape Medical News

Blogger, Public Library of Science

Mc
Graw
Hill
Education

HUMAN GENETICS: CONCEPTS AND APPLICATIONS, ELEVENTH EDITION

1 2 3 4 5 6 7 8 9 0 DOW/DOW 1 0 9 8 7 6 5 4

ISBN 978-1-259-09563-4
MHID 1-259-09563-0

About the Author

Ricki Lewis has built an eclectic career in communicating the excitement of genetics and genomics. She earned her Ph.D. in genetics in 1980 from Indiana University. It was the dawn of the modern biotechnology era, which Ricki chronicled in many magazines and journals. She published one of the first articles on DNA fingerprinting in *Discover* magazine in 1988, and a decade later one of the first articles on human stem cells in *The Scientist*.

Ricki has taught a variety of life science courses at Miami University, the University at Albany, Empire State College, and community colleges. She has authored or co-authored several university-level textbooks and is the author of *The Forever Fix: Gene Therapy and the Boy Who Saved It,* as well as an essay collection and a novel. Ricki has been a genetic counselor for a private medical practice since 1984 and is a frequent public speaker. Since 2012, Ricki has written hundreds of news stories for *Medscape Medical News*, articles for *Scientific American* and for several genetic disease organizations, and originated and writes the popular weekly DNA Science blog at *Public Library of Science*.

Ricki teaches an online course on "Genethics" for the Alden March Bioethics Institute of Albany Medical College. She lives in upstate New York and sometimes Martha's Vineyard, with husband Larry and several felines. Contact Ricki at rickilewis54@gmail.com, or join the discussion on DNA Science at http://blogs.plos.org/dnascience/.

Dedicated to the

families who live with genetic diseases, the

health care providers who help them, and

the researchers who develop new tests

and treatments.

Brief Contents

PART 1

Introduction 1

CHAPTER 1
What Is in a Human Genome? 1

CHAPTER 2
Cells 15

CHAPTER 3
Meiosis, Development, and Aging 42

PART 2

Transmission Genetics 68

CHAPTER 4
Single-Gene Inheritance 68

CHAPTER 5
Beyond Mendel's Laws 89

CHAPTER 6
Matters of Sex 110

CHAPTER 7
Multifactorial Traits 130

CHAPTER 8
Genetics of Behavior 148

PART 3

DNA and Chromosomes 163

CHAPTER 9
DNA Structure and Replication 163

CHAPTER 10
Gene Action: From DNA to Protein 180

CHAPTER 11
Gene Expression and Epigenetics 199

CHAPTER 12
Gene Mutation 212

CHAPTER 13
Chromosomes 237

PART 4

Population Genetics 263

CHAPTER 14
Constant Allele Frequencies 263

CHAPTER 15
Changing Allele Frequencies 279

CHAPTER 16
Human Ancestry and Evolution 302

PART 5

Immunity and Cancer 326

CHAPTER 17
Genetics of Immunity 326

CHAPTER 18
Cancer Genetics and Genomics 351

PART 6

Genetic Technology 374

CHAPTER 19
Genetic Technologies: Patenting, Modifying, and Monitoring DNA 374

CHAPTER 20
Genetic Testing and Treatment 389

CHAPTER 21
Reproductive Technologies 407

CHAPTER 22
Genomics 425

Contents

About the Author iii

Preface ix

Applying Human Genetics xiii

The Human Touch xiv

The Lewis Guided Learning System xv

PART 1 Introduction 1

CHAPTER 1

What Is in a Human Genome? 1

1.1 Introducing Genes and Genomes 2

1.2 Levels of Genetics and Genomics 3

1.3 Applications of Genetics and Genomics 7

1.4 A Global Perspective on Genomes 9

CHAPTER 2

Cells 15

2.1 Introducing Cells 16

2.2 Cell Components 16

2.3 Cell Division and Death 28

2.4 Stem Cells 33

2.5 The Human Microbiome 37

CHAPTER 3

Meiosis, Development, and Aging 42

3.1 The Reproductive System 43

3.2 Meiosis 44

3.3 Gametes Mature 47

3.4 Prenatal Development 51

3.5 Birth Defects 59

3.6 Maturation and Aging 62

PART 2 Transmission Genetics 68

CHAPTER 4

Single-Gene Inheritance 68

4.1 Following the Inheritance of One Gene 69

4.2 Single-Gene Inheritance Is Rare 72

4.3 Following the Inheritance of More Than One Gene 77

4.4 Pedigree Analysis 79

4.5 Family Exome Analysis 83

CHAPTER 5

Beyond Mendel's Laws 89

5.1 When Gene Expression Appears to Alter Mendelian Ratios 90

5.2 Mitochondrial Genes 98

5.3 Linkage 100

CHAPTER 6

Matters of Sex 110

6.1 Our Sexual Selves 111

6.2 Traits Inherited on Sex Chromosomes 117

6.3 Sex-Limited and Sex-Influenced Traits 122

6.4 X Inactivation 122

6.5 Parent-of-Origin Effects 124

CHAPTER 7

Multifactorial Traits 130

7.1 Genes and the Environment Mold Traits 131

7.2 Polygenic Traits Are Continuously Varying 133

7.3 Methods to Investigate Multifactorial Traits 135

7.4 A Closer Look: Body Weight 142

CHAPTER 8

Genetics of Behavior 148

8.1 Genes and Behavior 149

8.2 Sleep 150

8.3 Intelligence and Intellectual Disability 151

8.4 Drug Addiction 152

8.5 Mood Disorders 154

8.6 Schizophrenia 155

8.7 Autism 157

PART 3 DNA and Chromosomes 163

CHAPTER 9

DNA Structure and Replication 163

9.1 Experiments Identify and Describe the Genetic Material 164

9.2 DNA Structure 168

9.3 DNA Replication—Maintaining Genetic Information 170

9.4 Sequencing DNA 176

CHAPTER 10

Gene Action: From DNA to Protein 180

10.1 Transcription Copies the Information in DNA 181

10.2 Translation of a Protein 186

10.3 Processing a Protein 192

C H A P T E R 11

Gene Expression
and Epigenetics 199

11.1 Gene Expression Through Time
      and Tissue 200

11.2 Control of Gene Expression 203

11.3 Maximizing Genetic Information 205

11.4 Most of the Human Genome Does *Not*
      Encode Protein 206

C H A P T E R 12

Gene Mutation 212

12.1 The Nature of Mutations 213

12.2 A Closer Look at Two Mutations 214

12.3 Allelic Disorders 217

12.4 Causes of Mutation 218

12.5 Types of Mutations 221

12.6 The Importance of Position 227

12.7 DNA Repair 229

C H A P T E R 13

Chromosomes 237

13.1 Portrait of a Chromosome 238

13.2 Detecting Chromosomes 240

13.3 Atypical Chromosome Number 244

13.4 Atypical Chromosome Structure 253

13.5 Uniparental Disomy—A Double Dose from
      One Parent 258

PART 4 Population Genetics 263

C H A P T E R 14

Constant Allele
Frequencies 263

14.1 Population Genetics Underlies
      Evolution 264

14.2 Constant Allele Frequencies 265

14.3 Applying Hardy-Weinberg Equilibrium 267

14.4 DNA Profiling Uses Hardy-Weinberg
      Assumptions 268

C H A P T E R 15

Changing Allele
Frequencies 279

15.1 Nonrandom Mating 280

15.2 Migration 281

15.3 Genetic Drift 282

15.4 Mutation 286

15.5 Natural Selection 287

15.6 Eugenics 295

C H A P T E R 16

Human Ancestry
and Evolution 302

16.1 Human Origins 303

16.2 Methods to Study Molecular Evolution 311

16.3 The Peopling of the Planet 314

16.4 What Makes Us Human? 318

PART 5 Immunity and Cancer 326

CHAPTER 17

Genetics of
Immunity 326

17.1 The Importance of Cell Surfaces 327

17.2 The Human Immune System 330

17.3 Abnormal Immunity 335

17.4 Altering Immunity 341

17.5 Using Genomics to Fight Infection 345

CHAPTER 18

Cancer Genetics
and Genomics 351

18.1 Cancer Is an Abnormal Growth That Invades
and Spreads 352

18.2 Cancer at the Cellular Level 356

18.3 Cancer Genes and Genomes 360

18.4 The Challenges of Diagnosing and Treating
Cancer 369

PART 6 Genetic Technology 374

CHAPTER 19

Genetic Technologies:
Patenting, Modifying,
and Monitoring DNA 374

19.1 Patenting DNA 375

19.2 Modifying DNA 376

19.3 Monitoring Gene Function 382

19.4 Gene Silencing and Genome
Editing 384

CHAPTER 20

Genetic Testing and
Treatment 389

20.1 Genetic Counseling 390

20.2 Genetic Testing 392

20.3 Treating Genetic Disease 397

CHAPTER 21

Reproductive
Technologies 407

21.1 Savior Siblings and More 408

21.2 Infertility and Subfertility 409

21.3 Assisted Reproductive Technologies 411

21.4 Extra Embryos 419

CHAPTER 22

Genomics 425

22.1 From Genetics to Genomics 426

22.2 Analysis of the Human Genome 430

22.3 Personal Genome Sequencing 435

Glossary G-1

Credits C-1

Index I-1

Preface

Human Genetics Touches Us All

When I wrote the first edition of this book, in 1992, I could never have imagined that today, thousands of people would have had their genomes sequenced. Nor could I have imagined, when the first genomes were sequenced a decade later, that the process could take under a day, for less than $1,000. Of course, understanding all the information in a human genome will take much longer.

Each subsequent edition opened with a scenario of two students taking genetic tests, which grew less hypothetical and more real over time, even reaching the direct-to-consumer level. This new edition reflects the translation of gene and genome testing and manipulation from the research lab to the clinic.

The eleventh edition opens with "Eve's Genome" and ends with "Do You Want Your Genome Sequenced?" In between, the text touches on what exome and genome sequencing have revealed about single-gene diseases so rare that they affect only a single family, to clues to such common and complex conditions as intellectual disability and autism. Exome and genome sequencing are also important in such varied areas as understanding our origins, solving crimes, and tracking epidemics. In short, DNA sequencing will affect most of us.

As the cost of genome sequencing plummets, we all may be able to look to our genomes for echoes of our pasts and hints of our futures—if we so choose. We may also learn what we can do to counter our inherited tendencies and susceptibilities. Genetic knowledge is informative and empowering. This book shows you how and why this is true.

Ricki Lewis

Today, human genetics is for everyone. It is about our variation more than about our illnesses, and about the common as well as the rare. Once an obscure science or an explanation for an odd collection of symptoms, human genetics is now part of everyday conversation. At the same time, it is finally being recognized as the basis of medical science, and health care professionals must be fluent in the field's language and concepts. Despite the popular tendency to talk of "a gene for" this or that, we now know that for most traits and illnesses, several genes interact with each other and environmental influences to mold who we are.

What Sets This Book Apart

Current Content

The exciting narrative writing style, with clear explanations of concepts and mechanisms propelled by stories, reflects Dr. Lewis's eclectic experience as a medical news writer, blogger, professor, and genetic counselor, along with her expertise in genetics. Updates to this edition include

- Genetic tests, from preconception to old age
- Disease-in-a-dish stem cell technology
- From Mendel to molecules: family exome analysis
- Allelic diseases: one gene, more than one disease
- Admixture of archaic and modern humans
- Gene silencing and genome editing
- Cancer genomes guide treatment
- The reemergence of gene therapy
- Personal genome sequencing: promises and limitations

The transition of genetics to genomics catalyzed slight reorganization of the book. The order of topics remains, but material that had been boxed or discussed in later chapters because it was once new technology has been moved up as the "applications" become more integrated with the "concepts." The book has evolved with the science.

The Human Touch

Human genetics is about people, and their voices echo throughout these pages. They speak in the narrative as well as in many new chapter introductions, boxes, stories, and end-of-chapter questions and cases.

Compelling Stories and Cases When the parents of children with visual loss stood up at a conference to meet other families with the same very rare inherited disease, Dr. Lewis was there, already composing the opening essay to chapter 5. She knows the little girl in the "*In Their Own Words*" essay in chapter 2 and on the cover with her dog, who is 1 of about 70 people in the world with giant axonal neuropathy. Perhaps there is no more heart-wrenching image of Mendelian inheritance than the chapter 4 opening photo of a daughter and father, who died from Huntington disease within weeks of each other.

Clinical Application of Human Genetics A working knowledge of the principles and applications of human genetics is critical to being an informed citizen and health care consumer. Broad topics of particular interest include

- The roles that genes play in disease risk, physical characteristics, and behavior, with an eye toward the dangers of genetic determinism

- Biotechnologies, including next-generation DNA sequencing, genetic testing, stem cell technology, archaic human genome sequencing, gene therapy, familial DNA searches, exome sequencing, cell-free fetal DNA testing, and personal genome sequencing
- Ethical concerns that arise from the interface of genetic and genomic information and privacy.

The Lewis Guided Learning System

Each chapter begins with two views of the content. "*Learning Outcomes*" embedded in the table of contents guide the student in mastering material. "*The Big Picture*" encapsulates the overall theme of the chapter. The chapter opening essay and figure grab attention. Content flows logically through three to five major sections per chapter that are peppered with high-interest boxed readings ("*In Their Own Words,*" "*Clinical Connections,*" "*Bioethics: Choices for the Future,*" "*A Glimpse of History,*" and "*Technology Timelines*"). End-of-chapter pedagogy progresses from straight recall to applied and creative questions and challenges.

Dynamic Art

Outstanding photographs and dimensional illustrations, vibrantly colored, are featured throughout *Human Genetics: Concepts and Applications*. Figure types include process figures with numbered steps, micro to macro representations, and the combination of art and photos to relate stylized drawings to real-life structures.

New to This Edition!

The genomics of today evolved from the genetics of the twentieth century. *A Glimpse of History* features throughout the book capture key moments in time. *Clinical Connections* bring chapter concepts to patients and health care providers, with thought-provoking questions for discussion. *Key Concepts* after all major sections are now questions.

Highlights in the new edition include the following:

Chapter 1   What Is in a Human Genome?
- The story of young Nicholas Volker, near death when exome sequencing led to a diagnosis—and a treatment

Chapter 2   Cells
- The human microbiome

Chapter 3   Meiosis, Development, and Aging
- Progress for progeria
- Maternal and paternal age effects on gametes

Chapter 4   Single-Gene Inheritance
- Family exome analysis solves a medical mystery

Chapter 7   Multifactorial Traits
- Blond hair among the Melanesians
- Smoking-related lung cancer

Chapter 8   Genetics of Behavior
- Genetic risks for posttraumatic stress disorder, depression, autism
- Heritability of intelligence at different ages

Chapter 11   Gene Expression and Epigenetics
- Long noncoding RNAs

Chapter 12   Gene Mutation
- Gonadal mosaicism
- Allelic disease—more common than we thought
- Exon skipping causes and treats disease

Chapter 13   Chromosomes
- Harnessing XIST to silence trisomy 21
- Cell-free fetal DNA for noninvasive prenatal diagnosis

Chapter 15   Changing Allele Frequencies
- The Clinic for Special Children treats the Amish

Chapter 16   Human Ancestry and Evolution
- Updated terminology and evolutionary trees
- Admixture, the Neanderthals, Denisovans, and us
- What makes us human?

Chapter 17   Genetics of Immunity
- Genomic epidemiology tracks an outbreak
- Reverse vaccinology
- Mimicking *CCR5* mutations to prevent HIV infection

Chapter 18   Cancer Genetics and Genomics
- Summary figure of cancer at different levels

- Driver and passenger mutations
- Cancer genomes
- Cell-free tumor DNA
- How *BRCA1* causes cancer

Chapter 19   Genetic Technologies: Patenting, Modifying, and Monitoring DNA
- The Supreme court and DNA patents
- Gene silencing and genome editing

Chapter 22   Genomics
- Genome sequencing and annotation
- Practical medical matters
- Types of information in human genomes
- A gallery of genomes
- Comparative genomics
- Do you want your genome sequenced?

NEW FIGURES

- 4.6 Eye color
- 4.8 Loss-of-function and gain-of-function mutations
- 7.10 Copy number variants
- 8.6 Nicotine's effects at the cellular level
- 8.9 Exome sequencing and autism
- 9.17 Replication bubbles
- 12.5 Allelic disease of connective tissue
- 12.10 Exon skipping and Duchenne muscular dystrophy
- 13.14 XIST silences trisomy 21
- 14.11 Several steps identify STRs
- 15.13 Antibiotic resistance
- 16.13 Admixture of haplotypes
- 16.18 What makes us human?
- 17.14 Filaggrin and allergy
- 17.18 Genome sequencing to track outbreaks
- 18.1 Levels of cancer
- 18.12 Evolution of a cancer
- 18.13 Cancer chromosomes
- 19.7 Gene silencing and genome editing

NEW TABLES

- 2.2 Stem Cell Sources
- 3.4 Longevity Genes
- 7.6 Study Designs for Multifactorial Traits
- 13.2 Maternal Serum Markers
- 15.1 Clinical Connection: Genetic Disorders among the Amish
- 19.2 Genetically Modified Foods
- 22.1 Selected Projects to Analyze Human Genomes
- 22.2 Cost of Sequencing Human Genomes
- 22.3 A Gallery of Genomes

ACKNOWLEDGMENTS

Human Genetics: Concepts and Applications, Eleventh Edition, would not have been possible without the editorial and production dream team: senior brand manager Rebecca Olson, product development director Rose Koos, executive marketing manager Patrick Reidy, lead content licensing specialist Carrie Burger, designer Tara McDermott, developmental editors Anne Winch, Erin Guendelsberger, and Emily Nesheim, project manager Sheila Frank, copyeditor Beatrice Sussman, and photo editor extraordinaire, Toni Michaels. Many thanks to the fabulous reviewers. Special thanks to my friends in the rare disease community who have shared their stories, and to Jonathan Monkemeyer and David Bachinsky for helpful Facebook posts. As always, many thanks to my wonderful husband Larry for his support and encouragement and to my three daughters, my cats, and Cliff the hippo.

Eleventh Edition Reviewers

Andy Andres
Boston University
Elizabeth Alter
York College
Ann Blakey
Ball State University
Bruce Bowerman
University of Oregon
James Bradshaw
Utah Valley University
Dean Bratis
Villanova University
Susan Brown
Kansas State University
Michelle Coach
Asnuntuck Community College
Jonathon S. Coren
Elizabethtown College
Tracie Delgado
Northwest University

Dan Dixon
University of Kansas Medical Center
Jennifer Drew
University of Florida
Gregory Filatov
University of California Riverside
Yvette Gardner
Clayton State University
Ricki Glaser
Stevenson University
Debra Han
Palomar Community College
Bradley J. Isler
Ferris State University
Bridget Joubert
Northwestern State University
Patricia Matthews
Grand Valley State University
Gemma Niermann
University of California, Berkeley and Saint Mary's College
Ruth S. Phillips
North Carolina Central University
Mabel O. Royal
North Carolina Central University
Mark Sanders
University of California, Davis
Jennifer Smith
Triton College
Michael Torres
Warren Wilson College
Jo Ann Wilson
Florida Gulf Coast University
Erin Zimmer
Lewis University

This book continually evolves thanks to input from instructors and students. Please let me know your thoughts and suggestions for improvement. (rickilewis54@gmail.com)

Applying Human Genetics

Chapter Openers

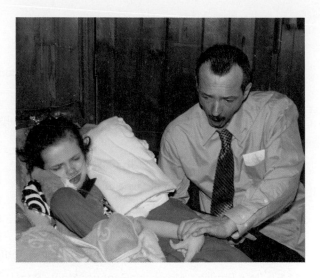

1	Eve's Genome	1
2	Diagnosis From a Tooth	15
3	Progress for Progeria	42
4	Juvenile Huntington Disease: The Cruel Mutation	68
5	Mutations in Different Genes Cause Blindness	89
6	Stem Cell and Gene Therapies Save Boys' Lives	110
7	The Complex Genetics of Athletics	130
8	Genetic Predisposition to Posttraumatic Stress Disorder	148
9	On the Meaning of Gene	163
10	An Inborn Error of Arginine Production	180
11	The Dutch Hunger Winter	199
12	One Mutation, Multiple Effects: Osteogenesis Imperfecta	212
13	The Curious Chromosomes of Werewolves	237
14	Postconviction DNA Testing	263
15	The Evolution of Lactose Tolerance	279
16	The Little Lady of Flores	302
17	Mimicking a Mutation to Protect Against HIV	326
18	A Genetic Journey to a Blockbuster Cancer Drug	351
19	Improving Pig Manure	374
20	Fighting Canavan Disease	389
21	Replacing Mitochondria	407
22	100,000 Genomes and Counting	425

A GLIMPSE OF HISTORY

Chapter 3 The first view of sperm
Chapter 4 Gregor Mendel
Chapter 5 The murdered Romanovs and mitochondrial DNA
Chapter 9 Kary Mullis invents PCR
Chapter 10 The RNA tie club
Chapter 12 The discovery of sickled cells
Chapter 13 Determining the human chromosome number
Chapter 14 Famous forensics cases
Chapter 15 Malaria in the United States
Chapter 18 Retinoblastoma
Chapter 20 Treating PKU
Chapter 22 Comparative genomics

The Human Touch

Clinical Connections

1.1	Exome Sequencing Saves a Boy's Life	10
2.1	Inborn Errors of Metabolism Affect the Major Biomolecules	19
2.2	Faulty Ion Channels Cause Inherited Disease	26
3.1	When an Arm Is Really a Leg: Homeotic Mutations	55
4.1	It's All in the Genes	73
4.2	"65 Roses": Progress in Treating Cystic Fibrosis	76
5.1	The Genetic Roots of Alzheimer Disease	96
6.1	Colorblindness	119
7.1	Many Genes Control Heart Health	132
12.1	Fragile X Mutations Affect Boys and Their Grandfathers	226
14.1	DNA Profiling: Molecular Genetics Meets Population Genetics	270
15.1	The Clinic for Special Children: The Founder Effect and "Plain" Populations	284
17.1	Viruses	328
17.2	A Special Immunological Relationship: Mother-to-Be and Fetus	339
18.1	The Story of Gleevec	364
21.1	The Case of the Round-Headed Sperm	410

In Their Own Words

A Little Girl with Giant Axons	28
Growing Human Mini-Brains	58
The Y Wars	113
Familial Dysautonomia: Rebekah's Story	223
Some Individuals With Trisomies Survive Childhood	251

Bioethics: Choices for the Future

Genetic Testing and Privacy	11
Banking Stem Cells: When Is It Necessary?	37
Why a Clone Is Not an Exact Duplicate	52
Infidelity Testing	175
Will Trisomy 21 Down Syndrome Disappear?	250
Should DNA Collected Today Be Used to Solve a Past Crime?	275
Two Views of Neural Tube Defects	297
Genetic Privacy: A Compromised Genealogy Database	318
Pig Parts	345
EPO: Built-in Blood Cell Booster or Performance-Enhancing Drug?	379
Incidental Findings: Does Sequencing Provide Too Much Information?	395
Removing and Using Gametes After Death	416

The Lewis Guided Learning System

Learning Outcomes preview major chapter topics in an inquiry-based format according to numbered sections.

The Big Picture encapsulates chapter content at the start.

Chapter Openers vividly relate content to real life.

Key Concepts Questions follow each numbered section.

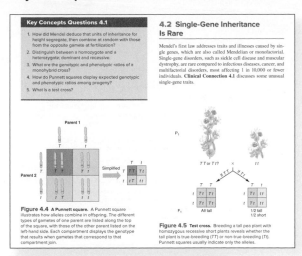

Key Concepts Questions 4.1

1. How did Mendel deduce that units of inheritance for height segregate, then combine at random with those from the opposite gamete at fertilization?
2. Distinguish between a homozygote and a heterozygote; dominant and recessive.
3. What are the genotypic and phenotypic ratios of a monohybrid cross?
4. How do Punnett squares display expected genotypic and phenotypic ratios among progeny?
5. What is a test cross?

4.2 Single-Gene Inheritance Is Rare

Mendel's first law addresses traits and illnesses caused by single genes, which are also called Mendelian or monofactorial. Single-gene disorders, such as sickle cell disease and muscular dystrophy, are rare compared to infectious diseases, cancer, and multifactorial disorders, most affecting 1 in 10,000 or fewer individuals. **Clinical Connection 4.1** discusses some unusual single-gene traits.

Figure 4.4 A Punnett square. A Punnett square illustrates how alleles combine in offspring. The different types of gametes of one parent are listed along the top of the square, with those of the other parent listed on the left-hand side. Each compartment displays the genotype that results when gametes that correspond to that compartment join.

Figure 4.5 Test cross. Breeding a tall pea plant with homozygous recessive short plants reveals whether the tall plant is true-breeding (*TT*) or non-true-breeding (*Tt*). Punnett squares usually indicate only the alleles.

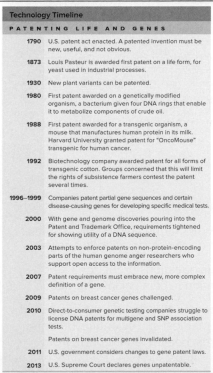

Technology Timeline

PATENTING LIFE AND GENES

1790	U.S. patent act enacted. A patented invention must be new, useful, and not obvious.
1873	Louis Pasteur is awarded first patent on a life form, for yeast used in industrial processes.
1930	New plant variants can be patented.
1980	First patent awarded on a genetically modified organism, a bacterium given four DNA rings that enable it to metabolize components of crude oil.
1988	First patent awarded for a transgenic organism, a mouse that manufactures human protein in its milk. Harvard University granted patent for "OncoMouse" transgenic for human cancer.
1992	Biotechnology company awarded patent for all forms of transgenic cotton. Groups concerned that this will limit the rights of subsistence farmers contest the patent several times.
1996–1999	Companies patent partial gene sequences and certain disease-causing genes for developing specific medical tests.
2000	With gene and genome discoveries pouring into the Patent and Trademark Office, requirements tightened for showing utility of a DNA sequence.
2003	Attempts to enforce patents on non-protein-encoding parts of the human genome anger researchers who support open access to the information.
2007	Patent requirements must embrace new, more complex definition of a gene.
2009	Patents on breast cancer genes challenged.
2010	Direct-to-consumer genetic testing companies struggle to license DNA patents for multigene and SNP association tests.
	Patents on breast cancer genes invalidated.
2011	U.S. government considers changes to gene patent laws.
2013	U.S. Supreme Court declares genes unpatentable.

In-Chapter Review Tools, such as Key Concepts Questions, summary tables, and timelines of major discoveries, are handy tools for reference and study. Most boldfaced terms are consistent in the chapters, summaries, and glossary.

Bioethics: Choices for the Future and **Clinical Connection** boxes include Questions for Discussion.

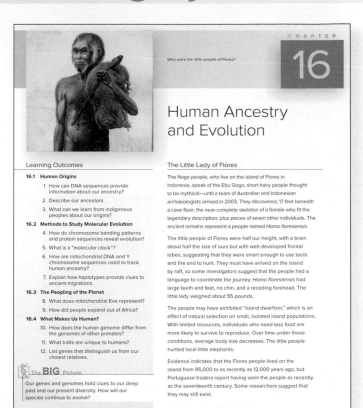

16

Who were the little people of Flores?

Human Ancestry and Evolution

Learning Outcomes

16.1 Human Origins
1. How can DNA sequences provide information about our ancestry?
2. Describe our ancestors.
3. What can we learn from indigenous peoples about our origins?

16.2 Methods to Study Molecular Evolution
4. How do chromosome banding patterns and protein sequences reveal evolution?
5. What is a "molecular clock"?
6. How are mitochondrial DNA and Y chromosome sequences used to track human ancestry?
7. Explain how haplotypes provide clues to ancient migrations.

16.3 The Peopling of the Planet
8. What does mitochondrial Eve represent?
9. How did people expand out of Africa?

16.4 What Makes Us Human?
10. How does the human genome differ from the genomes of other primates?
11. What traits are unique to humans?
12. List genes that distinguish us from our closest relatives.

The BIG Picture

Our genes and genomes hold clues to our deep past and our present diversity. How will our species continue to evolve?

The Little Lady of Flores

The Nage people, who live on the island of Flores in Indonesia, speak of the Ebu Gogo, short hairy people thought to be mythical—until a team of Australian and Indonesian archaeologists arrived in 2003. They discovered, 17 feet beneath a cave floor, the near-complete skeleton of a female who fit the legendary description, plus pieces of seven other individuals. The ancient remains represent a people named *Homo floresiensis*.

The little people of Flores were half our height, with a brain about half the size of ours but with well-developed frontal lobes, suggesting that they were smart enough to use tools and fire and to hunt. They must have arrived on the island by raft, so some investigators suggest that the people had a language to coordinate the journey. *Homo floresiensis* had large teeth and feet, no chin, and a receding forehead. The little lady weighed about 55 pounds.

The people may have exhibited "island dwarfism," which is an effect of natural selection on small, isolated island populations. With limited resources, individuals who need less food are more likely to survive to reproduce. Over time under these conditions, average body size decreases. The little people hunted local little elephants.

Evidence indicates that the Flores people lived on the island from 95,000 to as recently as 12,000 years ago, but Portuguese traders report having seen the people as recently as the seventeenth century. Some researchers suggest that they may still exist.

Bioethics: Choices for the Future

Banking Stem Cells: When Is It Necessary?

The parents-to-be were very excited by the company's promise: "Bank your baby's cord blood stem cells and benefit from breakthroughs. Be prepared for the unknowns in life."

The website profiled children saved from certain diseases using stored umbilical cord blood. The statistics were persuasive: More than 70 diseases are currently treatable with cord blood transplants, and 10,000 procedures have already been done.

With testimonials like that, it is little wonder that parents collectively spend more than $100 million per year to store cord blood. The ads and statistics are accurate but misleading, because of what they *don't* say. Most people never actually use the umbilical cord blood stem cells that they store. The scientific reasons go beyond the fact that treatable diseases are very rare. In addition, cord blood stem cells are not nearly as pluripotent as some other stem cells, limiting their applicability. Perhaps the most compelling reason that stem cell banks are rarely used is based on logic: For a person with an inherited disease, *healthy* stem cells are required—not his or her own, which could cause the disease all over again because the mutation is in every cell. The patient needs a well-matched donor, such as a healthy sibling.

Commercial cord blood banks may charge more than $1,000 for the initial collection plus an annual fee. However, the U.S. National Institutes of Health and organizations in many other nations have supported not-for-profit banks for years, and may not charge fees. Donations of cord blood to these facilities are not to help the donors directly, but to help whoever can use the cells.

Commercial stem cell banks are not just for newborns. One company, for example, offers to bank "very small embryonic-like stem cells" for an initial charge of $7,500 and a $750 annual fee, "enabling people to donate and store their own stem cells when they are young and healthy for their personal use in times of future medical need." The cells come from a person's blood and, in fact, one day may be very useful, but the research has yet to be done supporting any use of the cells in treatments.

Questions for Discussion

1. Storing stem cells is not regulated by the U.S. government the way that a drug or a surgical procedure is because it is a service that will be helpful for treatments not yet invented. Do you think such banks should be regulated, and if so, by whom and how?
2. What information do you think that companies offering to store stem cells should present on their websites?
3. Do you think that advertisements for cord blood storage services that have quotes and anecdotal reports, but do not mention that most people who receive stem cell transplants do not in fact receive their own cells, are deceptive? Or do you think it is the responsibility of the consumer to research and discover this information?
4. Several companies store stem cells extracted from baby teeth, although a use for such stem cells has not yet been found. Suggest a different way to obtain stem cells that have the genome of a particular child.

When an Arm Is Really a Leg: Homeotic Mutations

Flipping the X ray showed Stefan Mundlos, MD, that his hunch was right—the patient's arms were odd-looking and stiff because the elbows were actually knees! The condition, Liebenberg syndrome (OMIM 186550), had been described in 1973 among members of a five-generation white South African family (figure 1). Four males and six females had stiff elbows and wrists, and short fingers that looked strangely out of place. A trait that affects both sexes in every generation displays classic autosomal dominant inheritance—each child of a person with strange limbs had a 50:50 chance of having the condition too.

In 2000, a medical journal described a second family with Liebenberg syndrome. Several members had restricted movements because they couldn't bend their huge, misshapen elbows. Then in 2010, a report appeared on identical twin girls with the curious stiff elbows and long arms, with fingers that looked like toes.

In 2012, Dr. Mundlos noted that the muscles and tendons of the elbows, as well as the bones of the arms, weren't quite right in his patient. The doctor, an expert in the comparative anatomy of limb bones of different animals, realized that the stiff elbows were acting like knees. The human elbow joint hinges and rotates, but the knee extends the lower leg straight out. Then an X-ray scan of the patient's arm fell to the floor. "I realized that the entire limb had the appearance of a leg. Normally you would look at the upper limb X ray with the hand up, whereas the lower limb is looked at foot down. If you turn the X ray around, it looks just like a leg," Dr. Mundlos said.

Genes that switch body parts are termed homeotic. They are well studied in experimental organisms as evolutionarily diverse as fruit flies, flowering plants, and mice, affecting the positions of larval segments, petals, legs, and much more. Assignment of body parts begins in the early embryo, when cells look alike but are already fated to become specific structures. Gradients (increasing or decreasing concentrations) of "morphogen" proteins in an embryo program a particular region to develop a certain way. Mix up the messages, and an antenna becomes a leg, or an elbow a knee.

Homeotic genes include a 180-base-long DNA sequence, called the homeobox, which enables the encoded protein to bind other proteins that turn on sets of other genes, crafting an embryo, section by section. Homeotic genes line up on their chromosomes in the precise order in which they're deployed in development, like chapters in an instruction manual to build a body.

The human genome has four clusters of homeotic genes, and mutations in them cause disease. In certain lymphomas, a homeotic mutation sends white blood cells along the wrong developmental pathway, resulting in too many of some blood cell types and too few of others. The abnormal ears, nose, mouth, and throat of DiGeorge syndrome (OMIM 188400) echo the abnormalities in Antennapedia, a fruit fly mutant that has legs on its head. Extra and fused fingers and various bony alterations also stem from homeotic mutations.

The search for the mutation behind the arm-to-leg Liebenberg phenotype began with abnormal chromosomes. Affected members of the three known families were each missing 134 DNA bases in the same part of the fifth largest chromosome. The researchers zeroed in on a gene called PITX1 that controls other genes that in turn oversee limb development. In the Liebenberg families, the missing DNA places an "enhancer" gene near PITX1, altering its expression in a way that mixes up developmental signals so that the forming arm instead becomes a leg. Fortunately the condition appears more an annoying oddity than a disease.

Questions for Discussion

1. What is the genotype and phenotype of Liebenberg syndrome?
2. How can homeotic mutations be seen in such different species as humans, mice, fruit flies, and flowering plants?
3. Explain the molecular basis of a homeotic mutation and the resulting phenotype.
4. Name another human disease that results from a homeotic mutation.

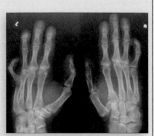

Figure 1 The hands of a person with Liebenberg syndrome resemble feet; the arms resemble legs.

Clinical Connection boxes discuss how genetics and genomics impact health and health care.

Summary

11.1 Gene Expression Through Time and Tissue

1. Changes in gene expression occur over time at the molecular and organ levels. **Epigenetic** changes to DNA alter gene expression, but do not change the DNA sequence.
2. **Proteomics** catalogs the types of proteins in particular cells, tissues, organs, or entire organisms under specified conditions.

11.2 Control of Gene Expression

3. Acetylation of certain histone proteins enables the transcription of associated genes, whereas phosphorylation and methylation prevent transcription. The effect of these three molecules is called **chromatin remodeling.**
4. **MicroRNAs** bind to certain mRNAs, blocking translation.

11.3 Maximizing Genetic Information

5. A small part of the genome encodes protein, but the number of proteins is much greater than the number of genes.
6. Alternate splicing, use of introns, protein modification, and cutting proteins translated from a single gene contribute to protein diversity.

11.4 Most of the Human Genome Does *Not* Encode Protein

7. The non-protein-encoding part of the genome includes viral sequences, noncoding RNAs, **pseudogenes,** introns, **transposons,** promoters and other controls, and repeats.
8. **Long noncoding RNAs** control gene expression.

www.mhhe.com/lewisgenetics11

Answers to all end-of-chapter questions can be found at **www.mhhe.com/lewisgenetics11**. You will also find additional practice quizzes, animations, videos, and vocabulary flashcards to help you master the material in this chapter.

Review Questions

1. Why is control of gene expression necessary?
2. Define *epigenetics*.
3. Distinguish between the type of information that epigenetics provides and the information in the DNA sequence of a protein-encoding gene.
4. Describe three types of cells and how they differ in gene expression from each other.
5. What is the environmental signal that stimulates globin switching?
6. How does development of the pancreas illustrate differential gene expression?
7. Explain how a mutation in a promoter can affect gene expression.
8. How do histones control gene expression, yet genes also control histones?
9. What controls whether histones enable DNA wrapped around them to be transcribed?
10. State two ways that methyl groups control gene expression.
11. Name a mechanism that silences transcription of a gene and a mechanism that blocks translation of an mRNA.
12. Why might a computational algorithm be necessary to evaluate microRNA function in the human genome?
13. Describe three ways that the number of proteins exceeds the number of protein-encoding genes in the human genome.
14. How can alternate splicing generate more than one type of protein from the information in a gene?
15. In the 1960s, a gene was defined as a continuous sequence of DNA, located permanently at one place on a chromosome, that specifies a sequence of amino acids from one strand. List three ways this definition has changed.
16. Give an example of a discovery mentioned in the chapter that changed the way we think about the genome.
17. What is the evidence that some long noncoding RNAs may hold clues to human evolution?

Applied Questions

1. The World Anti-Doping Agency warns against gene doping, which it defines as "the non-therapeutic use of cells, genes, genetic elements, or of the modulation of gene expression, having the capacity to improve athletic performance." The organization lists the following genes as candidates for gene doping when overexpressed:
   Insulin-like growth factor (IGF-1)
   Growth hormone (GH)

Each chapter ends with a point-by-point **Chapter Summary**.

Review Questions assess content knowledge.

Applied Questions help students develop problem-solving skills.

Web Activities

1. Gene expression profiling tests began to be marketed several years ago. Search for "Oncotype DX," "MammaPrint," or "gene expression profiling in cancer" and describe how classifying a cancer this way can improve diagnosis and/or treatment. (Or apply this question to a different type of disease.)
2. The government's Genotype-Tissue Expression (GTEx; https://commonfund.nih.gov/GTEx/) project is a database of gene expression profiles of 24 tissues (parts of organs) from 190 people who died while healthy.
   a. What type of data are compared?
   b. Suggest a way that a researcher can use this type of information.
3. Look up each of the following conditions using OMIM or another source, and describe how they arise from altered chromatin: alpha-thalassemia, ICF syndrome, Rett syndrome, Rubinstein-Taybi syndrome.

Web Activities encourage students to use the latest tools and databases in genetic analysis.

Forensics Focus questions probe the use of genetic information in criminal investigations.

Cases and Research Results use stories based on accounts in medical and scientific journals; real clinical cases; posters and reports from professional meetings; interviews with researchers; and fiction to ask students to analyze data and predict results.

Forensics Focus

1. Establishing time of death is critical information in a murder investigation. Forensic entomologists can estimate the "postmortem interval" (PMI), or the time at which insects began to deposit eggs on the corpse, by sampling larvae of specific insect species and consulting developmental charts to determine the stage. The investigators then count the hours backwards to estimate the PMI. Blowflies are often used for this purpose, but their three larval stages look remarkably alike in shape and color, and development rate varies with environmental conditions. With luck, researchers can count back 6 hours from the developmental time for the largest larvae to estimate the time of death.

   In many cases, a window of 6 hours is not precise enough to narrow down suspects when the victim visited several places and interacted with many people in the hours before death. Suggest a way that gene expression profiling might be used to more precisely define the PMI and extrapolate a probable time of death.

Case Studies and Research Results

1. To make a "reprogrammed" induced pluripotent stem (iPS) cell (see figure 2.22), researchers expose fibroblasts taken from skin to "cocktails" that include transcription factors. The fibroblasts divide and give rise to iPS cells, which, when exposed to other transcription factors, divide and yield daughter cells that specialize in distinctive ways that make them different from the original fibroblasts. How do transcription factors orchestrate these changes in cell type?
2. A study investigated "genomic signatures of global fitness" to identify gene expression patterns that indicate that a course of exercise is beneficial. In the study, sixty sedentary women representing different ethnic groups

Dynamic Art Program

Multilevel Perspective

Illustrations depicting complex structures show macroscopic and microscopic views to help students see relationships among increasingly detailed drawings.

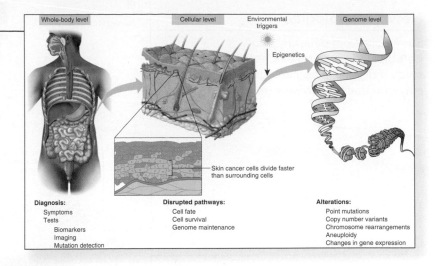

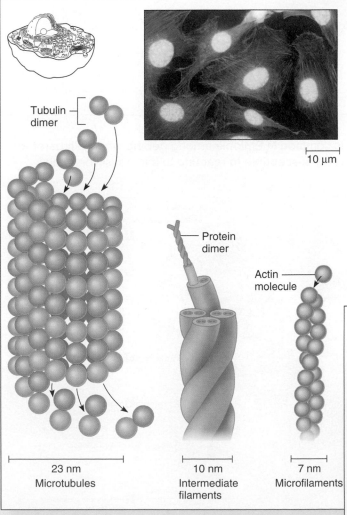

Tubulin dimer

Protein dimer

Actin molecule

23 nm
Microtubules

10 nm
Intermediate filaments

7 nm
Microfilaments

10 μm

Combination Art

Drawings of structures are paired with micrographs to provide the best of both perspectives: the realism of photos and the explanatory clarity of line drawings.

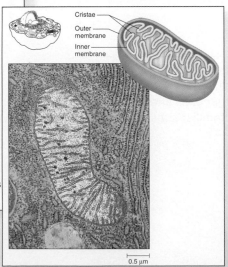

Cristae

Outer membrane

Inner membrane

0.5 μm

New Technologies

Stem cells from patients' skin fibroblasts enable researchers to study a disease's beginnings, and may one day lead to new treatments.

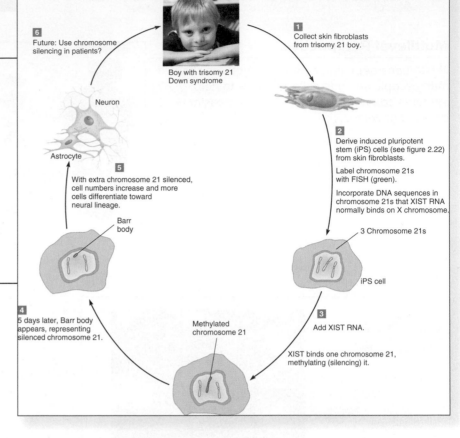

Clinical Coverage

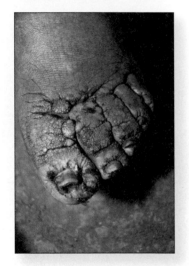

"Mossy foot," or podoconiosis, is common in Ethiopia among people who walk barefoot on volcanic rock and are genetically susceptible to reacting to mineral slivers. The treatment: *shoes*.

Process Figures

Complex processes are broken down into a series of numbered smaller steps that are easy to follow. Here, organelles interact to produce and secrete a familiar substance—milk (figure 2.6).

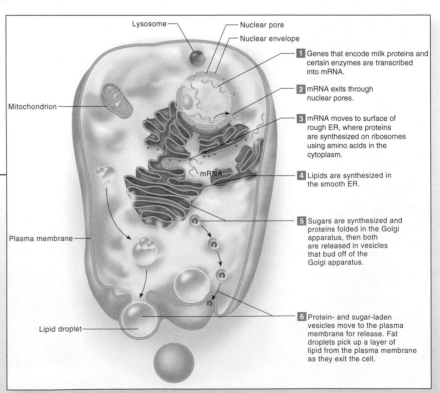

Teaching and Learning Tools

McGraw-Hill offers various tools and technology products to accompany *Human Genetics: Concepts and Applications,* Eleventh Edition.

McGraw-Hill Connect® Genetics

McGraw-Hill Connect Genetics provides online presentation, assignment, and assessment solutions. It connects your students with the tools and resources they'll need to achieve success.

With Connect Genetics you can deliver assignments, quizzes, and tests online. A robust set of questions and activities is presented and aligned with the textbook's learning outcomes. As an instructor, you can edit existing questions and author entirely new problems. Track individual student performance—by question, assignment, or in relation to the class overall—with detailed grade reports. Integrate grade reports easily with Learning Management Systems (LMS), such as WebCT and Blackboard®. And much more.

ConnectPlus® Genetics provides students with all the advantages of Connect Genetics, plus 24/7 online access to an eBook. This media-rich version of the book is available through the McGraw-Hill Connect platform and allows seamless integration of text, media, and assessments.

To learn more, visit
www.mcgrawhillconnect.com

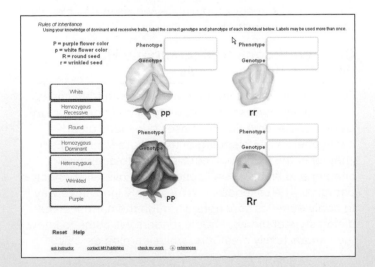

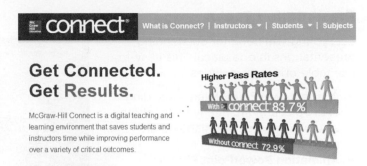

Get Connected. Get Results.

McGraw-Hill Connect is a digital teaching and learning environment that saves students and instructors time while improving performance over a variety of critical outcomes.

SEE FOR YOURSELF

LEARNSMART®

Integrated within Connect and available as a standalone, **McGraw-Hill LearnSmart**™ is the premier learning system designed to effectively assess a student's knowledge of course content. Through a series of adaptive questions, LearnSmart intelligently pinpoints concepts the student does not understand and maps out a personalized study plan for success. LearnSmart prepares students with a base of knowledge, allowing instructors to focus valuable class time on higher-level concepts.

SMARTBOOK™

New SmartBook™ facilitates the reading process by identifying what content a student knows and does't know through adaptive assessments. As the student reads, the reading material constantly adapts to ensure the student is focused on the content he or she needs the most to close any knowledge gaps.

Presentation Tools

Everything you need for outstanding presentations in one place! This easy-to-use table of assets includes

- **Image PowerPoint® Files**—Including every piece of art, nearly every photo, all tables, as well as unlabeled art pieces.
- **Animation PowerPoint Files**—Numerous full-color animations illustrating important processes are also provided. Harness the visual impact of concepts in motion by importing these files into classroom presentations or online course materials.
- **Lecture PowerPoint Files**—with animations fully embedded!
- **Labeled and Unlabeled JPEG Images**—Full-color digital files of all illustrations ready to incorporate into presentations, exams, or custom-made classroom materials.

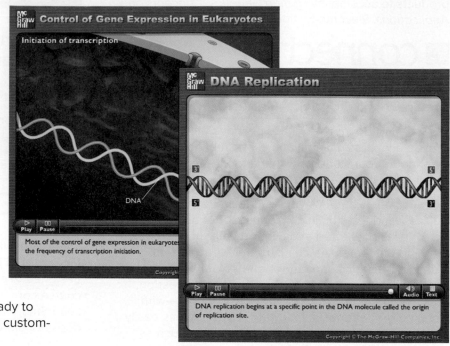

Computerized Test Bank written by Ricki Lewis!

The author has rewritten and expanded the test bank to include many more cases and problems. Terms match those used in the text, and the questions follow the order of topics within the chapters. This comprehensive bank of questions is provided within a computerized test bank powered by McGraw-Hill's flexible electronic testing program EZ Test Online. EZ Test Online allows you to create paper and online tests or quizzes in this easy-to-use program!

Imagine being able to create and access your test or quiz anywhere, at any time without installing the testing software. Now, with EZ Test Online, instructors can select questions from multiple McGraw-Hill test banks or author their own, and then either print the test for paper distribution or give it online.

Access the computerized test bank for Lewis, *Human Genetics* by going to www.mhhe.com/lewisgenetics11 and clicking on Instructor Resources.

My Lectures-Tegrity

McGraw-Hill Tegrity Campus™ records and distributes your class lecture, with just a click of a button. Students can view anytime/anywhere via computer, iPod, or mobile device. It indexes as it records your PowerPoint presentations and anything shown on your computer so students can use keywords to find exactly what they want to study.

Case Workbook to Accompany Human Genetics by Ricki Lewis

For those who enjoy learning and teaching from cases, *In the Family: A Case Workbook to Accompany Human Genetics, Eleventh Edition,* bases questions on a multigenerational blending of three core families. Each chapter in the workbook corresponds to a textbook chapter and highlights a section of the overall connected pedigree. The casebook is a fun, highly innovative way to apply genetics concepts. Through the narrative and dialog style of the workbook, readers will come to know the various family members, while learning genetics.

A child's genome holds information on where she came from and some of what she might experience and achieve—but the environment is very important too in guiding who she is and will become.

What Is in a Human Genome?

Learning Outcomes

1.1 Introducing Genes and Genomes

1. Explain what genetics is, and what it is not.

2. Distinguish between gene and genome.

3. Define *bioethics*.

1.2 Levels of Genetics and Genomics

4. Describe the levels of genetics, from nucleic acids to chromosomes, to cells, body parts, families, and populations.

5. Explain how genetics underlies evolution.

1.3 Applications of Genetics and Genomics

6. Provide examples of how genetics is used in identification of people and in health care.

1.4 A Global Perspective on Genomes

7. How can investigating genomes extend beyond interest in ourselves?

The BIG Picture

The human genome is a vast store of information encoded in the sequence of building blocks of the molecule deoxyribonucleic acid (DNA). Genetic information affects our health and traits, and reflects how we are biologically related to one another.

Eve's Genome

A baby is born. A few drops of blood from her heel are placed into a small device that within hours sends personal information into her electronic medical record. The device deciphers the entire sequence of DNA building blocks wound into the nucleus of a white blood cell. This is Eve's genome.

Sequencing the first human genome took 15 years; now it takes hours. Eve's genome sequence holds clues to her current and future health, as well as to her ancestry. Past, present, and future are encoded in nature's informational molecule, deoxyribonucleic acid, or DNA—with room for environmental influences.

Eve's genome indicates overall good genetic health. She has a mild clotting disorder that the nurse suspected when two gauze patches were needed to stop the bleeding from the heel stick. Two rare variants of the gene that causes cystic fibrosis (CF) mean that Eve is susceptible to certain respiratory infections and sensitive to irritants, but her parents knew that from prenatal testing. Fortunately the family lives in a rural area far from pollution, and Eve will have to avoid irritants such as smoke and dust.

The inherited traits that will emerge as Eve grows and develops range from interesting to important. Her hair will darken and curl, and genes that contribute to bone development indicate that she'll have a small nose, broad forehead, and chiseled cheekbones. If she follows a healthy diet, she'll be as tall as her parents. On the serious side, Eve has inherited a mutation in a gene that greatly raises her risk of one day developing certain types of cancers. Her genes predict a healthy heart, but she might develop diabetes unless she exercises regularly and limits carbohydrates in her diet.

Many traits are difficult to predict because of environmental influences. What will Eve's personality be like? How intelligent will she be? How will she react to stress? What will be her passions?

Genome sequencing also reveals clues to Eve's past, which is of special interest to her father, who was adopted. She has gene variants common among the Eastern European population her mother comes from, and others that match people from Morocco. Is that her father's heritage? Eve is the beautiful consequence of a mix of her parents' genomes, receiving half of her genetic material from each.

Over the next few years sequencing of our genomes, or perhaps relevant parts of them, will become routine in health care. Do you want to know the information in your genome?

1.1 Introducing Genes and Genomes

Genetics is the study of inherited traits and their variation. Sometimes people confuse genetics with genealogy, which considers relationships but not traits. Because some genetic tests can predict illness, genetics has also been compared to fortune-telling. However, genetics is a life science. Heredity is the transmission of traits and biological information between generations, and genetics is the study of how traits are transmitted.

Inherited traits range from obvious physical characteristics, such as the freckles and red hair of the young man in **figure 1.1**, to many aspects of health, including disease. Talents, quirks, personality traits, and other difficult-to-define characteristics might appear to be inherited if they affect several family members, but may reflect a combination of genetic and environmental influences. Attributing some behavioral traits to genetics, such as sense of humor, fondness for sports, and whether or not one votes, are oversimplifications.

Over the past few years, genetics has exploded from a mostly academic discipline and a minor medical specialty dealing mostly with very rare diseases, to a part of everyday discussion. Personal genetic information is accessible and we are learning the contribution of genes to the most common traits and disorders. Many physicians are taking continuing medical education courses to learn how to integrate genetic and genomic testing into clinical practice.

Like all sciences, genetics has its own vocabulary. Many terms may be familiar, but actually have precise technical

Figure 1.1 Inherited traits. This young man owes his red hair, fair skin, and freckles to a variant of a gene that encodes a protein (the melanocortin 1 receptor) that controls the balance of pigments in his skin.

definitions. "It's in her DNA," for example, usually means an inborn trait, not a specific DNA sequence. This chapter introduces terms and concepts that are explained in detail in subsequent chapters.

Genes are the units of heredity. Genes are biochemical instructions that tell **cells,** the basic units of life, how to manufacture certain proteins. These proteins, in turn, impart or control the characteristics that create much of our individuality. A gene consists of the long molecule **deoxyribonucleic acid (DNA).** The DNA transmits information in its sequence of four types of building blocks.

The complete set of genetic instructions characteristic of an organism, including protein-encoding genes and other DNA sequences, constitutes a **genome.** Researchers concluded sequencing the human genome in 2003. Nearly all of our cells contain two copies of the genome. Researchers are still analyzing what all of our genes do, and how genes interact and respond to environmental stimuli. Only a tiny fraction of the 3.2 billion building blocks of our genetic instructions determines the most interesting parts of ourselves—our differences. Comparing and analyzing genomes, which constitute the field of **genomics,** reveals how closely related we are to each other and to other species.

Genetics directly affects our lives and those of our relatives, including our descendants. Principles of genetics also touch history, politics, economics, sociology, anthropology, art, and psychology. Genetic questions force us to wrestle with concepts of benefit and risk, even tapping our deepest feelings about right and wrong. A field of study called **bioethics** was founded in the 1970s to address moral issues and controversies that arise in applying medical technology. Bioethicists today confront concerns that arise from new genetic technology, such as privacy, use of genetic information, and discrimination. Essays throughout this book address bioethical issues, beginning with the story of how DNA sequencing saved a boy's life, on page 10.

Key Concepts Questions 1.1

1. Distinguish between genetics and heredity.
2. Distinguish between a gene and a genome.
3. What is bioethics?

1.2 Levels of Genetics and Genomics

Genetics considers the transmission of information at several levels. It begins with the molecular level and broadens through cells, tissues and organs, individuals, families, and finally to populations and the evolution of species (**figure 1.2**).

The Instructions: DNA, Genes, Chromosomes, and Genomes

DNA resembles a spiral staircase or double helix. The "rails," or backbone, consist of alternating chemical groups (sugars and phosphates) that are the same in all DNA molecules. The "steps" of the DNA double helix hold the information because they are pairs of four types of building blocks, or bases, whose sequence varies from molecule to molecule (**figure 1.3**). The bases are adenine (A) and thymine (T), which attract each other, and cytosine (C) and guanine (G), which attract each other. The information is in the sequences of A, T, C, and G. The two strands of the double helix are oriented in opposite directions, like two snakes biting each other's tails.

The chemical structure of DNA enables it both to perpetuate itself when a cell divides and to provide the cell with information used to manufacture proteins. Each consecutive three DNA bases is a code for a particular amino acid, and amino acids are the building blocks of proteins.

In DNA replication, the chains of the helix part and each half builds a new partner chain by pulling in free DNA bases—A and T attracting and C and G attracting. To produce protein, a process called transcription copies the sequence of part of one strand of a DNA molecule into a related molecule, messenger **ribonucleic acid** (**RNA**). Each three RNA bases in a row then attract another type of RNA that functions as a connector, bringing in a particular amino acid. The amino acids align, forming a protein. Building a protein is called translation. Proteins provide the traits associated with genes, such as blood clotting factors. **Figure 1.4** is a conceptual look ahead to chapter 10, which presents these complex processes in detail.

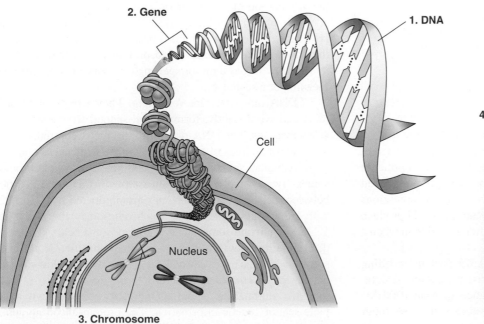

2. Gene

1. DNA

Cell

Nucleus

3. Chromosome

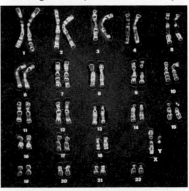

4. Human genome (23 chromosome pairs)

Figure 1.2 **Levels of genetics.** Genetics can be considered at several levels, from DNA, to genes, to chromosomes, to genomes, to the more familiar individuals, families, and populations. (A gene is actually several hundred or thousand DNA bases long.)

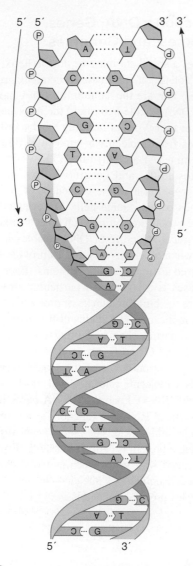

Figure 1.3 **The DNA double helix.** The 5′ and 3′ labels indicate the head-to-tail organization of the DNA double helix. A, C, T, and G are bases. S stands for sugar and P for phosphate.

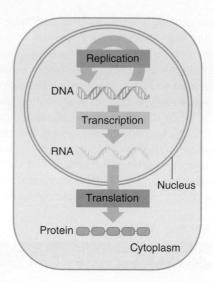

Figure 1.4 **The language of life:** DNA to RNA to protein.

A genome's worth of DNA is like a database that is accessed to run the cell. Different types of cells have different protein requirements. A muscle cell has abundant contractile proteins, but a skin cell contains mostly scaly proteins called keratins. A cell's protein production can change as conditions change. A cell lining the stomach, for example, would produce more protein-based digestive enzymes after a meal than when a person hasn't eaten in several hours.

The human genome has about 20,325 protein-encoding genes, and these DNA sequences comprise the **exome**. A database called Online Mendelian Inheritance in Man (OMIM) (http://www.ncbi.nlm.nih.gov/omim) describes the few thousand genes known to cause disorders or traits.

Protein-encoding genes account for only about 1.5 percent of the human genome. The rest includes many DNA sequences that assist in protein synthesis or turn protein-encoding genes

on or off. The ongoing effort to understand what individual genes do is termed annotation.

The same protein-encoding gene may vary slightly in base sequence from person to person. These gene variants are called **alleles**. The changes in DNA sequence that distinguish alleles arise by a process called **mutation**. A "mutation" is also used as a noun to refer to the changed gene. Once a gene mutates, the change is passed on when the cell that contains it divides. If the change is in a sperm or egg cell that becomes a fertilized egg, it is passed to the next generation.

Some mutations cause disease, and others provide variation, such as freckled skin. Mutations can also help. For example, a mutation makes a person's cells unable to manufacture a surface protein that binds HIV. These people are resistant to HIV infection. Many mutations have no visible effect because they do not change the encoded protein in a way that affects its function, just as a minor spelling error does not obscure the meaning of a sentence.

DNA molecules are very long. They wrap around proteins and wind tightly, forming rod-shaped structures called **chromosomes**. The DNA of a chromosome is continuous, but it includes hundreds of genes among other sequences.

A human somatic (non-sex) cell has 23 pairs of chromosomes. Twenty-two pairs are **autosomes,** which do not differ between the sexes. The autosomes are numbered from 1 to 22, with 1 the largest. The other two chromosomes, the X and the Y, are **sex chromosomes**. The Y chromosome bears genes that determine maleness. In humans, a female has two X chromosomes and a male has one X and one Y. Charts called **karyotypes** display the chromosome pairs from largest to smallest.

To summarize, a human somatic cell has two complete sets of genetic information. The protein-encoding genes are scattered among 3.2 billion DNA bases in each set of 23 chromosomes.

A trait caused by a single gene is termed Mendelian. Most traits are **multifactorial traits,** which means that they

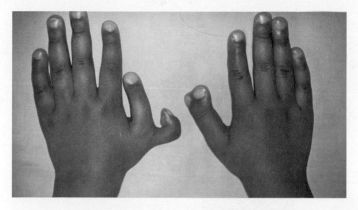 

Figure 1.5 **Mendelian versus multifactorial traits.** **(a)** Polydactyly—extra fingers and/or toes—is a Mendelian trait (single-gene). **(b)** Hair color is multifactorial, controlled by at least three genes plus environmental factors, such as the bleaching effects of sun exposure.

are determined by one or more genes and environmental factors (**figure 1.5**). The more factors that contribute to a trait or illness—inherited or environmental—the more difficult it is to predict the risk of occurrence in a particular family member. The bone-thinning condition osteoporosis illustrates the various factors that can contribute to a disease. Several genes confer susceptibility to fractures, as do smoking, lack of weight-bearing exercise, and a calcium-poor diet.

Environmental effects on gene action counter the idea of **"genetic determinism,"** that "we are our genes." Because of the role of the environment, some genetic test results given before symptoms are present indicate risks, not a diagnosis. A doctor might discuss the results of a test finding an inherited susceptibility to a form of breast cancer as, "You have a 45 percent chance of developing this form of cancer," not "You will get cancer."

Genetic determinism may be harmful or helpful, depending on circumstance. As part of social policy, genetic determinism can be disastrous. An assumption that one group of people is genetically less intelligent than another can lead to lowered expectations and/or fewer educational opportunities for people perceived as inferior. Environment, in fact, has a large impact on intellectual development. On the other hand, knowing the genetic contribution to a trait can give us more control over health outcomes by influencing noninherited factors, such as diet and exercise habits.

The Body: Cells, Tissues, and Organs

A human body consists of approximately 37 trillion cells. All cells except red blood cells contain the entire genome, but cells differ in appearance and activities because they use only some of their genes. Which genes a cell uses at any given time depends upon environmental conditions both inside and outside the body.

Like the Internet, a genome contains a wealth of information, but only some of it need be accessed. The use, or "expression," of different subsets of genes to manufacture proteins drives the **differentiation,** or specialization, of distinctive cell types. An adipose cell is filled with fat, but not the contractile proteins of muscle cells. Both cell types, however, have complete genomes. Groups of differentiated cells assemble and interact with each other and the nonliving material that they secrete to form aggregates called tissues.

The body has four basic tissue types, composed of more than 290 types of cells (see figure 2.2). Tissues intertwine and layer to form the organs of the body, which in turn connect into organ systems. The stomach shown at the center of **figure 1.6**, for example, is a sac made of muscle that also has a lining of epithelial tissue, nervous tissue, and a supply of blood, which is a type of connective tissue. **Table 1.1** describes tissue types.

Many organs include rare, unspecialized **stem cells**. A stem cell can divide to yield another stem cell and a cell that differentiates. Stem cells provide a reserve supply of cells that enable an organ to grow and repair damage.

Relationships: From Individuals to Families

Two terms distinguish the alleles that are *present* in an individual from the alleles that are *expressed*. The **genotype** refers to the underlying instructions (alleles present), whereas the **phenotype** is the visible trait, biochemical change, or effect on health (alleles expressed). Alleles are further distinguished by how many copies are necessary to affect the phenotype. A **dominant** allele has an effect when present in just one copy (on one chromosome), whereas a **recessive** allele must be present on both chromosomes of a pair to be expressed.

Individuals are genetically connected into families. A person has half of his or her gene variants in common with each parent and each sibling, and one-quarter with each grandparent. First cousins share one-eighth of their gene variants.

Charts called **pedigrees** depict the members of a family and indicate which individuals have particular inherited traits.

The Bigger Picture: From Populations to Evolution

Above the family level of genetic organization is the population. In a strict biological sense, a population is a group of individuals that can have healthy offspring together. In a genetic

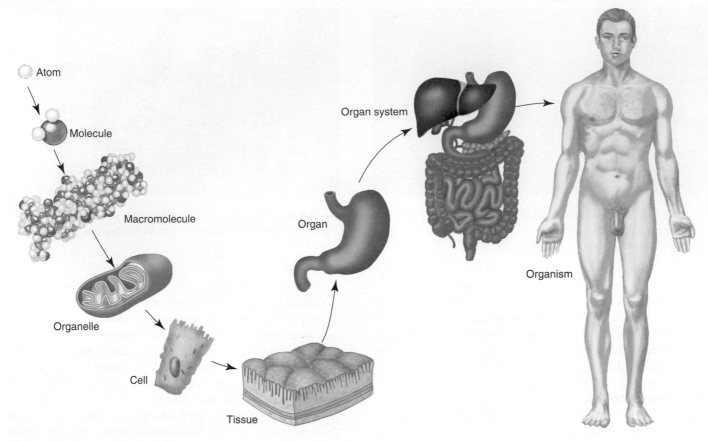

Figure 1.6 **Levels of biological organization.**

sense, a population is a large collection of alleles, distinguished by their frequencies. People from a Swedish population, for example, would have a greater frequency of alleles that specify light hair and skin than people from a population in Nigeria, who tend to have dark hair and skin. All the alleles in a population constitute the **gene pool**. (An individual does not have a gene pool.)

Population genetics is applied in health care, forensics, and other fields. It is also the basis of evolution, which is defined as changing allele frequencies in populations. These small-scale genetic changes underlie the more obvious species distinctions we most often associate with evolution.

Comparing DNA sequences for individual genes, or the amino acid sequences of the proteins that the genes encode, can reveal how closely related different types of organisms are. The assumption is that the more similar the sequences are, the more recently two species diverged from a shared ancestor, and the more closely related they are. This is a more plausible explanation than two species having evolved similar or identical gene sequences coincidentally. All life is related, and different species use the same basic set of genes that makes life possible.

Both the evolution of species and family patterns of inherited traits show divergence from shared ancestors. This idea is based on logic. It is more likely that a brother and sister share approximately half of their gene variants because they have the same parents than that half of their genetic material is identical by chance.

Genome sequence comparisons reveal more about evolutionary relationships than comparing single genes. Humans, for example, share more than 98 percent of the DNA sequence with chimpanzees. Our genomes differ from theirs more in gene organization and in the number of copies of genes than in the overall sequence. Learning the functions of the human-specific genes may explain the differences between us and them—such as our lack of hair and use of spoken language. Figure 16.8 highlights some of our distinctively human traits.

Table 1.1	Tissue Types
Tissue	**Function/Location/Description**
Connective tissues	A variety of cell types and materials around them that protect, support, bind to cells, and fill spaces throughout the body; include cartilage, bone, blood, and fat
Epithelium	Tight cell layers that form linings that protect, secrete, absorb, and excrete
Muscle	Cells that contract, providing movement
Nervous	Neurons transmit information as electrochemical impulses that coordinate movement and sense and respond to environmental stimuli; neuroglia are cells that support and nourish neurons

Comparisons of people at the genome level reveal that we are much more like each other genetically than are other mammals. Chimpanzees are more distinct from each other than we are! The most genetically diverse modern people are from Africa, where humanity arose. The gene variants among different modern ethnic groups include subsets of our ancestral African gene pool.

1.3 Applications of Genetics and Genomics

Genetics is impacting many areas of our lives, from health care choices, to what we eat and wear, to unraveling our pasts and guiding our futures. "Citizen scientists" are discovering genetic information about themselves while helping researchers compile databases that will help many. A direct-to-consumer genetic testing company used its clients' information to discover a new gene that causes Parkinson disease, for example. Thinking about genetics evokes fear, hope, anger, and wonder, depending upon context and circumstance. Following are glimpses of applications of genetics that are explored more fully in subsequent chapters.

Establishing Identity

A technique called DNA profiling compares DNA sequences among individuals to establish or rule out identity, relationships, or ancestry. DNA profiling has varied applications, in humans and other species.

Forensic science is the collecting of physical evidence of a crime. Comparing DNA samples from evidence at crime scenes to samples from suspects often leads to convictions, and also to reversing convictions erroneously made using other forms of evidence.

DNA profiling is useful in identifying victims of natural disasters, such as violent storms and earthquakes. In happier circumstances, DNA profiles maintained in databases assist adopted individuals in locating blood relatives and children of sperm donors in finding their biological fathers and half-siblings.

Another use of DNA profiling is to analyze food, because foods were once organisms, which have species-specific DNA sequences. For example, analyzing DNA sequences revealed

Figure 1.7 DNA reveals and clarifies history. After DNA evidence showed that Thomas Jefferson likely fathered a son of his slave, descendants of both sides of the family met.

horsemeat in meatballs sold at a restaurant chain, cheap fish sold as gourmet varieties, and worms in cans of sardines.

DNA analysis can clarify details of history. A famous case confirmed that Thomas Jefferson had children with his slave Sally Hemings. The president was near Hemings 9 months before each of her seven children were born, and the children looked like him. Male descendants of Sally Hemings share an unusual Y chromosome sequence with the president's male relatives. His only son with his wife died in infancy, so researchers deduced the sequence of the president's Y chromosome from descendants of his uncle. Today the extended family holds reunions (**figure 1.7**).

DNA testing can provide views into past epidemics of infectious diseases by detecting genes of the pathogens. For example, analysis of DNA in the mummy of the Egyptian king Tutankhamun revealed the presence of the microorganism that causes malaria. The child king likely died from complications of malaria following a leg fracture from weakened bones rather than from intricate murder plots, a kick from a horse, or fall from a chariot, as had been thought. His tomb included a cane and drugs, supporting the diagnosis based on DNA evidence.

Health Care

Looking at diseases from a genetic point of view is changing health care. Many diseases, not just inherited ones, are now viewed as the consequence of complex interactions among genes and environmental factors. Even the classic single-gene diseases are sensitive to the environment. A child with cystic fibrosis (OMIM 219700), for example, is more likely to suffer frequent respiratory infections if she regularly breathes second-hand smoke. In the opposite situation, whether or not a person is susceptible to a mostly environmentally caused condition depends upon genetics. **Figure 1.8** shows the feet of a young person with podoconiosis, also known as "mossy foot." The swollen, itchy, painful bumps result from walking barefoot on damp, red volcanic rock that contains microscopic slivers

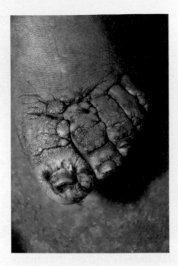

Figure 1.8 **A gene-environment interaction.**
Podoconiosis ("mossy foot") is a painful foot condition that develops in genetically susceptible individuals who walk barefoot on hard, volcanic rock that contains slivers of minerals. It is common in parts of Ethiopia.

of minerals. In the 15 countries that have the volcanic rock, 10 to 20 percent of the people have the painful foot disease. They share a pattern of gene variants that affects the immune response, causing extreme inflammation in response to irritation. In general, inherited differences in immunity are one of several reasons why some people are more susceptible to contracting certain infections than others, discussed in chapter 17.

Because genes instruct cells to manufacture specific proteins, inherited diseases can result from altered proteins or too little or too much of a protein, or proteins made at the wrong place or time. Genes also affect how people respond to particular drugs. For example, inheriting certain gene variants can make a person's body very slow at breaking down an anti-clotting drug, or extra sensitive to the drug. The person bleeds profusely at the same dose that most patients tolerate. Identifying individual drug reactions based on genetics is a growing field called pharmacogenetics. Tests based on pharmacogenetics can prevent adverse reactions or indicate that a particular drug will or will not work in a particular individual. Some physicians use these tests to prescribe antidepressants, anti-clotting drugs, chemotherapies, and cholesterol-lowering drugs.

Single-Gene Diseases

Inherited illness caused by a single gene differs from other types of illnesses in several ways (**table 1.2**). In families, we can predict inheritance of a single-gene disease by knowing exactly how a person is related to an affected relative. In contrast, an infectious disease requires that a pathogen pass from one person to another, which is much less predictable.

A second distinction of single-gene disorders is that tests can sometimes predict the risk of developing symptoms. This is possible because all cells harbor the mutation. A person with a family history of Huntington disease (HD; OMIM 143100), for example, can have a blood test that detects the mutation at any

age, even though symptoms typically do not occur until early middle age, and the disease affects the brain, not the blood. Inheriting the HD mutation predicts illness with near certainty. For many conditions, predictive power is much lower. For example, inheriting one copy of a particular variant of a gene called *APOE* raises risk of developing Alzheimer disease by threefold, and inheriting two copies raises it 15-fold. But without absolute risk estimates and no treatments for this disease, would you want to know if you have a high risk of developing it?

A third feature of single-gene diseases is that they may be much more common in some populations than others. Genes do not like or dislike certain types of people; rather, mutations stay in certain populations because we have children with people like ourselves. While it might not seem politically correct to offer a "Jewish genetic disease" screen, it makes biological and economic sense—several disorders are much more common in this population. A fourth characteristic of a genetic disease is that it may be "fixable" by altering the abnormal instructions.

A Genomic View Connects Diseases

"Gene expression" refers to whether a gene is "turned on" or "turned off" from being transcribed and translated into protein. Tracking gene expression in cells can reveal new information about diseases. It can show that diseases with different symptoms actually share the same underlying genetic defect, or that conditions with similar symptoms have different causes at the molecular level.

Figure 1.9 shows part of a huge disease map called the "diseasome." It connects diseases that share genes that have altered expression. Some of the links and clusters are well known, such as obesity, hypertension, and diabetes. Others are surprises, such as Duchenne muscular dystrophy (DMD; see figure 2.1) and heart attacks. The muscle disorder has no treatment, but heart attack does—researchers are now testing cardiac drugs on boys with DMD. In other cases, the association of a disease with genes whose expression goes up or down can suggest targets for new drugs.

The diseases in figure 1.9 are well known. Thousands of people have illnesses that their health care providers cannot diagnose, because the collection of symptoms does not match any known disease. A technology called exome sequencing is helping many of these patients finally learn the cause of their conditions.

Table 1.2	How Single-Gene Diseases Differ from Other Diseases

1. Risk can be predicted for family members.

2. Predictive (presymptomatic) testing may be possible.

3. Different populations may have different characteristic disease frequencies.

4. Correction of the underlying genetic abnormality may be possible.

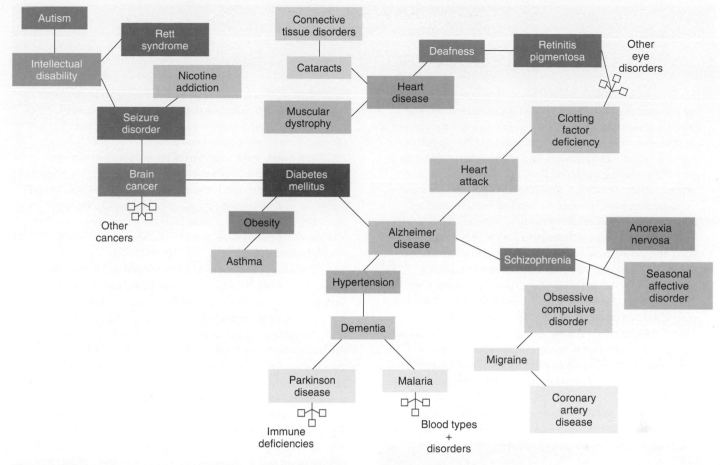

Figure 1.9 Part of the diseasome. This tool links diseases by shared gene expression. That is, a particular gene may be consistently overexpressed or underexpressed in two diseases, compared to the healthy condition. The lines refer to at least one gene connecting the disorders depicted in the squares. The conditions are not necessarily inherited because gene expression changes in all situations. For example, Alzheimer disease is linked to heart attack because in both conditions cholesterol builds up. Finding diseasome links can suggest existing drugs to "repurpose" to treat different illnesses. The cholesterol-lowering statin drugs are being tested on Alzheimer patients, for example. (Based on the work of A-L Barabási and colleagues.)

Exome sequencing reveals mutations in the protein-encoding part of an individual's genome. Powerful algorithms search the data for variants in genes that might explain the symptoms. **Clinical Connection 1.1** on page 10 describes the case of Nicholas Volker, one of the first patients to receive a diagnosis thanks to exome sequencing. It saved his life.

Key Concepts Questions 1.3

1. What are some uses of DNA profiling?
2. How can inheriting certain gene variants raise the risk of developing an illness without actually causing it?
3. How are single-gene diseases different from other types of diseases?
4. How do gene expression profiling and exome sequencing help us to better understand health and diseases?

1.4 A Global Perspective on Genomes

We share the planet with many thousands of other species. We aren't familiar with many of Earth's residents because we can't observe their habitats, or we can't grow them in laboratories. "Metagenomics" is a field that is revealing and describing much of the invisible living world by sequencing all of the DNA in a particular habitat. Such areas range from soil, to an insect's gut, to garbage, to a volume of captured air over a polluted city. Metagenomics studies are showing how species interact, and may yield new drugs and reveal novel energy sources.

Metagenomics researchers collect and sequence DNA, then consult databases of known genomes to imagine what the organisms might be like. The first metagenomics project described life in the Sargasso Sea. This 2-million-square-mile oval area off the coast of Bermuda has been thought to lack life beneath its thick cover of seaweed, which is so abundant that Christopher Columbus thought he'd reached land when

Exome Sequencing Saves a Boy's Life

Nicholas Volker was born October 26, 2004, in Monona, Wisconsin, the youngest of four and the only boy. He seemed fine until age 2, when his growth slowed. He developed an abscess in his rectum that burst, leaving two holes that enlarged instead of healing. Nicholas's digestive system was developing fistulas—holes that connect two parts that are normally separate. Food and feces leaked, and he became repeatedly infected. A feeding tube kept him from starving.

Doctors were stumped. No one had ever reported the shredding digestive tract that was killing Nicholas. Genetic tests were negative. Whatever Nicholas had, it was much more severe than inflammatory bowel disease. A clue was that he'd improve after a blood transfusion. He worsened, and by early 2008, doctors from all over the country were trying to diagnose him.

By January 2009, Nicholas weighed less than 20 pounds, as holes riddled his digestive tract, opening to the outside. The wounds needed cleaning daily, requiring anesthesia. Surgeons removed much of his large intestine. Whenever he'd gain a little weight, more holes would form. By the age of 4, he'd had more than 100 surgeries and by the summer of 2009 he'd been hospitalized for more than 300 days.

Nicholas's pediatrician knew that the Medical College of Wisconsin had been planning an exome sequencing experiment

Figure 1 An exome is to a genome like an orange slice is to an orange—a small part of a whole. However in a genome, the larger part controls the smaller.

for 2014. Could they move it up to search the boy's exome? The results might reveal mutations that could explain Nicholas's symptoms. The medical staff raised the funds, and a DNA sequencing company donated equipment.

The exome is the small part of the genome that encodes protein (**figure 1**). Everyone's exome has thousands of rare or even unique gene variants, and Nicholas's has more than 16,000. Using algorithms to search among the variants, researchers identified eight likely genes. Then an article provided a pivotal clue: Nicholas might have a never-before-seen version of an already known disease.

The paper described one of the genes found in the exome screen: *XIAP*. It had been known to affect the immune system, not digestion. But it does both. The gene encodes a protein that normally keeps the immune system from attacking the intestines when it is fighting viruses. When the gene is mutant, fighting a viral infection kills intestinal cells, explaining Nicholas's leaking digestive tract. His version of the gene has a one-base difference from the healthy version. The gene is on the X chromosome, passed from carrier mothers to affected sons. The mutation was known to affect only one in a million boys, and no one had ever seen a boy with leaky intestines. The known *XIAP* disease is lethal by age 10, but is curable with a bone marrow or stem cell transplant that would replace the immune system. Would a transplant help Nicholas? He had nothing to lose.

Nic had a cord blood stem cell transplant from an anonymous donor in June 2010. Within 2 years Nic was back at school and able to eat a variety of foods. He still has medical issues, including the colostomy bag that replaced his large intestine, and seizures that developed following the transplant.

This case was the first to demonstrate the value of exome sequencing in diagnosing rare diseases. It has now been used in thousands of cases, and may someday become routine.

Questions for Discussion

1. Explain how the environment interacts with Nic's mutation to cause disease.

2. Find a case of a person with unusual symptoms who was diagnosed using exome sequencing.

3. Why did initial single-gene tests that Nicholas took not reveal the cause of his illness?

---

his ships came upon it. Boats have been lost in the Sargasso Sea, which includes the area known as the Bermuda Triangle. Researchers collected more than a billion DNA bases from the depths, representing about 1,800 microbial species and including more than a million previously unknown genes.

In parallel to metagenomics, several projects are exploring biodiversity with DNA tags to "bar-code" species, rather than

sequencing entire genomes. DNA sequences that vary reveal more about ancestries, because they are informational, than do comparisons of physical features, such as body shape or size, which formed the basis of traditional taxonomy (biological classification).

Perhaps the most interesting subject of metagenomics is the human body, which is home to trillions of bacteria. Section 2.5 discusses the human microbiome, the collection of life within us.

Genetics is more than a branch of life science, because it affects us intimately. Equal access to genetic tests and treatments, misuse of genetic information, and abuse of genetics to intentionally cause harm are compelling social issues that parallel scientific progress.

Genetics and genomics are spawning technologies that may vastly improve quality of life. But at first, tests and treatments are costly and not widely available. While some advantaged people in economically and politically stable nations may be able to afford genetic tests or exome or genome sequencing, or even take genetic tests for curiosity or recreation, poor people in other nations just try to survive, often lacking basic vaccines and medicines. In an African nation where two out of five children suffer from AIDS and many die from other infectious diseases, newborn screening for rare single-gene defects hardly seems practical. However, genetic disorders weaken people so that they become more susceptible to infectious diseases, which they can pass to others.

Recognizing that human genome information can ultimately benefit everyone, organizations such as the United Nations, World Health Organization, and the World Bank, are discussing how nations can share new diagnostic tests and therapeutics that arise from genome information about ourselves and the microbes that make us sick.

Individual nations are adopting guidelines for how to use genetic information to suit their particular strengths. India, for example, has many highly inbred populations with excellent genealogical records, and is home to one-fifth of the world's population. Studies of genetic variation in East Africa are especially important because this region is the cradle of humanity—home of our forebears. The human genome belongs to us all, but efforts from around the world will tell us what our differences are and how they arose. *Bioethics: Choices for the Future* discusses instances when genetic testing can be intrusive.

Key Concepts Questions 1.4

1. What is metagenomics?
2. How is discovery of bacterial genes and genomes pertinent to human health?

Bioethics: Choices for the Future

Genetic Testing and Privacy

The field of bioethics began in the 1950s to address issues raised by medical experimentation during World War II. Bioethics initially centered on informed consent, paternalism, autonomy, allocation of scarce medical resources, justice, and definitions of life and death. Today, the field covers medical and biotechnologies and the dilemmas they present. Genetic testing is a key issue in current bioethics because its informational nature affects privacy (**figure 1**). Consider these situations.

Testing Incoming Freshmen

Today students in genetics classes at some colleges or in medical school take genetic tests or even have their exomes sequenced. In summer 2010, incoming freshmen at the University of California, Berkeley, were among the first students to be asked to take genetic tests. In addition to class schedules and dorm assignments, they also received kits to send in DNA samples to test for three genes that control three supposedly harmless traits. Participation was voluntary, and because the intent was to gather data, informed consent was not required. However, after genetics groups, bioethicists, policy analysts, and consumer groups protested, the Department of Public Health ruled that the tests provided personal medical information, and should be conducted by licensed medical labs. Because this quintupled the cost, the university changed the program to collect aggregate data, rather than individually identified results.

The three genetic tests were to detect lactose intolerance, alcohol metabolism, and folic acid metabolism. The alcohol test detects variants of a gene that cause a facial flush, nausea, and heart palpitations after drinking, particularly in East Asians—who make up a significant part of the freshman class. Certain mutations in this gene raise the risk of developing esophageal cancer, and so test results may be useful, but they could also encourage drinking.

The Military

A new recruit hopes that the DNA sample given when military service begins is never used—it is stored to identify remains. Until recently, genetic tests have only been performed for two specific illnesses that could endanger soldiers under certain environmental conditions. Carriers of sickle cell disease (OMIM 603903) can develop painful blocked circulation at high altitudes, and carriers of G6PD deficiency (OMIM 305900) react badly to anti-malaria medication. Carriers wear red bands on their arms to alert officers to keep the soldiers from harmful situations. In the future, the military may use genetic information to identify soldiers at risk for such conditions as depression and post-traumatic stress disorder. Deployments can be tailored to personal risks, minimizing suffering.

Research Study Participants

When genetic studies considered only a few hundred sites in the human genome, peoples' identities were protected, because there were many more people than genotypes. A person was unlikely to be the only one to have a particular genotype. That is no longer true. Because studies now probe a million or more pieces of genetic information, an individual's genotype can be traced to a

(Continued)

(Continued)

particular group being investigated. That is, the more ways that we can detect that people vary, the easier it is to identify any one of them. It is like adding digits to a zip code or a new area code to a phone number to increase the pool of identifiers.

Clever consulting of information other than genotypes can identify individuals. If a child's DNA information is in a study and his or her name is in a database that includes a rare disease name and a hometown, comparing these sources can match a name to a DNA sequence. *Bioethics: Choices for the Future* in chapter 16 describes a clever combination of information from a Google search and from a genealogical database to identify participants in an experiment.

Questions for Discussion

1. If a genetic test reveals a mutation that could harm a blood relative, should the first person's privacy be sacrificed to inform the second person?

2. If governments require that everyone's exome sequence be maintained in a database, what privacy protections should be instituted?

3. Some student athletes have died of complications from being carriers of sickle cell disease. What are the risks and benefits of testing student athletes for sickle cell disease carrier status?

4. Do you think that passenger screening at airports should include quick DNA scans, as at least one company is offering?

5. Exome sequencing to identify a mutation that could cause a particular set of symptoms in a patient can reveal another genetic condition that has not yet produced symptoms or been detected in another way. Under what circumstances, if any, do you think patients should receive such "unsolicited findings"?

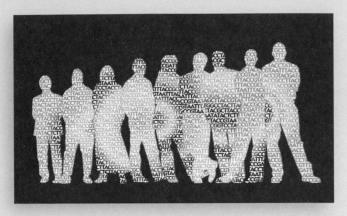

Figure 1 **Genetic privacy.** How can we protect DNA sequence data?

Summary

1.1 Introducing Genes and Genomes

1. **Genes** are the instructions to manufacture proteins, which determine inherited traits.

2. A **genome** is a complete set of genetic information. A **cell,** the unit of life, contains two genomes of **deoxyribonucleic acid (DNA)**. **Genomics** compares and analyzes the functions of many genes. **Bioethics** addresses controversies that arise from misuse of genetic information and other situations.

1.2 Levels of Genetics and Genomics

3. Genes encode proteins and the **ribonucleic acid (RNA)** molecules that synthesize proteins. RNA carries the gene sequence information so that it can be utilized, while the DNA is transmitted when the cell divides.

4. Much of the genome does not encode protein. The part that does is called the **exome**.

5. Variants of a gene, called **alleles,** arise by **mutation**. Alleles may differ slightly from one another, but encode the same product.

6. **Chromosomes** consist of DNA and protein. The 22 types of **autosomes** do not include genes that specify sex. The X and Y **sex chromosomes** bear genes that determine sex.

7. Cells undergo **differentiation** by expressing subsets of genes. **Stem cells** divide to yield other stem cells and cells that differentiate.

8. Single genes determine Mendelian traits. **Multifactorial traits** reflect the influence of one or more genes and the environment. Recurrence of a Mendelian trait is predicted based on Mendel's laws; predicting the recurrence of a multifactorial trait is more difficult.

9. **Genetic determinism** is the idea that the expression of an inherited trait cannot be changed.

10. The **phenotype** is the gene's expression. An allele combination constitutes the **genotype**. Alleles may be **dominant** (exerting an effect in a single copy) or **recessive** (requiring two copies for expression).

11. **Pedigrees** are diagrams used to study traits in families.

12. Genetic populations are defined by their collections of alleles, termed the **gene pool**. Genome comparisons among species reveal evolutionary relationships.

1.3 Applications of Genetics and Genomics

13. DNA profiling can provide information on identity, relationships, and history.

14. Most diseases arise from interactions between genes and environmental factors.

15. In health care, single-gene diseases are more predictable than other disease. Gene expression profiling is revealing how many types of diseases are related.

1.4 A Global Perspective on Genomes

16. In metagenomics, DNA collected from habitats, including the human body, is used to reconstruct ecosystems.

Review Questions

1. Place the following terms in size order, from largest to smallest, based on the structures or concepts they represent:
   a. chromosome
   b. gene pool
   c. gene
   d. DNA
   e. genome

2. Distinguish between
   a. an autosome and a sex chromosome.
   b. genotype and phenotype.
   c. DNA and RNA.
   d. recessive and dominant traits.
   e. pedigrees and karyotypes.
   f. gene and genome.
   g. exome and genome.

3. Explain how DNA encodes information.

4. Explain how all humans have the same genes, but vary genetically.

5. Explain how all cells in a person's body have the same genome, but are of hundreds of different types that look and function differently.

6. What is the assumption behind the comparison of DNA sequences to deduce that two individuals, or two species, are related?

7. Cite four ways that single-gene diseases differ from other types of illnesses.

Applied Questions

1. In 2012, Peter Kent was 62 years old and the director of a small biotech company. He was trying to decide whether to retire then, or after age 65, when his Social Security payment would be higher. He took a test offered by a direct-to-consumer genetic testing company that genotyped 260 single-gene conditions and susceptibilities. Based on the results, Peter concluded that he was healthy enough to wait until at least age 65 to retire. (Peter is still doing well but the Food and Drug Administration stopped the company from selling health tests as "information.") Do you agree or disagree with his decision? Cite a reason for your answer.

2. Describe how an environmental factor can influence an inherited disease, and how genes can influence an environmentally caused illness.

3. An artist travels around New York City collecting hairs, fingernails, discarded paper cups, gum, cigarette butts, and other trash, extracts the DNA, and uses a computer program that constructs a face from the genetic information. She then prints the results in three dimensions, producing a sculpture that she displays in a gallery along with the original DNA-bearing evidence. The algorithm considers 50 inherited traits that affect facial features, including information on ancestry.
   What would you do if you wandered into a gallery and found a sculpture of yourself or someone you know? How private is the DNA on a wad of discarded chewing gum?

Web Activities

1. Consult a website for a direct-to-consumer DNA testing company. Choose a test and explain why you would take it, and also discuss a genetic test that you would not take, and explain why not.

2. Describe a case of a person with unusual symptoms who was diagnosed using exome sequencing.

3. Research news coverage of a natural disaster or act of terrorism or war, or other form of violence, and describe the role of DNA profiling to identify victims.

4. The Gopher Kids Study collected DNA from children aged 1 to 11 at the Minnesota State Fair in 2010, in a saliva sample. The goal is to identify genes that contribute to normal health and development as the children grow up. The website is www.peds.umn.edu/gopherkids/. If you were a parent of a child at the fair, what questions would you have asked before donating his or her DNA?

5. Describe a program that is sequencing the exomes or genomes of a large number of people. What does the program hope to accomplish?

Forensics Focus

1. DNA samples are taken from inside the cheek of people arrested on suspicion of having committed serious crimes. Only the numbers of 13 repeated sequences used to identify individuals (see chapter 14) are stored—the DNA itself is discarded and no other sequences recorded. The repeats are not part of the exome. Explain why sampling this part of the genome does not indicate anything about the person's health, ancestry, or traits.

2. Researchers in Canada published a report on 468 legal cases that cited "genetic predisposition" to explain physical injury or criminal behavior due to mental illness or substance abuse. (In 86 cases the genetic information didn't harm the plaintiff, in 134 cases it did, and in 248 cases the legal significance of the genetic findings was unclear.) Under what circumstances, if any, do you think genetic information might be useful in a legal case, and from whose point of view?

3. Consult the websites for a television program that uses, or is based on, forensics (*CSI* or *Law and Order,* for example), and find an episode in which species other than humans are critical to the case. Explain how DNA analysis could help to solve the crime.

4. On an episode of the television program *House* from a few years ago, the main character, Dr. House, knew from age 12 that his biological father was a family friend, not the man who raised him. At his supposed father's funeral, the good doctor knelt over the body in the casket and sneakily snipped a bit of skin from the corpse's earlobe—for a DNA test.

   a. Do you think that this action was an invasion of anyone's privacy? Was Dr. House justified?

   b. Dr. House often ordered treatments for patients based on observing symptoms. Suggest a way that he could have used DNA testing to refine his diagnoses.

DNA from a saved baby tooth helped diagnose a little girl years after her death. Her parents learned that she had had Rett syndrome.

Cells

Learning Outcomes

2.1 Introducing Cells

1. Explain why it is important to know the cellular basis of a disease.

2. Define *differentiated cell.*

2.2 Cell Components

3. List the four major chemicals in cells.

4. Describe how organelles interact.

5. Describe the structure and function of a biological membrane.

6. List the components of the cytoskeleton.

2.3 Cell Division and Death

7. Distinguish between mitosis and apoptosis.

8. Describe the events and control of the cell cycle.

2.4 Stem Cells

9. List the characteristics of a stem cell.

10. Define stem and progenitor cell.

2.5 The Human Microbiome

11. How can the bacteria that live in and on our bodies affect our health?

The **BIG** Picture

Our bodies are built of trillions of cells that interact in complex ways to keep us alive. All cells in a body use the same genome, but have different structures and functions because they access different parts of the genome.

Diagnosis From a Tooth

The genome is like a database containing clues to health. One mother in Australia provided her daughter's saved baby tooth, from which researchers extracted DNA and diagnosed the disease behind the child's symptoms, years after she had died.

The little girl had seemed healthy until 16 months, when she still wasn't walking, and she began to lose some of the words she'd used frequently. A month later she fell down stairs, hitting her head. By age 2, she could no longer manipulate small objects, and developed odd repetitive hand-wringing movements. By age 3 she began to have frequent seizures. She became increasingly disabled, and died in 1991, undiagnosed.

In 2004, the mother read an article about Rett syndrome (OMIM 312750), and recognized her daughter's symptoms. The mother had kept her daughter's baby teeth. Could researchers test DNA from a tooth for the mutation in the *MECP2* gene that was behind the syndrome? She read up on teeth. Every tooth contains dental pulp, which is like a soup of cells, including connective tissue cells (fibroblasts), tooth precursors called odontoblasts, immune system cells, and stem cells. Within any one of them, lay a copy of her deceased daughter's genome. So she contacted the Australian Rett Syndrome project, which helped her connect with researchers who diagnosed the condition.

Diagnosis after death proved useful in a few ways. It dispelled guilt—the father had blamed himself when his daughter fell down the stairs, and the mother had blamed a vaccination. Testing of the mother revealed that she didn't have Rett syndrome—her daughter was a new mutation, and therefore other relatives needn't be tested. Both parents were reassured, once they learned that the mutation is on the X chromosome and dominant, that their son did not have the condition, nor could he pass it to his children.

2.1 Introducing Cells

Our inherited traits, quirks, and illnesses arise from the activities of cells. Understanding cell function reveals how a healthy body works, and how it develops from one cell to trillions. Understanding what goes wrong in certain cells to cause pain or other symptoms can suggest ways to treat the condition, because we learn what to repair or replace. For example, genes tell cells how to make the proteins that align to form the contractile apparatus of muscles. In Duchenne muscular dystrophy (OMIM 310200), one type of muscle protein is missing, and as a result muscle cells collapse under forceful contraction. Certain muscles become very weak in early childhood, and gradually the person becomes paralyzed. This form of muscular dystrophy is lethal by early adulthood. The little boy in **figure 2.1** shows the characteristic overdeveloped calf muscles from being unable to stand normally and having to use these muscles to do so.

Our bodies include more than 290 specialized, or differentiated, cell types that aggregate and interact to form the four basic tissue types: epithelial, connective, muscle, and nerve (**figure 2.2**). Most cells are **somatic cells,** also called body cells. Somatic cells have two copies of the genome and are said to be **diploid**. In contrast, the germ cells, which are sperm and egg cells, have one copy of the genome and are **haploid**. The meeting of sperm and egg restores the diploid state. **Stem cells** are diploid cells that divide to give rise to differentiated cells and replicate themselves in a process called self-renewal. Stem cells enable a body to develop, grow, and repair damage.

Cells interact. They send, receive, and respond to information. Some cells aggregate with others of like function, forming tissues, which in turn interact to form organs and organ systems. Other cells move about the body. Cell numbers are important, too—they are critical to development, growth, and healing. Staying healthy reflects a precise balance between cell division, which adds cells, and cell death, which takes them away.

Key Concepts Questions 2.1

1. Why is it important to understand the functions of cells?

2. How do diploid cells differ from haploid cells?

3. Distinguish between a somatic cell and a germ cell.

2.2 Cell Components

All cells share features that enable them to perform the basic life functions of reproduction, growth, response to stimuli, and energy use. Specialized characteristics emerge as cells express different subsets of the thousands of protein-encoding genes.

All multicellular organisms, including other animals, fungi, and plants, have differentiated cells. Some single-celled organisms, such as the familiar paramecium and amoeba, have cells as complex as our own. The most abundant organisms on the planet, however, are single-celled, such as bacteria. These microorganisms have occupied Earth much longer than we have, and so in an evolutionary sense have been more successful than us.

Biologists recognize three basic types of cells that define three major "domains" of life: the Archaea, the Bacteria, and the Eukarya. A domain is a broader classification than the familiar kingdom.

Members of the Archaea and Bacteria are single-celled, but they differ from each other in the sequences of many of their genes and in the types of molecules in their membranes. Archaea and Bacteria are prokaryotes. A **prokaryotic cell** does not have a **nucleus,** the structure that contains DNA in the cells of other types of organisms, which comprise the third domain of life, the Eukarya. Also known as eukaryotes, this group includes single-celled organisms that have nuclei, as well as all multicellular organisms (**figure 2.3**). A **eukaryotic cell** is also distinguished from a prokaryotic cell by structures called **organelles,** which perform specific functions. The cells of all three domains contain globular assemblies of RNA and protein called **ribosomes** that are essential for protein synthesis. The eukaryotes may have arisen from an ancient fusion of a bacterium with an archaean.

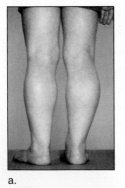
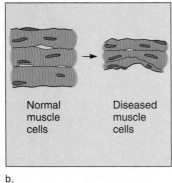

Normal muscle cells

Diseased muscle cells

a.

b.

Figure 2.1 **Genetic disease at the whole-person and cellular levels.** **(a)** An early sign of Duchenne muscular dystrophy is overdeveloped calf muscles that result from inability to rise from a sitting position the usual way. **(b)** Lack of the protein dystrophin causes skeletal muscle cells to collapse when they contract.

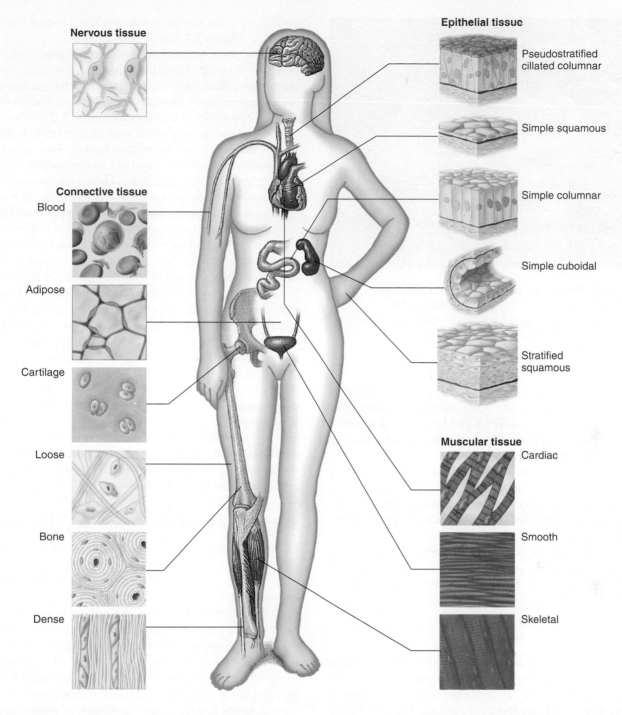

Figure 2.2 Tissue types. Cells interact to form four major types of tissues: nervous, epithelial, connective, and muscular.

Chemical Constituents

Cells are composed of molecules. Some of the chemicals of life (biochemicals) are so large that they are called macromolecules.

The major macromolecules that make up cells and that cells use as fuel are **carbohydrates** (sugars and starches), **lipids** (fats and oils), **proteins,** and **nucleic acids** (DNA and RNA). Cells require vitamins and minerals in much smaller amounts.

Carbohydrates provide energy and contribute to cell structure. Lipids form the basis of some hormones, form membranes, provide insulation, and store energy. Proteins have many diverse functions in the human body. They enable blood to clot, form the contractile fibers of muscle cells, and form the bulk of the body's connective tissue. Antibodies that fight infection are proteins. **Enzymes** are especially important proteins because

Figure 2.3 Eukaryotic and prokaryotic cells. The large purplish structure is a macrophage, part of the human immune system. Here it engulfs much smaller cells stained pink, the bacteria that cause tuberculosis.

they facilitate, or catalyze, biochemical reactions so that they occur fast enough to sustain life. Most important to the study of genetics are the nucleic acids DNA and RNA, which translate information from past generations into specific collections of proteins that give a cell its characteristics.

Macromolecules can combine in cells, forming larger structures. For example, the membranes that surround cells and compartmentalize their interiors consist of double layers (bilayers) of lipids embedded with carbohydrates, proteins, and other lipids.

Life is based on the chemical principles that govern all matter; genetics is based on a highly organized subset of the chemical reactions of life. **Clinical Connection 2.1** describes diseases that affect these major types of biological molecules.

Organelles

A typical eukaryotic cell holds a thousand times the volume of a bacterial or archaeal cell. To carry out the activities of life in such a large cell, organelles divide the labor by partitioning off certain areas or serving specific functions. The coordinated functioning of the organelles in a eukaryotic cell is much like the organization of departments in a big-box store, compared to the prokaryote-like simplicity of a small grocery store. In general, organelles keep related biochemicals and structures close enough to one another to interact efficiently. This eliminates the need to maintain a high concentration of a particular biochemical throughout the cell.

Organelles have a variety of functions. They enable a cell to retain as well as to use its genetic instructions, acquire energy, secrete substances, and dismantle debris. Saclike organelles sequester biochemicals that might harm other cellular constituents. Some organelles consist of membranes studded with enzymes embedded in the order in which they

participate in the chemical reactions that produce a particular molecule. **Figure 2.4** depicts organelles.

The most prominent organelle of most cells is the nucleus. A layer called the nuclear envelope surrounds the nucleus. Biochemicals can exit or enter the nucleus through nuclear pores, which are rings of proteins around an opening, like lined portholes in a ship's side (**figure 2.5**).

On the inner face of the nuclear membrane is a layer of fibrous material called the nuclear lamina. This layer has several important functions. The DNA in the nucleus touches the nuclear lamina as the cell divides. The nuclear lamina also provides mechanical support and holds in place the nuclear pores. The opener to chapter 3 discusses very rare, accelerated aging disorders that result from an abnormal nuclear lamina.

Inside the nucleus is an area that appears darkened under a microscope, called the **nucleolus** ("little nucleus"). Here, ribosomes are produced. The nucleus is filled with DNA complexed with many proteins to form chromosomes. Other proteins form fibers that fill out the nucleus, giving it a roughly spherical shape. RNA is abundant too, as are enzymes and proteins required to synthesize RNA from DNA. The fluid in the nucleus, minus these contents, is called nucleoplasm.

The remainder of the cell—that is, everything but the nucleus, organelles, and the outer boundary, or **plasma membrane**—is **cytoplasm**. Other cellular components include stored proteins, carbohydrates, and lipids; pigment molecules; and various other small chemicals. (The cytoplasm is called cytosol when these other parts are removed.) We now take a closer look at three cellular functions: secretion, digestion inside cells, and energy production.

Secretion—The Eukaryotic Production Line

Organelles interact in ways that coordinate basic life functions and provide the characteristics of specialized cell types. Secretion, which is the release of a substance from a cell, illustrates one way that organelles function together.

Secretion begins when the body sends a biochemical message to a cell to begin producing a particular substance. For example, when a newborn suckles the mother's breast, her brain releases hormones that signal cells in her breast to rapidly increase the production of the complex mixture that makes up milk (**figure 2.6**). In response, information in certain genes is copied into molecules of **messenger RNA** (**mRNA**), which then exit the nucleus (see steps 1 and 2 in figure 2.6). In the cytoplasm, the mRNAs, with the help of ribosomes and another type of RNA called **transfer RNA** (**tRNA**), direct the manufacture of milk proteins. These include nutritive proteins called caseins, antibodies that protect against infection, and enzymes.

Most protein synthesis occurs on a maze of interconnected membranous tubules and sacs called the **endoplasmic reticulum** (**ER**) (see step 3 in figure 2.6). The ER winds from the nuclear envelope outward to the plasma membrane. The ER nearest the nucleus, which is flattened and studded with ribosomes, is called rough ER, because the ribosomes make it appear fuzzy when viewed under an electron microscope.

Inborn Errors of Metabolism Affect the Major Biomolecules

Enzymes, by speeding specific chemical reactions, control a cell's production of all types of macromolecules. When an enzyme is not produced or cannot function, too much or too little of the product of the biochemical reaction may be made. These biochemical buildups and breakdowns may cause symptoms. Genetic disorders that result from deficient or absent enzymes are called "inborn errors of metabolism." Following are some examples.

Carbohydrates

The newborn yelled and pulled up her chubby legs in pain a few hours after each feeding, and had watery diarrhea. She had *lactase deficiency* (OMIM 223000). This is lack of the enzyme lactase, which enables the digestive system to break down the carbohydrate lactose. Bacteria multiplied in the undigested lactose in the child's intestines, producing gas, cramps, and bloating. Switching to a soybean-based, lactose-free infant formula helped. A different condition with milder symptoms is lactose intolerance (OMIM 150200), common in adults (see the opening essay to chapter 15).

Lipids

A sudden sharp pain began in the man's arm and spread to his chest. At age 36, he was younger than most people who suffer heart attacks, but he had inherited a mutation that halved the number of protein receptors for cholesterol on his liver cells. Because cholesterol could not enter liver cells efficiently, it built up in his arteries, constricting blood flow in his heart and causing the mild heart attack. He inherited *familial hypercholesterolemia* (OMIM 143890). Taking a cholesterol-lowering drug and exercising will lower his risk of suffering future heart attacks.

Proteins

Newborn Tim slept most of the time, and he vomited so often that he hardly grew. A blood test a day after his birth revealed *maple syrup urine disease* (OMIM 248600), which makes urine smell like maple syrup. Tim could not digest three types of amino acids (protein building blocks), which accumulated in his bloodstream. A diet very low in these amino acids controlled the symptoms.

Nucleic Acids

From birth, Troy's wet diapers held orange, sandlike particles, but otherwise he seemed healthy. By 6 months of age, urinating had become painful. A physician noted that Troy's writhing movements were involuntary rather than normal crawling. The orange particles

in Troy's diaper indicated *Lesch-Nyhan syndrome* (OMIM 300322), caused by deficiency of an enzyme called HGPRT. Troy's body could not recycle two of the four types of DNA building blocks (purines), instead converting them into uric acid, which crystallizes in urine. Later he developed severe intellectual disability, seizures, and aggressive and self-destructive behavior. By age 3, he uncontrollably bit his fingers and lips. His teeth were removed to keep him from harming himself, and he was kept in restraints. Troy would die before the age of 30 of kidney failure or infection.

Vitamins

Vitamins enable the body to use dietary carbohydrates, lipids, and proteins. Kaneesha inherited *biotinidase deficiency* (OMIM 253260), which greatly slows her body's use of the vitamin biotin. If Kaneesha hadn't been diagnosed as a newborn and quickly started on biotin supplements, she would have developed intellectual disability, seizures, skin rash, and loss of hearing, vision, and hair. Her slow growth, caused by her body's inability to extract energy from nutrients, would have eventually proved lethal.

Minerals

Ingrid, in her thirties, lived in an institution, unable to talk or walk. She grinned and drooled, but was alert and communicated using a computer. When she was a healthy high-school senior, symptoms of *Wilson disease* (OMIM 277900) began as her liver could no longer control the excess copper her digestive tract absorbed from food. The initial symptoms were stomachaches, headaches, and an inflamed liver (hepatitis). Then other changes began—slurred speech; loss of balance; a gravelly, low-pitched voice; and altered handwriting. A psychiatrist noted the telltale greenish rings around her irises, caused by copper buildup, and diagnosed Wilson disease (**figure 1**). A drug called penicillamine enabled Ingrid to excrete excess copper in her urine. The treatment halted the course of the illness, saving her life, but she did not improve.

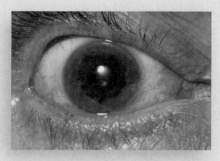

Figure 1 Wilson disease. A greenish ring around the brownish iris is one sign of the copper buildup of Wilson disease.

Questions for Discussion

1. Explain how inherited diseases can affect molecules in the body other than proteins.

2. Explain how absence of or abnormal forms of the same type of biological molecule—an enzyme—can cause very different diseases.

3. Describe how treatments for lactase deficiency, familial hypercholesterolemia, maple syrup urine disease, biotinidase deficiency, and Wilson disease help to alleviate symptoms.

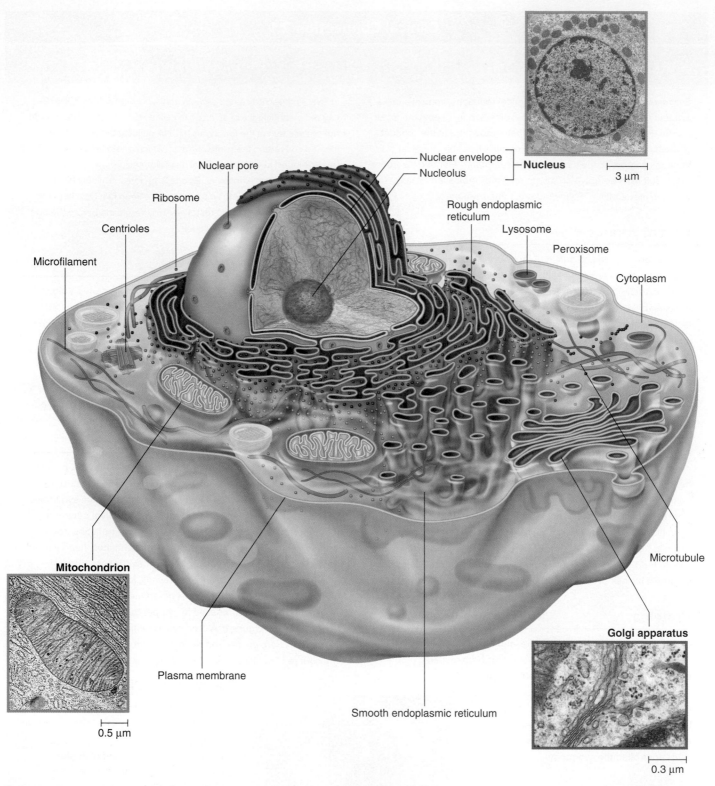

Figure 2.4 **Generalized animal cell.** Organelles provide specialized functions for the cell. Most of these structures are transparent; colors are used here to distinguish them. Different cell types have different numbers of organelles. All cell types have a single nucleus, except for red blood cells, which expel their nuclei as they mature.

Protein synthesis begins on the rough ER when messenger RNA attaches to the ribosomes. Amino acids from the cytoplasm are then linked, following the instructions in the mRNA's sequence, to form particular proteins that will either exit the cell or become part of membranes (step 3, figure 2.6). Sugars are added to certain proteins in the ER. Proteins are also synthesized on ribosomes not associated with the ER. These proteins remain in the cytoplasm.

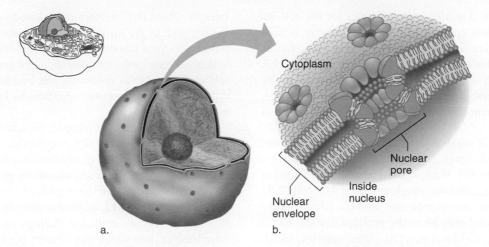

Figure 2.5 **The nucleus is the genetic headquarters. (a)** The largest structure in a typical human cell, the nucleus lies within two membrane layers that make up the nuclear envelope. **(b)** Nuclear pores allow specific molecules to move in and out of the nucleus through the envelope.

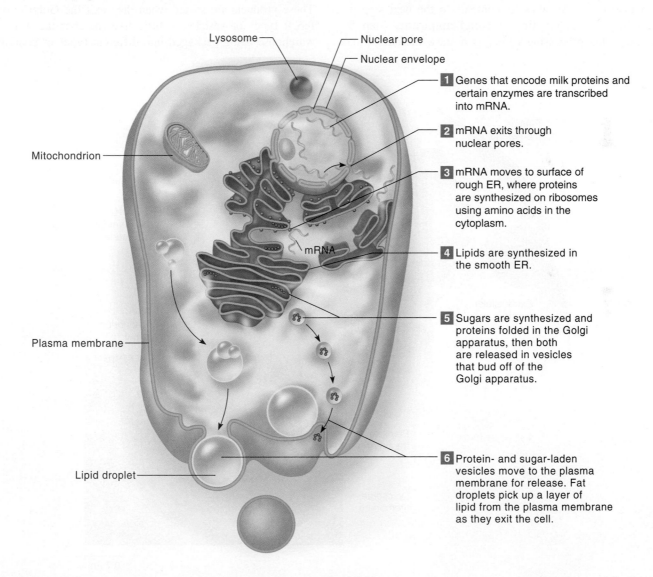

Lysosome — Nuclear pore

— Nuclear envelope

1 Genes that encode milk proteins and certain enzymes are transcribed into mRNA.

2 mRNA exits through nuclear pores.

3 mRNA moves to surface of rough ER, where proteins are synthesized on ribosomes using amino acids in the cytoplasm.

4 Lipids are synthesized in the smooth ER.

5 Sugars are synthesized and proteins folded in the Golgi apparatus, then both are released in vesicles that bud off of the Golgi apparatus.

6 Protein- and sugar-laden vesicles move to the plasma membrane for release. Fat droplets pick up a layer of lipid from the plasma membrane as they exit the cell.

Mitochondrion

mRNA

Plasma membrane

Lipid droplet

Figure 2.6 **Secretion: making milk.** Milk production and secretion illustrate organelle functions and interactions in a cell from a mammary gland: (1) through (6) indicate the order in which organelles participate in this process. Lipids are secreted in separate droplets from proteins and their attached sugars. This cell is highly simplified.

The ER acts as a quality control center for the cell. Its chemical environment enables the forming protein to start folding into the three-dimensional shape necessary for its specific function. Misfolded proteins are pulled out of the ER and degraded, much as an obviously defective toy might be pulled from an assembly line at a toy factory and discarded. Misfolded proteins can cause disease, as discussed further in chapter 10.

As the rough ER winds out toward the plasma membrane, the ribosomes become fewer, and the tubules widen, forming a section called smooth ER. Here, lipids are made and added to the proteins arriving from the rough ER (step 4, figure 2.6). The lipids and proteins are transported until the tubules of the smooth ER narrow and end. Then the proteins exit the ER in membrane-bounded, saclike organelles called **vesicles,** which pinch off from the tubular endings of the membrane. Lipids are exported without a vesicle, because a vesicle is itself made of lipid.

A loaded vesicle takes its contents to the next stop in the secretory production line, a **Golgi apparatus** (step 5, figure 2.6). This processing center is a stack of four to six interconnected flat, membrane-enclosed sacs. Here, additional sugars, such as the milk sugar lactose, are made that attach to proteins to form glycoproteins or to lipids to form glycolipids, which become parts of plasma membranes. Proteins finish folding in the Golgi apparatus and become active. Some cell types have just a few Golgi apparatuses, but those that secrete may have hundreds.

The components of complex secretions, such as milk, are temporarily stored in the Golgi apparatus. Droplets of proteins and sugars then bud off in vesicles that move outward to the plasma membrane, fleetingly becoming part of it until they are secreted to the cell's exterior. Lipids exit the plasma membrane directly, taking bits of it with them (step 6, figure 2.6).

In the breast, epithelial (lining) cells called lactocytes form tubules, into which they secrete the components of milk. When the baby suckles, contractile cells squeeze the milk through the tubules and out of holes in the nipples.

A Golgi apparatus processes many types of proteins. These products are sorted when they exit the Golgi into different types of vesicles, a little like merchandise leaving a warehouse being packaged into different types of containers,

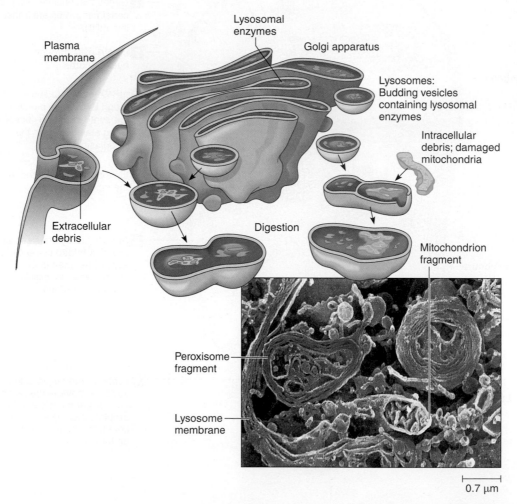

Figure 2.7 Lysosomes are trash centers. Lysosomes fuse with vesicles or damaged organelles, activating the enzymes within to recycle the molecules. Lysosomal enzymes also dismantle bacterial remnants. These enzymes require a very acidic environment to function.

depending upon the destination. Molecular tags target vesicles to specific locations in the cell, or indicate that the contents are to be released to outside the cell.

Intracellular Digestion—Lysosomes and Peroxisomes

Just as clutter and garbage accumulate in an apartment, debris builds up in cells. Organelles called **lysosomes** ("bodies that cut") handle the garbage. Lysosomes are membrane-bounded sacs that contain enzymes that dismantle bacterial remnants, worn-out organelles, and other material such as excess cholesterol (**figure 2.7**). The enzymes also break down some digested nutrients into forms that the cell can use.

Lysosomes fuse with vesicles carrying debris from outside or within the cell, and the lysosomal enzymes then degrade the contents. This process of the cell's disposing of its own trash is called **autophagy,** which means "eating self." For example, a type of vesicle that forms from the plasma membrane, called an endosome, ferries extra low-density lipoprotein (LDL) cholesterol to lysosomes. A loaded lysosome moves toward the plasma membrane and fuses with it, releasing its contents to the outside. Lysosomes maintain the very acidic environment that their enzymes require to function, without harming other cell parts that acids could destroy.

Cells differ in the number of lysosomes they contain. Certain types of blood cells, and macrophages that move about and engulf bacteria, are loaded with lysosomes. Liver cells require many lysosomes to break down cholesterol, toxins, and drugs.

All lysosomes contain 43 types of digestive enzymes, which must be maintained in a correct balance. Absence or malfunction of an enzyme causes a "lysosomal storage disease." In these inherited disorders, the molecule that the missing or abnormal enzyme normally degrades accumulates. The lysosomes swell, crowding other organelles and interfering with the cell's functions. In Tay-Sachs disease (OMIM 272800), for example, an enzyme is deficient that normally breaks down lipids in the cells that surround nerve cells. As the nervous system becomes buried in lipid, the infant begins to lose skills, such as sight, hearing, and the ability to move. Death is typically within 3 years. Even before birth, the lysosomes of affected cells swell. Lysosomal storage diseases are rare, but affect about 10,000 people worldwide.

Peroxisomes are sacs with outer membranes that are studded with several types of enzymes. These enzymes perform a variety of functions. They break down certain lipids and rare biochemicals, synthesize bile acids used in fat digestion, and detoxify compounds that result from exposure to oxygen free radicals. Peroxisomes are large and abundant in liver and kidney cells, which handle toxins.

The 1992 film *Lorenzo's Oil* recounted the true story of a child with an inborn error of metabolism caused by an absent peroxisomal enzyme. Lorenzo had adrenoleukodystrophy (ALD; OMIM 202370), in which a type of lipid called a very-long-chain fatty acid builds up in the brain and spinal cord. Early symptoms include low blood sugar, skin darkening, muscle weakness, altered behavior, and irregular heartbeat. The patient eventually loses control over the limbs and usually dies within a few years. Lorenzo's parents devised a combination of edible oils that diverts another enzyme to compensate for the missing one. Lorenzo lived 30 years, which may have been due to the oil, or to the excellent supportive care that he received. Providing a functional copy of the gene, called gene therapy, can halt progression of the disease. Gene therapy for ALD is being tested on newborns.

Energy Production—Mitochondria

Cells require continual energy to power the chemical reactions of life. Organelles called **mitochondria** provide energy by breaking the chemical bonds that hold together the nutrient molecules in food.

A mitochondrion has an outer membrane similar to those in the ER and Golgi apparatus and an inner membrane that forms folds called cristae (**figure 2.8**). These folds hold enzymes that catalyze the biochemical reactions that release energy from nutrient molecules. The freed energy is captured and stored in the bonds that hold together a molecule called adenosine triphosphate (ATP). In this way, ATP functions a little like an energy debit card.

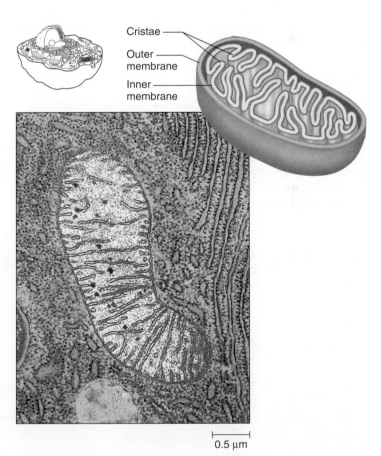

Cristae

Outer membrane

Inner membrane

0.5 µm

Figure 2.8 A mitochondrion extracts energy. Cristae, infoldings of the inner membrane, increase the available surface area containing enzymes for energy reactions in a mitochondrion.

The number of mitochondria in a cell varies from a few hundred to tens of thousands, depending upon the cell's activity level. A typical liver cell, for example, has about 1,700 mitochondria, but a muscle cell, with its very high energy requirements, may have up to 10,000. A major symptom of diseases that affect mitochondria is fatigue. Mitochondria contain a small amount of DNA different from the DNA in the nucleus (see figure 5.10). Chapter 5 discusses mitochondrial inheritance and disease, and chapter 16 describes how mitochondrial genes provide insights into early human migrations.

Table 2.1 summarizes the structures and functions of organelles.

Biological Membranes

Just as the character of a community is molded by the people who enter and leave it, the special characteristics of different cell types are shaped in part by the substances that enter and leave. The plasma membrane completely surrounds the cell and monitors the movements of molecules in and out. How the chemicals that comprise the plasma membrane associate with each other determines which substances can enter or leave the cell. Membranes similar to the plasma membrane form the outer boundaries of several organelles, such as the mitochondrion. Some organelles consist entirely of membranes, such as the endoplasmic reticulum and Golgi apparatus. A cell's membranes are more than mere coverings, because some of their molecules have specific functions.

Membrane Structure

A biological membrane has a distinctive structure. It is a double layer (bilayer) of molecules called phospholipids. A phospholipid is a fat molecule with attached phosphate groups. A phosphate group [PO_4] is a phosphorus atom bonded to four oxygen atoms. A phospholipid is depicted as a head with two parallel tails.

Membranes form because phospholipid molecules self-assemble into sheets (**figure 2.9**). The molecules do this because their ends react oppositely to water: The phosphate end of a phospholipid is attracted to water, and thus is hydrophilic ("water-loving"); the other end, which consists of two chains of fatty acids, moves away from water, and is therefore hydrophobic ("water-fearing"). Because of these forces, phospholipid molecules in water spontaneously form bilayers. Their hydrophilic surfaces are exposed to the watery exterior and interior of the cell, and their hydrophobic surfaces face each other on the inside of the bilayer, away from water.

A phospholipid bilayer forms the structural backbone of a biological membrane. Proteins are embedded in the bilayer. Some proteins traverse the entire structure, while others extend from one or both faces (**figure 2.10**).

The phospholipid bilayer is oily, and some proteins move within it like ships on a sea. Proteins with related functions may cluster on "lipid rafts" that float on the phospholipid bilayer, easing their interaction. The rafts are rich in cholesterol and other types of lipids. Proteins aboard lipid rafts contribute to the cell's identity, transport molecules into the cell, and keep out certain toxins and pathogens. HIV, for example, enters a cell by breaking a lipid raft.

The inner hydrophobic region of the phospholipid bilayer blocks entry and exit to most substances that dissolve in water. However, certain molecules can cross the membrane through proteins that form passageways, or when a "carrier" protein escorts the molecules. Some membrane proteins form channels for ions, which are atoms or molecules with electrical charge. **Clinical Connection 2.2** describes "channelopathies"—diseases that stem from faulty ion channels.

The Plasma Membrane Enables Cell-to-Cell Communication

The proteins, glycoproteins, and glycolipids that extend from a plasma membrane create surface topographies that are important in a cell's interactions with other cells. The surfaces of

Table 2.1	Structures and Functions of Organelles	
Organelle	**Structure**	**Function**
Endoplasmic reticulum	Membrane network; rough ER has ribosomes, smooth ER does not	Site of protein synthesis and folding; lipid synthesis
Golgi apparatus	Stacks of membrane-enclosed sacs	Site where sugars are made and linked into starches or joined to lipids or proteins; proteins finish folding; secretions stored
Lysosome	Sac containing digestive enzymes	Degrades debris; recycles cell contents
Mitochondrion	Two membranes; inner membrane enzyme-studded	Releases energy from nutrients, participates in cell death
Nucleus	Porous sac containing DNA	Separates DNA within cell
Peroxisome	Sac containing enzymes	Breaks down and detoxifies various molecules
Ribosome	Two associated globular subunits of RNA and protein	Scaffold and catalyst for protein synthesis
Vesicle	Membrane-bounded sac	Temporarily stores or transports substances

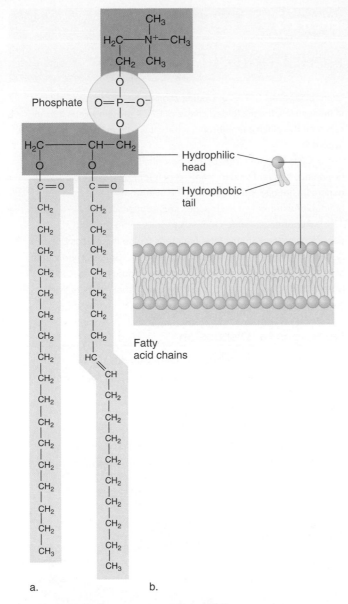

Phosphate

Hydrophilic head

Hydrophobic tail

C=O

Fatty acid chains

a.

b.

Figure 2.9 **The two faces of membrane phospholipids.**
(a) A phospholipid has one end attracted to water and
the other repelled by it. (b) A membrane phospholipid is
depicted as a circle with two tails.

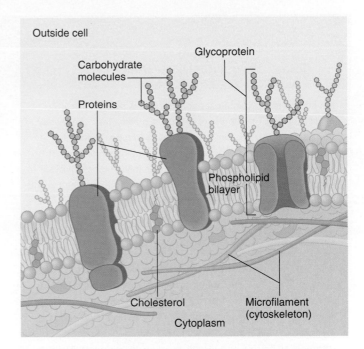

Outside cell

Carbohydrate molecules

Glycoprotein

Proteins

Phospholipid bilayer

Cholesterol

Microfilament (cytoskeleton)

Cytoplasm

Figure 2.10 **Anatomy of a plasma membrane.** Mobile
proteins are embedded throughout a phospholipid bilayer.
Carbohydrates jut from the membrane's outer face.

your cells indicate not only that they are part of your body, but
also that they are part of a particular organ and a particular tissue type.

Many molecules that extend from the plasma membrane
are receptors, which are structures that have indentations or
other shapes that fit and hold molecules outside the cell. The
molecule that binds to the receptor, called the ligand, may set
into motion a cascade of chemical reactions inside the cell that
carries out a particular activity, such as dividing.

In a cellular communication process called **signal
transduction,** a series of molecules that are part of the plasma
membrane form pathways that detect signals from outside
the cell and transmit them inward, where yet other molecules
orchestrate the cell's response. In a process called **cellular**

adhesion, the plasma membrane helps cells attach to certain
other cells. These cell-to-cell connections are important in
forming tissues.

Faulty signal transduction or cellular adhesion can
harm health. Cancer is due to cells' failure to recognize or
react to signals to cease dividing. Cancer cells also have
abnormal cellular adhesion, which enables them to invade
healthy tissue.

The Cytoskeleton

The **cytoskeleton** is a meshwork of protein rods and tubules
that serves as the cell's architecture, positioning organelles and
providing overall three-dimensional shapes. The proteins of
the cytoskeleton are continually broken down and built up as
a cell performs specific activities. Some cytoskeletal elements
function as rails that transport cellular contents; other parts,
called motor molecules, power the movement of organelles
along these rails as they convert chemical energy into mechanical energy.

The cytoskeleton includes three major types of elements—
microtubules, microfilaments, and **intermediate filaments**
(**figure 2.11**). They are distinguished by protein type, diameter, and how they aggregate into larger structures. Other
proteins connect these components, creating a framework
that provides the cell's strength and ability to resist force and
maintain shape.

Long, hollow microtubules provide many cellular movements. A microtubule is composed of pairs (dimers) of a protein,
called tubulin, assembled into a hollow tube. Adding or removing
tubulin molecules changes the length of the microtubule.

Faulty Ion Channels Cause Inherited Disease

Ion channels are specific for calcium (Ca^{2+}), sodium (Na^+), potassium (K^+), or chloride (Cl^-) ions. A plasma membrane may have a few thousand ion channels of each type. To function normally, a cell must allow certain types of ions in and out at specific rates through ion channels. Ten million ions can pass through an ion channel in one second! Abnormal ion channels lie behind the following "channelopathies."

Unusual Pain and Sodium Channels

A 10-year-old boy who lived in northern Pakistan could not feel pain. He became a performer, stabbing knives through his arms and walking on hot coals to entertain crowds. Several other people in this community where relatives often married relatives were also unable to feel pain. They share a mutation that alters sodium channels on certain nerve cells, blocking pain signals. The boy died at age 13 from jumping off a roof.

A different mutation affecting the same type of sodium channel causes very different symptoms. In "burning man syndrome," the channels become extra sensitive, opening and flooding the body with pain easily, in response to exercise, an increase in room temperature, or just putting on socks. In another condition, "paroxysmal extreme pain disorder," sodium channels stay open too long, causing excruciating pain in the rectum, jaw, and eyes. Researchers are using the information from studies of these genetic disorders to develop new painkillers.

Long-QT Syndrome and Potassium Channels

Four children in a Norwegian family were born deaf, and three of them died at ages 4, 5, and 9. All of the children had inherited from unaffected carrier parents "long-QT syndrome associated with deafness" (OMIM 176261). ("QT" refers to part of a normal heart rhythm.) These children had abnormal potassium channels in the cells of the heart muscle and in the inner ear. In the heart cells, the malfunctioning ion channels disrupted electrical activity, fatally disturbing heart rhythm. In the cells of the inner ear, the abnormal ion channels increased the extracellular concentration of potassium ions, impairing hearing.

Cystic Fibrosis and Chloride Channels

A seventeenth-century English saying, "A child that is salty to taste will die shortly after birth," described the consequence of abnormal chloride channels in cystic fibrosis (CF). The chloride channel is called CFTR, for cystic fibrosis transductance regulator. In most cases, the misfolded shape of CFTR protein keeps it from reaching the plasma membrane, where it would normally function (**figure 1**).

CF is inherited from carrier parents. It causes difficulty breathing, frequent severe respiratory infections, and a clogged pancreas that disrupts digestion. The symptoms result from buildup of extremely thick mucous secretions.

Abnormal chloride channels in cells lining the lung passageways and ducts of the pancreas cause the mucus build-up. Sodium channels are abnormal too. The result: Salt trapped inside cells draws moisture in and thickens surrounding mucus. Clinical Connection 4.2 discusses a new drug for CF that corrects the protein misfolding.

Questions for Discussion

1. What part of the cell does a channelopathy affect?
2. Explain how abnormal sodium channels can cause either lack of pain or too-intense pain.
3. How do abnormal potassium channels cause such diverse symptoms as abnormal heart rhythm and deafness?
4. What is one explanation for why chloride channels malfunction in cystic fibrosis?

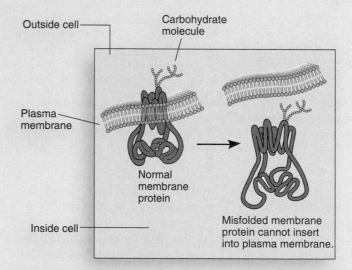

Figure 1 **In some cases of cystic fibrosis, CFTR protein remains in the cytoplasm, rather than anchoring in the plasma membrane.** The protein misfolds, which prevents normal chloride channel function.

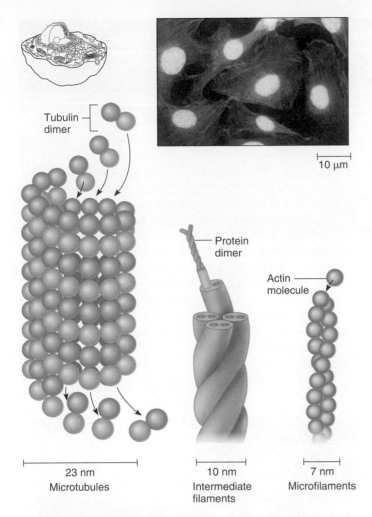

Figure 2.11 **The cytoskeleton is made of protein rods and tubules.** The three major components of the cytoskeleton are microtubules, intermediate filaments, and microfilaments. Using special stains, the cytoskeletons in the cells in the inset appear orange under the microscope. (nm stands for nanometer, which is a billionth of a meter.)

Cells contain both formed microtubules and individual tubulin molecules. When the cell requires microtubules to carry out a specific function—cell division, for example—free tubulin dimers self-assemble into more microtubules. After the cell divides, some of the microtubules fall apart into individual tubulin dimers, replenishing the cell's supply of building blocks. Cells perpetually build up and break down microtubules.

Microtubules form hairlike structures called cilia, from the Latin meaning "cells' eyelashes" (**figure 2.12**). Cilia are of two types: motile cilia that move, and primary cilia that do not move but serve a sensory function.

Motile cilia have one more pair of microtubules than primary microtubules. Coordinated movement of motile cilia

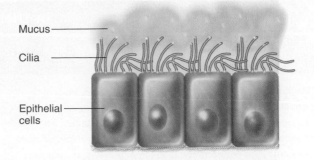

Figure 2.12 **Cilia are built of microtubules.** Motile cilia move secretions such as mucus on the cell surfaces of the respiratory tubes. In contrast, primary cilia do not move, functioning like antennae to sense changes. This figure shows only motile cilia.

generates a wave that moves the cell or propels substances along its surface. Motile cilia beat inhaled particles up and out of respiratory tubules and move egg cells in the female reproductive tract. Because motile cilia are so widespread, defects in them can cause multiple symptoms. One such ciliopathy—"sick cilia disease"—is Bardet-Biedl syndrome (OMIM 209900), which causes obesity, visual loss, diabetes, cognitive impairment, and extra fingers and/or toes. There are several types of Bardet-Biedl syndrome, and the condition affects both motile and primary cilia.

Many types of cells have primary cilia, which do not move and serve as antennae, sensing signals from outside the cell and passing them to particular locations inside cells. Primary cilia sense light entering the eyes, urine leaving the kidney tubules, blood flowing from vessels in the heart, and pressure on cartilage. Although these cilia do not move, they stimulate some cells to move, such as those that form organs in an embryo, and cells that help wounds to heal. Absence of primary cilia can harm health, such as in polycystic kidney disease (OMIM 173900). Cells may have many motile cilia but usually have only one cilium that does not move.

Microfilaments are long, thin rods composed of many molecules of the protein actin. They are solid and narrower than microtubules, enable cells to withstand stretching and compression, and help anchor one cell to another. Microfilaments provide many other functions in the cell through proteins that interact with actin. When any of these proteins is absent or abnormal, a genetic disease results.

Intermediate filaments have diameters intermediate between those of microtubules and microfilaments. Unlike microtubules and microfilaments, intermediate filaments are composed of different proteins in different cell types. However, all intermediate filaments consist of paired proteins entwined into nested coiled rods. Intermediate filaments are scarce in many cell types but are very abundant in nerve cells and skin cells. The *In Their Own Words* essay describes how abnormal intermediate filaments affect several cell types in a little girl, who has a very rare disease called giant axonal neuropathy (OMIM 256850).

A Little Girl with Giant Axons

Hannah Sames lacks a protein, called gigaxonin, which normally breaks down intermediate filaments and recycles their components. In Hannah's hair cells, intermediate filaments made of keratin proteins build up, kinking the strands. In her neurons, different types of intermediate filament proteins accumulate, swelling the long nerve extensions (axons) that send messages to muscles. She is slowly losing the ability to move and see. Hannah's condition is called giant axonal neuropathy (GAN). Her mother, Lori, recalls how hard it was to find a doctor familiar with this very rare disease, known to affect only a few dozen people in the world.

Hannah Sarah Sames is a beautiful little girl who was born on March 5, 2004. She has extremely curly blonde hair, a slight build, a precocious smile, and a charming personality. She loves to sing and dance, and play outdoors. Hannah is a beaming light of love.

When Hannah was 2 years, 5 months old, her grandmother noticed her left arch seemed to be rolling inward. I took Hannah to an orthopedist and a podiatrist, and was told Hannah would be fine. But by her third birthday, we suspected something was wrong—both arches were now involved, and her gait had become awkward. Her pediatrician gave her a rigorous physical exam and agreed she had an awkward gait, but felt that was just how Hannah walks.

Two months later, I took Hannah to another orthopedist, who told me to just let her live her life, she would be fine. Convinced otherwise, my sister showed cell phone video of Hannah walking to a physical therapist she works with, who thought Hannah's gait was like that of a child with muscular dystrophy. Our pediatrician referred us to a pediatric neurologist and a pediatric geneticist, and 6 months of testing for various diseases began. Results: all normal. During another visit with the pediatric neurologist, he took out a huge textbook and showed us a photo of a skinny little boy with kinky hair and a high forehead and braces that went just below the knee—he had GAN. He looked exactly like Hannah. So off we went to a children's hospital in New York City for more tests, and the diagnosis of GAN.

Meeting with a genetic counselor 3 days later brought devastation. Matt and I are each carriers, and we passed the disease to Hannah. Each of our two other daughters has a 2 in 3 chance of being a carrier. We learned GAN is a rare "orphan genetic disorder" for which there is no cure, no treatment, no clinical trial and no ongoing research. "So you are telling us this is a death sentence?" I asked. And, we were told, "Yes."

Matt and I walked around in a state of shock, anger, disbelief, and grief for 2 days. Then, we realized, as with any disease, someone has to be the first to be cured. Some family has to be the first to raise funds and awareness and pull the medical community together to find treatment. This is how Hannah's Hope Foundation was born! As a result, we held the world's first symposium for GAN, where clinicians and scientists brainstormed. Our foundation is now funding a number of projects aimed at treating GAN.

Lori Sames

http://www.hannahshopefund.org/

Hannah Sames has giant axonal neuropathy, a disorder that affects intermediate filaments in nerve cells. Her beautiful curls are one of the symptoms, but she prefers today to straighten her hair.

Key Concepts Questions 2.2

1. What are the components of cells?
2. How do organelles subdivide specific cell functions?
3. How is the plasma membrane more than just a covering for the cell?
4. What is the cytoskeleton?

2.3 Cell Division and Death

In a human body, new cells form as old ones die, at different rates in different tissues. Growth, development, maintaining health, and healing from disease or injury require an intricate interplay between the rates of **mitosis** and **cytokinesis,** which divide the DNA and the rest of the cell, respectively, and **apoptosis,** a form of cell death.

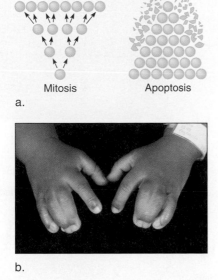

Figure 2.13 **Mitosis and apoptosis mold a body.**
(a) Cell numbers increase from mitosis and decrease from apoptosis. **(b)** In the embryo, apoptosis carves fingers and toes from webbed structures. In syndactyly, apoptosis fails to carve digits, and webbing persists, as it does in these hands.

About 10 trillion of a human body's 37 trillion cells are replaced daily. Yet cell death must happen to mold certain organs, just as a sculptor must remove some clay to shape the desired object. Apoptosis carves fingers and toes, for example, from weblike structures that telescope out from an embryo's developing form (**figure 2.13**). "Apoptosis" is Greek for "leaves falling from a tree." It is a precise, genetically programmed sequence of events that is a normal part of development.

The Cell Cycle

Many cell divisions enable a fertilized egg to develop into a many-trillion-celled organism. A series of events called the **cell cycle** describes the sequence of activities as a cell prepares to divide and then does.

Cell cycle rate varies in different tissues at different times. A cell lining the small intestine's inner wall may divide throughout life, whereas a neuron in the brain may never divide. A cell in the deepest skin layer may divide as long as a person lives. Some cells continue dividing even after a person dies! Frequent mitosis enables the embryo and fetus to grow rapidly. By birth, the mitotic rate slows dramatically. Later, mitosis maintains the numbers and positions of specialized cells in tissues and organs.

The cell cycle is continual, but we describe it with stages. The two major stages are **interphase** (not dividing) and mitosis (dividing) (**figure 2.14**). In mitosis, a cell duplicates its chromosomes, then in cytokinesis it apportions one set of chromosomes into each of two resulting cells, called daughter cells. This division maintains the set of 23 chromosome pairs

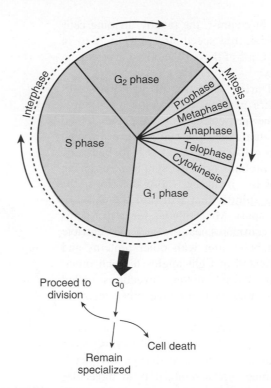

Figure 2.14 **The cell cycle.** The cell cycle is divided into interphase, when cellular components are replicated, and mitosis, when the cell distributes its contents into two daughter cells. Interphase is divided into G_1 and G_2, when the cell duplicates specific molecules and structures, and S phase, when it replicates DNA. Mitosis is divided into four stages plus cytokinesis, when the cells separate. G_0 is a "time-out" when a cell "decides" which course of action to follow.

characteristic of a human somatic cell. Another form of cell division, meiosis, produces sperm or eggs, which have half the amount of genetic material in somatic cells, as 23 single chromosomes. Chapter 3 discusses meiosis.

Interphase—A Time of Great Activity

During interphase a cell continues the basic biochemical functions of life, while also replicating its DNA and other subcellular structures. Interphase is divided into two gap (G_1 and G_2) **phases** and one synthesis (**S**) **phase**. In addition, a cell can exit the cell cycle at G_1 to enter a quiet phase called G_0. A cell in G_0 maintains its specialized characteristics but does not replicate its DNA or divide. From G_0, a cell may also proceed to mitosis and divide, or die. Apoptosis may ensue if the cell's DNA is so damaged that cancer might result. G_0, then, is when a cell's fate is either decided or put on hold.

During G_1, which follows mitosis, the cell resumes synthesis of proteins, lipids, and carbohydrates. These molecules will contribute to building the extra plasma membrane required to surround the two new cells that form from the original one. G_1 is the period of the cell cycle that varies the most in duration among different cell types. Slowly dividing cells, such as those in the liver, may exit at G_1 and enter G_0, where they remain for years. In contrast, the rapidly dividing cells in bone marrow speed through G_1 in 16 to 24 hours. Cells of the early embryo may skip G_1 entirely.

During S phase, the cell replicates its entire genome. As a result, each chromosome then consists of two copies joined at an area called the **centromere**. In most human cells, S phase takes 8 to 10 hours. Many proteins are also synthesized during this phase, including those that form the mitotic **spindle** that will pull the chromosomes apart. Microtubules form structures called **centrioles** near the nucleus. Centriole microtubules join with other proteins and are oriented at right angles to each other, forming paired, oblong structures called **centrosomes** that organize other microtubules into the spindle.

G_2 occurs after the DNA has been replicated but before mitosis begins. More proteins are synthesized during this phase. Membranes are assembled from molecules made during G_1 and are stored as small, empty vesicles beneath the plasma membrane. These vesicles will merge with the plasma membrane to enclose the two daughter cells.

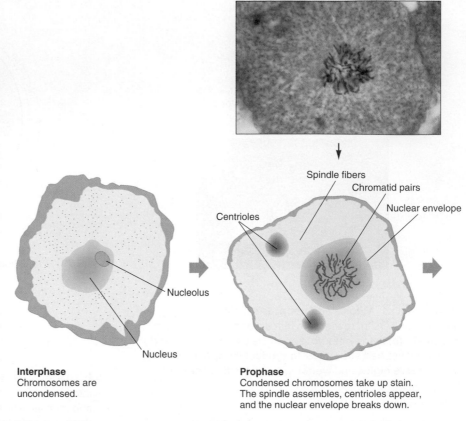

Interphase
Chromosomes are uncondensed.

Prophase
Condensed chromosomes take up stain. The spindle assembles, centrioles appear, and the nuclear envelope breaks down.

Figure 2.16 **Mitosis in a human cell.** Replicated chromosomes separate and are distributed into two cells from one. In a separate process called cytokinesis, the cytoplasm and other cellular structures distribute and pinch off into two daughter cells. (Not all chromosome pairs are depicted.)

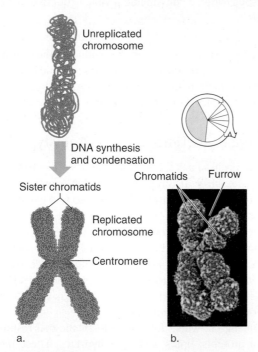

Figure 2.15 **Replicated and unreplicated chromosomes.** Chromosomes are replicated during S phase, before mitosis begins. Two genetically identical chromatids of a replicated chromosome join at the centromere **(a)**. The photomicrograph **(b)** shows a replicated human chromosome.

Mitosis—The Cell Divides

As mitosis begins, the replicated chromosomes are condensed enough to be visible, when stained, under a microscope. The two long strands of identical chromosomal material in a replicated chromosome are called **chromatids**, and the space between them is a furrow. (**figure 2.15**). At a certain point during mitosis, a replicated chromosome's centromere splits. This allows its chromatid pair to separate into two individual chromosomes that have all of the genetic material of that chromosome. Although the centromere of a replicated chromosome appears as a constriction, its DNA is replicated.

During **prophase,** the first stage of mitosis, DNA coils tightly. This shortens and thickens the chromosomes, which enables them to more easily separate (**figure 2.16**). Microtubules assemble from tubulin building blocks in the cytoplasm to form the spindles. Toward the end of prophase, the nuclear membrane breaks down. The nucleolus is no longer visible.

Metaphase follows prophase. Chromosomes attach to the spindle at their centromeres and align along the center of the cell, which is called the equator. Metaphase chromosomes are under great tension, but they appear motionless because they are pulled with equal force on both sides, like a tug-of-war rope pulled taut.

Next, during **anaphase,** the plasma membrane indents at the center, where the metaphase chromosomes line up. A band of microfilaments forms on the inside face of the

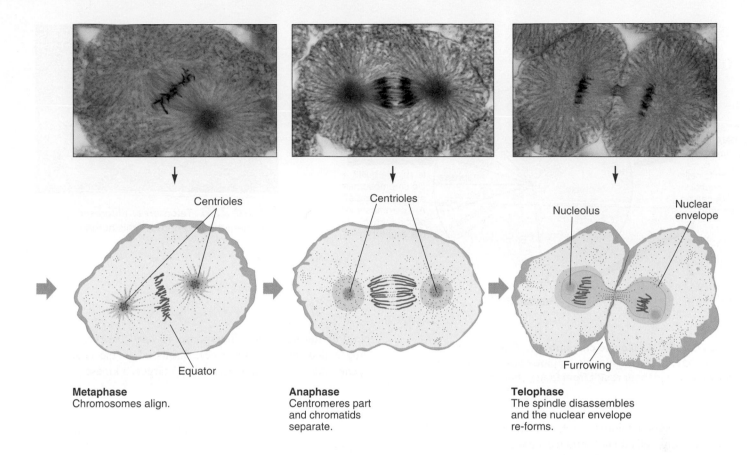

Metaphase
Chromosomes align.

Anaphase
Centromeres part and chromatids separate.

Telophase
The spindle disassembles and the nuclear envelope re-forms.

plasma membrane, constricting the cell down the middle. Then the centromeres part, which relieves the tension and releases one chromatid from each pair to move to opposite ends of the cell—like a tug-of-war rope breaking in the middle and the participants falling into two groups. Microtubule movements stretch the dividing cell. During the very brief anaphase, a cell fleetingly contains twice the normal number of chromosomes because each chromatid becomes an independently moving chromosome, but the cell has not yet physically divided.

In **telophase,** the final stage of mitosis, the cell looks like a dumbbell with a set of chromosomes at each end. The spindle falls apart, and nucleoli and the membranes around the nuclei re-form at each end of the elongated cell. Division of the genetic material is now complete. Next, during **cytokinesis,** organelles and macromolecules are distributed between the two daughter cells. Finally, the microfilament band contracts like a drawstring, separating the newly formed cells.

Control of the Cell Cycle

When and where a somatic cell divides is crucial to health. Illness can result from abnormally regulated mitosis. Control of mitosis is a daunting task. Quadrillions of mitoses occur in a lifetime, and not at random. Too little mitosis, and an injury goes unrepaired; too much, and an abnormal growth forms.

Groups of interacting proteins function at specific times in the cell cycle called checkpoints to ensure that chromosomes are faithfully replicated and apportioned into daughter cells (**figure 2.17**). A "DNA damage checkpoint" temporarily pauses the cell cycle while special proteins repair damaged DNA. An "apoptosis checkpoint" turns on as mitosis begins. During this checkpoint, proteins called survivins override signals telling the cell to die, ensuring that mitosis (division) rather than apoptosis (death) occurs. Later during mitosis, the "spindle assembly checkpoint" oversees construction of the spindle and the binding of chromosomes to it.

Cells obey an internal "clock" that tells them approximately how many times to divide. Mammalian cells grown (cultured) in a dish divide about 40 to 60 times. The mitotic clock ticks down with time. A connective tissue cell from a fetus, for example, will divide about 50 more times. A similar cell from an adult divides only 14 to 29 more times.

How can a cell "know" how many divisions remain? The answer lies in the chromosome tips, called **telomeres (figure 2.18).** Telomeres function like cellular fuses that burn down as pieces are lost from the ends. Telomeres consist of hundreds to thousands of repeats of a specific six DNA-base sequence. At each mitosis, the telomeres lose 50 to 200 endmost bases, gradually shortening the chromosome. After about 50 divisions, a critical length of telomere DNA is lost, which signals mitosis to stop. The cell may remain alive but not divide again, or it may die.

Not all cells have shortening telomeres. In eggs and sperm, in cancer cells, and in a few types of normal cells that must continually supply new cells (such as bone marrow cells),

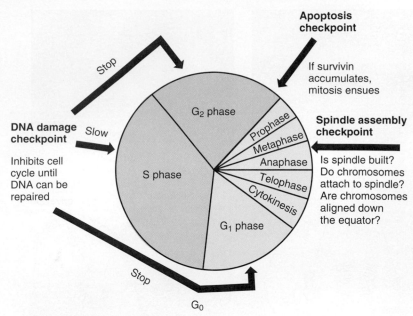

Figure 2.17 **Cell cycle checkpoints.** Checkpoints ensure that mitotic events occur in the correct sequence. Many types of cancer result from faulty checkpoints.

Labels on figure:

Stop

Apoptosis checkpoint

If survivin accumulates, mitosis ensues

DNA damage checkpoint

Slow

Inhibits cell cycle until DNA can be repaired

G₂ phase

Prophase
Metaphase
Anaphase
Telophase
Cytokinesis

Spindle assembly checkpoint

Is spindle built? Do chromosomes attach to spindle? Are chromosomes aligned down the equator?

S phase

G₁ phase

Stop

G₀

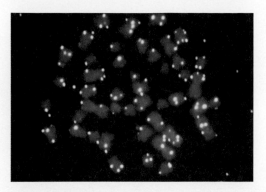

Figure 2.18 **Telomeres.** Fluorescent tags mark the telomeres in this human cell.

an enzyme called telomerase keeps chromosome tips long. However, most cells do not produce telomerase, and their chromosomes gradually shrink.

The rate of telomere shortening provides a "clock" for a cell's existence. The telomere clock not only counts down a cell's remaining lifespan, but may also sense environmental stimuli. Chronic stress, obesity, lack of exercise, and elevated blood sugar seem to accelerate telomere shortening.

Factors from outside the cell can affect a cell's mitotic clock. Crowding can slow or halt mitosis. Normal cells growing in culture stop dividing when they form a one-cell-thick layer lining the container. This limitation to division is called contact inhibition. If the layer tears, the cells that border the tear grow and divide, filling in the gap. The cells stop dividing once the space is filled. Perhaps a similar mechanism in the body limits mitosis.

Hormones and growth factors are chemical signals that control the cell cycle from outside. A **hormone** is made in a gland and transported in the bloodstream to another part of the body, where it exerts a specific effect. Hormones secreted in the brain, for example, signal the cells lining a woman's uterus to build up each month by mitosis in preparation for possible pregnancy. A growth factor acts more locally. Epidermal growth factor (EGF), for example, stimulates cell division in the skin beneath a scab. Certain cancer drugs work by plugging growth factor receptors on cancer cells, blocking the signals to divide.

Two types of proteins, cyclins and kinases, interact inside cells to activate the genes whose products carry out mitosis. The two types of proteins form pairs. Cyclin levels fluctuate regularly throughout the cell cycle, while kinase levels stay the same. A certain number of cyclin-kinase pairs turn on the genes that trigger mitosis. Then, as mitosis begins, enzymes degrade the cyclin. The cycle starts again as cyclin begins to build up during the next interphase. A very successful cancer drug called imatinib (Gleevec), discussed in the chapter 18 opener and Clinical Connection, 18.1 targets a kinase.

Apoptosis

Apoptosis rapidly and neatly dismantles a cell into membrane-enclosed pieces that a phagocyte (a cell that engulfs and destroys another) can mop up. It is a little like packing the contents of a messy room into garbage bags, then disposing of it all. In contrast is necrosis, a form of cell death associated with inflammation and damage, rather than an orderly, contained destruction.

Like mitosis, apoptosis is a continuous process. It begins when a "death receptor" on the cell's plasma membrane receives a signal to die. Within seconds, enzymes called caspases are activated inside the doomed cell, stimulating each other and snipping apart various cell components. These killer enzymes:

- destroy enzymes that replicate and repair DNA;
- activate enzymes that cut DNA into similarly sized pieces;
- tear apart the cytoskeleton, collapsing the nucleus and condensing the DNA within;
- cause mitochondria to release molecules that increase caspase activity and end the energy supply;
- abolish the cell's ability to adhere to other cells; and
- attract phagocytes that dismantle the cell remnants.

A dying cell has a characteristic appearance (**figure 2.19**). It rounds up as contacts with other cells are cut off, and the plasma membrane undulates, forming bulges called blebs. The nucleus bursts, releasing DNA pieces, as mitochondria decompose. Then the cell shatters. Almost instantly, pieces of membrane encapsulate the cell fragments, which prevents inflammation. Within an hour, the cell is gone.

From the embryo onward through development, mitosis and apoptosis are synchronized, so that tissue neither overgrows nor shrinks. In this way, a child's liver retains its shape as she grows into adulthood, yet enlarges. During early development, mitosis and apoptosis orchestrate the ebb and flow of cell number as new structures form. Later, these processes protect—mitosis produces new skin to heal a scraped knee; apoptosis peels away sunburnt skin cells that might otherwise become cancerous. Cancer is a profound derangement of the balance between cell division and cell death. In cancer, mitosis occurs too frequently or too many times, or apoptosis happens too infrequently. Chapter 18 discusses cancer in detail.

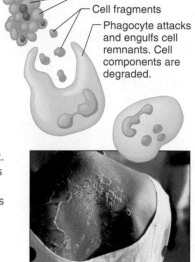

Death receptor on doomed cell binds signal molecule. Caspases are activated within.

Caspases destroy various proteins and other cell components. Cell undulates.

Blebs

Cell fragments

Phagocyte attacks and engulfs cell remnants. Cell components are degraded.

Key Concepts Questions 2.3

1. What are mitosis and apoptosis?
2. Why must mitosis and apoptosis be in balance?
3. Describe the phases of the cell cycle and of mitosis.
4. How is the cell cycle controlled?
5. What are the stages of apoptosis?

Figure 2.19 Death of a cell. A cell undergoing apoptosis loses its shape, forms blebs, and falls apart. Caspases destroy the cell's insides. Phagocytes digest the remains. Note the blebs on the dying liver cells in the first photograph. Sunburn peeling is one example of apoptosis.

2.4 Stem Cells

Bodies grow and heal thanks to cells that retain the ability to divide, generating new cells like themselves and cells that will specialize. **Stem cells** and **progenitor cells** renew tissues so that as the body grows, or loses cells to apoptosis, injury, and disease, other cells are produced that take their places.

Cell Lineages

A stem cell divides by mitosis to yield either two daughter cells that are stem cells like itself, or one that is a stem cell and one that is a progenitor cell, which may be partially specialized (**figure 2.20**). The characteristic of **self-renewal** is what makes a stem cell a stem cell—its ability to continue the lineage of cells that can divide to give rise to another cell like itself. A progenitor cell cannot self-renew, and its daughters specialize as any of a restricted number of cell types. A fully differentiated cell, such as a mature blood cell, descends from a sequence of increasingly specialized progenitor cell intermediates, each one less like a stem cell and more like a blood cell. Our more than 290 differentiated cell types develop from sequences, called lineages, of stem, progenitor, and increasingly differentiated cells. **Figure 2.21** shows parts of a few lineages.

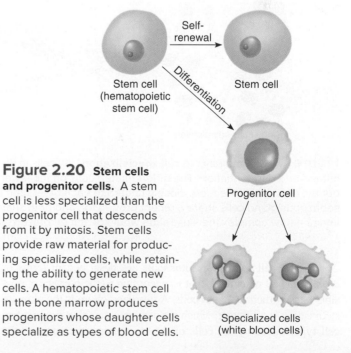

Self-renewal

Stem cell (hematopoietic stem cell)

Differentiation

Stem cell

Progenitor cell

Specialized cells (white blood cells)

Figure 2.20 Stem cells and progenitor cells. A stem cell is less specialized than the progenitor cell that descends from it by mitosis. Stem cells provide raw material for producing specialized cells, while retaining the ability to generate new cells. A hematopoietic stem cell in the bone marrow produces progenitors whose daughter cells specialize as types of blood cells.

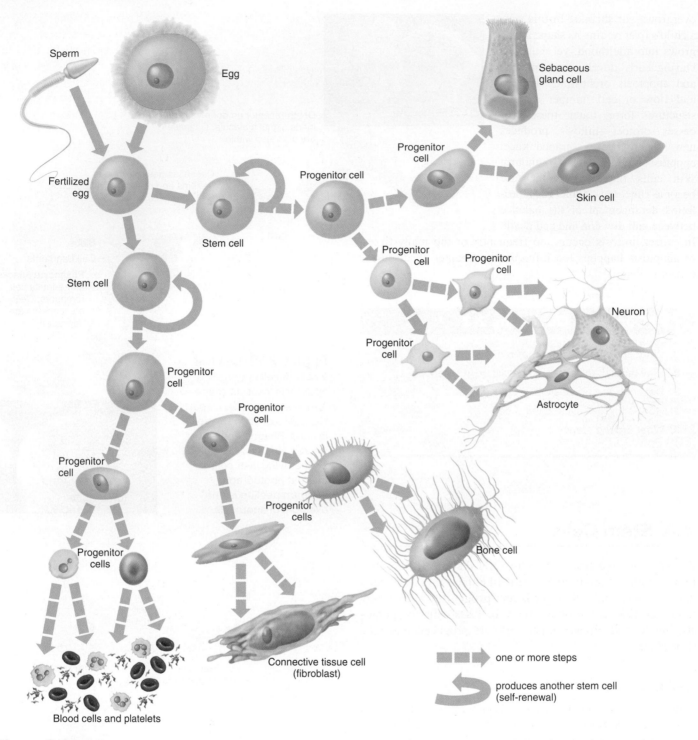

Figure 2.21 **Pathways to cell specialization.** All cells in the human body descend from stem cells, through the processes of mitosis and differentiation. The differentiated cells on the lower left are all connective tissues (blood, connective tissue, and bone), but the blood cells are more closely related to each other than they are to the other two cell types. On the upper right, the skin and sebaceous gland cells share a recent progenitor, and both share a more distant progenitor with neurons and supportive astrocytes. Imagine how complex the illustration would be if it embraced all 290-plus types of cells in a human body!

Stem cells and progenitor cells are described in terms of developmental potential—that is, according to the number of possible fates of their daughter cells. A fertilized ovum is the ultimate stem cell. It is totipotent, which means that it can give rise to every cell type, including the cells of the membranes that support the embryo. Other stem cells and progenitor cells are pluripotent: Their daughter cells have fewer possible fates. Some are multipotent: Their daughter cells have only a few developmental "choices." This is a little like a college freshman's consideration of many majors, compared to a junior's more narrowed focus in selecting courses.

As stem cell descendants specialize, they express some genes and ignore others. An immature bone cell forms from a progenitor cell by manufacturing mineral-binding proteins and enzymes. In contrast, an immature muscle cell forms from a muscle progenitor cell that accumulates contractile proteins. The bone cell does not produce muscle proteins, nor does the muscle cell produce bone proteins. All cells, however, synthesize proteins for basic "housekeeping" functions, such as energy acquisition and protein synthesis.

Many, if not all, of the organs in an adult human body have stem or progenitor cells. These cells can divide when injury or illness occurs and generate new cells to replace damaged ones. Stem cells in the adult may have been set aside in the embryo or fetus in particular organs as repositories of future healing. Some stem cells, such as those from bone marrow, can travel to and replace damaged or dead cells elsewhere in the body, in response to signals that are released in injury or disease. Because every cell except red blood cells contains all of an individual's genetic material, any cell type, given appropriate signals, can in theory become any other. This concept is the basis of much of stem cell technology.

Researchers are investigating stem cells to learn more about basic biology and to develop treatments for a great variety of diseases and injuries—not just inherited conditions. Clinical trials are currently testing stem cell–based treatments. These cells come from donors as well as from patients' own bodies, as **figure 2.22** illustrates. The cells can be mass-produced in laboratory glassware, and if they originate with a patient's cell or nucleus, they are a genetic match.

Stem Cell Sources

There are three general sources of human stem cells: **embryonic stem (ES) cells, induced pluripotent stem (iPS) cells,** and "adult" stem cells (**Table 2.2**).

Embryonic stem (ES) cells are not actually cells from an embryo, but are created in a laboratory dish using certain cells from a region of a very early embryo called the inner cell mass (ICM) (see figure 3.14). Some ICM cells, under certain conditions, become pluripotent and can self-renew—they are stem cells. The ICM cells used to derive ES cells can come from two sources: "leftover" embryos from fertility clinics that

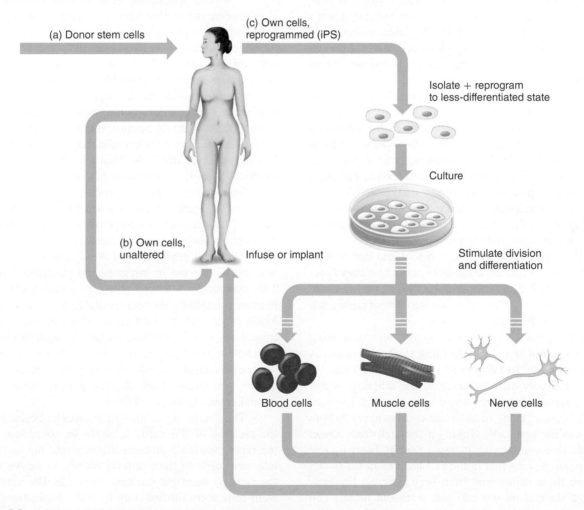

Figure 2.22 Using stem cells to heal. (a) Stem cells from donors (bone marrow or umbilical cord blood) are already in use. **(b)** A person's cells may be used, unaltered, to replace damaged tissue, such as bone marrow. **(c)** It is possible to "reprogram" a person's cells in culture, taking them back to a less specialized state and then nurturing them to differentiate as a needed cell type.

Table 2.2	Stem Cell Sources
Stem Cell Type	**Source**
Embryonic stem cell	Inner cell mass of embryo; somatic cell nuclear transfer into egg cell
Induced pluripotent stem cell	Genes or other chemicals reprogram somatic cell nucleus; no embryos required
"Adult" stem cell	Somatic cells that normally function as stem cells, from any stage of development from fertilized ovum through elderly

would otherwise be destroyed, and from somatic cell nuclear transfer, in which a nucleus from a somatic cell (such as a skin fibroblast) is transferred to an egg cell that has had its own nucleus removed. An embryo is grown and its inner cell mass cells used to make ES cells. Researchers can derive ES cells from somatic cell nuclear transfer into egg cells directly, without using embryos. Somatic cell nuclear transfer is popularly called "cloning" because it copies the nucleus donor's genome. It was made famous by Dolly the sheep, whose original nucleus came from a mammary gland cell of a six-year-old ewe. The age of the donor may be why Dolly did not live as long as expected.

Induced pluripotent stem cells are somatic cells that are "reprogrammed" to differentiate into any of several cell types. Reprogramming a cell may take it back through developmental time to an ES cell–like state. Then the cell divides and gives rise to cells that specialize as a different, desired cell type. Or, cells can be reprogrammed directly into another cell type. Reprogramming instructions are genes, certain RNA molecules, or other chemical factors. Induced pluripotent stem (iPS) cells do not require cells from an embryo. They are a genetic match to the person they come from.

For now, iPS cells are a research tool and not used to treat disease until researchers learn more about how they function in a human body. The *In Their Own Words* essay in chapter 3 describes how researchers are growing parts of embryonic human brains from iPS cells.

To return to the college major analogy, reprogramming a cell is like a senior in college deciding to change major. A French major wanting to become an engineer would have to start over, taking very different courses. But a biology major wanting to become a chemistry major would not need to start from scratch because many of the same courses apply to both majors. So it is for stem cells. Taking a skin cell from a man with heart disease and turning it into a healthy heart muscle cell might require taking that initial cell back to an ES or iPS state, because these cells come from very different lineages. But turning a skeletal muscle cell into a smooth muscle cell requires fewer steps backwards because the two cell types differentiate from the same type of progenitor cell.

A third source of stem cells are those that naturally are part of the body, called "adult" stem cells. They are more accurately called tissue-specific or somatic stem cells because they are found in the tissues of embryos, fetuses, and children, and not just in adult bodies. Adult stem cells self-renew, but most give rise only to a few types of specialized daughter cells. Researchers are still discovering niches of adult stem cells in the body. Many potentially valuable adult stem cells are discarded as medical waste, such as fat and brain tissue.

Stem Cell Applications

Stem cells are being used in four basic ways. In drug discovery and development, stem cell cultures supply the human cells that are affected in a particular disease, which may be difficult or impossible to culture in a laboratory. Drugs are tested on these cells. Liver and heart cells derived from stem cells are particularly useful in testing drugs for side effects, because the liver detoxifies many drugs, and many drugs harm the heart. Using stem cells in drug development can minimize the need to experiment on animals and can eliminate drugs with adverse effects before they are tested on people.

A second application of stem cells growing in culture is to observe the earliest signs of a disease, which may begin long before symptoms appear in a person. The first disease for which human iPS cells were derived was amyotrophic lateral sclerosis (ALS), also known as Lou Gehrig's disease. In ALS, motor neurons that enable a person to move gradually fail. A few years after the first signs of weakness or stumbling, death comes from failure of the respiratory muscles. ALS had been difficult to study because motor neurons do not survive for long in the laboratory because they do not divide. However, iPS cells derived from fibroblasts in patients' skin are reprogrammed in culture to become ALS motor neurons, providing an endless source of the hard-to-culture cells. Thanks to these and other iPS cells, researchers are now observing the beginnings of hundreds of diseases, many of them inherited—and discovering new ways to treat them.

The third application of stem cells is to create tissues and organs, for use in implants and transplants, or to study. This approach is not new—the oldest such treatment, a bone marrow transplant, has been available for more than 60 years. Many other uses of adult stem cells, delivered as implants, transplants, or infusions into the bloodstream, are being tested. A patient's own bone marrow stem cells, for example, can be removed, isolated, perhaps altered, grown, bathed in selected factors, and infused back into the patient, where they follow natural signals to damaged tissues.

The fourth application of stem cells became clear with the creation of iPS cells. It might be possible to introduce the reprogramming proteins directly into the body to stimulate stem cells in their natural niches. Once we understand the signals, we might not need the cells. The applications of stem cells seem limited only by our imaginations, but some companies take advantage of what the public does not know about the science. Bioethicists call this practice "stem cell tourism." *Bioethics: Choices for the Future* discusses stem cell banking.

Banking Stem Cells: When Is It Necessary?

The parents-to-be were very excited by the company's promise: "Bank your baby's cord blood stem cells and benefit from breakthroughs. Be prepared for the unknowns in life."

The website profiled children saved from certain diseases using stored umbilical cord blood. The statistics were persuasive: More than 70 diseases are currently treatable with cord blood transplants, and 10,000 procedures have already been done.

With testimonials like that, it is little wonder that parents collectively spend more than $100 million per year to store cord blood. The ads and statistics are accurate but misleading, because of what they *don't* say. Most people never actually use the umbilical cord blood stem cells that they store. The scientific reasons go beyond the fact that treatable diseases are very rare. In addition, cord blood stem cells are not nearly as pluripotent as some other stem cells, limiting their applicability. Perhaps the most compelling reason that stem cell banks are rarely used is based on logic: For a person with an inherited disease, *healthy* stem cells are required—not his or her own, which could cause the disease all over again because the mutation is in every cell. The patient needs a well-matched donor, such as a healthy sibling.

Commercial cord blood banks may charge more than $1,000 for the initial collection plus an annual fee. However, the U.S. National Institutes of Health and organizations in many other nations have supported not-for-profit banks for years, and may not charge fees. Donations of cord blood to these facilities are not to help the donors directly, but to help whoever can use the cells.

Commercial stem cell banks are not just for newborns. One company, for example, offers to bank "very small embryonic-like stem cells" for an initial charge of $7,500 and a $750 annual fee, "enabling people to donate and store their own stem cells when they are young and healthy for their personal use in times of future medical need." The cells come from a person's blood and, in fact, one day may be very useful, but the research has yet to be done supporting any use of the cells in treatments.

Questions for Discussion

1. Storing stem cells is not regulated by the U.S. government the way that a drug or a surgical procedure is because it is a service that will be helpful for treatments not yet invented. Do you think such banks should be regulated, and if so, by whom and how?

2. What information do you think that companies offering to store stem cells should present on their websites?

3. Do you think that advertisements for cord blood storage services that have quotes and anecdotal reports, but do not mention that most people who receive stem cell transplants do not in fact receive their own cells, are deceptive? Or do you think it is the responsibility of the consumer to research and discover this information?

4. Several companies store stem cells extracted from baby teeth, although a use for such stem cells has not yet been found. Suggest a different way to obtain stem cells that have the genome of a particular child.

Key Concepts Questions 2.4

1. What are the two general characteristics of stem cells?
2. How do stem cells differ from progenitor cells?
3. How do cells differentiate down cell lineages?
4. What are the functions of stem cells throughout life?
5. Distinguish among embryonic stem cells, induced pluripotent stem cells, and adult stem cells.
6. How might stem cells be used in health care?

2.5 The Human Microbiome

Ninety percent of the cells within a human body are not actually human—they are bacterial. Our bodies are vast ecosystems for microscopic life. The cells within and on us that are not actually of us constitute the human **microbiome**. The term "microbiome" borrows from the ecological term "biome" to indicate all of the species in an area. Many trillions of bacteria can fit into a human body because their cells, which are prokaryotic, are much smaller than ours. The Human Microbiome Project has revealed our bacterial residents. What we learn about the human microbiome will be increasingly applied to improving health, once researchers determine whether a particular collection of microbes causes or contributes to an illness or is a result of it.

Each of us has a "core microbiome" of bacterial species, but also many other species that reflect our differing genomes, environments, habits, ages, diets, and health. Different body parts house distinctive communities of microbes, altered by experience. A circumcised penis has a different microbiome than an uncircumcised one, for example, and the vaginal microbiome of a mother differs from that of a woman who hasn't had children. Different communities of bacteria inhabit our armpits, groins, navels, the bottoms of our feet, between our fingers, and between our buttocks.

The 10 trillion bacteria that make up the human "gut microbiome" have long been known to help us digest certain foods. The mouth alone houses more than 600 species of bacteria. Analysis of their genomes yields practical information. For example, the genome sequence of the bacterium *Treponema denticola* reveals how it survives amid the films other bacteria form in the mouth, and causes gum disease.

The more distant end of our digestive tract is easy to study in feces. The large intestine alone houses 6,800 bacterial species. One of the first microbiome studies examined soiled diapers from babies during their first year—one the child of the chief investigator—chronicling the establishment of the gut bacterial community. Newborns start out with clean intestines. After various bacteria come and go, very similar species remain from baby to baby by the first birthday.

We can use knowledge of our gut microbiome to improve health, because illness can alter the bacterial populations within us. Probiotics are live microorganisms, such as bacteria and yeasts, that, when ingested, confer a health benefit. For example, certain *Lactobacillus* strains added to yogurt help protect against *Salmonella* foodborne infection.

Probiotics typically consist of one or several microbial species. In contrast, "fecal transplantation" is a treatment based on altering the microbiome that replaces hundreds of bacterial species at once. In the procedure, people with recurrent infection from *Clostridium difficile,* which causes severe diarrhea, receive feces from a healthy donor. The bacteria transferred in the donated feces reconstitute a healthy gut microbiome, treating the infection. Fecal transplantation has been performed by enema in cattle for a century, and since 1958 in humans. A recent clinical trial introduced the fecal material in a tube through the nose to the small intestine, but may soon be delivered in a capsule. Do not try this yourself!

Researchers have been studying the human microbiome for a short time, but already have learned a lot. Facts learned or confirmed through studying the human microbiome include:

1. Certain skin bacteria cause acne, but others keep skin clear.
2. Circumcision protects against certain viral infections, including HIV.
3. Lowered blood sugar following weight-loss surgery is partly due to a changed gut microbiome.
4. An altered microbiome hastens starvation in malnourished children.
5. Antibiotics temporarily alter the gut microbiome.
6. The birth of agriculture 10,000 years ago introduced the bacteria that cause dental caries.
7. Microbiome imbalances may contribute to or cause asthma, cancers, obesity, psoriasis, Crohn's disease, gum disease, and other conditions.

Investigation of the human microbiome is part of metagenomics. Recall from chapter 1 that metagenomics sequences all DNA in a particular habitat. It is intriguing to think of the human body as an ecosystem!

Key Concepts Questions 2.5

1. What is the human microbiome?
2. Give an example of a body part that houses a unique microbiome.
3. Cite one way that the microbiome affects health.
4. What is metagenomics?

Summary

2.1 Introducing Cells

1. Cells are the units of life and comprise the human body. Inherited traits and illnesses can be understood at the cellular and molecular levels.
2. All cells share certain features, but they are also specialized because they express different subsets of genes.
3. **Somatic** (body) **cells** are **diploid** and sperm and egg cells are **haploid**. **Stem cells** produce new cells.

2.2 Cell Components

4. A **prokaryotic cell** is small and does not have a **nucleus** or other **organelles**. A **eukaryotic cell** has organelles, including a nucleus.
5. Cells consist primarily of water and several types of macromolecules: **carbohydrates, lipids, proteins,** and **nucleic acids**.
6. Organelles sequester related biochemical reactions, improving the efficiency of life functions and protecting the cell. The cell also consists of **cytoplasm** and other chemicals.

7. The nucleus contains DNA and a **nucleolus**, which is a site of ribosome synthesis. **Ribosomes** provide scaffolds for protein synthesis; they exist free in the cytoplasm or complexed with the rough **endoplasmic reticulum (ER)**.
8. In secretion, the rough ER is where protein synthesis starts. Smooth ER is the site of lipid synthesis, transport, and packaging, and the **Golgi apparatus** packages secretions into **vesicles,** which exit through the **plasma membrane**. **Lysosomes** contain enzymes that dismantle debris (**autophagy**), and **peroxisomes** house enzymes that perform a variety of functions. Enzymes in **mitochondria** extract energy from nutrients.
9. The plasma membrane is a protein-studded phospholipid bilayer. It controls which substances exit and enter the cell (**signal transduction**), and how the cell interacts with other cells (**cell adhesion**).
10. The **cytoskeleton** is a protein framework of hollow **microtubules,** made of tubulin, and solid **microfilaments,**

which consist of actin. **Intermediate filaments** are made of more than one protein type.

2.3 Cell Division and Death

11. Coordination of cell division (**mitosis**) and cell death (**apoptosis**) maintains cell numbers, enabling structures to enlarge during growth and development but preventing abnormal growth.

12. The **cell cycle** describes whether a cell is dividing (mitosis) or not (**interphase**). Interphase consists of two gap phases, when proteins and lipids are produced, and a synthesis phase, when DNA is replicated.

13. Mitosis proceeds in four stages. In **prophase,** replicated chromosomes consisting of two **chromatids** condense, the **spindle** assembles, the nuclear membrane breaks down, and the nucleolus is no longer visible. In **metaphase,** replicated chromosomes align along the center of the cell. In **anaphase,** the **centromeres** part, equally dividing the now unreplicated chromosomes into two daughter cells. In **telophase,** the new cells separate. **Cytokinesis** apportions other components into daughter cells.

14. Internal and external factors control the cell cycle. Checkpoints are times when proteins regulate the cell cycle. **Telomere** (chromosome tip) length determines how many more mitoses will occur. Crowding, **hormones,** and growth factors signal cells from the outside; the interactions of cyclins and kinases trigger mitosis from inside.

15. In apoptosis, a receptor on the plasma membrane receives a death signal, which activates caspases that tear apart the cell in an orderly fashion. Membrane surrounds the pieces, preventing inflammation.

2.4 Stem Cells

16. **Stem cells** self-renew, producing daughter cells that retain the ability to divide and daughter cells that specialize. **Progenitor cells** give rise to more specialized daughter cells but do not **self-renew**.

17. A fertilized ovum is totipotent. Some stem cells are pluripotent, and some are multipotent. Cells are connected through lineages.

18. The three sources of stem cells are **embryonic stem (ES) cells, induced pluripotent stem (iPS) cells,** and adult stem cells.

19. Stem cell technology enables researchers to observe the origins of diseases and to devise new types of treatments.

2.5 The Human Microbiome

20. Ninety percent of the cells in a human body are microorganisms. They are our **microbiome**.

21. Different body parts house different communities of microbes.

22. Genes interact with the microbiome and with environmental factors to mold most traits.

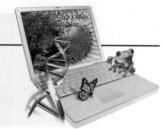

www.mhhe.com/lewisgenetics11

Answers to all end-of-chapter questions can be found at **www.mhhe.com/lewisgenetics11**. You will also find additional practice quizzes, animations, videos, and vocabulary flashcards to help you master the material in this chapter.

Review Questions

1. Match each organelle to its function.

Organelle	Function
a. lysosome	1. lipid synthesis
b. rough ER	2. houses DNA
c. nucleus	3. energy extraction
d. smooth ER	4. dismantles debris
e. Golgi apparatus	5. detoxification
f. mitochondrion	6. protein synthesis
g. peroxisome	7. processes secretions

2. List and describe a disease that affects each of the following organelles or structures, using information from the text or elsewhere.

   a. lysosome
   b. peroxisome
   c. mitochondrion
   d. cytoskeleton
   e. ion channel

3. What advantage does compartmentalization provide to a large and complex cell?

4. Give two examples of how the plasma membrane is more than just a covering of the cell's insides.

5. List four types of controls on cell cycle rate.

6. How can all of a person's cells contain exactly the same genetic material, yet specialize as bone cells, nerve cells, muscle cells, and connective tissue cells?

7. Distinguish between

   a. a bacterial cell and a eukaryotic cell.
   b. interphase and mitosis.
   c. mitosis and apoptosis.
   d. rough ER and smooth ER.
   e. microtubules and microfilaments.
   f. a stem cell and a progenitor cell.
   g. totipotent and pluripotent.

8. How are intermediate filaments similar to microtubules and microfilaments, and how are they different?

9. Find an example of an article that defines stem cells as "turning into any cell type" and explain how this definition is not quite correct.

10. Distinguish among ES cells, iPS cells, and adult stem cells, and state the pros and cons of working with each to develop a therapy.

11. Define *microbiome*.

12. "I heard about a great new organic, gluten-free, nonfat, low-calorie, high-energy way to remove all of the bacteria in our intestines!" enthused the young woman in the health food store to her friend. Is this a good idea or not? Cite a reason for your answer.

13. State one or more ways that the human genome and human microbiome are similar, and one or more ways that they are different.

Applied Questions

1. How might abnormalities in each of the following contribute to cancer?
   a. cellular adhesion
   b. signal transduction
   c. balance between mitosis and apoptosis
   d. cell cycle control
   e. telomerase activity
   f. an activated stem cell

2. In which organelle would a defect cause fatigue?

3. If you wanted to create a synthetic organelle to test new drugs for toxicity, which natural organelle's function would you try to replicate?

4. An inherited form of migraine is caused by a mutation in a gene *(SCN1A)* that encodes a sodium channel in neurons. What is a sodium channel, and in which cell structure is it located?

5. In a form of Parkinson disease, a protein called alpha synuclein accumulates because of impaired autophagy. Which organelle is implicated in this form of the disease?

6. Why wouldn't a cell in an embryo likely be in phase G_0?

7. Explain how replacing an enzyme or implanting umbilical cord stem cells from a donor can treat a lysosomal storage disease.

8. Describe two ways to derive stem cells without using human embryos.

9. Researchers isolated stem cells from fat removed from people undergoing liposuction, a procedure to remove fat. The stem cells can give rise to muscle, fat, bone, and cartilage cells.
   a. Are the stem cells totipotent or pluripotent?
   b. Are these stem cells ES cells, iPS cells, or adult stem cells?

10. Two roommates purchase several 100-calorie packages of cookies. After a semester of each roommate eating one package a night, and following similar diet and exercise plans otherwise, one roommate has gained 12 pounds, but the other hasn't. If genetics does not account for the difference in weight gain, what other factor discussed in the chapter might?

Web Activities

1. Several companies offer expensive "banking" of stem cells, even if treatments using the cells have not yet been invented. Find a website offering to bank the cells, and discuss whether or not the company provides enough information for you to make an informed choice as to whether or not to use the service.

2. Consult Online Mendelian Inheritance in Man for the following disorders, and state what they have in common at the cellular level: Meckel syndrome (OMIM 249000), Joubert syndrome (OMIM 213300), Ellis van Creveld disease (OMIM 225500), and Senior-Loken disease (OMIM 266900).

Forensics Focus

1. Michael Mastromarino was sentenced to serve many years in prison for trafficking in body parts. As the owner of Biomedical Tissue Services in Fort Lee, NJ, Mastromarino and his cohorts dismembered corpses taken, without consent, from funeral homes in Pennsylvania, New Jersey, and New York. Thousands of parts from hundreds of bodies were used in surgical procedures, including hip replacements and dental implants. The most commonly used product was a paste made of bone. Many family members testified at the trials. Explain why cells from bone tissue can be matched to blood or cheek lining cells from blood relatives.

Case Studies and Research Results

1. Taylor King has a form of neuronal ceroid lipofuscinosis (OMIM 610127). She suffers from seizures, loss of vision, and lack of coordination. The condition is a lysosomal storage disease. State what her cells cannot do. (Read about Taylor at www.taylorstale.com).

2. Studies show that women experiencing chronic stress, such as from caring for a severely disabled child, have telomeres that shorten at an accelerated rate. Suggest a study that would address the question of whether men have a similar reaction to chronic stress.

3. Hannah Sames is having gene therapy to treat her giant axonal neuropathy, described in the *In Their Own Words* essay on page 28. First, she must have a transplant of cord blood stem cells to replace her immune system so her body will not reject the protein that the gene therapy will enable her cells to make. Hannah is getting the stem cells from her sister, who is healthy. Why wouldn't Hannah's own banked cord blood stem cells be helpful?

4. Researchers investigated all of the bacteria found on various surfaces in public restrooms, such as sinks, toilet seats, and toilet paper dispensaries. They identified bacterial species normally found on human skin, the anus, nostrils, and vagina. Is this an investigation of the human microbiome, a metagenomics study, or both? Cite a reason for your answer.

5. Malnutrition is common among children in the African nation of Malawi. Researchers hypothesized that the microbiome may play a role in starvation because in some families, some children are malnourished and their siblings not, even though they eat the same diet. Even identical twins may differ in nutritional status.

    Researchers followed 317 sets of twins in Malawi, from birth until age 3. In half of the pairs, one or both twins developed kwashiorkor, the type of protein malnutrition that swells bellies. The researchers focused on twin pairs in which only one was starving, including both identical and fraternal pairs. At the first sign that one twin was malnourished, both were placed on a diet of healthy "therapeutic food." Four weeks later, the pair returned to the nutrient-poor village diet. If the malnourished twin became so again, then the researchers compared his or her microbiome to that of the healthy sibling. The goal was to identify bacterial species that impair the ability of a child to extract nutrients from the native diet.

    a. How might the findings from this study be applied to help prevent or treat malnutrition?

    b. Do you think that the study was conducted ethically? Why or why not?

    c. Explain how identical twins who follow the same diet can differ in nutritional status.

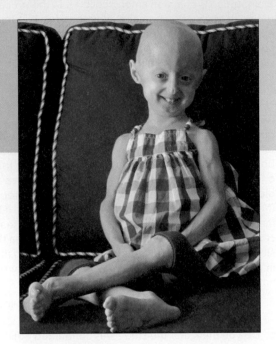

Lindsay has progeria, which greatly accelerates some signs of aging. She is 11 years old in this photograph. A drug being tested in children with progeria not only improved symptoms, but extended life by 1.6 years—significant because average survival is only 14.

Meiosis, Development, and Aging

Learning Outcomes

3.1 The Reproductive System

1. Describe the structures of the male and female reproductive systems.

3.2 Meiosis

2. Explain why meiosis is necessary to reproduce.

3. Summarize the events of meiosis.

3.3 Gametes Mature

4. List the steps in sperm and oocyte formation.

3.4 Prenatal Development

5. Describe early prenatal development.

6. Explain how the embryo differs from the fetus.

3.5 Birth Defects

7. Define *critical period*.

8. List some teratogens.

3.6 Maturation and Aging

9. Describe common disorders that begin in adulthood.

10. Explain how rapid aging disorders occur.

 The **BIG** Picture

Our reproductive systems enable us to start a new generation. First our genetic material must be halved, so it can combine with that of a partner to reconstitute a full diploid genome. Then genetic programs operate as the initial cell divides and its daughter cells specialize. The forming tissues fold into organs and the organs interact, slowly building a new human body.

Progress for Progeria

A child with Hutchinson-Gilford progeria syndrome (HGPS) appears old. Shortly after birth, weight gain begins to slow and hair to thin. The gums remain smooth, as teeth do not erupt. Joints stiffen and bones weaken. Skin wrinkles as the child's chubbiness melts away too quickly, and a cherubic toddler becomes increasingly birdlike in appearance.

Beneath the child's toughening skin, blood vessels stiffen with premature atherosclerosis, fat pockets shrink, and connective tissue hardens. Inside cells, chromosome tips whittle down at an accelerated pace, marking time too quickly. However, some organs remain healthy, and intellect is spared. The child usually dies during adolescence, from a heart attack or stroke.

Fewer than 150 children worldwide are known to have this strange illness, first described in 1886. When researchers in 2003 identified the mutant gene that causes HGPS, they realized, from the nature of the molecular defect, that several existing drugs might alleviate certain symptoms.

Researchers tested a failed cancer drug, tracking children's weight to assess whether the drug had an effect. These children usually stop growing at age 3, but 9 of 25 children in the initial clinical trial had a greater than 50 percent increase in the rate of weight gain. The artery stiffening and blockages of atherosclerosis not only stopped progressing, but reversed! For some children, hearing improved and bones became stronger.

The Progeria Research Foundation is trying to locate all of the children with progeria in the world, and get them involved in testing several promising drugs, both old and new. Understanding how a mutation causes a disease is an important first step in finding a treatment. The drugs discovered to help in progeria might also be useful in those who develop the symptoms as part of advancing age.

3.1 The Reproductive System

Genes orchestrate our physiology from a few days after conception through adulthood. Expression of specific sets of genes sculpts the differentiated cells that interact, aggregate, and fold, forming the organs of the body. Abnormal gene functioning can affect health at all stages of development. Certain single-gene mutations act before birth, causing broken bones, dwarfism, cancer, and other conditions. Some mutant genes exert their effects during childhood, appearing as early developmental delays or loss of skills. Inherited forms of heart disease and breast cancer can appear in early or middle adulthood, which is earlier than multifactorial forms of these conditions. Pattern baldness is a very common single-gene trait that may not become obvious until well into adulthood. This chapter explores the stages of the human life cycle that form the backdrop against which genes function.

The first cell that leads to development of a new individual forms when a **sperm** from a male and an **oocyte** (also called an egg) from a female join. Sperm and oocytes are **gametes,** or sex cells. They mix genetic material from past generations. Because we have thousands of genes, some with many variants, each person (except for identical twins) has a unique combination of inherited traits.

Sperm and oocytes are produced in the reproductive system. The reproductive organs are organized similarly in the male and female. Each system has

- paired structures, called **gonads,** where the sperm and oocytes are manufactured;
- tubular structures that transport these cells; and
- hormones and secretions that control reproduction.

The Male

Sperm cells develop within a 125-meter-long network of seminiferous tubules, which are packed into paired, oval organs called **testes** (testicles) (**figure 3.1**). The testes are the male gonads. They lie outside the abdomen within a sac, the scrotum. This location keeps the testes cooler than the rest of the body, which is necessary for sperm to develop. Leading from each testis is a tightly coiled tube, the epididymis, in which sperm cells mature and are stored. Each epididymis continues into another tube, the ductus deferens. Each ductus deferens bends behind the bladder and joins the urethra, which is the tube that carries sperm and urine out through the penis.

Along the sperm's path, three glands add secretions. The ductus deferentia pass through the prostate gland, which produces a thin, milky, acidic fluid that activates the sperm to swim. Opening into the ductus deferens is a duct from the seminal vesicles, which secrete fructose (an energy-rich sugar) and hormonelike prostaglandins, which may stimulate contractions in the female that help sperm and oocyte meet. The bulbourethral glands, each about the size of a pea, join the urethra where it passes through the body wall. They secrete an alkaline mucus that coats the urethra before sperm are released. All of these secretions combine to form the seminal fluid that carries sperm. The fluid is alkaline.

During sexual arousal, the penis becomes erect, which enables it to penetrate and deposit sperm in the female reproductive tract. At the peak of sexual stimulation, a pleasurable sensation called orgasm occurs, accompanied by rhythmic muscular contractions that eject the sperm from each ductus deferens through the urethra and out the penis. The discharge of sperm from the penis, called ejaculation, delivers 200 to 600 million sperm cells.

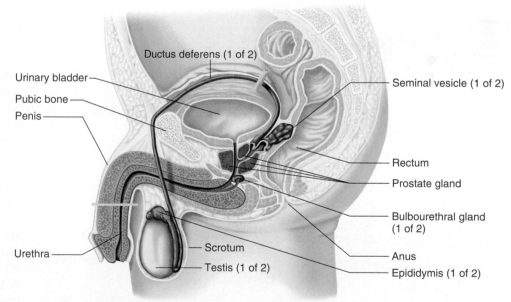

Figure 3.1 The human male reproductive system. Sperm cells are manufactured in the seminiferous tubules, which wind tightly within the testes, which descend into the scrotum. The prostate gland, seminal vesicles, and bulbourethral glands add secretions to the sperm cells to form seminal fluid. Sperm mature and are stored in the epididymis and exit through the ductus deferens. The paired ductus deferentia join in the urethra, which transports seminal fluid from the body.

The Female

The female sex cells develop in paired organs in the abdomen called **ovaries** (**figure 3.2**), which are the female gonads. Within each ovary of a newborn girl are about a million immature oocytes. Each individual oocyte nestles within nourishing follicle cells, and each ovary houses oocytes in different stages of development. After puberty, about once a month, one ovary releases the most mature oocyte. Beating cilia sweep the mature oocyte into the fingerlike projections of one of two uterine (also called fallopian) tubes. The tube carries the oocyte into a muscular, saclike organ called the uterus, or womb.

The oocyte released from the ovary may encounter a sperm, usually in a uterine tube. If the sperm enters the oocyte and the DNA of the two gametes merges into a new nucleus, the result is a fertilized ovum. After a day, this first cell divides while moving through the uterine tube. It settles into the lining of the uterus, where it may continue to divide and an embryo develops. If fertilization does not occur, the oocyte, along with much of the uterine lining, is shed as the menstrual flow. Hormones coordinate the monthly menstrual cycle.

The lower end of the uterus narrows and leads to the cervix, which opens into the tubelike vagina. The vaginal opening is protected on the outside by two pairs of fleshy folds. At the upper juncture of both pairs is the 2-centimeter-long clitoris, which is anatomically similar to the penis. Rubbing the clitoris triggers female orgasm. Hormones control the cycle of oocyte maturation and the preparation of the uterus to nurture a fertilized ovum.

Key Concepts Questions 3.1

1. Describe where sperm develop.
2. Describe where oocytes develop.

3.2 Meiosis

Gametes form from special cells, called germline cells, in a type of cell division called **meiosis** that halves the chromosome number. A further process, maturation, sculpts the distinctive characteristics of sperm and oocyte. The organelle-packed oocyte has 90,000 times the volume of the streamlined sperm.

Meiosis halves the amount of genetic material; the full amount is restored when sperm meets oocyte. Gametes contain 23 different chromosomes, constituting one copy of the genome. In contrast, somatic cells contain 23 *pairs,* or 46 chromosomes. One member of each pair comes from the person's mother and one comes from the father. The chromosome pairs are called **homologous pairs,** or *homologs* for short. Homologs have the same genes in the same order but may carry different alleles, or variants, of the same gene. Recall from chapter 2 that gametes are **haploid** (1*n*), which means that they have only one of each type of chromosome, and somatic cells are **diploid** (2*n*), with two copies of each chromosome type.

Without meiosis, the sperm and oocyte would each contain 46 chromosomes, and the fertilized ovum would have twice the normal number of chromosomes, or 92. Such a genetically overloaded cell, called a polyploid, usually does not develop far enough to be born.

In addition to producing gametes, meiosis mixes up trait combinations. For example, a person might produce one gamete containing alleles encoding green eyes and freckles, yet another gamete with alleles encoding brown eyes and no freckles. Meiosis explains why siblings differ genetically from each other and from their parents.

In a much broader sense, meiosis, as the mechanism of sexual reproduction, provides genetic diversity. A population of sexually reproducing organisms is made up of individuals with different genotypes and phenotypes. Genetic diversity may enable a population to survive an environmental challenge. A population of asexually reproducing organisms, such as bacteria or genetically identical crops, consists of individuals with the same genome sequence. Should a new threat arise, such as an infectious disease that kills only organisms with a certain genotype, then the entire asexual population could be wiped out. However,

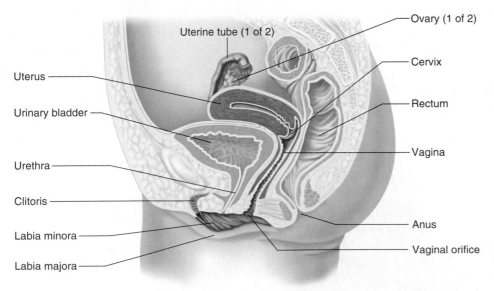

Figure 3.2 **The human female reproductive system.** Oocytes mature in the paired ovaries. Once a month after puberty, an ovary releases one oocyte, which is drawn into a nearby uterine tube. If a sperm fertilizes the oocyte in the uterine tube, the fertilized ovum continues into the uterus, where for 9 months it divides and develops. If the oocyte is not fertilized, the body expels it, along with the built-up uterine lining, as the menstrual flow.

in a sexually reproducing population, individuals that inherited a certain combination of genes might survive. This differential survival of certain genotypes is the basis of evolution by natural selection, discussed in chapter 15.

Meiosis is actually two divisions of the genetic material. The first division is called a **reduction division** (or meiosis I) because it reduces the number of replicated chromosomes from 46 to 23. The second division, called an **equational division** (or meiosis II), produces four cells from the two cells formed in the first division by splitting the replicated chromosomes. **Figure 3.3** shows an overview of the process, and **figure 3.4** depicts the major events of each stage.

As in mitosis, meiosis occurs after an interphase period when DNA is replicated (doubled) (**table 3.1**). For each chromosome pair in the cell undergoing meiosis, one homolog comes from the person's mother, and one from the father. In figures 3.3 and 3.4, the colors represent the contributions of the two parents, whereas size indicates different chromosomes.

After interphase, prophase I (so called because it is the prophase of meiosis I) begins as the replicated chromosomes condense and become visible when stained. A spindle forms. Toward the middle of prophase I, the homologs line up next to one another, gene by gene, in an event called synapsis. A mixture of RNA and protein holds the chromosome pairs together. At this time, the homologs exchange parts, or **cross over**

(**figure 3.5**). All four chromatids that comprise each homologous chromosome pair are pressed together as exchanges occur. After crossing over, each homolog bears some genes from each parent. Prior to this, all of the genes on a homolog were derived from one parent. New gene combinations arise from crossing over when the parents carry different alleles. Toward the end of prophase I, the synapsed chromosomes separate but remain attached at a few points along their lengths.

To understand how crossing over mixes trait combinations, consider a simplified example. Suppose that homologs carry genes for hair color, eye color, and finger length. One of

Table 3.1	Comparison of Mitosis and Meiosis
Mitosis	**Meiosis**
One division	Two divisions
Two daughter cells per cycle	Four daughter cells per cycle
Daughter cells genetically identical	Daughter cells genetically different
Chromosome number of daughter cells same as that of parent cell (2n)	Chromosome number of daughter cells half that of parent cell (1n)
Occurs in somatic cells	Occurs in germline cells
Occurs throughout life cycle	In humans, completes after sexual maturity
Used for growth, repair, and asexual reproduction	Used for sexual reproduction, producing new gene combinations

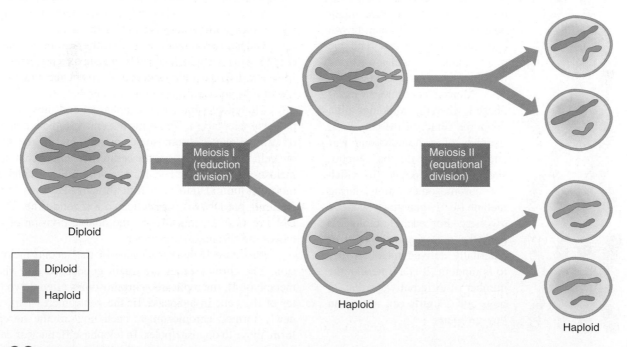

Diploid

Meiosis I (reduction division)

Meiosis II (equational division)

Diploid

Haploid

Haploid

Haploid

Figure 3.3 Overview of meiosis. Meiosis is a form of cell division in which certain cells are set aside and give rise to haploid gametes. This simplified illustration follows the fate of two chromosome pairs rather than the true 23 pairs. In actuality, the first meiotic division reduces the number of chromosomes to 23, all in the replicated form. In the second meiotic division, the cells essentially undergo mitosis. The result of the two meiotic divisions (in this illustration and in reality) is four haploid cells. In this illustration, homologous pairs of chromosomes are indicated by size, and parental origin of chromosomes by color.

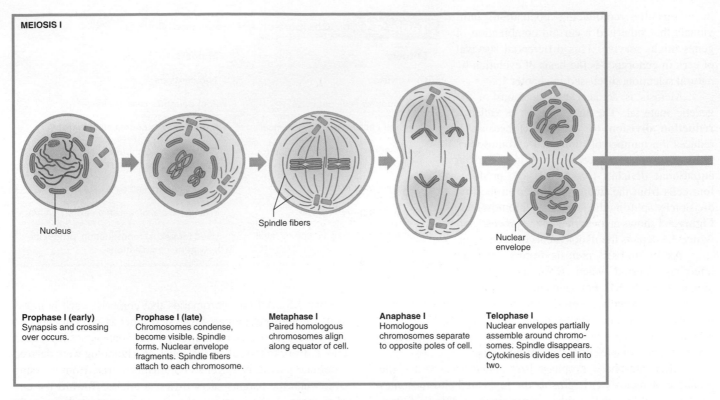

MEIOSIS I

Nucleus

Spindle fibers

Nuclear envelope

Prophase I (early)
Synapsis and crossing over occurs.

Prophase I (late)
Chromosomes condense, become visible. Spindle forms. Nuclear envelope fragments. Spindle fibers attach to each chromosome.

Metaphase I
Paired homologous chromosomes align along equator of cell.

Anaphase I
Homologous chromosomes separate to opposite poles of cell.

Telophase I
Nuclear envelopes partially assemble around chromosomes. Spindle disappears. Cytokinesis divides cell into two.

Figure 3.4 Meiosis. An actual human cell undergoing meiosis has 23 chromosome pairs.

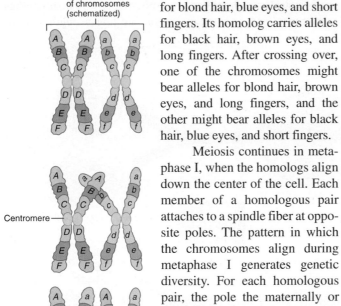

Homologous pair of chromosomes (schematized)

Centromere

Figure 3.5 Crossing over recombines genes. Crossing over generates genetic diversity by recombining genes, mixing parental traits. The capital and lowercase forms of the same letter represent different variants (alleles) of the same gene. A chromosome has hundreds to thousands of genes.

the chromosomes carries alleles for blond hair, blue eyes, and short fingers. Its homolog carries alleles for black hair, brown eyes, and long fingers. After crossing over, one of the chromosomes might bear alleles for blond hair, brown eyes, and long fingers, and the other might bear alleles for black hair, blue eyes, and short fingers.

Meiosis continues in metaphase I, when the homologs align down the center of the cell. Each member of a homologous pair attaches to a spindle fiber at opposite poles. The pattern in which the chromosomes align during metaphase I generates genetic diversity. For each homologous pair, the pole the maternally or paternally derived member goes to is random. It is a little like the number of different ways that 23 boys and 23 girls can line up in boy-girl pairs.

The greater the number of chromosomes, the greater the genetic diversity generated in metaphase I. For two pairs of homologs, four (2^2) different metaphase alignments are possible. For three pairs of homologs, eight (2^3) different alignments can occur. Our 23 chromosome pairs can line up in 8,388,608 (2^{23}) different ways. This random alignment of chromosomes causes **independent assortment** of the genes that they carry. Independent assortment means that the fate of a gene on one chromosome is not influenced by a gene on a different chromosome (**figure 3.6**). Independent assortment accounts for a basic law of inheritance discussed in chapter 4.

Homologs separate in anaphase I and move to opposite poles by telophase I. These movements establish a haploid set of still-replicated chromosomes at each end of the stretched-out cell. Unlike in mitosis, the centromeres of each homolog in meiosis I remain together. During a second interphase, chromosomes unfold into very thin threads. Proteins are manufactured, but DNA is not replicated a second time. The single DNA replication, followed by the double division of meiosis, halves the chromosome number.

Prophase II marks the start of the second meiotic division. The chromosomes are again condensed and visible. In metaphase II, the replicated chromosomes align down the center of the cell. In anaphase II, the centromeres part, and the newly formed chromosomes, each now in the unreplicated form, move to opposite poles. In telophase II, nuclear envelopes form around the four nuclei, which then separate into individual cells. The net result of meiosis is four haploid cells, each carrying a new assortment of genes and chromosomes that hold a single copy of the genome.

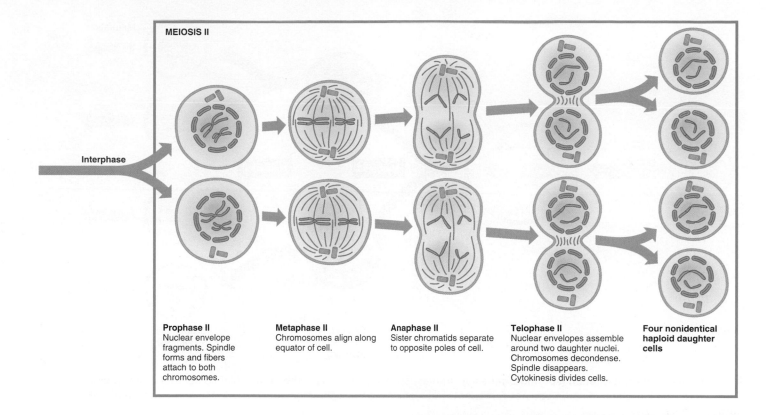

MEIOSIS II

Interphase

Prophase II
Nuclear envelope fragments. Spindle forms and fibers attach to both chromosomes.

Metaphase II
Chromosomes align along equator of cell.

Anaphase II
Sister chromatids separate to opposite poles of cell.

Telophase II
Nuclear envelopes assemble around two daughter nuclei. Chromosomes decondense. Spindle disappears. Cytokinesis divides cells.

Four nonidentical haploid daughter cells

Meiosis generates astounding genetic variety. Any of a person's more than 8 million possible combinations of chromosomes can meet with any of the more than 8 million combinations of a partner, raising potential variability to more than 70 trillion ($8,388,608^2$) genetically unique individuals! Crossing over contributes almost limitless genetic diversity.

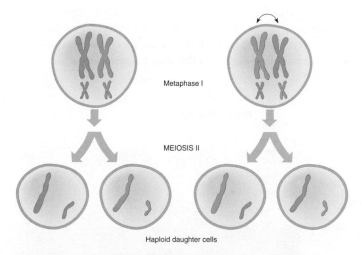

Metaphase I

MEIOSIS II

Haploid daughter cells

Figure 3.6 Independent assortment. The pattern in which homologs randomly align during metaphase I determines the combination of maternally and paternally derived chromosomes in the daughter cells. Two pairs of chromosomes can align in two ways to produce four possibilities in the daughter cells. The potential variability that meiosis generates skyrockets when one considers all 23 chromosome pairs and the effects of crossing over.

Key Concepts Questions 3.2

1. Distinguish between haploid and diploid.
2. Explain how meiosis maintains the chromosome number over generations and mixes gene combinations.
3. Discuss the two mechanisms that generate genotypic diversity.

3.3 Gametes Mature

Meiosis happens in both sexes, but further steps elaborate the very different-looking sperm and oocyte. Different distributions of cell components create the distinctions between sperm and oocytes. The forming gametes of the maturing male and female proceed through similar stages, but with sex-specific terminology and different timetables. A male begins manufacturing sperm at puberty and continues throughout life, whereas a female begins meiosis when she is a fetus. Meiosis in the female completes only if a sperm fertilizes an oocyte.

Sperm Form

Spermatogenesis, the formation of sperm cells, begins in a diploid stem cell called a **spermatogonium (figure 3.7).** This cell divides mitotically, yielding two daughter cells. One continues to specialize into a mature sperm. The other daughter cell remains a stem cell, able to self-renew and continually produce more sperm.

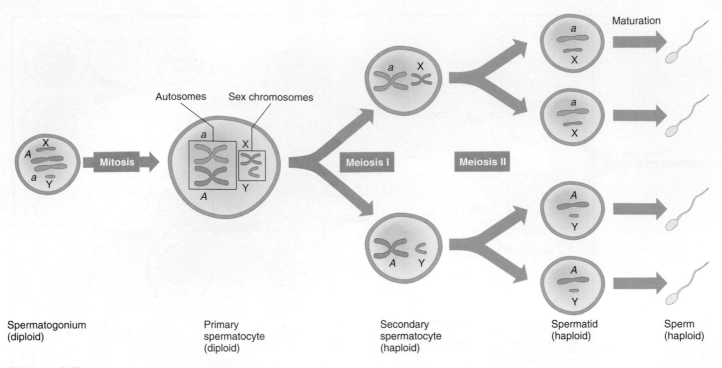

| Spermatogonium (diploid) | Primary spermatocyte (diploid) | Secondary spermatocyte (haploid) | Spermatid (haploid) | Sperm (haploid) |

Figure 3.7 **Sperm formation (spermatogenesis).** Primary spermatocytes have the normal diploid number of 23 chromosome pairs. The large pair of chromosomes represents autosomes (non-sex chromosomes). The X and Y chromosomes are sex chromosomes.

Bridges of cytoplasm attach several spermatogonia, and their daughter cells enter meiosis together. As these spermatogonia mature, they accumulate cytoplasm and replicate their DNA, becoming primary spermatocytes.

During reduction division (meiosis I), each primary spermatocyte divides, forming two equal-sized haploid cells called secondary spermatocytes. In meiosis II, each secondary spermatocyte divides to yield two equal-sized spermatids. Each spermatid then develops the characteristic sperm tail, or flagellum. The base of the tail has many mitochondria, which will split ATP molecules to release energy that will propel the sperm inside the female reproductive tract. After spermatid differentiation, some of the cytoplasm connecting the cells falls away, leaving mature, tadpole-shaped spermatozoa (singular, *spermatozoon*), or sperm. **Figure 3.8** presents an anatomical view showing the stages of spermatogenesis within the seminiferous tubules.

A sperm, which is a mere 0.006 centimeter (0.0023 inch) long, must travel about 18 centimeters (7 inches) to reach an oocyte. Each sperm cell consists of a tail, body or midpiece, and a head region (**figure 3.9**). A membrane-covered area on the front end, the acrosome, contains enzymes that help the sperm cell penetrate the protective layers around the oocyte. Within the large sperm head, DNA is wrapped around proteins. The sperm's DNA at this time is genetically inactive. A male manufactures trillions of sperm in his lifetime. Although many of these will come close to an oocyte, very few will actually touch one.

Meiosis in the male has built-in protections that help prevent sperm from causing some birth defects. Spermatogonia that are exposed to toxins tend to be so damaged that they never mature into sperm. More mature sperm cells exposed to toxins are often so damaged that they cannot swim.

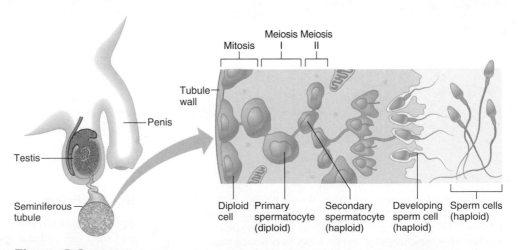

Figure 3.8 **Meiosis produces sperm cells.** Diploid cells divide through mitosis in the linings of the seminiferous tubules. Some daughter cells then undergo meiosis, producing haploid secondary spermatocytes, which differentiate into mature sperm cells.

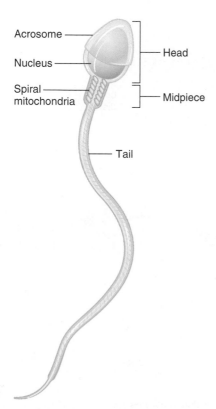

Figure 3.9 Sperm. A sperm has distinct regions that assist in delivering DNA to an oocyte.

Oocytes Form

Meiosis in the female, called **oogenesis** (egg making), begins with a diploid cell, an **oogonium**. Unlike male cells, oogonia are not attached. Instead, follicle cells surround each oogonium. As each oogonium grows, cytoplasm accumulates, DNA replicates, and the cell becomes a primary oocyte. The ensuing meiotic division in oogenesis, unlike the male pathway, produces cells of different sizes.

In meiosis I, the primary oocyte divides into two cells: a small cell with very little cytoplasm, called a first **polar body,** and a much larger cell called a secondary oocyte (**figure 3.10**). Each cell is haploid, with the chromosomes in replicated form. In meiosis II, the tiny first polar body may divide to yield two polar bodies of equal size, with unreplicated chromosomes; or the first polar body may decompose. The secondary oocyte, however, divides unequally in meiosis II to produce another small polar body, with unreplicated chromosomes, and the mature egg cell, or ovum, which contains a large volume of cytoplasm. **Figure 3.11** summarizes meiosis in the female, and **figure 3.12** provides an anatomical view of the process.

Most of the cytoplasm among the four meiotic products in the female ends up in only one cell, the ovum. The woman's body absorbs the polar bodies, which normally play no further role in development. Rarely, a sperm fertilizes a polar body. When this happens, the woman's hormones respond as if she is pregnant, but a disorganized clump of cells that is not an embryo grows for a few weeks, and then leaves the woman's body. This event is a type of miscarriage called a "blighted ovum."

Before birth, a female has a million or so oocytes arrested in prophase I. (This means that when your grandmother was pregnant with your mother, the oocyte that would be fertilized and eventually become you was already there.) By puberty, about 400,000 oocytes remain. After puberty, meiosis I continues in one or several oocytes each month, but halts again at metaphase II. In response to specific hormonal cues each month, one ovary releases a secondary oocyte; this event is ovulation. The oocyte drops into a uterine tube, where waving cilia move it toward the uterus. Along the way, if a sperm penetrates the oocyte membrane, then female meiosis completes, and a fertilized ovum forms. If the secondary oocyte is not fertilized, it degenerates and leaves the body in the menstrual flow, meiosis never completed.

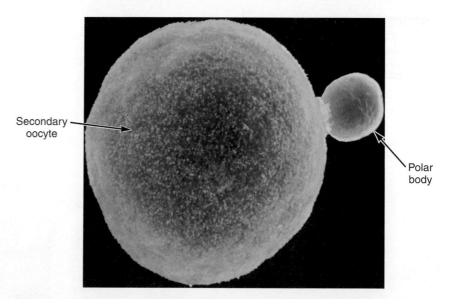

Figure 3.10 Meiosis in a female produces a secondary oocyte and a polar body. Unequal division and apportioning of cell parts enable the cell destined to become a fertilized ovum to accumulate most of the cytoplasm and organelles from the primary oocyte, but with only one genome copy. The oocyte accumulates abundant cytoplasm that would have gone into the meiotic product that became the polar body if the division had been equal.

A female ovulates about 400 oocytes between puberty and menopause. Most oocytes degrade, because fertilization is so rare. Furthermore, only one in three of the oocytes that do meet and merge with a sperm cell will continue to grow, divide, and specialize to eventually form a new individual.

The diminishing number of oocytes present as a female ages was thought to be due to a dwindling original supply of the cells. However, researchers recently discovered that human ovaries contain oocyte-producing stem cells.

Meiosis and Mutations

The gametes of older people are more likely to have new mutations (e.g., not inherited mutations) than the gametes of younger people. Older women are at higher risk of producing oocytes that have an extra or missing chromosome. If fertilized, such oocytes lead to offspring with the chromosomal conditions described in chapter 13. Older men are also more likely to produce gametes that have genetic errors, but the "paternal age effect" usually causes dominant single-gene diseases. That is, only one copy of the mutant gene causes the condition.

Men introduce single-gene conditions, whereas women originate chromosome imbalances because of different timetables of meiosis. In females, oocytes exist on the brink of meiosis I for years, and the cells complete meiosis II only if they are fertilized. Mistakes occur when gametes are active, and that is when they are distributing their chromosomes. If a homologous pair doesn't separate, the result could be an oocyte with an extra or missing chromosome.

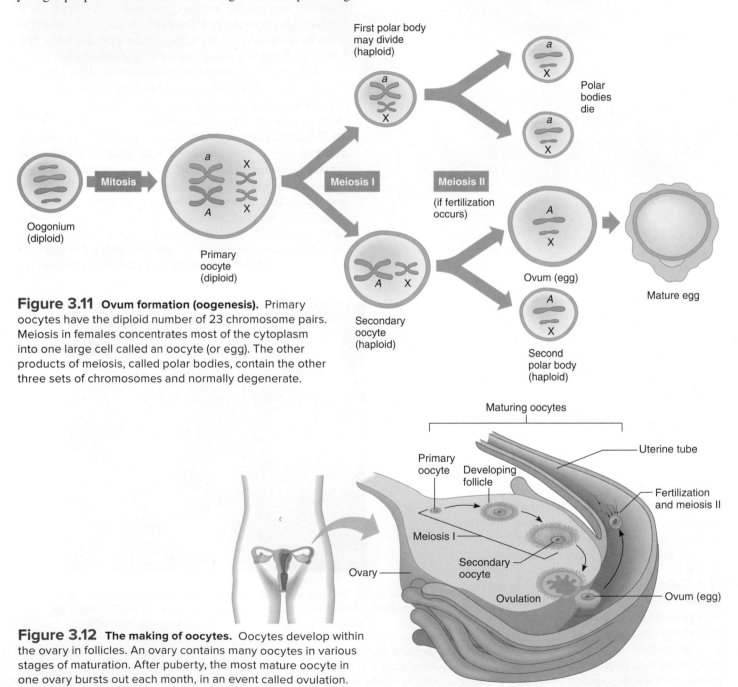

Figure 3.11 Ovum formation (oogenesis). Primary oocytes have the diploid number of 23 chromosome pairs. Meiosis in females concentrates most of the cytoplasm into one large cell called an oocyte (or egg). The other products of meiosis, called polar bodies, contain the other three sets of chromosomes and normally degenerate.

Figure 3.12 The making of oocytes. Oocytes develop within the ovary in follicles. An ovary contains many oocytes in various stages of maturation. After puberty, the most mature oocyte in one ovary bursts out each month, in an event called ovulation.

In contrast to oocytes, sperm develop in only 74 days. "Paternal age effect" conditions arise from stem cells in the testis that divide every 16 days, from puberty on, offering many opportunities for DNA replication (discussed in chapter 9) to make a mistake, generating a dominant mutation. Such paternal age effect conditions include a form of dwarfism; premature fusion of an infant's skull bones; several birth defect syndromes; and multiple endocrine neoplasia (OMIM 131100), which causes glandular cancers.

The maternal age effect in causing chromosomal imbalances has been recognized since the nineteenth century, when physicians noticed that babies with trisomy 21 Down syndrome tended to be the youngest children in large families. The paternal age effect has only recently been recognized. Researchers examined testes from elderly men who had died and donated their bodies to science. Each testis was divided into 6 slices of 32 pieces, and probed for mutations that cause multiple endocrine neoplasia. Mutations occurred in discrete sections of the testes, indicating the involvement of stem cells perpetuating the mutation as they divided. The older the man, the more testis cells had the mutations.

Key Concepts Questions 3.3

1. Explain the steps of sperm and oocyte formation and specialization.

2. How do the timetables of spermatogenesis and oogenesis differ?

3.4 Prenatal Development

A prenatal human is considered an **embryo** for the first 8 weeks. During this time, rudiments of all body parts form. The embryo in the first week is in a "preimplantation" stage because it has not yet settled into the uterine lining. Some biologists do not consider a prenatal human an embryo until it begins to develop tissue layers, at about 2 weeks.

Prenatal development after the eighth week is the fetal period, when structures grow and specialize. From the start of the ninth week until birth, the prenatal human organism is a **fetus**.

Sperm and Oocyte Meet at Fertilization

Hundreds of millions of sperm cells are deposited in the vagina during sexual intercourse. A sperm cell can survive in a woman's body for up to 3 days, but the oocyte can only be fertilized in the 12 to 24 hours after ovulation.

The woman's body helps sperm reach an oocyte. A process in the female called capacitation chemically activates sperm, and the oocyte secretes a chemical that attracts sperm. Contractions of the female's muscles and moving of sperm tails propel the sperm. Still, only 200 or so sperm get near the oocyte.

A sperm first contacts a covering of follicle cells, called the corona radiata, that guards a secondary oocyte. The sperm's acrosome then bursts, releasing enzymes that bore through a protective layer of glycoprotein (the zona pellucida) beneath the corona radiata. Fertilization, or conception, begins when the outer membranes of the sperm and secondary oocyte meet and a protein on the sperm head contacts a different protein on the oocyte (**figure 3.13**). A wave of electricity spreads physical and chemical changes across the oocyte surface, which keep other sperm out. More than one sperm can enter an oocyte, but the resulting cell has too much genetic material for development to follow.

Usually only the sperm's head enters the oocyte. Within 12 hours of the sperm's penetration, the ovum's nuclear membrane disappears, and the two sets of chromosomes, called pronuclei, approach one another. Within each pronucleus, DNA replicates. Fertilization completes when the two genetic packages meet and merge, forming the genetic instructions for a new individual. The fertilized ovum is called a **zygote**. The *Bioethics: Choices for the Future* reading discusses cloning, which is a way to start development without a fertilized egg.

The Embryo Cleaves and Implants

A day after fertilization, the zygote divides by mitosis, beginning a period of frequent cell division called **cleavage** (**figure 3.14**). The resulting early cells are called **blastomeres**. When the blastomeres form a solid ball of sixteen or more cells, the embryo is called a **morula** (Latin for "mulberry," which it resembles).

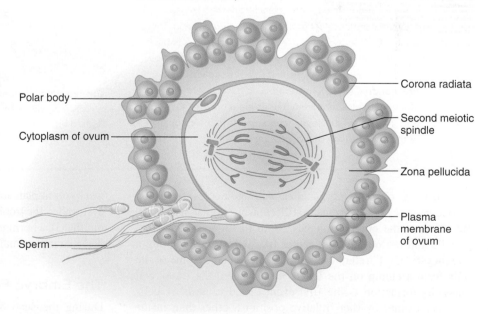

Polar body

Cytoplasm of ovum

Sperm

Corona radiata

Second meiotic spindle

Zona pellucida

Plasma membrane of ovum

Figure 3.13 **Fertilization.** Fertilization by a sperm cell induces the oocyte (arrested in metaphase II) to complete meiosis. Before fertilization occurs, the sperm's acrosome bursts, spilling enzymes that help the sperm's nucleus enter the oocyte.

Why a Clone Is Not an Exact Duplicate

Cloning creates a genetic replica of an individual. In contrast, normal reproduction and development combine genetic material from two individuals. In fiction, scientists have cloned Nazis, politicians, dinosaurs, children, and organ donors. Real scientists have cloned sheep, mice, cats, pigs, dogs, monkeys, and amphibians.

Cloning transfers a nucleus from a somatic cell into an oocyte whose nucleus has been removed. The technique is more accurately called "somatic cell nuclear transfer" or just "nuclear transfer." Cloning cannot produce an exact replica of a person, for several reasons:

- **Premature cellular aging.** In some species, telomeres of chromosomes in the donor nucleus are shorter than those in the recipient cell (see figure 2.18).
- **Altered gene expression.** In normal development, for some genes, one copy is turned off, depending upon which parent transmits it. This phenomenon is called genomic imprinting (see section 6.5). In cloning, genes in a donor nucleus skip passing through a sperm or oocyte, and thus are not imprinted. Lack of imprinting somehow causes cloned animals to be unusually large. Regulation of gene expression is abnormal at many times during prenatal development of a clone.
- **More mutations.** DNA from a donor cell has had years to accumulate mutations. A mutation might not be noticeable in one of millions of somatic cells in a body but it could be devastating if that nucleus is used to program development of a new individual.
- **X inactivation.** At a certain time in early prenatal development in female mammals, one X chromosome is inactivated. Whether that X chromosome is from the mother or the father occurs at random in each cell, creating an overall mosaic pattern of expression for genes on the X chromosome. The pattern of X inactivation of a female clone would probably not match that of her nucleus donor, because X inactivation occurs in the embryo, not the first cell (see section 6.4).
- **Mitochondrial DNA.** Mitochondria contain DNA. A clone's mitochondria descend from the recipient oocyte, not from the donor cell, because mitochondria are in the cytoplasm, not the nucleus (see figure 5.10).

The environment is another powerful factor in why a clone isn't an identical copy. For example, coat color patterns differ in cloned calves and cats. When the animals were embryos, cells destined to produce pigment moved in a unique way in each individual, producing different color patterns. In humans, experience, nutrition, stress, exposure to infectious disease, and many other factors join our genes in molding who we are.

Questions For Discussion

1. Which of your characteristics do you think could not be duplicated in a clone, and why?
2. What might be a reason to clone humans?
3. What are potential dangers of cloning humans?
4. Do human clones exist naturally?

Cloned cats. A company that tried to sell cloned cats for $50,000 quickly went out of business. The author has not figured out how to clone her cat Juice, but with nine lives, cloning might not be necessary.

During cleavage, organelles and molecules from the secondary oocyte's cytoplasm still control cellular activities, but some of the embryo's genes begin to function. The ball of cells hollows out, and its center fills with fluid, creating a **blastocyst**. "Cyst" refers to the fluid-filled center. Some of the cells form a clump on the inside lining called the **inner cell mass**. Its formation is the first event that distinguishes cells from each other by their relative positions, other than inside and outside the morula. The inner cell mass continues developing, forming the embryo.

A week after conception, the blastocyst nestles into the uterine lining. This event, called implantation, takes about a week. As implantation begins, the outermost cells of the blastocyst, called the trophoblast, secrete human chorionic gonadotropin (hCG), a hormone that prevents menstruation. This hormone detected in a woman's urine or blood is one sign of pregnancy.

The Embryo Forms

During the second week of prenatal development, a space called the amniotic cavity forms between the inner cell mass and the outer cells anchored to the uterine lining. Then the inner cell mass flattens into a two-layered embryonic disc. The layer nearest the amniotic cavity is the **ectoderm;** the inner

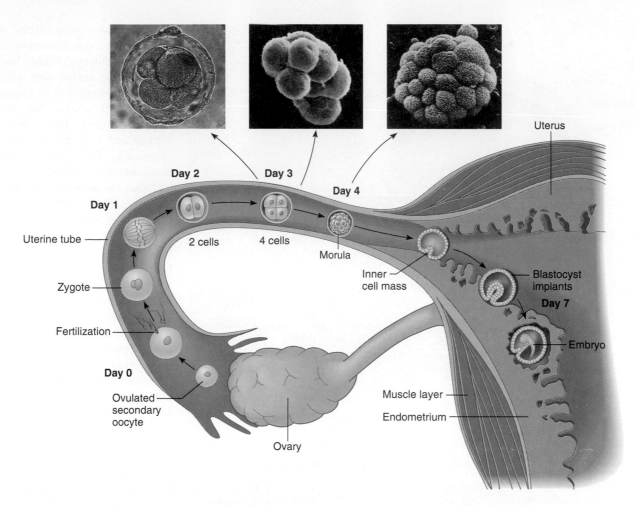

Figure 3.14 **Cleavage: From ovulation to implantation.** The zygote forms in the uterine tube when a sperm nucleus fuses with the nucleus of an oocyte. The first divisions proceed while the zygote moves toward the uterus. By day 7, the zygote, now called a blastocyst, begins to implant in the uterine lining.

layer, closer to the blastocyst cavity, is the **endoderm**. Shortly after, a third layer, the **mesoderm,** forms in the middle. This three-layered structure is called the primordial embryo, or the **gastrula (figure 3.15)**.

The cells that make up the layers of the primordial embryo, called **primary germ layers,** become "determined," or fated, to develop as specific cell types. The cells specialize by the progressive switching off of the expression of genes important and active in the early embryo as other genes begin to be expressed. Gene expression shuts off when a small molecule called a methyl (CH_3; a carbon bonded to three hydrogen atoms), binds certain genes and proteins. In the early embryo, methyl groups bind the proteins (called histones) that hold the long DNA molecules in a coil (see figure 9.13). In the later embryo, methyl groups bind to and silence specific sets of genes. The pattern of methyl group binding that guides the specialization of the embryo establishes the epigenome. "Epigenetics" refers to changes to DNA other than changes in DNA nucleotide base sequence.

Each primary germ layer gives rise to certain structures. Cells in the ectoderm become skin, nervous tissue, or parts of certain glands. Endoderm cells form parts of the liver and

pancreas and the linings of many organs. The middle layer of the embryo, the mesoderm, forms many structures, including muscle, connective tissues, the reproductive organs, and the kidneys.

Genes called homeotics control how the embryo develops parts in the right places. Mutations in these genes cause some very interesting conditions, including forms of intellectual disability, autism, blood cancer, and blindness. The homeotic mutations were originally studied in fruit flies that had legs growing where their antennae should be.

Table 3.2 summarizes the stages of early human prenatal development. **Clinical Connection 3.1** describes two human diseases that result from mutations in homeotic genes.

Supportive Structures Form

As an embryo develops, structures form that support and protect it. These include chorionic villi, the placenta, the yolk sac, the allantois, the umbilical cord, and the amniotic sac.

By the third week after conception, fingerlike outgrowths called chorionic villi extend from the area of the embryonic disc close to the uterine wall. The villi project into pools of the woman's blood. Her blood system and the embryo's are separate, but

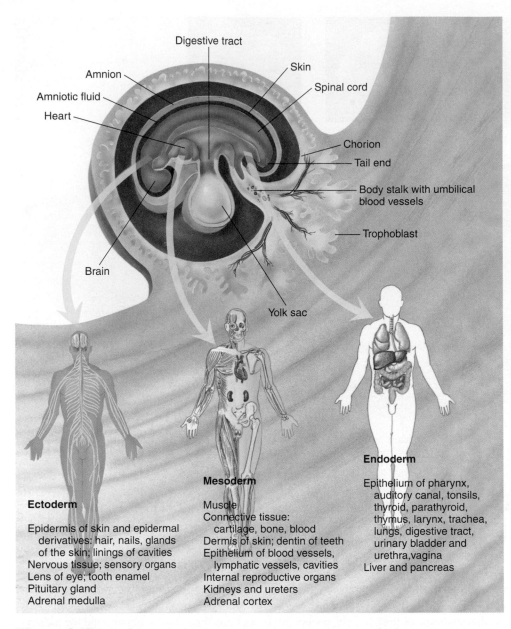

Figure 3.15 **The primordial embryo.** When the three primary germ layers of the embryo form at gastrulation, many cells become "fated" to follow a specific developmental pathway. Each layer retains stem cells as the organism develops. Under certain conditions, stem cells may produce daughter cells that can specialize as many cell types.

Ectoderm

Epidermis of skin and epidermal
  derivatives; hair, nails, glands
  of the skin; linings of cavities
Nervous tissue; sensory organs
Lens of eye; tooth enamel
Pituitary gland
Adrenal medulla

Mesoderm

Muscle
Connective tissue:
  cartilage, bone, blood
Dermis of skin; dentin of teeth
Epithelium of blood vessels,
  lymphatic vessels, cavities
Internal reproductive organs
Kidneys and ureters
Adrenal cortex

Endoderm

Epithelium of pharynx,
  auditory canal, tonsils,
  thyroid, parathyroid,
  thymus, larynx, trachea,
  lungs, digestive tract,
  urinary bladder and
  urethra,vagina
Liver and pancreas

Table 3.2	Stages and Events of Early Human Prenatal Development	
Stage	**Time Period**	**Principal Events**
Fertilized ovum	12–24 hours following ovulation	Oocyte fertilized; zygote has 23 pairs of chromosomes and is genetically distinct
Cleavage	30 hours to third day	Mitosis increases cell number
Morula	Third to fourth day	Solid ball of cells
Blastocyst	Fifth day through second week	Hollowed ball forms trophoblast (outside) and inner cell mass, which implants and flattens to form embryonic disc
Gastrula	End of second week	Primary germ layers form

nutrients and oxygen diffuse across the chorionic villi from her circulation to the embryo. Wastes leave the embryo's circulation and enter the woman's circulation to be excreted.

By 10 weeks, the placenta is fully formed. This organ links woman and fetus for the rest of the pregnancy. The placenta secretes hormones that maintain pregnancy and alter the woman's metabolism to send nutrients to the fetus.

Other structures nurture the developing embryo. The yolk sac manufactures blood cells, as does the allantois, a membrane surrounding the embryo that gives rise to the umbilical blood vessels. The umbilical cord forms around these vessels and attaches to the center of the placenta. Toward the end of the embryonic period, the yolk sac shrinks, and the amniotic sac swells with fluid that cushions the embryo and maintains a constant temperature and pressure. The amniotic fluid contains fetal urine and cells.

Two of the supportive structures that develop during pregnancy provide material for prenatal tests discussed in chapters 13 and 20. Chorionic villus sampling examines chromosomes from cells snipped off the chorionic villi at 10 weeks. Because the villi cells and the embryo's cells come from the same fertilized ovum, an abnormal chromosome in villi cells should also be in the embryo. In amniocentesis, a sample of amniotic fluid is taken and fetal cells in it are examined for biochemical, genetic, and chromosomal anomalies.

The umbilical cord is another prenatal structure that has medical applications. In addition to blood stem cells mentioned in *Bioethics: Choices for the Future* on page 37, the cord yields bone, fat, nerve, and cartilage cells. Stem cells from the cord itself are used to treat a respiratory disease of newborns that scars and inflames the lungs. The stem cells become lung cells that secrete surfactant, which is the chemical that inflates the microscopic air sacs, and the cell type that

When an Arm Is Really a Leg: Homeotic Mutations

Flipping the X ray showed Stefan Mundlos, MD, that his hunch was right—the patient's arms were odd-looking and stiff because the elbows were actually knees! The condition, Liebenberg syndrome (OMIM 186550), had been described in 1973 among members of a five-generation white South African family (**figure 1**). Four males and six females had stiff elbows and wrists, and short fingers that looked strangely out of place. A trait that affects both sexes in every generation displays classic autosomal dominant inheritance—each child of a person with strange limbs had a 50:50 chance of having the condition too.

In 2000, a medical journal described a second family with Liebenberg syndrome. Several members had restricted movements because they couldn't bend their huge, misshapen elbows. Then in 2010, a report appeared on identical twin girls with the curious stiff elbows and long arms, with fingers that looked like toes.

In 2012, Dr. Mundlos noted that the muscles and tendons of the elbows, as well as the bones of the arms, weren't quite right in his patient. The doctor, an expert in the comparative anatomy of limb bones of different animals, realized that the stiff elbows were acting like knees. The human elbow joint hinges and rotates, but the knee extends the lower leg straight out. Then an X-ray scan of the patient's arm fell to the floor. "I realized that the entire limb had the appearance of a leg. Normally you would look at the upper limb X ray with the hand up, whereas the lower limb is looked at foot down. If you turn the X ray around, it looks just like a leg," Dr. Mundlos said.

Genes that switch body parts are termed *homeotic*. They are well studied in experimental organisms as evolutionarily diverse as fruit flies, flowering plants, and mice, affecting the positions of larval segments, petals, legs, and much more. Assignment of body parts begins in the early embryo, when cells look alike but are already fated to become specific structures. Gradients (increasing or decreasing concentrations) of "morphogen" proteins in an embryo program a particular region to develop a certain way. Mix up the messages, and an antenna becomes a leg, or an elbow a knee.

Homeotic genes include a 180-base-long DNA sequence, called the homeobox, which enables the encoded protein to bind other proteins that turn on sets of other genes, crafting an embryo, section by section. Homeotic genes line up on their chromosomes in the precise order in which they're deployed in development, like chapters in an instruction manual to build a body.

The human genome has four clusters of homeotic genes, and mutations in them cause disease. In certain lymphomas, a homeotic mutation sends white blood cells along the wrong developmental pathway, resulting in too many of some blood cell types and too few of others. The abnormal ears, nose, mouth,

and throat of DiGeorge syndrome (OMIM 188400) echo the abnormalities in *Antennapedia,* a fruit fly mutant that has legs on its head. Extra and fused fingers and various bony alterations also stem from homeotic mutations.

The search for the mutation behind the arm-to-leg Liebenberg phenotype began with abnormal chromosomes. Affected members of the three known families were each missing 134 DNA bases in the same part of the fifth largest chromosome. The researchers zeroed in on a gene called *PITX1* that controls other genes that in turn oversee limb development. In the Liebenberg families, the missing DNA places an "enhancer" gene near *PITX1,* altering its expression in a way that mixes up developmental signals so that the forming arm instead becomes a leg. Fortunately the condition appears more an annoying oddity than a disease.

Questions for Discussion

1. What is the genotype and phenotype of Liebenberg syndrome?

2. How can homeotic mutations be seen in such different species as humans, mice, fruit flies, and flowering plants?

3. Explain the molecular basis of a homeotic mutation and the resulting phenotype.

4. Name another human disease that results from a homeotic mutation.

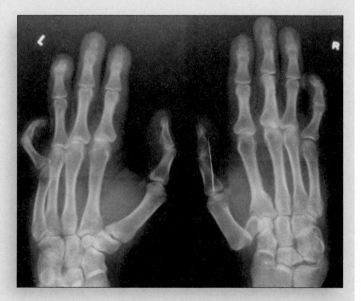

Figure 1 The hands of a person with Liebenberg syndrome resemble feet; the arms resemble legs.

exchanges oxygen for carbon dioxide. Stem cells from umbilical cords are abundant, easy to obtain and manipulate, and can give rise to almost any cell type.

Multiples

Twins and other multiples arise early in development. Fraternal, or **dizygotic (DZ)**, **twins** result when two sperm fertilize two oocytes. This can happen if ovulation occurs in two ovaries in the same month, or if two oocytes leave the same ovary and are both fertilized. DZ twins are no more alike than any two siblings, although they share a very early environment in the uterus. The tendency to ovulate two oocytes in a month, leading to conception of DZ twins, can run in families.

Identical, or **monozygotic (MZ)**, **twins** descend from a single fertilized ovum and therefore are genetically identical. They are natural clones. Three types of MZ twins can form, depending upon when the fertilized ovum or very early embryo splits (**figure 3.16**). This difference in timing determines which supportive structures the twins share. About a third of all MZ twins have completely separate chorions and amnions, and about two-thirds share a chorion but have separate amnions. Slightly fewer than 1 percent of MZ twins share both amnion and chorion. (The amnion is the fluid-filled sac that surrounds the fetus. The chorion develops into the placenta.) These differences may expose the different types of MZ twins to slightly different uterine environments. For example, if one chorion develops more attachment sites to the maternal circulation, one twin may receive more nutrients and gain more weight than the other.

In 1 in 50,000 to 100,000 pregnancies, an embryo divides into twins after the point at which the two groups of cells can develop as two individuals, between days 13 and 15. The result is conjoined or "Siamese" twins. The latter name comes from

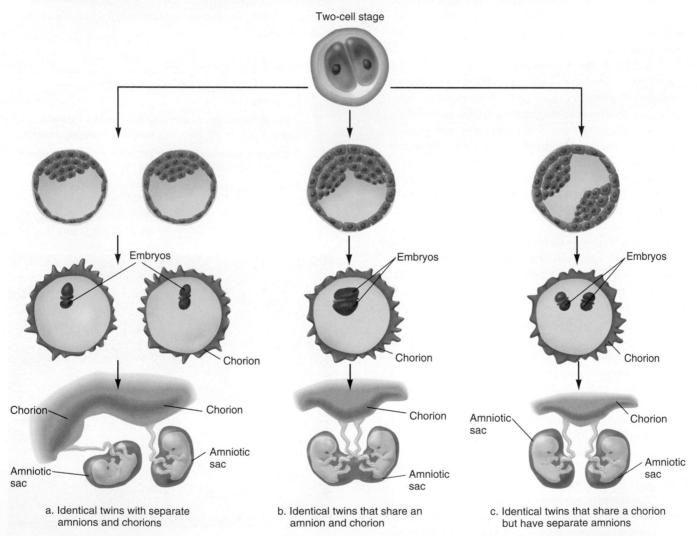

a. Identical twins with separate amnions and chorions

b. Identical twins that share an amnion and chorion

c. Identical twins that share a chorion but have separate amnions

Figure 3.16 **Types of identical twins.** Identical twins originate at three points in development. **(a)** In about one-third of identical twins, separation of cells into two groups occurs before the trophoblast forms on day 5. These twins have separate chorions and amnions. **(b)** About 1 percent of identical twins share a single amnion and chorion, because the tissue splits into two groups after these structures have already formed. **(c)** In about two-thirds of identical twins, the split occurs after day 5 but before day 9. These twins share a chorion but have separate amnions. Fraternal twins result from two sperm fertilizing two secondary oocytes. These twins develop their own amniotic sacs, yolk sacs, allantois, placentas, and umbilical cords.

Chang and Eng Bunker, who were born in Thailand, then called Siam, in 1811. They were joined by a band of tissue from the navel to the breastbone, and could easily have been separated today. Chang and Eng lived for 63 years, attached. They fathered 22 children and divided each week between their wives.

For Abigail and Brittany Hensel, shown in **figure 3.17**, the separation occurred after day 9 of development, but before day 14. Biologists know this because the shared organs have derivatives of ectoderm, mesoderm, and endoderm; that is, when the lump of cells divided incompletely, the three primary germ layers had not yet completely sorted themselves out. The Hensels are extremely rare "incomplete twins." They are "dicephalic," which means that they have two heads. They are very much individuals.

Each twin has her own neck, head, heart, stomach, gallbladder, and lungs. Each has one leg and one arm, and a third arm between their heads was surgically removed. Each twin also has her own nervous system. The twins share a large liver, a single bloodstream, and all organs below the navel. They have three kidneys. Because at birth Abigail and Brittany were strong and healthy, doctors suggested surgery to separate them. But their parents, aware of other cases where only one child survived separation, declined surgery.

As teens, Abigail and Brittany enjoyed kickball, volleyball, basketball, and cycling. Like any teen girls, they developed distinctive tastes in clothing and in food. They graduated from college, had their own reality TV series, and posted a YouTube video of themselves driving. They are elementary school teachers. Clearly, these young women have not let their unusual development stop them from enjoying life and achieving some of their dreams.

MZ twins occur in 3 to 4 pregnancies per 1,000 births worldwide. In North America, twins occur in about 1 in 81 pregnancies, which means that 1 in 40 of us is a twin. However, not all twins survive to be born. One study of twins detected early in pregnancy showed that up to 70 percent of the eventual births are of a single child. This is called the "vanishing twin" phenomenon.

Figure 3.17 Conjoined twins. Abigail and Brittany Hensel are the result of incomplete twinning during the first 2 weeks of prenatal development.

The Embryo Develops

As prenatal development proceeds, different rates of cell division in different parts of the embryo fold the forming tissues into intricate patterns. In a process called embryonic induction, the specialization of one group of cells causes adjacent groups of cells to specialize. Gradually, these changes mold the three primary germ layers into organs and organ systems. Organogenesis is the transformation of the simple three layers of the embryo into distinct organs. During the weeks of organogenesis, the developing embryo is particularly sensitive to environmental influences such as chemicals and viruses.

During the third week of prenatal development, a band called the primitive streak appears along the back of the embryo. Some nations designate day 14 of prenatal development and primitive streak formation as the point beyond which they ban research on the human embryo. The reasoning is that the primitive streak is the first sign of a nervous system and day 14 is when implantation completes.

The primitive streak gradually elongates to form an axis that other structures organize around as they develop. The primitive streak eventually gives rise to connective tissue progenitor cells and the notochord, which is a structure that forms the basic framework of the skeleton. The notochord induces a sheet of overlying ectoderm to fold into the hollow **neural tube,** which develops into the brain and spinal cord (central nervous system). If the neural tube does not completely zip up by day 28, a birth defect called a neural tube defect (NTD) occurs. Parts of the brain or spinal cord protrude from the open head or spine, and body parts below the defect cannot move. Surgery can correct some NTDs (see *Bioethics: Choices for the Future* on page 297). Lack of the B vitamin folic acid can cause NTDs in embryos with a genetic susceptibility. For this reason, the U.S. government adds the vitamin to grains, and pregnant women take supplements. A blood test during the 15th week of pregnancy detects a substance from the fetus's liver called alpha fetoprotein (AFP) that moves at an abnormally rapid rate into the woman's circulation if there is an open NTD. Using hints from more than 200 genes associated with NTDs in other vertebrates, investigators are searching the genome sequences of people with NTDs to identify patterns of gene variants that could contribute to or cause these conditions.

Appearance of the neural tube marks the beginning of organ development. Shortly after, a reddish bulge containing the heart appears. The heart begins to beat around day 18, and this is easily detectable by day 22. Soon the central nervous system starts to form.

The fourth week of embryonic existence is one of spectacularly rapid growth and differentiation (**figure 3.18**). Arms and legs begin to extend from small buds on the torso. Blood cells form and fill primitive blood vessels. Immature lungs and kidneys begin to develop.

By the fifth and sixth weeks, the embryo's head appears too large for the rest of its body. Limbs end in platelike structures with tiny ridges, and gradually apoptosis sculpts the fingers and toes. The eyes are open, but they do not yet have lids or irises. By the seventh and eighth weeks, a skeleton composed

Growing Human Mini-Brains

Most of what we know about human development comes from observing what happens when the process goes awry, in birth defects or in model organisms. A new tool to watch human organs form comes from induced pluripotent stem (iPS) cell technology. Recall from figure 2.22 that iPS cells are "reprogrammed" from somatic cell nuclei.

Researchers Juergen Knoblich and Madeline Lancaster at the Institute of Molecular Biotechnology in Austria use iPS cells to culture the beginnings of human brains in glass vessels called bioreactors. The "cerebral organoids," layers of neurons and the progenitor cells that give rise to them (**figure 1**), provide peeks into how the brain develops. Here the researchers discuss their experiments.

Dr. Lancaster:

We begin with human iPS cells generated from many cell types. We use those cells to generate [structures called] embryoid bodies that show the first specification of cell types. Then we transfer cells to special culture conditions in which most types of embryonic tissues stop developing, and neuroectoderm is allowed, which gives rise to the central nervous system. We transfer neuroectoderm to an extracellular matrix scaffold that allows it to develop into more complex tissues, then to a spinning bioreactor that constantly agitates the medium for better nutrient and oxygen exchange. This treatment results in three-dimensional tissues with characteristic human brain development, cerebral organoids. They have discrete regions that resemble different areas of the early developing human brain and immature retina.

The organoids develop to just past the transition from embryo to fetus. The researchers compared cerebral organoids from healthy people to those made from a patient with microcephaly (a very small head). Mutations in several different genes cause or contribute to microcephaly, which often causes great intellectual disability because the brain stops growing. The patient's microcephaly stemmed from mutations in both copies of the gene *CDK5RAP2* (OMIM 608201).

Dr. Lancaster:

Organoids derived from a patient were much smaller than control healthy organoids. At this point a neural stem cell population undergoes rounds of division to make more stem cells, but the patient-derived organoids start making neurons too early. Stem cells not expanding deplete stem cells, which leads to the overall size decrease. We think this is the basis of the microcephaly in this patient.

The human brain has a layer of neural stem cells hugging the ventricles (fluid-filled spaces) that mice do not have.

Dr. Knoblich:

Our system allows us to study the human-specific aspects of brain development and the several genes that cause microcephaly. We'd like to move to more common diseases, like schizophrenia or autism, to study the underlying defects in brain development. Organoid cultures may be very useful for pharmacology. Drugs are tested in model organisms and isolated human cells in culture, but organoids will make it possible to test drugs directly in a human setting and avoid animal experiments. We hope our organoids will contribute to the transfer of knowledge from model organisms to cure disease in humans.

Cerebral organoids do not recapitulate development of a complete human brain. For now, they are a research tool used to understand how development can fail. Perhaps someday they will serve as replacement parts.

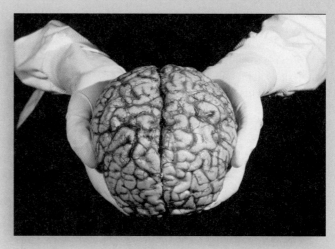

Figure 1 **Watching a human brain develop.** Researchers use induced pluripotent stem (iPS) cells to grow "cerebral organoids," which look more like layers of cells than this fully-formed brain.

of cartilage forms. The embryo is now about the length and weight of a paper clip. At 8 weeks of gestation, the prenatal human has tiny versions of all of the structures that will be present at birth. It is now a fetus. Researchers study how organs form in the embryo using stem cells to give rise to "organoids" that grow and develop in laboratory glassware, as described in *In Their Own Words.*

The Fetus Grows

During the fetal period, body proportions approach those of a newborn. Initially, the ears lie low, and the eyes are widely spaced. Bone begins to replace the softer cartilage. As nerve and muscle functions become coordinated, the fetus moves.

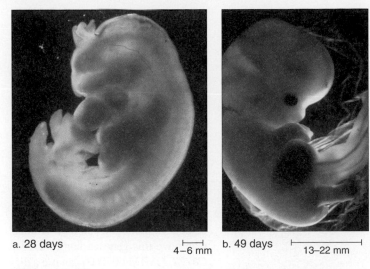

a. 28 days |—| 4–6 mm   b. 49 days |——| 13–22 mm

Figure 3.18 Human embryos at (**a**) 28 days, and (**b**) 49 days.

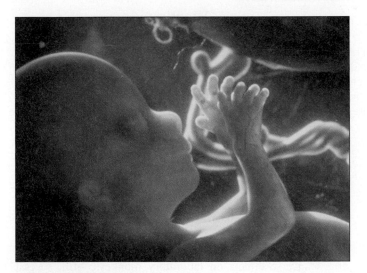

Figure 3.19 **A 16-week fetus.** A fetus at this stage has hair, eyebrows, lashes, and other features that look human.

Sex is determined at conception, when a sperm bearing an X or Y chromosome meets an oocyte, which always carries an X chromosome. An individual with two X chromosomes is a female, and one with an X and a Y is a male. A gene on the Y chromosome, called *SRY* (for "sex-determining region of the Y"), determines maleness.

Anatomical differences between the sexes appear at week 6, after the *SRY* gene is expressed in males. Male hormones stimulate male reproductive organs and glands to differentiate from existing, indifferent structures. In a female, the indifferent structures of the early embryo develop as female organs and glands, under the control of other genes. Differences may be noticeable on ultrasound scans by 12 to 15 weeks. Chapter 6 discusses sexual development further.

By week 12, the fetus sucks its thumb, kicks, makes fists and faces, and has the beginnings of teeth. It breathes amniotic fluid in and out, and urinates and defecates into it. The first trimester (3 months) of pregnancy ends.

By the fourth month, the fetus has hair, eyebrows, lashes, nipples, and nails (**figure 3.19**). By 18 weeks, the vocal cords have formed, but the fetus makes no sound because it doesn't breathe air. By the end of the fifth month, the fetus curls into a head-to-knees position. It weighs about 454 grams (1 pound). During the sixth month, the skin appears wrinkled because there isn't much fat beneath it, and turns pink as capillaries fill with blood. By the end of the second trimester, the woman feels distinct kicks and jabs and may even detect a fetal hiccup. The fetus is now about 23 centimeters (9 inches) long.

In the final trimester, fetal brain cells rapidly link into networks as organs elaborate and grow. A layer of fat forms beneath the skin. The digestive and respiratory systems mature last, which is why infants born prematurely often have difficulty digesting milk and breathing. Approximately 266 days after a single sperm burrowed its way into an oocyte, a baby is ready to be born.

The birth of a healthy baby is against the odds. Of every 100 secondary oocytes exposed to sperm, 84 are fertilized. Of these 84, 69 implant in the uterus, 42 survive one week or longer, 37 survive 6 weeks or longer, and only 31 are born alive. Of the fertilized ova that do not survive, about half have chromosomal abnormalities that cause problems too severe for development to proceed.

Key Concepts Questions 3.4

1. Describe the events of fertilization.
2. Distinguish among the zygote, morula, and blastocyst.
3. What happens to the inner cell mass?
4. How are some genes silenced as development proceeds?
5. What is the significance of the formation of primary germ layers?
6. What supportive structures enable the embryo to develop?
7. How do twins arise?
8. List the events of the embryonic and fetal periods.

3.5 Birth Defects

Certain genetic abnormalities or toxic exposures can affect development in an embryo or fetus, causing birth defects. Only a genetic birth defect can be passed to future generations. Although development can be derailed in many ways, about 97 percent of newborns appear healthy at birth.

The Critical Period

The specific nature of a birth defect reflects the structures developing when the damage occurs. The time when genetic abnormalities, toxic substances, or viruses can alter a specific structure is its **critical period** (**figure 3.20**). Some body parts, such as fingers and toes, are sensitive for short periods of time.

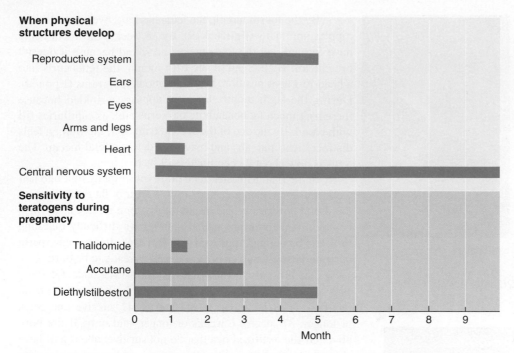

When physical structures develop

Sensitivity to teratogens during pregnancy

Figure 3.20 Critical periods of development. The nature of a birth defect resulting from drug exposure depends upon which structures were developing at the time of exposure. Isotretinoin (Accutane) is an acne medication that causes cleft palate and eye, brain, and heart defects. Diethylstilbestrol (DES) was used in the 1950s to prevent miscarriage. It caused vaginal cancer in some "DES daughters." Thalidomide was used to prevent morning sickness.

In contrast, the brain is sensitive throughout prenatal development, and connections between nerve cells continue to change throughout life. Because of the brain's continuous critical period, many birth defect syndromes include learning disabilities or intellectual disability.

About two-thirds of all birth defects arise from a disruption during the embryonic period. More subtle defects, such as learning disabilities, that become noticeable only after infancy, are often caused by interventions during the fetal period. A disruption in the first trimester might cause severe intellectual disability; in the seventh month of pregnancy, it might cause difficulty in learning to read.

Some birth defects can be attributed to an abnormal gene that acts at a specific point in prenatal development. In a rare inherited condition called phocomelia (OMIM 276826), for example, a mutation halts limb development from the third to the fifth week of the embryonic period, severely stunting arms and legs. The risk that a genetically caused birth defect will affect a particular family member can be calculated.

Many birth defects are caused by toxic substances the pregnant woman encounters. These environmentally caused problems will not affect other family members unless they, too, are exposed to the environmental trigger. Chemicals or other agents that cause birth defects are called **teratogens** (Greek for "monster-causing"). While it is best to avoid teratogens while pregnant, some women may need to continue to take potentially teratogenic drugs to maintain their own health.

Teratogens

Most drugs are not teratogens. Whether or not exposure to a particular drug causes birth defects may depend upon a woman's genes. For example, certain variants of a gene that control the body's use of an amino acid called homocysteine affect whether or not the medication valproic acid causes birth defects. Valproic acid is used to prevent seizures and symptoms of bipolar disorder. Rarely, it can cause NTDs, heart defects, hernias, and clubfoot. Women who have this gene variant (*MTHFR C677T,* OMIM 607093) can use a different drug when they try to conceive.

Thalidomide

The idea that the placenta protects the embryo and fetus was tragically disproven between 1957 and 1961, when 10,000 children were born in Europe with what seemed, at first, to be phocomelia. This genetic disorder is very rare, and therefore couldn't be the cause of the sudden problem. Instead, the mothers had all taken a mild tranquilizer, thalidomide, to alleviate nausea early in pregnancy, during the critical period for limb formation. Many "thalidomide babies" were born with incomplete or missing legs and arms.

The United States was spared from the thalidomide disaster because an astute government physician noted the drug's adverse effects on laboratory monkeys. Still, several "thalidomide babies" were born in South America in 1994, where pregnant women were given the drug. In spite of its teratogenic effects, thalidomide is used to treat leprosy, AIDS, and certain blood and bone marrow cancers.

Cigarettes

Chemicals in cigarette smoke stress a fetus. Carbon monoxide crosses the placenta and prevents the fetus's hemoglobin molecules from adequately binding oxygen. Other chemicals in smoke block nutrients. Smoke-exposed placentas lack important growth factors, causing poor growth before and after birth. Cigarette smoking during pregnancy increases the risk of spontaneous abortion, stillbirth, prematurity, and low birth weight.

Alcohol

A pregnant woman who has just one or two alcoholic drinks a day, or perhaps a large amount at a single crucial time, risks fetal alcohol syndrome (FAS) in her unborn child. Tests for variants of genes that encode proteins that regulate alcohol metabolism may be able to predict which women and fetuses are at elevated risk for developing FAS, but until these tests are validated, pregnant women are advised to avoid all alcohol.

Most children with FAS have small heads and flat faces (**figure 3.21**). Growth is slow before and after birth. Teens and young adults who have FAS are short and have small heads. More than 80 percent of them retain the facial characteristics of a young child with FAS.

The long-term cognitive effects of prenatal alcohol exposure are more severe than the physical vestiges. Intellectual impairment ranges from minor learning disabilities to intellectual disability. Many adults with FAS function at early grade-school level. They lack social and communication skills and cannot understand the consequences of actions, form friendships, take initiative, and interpret social cues. A person with FAS may have the cognitive symptoms without the facial characteristics, so considering alcohol consumption is important for diagnosis.

Greek philosopher Aristotle noticed problems in children of alcoholic mothers more than 23 centuries ago. In the United States today, 1 to 3 of every 1,000 infants has the syndrome, meaning 2,000 to 6,000 affected children are born each year. Many more children have milder "alcohol-related effects." A fetus of a woman with active alcoholism has a 30 to 45 percent chance of harm from prenatal alcohol exposure.

Nutrients

Certain nutrients ingested in large amounts, particularly vitamins, act as drugs. The acne medicine isotretinoin (Accutane) is a vitamin A derivative that causes spontaneous abortion and defects of the heart, nervous system, and face in exposed embryos. Physicians first noted the tragic effects of this drug 9 months after dermatologists began prescribing it to young women in the early 1980s. Another vitamin A–based drug, used to treat psoriasis, as well as excesses of vitamin A itself, also cause birth defects. Some forms of vitamin A are stored in body fat for up to 3 years.

Excess vitamin C can harm a fetus that becomes accustomed to the large amounts the woman takes. After birth, when the vitamin supply suddenly plummets, the baby may develop symptoms of vitamin C deficiency (scurvy), such as bruising and becoming infected easily.

Malnutrition threatens a fetus. The opening essay to chapter 11 describes the effects of starvation on embryos during the bleak "Dutch Hunger Winter" of 1944–1945. Poor nutrition later in pregnancy affects the development of the placenta and can cause low birth weight, short stature, tooth decay, delayed sexual development, and learning disabilities.

Occupational Hazards

Teratogens are present in some workplaces. Researchers note increased rates of spontaneous abortion and children born with birth defects among women who work with textile dyes, lead, certain photographic chemicals, semiconductor materials, mercury, and cadmium. Men whose jobs expose them to sustained heat, such as smelter workers, glass manufacturers, and bakers, may produce sperm that can fertilize an oocyte and then cause spontaneous abortion or a birth defect. A virus or a toxic chemical carried in semen may also cause a birth defect.

Viral Infection

Viruses are small enough to cross the placenta and reach a fetus. Some viruses that cause mild symptoms in an adult, such as the chickenpox virus, may devastate a fetus. Men can transmit viral infections to an embryo or fetus during sexual intercourse.

HIV can reach a fetus through the placenta or infect a newborn via blood contact during birth. The risk of transmission is significantly reduced if a pregnant woman takes anti-HIV drugs. Fetuses of HIV-infected women are at higher risk for low birth weight, prematurity, and stillbirth if the woman's health is failing.

German measles (rubella) is a well-known viral teratogen. In the United States, in the early 1960s, an epidemic of the usually mild illness caused 20,000 birth defects and 30,000 stillbirths. Exposure during the first trimester of pregnancy caused cataracts, deafness, and heart defects. Exposure during the second or third trimester of pregnancy caused learning disabilities, speech and hearing problems, and type 1 diabetes mellitus. Incidence of these problems, called congenital rubella syndrome, has dropped markedly since vaccination eliminated the disease in the United States.

Herpes simplex virus can harm a fetus or newborn. Forty percent of babies exposed to active vaginal herpes lesions become infected, and half of them die. Of the survivors, 25 percent sustain severe nervous system damage, and another 25 percent have skin sores. A surgical (caesarean) delivery can protect the child.

Pregnant women are routinely checked for hepatitis B infection, which in adults causes liver inflammation, great fatigue, and other symptoms. Each year in the United States, 22,000 infants are infected with this virus during birth. These babies are healthy, but at high risk for developing serious liver problems as adults. When infected women are identified, a vaccine can be given to their newborns to help prevent complications.

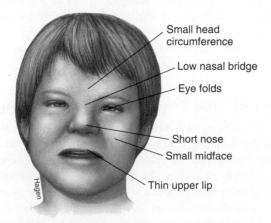

Small head circumference

Low nasal bridge

Eye folds

Short nose

Small midface

Thin upper lip

Figure 3.21 **Fetal alcohol syndrome.** Some children whose mothers drank alcohol during pregnancy have characteristic flat faces.

Key Concepts Questions 3.5

1. What is the critical period?
2. Why are most birth defects that develop during the embryonic period more severe than problems that arise during fetal development?
3. What are teratogens?

3.6 Maturation and Aging

"Aging" means moving through the life cycle, despite advertisements for products that promise to reverse it. As we age, the limited life spans of cells are reflected in the waxing and waning of biological structures and functions. Although some aspects of our anatomy and physiology peak very early—such as the number of brain cells or hearing acuity, which do so in childhood—age 30 seems to be a turning point for decline. Some researchers estimate that, after this age, the human body becomes functionally less efficient by about 0.8 percent each year.

Can we slow aging? In the quest to extend life, people have sampled everything from turtle soup to owl meat to human blood. More recently, attention has turned to a component of red wine called resveratrol. It is a member of a class of enzymes called **sirtuins** that regulate energy use in cells by altering the expression of certain sets of genes. Through their effect on energy metabolism, the sirtuins seem to prevent or delay several diseases that are more common in the aged, such as heart disease and neurodegenerative disorders. Drug companies are exploring compounds that are 1,000 times as active as resveratrol to fight particular diseases.

Many diseases that begin in adulthood, or are associated with aging, have genetic components. These disorders tend to be multifactorial, because it takes many years for environmental exposures to alter gene expression in ways that noticeably affect health. Following is a closer look at how genes may impact health throughout life.

Adult-Onset Inherited Disorders

Human prenatal development is a highly regulated program of genetic switches that are turned on in specific body parts at specific times. Environmental factors can affect how certain genes are expressed before birth in ways that create risks that appear much later. For example, adaptations that enable a fetus to grow despite near-starvation become risk factors for certain common conditions of adulthood, when conserving calories is no longer needed. Such disorders include coronary artery disease, obesity, stroke, hypertension, schizophrenia, and type 2 diabetes mellitus. A malnourished fetus has intrauterine growth retardation (IUGR), and though born on time, is very small. Premature infants, in contrast, are small but are born early, and are not predisposed to conditions resulting from IUGR.

More than 100 studies correlate low birth weight due to IUGR with increased incidence of cardiovascular disease and diabetes later in life. Much of the data come from war records because enough time has elapsed to study the effects of prenatal malnutrition as people age. How can poor nutrition before birth cause disease decades later? Perhaps to survive, the starving fetus redirects its circulation to protect vital organs, as muscle mass and hormone production change to conserve energy. Growth-retarded babies have too little muscle tissue, and because muscle is the primary site of insulin action, glucose metabolism changes. Thinness at birth, and the accelerated weight gain in childhood that often compensates, sets the stage for coronary heart disease and type 2 diabetes much later.

In contrast to the delayed effects of fetal malnutrition, symptoms of single-gene disorders can begin anytime. Most inherited conditions that affect children are recessive. Even a fetus can have symptoms of inherited disease, such as osteogenesis imperfecta ("brittle bone disease"), described in the opener to Chapter 12.

Most dominantly inherited conditions start to affect health in early to middle adulthood. In polycystic kidney disease (OMIM 173900), cysts that may have been present but undetected in the kidneys during one's twenties begin causing bloody urine, high blood pressure, and abdominal pain in the thirties. The joint destruction of osteoarthritis may begin in one's thirties, but not be painful for 20 years.

Mutations in single genes cause from 5 to 10 percent of Alzheimer disease cases, producing initial symptoms in the forties and fifties. German neurologist Alois Alzheimer first identified the condition in 1907 as affecting people in mid-adulthood. Noninherited Alzheimer disease typically begins later in life.

Whatever the age of onset, Alzheimer disease starts gradually. Mental function declines steadily for 3 to 10 years following the first symptoms of depression and short-term memory loss. Cognitive skills gradually decline. A person may become unable to walk by forgetting how to put one foot in front of the other. Confused and forgetful, Alzheimer patients may wander away from family and friends. Finally, the patient cannot recognize loved ones and can no longer perform basic functions such as speaking or eating.

The brains of Alzheimer disease patients contain deposits of a protein, called beta amyloid, in learning and memory centers. Alzheimer brains also contain structures called neurofibrillary tangles, which consist of a protein called tau. Tau binds to and disrupts microtubules in nerve cell branches, destroying the shape of the cell, impairing its ability to communicate. Clinical Connection 5.1 discusses Alzheimer disease further.

Disorders That Resemble Accelerated Aging

Genes control aging both passively (as structures break down) and actively (by initiating new activities). A group of "rapid aging" inherited disorders are very rare, but may hold clues to how genes control aging in all of us, as discussed in the chapter opener.

The most severe rapid aging disorders are the segmental progeroid syndromes, sometimes considered together as "progeria." Most of these disorders are caused by cells' inability to repair DNA, which is discussed in section 12.7. Mutations that would ordinarily be corrected persist. Over time, the accumulation of mutations destabilizes the entire genome, and even

Table 3.3	Rapid Aging Syndromes			
Disorder	**Incidence**	**Average Life Span**	**OMIM Number**	
Ataxia telangiectasia	1/60,000	20	208900	
Cockayne syndrome	1/100,000	20	216400	
Hutchinson-Gilford syndrome	<1/1,000,000	13	176670	
Rothmund-Thomson syndrome	<1/100,000	Normal	268400	
Trichothiodystrophy	<1/100,000	10	601675	
Werner syndrome	<1/100,000	50	277700	

more mutations occur in somatic cells. The various changes that we associate with aging ensue.

Table 3.3 lists the more common segmental progeroid syndromes. They vary in severity. People with Rothmund-Thomson syndrome, for example, may have a normal life span, but develop gray hair or baldness, cataracts, cancers, and osteoporosis at young ages. Werner syndrome becomes apparent before age 20, causing death before age 50 from diseases associated with aging. Young adults with Werner syndrome develop atherosclerosis, type 2 diabetes mellitus, hair graying and loss, osteoporosis, cataracts, and wrinkled skin. They are short because they skip the growth spurt of adolescence.

Hutchinson-Gilford progeria syndrome is the most severe rapid aging disorder. In addition to the symptoms the chapter opener describes are abnormalities at the cellular level. Normal cells growing in culture divide about 50 times before dying. Cells from Hutchinson-Gilford progeria syndrome patients die in culture after only 10 to 30 divisions.

Hutchinson-Gilford progeria is caused by a single DNA base change in the gene that encodes a protein called lamin A. That base is a site that determines how parts of the protein are cut and joined, and when it is altered, the protein lacks 50 amino acids. The shortened protein is called progerin. It remains stuck to the endoplasmic reticulum, instead of being transported into the nucleus through the nuclear pores, as happens to normal lamin A protein. Progerin diffuses into the tubules of the ER and travels within them to the nuclear membrane. This route of entry stresses the nuclear membrane, causing it to bubble or "bleb" inward, altering the way that the nuclear lamina (the layer on the inside face) binds the chromatin (DNA complexed with protein) within. Somehow, disturbing the chromatin hampers DNA repair, allowing mutations associated with the signs of aging to occur. Several drugs block the molecule that holds progerin to the ER.

Studies on stem cells from bone marrow of patients add further evidence: Progerin shifts the activities of certain genes in ways that promote bone formation and suppress fat deposition. This skewed development perhaps explains the failure to thrive and skeletal appearance of affected individuals. The molecular view of progeria suggests that DNA repair is what enables us to live many years. These children lack that

protection, and the mutations that age us all accumulate much faster.

Genes and Longevity

Aging reflects gene activity plus a lifetime of environmental influences (**figure 3.22**). Families and genetically isolated populations with many very aged members have a lucky collection of gene variants plus shared environmental influences such as good nutrition, excellent health care, and devoted relatives. Genome comparisons among people who've passed their 100th birthdays to those who have died of the common illnesses of older age are revealing genes that influence longevity.

People who have lived past 100 years are called centenarians. Most enjoy excellent health and are socially active, then succumb rapidly to diseases that typically claim people decades earlier. Centenarians fall into three broad groups—about 20 percent of them never get the diseases that kill most people; 40 percent get these diseases, but at a much older age; and the other 40 percent live with and survive the more common disorders of aging: heart disease, stroke, cancers, type 2 diabetes mellitus, and dementias. Those past 110 are supercentenarians. About 70 are alive worldwide.

While the environment seems to play an important role in the deaths of people ages 60 to 85, past that age, genetic effects predominate. That is, someone who dies at age 68 of lung cancer can probably blame a lifetime of cigarette smoking. But a smoker who dies at age 101 of the same disease probably had gene variants that protected against lung cancer. Centenarians have higher levels of large lipoproteins that carry cholesterol (high-density lipoprotein [HDL]) than other people, which researchers estimate adds 20 years of life.

Figure 3.22 Analyzing the genomes of older people can provide clues to health for everyone.

Children and siblings of centenarians tend to be long-lived as well, supporting the idea that longevity is inherited. Brothers of centenarians are 17 times as likely to live past age 100 as the average man, and sisters are 8.5 times as likely. Unhealthy habits among some centenarians hint that genes are protecting them. One researcher suggests that the saying, "The older you get, the sicker you get" be replaced with "The older you get, the healthier you've been."

Centenarians have inherited two types of gene variants—those that directly protect them, and wild type alleles of genes that, when mutant, cause disease.

Types of genes that affect longevity control:

- immune system functioning;
- insulin secretion and glucose metabolism;
- response to stress;
- the cell cycle;
- DNA repair;
- lipid (including cholesterol) metabolism;
- nutrient metabolism; and
- production of antioxidant enzymes.

Researchers estimate that at least 130 genes have variants that influence how long a person is likely to live.

One well-studied aging gene encodes the growth hormone receptor, which is a protein that enables a cell to respond to signals to enlarge, a key part of early development. People who inherit two recessive mutations in the gene have Laron syndrome (OMIM 262500). They are very short—adults stand under 4 feet tall—but they do not develop cancer or diabetes. Because of the near-absence of these common diseases of aging, people with Laron syndrome live very long. In El Oro Province in southern Ecuador, more than 100 people have Laron syndrome. They descend from Eastern European Jewish people who fled to the area to escape the Spanish Inquisition in the fifteenth century, and avoiding churches led them to the unoccupied, rural region.

An isolated population where people have children with relatives creates a situation in which mutations are overrepresented, but the genomes are relatively uniform, making gene variants easier to identify than in more outbred populations. Something about inheriting Laron syndrome protects these people from two major diseases of older age. Table 3.4 lists other single genes that are associated with longevity in isolated populations. Environmental factors are also important in longevity. The very-long-lived people of Calabria, an area of southern Italy, for example, eat mostly fruits and vegetables, which may contribute to their excellent health.

Several research projects are collecting and curating genome sequences from specific populations, many of them genetically more uniform than most, like the "little people" of southern Ecuador. The longest study is the New England Centenarian Study, which began in 1988 to amass information on the families of the oldest citizens then known in the United States. Another study is investigating genomes of nursing home residents, and a program in California is probing the genomes of the "wellderly." So far, these people share never having had heart disease and never having smoked. Several very-long-lived people have had cancer, indicating that cancers are survivable.

Researchers at the University of Pittsburgh have identified places in the genome that harbor "successful aging genes" that have variants that preserve cognition. Other studies are looking at genes implicated in the diseases that kill most of us for variants that long-lived siblings share. Considered together, perhaps these studies will provide information that will help the majority of us who have not been fortunate enough to have inherited longevity gene variants.

Key Concepts Questions 3.6

1. How can starvation before birth set the stage for later disease?
2. How do recessive and dominant conditions differ in terms of the human life cycle?
3. Describe a single-gene disorder that speeds aspects of aging.
4. How do genes control aging?

Table 3.4	Longevity Genes	
Gene	**Protects Against**	**Population Studied**
Apolipoprotein C3 (*APOC-3*)	Hypertension, diabetes	Ashkenazi Jews
	Cardiovascular disease	Amish
Bitter taste receptor (*TAW2R16*)	Toxin detection, digestive problems	Calabria, Italy
Cholesteryl ester transfer protein (*CETP*)	Cardiovascular disease	Ashkenazi Jews
Forkhead box O3 (*FOXO3*)	Cancer, cardiovascular disease	Japanese-Americans
Growth hormone receptor (*GHR*)	Diabetes, cancer	Ecuador, Israel
Uncoupling proteins (*UCP 2, 3, 4*)	Oxidative damage, poor energy use	Calabria, Italy

Summary

3.1 The Reproductive System

1. The male and female reproductive systems include paired **gonads** and networks of tubes in which **sperm** and **oocytes** are made.

2. Male **gametes** originate in seminiferous tubules within the **testes**, then pass through the epididymis and ductus deferentia, where they mature before exiting the body through the urethra during sexual intercourse. The prostate gland, the seminal vesicles, and the bulbourethral glands add secretions.

3. Female gametes originate in the **ovaries**. Each month after puberty, one ovary releases an oocyte into a uterine tube. The oocyte then moves to the uterus for implantation (if fertilized) or expulsion.

3.2 Meiosis

4. **Meiosis** reduces the chromosome number in gametes from **diploid** to **haploid**, maintaining the chromosome number between generations. Meiosis ensures genetic variability by **independently assorting** combinations of genes into gametes as chromosomes randomly align and **cross over**.

5. Meiosis I, a **reduction division,** halves the number of chromosomes. Meiosis II, an **equational division**, produces four cells from the two that result from meiosis I, without another DNA replication.

6. Crossing over occurs during prophase I. It mixes up paternally and maternally derived genes on **homologous pairs** of chromosomes.

7. Chromosomes segregate and independently assort in metaphase I, which determines the distribution of genes from each parent.

3.3 Gametes Mature

8. **Spermatogenesis** begins with **spermatogonia**, which accumulate cytoplasm and replicate their DNA, becoming primary spermatocytes. After meiosis I, the cells become haploid secondary spermatocytes. In meiosis II, the secondary spermatocytes divide, each yielding two spermatids, which then differentiate into spermatozoa.

9. In **oogenesis,** some **oogonia** grow and replicate their DNA, becoming primary oocytes. In meiosis I, the primary oocyte divides to yield one large secondary oocyte and a small **polar body**. In meiosis II, the secondary oocyte divides to yield the large ovum and another polar body. Female meiosis is completed at fertilization.

10. Genetic errors in sperm from older men are usually dominant single-gene mutations. Genetic errors in oocytes from older women are typically extra or absent chromosomes.

3.4 Prenatal Development

11. In the female, sperm are capacitated and drawn toward a secondary oocyte. One sperm burrows through the oocyte's protective layers with acrosomal enzymes. Fertilization occurs when the sperm and oocyte fuse and their DNA combines in one nucleus, forming the **zygote**. Electrochemical changes in the egg surface block additional sperm from entering. **Cleavage** begins and a 16-celled **morula** forms. Between days 3 and 6, the morula arrives at the uterus and hollows, forming a **blastocyst** made up of **blastomeres**. The trophoblast and **inner cell mass** form. Around day 6 or 7, the blastocyst implants, and trophoblast cells secrete hCG, which prevents menstruation.

12. During the second week, the amniotic cavity forms as the inner cell mass flattens. **Ectoderm** and **endoderm** form, and then **mesoderm** appears, establishing the **primary germ layers** of the **gastrula**. Epigenetic effects oversee cell differentiation as cells in each germ layer begin to develop into specific organs.

13. During the third week, the placenta, yolk sac, allantois, and umbilical cord begin to form as the amniotic cavity swells with fluid.

14. **Monozygotic (MZ) twins** result when one fertilized ovum splits. **Dizygotic (DZ) twins** result from two fertilized ova.

15. Organs form throughout the embryonic period. The primitive streak, notochord and **neural tube**, arm and leg buds, heart, facial features, and skeleton develop.

16. At the eighth week, the **embryo** becomes a **fetus**, with all structures present but not fully grown. Organ rudiments grow and specialize. The developing organism moves and reacts. In the last trimester, the brain develops rapidly, and fat is deposited beneath the skin. The digestive and respiratory systems mature last.

3.5 Birth Defects

17. Birth defects can result from a mutation or an environmental intervention.

18. A substance that causes birth defects is a **teratogen**. Environmentally caused birth defects are not transmitted to future generations.

19. The time when a structure is sensitive to damage from an abnormal gene or environmental intervention is its **critical period**.

3.6 Maturation and Aging

20. Genes cause or predispose us to illness throughout life. Single-gene disorders that strike early tend to be recessive; most adult-onset single-gene conditions are dominant.

21. Malnutrition before birth can alter gene expression in ways that cause illness later in life.

22. The segmental progeroid syndromes are single-gene disorders that speed aging-associated changes.

23. Long life is due to genetics and environmental influences. Researchers identify longevity genes by studying genetically isolated populations with many long-lived members.

Review Questions

1. How many sets of human chromosomes are in each of the following cell types?

   a. an oogonium
   b. a primary spermatocyte
   c. a spermatid
   d. a cell from either sex during anaphase of meiosis I
   e. a cell from either sex during anaphase of meiosis II
   f. a secondary oocyte
   g. a polar body derived from a primary oocyte

2. List the structures and functions of the male and female reproductive systems.

3. A dog has 39 pairs of chromosomes. Considering only the random alignment of chromosomes, how many genetically different puppies are possible when two dogs mate? Is this number an underestimate or overestimate of the actual total? Why?

4. How does meiosis differ from mitosis?

5. What do oogenesis and spermatogenesis have in common, and how do they differ?

6. How does gamete maturation differ in the male and female?

7. Why is it necessary for spermatogenesis and oogenesis to generate stem cells?

8. Describe the events of fertilization.

9. Define epigenome. How does the epigenome differ from the gene variants that make up the genome of an embryo?

10. Write the time sequence in which the following structures begin to develop: notochord, gastrula, inner cell mass, fetus, zygote, morula.

11. Why does exposure to teratogens produce more severe health effects in an embryo than in a fetus?

12. The same birth defect syndrome can be caused by a mutant gene or exposure to a teratogen. How do the consequences of each cause differ for future generations?

13. List three teratogens, and explain how they disrupt prenatal development.

14. How are sirtuins thought to extend life?

15. Describe two ways that children with Hutchinson-Gilford progeria syndrome age rapidly.

16. Cite two pieces of evidence that genes control aging.

Applied Questions

1. Up to what stage, if any, do you think it is ethical to experiment on a prenatal human? Cite reasons for your answer.

2. Under a microscope, a first and second polar body look alike. What structure would distinguish them?

3. Armadillos always give birth to identical quadruplets. Are the offspring clones?

4. The morning-after pill prevents pregnancy if taken up to 5 days after having unprotected sex. Based on the stages of prenatal development, how does the pill prevent pregnancy?

5. In about 1 in 200 pregnancies, a sperm fertilizes a polar body instead of an oocyte. A mass of tissue that is not an embryo develops. Why can't a polar body support the development of an entire embryo?

6. Should a woman be held legally responsible if she drinks alcohol, smokes, or abuses drugs during pregnancy and it harms her child? Should liability apply to all substances that can harm a fetus, or only to those that are illegal?

7. What are possible benefits and dangers of predicting how long a person will live from analyzing his or her genome sequence?

Web Activities

1. Look over the "Living to 100 Life Expectancy Calculator" at www.livingto100.com and list 10 ways that you can change your behavior to possibly live longer. What does this quiz suggest about the relative role of genes and the environment in determining longevity?

2. Go to a website that lists teratogens. Identify three drugs that are safe to take during pregnancy and three that are not safe, and list the associated medical problems.

Case Studies and Research Results

1. Human embryonic stem cells can be derived and cultured from an 8-celled cleavage embryo and from a cell of an inner cell mass. Explain the difference between these stages of human prenatal development.

2. Victor, a 34-year-old artist, was killed in a car accident. He and his wife Emma hadn't started a family yet, but planned to soon. The morning after the accident, Emma asked if some of her husband's sperm could be collected and frozen, for her to use to have a child. Do you think that this "post-mortem sperm retrieval" should be done? For further information skip ahead to "Bioethics: Choices for the Future" in Chapter 21, "Removing and Using Gametes After Death."

3. A pediatrician with training in anatomy and embryology was examining the jaw of a little girl with auriculocondylar syndrome (OMIM 602483). The child had tiny ears, a small head and mouth, and a very unusual jaw. The doctor realized from studying X rays that the lower jaw looked exactly like an upper jaw—and the girl's mother looked just like the child. What type of mutation does this family have? Describe in general terms how it causes the phenotype.

4. The television program "Orphan Black" is about a young woman and how she interacts with her clones—more than a dozen.

   a. Describe how the clones might have been created using somatic cell nuclear transfer.
   b. Wikipedia describes the clones as being made using *in vitro* fertilization, in which sperm fertilize oocytes in a laboratory dish. How would the process of meiosis make cloning via *in vitro* fertilization impossible?
   c. List 3 ways that clones are not genetically identical.

5. Hendrikje van Andel-Schipper died in 2005 at age 115. In 2014, researchers identified 450 mutations in white blood cells from a blood sample taken at her autopsy. "Henny" had lived a very healthy life, dying of a stomach cancer she didn't even know she had. She never had a blood disease. What is the significance of the discovery that her white blood cells had so many mutations?

6. A study of 202 individuals over the age of 90 living in Calabria in southern Italy is comparing the siblings of these people to their spouses. How could this strategy reveal the genetic underpinnings of longevity?

PART 2 Transmission Genetics

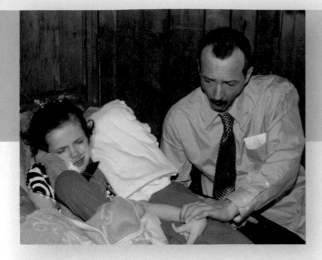

Karli Mervar and her father Karl suffered from Huntington disease at the same time. Karl's gaunt look results from the constant movements that are part of the disease—his cells used energy rapidly.

CHAPTER

4

Single-Gene Inheritance

Learning Outcomes

4.1 Following the Inheritance of One Gene

1. Describe how Mendel deduced that recessive traits seem to disappear in hybrids.

2. Define and distinguish *heterozygote* and *homozygote; dominant* and *recessive; phenotype* and *genotype*.

3. Explain how the law of segregation reflects the events of meiosis.

4. Describe a Punnett square.

4.2 Single-Gene Inheritance Is Rare

5. Explain how a gene alone usually does not solely determine a trait.

6. Distinguish between autosomal recessive and autosomal dominant inheritance.

4.3 Following the Inheritance of More Than One Gene

7. Explain how Mendel's experiments followed the inheritance of more than one gene.

8. Explain how the law of independent assortment reflects the events of meiosis.

4.4 Pedigree Analysis

9. Explain how pedigrees show single-gene transmission.

4.5 Family Exome Analysis

10. Explain how exome sequencing in a family can reveal Mendelian inheritance patterns.

The **BIG** Picture

Gregor Mendel deduced the basis of inheritance patterns using pea plants, but his insights apply to all diploid species, including our own. Mendel's laws even lie behind sequencing exomes and genomes to identify the genes behind mysterious medical conditions.

Juvenile Huntington Disease: The Cruel Mutation

Huntington disease (HD) causes uncontrollable movements and changes in behavior and thinking (cognition). It is dominant, which means that each child of an affected individual need inherit only one copy of the abnormal gene to develop the disease. Typically, symptoms begin in early adulthood, with death 15 to 20 years later. Only 10 percent of people with HD are under age 20.

HD in the Mervar family is very unusual. The first symptoms for Karl started when he and his wife Jane's youngest daughter, Karli, began to have trouble paying attention in school. Karl had become abusive, paranoid, and unstable on his feet. By age 5 Karli's left side stiffened and her movements slowed. Soon she could no longer skip, hop, or jump, and she developed a racing heartbeat, itching, pain, and unintelligible speech. Karl was diagnosed 6 weeks after Karli in 2002. He was 35, she just 6. Karli's disease progressed rapidly. By age 9 she required a feeding tube, suffered seizures, and barely slept. Meanwhile, Karl became so violent that he spent his last years in a nursing home. Father and daughter died within weeks of each other, in early 2010.

An inherited disease affects each child independently, and so the fact that Karli was sick did not spare her sisters. Jacey was diagnosed in 2004 at age 13, and Erica in 2007 at age 17. This family illustrates classic autosomal dominant inheritance— affecting every generation, and both sexes. Jane Mervar has cared for them all.

4.1 Following the Inheritance of One Gene

Single-gene diseases such as Huntington disease (HD) and cystic fibrosis (CF) affect families in patterns termed modes of inheritance. Knowing these patterns makes it possible to predict risks that people related in particular ways inherit the mutation. HD is **autosomal dominant,** which means that it affects both sexes and appears every generation. In contrast, CF is **autosomal recessive,** which means that the disease affects both sexes and can "skip" generations through carriers, who do not have symptoms.

Today, genetic tests, including exome and genome sequencing, reveal which single-gene health conditions we carry. Single-gene traits are also called "Mendelian," in honor of the researcher who first derived the two laws of inheritance that determine how these traits are transmitted from one generation to the next. Mendel's laws remain important in this age of sequencing because knowing who to sequence, why to sequence, and interpreting the data often depend on knowing family relationships. Mendel's tale begins with pea plants in a long-ago garden in what is now the Czech Republic.

Mendel's Experiments

Gregor Mendel was the first thinker to probe the underlying rules of logic that make it possible to predict inheritance of specific traits. He was an inquisitive man who saw patterns in the natural variation in plants. By breeding pea plants, Mendel described units of inheritance that pass traits from generation to generation. He called these units "elementen." He could not see them, but he inferred their existence from the appearances of his plants. Although Mendel knew nothing of DNA, chromosomes, or cells, his "laws" of inheritance have not only stood the test of time, but explain trait transmission in any diploid species.

From 1857 to 1863, Mendel crossed and cataloged traits in 24,034 plants, through several generations. He deduced that consistent ratios of traits in the offspring indicated that the plants transmitted distinct units. He derived two hypotheses to explain how inherited traits are transmitted. Mendel published his findings in 1866, but few people read it. In 1901, three botanists independently rediscovered the laws of inheritance. Once they read Mendel's paper, they credited him, and Mendel became known as the "father of genetics."

Peas are ideal for probing heredity because they are easy to grow, develop quickly, and have many traits that take one of two easily distinguishable forms. **Figure 4.1** illustrates the seven traits that Mendel followed through several pea generations. When analyzing genetic crosses, the first generation is the parental generation, or P_1; the second generation is the first filial generation, or F_1; the next generation is the second filial generation, or F_2, and so on.

Mendel's first experiments followed single traits with two expressions, such as "short" and "tall." He began with plants that had been cultivated over many generations to ensure that they always "bred true," producing the same expression of the trait. He set up all possible combinations of artificial pollinations, crossing tall with tall, short with short, and tall with short plants. This last combination, plants with one trait variant

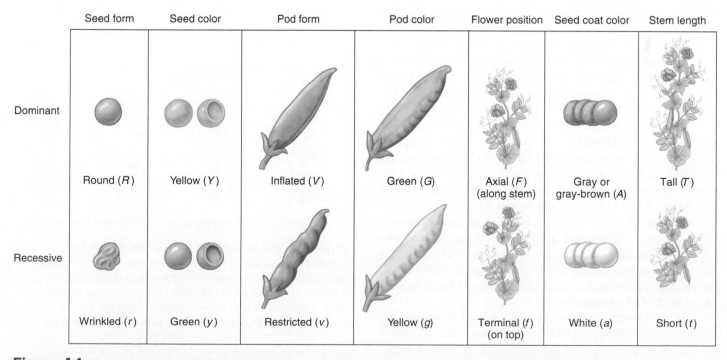

Figure 4.1 Traits Mendel studied. Gregor Mendel studied the transmission of seven traits in the pea plant. Each trait has two easily distinguished expressions, or phenotypes.

from each parent—to offspring. Both are inherited in random combinations. Chromosomes provided a physical mechanism for Mendel's hypotheses. In 1909, English embryologist William Bateson renamed Mendel's elementen *genes* (Greek for "give birth to"). In the 1940s, scientists began investigating the gene's chemical basis, discussed in chapter 9.

In the twentieth century, researchers discovered the molecular basis of some of the traits that Mendel studied. "Short" and "tall" plants reflect expression of a gene that enables a plant to produce the hormone gibberellin, which elongates the stem. One tiny change to the DNA, and a short plant results. Likewise, "round" and "wrinkled" peas arise from the R gene, whose encoded protein connects sugars into branching polysaccharide molecules. Seeds with a mutant R gene cannot attach the sugars. As a result, water exits the cells, and the peas wrinkle.

crossed to plants with the alternate, produces hybrids, which are offspring that inherit a different gene variant (allele) from each parent. He also self-crossed plants.

Mendel noted that short plants crossed to other short plants were "true-breeding," always producing the same phenotype, in this case short plants. The crosses of tall plants to each other were more confusing. Some tall plants were true-breeding, but others crossed with each other yielded short plants in about one-quarter of the next generation. In some plants, tallness appeared to mask shortness. One trait that masks another is **dominant;** the masked trait is **recessive.**

Mendel conducted up to 70 hybrid self-crosses for each of the seven traits. This experiment is called a **monohybrid cross** because it follows one trait and the self-crossed plants are hybrids. The progeny were in the ratio of one-quarter short to three-quarters tall plants (**figure 4.2**). In further crosses, two-thirds of the tall plants in the F_2 generation were non-true-breeding, and the remaining third were true-breeding. (Consider only the tall plants at the bottom of the figure.)

These experiments confirmed that hybrids hide one expression of a trait—short, in this case—which reappears when hybrids are self-crossed. Mendel tried to explain how this happened: Gametes distribute "elementen" because these cells physically link generations. Paired sets of elementen separate as gametes form. When gametes join at fertilization, the elementen combine anew. Mendel reasoned that each elementen was packaged in a separate gamete. If opposite-sex gametes combine at random, he could mathematically explain the different ratios of traits produced from his pea plant crosses. Mendel's idea that elementen separate in the gametes would later be called the **law of segregation.**

When Mendel's ratios were demonstrated in several species in the early 1900s, just when chromosomes were being discovered, it became apparent that elementen and chromosomes had much in common. Both paired elementen and pairs of chromosomes separate at each generation and are transmitted—one

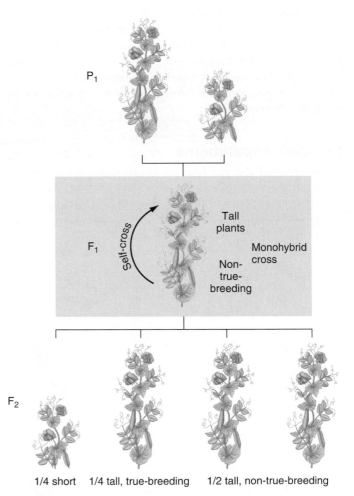

Figure 4.2 **A monohybrid cross.** When Mendel crossed true-breeding tall plants with short plants, the next generation plants were all tall. When he self-crossed the F_1 plants, one-quarter of the plants in the next generation, the F_2, were short, and three-quarters were tall. Of the tall plants in the F_2, one-third were true-breeding, and the other two-thirds were not true-breeding. He could tell this by conducting further crosses of the tall plants to short plants, to see which bred true. The monohybrid cross is the F_1 generation.

Terms and Tools to Follow Segregating Genes

The law of segregation reflects the actions of chromosomes and the genes they carry during meiosis. Because a gene is a long sequence of DNA, it can vary in many ways. An individual with two identical alleles for a gene is **homozygous** for that gene. An individual with two different alleles is **heterozygous**—what Mendel called "non-true-breeding" or "hybrid."

For a gene with two alleles, the dominant one is shown as a capital letter and the recessive with the corresponding small letter. If both alleles are recessive, the individual is homozygous recessive, shown with two small letters. An individual with two dominant alleles is homozygous dominant, and has two capital letters. One dominant and one recessive allele, such as *Tt* for non-true-breeding tall pea plants, indicates a heterozygote.

An organism's appearance does not always reveal its alleles. Both a *TT* and a *Tt* pea plant are tall, but *TT* is a homozygote and *Tt* a heterozygote. The **genotype** describes the organism's alleles, and the **phenotype** describes the outward expression of an allele combination. A **wild type** phenotype is the most common expression of a particular allele combination in a population. The wild type allele may be recessive or dominant. A **mutant** phenotype is a variant of a gene's expression that arises when the gene undergoes a change, or **mutation.**

Mendel was observing the events of meiosis. When a gamete is produced, the two copies of a gene separate with the homologs that carry them. In a plant of genotype *Tt,* for example, gametes carrying either *T* or *t* form in equal numbers during anaphase I and the ratio holds up when sister chromatids separate in anaphase II. Gametes combine at random. A *t*-bearing oocyte is neither more nor less attractive to a sperm than is a *T*-bearing oocyte. These two factors—equal allele distribution into gametes and random combinations of gametes—underlie Mendel's law of segregation (**figure 4.3**).

When Mendel crossed short plants (*tt*) with true-breeding tall plants (*TT*), the seeds grew into F_1 plants that were all tall (genotype *Tt*). Next, he self-crossed the F_1 plants. The progeny were *TT, tt,* and *Tt.* A *TT* individual resulted when a *T* sperm fertilized a *T* oocyte; a *tt* plant resulted when a *t* oocyte met a *t* sperm; and a *Tt* individual resulted when either a *t* sperm fertilized a *T* oocyte, or a *T* sperm fertilized a *t* oocyte.

Because two of the four possible gamete combinations produce a heterozygote, and each of the others produces a homozygote, the genotypic ratio expected of a monohybrid cross is 1*TT*: 2*Tt*: 1*tt*. The corresponding phenotypic ratio is three tall plants to one short plant, a 3:1 ratio. Mendel saw these results for all seven traits that he studied (**table 4.1**). A diagram called a **Punnett square** shows these ratios (**figure 4.4**). A Punnett square represents how genes in gametes join if they are on different chromosomes. Experimental crosses yielded numbers of offspring that approximate these ratios.

Mendel used additional crosses to distinguish the two genotypes resulting in tall progeny—*TT* from *Tt* (**figure 4.5**). He bred tall plants of unknown genotype with short (*tt*) plants. If a tall plant crossed with a *tt* plant produced both tall and short progeny, it was genotype *Tt;* if it produced only tall plants, it must be *TT.*

Crossing an individual of unknown genotype with a homozygous recessive individual is called a test cross. The logic is that the homozygous recessive is the only genotype that can be identified by its phenotype—that is, a short plant is always *tt*. The homozygous recessive is a "known" that can reveal the unknown genotype of another individual to which it is crossed.

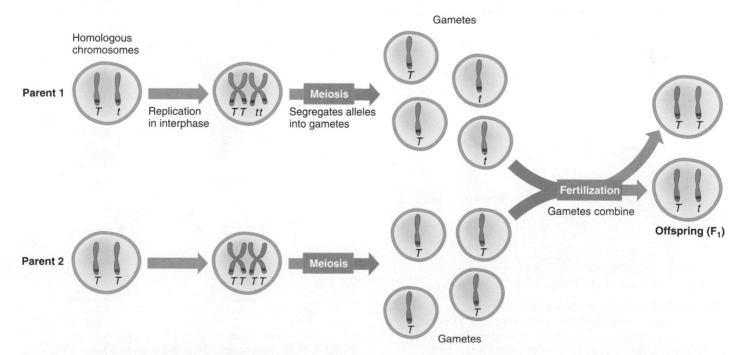

Figure 4.3 **Mendel's first law—gene segregation.** During meiosis, homologous pairs of chromosomes and their genes separate and are packaged into separate gametes. At fertilization, gametes combine at random. Green and blue denote different parental origins of the chromosomes. This cross yields offspring of genotypes *TT* and *Tt.*

Table 4.1	Mendel's Law of Segregation			
Experiment	Total	Dominant	Recessive	F$_2$ Phenotypic Ratios
1. Seed form	7,324	5,474	1,850	2.96:1
2. Seed color	8,023	6,022	2,001	3.01:1
3. Seed coat color	929	705	224	3.15:1
4. Pod form	1,181	882	299	2.95:1
5. Pod color	580	428	152	2.82:1
6. Flower position	858	651	207	3.14:1
7. Stem length	1,064	787	277	2.84:1
				Average = 2.98:1

4.2 Single-Gene Inheritance Is Rare

Mendel's first law addresses traits and illnesses caused by single genes, which are also called Mendelian or monofactorial. Single-gene disorders, such as sickle cell disease and muscular dystrophy, are rare compared to infectious diseases, cancer, and multifactorial disorders, most affecting 1 in 10,000 or fewer individuals. **Clinical Connection 4.1** discusses some unusual single-gene traits.

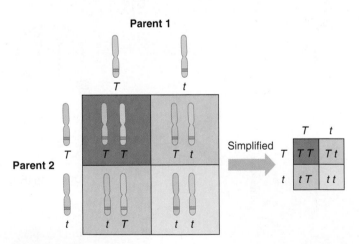

Figure 4.4 **A Punnett square.** A Punnett square illustrates how alleles combine in offspring. The different types of gametes of one parent are listed along the top of the square, with those of the other parent listed on the left-hand side. Each compartment displays the genotype that results when gametes that correspond to that compartment join.

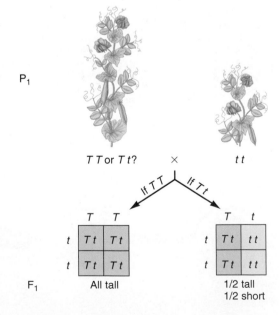

Figure 4.5 **Test cross.** Breeding a tall pea plant with homozygous recessive short plants reveals whether the tall plant is true-breeding (*TT*) or non-true-breeding (*Tt*). Punnett squares usually indicate only the alleles.

It's All in the Genes

Do you have uncombable hair, misshapen toes or teeth, or a pigmented tongue tip? Are you unable to smell a squashed skunk, or do you sneeze repeatedly in bright sunlight? Do you lack teeth, eyebrows, eyelashes, nasal bones, thumbnails, or fingerprints? If so, your unusual trait may be one of thousands described in Online Mendelian Inheritance in Man (OMIM).

Genes control whether hair is blond, brown, or black, has red highlights, and is straight, curly, or kinky. Widow's peaks, cowlicks, a whorl in the eyebrow, and white forelocks run in families; so do hairs with triangular cross-sections. Some people have multicolored hairs, like cats; others have hair in odd places, such as on the elbows, nose tip, knuckles, palms, or soles. Teeth can be missing or extra, protuberant or fused, present at birth, shovel-shaped, or "snowcapped." A person can have a grooved tongue, duckbill lips, flared ears, egg-shaped pupils, three rows of eyelashes, spotted nails, or "broad thumbs and great toes." Extra breasts are known in humans and guinea pigs, and one family's claim to genetic fame is a double nail on the littlest toe.

Unusual genetic variants can affect metabolism, producing either disease or harmless, yet noticeable, effects. Members of some families experience "urinary excretion of odoriferous component of asparagus" or "urinary excretion of beet pigment," producing a strange odor or dark pink urine after consuming the offending vegetable. In blue diaper syndrome, an infant's urine turns blue on contact with air, thanks to an inherited inability to break down the amino acid tryptophan.

The Jumping Frenchmen of Maine syndrome (OMIM 244100) is an exaggerated startle reflex first noted among French-Canadian lumberjacks from the Moosehead Lake area of Maine, whose ancestors were from the Beauce region of Quebec. Physicians first reported the condition at a medical conference in 1878. Geneticists videotaped the startle response in 1980, and the condition continues to appear in genetics journals. OMIM offers a vivid description:

> If given a short, sudden, quick command, the affected person would respond with the appropriate action, often echoing the words of command. . . . For example, if one of them was abruptly asked to strike another, he would do so without hesitation, even if it was his mother and he had an ax in his hand.

The Jumping Frenchmen of Maine syndrome may be an extreme variant of the more common Tourette syndrome, which causes tics and other uncontrollable movements. **Figure 1** illustrates other genetic variants.

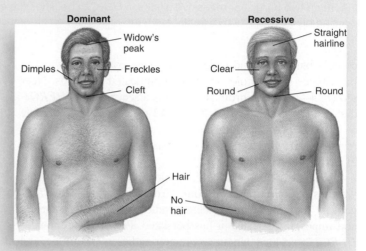

Figure 1 Inheritance of some common traits. Freckles, dimples, hairy arms, widow's peak, and a cleft chin are examples of dominant traits.

Single-gene inheritance is usually more complicated than a pea plant having green or yellow peas because many phenotypes associated with single genes are influenced by other genes as well as by environmental factors. That is, the single gene controls trait transmission, but other genes and the environment affect the degree of the trait or severity of the illness. Eye color is a good example of the view of single-gene traits that has evolved with increasing knowledge of our genomes.

Eye Color

Most people have brown eyes; blue and green eyes are almost exclusively in people of European ancestry. The color of the iris is due to melanin pigments, which come in two forms—the dark brown/black eumelanin, and the red-yellow pheomelanin. In the eye, cells called melanocytes produce melanin, which is stored in structures called melanosomes in the outermost layer of the iris. People differ in the amount of melanin and number of melanosomes, but have about the same number of melanocytes in their eyes.

Nuances of eye color—light versus dark brown, clear blue versus greenish or hazel—arise from the distinctive peaks and valleys at the back of the iris. Thicker regions darken appearance of the pigments, rendering brown eyes nearly black in some parts and blue eyes closer to purple. The bluest eyes have thin irises with very little pigment. The effect of the iris surface on color is a little like the visual effect of a rough-textured canvas on paint.

A single gene on chromosome 15, *OCA2* (OMIM 611409), confers eye color by controlling melanin synthesis. If this gene is missing, albinism results, causing very pale skin and red eyes (**figure 4.6**; see also figure 4.15). A recessive allele of this gene

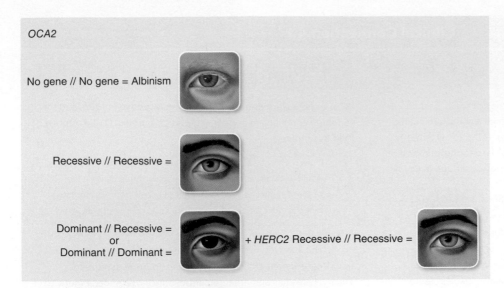

Figure 4.6 Eye color. At least two genes determine eye color in humans. Inheriting two particular recessive alleles in *HERC2* abolishes the effect of *OCA2*, causing blue eyes. The two genes are near each other on chromosome 15.

confers blue color and a dominant allele confers brown. But inheritance of eye color is more complicated than this. Near the *OCA2* gene on chromosome 15 is a second gene, *HERC2*, that controls expression of the *OCA2* gene. A recessive allele of *HERC2* abolishes the control over *OCA2*, and blue eyes result. A person must inherit two copies of the recessive allele in *HERC2* to have blue eyes.

If blue eye color is the disruption of a "normal" function, why has it persisted? A clue comes from evolution. The *HERC2* gene is found in many species, indicating that it is ancient and important, because it has persisted. Perhaps mutations in *HERC2* arose long ago among hunter-gatherers in Europe, and the unusual individuals with the pale eyes were, for whatever reason, more desirable as sexual partners. Over time, this sexual selection would have increased the proportion of the population with blue eyes. Other explanations are possible.

Modes of Inheritance

Modes of inheritance are rules that explain the common patterns of single-gene transmission. They are derived from Mendel's laws. Knowing mode of inheritance makes it possible to calculate the probability that a particular couple will have a child who inherits a particular condition. The way that a trait is passed depends on whether the gene that determines it is on an autosome or on a sex chromosome, and whether the allele is recessive or dominant.

In autosomal dominant inheritance, a trait can appear in either sex because an autosome carries the gene. If a child has the trait, at least one parent also has it. That is, autosomal dominant traits do not skip generations. If no offspring inherit the mutation in one generation, its transmission stops because the offspring can pass on only the recessive form of the gene. Huntington disease, discussed in the chapter opener, is an autosomal dominant condition. The Punnett square in

figure 4.7 depicts inheritance of an autosomal dominant trait or condition, and **table 4.2** summarizes the criteria. Many autosomal dominant diseases do not cause symptoms until adulthood.

In autosomal recessive inheritance, a trait can appear in either sex. Affected individuals have a homozygous recessive genotype, whereas in heterozygotes (carriers) the wild type allele masks expression of the mutant allele. A person with cystic fibrosis, for example, inherits a mutant allele from each carrier parent.

Mendel's first law can be used to calculate the probability that an individual will have either of two phenotypes. The probabilities of each possible genotype are added. For example, the chance that a child whose parents are both carriers of cystic fibrosis will *not* have the condition is the sum of the probability that she has inherited two normal alleles (1/4) plus the chance that she herself is a heterozygote (1/2), or 3/4 in total. Note that this also equals

Table 4.2	Criteria for an Autosomal Dominant Trait

1. Males and females can be affected. Male-to-male transmission can occur.

2. Males and females transmit the trait with equal frequency.

3. Successive generations are affected.

4. Transmission stops after a generation in which no one inherits the mutation.

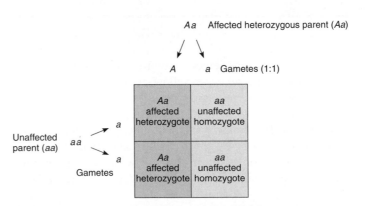

Figure 4.7 Autosomal dominant inheritance. When one parent has an autosomal dominant condition and the other does not, each offspring has a 50 percent probability of inheriting the mutant allele and the condition. In the family from the chapter opener, Karl was the "*Aa*" parent and Jane the "*aa*" parent. Each of their daughters faced the 1 in 2 probability of inheriting Huntington disease.

1 minus the probability that she is homozygous recessive and has the condition.

The ratios that Mendel's first law predicts for autosomal recessive inheritance apply to each offspring anew. If a couple has a child with an autosomal recessive illness, each of their next children faces the same 25 percent risk of inheriting the condition.

Most autosomal recessive conditions appear unexpectedly in families, because they are transmitted silently, through heterozygotes (carriers). **Clinical Connection 4.2** discusses one of them, cystic fibrosis. However, a situation in which an autosomal recessive condition is more likely to recur is when blood relatives have children together. The higher risk of having a child with a particular autosomal recessive condition is because the related parents may carry the same alleles inherited from an ancestor that they have in common, such as a great-grandparent. Marriage between relatives introduces **consanguinity,** which means "shared blood"—a figurative description, because genes are not passed in blood. Alleles inherited from shared ancestors are said to be "identical by descent."

Consanguinity is part of many cultures. For example, marriage between first cousins occurs in about a third of the Pakistani community in England, a population that researchers have been following. In this group, consanguinity doubles the birth defect rate, but it is still low—less than 4 percent of births. Consanguinity may happen unknowingly in close-knit communities where people do not realize they are related, and among families using the same sperm donor.

Logic explains why consanguinity raises risk of inheriting autosomal recessive diseases. An unrelated man and woman have eight different grandparents, but first cousins have only six, because they share one pair through their parents, who are siblings (see figure 4.14c). That is, the probability of two relatives inheriting the same disease-causing recessive allele is greater than that of two unrelated people having the same allele by chance. However, genome-wide studies show that cousins tend to inherit different parts of the shared ancestor's genome, which explains why these populations are healthier than might be expected given the high frequency of consanguinity.

The nature of the phenotype is important when evaluating the transmission of single-gene traits. Some diseases are too severe for people to live long enough or feel well enough to have children. For example, each adult sibling of a person who is a known carrier of Tay-Sachs disease has a two-thirds chance of also being a carrier because only two genotypes are possible for an adult—homozygous for the wild type allele or a carrier who inherits the mutant allele from either parent. A homozygous recessive individual for this brain disease would not have survived childhood.

Geneticists who study human traits and illnesses can hardly set up crosses as Mendel did to demonstrate modes of inheritance, but they can pool information from families whose members have the same trait or illness. Consider a simplified example of 50 couples in whom both partners are carriers of sickle cell disease. If 100 children are born, about 25 of them would be expected to have sickle cell disease. Of the remaining 75, theoretically 50 would be carriers like their parents, and the remaining 25 would have two wild type alleles. **Table 4.3** lists criteria for an autosomal recessive trait.

Solving a Problem in Following a Single Gene

Using Mendel's laws to predict phenotypes and genotypes requires a careful reading of the problem to identify and organize relevant information. Sometimes common sense is useful, too. The following general steps can help to solve a problem based on the inheritance of a single-gene trait:

1. List all possible genotypes and phenotypes for the trait.
2. Determine the genotypes of the individuals in the first (P_1) generation. Use information about those individuals' parents.
3. After deducing genotypes, derive the possible alleles in gametes each individual produces.
4. Unite these gametes in all combinations to reveal all possible genotypes. Calculate ratios for the F_1 generation.
5. To extend predictions to the F_2 generation, use the genotypes of the specified F_1 individuals and repeat steps 3 and 4.

As an example, consider curly hair. If C is the dominant allele, conferring curliness, and c is the recessive allele, then CC and Cc genotypes confer curly hair. A person with a cc genotype has straight hair.

Wendy has beautiful curls, and her husband Rick has straight hair. Wendy's father is bald, but once had curly hair, and her mother has stick-straight hair. What is the probability that Wendy and Rick's child will have straight hair? Steps 1 through 5 solve the problem:

1. State possible genotypes: CC, Cc = curly; cc = straight.
2. Determine genotypes: Rick must be cc, because his hair is straight. Wendy must be Cc, because her mother has straight hair and therefore gave her a c allele.
3. Determine gametes: Rick's sperm carry only c. Half of Wendy's oocytes carry C, and half carry c.
4. Unite the gametes:

		Wendy	
		C	c
Rick	c	Cc	cc
	c	Cc	cc

5. Conclusion: Each child of Wendy and Rick's has a 50 percent chance of having curly hair (Cc) and a 50 percent chance of having straight hair (cc).

Table 4.3	Criteria for an Autosomal Recessive Trait

1. Males and females can be affected.

2. Affected males and females can transmit the gene, unless it causes death before reproductive age.

3. The trait can skip generations.

4. Parents of an affected individual are heterozygous or have the trait.

"65 Roses": Progress in Treating Cystic Fibrosis

Young children who cannot pronounce the name of their disease call it "65 Roses." Cystic fibrosis (CF), although still a very serious illness, is a genetic disease success story thanks to a new class of drugs that correct the underlying problem in protein folding that causes many cases.

CF affects ion channels that control chloride movement out of cells in certain organs (see Clinical Connection 2.2, figure 1). In the lungs, thick, sticky mucus accumulates and creates an environment hospitable to certain bacteria that are uncommon in healthy lungs. A clogged pancreas prevents digestive secretions from reaching the intestines, impairing nutrient absorption. A child with CF has trouble breathing and maintaining weight.

Life with severe CF is challenging. In summertime, a child must avoid water from hoses, which harbor lung-loving *Pseudomonas* bacteria. A bacterium called *Burkholderia cepacia* easily spreads in summer camps. Cookouts spew lung-irritating particulates. Too much chlorine in pools irritates lungs whereas too little invites bacterial infection. New infections arise too. In the past few years, multidrug-resistant *Mycobacterium abscessus,* related to the pathogen that causes tuberculosis, has affected 3 to 10 percent of CF patients in the United States and Europe. Other bacteria that can't grow in a laboratory infect CF patients. Sequenced bacterial DNA in lung fluid from children with CF reveals more than 60 species of bacteria.

CF is inherited from two carrier parents, and affects about 30,000 people in the United States and about 70,000 worldwide. Many other people may have cases so mild that they are not diagnosed with CF but instead with frequent respiratory infections. More than 1,600 mutations are known in the cystic fibrosis transmembrane regulator (*CFTR*) gene, which encodes the chloride channel protein. Today pregnant women are tested for the more common mutations to see if they are carriers, and if they are, their partners are tested. If both parents-to-be are carriers, then the fetus has a 25% chance of having inherited CF and may be tested as well. In most states all newborns receive a genetic test for CF. Before genetic testing became available, diagnosis typically took months and was based on "failure to thrive," salty sweat, and foul-smelling stools.

When CF was recognized in 1938, life expectancy was only 5 years; today median survival is about age 40. Many patients are living longer, thanks to treatments. Inhaled antibiotics control respiratory infections and daily "bronchial drainage" exercises shake mucus from the lungs (**figure 1**). A vibrating vest worn for half-hour periods two to four times a day loosens mucus. Digestive enzymes mixed into soft foods enhance nutrient absorption, although some patients require feeding tubes.

Discovery of the most common *CFTR* mutation (delta F508) in 1989 enabled development of more targeted treatments. The new drugs work in various ways: correcting misfolded CFTR protein, restoring liquid on airway surfaces, breaking up mucus, improving nutrition, and fighting inflammation and infection. A drug called ivacaftor (Kalydeco) that became available in 2012 at first helped only the 5 percent of patients with a mutation called G551D, in whom the ion channel proteins reach the cell membrane but once there, remain closed, like locked gates. The drug binds the proteins in a way that opens the channels. Kalydeco given with a second drug helps the 70 percent of CF patients who have the common mutation. A third new drug enables protein synthesis to ignore mutations that shorten the protein. Patients report better lung function

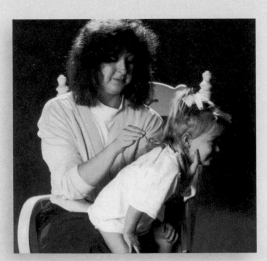

Figure 1 **Treating cystic fibrosis.** In cystic fibrosis, the thick, sticky mucus that clogs airways must be coughed up at least twice every day. Bronchial drainage treatments—tapping hard on the chest and wearing a vibrating vest—help to shake free the mucus.

Questions for Discussion

1. Which parts of the cell are affected in CF?

2. If a child with CF has two parents who do not have the disease, what is the risk that a future sibling will inherit CF?

3. How does the drug Kalydeco work?

4. For some people who have taken Kalydeco, the first sign that the drug is working is that their flatulence no longer smells foul. What characteristic of the disease might this observation indicate is improving?

On the Meaning of Dominance and Recessiveness

Knowing whether an allele is dominant or recessive is critical in determining the risk of inheriting a particular condition (phenotype). Dominance and recessiveness arise from the genotype, and reflect the characteristics or abundance of a protein.

Mendel based his definitions of dominance and recessiveness on what he could see—one allele masking the other. Today we can often add a cellular or molecular explanation.

The basis of an inborn error of metabolism is clear. These disorders are typically recessive because the half normal amount of the enzyme that a carrier produces is usually sufficient to maintain health. The one normal allele, therefore, compensates for the mutant one, to which it is dominant.

A recessive trait is said to arise from a "loss-of-function" because the recessive allele usually prevents the production or activity of the normal protein. In contrast, some dominantly inherited disorders are said to be due to a "gain-of-function," because they result from the action of an abnormal protein that interferes with the function of the normal protein. Huntington disease results from a gain-of-function in which the dominant mutant allele encodes an abnormally long protein that prevents the normal protein from functioning in certain brain cells. Huntington disease is a gain-of-function because individuals who are missing one copy of the gene do not have the illness. That is, the protein encoded by the mutant HD allele must be abnormal, not absent, to cause the disease. The gain-of-function nature of HD is why people with one mutant allele have the same phenotype as the rare individuals with two mutant alleles (**figure 4.8**).

Recessive disorders tend to be more severe, and produce symptoms earlier than dominant disorders. Disease-causing recessive alleles remain in populations because healthy heterozygotes pass them to future generations. In contrast, if a dominant mutation arises that harms health early in life, people who have the allele are either too ill or do not survive long enough to reproduce. The allele eventually becomes rare in the population unless it arises anew by mutation in the gametes of a person who does not have the disease. Dominant disorders whose symptoms do not appear until adulthood, or that do not severely disrupt health, remain in a population because they do not prevent a person from having children and passing on the mutation.

Under certain circumstances, for some genes, a heterozygous individual (a carrier) can develop symptoms. This is the case for sickle cell disease. Carriers can develop a life-threatening breakdown of muscle if exposed to the combination of environmental heat, intense physical activity, and dehydration. Several college athletes died from these symptoms, prompting sports authorities to begin testing athletes for sickle cell disease carrier status. Case Studies and Research Results 2 at the chapter's end discusses this problem.

Key Concepts Questions 4.2

1. State two factors that can influence single gene inheritance patterns.
2. What is a mode of inheritance?
3. Distinguish between autosomal dominant traits and autosomal recessive traits.
4. What is the effect of relatives having children with relatives on inherited trait transmission?
5. How are Mendel's first law and logic used to solve genetics problems?
6. Explain how recessive traits or illnesses can result from a loss-of-function, whereas dominant traits can result from a gain-of-function.

4.3 Following the Inheritance of More Than One Gene

In a second set of experiments, Mendel examined the inheritance of two traits at a time. Today, the ability of exome and genome sequencing to reveal gene interactions makes Mendel's experiments on the inheritance of more than one trait more relevant than ever because many genes are considered simultaneously.

Mendel's Second Law

The second law, the **law of independent assortment,** states that for two genes on different chromosomes, the inheritance of one gene does not influence the chance of inheriting the other gene. The two genes are said to "independently assort" because they are packaged into gametes at random (**figure 4.9**). Two genes that are far apart on the same chromosome also appear to independently assort, because so many crossovers take place between them that it is as if they are part of separate chromosomes (see figure 3.5.)

Mendel looked at seed shape, which was either round or wrinkled (determined by the *R* gene), and seed color, which was either yellow or green (determined by the *Y* gene). When he crossed true-breeding plants that had round, yellow seeds

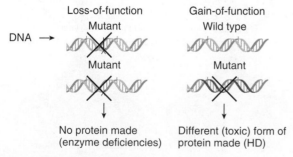

Figure 4.8 **A mutation can confer a loss or a gain of function.** A loss-of-function mutation is typically recessive and results in a deficit or absence of the gene's protein product. A gain-of-function mutation is typically dominant and alters the encoded protein, introducing a new activity.

to true-breeding plants that had wrinkled, green seeds, all the progeny had round, yellow seeds (**figure 4.10**). These offspring were double heterozygotes, or dihybrids, of genotype *RrYy*. From their appearance, Mendel deduced that round is dominant to wrinkled, and yellow to green.

Next, he self-crossed the dihybrid plants in a **dihybrid cross,** so named because two genes and traits are followed. Mendel found four types of seeds in the next, third generation: 315 plants with round, yellow seeds; 108 plants with round, green seeds; 101 plants with wrinkled, yellow seeds; and 32 plants with wrinkled, green seeds. These classes appeared in a ratio of 9:3:3:1.

Mendel then crossed each plant from the third generation to plants with wrinkled, green seeds (genotype *rryy*). These test crosses established whether each plant in the third generation was true-breeding for both genes (genotypes *RRYY* or *rryy*), true-breeding for one gene but heterozygous for the other (genotypes *RRYy, RrYY, rrYy,* or *Rryy*), or heterozygous for both genes (genotype *RrYy*). Mendel could explain the 9:3:3:1 proportion of progeny classes only if one gene does not influence transmission of the other. Each parent would produce equal numbers of four different types of gametes: *RY, Ry, rY,* and *ry*. Each of these combinations has one gene for each trait. A Punnett square for this cross shows that the four types of seeds:

1. round, yellow (*RRYY, RrYY, RRYy,* and *RrYy*),
2. round, green (*RRyy* and *Rryy*),
3. wrinkled, yellow (*rrYY* and *rrYy*), and
4. wrinkled, green (*rryy*)

are present in the ratio 9:3:3:1, just as Mendel found.

Solving a Problem in Following Multiple Genes

A Punnett square for three genes has 64 boxes; for four genes, 256 boxes. An easier way to predict genotypes and phenotypes in multigene crosses is to use the mathematical laws of probability that are the basis of Punnett squares. Probability predicts the likelihood of an event.

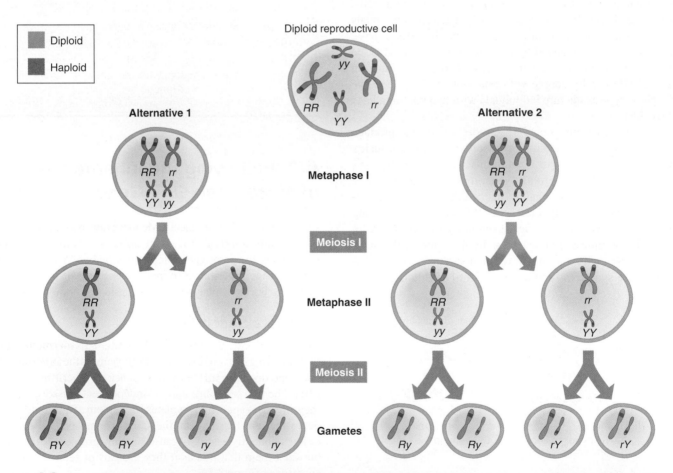

Figure 4.9 **Mendel's second law—independent assortment.** The independent assortment of genes carried on different chromosomes results from the random alignment of chromosome pairs during metaphase of meiosis I. An individual of genotype *RrYy,* for example, manufactures four types of gametes, containing the dominant alleles of both genes (*RY*), the recessive alleles of both genes (*ry*), and a dominant allele of one with a recessive allele of the other (*Ry* or *rY*). The allele combination depends upon which chromosomes are packaged together in a gamete—and this happens at random.

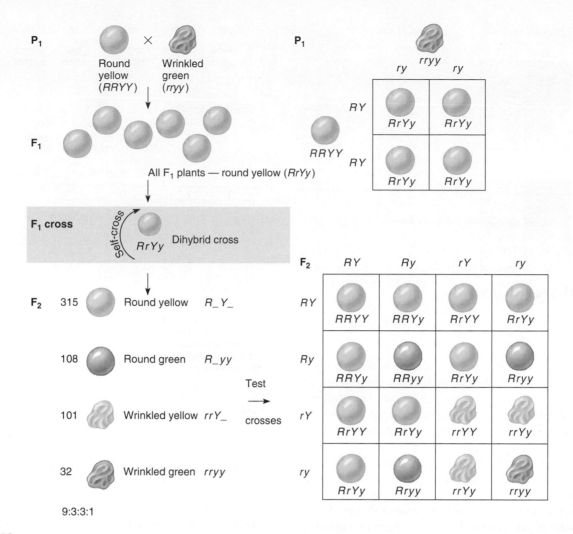

Figure 4.10 **Plotting a dihybrid cross.** A Punnett square can represent the random combinations of gametes produced by dihybrid individuals. An underline in a genotype (in the F$_2$ generation) indicates that either a dominant or a recessive allele is possible. The numbers in the F$_2$ generation are Mendel's experimental data. Test crosses with *rryy* plants revealed the genotypes of the F$_2$ generation, depicted in the 16-box Punnett square.

An application of probability theory called the product rule can predict the chance that parents with known genotypes can produce offspring of a particular genotype. The product rule states that the chance that two independent events will both occur equals the product of the chance that either event will occur alone. Consider the probability of obtaining a plant with wrinkled, green peas (genotype *rryy*) from dihybrid (*RrYy*) parents. Do the reasoning for one gene at a time, then multiply the results (**figure 4.11**).

A Punnett square depicting a cross of two *Rr* plants indicates that the probability of producing *rr* progeny is 25 percent, or 1/4. Similarly, the chance of two *Yy* plants producing a *yy* plant is 1/4. Therefore, the chance of dihybrid parents (*RrYy*) producing homozygous recessive (*rryy*) offspring is 1/4 multiplied by 1/4, or 1/16. Now consult the 16-box Punnett square for Mendel's dihybrid cross again (see figure 4.10). Only one of the 16 boxes is *rryy*, just as the product rule predicts. **Figure 4.12** shows how these tools can be used to predict offspring genotypes and phenotypes for three human traits simultaneously.

4.4 Pedigree Analysis

For genetics researchers and genetic counselors, families are tools, and the bigger the family the better—the more children in a generation, the easier it is to see a mode of inheritance. Charts called **pedigrees** display family relationships

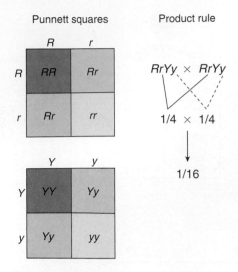

Figure 4.11 The product rule.

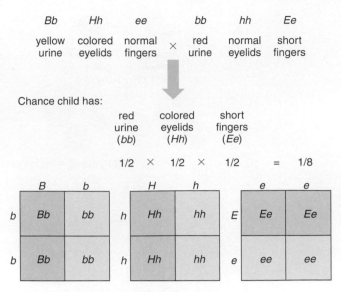

Bb yellow (normal) urine
bb beeturia (red urine after eating beets)
Hh colored eyelids
hh normal eyelids
Ee brachydactyly (short fingers)
ee normal fingers

Figure 4.12 **Using probability to track three traits in people.** A man has normal urine, colored eyelids, and normal fingers. His partner has red urine after she eats beets, and has normal eyelids and short fingers. The chance that their child will have red urine after eating beets, colored eyelids, and short fingers is 1/8.

and depict which relatives have specific phenotypes and, sometimes, genotypes. A human pedigree serves the same purpose as one for purebred dogs or cats or thoroughbred horses—it represents relationships. A pedigree in genetics differs from a family tree in genealogy, and from a genogram in social work, in that it indicates disorders or traits as well as relationships and ancestry. Pedigrees may also include molecular data, test results, and information on variants of multiple genes.

A pedigree consists of lines that connect shapes. Vertical lines represent generations; horizontal lines that connect two shapes at their centers depict partners; shapes connected by vertical lines that are joined horizontally represent siblings. Squares indicate males; circles, females; and diamonds, individuals of unspecified sex. Roman numerals designate generations. Arabic numerals or names indicate individuals. **Figure 4.13** shows these and other commonly used pedigree symbols. Colored or shaded shapes indicate individuals who express a trait, and half-filled shapes are known carriers.

Pedigrees Then and Now

The earliest pedigrees were strictly genealogical, not indicating traits. **Figure 4.14a** shows such a pedigree for a highly inbred branch of the ancient Egyptian royal family tree. The term *pedigree* arose in the fifteenth century, from the French *pie de grue,* which means "crane's foot." Pedigrees at that time, typically depicting large families, showed parents linked by curved lines to their many offspring. The overall diagram often resembled a bird's foot.

One of the first pedigrees to trace an inherited illness was an extensive family tree of several European royal families, indicating which members had the clotting

disorder hemophilia (see figure 6.7). The mutant gene probably originated in Queen Victoria of England in the nineteenth century. In 1845, a genealogist named Pliny Earle constructed a pedigree of a family with colorblindness using musical notation—half notes for unaffected females, quarter notes for colorblind females, and filled-in and squared-off notes to represent the many colorblind males. In the early twentieth century, eugenicists tried to use pedigrees to show that traits such as criminality, feeblemindedness, and promiscuity were the consequence of faulty genes.

Today, pedigrees are important both for helping families identify the risk of transmitting an inherited illness and as starting points for identifying and describing, or annotating, a gene from the human genome sequence. People who have kept meticulous family records, such as the Mormons and the Amish, are invaluable in helping researchers follow the inheritance of particular genes. (see Clinical Connection 15.1) Very large pedigrees can provide information on many individuals with a particular rare disorder. The researchers then search affected individuals' DNA to identify a particular sequence they have all inherited that is not found in healthy family members. Within this section of a chromosome lies the causative mutation. Discovery of a mutation that causes an early-onset form of Alzheimer disease, for

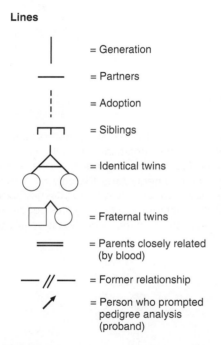

Symbols

○, □ = Normal female, male

●, ■ = Female, male who expresses trait

◐, ◧ = Female, male who carries an allele for the trait but does not express it (carrier)

Ø, ⊘ = Dead female, male

◇ = Sex unspecified

Ø, ⊘ = Stillbirth
SB SB

Ⓟ, Ⓟ, ⟨Ⓟ⟩ = Pregnancy

△ = Spontaneous abortion (miscarriage)

⚲ = Terminated pregnancy (shade if abnormal)

Lines

| = Generation

— = Partners

┊ = Adoption

┌┬┐ = Siblings

= Identical twins

= Fraternal twins

= = Parents closely related (by blood)

—//— = Former relationship

↗ = Person who prompted pedigree analysis (proband)

Numbers

Roman numerals = generations

Arabic numerals = individuals in a generation

Figure 4.13 Pedigree components. Symbols representing individuals connect to form pedigree charts, which display the inheritance patterns of traits.

example, took researchers to a remote village in Colombia, where the original mutation present today in a 1,000-plus-member family came from a Spanish settler who had arrived in the 17th century.

Pedigrees Display Mendel's Laws

Visual learners can easily "see" a mode of inheritance in a pedigree. Consider a pedigree for an autosomal recessive trait, albinism. Homozygous recessive individuals in the third (F_2) generation lack an enzyme necessary to manufacture the pigment melanin and, as a result, hair and skin are very pale (**figure 4.15**). Their parents are inferred to be heterozygotes (carriers). One partner from each pair of grandparents must also be a carrier. For some disorders, carriers have half the wild type amount of a key biochemical in a body fluid, such as blood or urine. Their carrier status is detectable with a blood or urine test.

An autosomal dominant trait does not skip generations and can affect both sexes. A typical pedigree for an autosomal dominant trait has some squares and circles filled in to indicate affected individuals in each generation. Figure 4.14*b* is a pedigree for an autosomal dominant trait, extra fingers and toes (polydactyly), which is shown in figure 1.5a.

A pedigree may be inconclusive, which means that either autosomal recessive or autosomal dominant inheritance can explain the pattern of filled-in symbols. **Figure 4.16** shows one such pedigree, for a type of hair loss called alopecia areata (OMIM 104000). According to the pedigree, this trait can be passed in an autosomal dominant mode because it affects both males and females and is present in every generation. However, the pedigree can also depict autosomal recessive inheritance if the individuals represented by unfilled symbols are carriers. Inconclusive pedigrees tend to arise when families are small and the trait is not severe enough to impair fertility.

Pedigrees may be difficult to construct or interpret for several reasons. People may hesitate to supply information because the symptoms embarrass them. Families with adoption, children born out of wedlock, serial relationships, blended families, and assisted reproductive technologies (see chapter 21) may not fit easily into the rules of pedigree construction. Many people cannot trace their families back far enough to reveal a mode of inheritance.

Solving a Problem in Conditional Probability

Pedigrees and Punnett squares can trace a conditional probability, which is when an offspring's genotype depends on the parents' genotypes, which may not be obvious from their phenotypes. For example, if a person has an autosomal recessive trait or condition, his or her parents are inferred to be carriers (unless a new mutation arises in the affected individual).

The family represented in **figure 4.17** illustrates conditional probability. Deshawn has sickle cell disease. His unaffected parents, Kizzy and Ike, must each be heterozygotes (carriers). Deshawn's sister, Taneesha, is also healthy, and she is expecting her first child. Taneesha's husband, Antoine, has no family history of sickle cell disease. What is the probability that Taneesha's child will inherit her mutant allele and be a carrier?

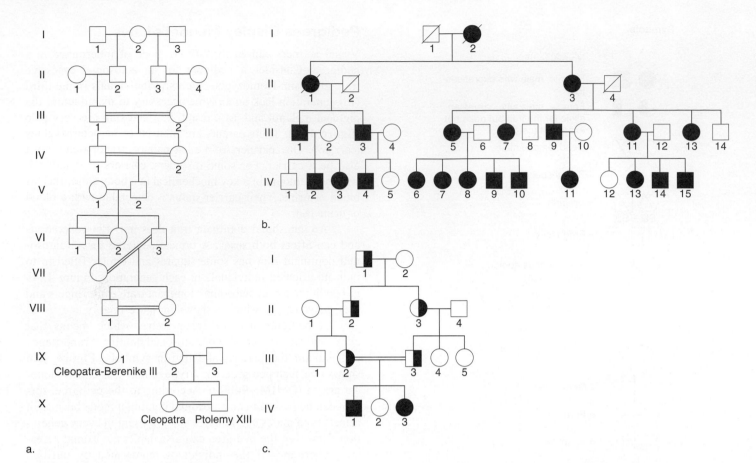

Figure 4.14 **Unusual pedigrees.** **(a)** A partial pedigree of Egypt's Ptolemy dynasty shows only genealogy, not traits. It appears almost ladderlike because of the extensive consanguinity. From 323 B.C. to Cleopatra's death in 30 B.C., the family experienced one pairing between cousins related through half-brothers (generation III), four brother-sister pairings (generations IV, VI, VIII, and X), and an uncle-niece relationship (generations VI and VII). Cleopatra married her brother, Ptolemy XIII, when he was 10 years old! These marriage patterns were an attempt to preserve the royal blood. **(b)** A family with polydactyly (extra fingers and toes) extends laterally, with many children because the phenotype doesn't affect reproduction. **(c)** This pedigree shows marriage of first cousins. They share one set of grandparents, and therefore risk passing on the same recessive alleles to offspring.

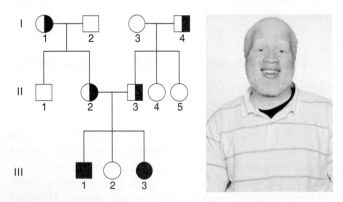

Figure 4.15 **Albinism is autosomal recessive.** Albinism affects males and females and can skip generations, as it does in generations I and II in this pedigree. Homozygous recessive individuals lack an enzyme needed to produce melanin, which colors the eyes, skin, and hair.

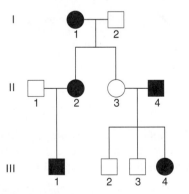

Figure 4.16 **An inconclusive pedigree.** This pedigree could account for an autosomal dominant trait or an autosomal recessive trait that does not prevent affected individuals from having children. (Unfilled symbols could represent carriers.)

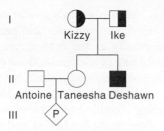

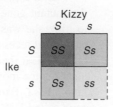

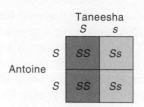

a. Taneesha's brother, Deshawn, has sickle cell disease.

b. Probability that Taneesha is a carrier: $^2/_3$

c. If Taneesha is a carrier, chance that fetus is a carrier: $^1/_2$

Total probability = $^2/_3 \times ^1/_2 = ^1/_3$

Figure 4.17 **Making predictions.** Taneesha's brother Deshawn has sickle cell disease **(a).** Taneesha wonders if her fetus has inherited the sickle cell allele. First, she must calculate the chance that she is a carrier. The Punnett square in **(b)** shows that this risk is 2 in 3. (She must be genotype *SS* or *Ss*, but cannot be *ss* because she does not have the disease.) The risk that the fetus is a carrier, assuming that the father is not a carrier, is half Taneesha's risk of being a carrier, or 1 in 3 **(c).**

Taneesha's concern raises two questions. First, what is the probability that she is a carrier? Because she is the product of a monohybrid cross, and she is not homozygous recessive (sick), she has a 2 in 3 chance of being a carrier. If so, the chance that she will transmit the mutant allele is 1 in 2, because she has two copies of the gene, and only one allele goes into each gamete. To calculate the overall risk to her child, multiply the probability that she is a carrier (2/3) by the chance that if she is, she will transmit the mutant allele (1/2). The result is 1/3.

4.5 Family Exome Analysis

About 7,000 single-gene disorders are recognized, but many are so rare that affected individuals are never diagnosed, for several reasons. Physicians might not recognize combinations of symptoms as reflecting a specific genetic syndrome, or families may be so small that only one person is affected, and therefore the condition does not appear to be inherited. Or, a person with an undiagnosed condition may have a spontaneous mutation, and so the illness is genetic, but not inherited.

Sequencing exomes (the protein-encoding portion of the genome) is filling in the blanks in our knowledge of Mendelian (single-gene) traits and illnesses, solving some diagnostic mysteries. In **family exome analysis,** researchers compare the DNA sequence of the exome of a relative who has unexplained symptoms or traits to the DNA sequences of the exomes of other family members. Then researchers search among the identified mutations for likely candidate genes whose functions, if disrupted or absent, might explain the symptoms. Two general results are possible. The affected family member inherited recessive mutations from either or both parents, or the affected individual is a "new mutation." The case of Nicholas Volker described in Clinical Connection 1.1 on page 10 is the first type of result, and the case of Bea Rienhoff is the second.

One of the first examples of family exome analysis was that of Bea's father's search for an explanation of her unusual physical characteristics. When Dr. Hugh Rienhoff, an internist and clinical geneticist, first saw Bea when she was born in 2003, he knew that her long feet, clenched fingers, poor muscle tone, widely spaced eyes, and facial birthmark might be due to a genetic syndrome (**figure 4.18**).

As Bea got older but did not gain much weight, appearing thin with birdlike legs, her father recognized signs of Marfan syndrome (OMIM 609192) and Loeys-Dietz syndrome

Figure 4.18 Dr. Hugh Rienhoff performed exome sequencing on his family to identify his daughter's connective tissue condition.

(OMIM 609192, 610380), both disorders of connective tissue. When pediatricians couldn't diagnose Bea, Dr. Rienhoff bought second-hand DNA sequencing equipment and searched for an answer himself, working at home. "It was eerie examining her DNA, as though I were peering through a powerful microscope looking deep into my daughter while she patiently lay on the microscope stage, looking up, hoping for answers," he wrote.

When probing genes that might be mutant and looking at which genes only Bea expressed proved fruitless, Dr. Rienhoff turned to exome sequencing. Did Bea's protein-encoding genes differ from those of her parents and two brothers?

Exome sequencing typically finds hundreds of gene variants. Most are dismissed because they don't change the encoded protein in a way that can affect the phenotype, or because they are so common that they couldn't cause a condition so rare than no doctor had recognized it. Exome sequencing revealed that Bea has a mutation in the gene that encodes transforming growth factor β-3, blocking her cells from receiving certain growth signals. The gene is similar to the genes behind the two conditions that Dr. Rienhoff had suspected. The parents and brothers are wild type.

Discovering mutations using DNA sequencing, of the exome or entire genome, is a very recent technology, but interpretation of the results is based on Mendel's classic laws of single-gene inheritance. Because Bea's mutation appears only in her, and in one copy, it is dominant and spontaneous —she did not inherit it from her parents. Origin of the mutation in Bea explains why there was no family history from which to predict a mode of inheritance.

In the case of Nicholas Volker, the boy's mother carried the mutant gene on one of her X chromosomes and passed it to him; she was unaffected. Family exome analysis can track autosomal recessive inheritance too. In one family, three of four children lost their sight as they entered adolescence, but tested negative for more than 50 mutations known to cause retinitis pigmentosa. Exome sequencing identified two mutations in a gene called *DHDDS* in two sisters and a brother, as well as one copy in the heterozygous parents. Therefore, exome sequencing can confirm a mode of inheritance, or reveal a genotype that is either due to a spontaneous mutation or hidden in heterozygotes. The genetic mysteries being solved with DNA sequencing tell a bigger tale—that understanding heredity requires both observations at the person and family levels, as well as analyzing genes.

Key Concepts Questions 4.5

1. Explain how exome sequencing can be used to identify a mutation that causes a phenotype in a family in which only one person is affected.

2. Explain how a dominant trait or condition might appear differently in a family from a recessive condition.

3. Explain the differences in the findings of exome sequencing in the cases of Nicholas Volker and Bea Rienhoff.

Summary

4.1 Following the Inheritance of One Gene

1. Mendel's laws, based on pea plant crosses, derive from how chromosomes act during meiosis. The laws apply to all diploid organisms.

2. Mendel used statistics to investigate why some traits vanish in hybrids. The **law of segregation** states that alleles of a gene go into separate gametes during meiosis. Mendel demonstrated segregation of seven traits in pea plants using **monohybrid crosses**.

3. A diploid individual with two identical alleles of a gene is a **homozygote.** A **heterozygote** has two different alleles of a gene. A gene may have many alleles.

4. A **dominant** allele masks the expression of a **recessive** allele. An individual may be homozygous dominant, homozygous recessive, or heterozygous. The alleles for a gene in an individual constitute the **genotype**, and their expression the **phenotype**. The most common allele in a population is **wild type**, and variants are **mutant**. **Mutation** is a change in a gene.

5. When Mendel crossed two true-breeding types, then bred the resulting hybrids to each other, the two variants of the trait appeared in a 3:1 phenotypic ratio. Crossing these progeny revealed a genotypic ratio of 1:2:1.

6. A **Punnett square** follows the transmission of alleles and is based on probability.

4.2 Single-Gene Inheritance Is Rare

7. Eye color illustrates how a single-gene trait can be affected by other genes.

8. An **autosomal dominant** trait affects males and females, and does not skip generations.

9. An **autosomal recessive** trait affects males or females and may skip generations. Autosomal recessive conditions are more likely in families with **consanguinity.** Recessive disorders tend to be more severe and cause symptoms earlier than dominant disorders.

10. Genetic problems can be solved by tracing alleles as gametes form and then combine in a new individual.

11. Dominance and recessiveness reflect how alleles affect the abundance or activity of the gene's protein product. Loss-of-function mutations lead to a missing protein. Gain-of-function mutations alter a protein's function.

4.3 Following the Inheritance of More Than One Gene

12. Mendel's second law, the **law of independent assortment,** follows transmission of two or more genes on different chromosomes. Mendel used **dihybrid crosses** to show that a random assortment of maternally and paternally derived chromosomes during meiosis yields gametes with different gene combinations.

13. The chance that two independent genetic events occur equals the product of the probabilities that each event will occur. This product rule is useful in calculating the risk that individuals will inherit a genotype and

in following the inheritance of genes on different chromosomes.

4.4 Pedigree Analysis

14. A **pedigree** is a chart that depicts family relationships and patterns of inheritance for particular traits. A pedigree can be inconclusive.

4.5 Family Exome Analysis

15. Comparing exome sequences of family members extends Mendelian analysis to include specific DNA sequences. Such family exome analysis is useful in identifying a disease-causing gene variant inherited from a parent, or one that has arisen in the child.

www.mhhe.com/lewisgenetics11

Answers to all end-of-chapter questions can be found at **www.mhhe.com/lewisgenetics11**. You will also find additional practice quizzes, animations, videos, and vocabulary flashcards to help you master the material in this chapter.

Review Questions

1. How does meiosis explain Mendel's laws?

2. Describe two ways to inherit blue eyes.

3. Compare how an autosomal recessive condition affects a family to how an autosomal dominant condition does so.

4. Discuss how Mendel derived the two laws of inheritance without knowing about chromosomes.

5. Distinguish between
   a. autosomal recessive and autosomal dominant inheritance.
   b. Mendel's first and second laws.
   c. a homozygote and a heterozygote.
   d. a monohybrid and a dihybrid cross.
   e. a loss-of-function mutation and a gain-of-function mutation.
   f. a Punnett square and a pedigree.

6. Explain how a dominant disease can occur if its symptoms prevent reproduction.

7. Why would Mendel's results for the dihybrid cross have been different if the genes for the traits he followed were near each other on the same chromosome?

8. Why are extremely rare autosomal recessive disorders more likely to appear in families in which blood relatives have children together?

9. How does the pedigree of the ancient Egyptian royal family in figure 4.14*a* differ from a pedigree a genetic counselor might use today?

10. What are possible genotypes of the parents of a person who has two HD mutations?

11. What is the probability that two individuals with an autosomal recessive trait, such as albinism, will have a child with the same genotype and phenotype as they have?

12. How does family exome sequencing and analysis apply Mendel's laws?

Applied Questions

1. Predict the phenotypic and genotypic ratios for crossing the following pea plants:
   a. short × short
   b. short × true-breeding tall
   c. true-breeding tall × true-breeding tall

2. What are the genotypes of the pea plants that would have to be bred to yield one plant with restricted pods for every three plants with inflated pods?

3. If pea plants with all white seed coats are crossed, what are the possible phenotypes of their progeny?

4. Pea plants with restricted yellow pods are crossed to plants that are true-breeding for inflated green pods and

the F$_1$ crossed. Derive the phenotypic and genotypic ratios for the F$_2$ generation.

5. More than 100 genes cause deafness when mutant. What is the most likely mode of inheritance in families in which all children and the parents were born deaf?

6. The MacDonalds raise Labrador retrievers. In one litter, two of eight puppies, a male and a female, have a condition called exercise-induced collapse. After about 15 minutes of intense exercise, the dogs wobble about, develop a fever, and their hind legs collapse. The parents are healthy. What is the mode of inheritance?

7. Draw a pedigree to depict the following family: One couple has a son and a daughter with normal skin

pigmentation. Another couple has one son and two daughters with normal skin pigmentation. The daughter from the first couple has three children with the son of the second couple. Their son and one daughter have albinism; their other daughter has normal skin pigmentation.

8. Chands syndrome (OMIM 214350) is autosomal recessive and causes very curly hair, underdeveloped nails, and abnormally shaped eyelids. In the following pedigree, which individuals must be carriers?

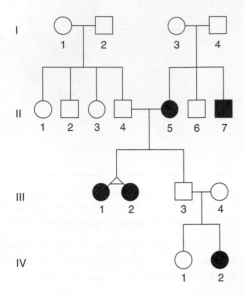

Chands syndrome

9. Lorenzo has a double row of eyelashes (OMIM 126300), which he inherited from his mother as a dominant trait. His maternal grandfather is the only other relative to have it. Fatima, who has normal eyelashes, marries Lorenzo. Their first child, Nicola, has normal eyelashes. Now Fatima is pregnant again and hopes for a child with double eyelashes. What chance does the child have of inheriting double eyelashes? Draw a pedigree of this family.

10. In peeling skin syndrome (OMIM 270300) the outer skin layers fall off on the hands and feet. The pedigrees depict three affected families. What do the families share that might explain the appearance of this otherwise rare condition?

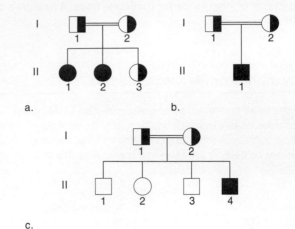

Peeling skin syndrome

11. In sclerosteosis (OMIM 269500) (see below), the skull and jaws overgrow, producing a characteristic face, gigantism, facial paralysis, hearing loss, severe headaches, and even sudden death. In the pedigree below for a family with sclerosteosis:

a. What is the relationship between the individuals who are connected by slanted double lines?

b. Which individuals in the pedigree must be carriers?

12. The child referenced in figure 4.12 who has red urine after eating beets, colored eyelids, and short fingers, is genotype *bbHhEe*. The genes for these traits are on different chromosomes. If he has children with a woman who is a trihybrid, what are the expected genotypic and phenotypic ratios for their offspring?

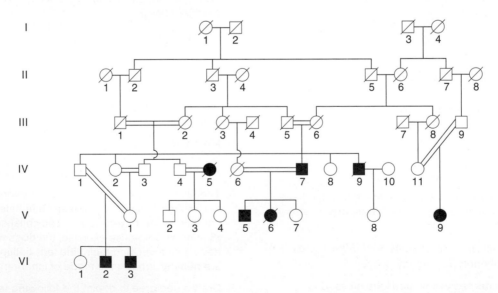

Sclerosteosis

Web Activities

1. Visit a website that features families who have rare diseases, such as Check Orphan (http://www.checkorphan.org/), the National Organization for Rare Diseases (http://www.rarediseases.org/), or the Office of Rare Diseases Research (http://rarediseases.info.nih.gov/resources/5), or a website for a particular disease. Describe the pattern of inheritance in a family, indicating whether the disease is inherited as an autosomal dominant, autosomal recessive, or X-linked trait. (X-linked traits are transmitted on the X chromosome and are discussed in chapter 6).

2. Jacey Mervar, whose family is described in the chapter opener, began jhdkids.com to help others with juvenile Huntington disease. Read the website, and explain how juvenile HD is like the disease in adults, and how it is different.

3. Look up Kalydeco Miracles on Facebook. List three ways that taking the drug has improved the lives of patients.

Forensics Focus

1. A woman, desperate to complete her family tree for an upcoming family reunion, cornered a stranger in a fast-food restaurant. Her genealogical research had identified him as a distant cousin, and she needed his DNA. He refused to cooperate, looked scared, and ran off. The woman took his discarded coffee cup and collected DNA from traces of saliva, which she sent on a swab to a DNA ancestry testing company. (This is a true story.)

   a. Do you think the woman was justified in her action? Why or why not?

   b. What are the strengths and limitations of using genealogical information (family records, word-of-mouth) versus DNA testing to construct a pedigree?

2. A young woman walking to her car in a parking lot late at night was attacked and raped. She recalled that the rapist wore a green silk shirt, a belt with a metal buckle that left a bruise on her abdomen, jeans, and running shoes. She also remembered that he had white skin, a shaved head, and startling blue eyes. She recalled footsteps approaching and then someone yanking the man off her, but her head was let go and hit the pavement, knocking her unconscious.

   A few weeks after the rape, a naked male body washed up in a nearby river, and police found a belt buckle farther downstream that matched the pattern of the woman's bruise. The body, however, was headless, as if cut. How can police determine the eye color of the corpse to help in identifying or ruling him out as the rapist?

Case Studies and Research Results

1. Calculate the probability that all three children in the Mervar family would inherit Huntington disease.

2. In 2006, a 19-year-old freshman at Rice University, Dale Lloyd II, collapsed during a strenuous football practice. He died the next morning following acute exertional rhabdomyolysis, a complication of being a carrier (heterozygote) for sickle cell disease in which muscles break down. The student's parents sued Rice University and the National Collegiate Athletic Association (NCAA). In response the NCAA mandated sickle cell trait testing for all Division 1 student-athletes. However, a student can sign a waiver to opt out of sickle cell testing, absolving the NCAA of liability. In some states students may already know their carrier status because they were tested as newborns.

   Dehydration and heat exposure are also risk factors for the deadly muscle condition, whether or not someone is a carrier of sickle cell disease. Drinking enough water and resting can help prevent the muscle problem. The genetic disease is more prevalent among people of African ancestry, but anyone can be a carrier. The NCAA estimates that requiring testing in all Division 1 schools would identify 300 to 400 carriers each year, and save one or two lives.

   a. Was the NCAA response to test all Division 1 students a good idea? State a reason for your answer.

   b. How might the NCAA improve its plan to test all Division 1 athletes?

   c. Suggest a way to prevent deaths from exertional rhabdomyolysis other than genetic testing.

3. When Peter and Martha were 24 and expecting their first child, they learned that Peter's mother, who was adopted, had early signs of Huntington disease (see chapter opener). A genetic counselor explained the mode of inheritance and said that Peter could take a "predictive" genetic test to find out if he had inherited the dominant mutation. Peter did not want that information, but Martha did not want to have a child who would inherit HD. Martha requested that the fetus be tested, but not Peter. The genetic counselor explained that people under age 18 were discouraged from having predictive testing. Peter was against testing the fetus, pointing out that symptoms do not begin until adulthood, and a treatment might be available 20 or so years in the future.

   a. Why do you think geneticists advise against testing people under 18 years of age? Do you agree or disagree with this practice?

b. What is the mode of inheritance of HD?

c. What is the risk that Peter's sister Kate, who is 19, inherited the mutation?

d. If the fetus could be tested, how might this pose a problem for Peter?

4. Dr. Hugh Rienhoff and his daughter Bea were featured in the opener to chapter 4 in the ninth edition of this textbook. They were not in the tenth edition, because at that time, Dr. Rienhoff had been unable to find a mutant gene that explained his daughter's symptoms. Reviewers of this book advised removal of the story, because Bea's disease might not be genetic.

a. What technology enabled the father to finally discover the genetic basis of his daughter's condition?

b. Why is Bea's condition genetic but not inherited?

c. What is the probability that a child of Bea's inherits her condition?

5. More than a dozen recessive illnesses that are very rare in most of the world are fairly common among the Bedouin people who live in the Negev Desert area of Israel. More than 65 percent of Bedouins marry their first or second cousins. This practice helped the group to survive a nomadic existence in the harsh environment in the past. Two physicians and a geneticist set up a service that enables people wishing to marry to take genetic tests to learn if they are carriers for the same diseases. Prenatal testing has also been introduced to provide the option of terminating pregnancies that would otherwise lead to the births of children who would die of a recessive disorder in early childhood. Discuss the pros and cons of introducing genetic testing in this community. Should researchers or clinicians interfere with a society's long-held cultural practices?

Gavin Stevens has a form of Leber congenital amaurosis that was the 18th to be discovered, using exome sequencing. He cannot see, but is an amazing musician. Gavin bounced to hiphop as a toddler, could play anything he heard on the piano before he turned two, and today attends the Academy of Music for the Blind in Los Angeles, where he sings opera.

Beyond Mendel's Laws

Learning Outcomes

5.1 When Gene Expression Appears to Alter Mendelian Ratios

1. Discuss phenomena that can appear to alter expected Mendelian ratios.

5.2 Mitochondrial Genes

2. Describe the mode of inheritance of a mitochondrial trait.

3. Explain how mitochondrial DNA differs from nuclear DNA.

5.3 Linkage

4. Explain how linked traits are inherited differently from Mendelian traits.

5. Discuss the basis of linkage in meiosis.

6. Explain how linkage is the basis of genetic maps and genome-wide association studies.

The **BIG** Picture

The modes of inheritance that Gregor Mendel's elegant experiments on peas revealed can be obscured when genes have many variants, interact with each other or the environment, are in mitochondria, or are linked on the same chromosome.

Mutations in Different Genes Cause Blindness

"Genetic heterogeneity" refers to mutations in different genes that cause the same symptoms. This technical concept came to life when, on a sunny summer Saturday in Philadelphia in 2010, families with a form of visual loss called Leber congenital amaurosis (LCA) gathered to hear researchers talk about gene therapy that treats, and possibly cures, one type of the condition. At least twenty-one genes cause LCA.

Jennifer Pletcher was at the conference. Her daughter Finley has a mutation in a gene called *RDH12*. Jennifer had met a few LCA families on Facebook, but they had mutations in different genes. At the conference, one of the speakers asked attendees to stand when their mutation was called. When she said *"RDH12,"* "we stood up, and soon three families were around us. It was awesome! Now we are friends and can keep in contact about changes in our kids," said Jennifer.

Not all were able to stand and be counted among those with known mutations. Jennifer and Troy Stevens were there to learn which mutation their then-2-year-old son Gavin had. The parents were to meet with the head of the lab testing their DNA, but the family's mutation was not among the known mutations. So they all had exome sequencing, because knowing the gene behind an illness is the first step in developing a gene therapy, which adds a functional gene. Two years later, exome sequencing discovered Gavin's mutation, and the foundation his parents started is funding gene therapy research. Gavin cannot see, but he is a gifted musician and a delight.

5.1 When Gene Expression Appears to Alter Mendelian Ratios

Single genes rarely completely control a phenotype in the way that Mendel's experiments with peas suggested. Genes interact with each other and with environmental influences. When transmission patterns of a visible trait do not exactly fit autosomal recessive or autosomal dominant inheritance, Mendel's laws are still operating. The underlying genotypic ratios persist, but other factors affect the phenotypes.

This chapter considers three general phenomena that seem to be exceptions to Mendel's laws, but are really not: gene expression, mitochondrial inheritance, and linkage. In several circumstances, phenotypic ratios appear to contradict Mendel's laws, but they do not.

Lethal Allele Combinations

A genotype (allele combination) that causes death is, by strict definition, lethal. Death from genetic disease can occur at any stage of development or life. Tay-Sachs disease is lethal by age 3 or 4, whereas Huntington disease may not be lethal until late middle age. In a population and evolutionary sense, a lethal genotype has a more specific meaning—it causes death before the individual can reproduce, which prevents passage of genes to the next generation.

In organisms used in experiments, such as fruit flies, pea plants, or mice, lethal allele combinations remove an expected progeny class following a specific cross. For example, in a cross of heterozygous flies carrying lethal alleles in the same gene, homozygous recessive progeny die as embryos, leaving only heterozygous and homozygous dominant adult fly offspring. In humans, early-acting lethal alleles cause spontaneous abortion. When both parents carry a recessive lethal allele for the same gene, each pregnancy has a 25 percent chance of spontaneously aborting—this is the homozygous recessive class.

An example of a lethal genotype in humans is achondroplastic dwarfism (OMIM 100800), which has a very distinct phenotype of a long trunk, very short limbs, and a large head with a flat face (**figure 5.1**). It is an autosomal dominant trait, but is most often the result of a spontaneous (new) mutation. Each child of two people with achondroplasia has a one in four chance of inheriting both mutant alleles. However, because such homozygotes are not seen, this genotype is presumed to be lethal. Each child therefore faces a 2/3 probability of having achondroplasia and a 1/3 probability of being of normal height. Homozygotes for achondroplasia mutations in other species cannot breathe because the lungs do not have room to inflate. The mutation is in the gene that encodes a receptor for a growth factor. Without the receptor, growth is severely stunted.

a.

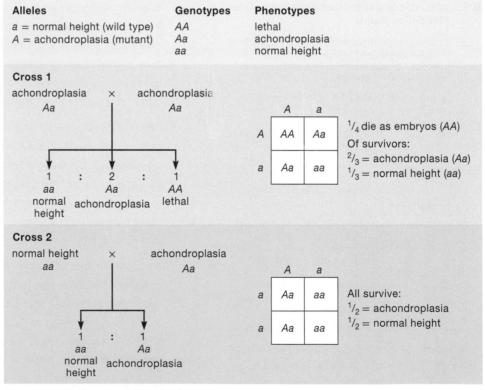

b.

Figure 5.1 Lethal alleles. (a) A person with achondroplasia has inherited a mutant dominant allele. Inheriting two such alleles is lethal to embryos. **(b)** If two people with achondroplasia have children, each child has a 2/3 chance of inheriting achondroplasia and 1/3 of being normal height. This is a conditional probability because one in four conceptions does not survive.

Multiple Alleles

An individual has two alleles for any autosomal gene, one on each homolog. However, a gene can exist in more than two allelic forms in a population because it can mutate in many ways. That is, the sequence of hundreds of DNA bases that makes up a gene can be altered in many ways, just as mistakes can occur anywhere in a written sentence. Different allele combinations can produce variations in the phenotype. The more alleles, the more variations of the phenotype are possible. An individual with two different mutant alleles for the same gene is called a *compound heterozygote.*

For some inherited diseases, knowing the genotype enables a physician to predict the general course of an illness, or which of several symptoms are likely to develop. This is the case for carrier testing for CF, which is done routinely in early pregnancy (see Clinical Connection 4.2). One mutant allele, *ΔF508,* when homozygous causes about 70 percent of cases. People who have this "double delta" genotype have frequent, severe respiratory infections, very congested lungs, and poor weight gain. A different CF genotype increases susceptibility to bronchitis and pneumonia, and another causes only male infertility.

Different Dominance Relationships

In complete dominance, one allele is expressed, while the other isn't. In **incomplete dominance,** the heterozygous phenotype is intermediate between that of either homozygote.

Enzyme deficiencies in which a threshold level is necessary for health illustrate both complete and incomplete dominance. For example, on a whole-body level, Tay-Sachs disease displays complete dominance because the heterozygote (carrier) is as healthy as a homozygous dominant individual. However, if phenotype is based on enzyme level, then the heterozygote has an intermediate level of enzyme between the homozygous dominant (full enzyme level) and homozygous recessive (no enzyme). Half the normal amount of enzyme is sufficient for health, which is why at the whole-person level, the wild type allele is completely dominant.

Familial hypercholesterolemia (FH; OMIM 143890) is an example of incomplete dominance that can be observed in carriers on both the molecular and whole-body levels. A person with two disease-causing alleles lacks receptors on liver cells that take up the low-density lipoprotein (LDL) form of cholesterol from the bloodstream, so it builds up. A person with one disease-causing allele has half the normal number of receptors. Someone with two wild type alleles has the normal number of receptors. **Figure 5.2** shows how measurement of plasma cholesterol reflects these three genotypes. The phenotypes parallel the number of receptors—individuals with two mutant alleles die in childhood of heart attacks, those with one mutant allele may suffer heart attacks in young adulthood, and those with two wild type alleles do not develop this inherited form of heart disease.

Different alleles that are both expressed in a heterozygote are **codominant**. The ABO blood group is based on the expression of codominant alleles.

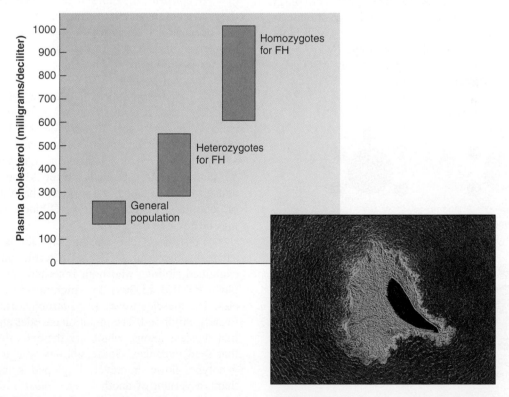

Figure 5.2 Incomplete dominance. A heterozygote for familial hypercholesterolemia (FH) has approximately half the normal number of cell surface receptors in the liver for LDL cholesterol. An individual with two mutant alleles has the severe form of FH, with liver cells that totally lack the receptors. As a result, serum cholesterol level is very high. The photograph shows an artery blocked with cholesterol-rich plaque. Cholesterol is also deposited in joints and many other body parts.

Blood types are determined by the patterns of molecules on the surfaces of red blood cells. Most of these molecules are proteins embedded in the plasma membrane with attached sugars that extend from the cell surface. The sugar is the antigen, which is the molecule that the immune system recognizes. People who belong to blood group A have an allele that encodes an enzyme that adds a final piece to a certain sugar to produce antigen A. In people with blood type B, the allele and its encoded enzyme are slightly different, which causes a different piece to attach to the sugar, producing antigen B. People in blood group AB have both antigen types. Blood group O reflects yet a third allele of this gene. It is missing just one DNA nucleotide, but this changes the encoded enzyme in a way that removes the sugar chain from its final piece (**figure 5.3**). Type O red blood cells lack both A and B antigens.

The *A* and *B* alleles are codominant, and both are completely dominant to O. Considering the genotypes reveals how these interactions occur. In the past, ABO blood types have been described as variants of a gene called "*I*," although OMIM now abbreviates the designations. The older *I* system is easier to understand. ("*I*" stands for isoagglutinin.) The three alleles are I^A, I^B, and *i*. People with blood type A have antigen A on the surfaces of their red blood cells, and may be of genotype $I^A I^A$ or $I^A i$. People with blood type B have antigen B on their red blood cell surfaces, and may be of genotype $I^B I^B$ or $I^B i$. People with the rare blood type AB have both antigens A and B on their cell surfaces, and are genotype $I^A I^B$. People with blood type O have neither antigen, and are genotype *ii*.

Fiction plots often misuse ABO blood type terminology, assuming that a child's ABO type must match that of one parent. This is not true, because a person with type A or B blood can be heterozygous. A person who is genotype $I^A i$ and a person who is $I^B i$ can jointly produce offspring of any ABO genotype or phenotype, as **figure 5.4** illustrates.

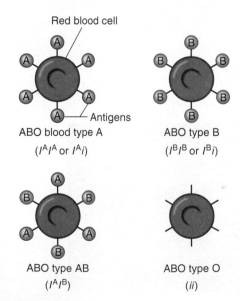

Figure 5.3 ABO blood types illustrate codominance. ABO blood types are based on antigens on red blood cell surfaces. This depiction greatly exaggerates the size of the A and B antigens. Genotypes are in parentheses.

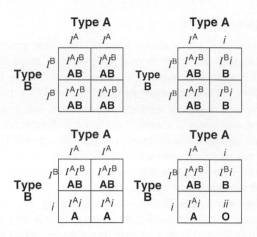

Figure 5.4 **Codominance.** The I^A and I^B alleles of the *I* gene are codominant, but they follow Mendel's law of segregation. These Punnett squares follow the genotypes that could result when a person with type A blood has children with a person with type B blood.

Epistasis

Mendel's laws can appear not to operate when one gene masks or otherwise affects the phenotype of another. This phenomenon is called **epistasis**. It refers to interaction between different genes, not between the alleles of the same gene. A gene that affects expression of another is called a modifier gene.

In epistasis, the blocked gene is expressed normally, but the product of the modifier gene inactivates it, removes a structure needed for it to contribute to the phenotype, or otherwise counteracts its effects. An epistatic interaction seen in many species is albinism, in which one gene blocks the action of genes that confer color.

A blood type called the Bombay phenotype also illustrates epistasis. It results from an interaction between a gene called *H* and the *I* gene that confers ABO blood type. The *H* gene controls the placement of a molecule to which antigens A and B attach on red blood cell surfaces. In a person of genotype *hh*, that molecule isn't made, so the A and B antigens cannot attach to the red blood cell surface. The A and B antigens fall off, and the person tests as type O blood. However, he or she may be any ABO genotype.

Epistasis is one reason why siblings who inherit the same disorder can suffer to differing degrees. One study examined siblings who both inherited spinal muscular atrophy 1 (OMIM 253300), in which nerves cannot signal muscles. The muscles weaken and atrophy, usually proving fatal in early childhood. The mutation encodes an abnormal protein that shortens axons, which are the extensions on nerve cells that send messages. Some siblings who inherited the SMA genotype, however, never developed symptoms. They can thank a variant of another gene, *plastin 3* (OMIM 300131), which increases production of the cytoskeletal protein actin that extends axons. Because the healthy siblings inherited the ability to make extra long axons, the axon-shortening effects of SMA were not harmful.

Penetrance and Expressivity

The same genotype can produce different degrees of a phenotype in different individuals because of influences of other genes, as well as environmental influences such as nutrition, exposure to toxins, and stress. This is why two individuals who have the same *CF* genotype may have different clinical experiences. One person may be much sicker because she also inherited gene variants predisposing her to develop asthma and respiratory allergies. Even identical twins with the same genetic disease may be affected to different degrees due to environmental influences.

The terms *penetrance* and *expressivity* describe degrees of expression of a single gene. **Penetrance** refers to the all-or-none expression of a genotype; **expressivity** refers to severity or extent.

An allele combination that produces a phenotype in everyone who inherits it is completely penetrant. Huntington disease (see the opener to chapter 4) is nearly completely penetrant. Almost all people who inherit the mutant allele will develop symptoms if they live long enough. Complete penetrance is very rare. Sometimes we do not know why a person with a particular genotype is ill or has a trait, yet another doesn't.

A genotype is incompletely penetrant if some individuals do not express the phenotype (have no symptoms). Polydactyly (see figure 1.5) is incompletely penetrant. Some people who inherit the dominant allele have more than five digits on a hand or foot. Yet others who we know have the allele because they have an affected parent and child have ten fingers and ten toes. Penetrance is described numerically. If 80 of 100 people who inherit the dominant polydactyly allele have extra digits, the genotype is 80 percent penetrant.

A phenotype is variably expressive if symptoms vary in intensity among different people. One person with polydactyly might have an extra digit on both hands and a foot, but another might have just one extra fingertip. Polydactyly is both incompletely penetrant and variably expressive.

Pleiotropy

A single-gene disorder with many symptoms, or a gene that controls several functions or has more than one effect, is termed **pleiotropic**. Such conditions can be difficult to trace through families because people with different subsets of symptoms may appear to have different disorders. A classic example of pleiotropy is porphyria variegata, an autosomal dominant disease that affected several members of the royal families of Europe (**figure 5.5**).

King George III ruled England during the American Revolution. At age 50, he first experienced abdominal pain and constipation, followed by weak limbs, fever, a fast pulse, hoarseness, and dark red urine. Next, nervous system signs and symptoms began, including insomnia, headaches, visual problems, restlessness, delirium, convulsions, and stupor. His confused and racing thoughts, combined with actions such as ripping off his wig and running about naked while at the peak of a fever, convinced court observers that the king was mad. Just as Parliament was debating his ability to rule, he recovered.

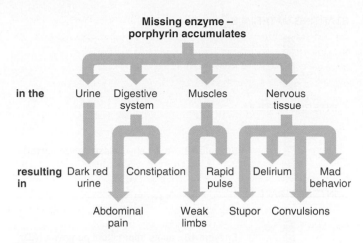

Figure 5.5 **Pleiotropy.** King George III suffered from the autosomal dominant disorder porphyria variegata, and so did several other family members. Because of pleiotropy, the family's varied illnesses and quirks appeared to be different, unrelated disorders. Symptoms appear every few years in a particular order.

But the king's ordeal was far from over. He relapsed 13 years later, then again 3 years after that. Always the symptoms appeared in the same order, beginning with abdominal pain, fever, and weakness, and progressing to nervous system symptoms. Finally, an attack in 1811 placed George in a prolonged stupor, and the Prince of Wales dethroned him. George III lived for several more years, experiencing further episodes.

In George III's time, doctors were permitted to do very little to the royal body, and their diagnoses were based on what the king told them. Twentieth-century researchers found that porphyria variegata caused George's red urine. It is one of several types of porphyrias, which result from deficiency of any of several enzymes required to manufacture heme. The king's disorder arises from lack of enzyme #7 in the heme pathway shown in **figure 5.6**. Heme is part of hemoglobin, the molecule that carries oxygen in the blood and imparts the red color. In the disease, a part of heme called a porphyrin ring is routed into the urine instead of being broken down and metabolized in cells. Porphyrin also builds up and attacks the nervous system.

Physicians' records of George's royal relatives reported the disorder as several different illnesses. Today, porphyria variegata remains rare, and is often misdiagnosed as a seizure disorder. Unfortunately, some seizure medications and anesthetics worsen symptoms. Ironically, treatment may have worsened King George III's disease. Medical records and hair analysis indicate that a medicine based on the element antimony was forced upon the king in the madhouse. Antimony was often contaminated with arsenic, and arsenic inactivates several of the enzymes in the heme biosynthetic pathway.

On a molecular level, pleiotropy occurs when a single protein affects different body parts, participates in more than one biochemical reaction, or has different effects in different amounts. Consider Marfan syndrome. The most common form of this autosomal dominant condition is a defect in an elastic connective tissue protein called fibrillin (OMIM 134797).

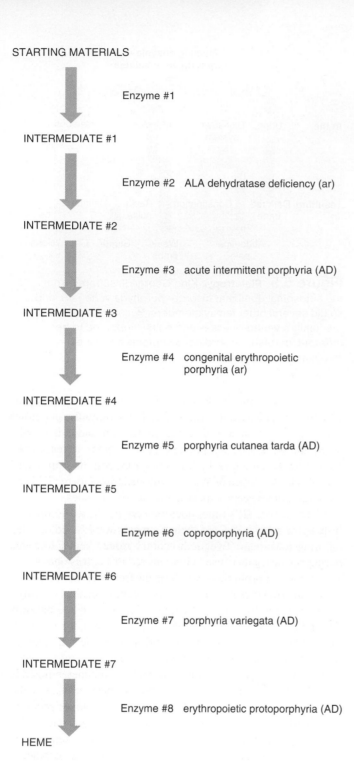

STARTING MATERIALS

Enzyme #1

INTERMEDIATE #1

Enzyme #2   ALA dehydratase deficiency (ar)

INTERMEDIATE #2

Enzyme #3   acute intermittent porphyria (AD)

INTERMEDIATE #3

Enzyme #4   congenital erythropoietic
porphyria (ar)

INTERMEDIATE #4

Enzyme #5   porphyria cutanea tarda (AD)

INTERMEDIATE #5

Enzyme #6   coproporphyria (AD)

INTERMEDIATE #6

Enzyme #7   porphyria variegata (AD)

INTERMEDIATE #7

Enzyme #8   erythropoietic protoporphyria (AD)

HEME

Figure 5.6 **The porphyrias.** Errors in the heme biosynthetic pathway cause seven related, yet distinct, diseases. In each disorder, the intermediate biochemical that a deficient enzyme would normally affect builds up. The excess is excreted in the urine or accumulates in blood, feces, or inside red blood cells, causing symptoms. People with various porphyria-related symptoms may have inspired the vampire and werewolf legends, including reddish teeth, pink urine, excess hair, and photosensitivity (avoidance of daylight).

The protein is abundant in the lens of the eye, in the aorta (the largest artery in the body, leading from the heart), and in the bones of the limbs, fingers, and ribs. The symptoms are lens dislocation, long limbs, spindly fingers, and a caved-in chest (**figure 5.7**). The most serious symptom is a life-threatening weakening in the aorta, which can suddenly burst. If the weakening is detected early, a synthetic graft can replace the section of artery wall.

Certain proteins that form a structure in the eye called lens crystallins beautifully illustrate pleiotropy. If in low abundance as single molecules, these proteins are metabolic enzymes, functioning in many cell types. At higher abundance, however, they join and form crystallins, which aggregate to create a transparent lens whose physical properties enable it to focus incoming light on the retina.

Genetic Heterogeneity

Mutations in different genes that produce the same phenotype is called **genetic heterogeneity**. The phenomenon can occur when genes encode enzymes or other proteins that are part

Figure 5.7 **Marfan syndrome is pleiotropic.** Jonathan Larson died just before the opening of the Broadway play that he wrote, *Rent,* when his aorta tore apart. He had gone to two New York City hospitals complaining of chest pain, but no one recognized the symptoms of Marfan syndrome.

of the same biochemical pathway, or when gene products affect the same specific body part, such as the vision disorders described in the chapter opener. Genetic heterogeneity may make it appear that Mendel's laws are not operating, even though they are. The different forms of Leber congenital amaurosis arise because there are many ways that a mutation can disrupt the functioning of the rods and cones, the cells that provide vision (**figure 5.8**). If a man who is homozygous recessive for a mutation in one of the genes that causes the condition has a child with a woman who is homozygous recessive for a different gene, then the child would not inherit Leber congenital amaurosis, because he or she would be heterozygous for both genes.

Discovering additional genes that can cause a known disorder is happening more as the human genome is analyzed, and can have practical repercussions. Consider osteogenesis imperfecta, in which abnormal collagen causes very fragile bones (see the Opener to chapter 12). Before a second causative gene was discovered, some parents of children brought to the hospital with frequent fractures, and who did not have a mutation in that gene, were accused of child abuse. Today eight genetically distinct forms of the disease are recognized. Misunderstandings based on genetic heterogeneity, such as accusations of child abuse, will become less common as exome sequencing reveals the functions of more genes.

Clinical Connection 5.1 describes a common genetically heterogeneic condition—Alzheimer disease.

Phenocopies

An environmentally caused trait that appears to be inherited is a **phenocopy**. Such a trait can either produce symptoms that resemble those of a known single-gene disorder or mimic inheritance patterns by affecting certain relatives. For example, the limb birth defect caused by the drug thalidomide, discussed

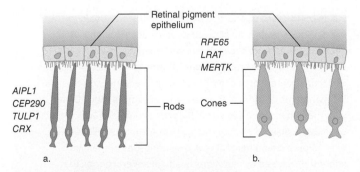

a.   b.

Figure 5.8 Many routes to blindness. Mutations in more than 100 genes cause degeneration of the retina, the multilayered structure at the back of the eye that includes the rods and cones, which are the photoreceptor cells that signal incoming light to the brain. This illustration indicates seven of these genes. Their encoded proteins have various functions: activating vitamin A, cleaning up debris, providing energy, and maintaining the functioning of the rods and cones.

in chapter 3, is a phenocopy of the inherited illness phocomelia. Physicians recognized the environmental disaster when they began seeing many children born with what looked like the very rare phocomelia. A birth defect caused by exposure to a teratogen was more likely than a sudden increase in incidence of a rare inherited disease.

An infection can be a phenocopy if it affects more than one family member. Children who have AIDS may have parents who also have the disease, but these children acquired AIDS by viral infection, not by inheriting a gene.

Sometimes, common symptoms may resemble those of an inherited condition until medical tests rule heredity out. For example, an underweight child who has frequent colds may show some signs of cystic fibrosis, but may instead suffer from malnutrition. Negative test results for several common CF alleles would alert a physician to look for another cause.

The Human Genome Sequence Adds Perspective

As researchers continue to describe the genes sequenced in the human genome project, it is becoming clear that phenomena once considered to only rarely complicate single-gene inheritance may be common. As a result, terms such as *epistasis* and *genetic heterogeneity* are beginning to overlap and blur. Consider Marfan syndrome. Most affected individuals have a mutation in the fibrillin gene. However, some people with the syndrome instead have a mutation in the gene that encodes the transforming growth factor beta receptor (TGFβR) (OMIM 190181). Fibrillin and TGFβR are part of the same biochemical pathway. The conditions fit the definition of genetic heterogeneity because mutations in different genes cause identical symptoms. Yet they are also epistatic because a mutation in TGFβR blocks the activity of fibrillin.

Gene interactions also underlie penetrance and expressivity, because even genes that do not directly interact, in space or time, can affect each other's expression. This is the case for Huntington disease, in which cells in a certain part of the brain die, usually beginning in young adulthood. Siblings who inherit the exact same HD mutation may differ in the number of cells that they have in the affected brain area, because of variants of other genes that affected the division rate of stem cells in the brain during embryonic development. As a result, an individual who inherits HD, but also extra brain cells, might develop symptoms much later in life than a brother or sister who does not have such a built-in reserve supply. If the delay is long enough that death comes from another cause, HD would then be nonpenetrant.

Technologies that reveal gene expression patterns in different tissues are painting detailed portraits of pleiotropy, showing that inherited disorders may affect more tissues or organs than are obvious as symptoms. Finally, more cases of genetic heterogeneity are being discovered as researchers identify genes with redundant or overlapping functions.

Table 5.1 summarizes phenomena that appear to alter single-gene inheritance. Our definitions and designations are changing as improving technology enables us to describe and differentiate disorders in greater detail. Phenomena such as variable expressivity, incomplete penetrance, epistasis, pleiotropy, and genetic heterogeneity, once considered unusual characteristics of single genes, may turn out to be the norm.

The Genetic Roots of Alzheimer Disease

"What is that thing for, that you put in your ear?" asked 72-year-old Ginny for the fifth time in half an hour.

"Mom, don't you know you've asked me that several times? It gets cell phone calls," answered her son, trying not to become annoyed.

"No, I've never asked you that before." She paused, looking puzzled. *"What did you say it is?"*

In the months following that conversation, Ginny's short-term memory declined further. She could rarely concentrate long enough to finish reading an article, or follow a conversation. In the grocery store, she couldn't find items she'd been buying for decades. Aware of her growing deficits, she became depressed. Finally, her son suggested she have a complete neurological exam. By the time she could see a physician, other signs had emerged. Ginny couldn't recall her zip code or the name of the small town where she grew up. Sometimes she couldn't remember where things belonged—she put a cantaloupe in the bathtub, and still had trouble with that contraption her son put in his ear to receive phone calls.

Ginny was showing signs of "mild cognitive impairment" (MCI). It could be the start of Alzheimer disease, which affects 26 million people worldwide. After ruling out a vitamin deficiency and lack of oxygen to the brain as causes of Ginny's forgetfulness, a neurologist started her on a drug to slow breakdown of the neurotransmitter acetylcholine in the brain. The doctor also prescribed an antidepressant, which revived Ginny enough so that she was more willing to leave her apartment. If the MCI progressed to Alzheimer disease, Ginny would continue to lose thinking, reasoning, learning, and communicating abilities, and one day would not recognize her loved ones. Eventually, she would no longer speak and smile, and would cease walking, not because her legs wouldn't function, but because she would forget how to walk. Yet the haze would sometimes seem to lift from her eyes and the old Ginny would return.

In Alzheimer disease, certain brain parts—the amygdala (seat of emotion) and hippocampus (the memory center)—become buried in two types of protein. Amyloid precursor protein normally converts iron into a safe form. In Alzheimer disease, the protein is cut into unusually sized amyloid beta peptides, which aggregate to form "plaques" outside brain cells because cells cannot remove the material fast enough. At the same time, unsafe iron accumulates

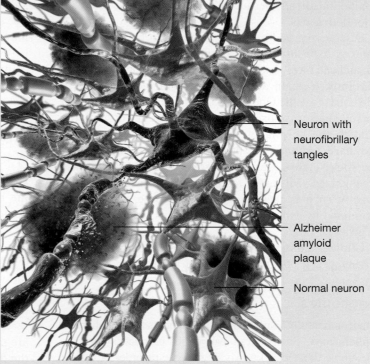

Neuron with neurofibrillary tangles

Alzheimer amyloid plaque

Normal neuron

Figure 1 In Alzheimer disease, amyloid plaques form outside neurons and neurofibrillary tangles form inside the cells.

inside brain neurons, and zinc builds up in the plaques outside them. Ginny's brain scan showed plaques accumulating. The second type of protein, called tau, was building up too, clumping into "tangles" inside brain neurons (**figure 1**). The telltale plaques and tangles, present to a lesser extent in everyone, could cause the symptoms of Alzheimer disease or result from them. Either way, their abundance in a spinal fluid test, combined with Ginny's cognitive problems, strongly suggested Alzheimer disease.

Fewer than 1 percent of Alzheimer cases are familial (inherited), caused by mutations in any of several genes that affect amyloid accumulation or clearance (**table 1**). However, some variants of genes associated with Alzheimer disease are actually protective. A variant of a gene called *APOE4* increases the risk of developing a late-onset form three-fold in a heterozygote and 15-fold in a homozygote. Other genes increase susceptibility but raise risk to such a small degree that they have little predictive value.

(Continued)

Table 1	Genes Associated with Alzheimer Disease			
Causative Gene	**OMIM**	**Chromosome**	**Mechanism**	
Amyloid precursor protein (*APP*)	104760	21	Unusually sized pieces aggregate outside brain cells.	
Calcium homeostasis modulator 1 (*CALHM1*)	612234	10	Controls amyloid cutting.	
Presenilin 1 (*PSEN1*)	607822 104311	14	Forms part of secretase (enzyme) that cuts APP.	
Presenilin 2 (*PSEN2*)	606889 600759	1	Forms part of secretase (enzyme) that cuts APP.	
Risk Gene				
Apolipoprotein E4 (*APOE4*)	107741	19	Unusually sized pieces add phosphates to tau protein, making it accumulate and impairing microtubule binding.	
Triggering receptor expressed on myeloid cells (*TREM*)	605086	6	Promotes brain inflammation.	

Questions for Discussion

1. Why is Alzheimer disease difficult to diagnose, especially in the early stage?

2. How is the genetics of Alzheimer disease complex?

3. Describe the protein abnormalities that occur in Alzheimer disease.

4. How common are inherited forms of Alzheimer disease?

Table 5.1	Factors That Alter Single-Gene Phenotypic Ratios	
Phenomenon	**Effect on Phenotype**	**Example**
Lethal alleles	A phenotypic class does not survive to reproduce.	Achondroplasia
Multiple alleles	Many variants or degrees of a phenotype are possible.	Cystic fibrosis
Incomplete dominance	A heterozygote's phenotype is intermediate between those of two homozygotes.	Familial hypercholesterolemia
Codominance	A heterozygote's phenotype is distinct from and not intermediate between those of the two homozygotes.	ABO blood types
Epistasis	One gene masks or otherwise affects another's phenotype.	Bombay phenotype
Penetrance	Some individuals with a particular genotype do not have the associated phenotype.	Polydactyly
Expressivity	A genotype is associated with a phenotype of varying intensity.	Polydactyly
Pleiotropy	The phenotype includes many symptoms, with different subsets in different individuals.	Porphyria variegata
Phenocopy	An environmentally caused condition has symptoms and a recurrence pattern similar to those of a known inherited trait.	Infection
Genetic heterogeneity	Different genotypes are associated with the same phenotype.	Leber congenital amaurosis

Gregor Mendel derived the two laws of inheritance working with traits conferred by genes located on different chromosomes in the nucleus. When genes are not on different chromosomes, however, the associated traits may not appear in Mendelian ratios. The rest of this chapter considers two types of gene transmission that are not Mendelian—mitochondrial inheritance and linkage.

5.2 Mitochondrial Genes

Mitochondria are the cellular organelles that house the reactions that derive energy from nutrients (see figure 2.8). Each of the hundreds to thousands of mitochondria in each human cell contains several copies of a "mini-chromosome" that carries just 37 genes. Mitochondrial DNA (mtDNA)-encoded genes act in the mitochondrion, but the organelle also requires the activities of certain genes from the nucleus.

The inheritance patterns and mutation rates for mitochondrial genes differ from those for genes in the nucleus. Rather than being transmitted equally from both parents, mitochondrial genes are maternally inherited. They are passed only from an individual's mother because the sperm head, which enters an oocyte at fertilization, does not include mitochondria, which instead are found in the sperm midsection, where they provide energy for movement of the tail. In the rare instances when mitochondria from sperm enter an oocyte, they are usually selectively destroyed early in development. Pedigrees that follow mitochondrial genes therefore show a woman passing the trait to all her children, while a male cannot pass the trait to any of his (**figure 5.9**).

DNA in the mitochondria differs functionally from DNA in the nucleus in several ways (**table 5.2** and **figure 5.10**). Mitochondrial DNA does not cross over. It mutates faster than DNA in the nucleus for two reasons: It has fewer ways to repair DNA (discussed in chapter 12), and the

mitochondrion is the site of energy reactions that produce oxygen free radicals that damage DNA. Also unlike nuclear DNA, mtDNA is not wrapped in proteins, nor are genes "interrupted" by DNA sequences that do not encode protein. Finally, a cell has one nucleus but many mitochondria, and each mitochondrion harbors several copies of its chromosome. Therefore, a mitochondrial gene may be present in many copies in a cell. Mitochondria with different alleles for the same gene can reside in the same cell.

Mitochondrial Disorders

Mitochondrial genes encode proteins that participate in protein synthesis and energy production. Twenty-four of the 37 genes encode RNA molecules (22 transfer RNAs and 2 ribosomal RNAs) that help assemble proteins. The other 13 mitochondrial genes encode proteins that function in cellular respiration, which is the process that uses energy from digested nutrients to synthesize ATP, the biological energy molecule.

A class of diseases results from mutations in mitochondrial genes. They are called mitochondrial myopathies and have specific names, but news reports often lump them together as "mitochondrial disease." Symptoms arise from tissues whose cells have many mitochondria, such as skeletal muscle. Symptoms are great fatigue, weak and flaccid muscles, and intolerance to exercise. Skeletal muscle fibers appear "red and ragged" when stained and viewed under a light microscope, their abundant abnormal mitochondria visible beneath the plasma membrane.

A mutation in a mitochondrial gene that encodes a tRNA or rRNA can be devastating because it impairs the organelle's general ability to manufacture proteins. Consider what happened to Lindzy S., a once active and articulate dental hygienist. In her forties, Lindzy gradually began to slow down at work. She heard a buzzing in her ears and developed difficulty talking and walking. Then her memory began to fade in and out, she became lost easily in familiar places, and her conversation made no sense. Her condition worsened, and she developed diabetes, seizures, and pneumonia and became deaf and demented. She was finally diagnosed with MELAS, which stands for "mitochondrial myopathy encephalopathy lactic acidosis syndrome" (OMIM 540000). Lindzy died. Her son and

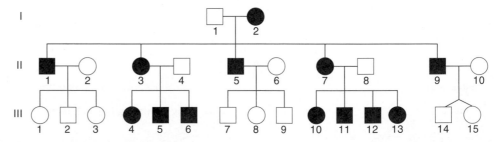

Figure 5.9 Inheritance of mitochondrial genes. Mothers pass mitochondrial genes to all offspring. Fathers do not transmit mitochondrial genes because sperm only very rarely contribute mitochondria to fertilized ova. If mitochondria from a male do enter, they are destroyed.

Table 5.2	Features of Mitochondrial DNA
No crossing over	
Fewer types of DNA repair	
Inherited from the mother only	
Many copies per mitochondrion and per cell	
High exposure to oxygen free radicals	
No histones (DNA-associated proteins)	
Genes not interrupted	

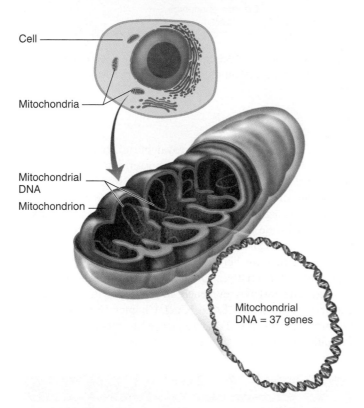

Cell

Mitochondria

Mitochondrial DNA

Mitochondrion

Mitochondrial DNA = 37 genes

Figure 5.10 **Mitochondrial DNA.** A mitochondrion contains several rings of DNA. Different alleles can reside on different copies of the mitochondrial chromosome. A typical cell has hundreds to thousands of mitochondria, each of which has many copies of its "mini-chromosome."

daughter will likely develop the condition because they inherited her mitochondria.

Theoretically, a woman with a mitochondrial disorder can avoid transmitting it to her children if her mitochondria can be replaced with healthy mitochondria from a donor. The opener to Chapter 21 describes two ways to do this. Such mitochondrial replacement is controversial for several reasons, including the fact that it creates a "three-parent" child because DNA is present from the mother, the father, and the donor of the mitochondria. However, very little DNA comes from the donor mitochondria.

About 1 in 200 people has a mutation in a mitochondrial gene that could cause disease. However, mitochondrial diseases are rare, affecting about 1 in 6,500 people, apparently because of a weeding-out process during egg formation. Such a mutation may disrupt energy acquisition so greatly in an oocyte that it cannot survive.

Heteroplasmy

The fact that a cell contains many mitochondria makes possible a condition called **heteroplasmy** (see *A Glimpse of History*). In this state, a mutation is in some mitochondrial chromosomes, but not others. At each cell division, the mitochondria are distributed at random into daughter cells. Over time, the chromosomes within a mitochondrion tend to be all wild type or all mutant for any particular gene, but different mitochondria can have different alleles predominating. As an oocyte matures, the number of mitochondria drops from about 100,000 to 100 or fewer. If the woman is heteroplasmic for a mutation, by chance, she can produce an oocyte that has mostly mitochondria that are wild type mostly mitochondria that have the mutation, or anything in between (**figure 5.11**). In this way, a woman who does not have a mitochondrial disorder, because the mitochondria bearing the mutation are either rare or not abundant in affected cell types, can nevertheless pass the associated condition to a child.

Heteroplasmy has several consequences for the inheritance of mitochondrial phenotypes. Expressivity may vary widely among siblings, depending upon how many mutation-bearing mitochondria were in the oocyte that became each individual. Severity of symptoms reflects which tissues have cells whose mitochondria bear the mutation. This is the case for a family with Leigh syndrome (OMIM 256000), which affects the

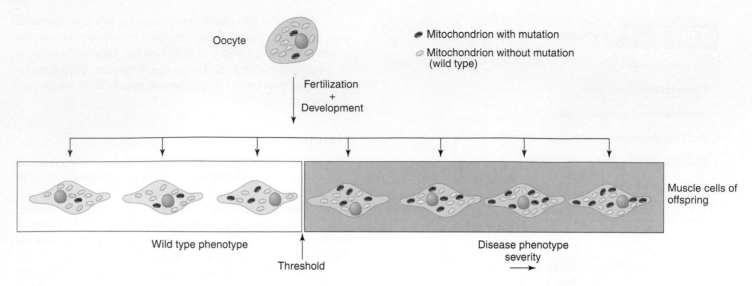

Oocyte

Fertilization
+
Development

● Mitochondrion with mutation
◐ Mitochondrion without mutation
(wild type)

Muscle cells of offspring

Wild type phenotype

Threshold

Disease phenotype severity →

Figure 5.11 **Mitochondrial inheritance.** Mitochondria and their genes are passed only from the mother. Cells have many mitochondria. If an oocyte is heteroplasmic, differing numbers of copies of a mitochondrial mutation may be transmitted. The phenotype reflects the proportion of mitochondria bearing the mutation.

enzyme that directly produces ATP. Two boys died of the severe form of the disorder because the brain regions that control movement rapidly degenerated. Another sibling was blind and had central nervous system degeneration. Several relatives, however, suffered only mild impairment of their peripheral vision. The more severely affected family members had more brain cells that received the mutation-bearing mitochondria.

The most severe mitochondrial illnesses are heteroplasmic. This is presumably because *homo*plasmy—when all mitochondria bear the mutant allele—too severely impairs protein synthesis or energy production for embryonic development to complete. Often, severe heteroplasmic mitochondrial disorders do not produce symptoms until adulthood because it takes many cell divisions, and therefore years, for a cell to receive enough mitochondria bearing mutant alleles to cause symptoms. In the case of the Russian royals, heteroplasmy did not cause illness, but provided a forensic clue.

About one in ten of the DNA bases in the mitochondrial genome are heteroplasmic—that is, they can differ within an individual. The mutations that generate these single-base variations in mtDNA probably occur all the time, but only those that originated early in development have a chance to accumulate enough to be detectable.

Mitochondrial DNA Reveals the Past

Interest in mtDNA extends beyond the medical. mtDNA provides a powerful forensic tool used to link suspects to crimes, identify war dead, and support or challenge historical records. mtDNA is used in forensics because it is more likely to remain after extensive damage, because cells have many copies of it.

Sequencing mtDNA identified the son of Marie Antoinette and Louis XVI, who supposedly died in prison at age 10.

In 1845, the boy was given a royal burial, but some people thought the buried child was an imposter. His heart had been stolen at the autopsy, and through a series of bizarre events, wound up, dried out, in the possession of the royal family. A few years ago, researchers compared mtDNA sequences from cells in the boy's heart to corresponding sequences in heart and hair cells from Marie Antoinette (her decapitated body identified by her fancy underwear), two of her sisters, and living relatives Queen Anne of Romania and her brother. The mtDNA evidence showed that the buried boy was indeed the prince, Louis XVII. Chapter 16 discusses how researchers consult mtDNA sequences to reconstruct ancient migration patterns.

Key Concepts Questions 5.2

1. How does the inheritance pattern of mitochondrial DNA differ from inheritance of single-gene traits?

2. Explain why mitochondrial DNA is more prone to mutation than DNA in the nucleus.

3. What is heteroplasmy?

5.3 Linkage

The genes responsible for the traits that Mendel studied in pea plants were on different chromosomes. When genes are close to each other on the same chromosome, they usually do not segregate at random during meiosis and therefore do not support Mendel's predictions. Instead, genes close on a chromosome are packaged into the same gametes and are said to be "linked" (**figure 5.12**). Linkage has this very precise meaning

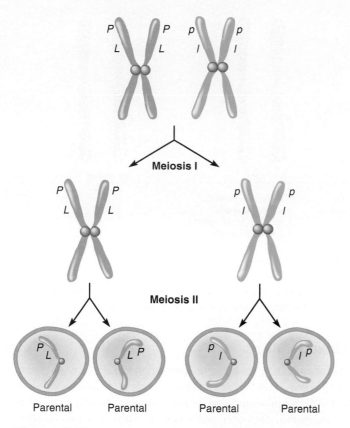

Figure 5.12 Inheritance of linked genes. Genes linked closely to one another are usually inherited together when the chromosome is packaged into a gamete.

in genetics. The term is popularly used to mean any association between two events or observations.

Linkage refers to the transmission of genes on the same chromosome. Linked genes do *not* assort independently and do *not* produce Mendelian ratios for crosses tracking two or more genes. Understanding and using linkage as a mapping tool helped to identify many disease-causing genes before genome sequencing became possible.

Discovery in Pea Plants

William Bateson and R. C. Punnett first observed the unexpected ratios indicating linkage in the early 1900s, again in pea plants. They crossed true-breeding plants with purple flowers and long pollen grains (genotype *PPLL*) to true-breeding plants with red flowers and round pollen grains (genotype *ppll*). The plants in the next generation, of genotype *PpLl*, were then self-crossed. But this dihybrid cross did not yield the expected 9:3:3:1 phenotypic ratio that Mendel's second law predicts (**figure 5.13**).

Bateson and Punnett noticed that two types of third-generation peas, those with the parental phenotypes *P_L_* and *ppll*, were more abundant than predicted, while the other two progeny classes, *ppL_* and *P_ll*, were less common (the blank indicates that the allele can be dominant or recessive). The more prevalent parental allele combinations, Bateson and

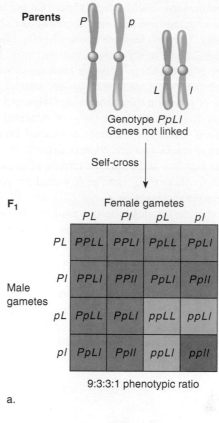

a.

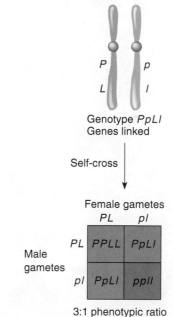

b.

Figure 5.13 Expected results of a dihybrid cross.
(a) Unlinked genes assort independently. The gametes represent all possible allele combinations. The expected phenotypic ratio of a dihybrid cross is 9:3:3:1. **(b)** If genes are linked, only two allele combinations are expected in the gametes. The phenotypic ratio is 3:1, the same as for a monohybrid cross.

Punnett hypothesized, could reflect genes that are transmitted on the same chromosome and that therefore do not separate during meiosis. The two less common offspring classes could also be explained by a meiotic event, crossing over. Recall that crossing over is an exchange between homologs that mixes up maternal and paternal gene combinations without disturbing the sequence of genes on the chromosome (**figure 5.14**).

Progeny that exhibit this mixing of maternal and paternal alleles on a single chromosome are called **recombinant**. *Parental* and *recombinant* are context dependent terms. Had the parents in Bateson and Punnett's crosses been of genotypes *ppL_* and *P_ll*, then *P_L_* and *ppll* would be recombinant rather than parental classes.

Two other terms describe the configurations of linked genes in dihybrids. Consider a pea plant with genotype *PpLl*. These alleles can be part of the chromosomes in either of two ways. If the two dominant alleles are on one chromosome and the two recessive alleles on the other, the genes are in *"cis."* In the opposite configuration, with one dominant and one recessive allele on each chromosome, the genes are in *"trans"* (**figure 5.15**). Whether alleles in a dihybrid are in *cis* or *trans* is important in distinguishing recombinant from parental progeny classes in specific crosses.

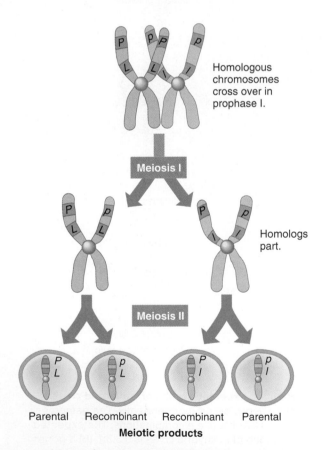

Figure 5.14 **Crossing over disrupts linkage.** The linkage between two genes may be interrupted if the chromosome they are on crosses over with its homolog between the two genes. Crossing over packages recombinant groupings of the genes into gametes.

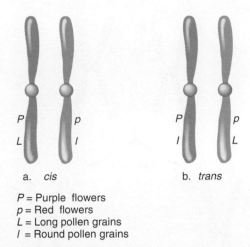

P = Purple flowers
p = Red flowers
L = Long pollen grains
l = Round pollen grains

Figure 5.15 **Allele configuration is important.** Parental chromosomes can be distinguished from recombinant chromosomes only if the allele configuration of the two genes is known—they are either in *cis* **(a)** or in *trans* **(b)**.

Linkage Maps

As Bateson and Punnett were discovering linkage in peas, geneticist Thomas Hunt Morgan and his coworkers in Columbia University's "fly room" were doing the same using the fruit fly *Drosophila melanogaster*. They assigned genes to relative positions on chromosomes and compared progeny class sizes to assess whether traits were linked. The pairs of traits fell into four groups. Within each group, crossed dihybrids did not produce offspring classes according to Mendel's second law. Also, the number of linkage groups—four—matched the number of chromosome pairs in the fly. Coincidence? No. The traits fell into four groups because their genes are inherited together on the same chromosome.

The genius of the work on linkage in fruit flies was twofold. First, the researchers used test crosses (see figure 4.5) to follow parental versus recombinant progeny. Second, the fly room investigators translated their data into actual maps depicting positions of genes on chromosomes. Morgan wondered why the sizes of the recombinant classes varied for different genes. In 1911 Morgan proposed that the farther apart two genes are on a chromosome, the more likely they are to cross over simply because more physical distance separates them (**figure 5.16**). He proposed that the size of a recombinant class is directly proportional to the distance between the genes on the chromosome they share. Then undergraduate student Alfred Sturtevant devised a way to represent the correlation between crossover frequency and the distance between genes as a **linkage map**. These diagrams showed the order of genes on chromosomes and the relative distances between them. The distance was represented using "map units" called centimorgans, where 1 centimorgan equals 1 percent recombination. These units are still used today to estimate genetic distance along a chromosome. Linkage maps of human genes were important in the initial sequencing of the human genome.

The frequency of a crossover between any two linked genes is inferred from the proportion of offspring from a cross that are recombinant. Frequency of recombination is based

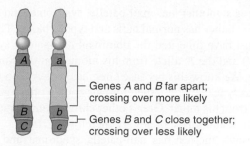

Figure 5.16 Breaking linkage. Crossing over is more likely between widely spaced linked genes *A* and *B*, or between *A* and *C*, than between more closely spaced linked genes *B* and *C*, because there is more room for an exchange to occur.

Genes *A* and *B* far apart; crossing over more likely

Genes *B* and *C* close together; crossing over less likely

on the percentage of meiotic divisions that break the linkage between—that is, separate—two parental alleles. Genes at opposite ends of the same chromosome cross over often, generating a large recombinant class. Genes lying very close on the chromosome would only rarely be separated by a crossover. The probability that genes on opposite ends of a chromosome cross over approaches the probability that, if on different chromosomes, they would independently assort—about 50 percent. **Figure 5.17** illustrates this distinction.

The situation with linked genes is like a street lined with stores on both sides. There are more places to cross the street between stores at opposite ends on opposite sides than between two stores in the middle of the block on opposite sides of the street. Similarly, more crossovers, or progeny with recombinant genotypes, are seen when two genes are farther apart on the same chromosome.

As the twentieth century progressed, geneticists in Columbia University's fly room mapped several genes on all four chromosomes, as researchers in other labs assigned many genes to the human X chromosome. Localizing genes on the X chromosome was easier than doing so on the autosomes, because X-linked traits follow an inheritance pattern that is distinct from the one all autosomal genes follow. In human males, with their single X chromosome, recessive alleles on the X are expressed and observable. Chapter 6 returns to this point.

By 1950, geneticists began to think about mapping genes on the 22 human autosomes. To start, a gene must be matched to its chromosome. This became possible when geneticists identified people with a particular inherited condition or trait and an unusual chromosome.

In 1968, researcher R. P. Donohue was looking at chromosomes in his own white blood cells when he noticed a dark area consistently located near the centromere of one member of his largest chromosome pair (chromosome 1). He examined chromosomes from several family members for the dark area, noting also whether each family member had a blood type called Duffy. (Recall that blood types refer to the patterns of sugars on red blood cell surfaces.) Donohue found that the Duffy blood type was linked to the chromosome variant. He could predict a relative's Duffy blood type by whether or not the chromosome had the dark area. This was the first assignment of a trait in humans to an autosome.

Finding a chromosomal variation linked to a family trait like Donohue did was unusual. More often, researchers mapped genes in experimental organisms, such as fruit flies, by calculating percent recombination (crossovers) between two genes with known locations on a chromosome. However, because human parents do not have hundreds of offspring, nor do they produce a new generation every 10 days, getting enough information to establish linkage relationships for us requires observing the same traits in many families and pooling the results. Today, even though we can sequence human genomes, linkage remains a powerful tool to track disease-associated genes.

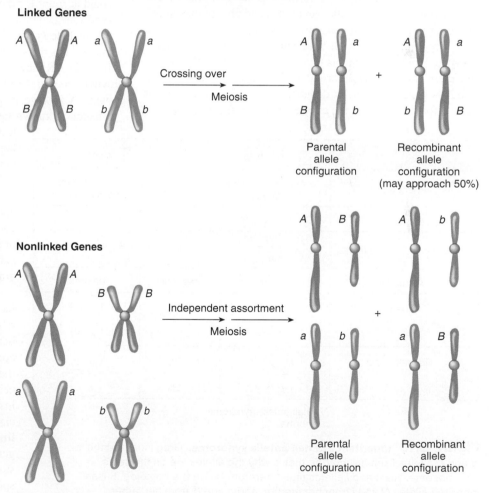

Linked Genes

Crossing over

Meiosis

Parental allele configuration

Recombinant allele configuration (may approach 50%)

Nonlinked Genes

Independent assortment

Meiosis

Parental allele configuration

Recombinant allele configuration

Figure 5.17 Linkage versus nonlinkage (independent assortment). When two genes are widely separated on a chromosome, the likelihood of a crossover is so great that the recombinant class may approach 50 percent—which may appear to be the result of independent assortment.

Solving Linkage Problems Uses Logic

To illustrate determining the degree of linkage by percent recombination using several families, consider the traits of Rh blood type (OMIM 111700) and a form of anemia called elliptocytosis (OMIM 130500). An Rh positive phenotype corresponds to genotypes *RR* or *Rr*. (This is simplified.) The anemia corresponds to genotypes *EE* or *Ee*.

In 100 one-child families, one parent is Rh negative with no anemia (*rree*), and the other parent is Rh positive with anemia (*RrEe*), and the *R* and *E* (or *r* and *e*) alleles are in *cis*. Of the 100 offspring, 96 have parental genotypes (*re/re* or *RE/re*) and four are recombinants for these two genes (*Re/re* or *rE/re*). Percent recombination is therefore 4 percent, and the two linked genes are 4 centimorgans apart.

Another pair of linked genes in humans is nail-patella syndrome (OMIM 161200), a rare autosomal dominant trait that causes absent or underdeveloped fingernails and toenails and painful arthritis in the knee and elbow joints, and the *I* gene that determines the ABO blood type, on chromosome 9. These two genes are 10 map units apart, which geneticists determined by pooling information from many families. It is the first example of linked autosomal genes in humans. The information is used to predict genotypes and phenotypes in offspring, as in the following example.

Greg and Susan each have nail-patella syndrome. Greg has type A blood. Susan has type B blood. What is the chance that their child inherits normal nails and knees and type O blood? A genetic counselor deduces their allele configurations using information about the couple's parents (**figure 5.18**).

Greg's mother has nail-patella syndrome and type A blood. His father has normal nails and type O blood. Therefore, Greg must have inherited the dominant nail-patella syndrome allele (*N*) and the I^A allele from his mother, on the same chromosome. We know this because Greg has type A blood and his father has type O blood—therefore, he couldn't have gotten the I^A allele from his father. Greg's other chromosome 9 must carry the alleles *n* and *i*. His alleles are therefore in *cis*.

Susan's mother has nail-patella syndrome and type O blood, and so Susan inherited *N* and *i* on the same chromosome. Because her father has normal nails and type B blood, her homolog from him bears alleles *n* and I^B. Her alleles are in *trans*.

Determining the probability that Susan and Greg's child could have normal nails and knees and type O blood is the easiest question the couple could ask. The only way this genotype can arise from theirs is if an *ni* sperm (which occurs with a frequency of 45 percent, based on pooled data) fertilizes an *ni* oocyte (which occurs 5 percent of the time). The result—according to the product rule—is a 2.25 percent chance of producing a child with the *nnii* genotype.

Calculating other genotypes for their offspring is more complicated, because more combinations of sperm and oocytes could account for them. For example, a child with nail-patella syndrome and type AB blood could arise from all combinations that include I^A and I^B as well as at least one *N* allele (assuming that *NN* has the same phenotype as *Nn*).

The Rh blood type and elliptocytosis, and nail-patella syndrome and ABO blood type, are examples of linked gene pairs. A linkage map begins to emerge when percent recombination is known between all possible pairs of three or more linked genes, just as a road map with more landmarks provides more information on distance and direction. Consider genes *x*, *y*, and *z* (**figure 5.19**). If the percent recombination between *x* and *y* is 10, between *x* and *z* is 4, and between *z* and *y* is 6, then the order of the genes on the chromosome is *x-z-y*, the only order that accounts for the percent recombination data. It is a little like deriving a geographical map from distances between cities.

Genetic maps derived from percent recombination between linked genes accurately reflect the order on the chromosome, but the distances are estimates because crossing over is not equally likely across the genome. Some DNA sequences are nearly always inherited together, like two inseparable friends. This nonrandom association between DNA sequences is called **linkage disequilibrium** (**LD**). The human genome consists of many "LD" blocks where stretches of alleles stick together, interspersed with areas where crossing over is prevalent. Chapter 7 discusses the use of LD blocks, called **haplotypes,** to track genes in populations.

	Greg		Susan
Phenotype	nail-patella syndrome, type A blood		nail-patella syndrome, type B blood
Genotype	NnI^Ai		$NnI^B\_\_$
Allele configuration	$\dfrac{N \quad I^A}{n \quad i}$		$\dfrac{N \quad i}{n \quad I^B}$
Gametes:	sperm	frequency	oocytes
Parental	$N\ I^A$	45%	$N\ i$
	$n\ i$	45%	$n\ I^B$
Recombinants	$N\ i$	5%	$N\ I^B$
	$n\ I^A$	5%	$n\ i$

N = nail-patella syndrome
n = normal

Figure 5.18 **Inheritance of nail-patella syndrome.** Greg inherited the *N* and I^A alleles from his mother; that is why the alleles are on the same chromosome. His *n* and *i* alleles must therefore be on the homolog. Susan inherited alleles *N* and *i* from her mother, and *n* and I^B from her father. Population-based probabilities are used to calculate the likelihood of phenotypes in the offspring of this couple. Note that in this figure, map distances are known and are used to predict outcomes.

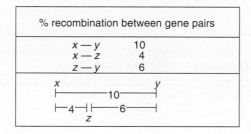

% recombination between gene pairs	
x — y	10
x — z	4
z — y	6

Figure 5.19 **Recombination mapping.** If we know the percent recombination between all possible pairs of three genes, we can determine their relative positions on the chromosome.

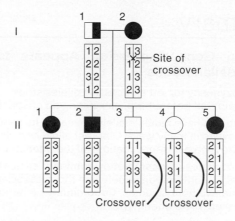

Figure 5.20 **Haplotypes.** The numbers in bars beneath pedigree symbols enable researchers to track specific chromosome segments with markers. Disruptions of a marker sequence indicate crossover sites.

From Linkage to Genome-Wide Associations

The first human genes mapped to their chromosomes encoded blood proteins, because these were easy to study. In 1980, researchers began using DNA sequences near genes of interest as landmarks called **genetic markers**. These markers need not encode proteins that cause a phenotype. They might be DNA sequence differences that alter where a DNA cutting enzyme cuts, differing numbers of short repeated sequences of DNA with no obvious function, or single sites where the base varies among individuals. The term "genetic marker" is used popularly to mean any DNA sequence that is associated with a phenotype, usually one affecting health.

Computers tally how often genes and their markers are inherited together. The "tightness" of linkage between a marker and a gene of interest is represented as a LOD score, which stands for "logarithm of the odds." A LOD score indicates the likelihood that particular crossover frequency data indicate linkage, rather than the inheritance of two alleles by chance. The higher the LOD score, the closer the two genes.

A LOD score of 3 or greater signifies linkage. It means that the observed data are 1,000 (10^3) times more likely to have occurred if the two DNA sequences (a disease-causing allele and its marker) are linked than if they reside on different chromosomes and just happen to often be inherited together by chance. It is somewhat like deciding whether two coins tossed together 1,000 times always come up both heads or both tails by chance, or because they are taped together side by side in that position, as are linked genes. If the coins land with the same sides up in all 1,000 trials, it indicates they are very likely taped.

Before sequencing of the human genome, genetic markers were used to predict which individuals in some families were most likely to have inherited a particular disorder, before

symptoms began. Such tests are no longer necessary, because tests directly detect disease-causing genes. However, genetic markers are still used to distinguish parts of chromosomes. In pedigrees, designations for markers linked into haplotypes are sometimes placed beneath the symbols to further describe chromosomes. In the family with cystic fibrosis depicted in **figure 5.20**, each set of numbers beneath a symbol represents a haplotype. Knowing the haplotype of individual II–2 reveals which chromosome in parent I–1 contributes the mutant allele. Because Mr. II–2 received haplotype 3233 from his affected mother, his other haplotype, 2222, comes from his father. Because Mr. II–2 is affected and his father is not, the father must be a heterozygote, and 2222 must be the haplotype linked to the mutant *CFTR* allele.

Today, entire human genomes can be sequenced within a day. But researchers are still filling in the orders of genes on chromosomes and determining gene functions.

Key Concepts Questions 5.3

1. Why are linked genes inherited in different patterns than unlinked genes?

2. What is the relationship between crossover frequency and relative positions of genes on chromosomes?

3. How were the first human linkage maps constructed?

4. What is linkage disequilibrium?

5. What are genetic markers, LOD scores, and haplotypes?

Summary

5.1 When Gene Expression Appears to Alter Mendelian Ratios

1. Homozygosity for lethal recessive alleles stops development before birth, eliminating an offspring class.

2. A gene can have multiple alleles because its sequence can deviate in many ways. Different allele combinations may produce different variations of the phenotype.

3. Heterozygotes of **incompletely dominant** alleles have phenotypes intermediate between those associated with the two homozygotes. **Codominant** alleles are both expressed in the phenotype.

4. In **epistasis,** one gene affects the phenotype of another.

5. An incompletely **penetrant** genotype is not expressed in all individuals who inherit it. Phenotypes that vary in intensity among individuals are **variably expressive**.

6. **Pleiotropic** genes have several expressions due to functioning in shared pathways, processes, or structures.

7. In **genetic heterogeneity,** two or more genes specify the same phenotype.

8. A **phenocopy** is a characteristic that appears to be inherited but is environmentally caused.

5.2 Mitochondrial Genes

9. Cells have many mitochondria and each mitochondrion has many copies of the mitochondrial genome.

10. Only females transmit mitochondrial genes; males can inherit such a trait but cannot pass it on.

11. Mitochondrial genes do not cross over, and they mutate more frequently than nuclear DNA.

12. The 37 mitochondrial genes encode tRNA, rRNA, or proteins involved in protein synthesis or energy reactions.

13. Many mitochondrial disorders are **heteroplasmic,** with mitochondria in a single cell harboring different alleles.

5.3 Linkage

14. Genes on the same chromosome are **linked** and, unlike genes that independently assort, produce many individuals with parental genotypes and a few with **recombinant** genotypes.

15. **Linkage maps** depict linked genes. Researchers can examine a group of known linked DNA sequences (a **haplotype**) to follow the inheritance of certain chromosomes.

16. We can predict the probabilities that certain genotypes will appear in progeny if we know crossover frequencies from pooled data and whether linked alleles are in *cis* or *trans*.

17. Genetic linkage maps assign distances to linked genes based on crossover frequencies.

18. **Genetic markers** are DNA sequences near a gene of interest that are inherited with it due to linkage.

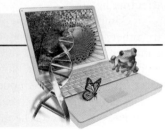

www.mhhe.com/lewisgenetics11

Answers to all end-of-chapter questions can be found at **www.mhhe.com/lewisgenetics11**. You will also find additional practice quizzes, animations, videos, and vocabulary flashcards to help you master the material in this chapter.

Review Questions

1. Explain how each of the following phenomena can disrupt Mendelian phenotypic ratios.

   a. lethal alleles
   b. multiple alleles
   c. incomplete dominance
   d. codominance
   e. epistasis
   f. incomplete penetrance
   g. variable expressivity
   h. pleiotropy
   i. a phenocopy
   j. genetic heterogeneity

2. How does the relationship between dominant and recessive alleles of a gene differ from epistasis?

3. Why can transmission of an autosomal dominant trait with incomplete penetrance look like autosomal recessive inheritance?

4. How does inheritance of ABO blood type exhibit both complete dominance and codominance?

5. How could two people with albinism have a child who has normal skin pigment?

6. How do the porphyrias exhibit variable expressivity, pleiotropy, and genetic heterogeneity?

7. How can epistasis explain incomplete penetrance?

8. The lung condition emphysema may be caused by lack of an enzyme, or by smoking. Which cause is a phenocopy?

9. List three ways that mtDNA differs from DNA in a cell's nucleus.

10. Describe why inheritance of mitochondrial DNA and linkage are exceptions to Mendel's laws.

11. How does a child conceived with donor mitochondria have three genetic parents? Do the parents contribute equally in a genetic sense? Explain your answer.

12. How does a pedigree for a maternally inherited trait differ from one for an autosomal dominant trait?

13. If researchers could study pairs of human genes as easily as they can study pairs of genes in fruit flies, how many linkage groups would they detect?

14. The popular media often use words that have precise meanings in genetics, but more general common meanings. Explain the two types of meanings of "linked" and "marker."

Applied Questions

1. For each of the diseases described in situations *a* through *i*, indicate which of the following phenomena (A–H) is at work. More than one may apply.

   A. lethal alleles

   B. multiple alleles

   C. epistasis

   D. incomplete penetrance

   E. variable expressivity

   F. pleiotropy

   G. a phenocopy

   H. genetic heterogeneity

   a. A woman with severe neurofibromatosis type 1 has brown spots on her skin and several large tumors beneath her skin. A genetic test shows that her son has the disease-causing autosomal dominant allele, but he has no symptoms.

   b. A man would have a widow's peak, if he wasn't bald.

   c. A man and woman have six children. They also had two stillbirths—fetuses that died shortly before birth.

   d. Mutations in a gene that encodes a muscle protein called titin cause 22 percent of cases of inherited dilated cardiomyopathy, a form of heart disease. Other single genes cause the remaining cases.

   e. A woman with dark brown skin uses a bleaching cream with a chemical that darkens her fingertips and ears, just like the inherited disease alkaptonuria.

   f. In Labrador retrievers, the *B* allele confers black coat color and the *b* allele brown coat color. The *E* gene controls the expression of the *B* gene. If a dog inherits the *E* allele, the coat is golden no matter what the *B* genotype is. A dog of genotype *ee* expresses the *B* (black) phenotype.

   g. Two parents are heterozygous for genes that cause albinism, but each gene specifies a different enzyme in the biochemical pathway for skin pigment synthesis. Their children thus do not face a 25 percent risk of having albinism.

   h. A woman with a form of osteogenesis imperfecta has brittle bones, and also glaucoma (high pressure in the eyeballs), hearing loss, muscle weakness, fatigue, and loose joints.

   i. Two young children in a family have very decayed teeth. Their parents think it is genetic, but the true cause is a babysitter who puts them to sleep with juice bottles in their mouths.

2. If many family studies for a particular autosomal recessive condition reveal fewer affected individuals than Mendel's law predicts, the explanation may be either incomplete penetrance or lethal alleles. How might you use haplotypes to determine which of these two possibilities is the cause?

3. A man who has type O blood has a child with a woman who has type A blood. The woman's mother has AB blood, and her father, type O. What is the probability that the child has the following blood types?

   a. O

   b. A

   c. B

   d. AB

4. Enzymes are used in blood banks to remove the A and B antigens from blood types A and B, making the blood type O. Does this process alter the phenotype or the genotype?

5. Ataxia-oculomotor apraxia syndrome (OMIM 208920), which impairs the ability to feel and move the limbs, usually begins in early adulthood. The molecular basis of the disease is impairment of ATP production in mitochondria, but the mutant gene is in the nucleus of the cells. Would this disorder be inherited in a Mendelian fashion? Explain your answer.

6. What is the chance that Greg and Susan, the couple with nail-patella syndrome, could have a child with normal nails and type AB blood?

7. A gene called secretor (OMIM 182100) is 1 map unit from the *H* gene that confers the Bombay phenotype on chromosome 19. Secretor is dominant, and a person of either genotype *SeSe* or *Sese* secretes the ABO and H blood type antigens in saliva and other body fluids. This secretion, which the person is unaware of, is the phenotype. A man has the Bombay phenotype and is not a secretor. A woman does not have the Bombay phenotype and is a secretor. She is a dihybrid whose alleles are in *cis*. What is the chance that their child will have the same genotype as the father?

8. In prosopagnosia (OMIM 610382), a person has "face blindness"—he or she cannot identify individuals by their faces. It is inherited as an autosomal dominant trait, and affects people to different degrees. Some individuals learn early in life to identify people by voice or style of dress, and so appear not to have the condition. Only a small percentage of cases are inherited; most are the result of stroke or brain injury. (www.faceblind.org offers tests to help you imagine what it is like not to be able to recognize faces.) Which

of the following does face blindness demonstrate? Explain your choices.

   a. incomplete penetrance
   b. variable expressivity
   c. pleiotropy
   d. phenocopy

9. Many people who have the "iron overload" disease hereditary hemochromatosis (OMIM 235200; see section 20.2) are homozygous for a variant of the *C282Y* gene. How would you determine the penetrance of this condition?

10. A Martian creature called a gazook has 17 chromosome pairs. On the largest chromosome are genes for three traits—round or square eyeballs (*R* or *r*); a hairy or smooth tail (*H* or *h*); and 9 or 11 toes (*T* or *t*). Round eyeballs, hairy tail, and 9 toes are dominant to square eyeballs, smooth tail, and 11 toes. A trihybrid male has offspring with a female who has square eyeballs, a smooth tail, and 11 toes on each of her three feet. She gives birth to 100 little gazooks, who have the following phenotypes:

   - 40 have round eyeballs, a hairy tail, and 9 toes
   - 40 have square eyeballs, a smooth tail, and 11 toes
   - 6 have round eyeballs, a hairy tail, and 11 toes
   - 6 have square eyeballs, a smooth tail, and 9 toes
   - 4 have round eyeballs, a smooth tail, and 11 toes
   - 4 have square eyeballs, a hairy tail, and 9 toes

   a. Draw the allele configurations of the parents.
   b. Identify the parental and recombinant progeny classes.
   c. What is the crossover frequency between the *R* and *T* genes?

Web Activities

1. Look at the gene descriptions at RetNet: http://www .sph.uth.tmc.edu/retnet/disease.htm. Select three retinal diseases and use Mendelian Inheritance in Man or other sources to describe how they affect vision differently. (Some researchers consider Leber congenital amaurosis to be a subtype of retinitis pigmentosa.)

2. Go to the websites for the Genetic Alliance (www .geneticalliance.org) or the National Institutes of Health's Office of Rare Diseases Research (http://rarediseases.info .nih.gov/). Learn about different inherited diseases, and identify one that exhibits pleiotropy.

3. Go to the United Mitochondrial Disease Foundation website and describe the phenotype of a mitochondrial disorder.

4. Browse the National Center for Biotechnology Information (NCBI) site, and list three sets of linked genes. Consult OMIM or other sources to describe the trait or disorder that each specifies.

5. Use OMIM or other sources to identify a genetically heterogeneic condition, and explain why this description applies.

6. For some of the porphyrias, attacks are precipitated by an environmental trigger. Using OMIM or other sources, describe factors that can trigger an attack of any of the following:

   a. acute intermittent porphyria
   b. porphyria cutanea tarda
   c. coproporphyria
   d. porphyria variegata
   e. erythropoietic protoporphyria

Forensics Focus

1. "Earthquake McGoon" was 32 years old when the plane he was piloting over North Vietnam was hit by groundfire on May 6, 1954. Of the five others aboard, only two survived. McGoon, actually named James B. McGovern, was well known for his flying in World War II, and for his jolliness. Remains of a man about his height and age at death were discovered in late 2002, but could not be identified by dental records. However, DNA sampled from a leg bone enabled forensic scientists to identify him. Describe the type of DNA likely analyzed, and what further information was needed to make the identification.

Case Studies and Research Results

1. Shiloh Winslow is deaf. In early childhood, she began having fainting spells, especially when she became excited. When she fainted while opening Christmas gifts, her parents took her to the hospital, where doctors said, again, that there wasn't a problem. As the spells continued, Shiloh became able to predict the attacks, telling her parents that her head hurt beforehand. Her parents took her to a neurologist, who checked Shiloh's heart and diagnosed long QT syndrome with deafness, a severe form of inherited heartbeat irregularity (see Clinical

Connection 2.2). Ten different genes can cause long QT syndrome. The doctor told them of a case from 1856: a young girl, called at school to face the headmaster for an infraction, became so agitated that she dropped dead. The parents were not surprised; they had lost two other children to great excitement.

The Winslows visited a medical geneticist, who discovered that each parent had a mild heartbeat irregularity that did not produce symptoms. Shiloh's parents had normal hearing. Shiloh's younger brother Pax was also hearing-impaired and suffered night terrors, but had so far not fainted during the day. Like Shiloh, he had the full syndrome. Vivienne, still a baby, was also tested. She did not have either form of the family's illness; her heartbeat was normal.

Today, Shiloh and Pax are treated with beta blocker drugs, and each has an implantable defibrillator to correct a potentially fatal heartbeat. Shiloh's diagnosis may have saved her brother's life.

a. Which of the following applies to the condition in this family?

   i. genetic heterogeneity

   ii. pleiotropy

   iii. variable expressivity

   iv. incomplete dominance

   v. a phenocopy

b. How is the inheritance pattern of this form of long QT syndrome similar to that of familial hypercholesterolemia?

c. How is it possible that Vivienne did not inherit either the serious or asymptomatic form of the illness?

d. Do the treatments for the condition affect the genotype or the phenotype?

2. Barnabas Collins has congenital erythropoietic porphyria, and his wife Angelique is a carrier of ALA dehydratase deficiency. What is the chance that, if they have a child, he or she will have a porphyria?

3. Max Watson, born in 2004, was well for the first few weeks of his life, but then he became difficult to awaken, stopped gaining weight and then began to lose it, and then developed seizures. His pediatrician suspected that Max had cobalamin C deficiency (OMIM 277400), an inability to use vitamin B12, also called cobalamin. But Max did not have a mutation in what at the time was the only known causative gene, on an autosome. When Max was 9, researchers sequenced his exome and discovered that he had a mutation in a gene on the X chromosome that controls a transcription factor that in turn controls use of the vitamin. What phenomenon described in section 5.1 does this case illustrate?

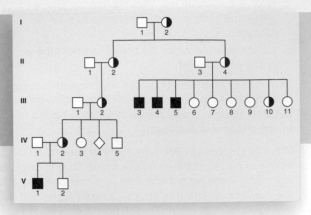

In 1937, Alfred Wiskott described the family depicted in this pedigree, which shows classic X-linked inheritance—males affected, female carriers. Wiskott-Aldrich syndrome is treatable today with a combination of stem cell and gene therapies.

Matters of Sex

Learning Outcomes

6.1. Our Sexual Selves

1. Describe the factors that contribute to whether we are and feel male or female.

2. Distinguish between the X and Y chromosomes.

3. Discuss how manipulating sex ratio can affect societies.

6.2. Traits Inherited on Sex Chromosomes

4. Distinguish between Y linkage and X linkage.

5. Compare and contrast X-linked recessive inheritance and X-linked dominant inheritance.

6.3 Sex-Limited and Sex-Influenced Traits

6. Discuss the inheritance pattern of a trait that appears in only one sex.

7. Define *sex-influenced trait*.

6.4 X Inactivation

8. Explain why X inactivation is necessary.

9. Explain how X inactivation is an epigenetic change.

10. Discuss how X inactivation affects the phenotype in female mammals.

6.5 Parent-of-Origin Effects

11. Explain the chemical basis of silencing the genetic contribution from one parent.

12. Explain how differences in the timetables of sperm and oocyte formation can lead to parent-of-origin effects.

The BIG Picture

Sex affects our lives in many ways. Which sex chromosomes we are dealt at conception sets the developmental program for maleness or femaleness, but gene expression before and after birth greatly influences how that program unfolds.

Stem Cell and Gene Therapies Save Boys' Lives

In 1937, Alfred Wiskott, a pediatrician in Germany, saw a family that had six healthy girls, but three baby boys who died of the same illness. They had bruises, a skin condition, bloody diarrhea, ear infections, and pneumonia. All died from bleeding in their digestive tracts and infection in the blood. Dr. Wiskott noted the inherited nature of the condition and that it affected only boys.

In 1954, Robert Aldrich described the disorder in a large Dutch family. Sick boys who survived childhood developed cancers of the blood or autoimmune disorders. The disease was named Wiskott-Aldrich syndrome (OMIM 301000), and the mutant gene is on the X chromosome. Only boys are affected because, unlike girls, they lack a second X chromosome to block the effects of the mutation.

In 2006, German researchers contacted living relatives of the original family with the three sick boys. Genetic tests revealed a mutation in the *WAS* gene. An affected little boy, the first cousin twice removed of the three original boys, was successfully treated with a stem cell transplant from a matched, unrelated donor. Today, patients can be successfully treated with an experimental combination of stem cell and gene therapy. They receive their own bone marrow stem cells given functional *WAS* genes.

6.1 Our Sexual Selves

Whether we are male or female is enormously important in our lives. It affects our relationships, how we think and act, and how others perceive us. Gender is, at one level, dictated by genes, but it is also layered with psychological and sociological components.

Maleness or femaleness is determined at conception, when he inherits an X and a Y chromosome, or she inherits two X chromosomes (**figure 6.1**). Another level of sexual identity comes from the control that hormones exert over the development of reproductive structures. Finally, both biological factors and social cues influence sexual feelings, including the strong sense of whether we are male or female.

Sexual Development

Gender differences become apparent around the ninth week of prenatal development. During the fifth week, all embryos develop two unspecialized gonads, which are organs that will develop as either testes or ovaries. Each such "indifferent" gonad forms near two sets of ducts that offer two developmental options. If one set of tubes, called the Müllerian ducts, continues to develop, female sexual structures form. If the other set, the Wolffian ducts, persist, male sexual structures form.

The choice to follow either developmental pathway occurs during the sixth week, depending upon the sex chromosome constitution and actions of certain genes. If a gene on the Y chromosome called *SRY* (for "sex-determining region of the Y") is activated, hormones steer development along a male route. In the absence of *SRY* activation, a female develops (**figure 6.2**).

Femaleness was once considered a "default" option in human development, defined as the absence of maleness. Sex determination is more accurately described as a fate imposed on ambiguous precursor structures. Several genes in addition to *SRY* guide early development of reproductive structures. A mutation in a gene called *Wnt4*, for example, causes an XX female to have high levels of male sex hormones and lack a vaginal canal and uterus. Ovaries do not develop properly and secondary sex characteristics do not appear. Hence, *Wnt4* is essential for development and maturation as a female.

Sex Chromosomes

The sex with two different sex chromosomes is called the **heterogametic** sex, and the other, with two of the same sex chromosome, is the **homogametic** sex. In humans, males are

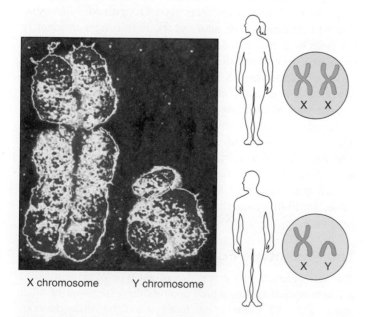

Figure 6.1 **The sex chromosomes.** A human male has one X chromosome and one Y chromosome. A female has two X chromosomes.

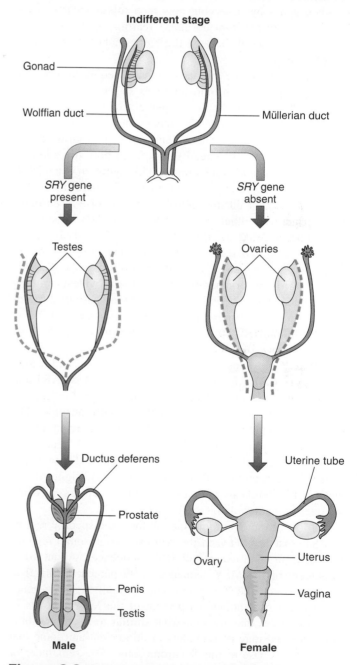

Figure 6.2 **Male or female?** The paired duct systems in the early human embryo may develop into male *or* female reproductive organs. The red tubes represent female structures and the blue tubes, male structures.

heterogametic (XY) and females are homogametic (XX). Some other species are different. In birds and snakes, for example, males are ZZ (homogametic) and females are ZW (heterogametic).

The sex chromosomes differ in size and capacity. The human X chromosome has more than 1,500 genes and is much larger than the Y chromosome, which has 231 protein-encoding genes. In meiosis in a male, the X and Y chromosomes act as if they are a pair of homologs.

Identifying genes on the human Y chromosome has been difficult. Before the human genome sequence became available, researchers inferred the functions and locations of Y-linked genes by observing how men missing parts of the chromosome differ from normal. Creating linkage maps for the Y was not possible because the Y does not cross over along all of its length.

Analysis of the human genome sequence revealed why mapping the Y chromosome was so hard: It has a very unusual organization. In the 95 percent of the chromosome that harbors 22 male-specific genes, many DNA segments are palindromes. In written languages, palindromes are sequences of letters that read the same in both directions— "Madam, I'm Adam," for example. This symmetry in a DNA sequence, compared to "a hall of mirrors," destabilizes DNA replication so that during meiosis, sections of a Y chromosome attract each other. This can loop out parts in between, which may account for infertility caused by missing parts of the Y, although the human Y chromosome has shrunk by about 3 percent over the past 25 million years. Yet this "hall of mirrors" organization may also provide a way for the chromosome to recombine with itself, essentially sustaining its structure. Two researchers—one an XX, one an XY— consider the curious structure of the human Y chromosome in *In Their Own Words* on page 113.

The Y chromosome has a very short arm and a long arm (**figure 6.3**). At both tips are regions, called PAR1 and PAR2, that are termed "pseudoautosomal" because they can cross over with counterparts on the X chromosome. The pseudoautosomal regions comprise only 5 percent of the chromosome and include 63 pseudoautosomal genes that encode proteins that function in both sexes. These genes control bone growth, cell division, immunity, signal transduction, the synthesis of hormones and receptors, fertility, and energy metabolism.

Most of the Y chromosome is the male-specific region, or MSY, that lies between the two pseudoautosomal regions. It consists of three classes of DNA sequences. About 10 to 15 percent of the MSY sequence is 99 percent identical to counterparts on the X chromosome. Protein-encoding genes are scarce here. Another 20 percent of the MSY consists of DNA sequences that are somewhat similar to X chromosome sequences and may be remnants of an ancient autosome that long ago gave rise to the X chromosome. The remainder of the MSY includes palindrome-ridden regions. The genes in the MSY include many repeats and specify protein segments that combine in different ways, which is one reason why counting the number of protein-encoding genes on the Y chromosome

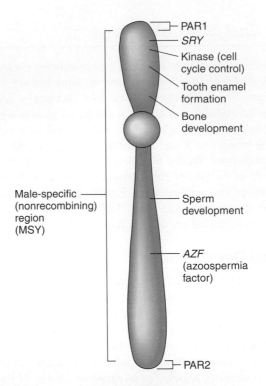

Figure 6.3 **Anatomy of the Y chromosome.** The Y chromosome has two pseudoautosomal regions (PAR1 and PAR2) and a large central area (MSY) that comprises about 95 percent of the chromosome and has only 22 genes. *SRY* determines sex. *AZF* encodes a protein essential to producing sperm; mutations in it cause infertility.

has been difficult. Most males have 22 such genes, but some have only 20. Many of the genes in the MSY are essential to fertility, including *SRY*.

The Y chromosome was first seen under a light microscope in 1923, and researchers soon recognized its association with maleness. For many years, they sought to identify the gene or genes that determine sex. Important clues came from two very interesting types of people—men who have two X chromosomes (XX male syndrome), and women who have one X and one Y chromosome (XY female syndrome). A close look at their sex chromosomes revealed that the XX males actually had a small piece of a Y chromosome, and the XY females lacked a small part of the Y chromosome. The part of the Y chromosome present in the XX males was the same part that was missing in the XY females. This critical area accounted for half a percent of the Y chromosome. In 1990, researchers isolated and identified the *SRY* gene here.

The Phenotype Forms

The *SRY* gene encodes a very important type of protein called a **transcription factor,** which controls the expression of other genes. The *SRY* transcription factor stimulates male development by sending signals to the indifferent gonads that destroy potential female structures while stimulating development of male structures.

The Y Wars

Researcher Jennifer Marshall-Graves predicts that the Y chromosome will "self-destruct" within the next 10 million years. Her comparison of Y chromosomes in a wide variety of mammals indicates that, gradually, important genes are being transferred to other chromosomes. David Page, who has led the mapping of the Y chromosome, has a more optimistic view. Each researcher spoke out, in jest, at two scientific conferences. Here is some of what they had to say:

The Rise and Fall of the Human Y Chromosome

Jennifer A. Marshall-Graves, Australian National University

The Y chromosome is unique in the human genome. It is small, gene-poor, prone to deletion and loss, variable among species, and useless. You can lack a Y and not be dead, just female. It is impossible to understand why this chromosome is so weird without understanding where it came from. It is a sad decline, and I predict its imminent loss.

The X is a decent sort of chromosome. It accounts for 4 percent of the genome, with about 1,500 perfectly normal genes. The Y is a pathetic little chromosome that has few genes interspersed with lots of junk. And those genes are a weird lot. They are particularly concerned with male sexual development. Some are quite bizarre and many inactive. The Y clearly diverged from the X.

There are several models of the Y (**figure 1**). The dominant Y model of a macho Y reflects the fact that the Y contains the male-determining *SRY* gene. The selfish Y model predicts that the Y kidnapped genes from elsewhere. The wimp Y model says that the Y is just a relic of the once glorious X chromosome. This model was first proposed by biologist Susumo Ohno in 1967 in the theory that the X and Y originated as a pair of autosomes. Then the Y acquired the male-determining locus, and other genes that are required for spermatogenesis gathered nearby. This led to suppressed recombination in this region of the Y, which allowed all sorts of horrible genetic accidents to occur that could not be

repaired. Mutations, deletions, and insertions accumulated until almost nothing was left, except bits at the top and bottom that still pair with the X. A few genes survived because they found a useful male-specific function, and many of these have made copies of themselves in a desperate race to stave off disappearing altogether.

The Y is degrading fast, losing genes at the rate of 5 per million years. I predict that it will be completely gone in 5 to 10 million years. Will we have males? The males in the audience can take comfort from the mole vole *Ellobius lutescens* (**figure 2**). It has no Y, but it does have males and females. It has no *SRY*, no Y chromosome at all. Both sexes are XO. How do they do it? We don't know. Clearly another gene takes over and new sex genes start evolving. Will there be new sex chromosome evolution in humans? Maybe it will happen in different ways in different populations, and we will split into two species.

Rethinking the Rotting Y Chromosome

David Page, Massachusetts Institute of Technology and Howard Hughes Medical Institute Investigator

The Y chromosome has had a public relations problem for a long time. For most of the last half of the past century, people thought that the Y chromosome was a junk heap. We can now update that model.

Back 300 million years ago, when we were reptiles, we had no sex chromosomes, only ordinary autosomes. Shortly after our ancestors parted company with the ancestors of birds, a mutation arose on one member of a pair of ordinary autosomes to give rise to *SRY*. The process of shutting down XY crossing over began, first in the vicinity of *SRY*, and then in an expanding region. Once a piece of the Y was no longer able to recombine with the X, its genes began to rot. The purpose of sex (recombination in meiosis)

(Continued)

Models of the Human Y

Dominant Y   Selfish Y   Wimp Y

Figure 1 **Models of the human Y chromosome.** Researcher Jennifer Marshall-Graves offers a tongue-in-cheek look at the Y chromosome, but her research findings are serious—the chromosome is shrinking.

Figure 2 **Life without a Y?** Males of all mammals, except two species of mole voles, have Y chromosomes. Birds and reptiles do not. The Y chromosome probably arose from an X chromosome about 310 million years ago. The X lost many genes and gained a few that set their carriers on the road to maleness. This animal is a Y-less male mole vole— it reproduces just fine.

(Continued)

is not just to generate new gene combinations, but to allow genes to rid themselves of mildly deleterious mutations that accumulate. Y genes are not protected because they have lots of areas of no crossing over. Genes decayed, except for *SRY* and the tips. It wasn't a very flattering model for the Y.

When Jennifer Marshall-Graves and John Aitken wrote that the Y would self-destruct in 10 million years, it truly frightened the people in my lab. We decided we needed to pick up the pace.

Based on the sequencing of the Y, we've been able to rethink its evolution, and realized that the chromosome may have found a way around its seemingly inevitable problems. We looked closely at the male-specific region of the Y, reanalyzing sequences

in a different way, chopped into smaller bits. And we found that each piece would find a match elsewhere on the Y. So segments on the Y are effectively functioning as alleles—30 percent have a perfect match elsewhere on the chromosome. These are not simple repeats, but highly complex sequences of tens to hundreds of kilobases. The region includes eight palindromes and one inverted repeat (**figure 3**). We propose that there is intense recombination within the palindromes. And so the Y has two forms of productive recombination: conventional routine recombination of crossing over with the X at pseudoautosomal regions, and recombination within the Y. It's not that the Y doesn't recombine, it just does it its own way. The Y does copying that preserves its identity.

cen ... u1 b1 t1 u2 t2 b2 u3 g1 r1 r2 b3 g2 r3 r4 g3 b4 ... qter

1 Mb

Figure 3 **The Y chromosome DNA sequence is highly repetitive.** Part of the Y chromosome that David Page studies, called *AZFc* (for azoospermia factor c), consists of DNA sequences that read the same in either direction, an organization that can lead to instability as well as provide a mechanism to generate new alleles. Other parts of the chromosome house similar repeats. Matching colors represent identical sequences. Same-color arrows that point in opposite directions indicate inverted repeats.

Prenatal sexual development is a multistep process, and mutations can intervene at several points (**figure 6.4**). The result may be an XY individual who looks female because of a block in the development of male structures. For example, in androgen insensitivity syndrome (OMIM 300068), caused by a mutation in a gene on the X chromosome, the absence of receptors for androgens (the male sex hormones testosterone and dihydrotestosterone [DHT]) stops cells in early reproductive structures from receiving the signal to develop as male. The person looks female, but is XY.

Several terms are used to describe individuals whose genetic/chromosomal sex and physical structures, both internal and external, are not consistent with one gender. *Hermaphroditism* is an older and more general term for an individual with both male and female sexual structures. *Intersex* refers to individuals whose internal structures are inconsistent with external structures, or whose genitalia are ambiguous. It is the preferred term outside of medical circles. *Pseudohermaphroditism* refers to the presence of both female and male structures but at different life stages.

Living with pseudohermaphroditism can be confusing. Consider 5-alpha reductase deficiency (OMIM 264600), which is autosomal recessive. Affected individuals have a normal Y chromosome, a wild type *SRY* gene, and testes. The internal male reproductive tract develops and internal female structures do not, so the male anatomy is present on the inside. But the child is unaware of the insides, and on the outside looks like a girl. Without 5-alpha reductase, which normally catalyzes the reaction of testosterone to form DHT, a penis cannot form. At puberty, when the adrenal glands, which sit atop the kidneys, start to produce testosterone, this XY individual,

who thought she was female, starts to experience the signs of maleness—deepening voice, growth of facial hair, and the sculpting of muscles into a masculine physique. Instead of developing breasts and menstruating, the clitoris enlarges into a penis. Usually sperm production is normal. XX individuals with 5-alpha reductase deficiency look female.

The degree to which pseudohermaphroditism disturbs the individual depends as much on society as it does on genetics. In the Dominican Republic in the 1970s, 22 young girls reached the age of puberty and began to transform into boys. They had a form of 5-alpha reductase deficiency that was fairly common in the population due to consanguinity (relatives having children with relatives). The parents were happy that they had sons after all, and so these special adolescents were given their own gender name—*guevedoces,* for "penis at age 12." They were fully accepted as whatever they wanted to be. This isn't always the case. A very realistic novel, *Middlesex,* tells the story of a young man with this condition who grew up as a female.

In a more common form of pseudohermaphroditism, congenital adrenal hyperplasia (CAH) due to 21-hydroxylase deficiency (OMIM 201910), an enzyme block causes testosterone and DHT to accumulate. It is autosomal recessive, and both males and females are affected. The higher levels of androgens cause precocious puberty in males or male secondary sex characteristics to develop in females. Boys may enter puberty as early as 3 years old, with well-developed musculature, small testes, and enlarged penises. At birth, girls may have swollen clitorises that appear to be small penises. They are female internally, but as they reach puberty, their voices deepen, facial hair grows, and they do not menstruate.

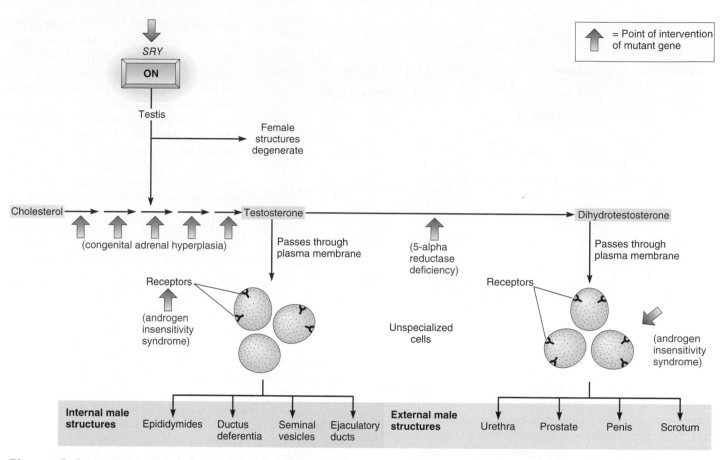

Figure 6.4 **Mutations that affect male sexual development.** In normal male prenatal development, activation of a set of genes beginning with *SRY* sends signals that destroy female rudiments, while activating the biochemical pathway that produces testosterone and dihydrotestosterone, which promote the development of male structures. The green arrows indicate where mutations disrupt normal sexual development. Diseases are indicated in red.

Prenatal tests that detect chromosomal sex have changed the way that pseudohermaphroditism is diagnosed. Before, the condition was detected only after puberty, when a girl began to look like a boy. Today pseudohermaphroditism is suspected when a prenatal chromosome check reveals an X and a Y chromosome, but the newborn looks like a girl.

Transgender is a poorly understood condition related to sexual identity. A transgendered individual has the phenotype and sex chromosomes of one gender, but identifies extremely strongly with the opposite gender. It is a much more profound condition than transvestitism, which refers to a male who prefers women's clothing. The genetic or physical basis of transgender is not known. Some affected individuals have surgery to better match their physical selves with the gender that they feel certain they are.

Is Homosexuality Inherited?

No one really knows why we have feelings of belonging to one gender or the other, or of being attracted to the same or opposite gender, but these feelings are intense. In homosexuality, a person's phenotype and genotype are consistent, and physical attraction is toward members of the same sex. Homosexuality is seen in all human cultures and has been observed for thousands of years. It has been documented in more than 500 animal species.

Homosexuality reflects complex input from genes and the environment. The genetic influence may be seen in the strong feelings that homosexual individuals have as young children, long before they know of the existence or meaning of the term. Other evidence comes from identical twins, who are more likely to both be homosexual than are both members of fraternal twin pairs.

Experiments in the 1990s identified genetic markers on the X chromosome that were more often identical among pairs of homosexual brothers than among other pairs of brothers. This finding led to the idea that a single gene, or a few genes, dictates sexual preference but these results could not be confirmed. Further studies on twins have indicated what many people have long suspected—that the roots of homosexuality are not simple.

Twin studies compare a trait between identical and fraternal twin pairs, to estimate the rough proportion of a trait that can be attributed to genes. Chapter 7 discusses this approach further. Such a study done on all of the adult twins in Sweden found that in males, genetics contributes about 35 percent to homosexuality, whereas among females the genetic contribution is about 18 percent. Clearly, homosexuality reflects the input of many genes and environmental factors, and may in fact arise in a variety of ways.

Table 6.1 summarizes the components of sexual identity.

Table 6.1   Sexual Identity

Level	Events	Timing
Chromosomal/genetic	XY = male XX = female	Fertilization
Gonadal sex	Undifferentiated structure begins to develop as testis or ovary	6 weeks after fertilization
Phenotypic sex	Development of external and internal reproductive structures continues as male or female in response to hormones	8 weeks after fertilization, puberty
Gender identity	Strong feelings of being male or female develop	From childhood, possibly earlier
Sexual orientation	Attraction to same or opposite sex	From childhood

Sex Ratio

Mendel's law of segregation predicts that populations should have approximately equal numbers of male and female newborns. That is, male meiosis should yield equal numbers of X-bearing and Y-bearing sperm. After birth, societal and environmental factors may favor survival of one gender over the other.

The proportion of males to females in a human population is called the **sex ratio**. Sex ratio is calculated as the number of males divided by the number of females multiplied by 1,000, for people of a particular age. (Some organizations describe sex ratio based on 1.0.) A sex ratio of equal numbers of males and females would be designated 1,000. The sex ratio at conception is called the primary sex ratio. In the United States for the past six decades, newborn boys have slightly outnumbered newborn girls, with the primary sex ratio averaging 1,050. The slight excess of boys may reflect the fact that Y-bearing sperm weigh slightly less than X-bearing sperm, and so they may reach the oocyte faster and more of them enter.

Sex ratio at birth is termed secondary and at maturity is called tertiary. Sex ratio can change markedly with age. This reflects medical conditions that affect the sexes differently, as well as environmental factors that affect one sex more than the other, such as participation in combat or engaging in other dangerous behaviors.

When a society intentionally alters the sex ratio, results can be drastic. In India and China, for example, researchers have inferred the existence of many "missing females." In these societies, prenatal diagnostic techniques were used to identify XX fetuses. Termination of XX fetuses, underreporting of female births, and, rarely, selective infanticide of girl babies all contributed to a very unnaturally skewed sex ratio favoring males.

The effects of altering sex ratio echo for years. In China, by the year 2020, 20 million men will lack female partners as a long-term consequence of that nation's "one-child policy." It began in 1979, with financial incentives to control runaway population growth. If a couple had a second child, the government revoked benefits. Some families, wanting their only child to be a boy, failed to continue or report female pregnancies. The reasoning was societal: A son would care for his aging parents, but a daughter would care for her in-laws.

China's one-child policy prevented hundreds of millions of births. The average number of births per woman fell from 5.4 in 1971 to 1.8 in 2001. By the turn of the century, 117 boys were being born for every 100 girls. In 2007, a committee of experts asked the government to stop the policy because it created "social problems and personality disorders." Today, many children in China have few siblings, cousins, aunts, or uncles. Young women, now rare, have become valued once again. Coddled only-children, called "little emperors," now dominate the younger generation, and studies have shown them to be more self-centered and less cooperative than people who grew up in other times and with siblings. Some job ads indicate "no single children." Being an only child, once desired, is becoming a stigma. The Chinese government now awards housing subsidies and scholarships to families that have girls, in the hope that a more natural sex ratio will return.

India is experiencing male bias similar to the situation in China. A survey of 1.1 million families revealed that the ratio of boys to girls was about equal when the first or second child was a boy, but if the first child and especially if the first two children were girls, then the secondary sex ratio fell to about 750 girls to every 1,000 boys. Families were using prenatal diagnosis to detect female pregnancies and were terminating about a fourth of them. Researchers estimate that India has about 100 million "missing females."

At the other end of the human life cycle, sex ratio favors females in most populations. For people over the age of 65 in the United States, for example, the sex ratio is 720, meaning that there are 72 men for every 100 women. The ratio among older people is the result of disorders that are more likely to be fatal in males as well as behaviors that may shorten their life spans compared to women.

Key Concepts Questions 6.1

1. What are the sex chromosome constitutions of human males and females?
2. Describe male and female development before birth.
3. How is the structure of the Y chromosome unusual?
4. Describe conditions that disrupt sexual development.
5. How do genes and the environment contribute to homosexuality?
6. Define *sex ratio*.

6.2 Traits Inherited on Sex Chromosomes

Genes on the Y chromosome are **Y-linked** and genes on the X chromosome are **X-linked**. Y-linked traits are rare because the chromosome has few genes, and many have counterparts on the X chromosome. Y-linked traits are passed from male to male, because a female does not have a Y chromosome. No other Y-linked traits besides infertility (which obviously can't be passed on) are yet clearly defined. Some traits at first attributed to the Y chromosome are actually genes that inserted into the chromosome from other chromosomes, such as a deafness gene.

A disproportionate number of X-linked genes cause illness when mutant. The chromosome includes 4 percent of all the genes, but accounts for about 10 percent of Mendelian (single-gene) diseases.

Genes on the X chromosome have different patterns of expression in females and males, because a female has two X chromosomes and a male just one. In females, X-linked traits are passed just like autosomal traits—that is, two copies are required for expression of a recessive allele, and one copy for a dominant allele. In males, however, a single copy of an X-linked allele causes expression of the trait or illness because there is no copy of the gene on a second X chromosome to mask the effect. A man inherits an X-linked trait only from his mother. The human male is considered **hemizygous** for X-linked traits, because he has only one set of X-linked genes.

Understanding how sex chromosomes are inherited is important in predicting phenotypes and genotypes in offspring. A male inherits his Y chromosome from his father and his X

chromosome from his mother (**figure 6.5**). A female inherits one X chromosome from each parent. If a mother is heterozygous for a particular X-linked gene, her son or daughter has a 50 percent chance of inheriting either allele from her. X-linked traits are always passed on the X chromosome from mother to son or from either parent to daughter, but there is no male-to-male transmission of X-linked traits.

X-Linked Recessive Inheritance

An X-linked recessive trait is expressed in females if the causative allele is present in two copies. Many times, an X-linked trait passes from an unaffected heterozygous mother to an affected son. **Table 6.2** summarizes the transmission of an X-linked recessive trait.

If an X-linked condition is not lethal, a man may be healthy enough to transmit it to offspring. Consider the small family depicted in **figure 6.6**. A middle-aged man who had

Table 6.2	Criteria for an X-Linked Recessive Trait

1. Always expressed in the male.

2. Expressed in a female homozygote and very rarely in a heterozygote.

3. Affected male inherits trait from heterozygote or homozygote mother.

4. Affected female inherits trait from affected father and affected or heterozygote mother.

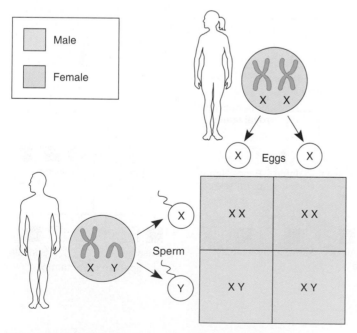

Figure 6.5 **Sex determination in humans.** An oocyte has a single X chromosome. A sperm cell has either an X or a Y chromosome. If a Y-bearing sperm cell with a functional *SRY* gene fertilizes an oocyte, the zygote is a male (XY). If an X-bearing sperm cell fertilizes an oocyte, then the zygote is a female (XX).

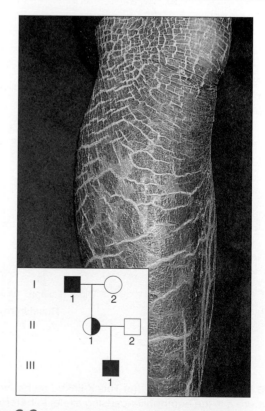

Figure 6.6 **An X-linked recessive trait.** Ichthyosis is transmitted as an X-linked recessive trait. A grandfather and grandson were affected in this family.

rough, brown, scaly skin did not realize his condition was inherited until his daughter had a son. By a year of age, the boy's skin resembled his grandfather's. In the condition, called ichthyosis (OMIM 308100), an enzyme deficiency blocks removal of cholesterol from skin cells. The upper skin layer cannot peel off as it normally does, and appears brown and scaly. A test of the daughter's skin cells revealed that she produced half the normal amount of the enzyme, indicating that she was a carrier.

Colorblindness is another X-linked recessive trait that does not hamper the ability of a man to have children. About 8 percent of males of European ancestry are colorblind, as are 4 percent of males of African descent. Only 0.4 percent of females in both groups are colorblind. **Clinical Connection 6.1,** on page 119, takes a closer look at this interesting trait.

Figure 6.7 shows part of an extensive pedigree for another X-linked recessive trait, the blood-clotting disorder hemophilia B (OMIM 306900), also known as Christmas disease. Note the combination of pedigree symbols and a Punnett square to trace transmission of the trait. Dominant and recessive alleles are indicated by superscripts to the X and Y chromosomes. In the royal families of England, Germany, Spain, and Russia, the mutant allele arose in one of Queen Victoria's X chromosomes.

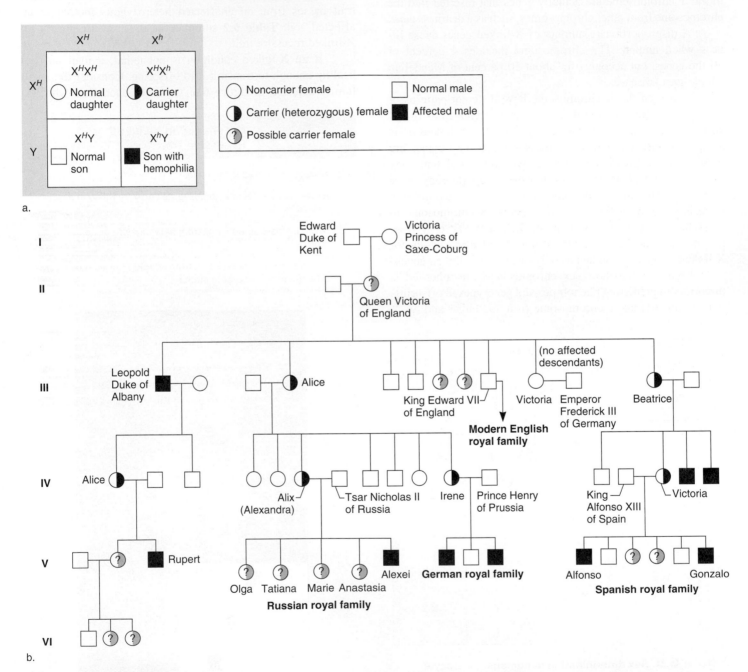

Figure 6.7 Hemophilia B. (a) This X-linked recessive disease usually passes from a heterozygous woman (designated X^HX^h, where h is the hemophilia-causing allele) to heterozygous daughters or hemizygous sons. The father is normal. **(b)** The disorder has appeared in the royal families of England, Germany, Spain, and Russia. The modern royal family in England does not carry hemophilia.

Colorblindness

English chemist John Dalton saw things differently from most people. Sealing wax that appeared red to other people was as green as a leaf to Dalton and his brother. Pink wildflowers were blue. The Dalton brothers had X-linked recessive colorblindness.

Curious about the cause of his colorblindness, Dalton asked his physician, Joseph Ransome, to dissect his eyes after he died. When that happened, Ransome snipped off the back of one eye, removing the retina, where the cone cells that provide color vision are nestled among the more abundant rod cells that impart black-and-white vision. Because Ransome could see red and green normally when he peered through the back of his friend's eyeball, he concluded that it was not an abnormal filter in front of the eye that altered color vision. He stored the eyes in dry air, enabling researchers at the London Institute of Ophthalmology to analyze DNA in Dalton's eyeballs in 1994. Dalton's remaining retina lacked one of the three types of photopigments that enable cone cells to capture certain wavelengths of light.

Color Vision Basics

Cone cells are of three types, defined by three types of photopigments. Hence humans are called "trichromats." Many other primate species are dichromats. An object appears colored because it reflects certain wavelengths of light, and each photopigment captures a particular range of wavelengths. The brain interprets the information as a visual perception, much as an artist mixes the three primary colors to create many hues.

Each photopigment has a vitamin A–derived portion called retinal and a protein portion called an opsin. The three types of opsins correspond to short, middle, and long wavelengths of light. Mutations in opsin genes cause three types of colorblindness. A gene on chromosome 7 encodes shortwave opsins, and mutations in it produce rare "blue" colorblindness (OMIM 190900). Dalton had deuteranopia (green colorblindness); his eyes lacked the middle-wavelength opsin. The third type, protanopia (red colorblindness), lacks long-wavelength opsin. Deuteranopia (OMIM 303800) and protanopia (OMIM 303900) are X-linked.

A Molecular View

Jeremy Nathans of Johns Hopkins University also personally contributed to understanding color vision. He used a cow version of a protein called rhodopsin that provides black-and-white vision to identify the human version of the gene. Then he used the human rhodopsin gene to search his own DNA and found three genes with similar sequences: One on chromosome 7, and two on the X chromosome.

Nathans can see colors, but his opsin genes provided a clue to how colorblindness arises and why it is so common. His X chromosome has one red opsin gene and two green genes, instead of the normal one of each. Because the red and green genes have similar sequences, Nathans reasoned, they can misalign during meiosis in the female (**figure 1**). The resulting oocytes then have either two or none of one opsin gene type. An oocyte lacking either a red or a green opsin gene would, when fertilized by a Y-bearing sperm, give rise to a colorblind male.

A mutation in an opsin gene can also be beneficial. A few women are known to have an opsin gene variant that gives them four types of opsins: blue, green, red, and the new variant. This means that they can detect an expanded spectrum of light wavelengths. Researchers are studying these special women, called "tetrachromats," to learn whether they can see more colors or have more nuanced color vision than the rest of us.

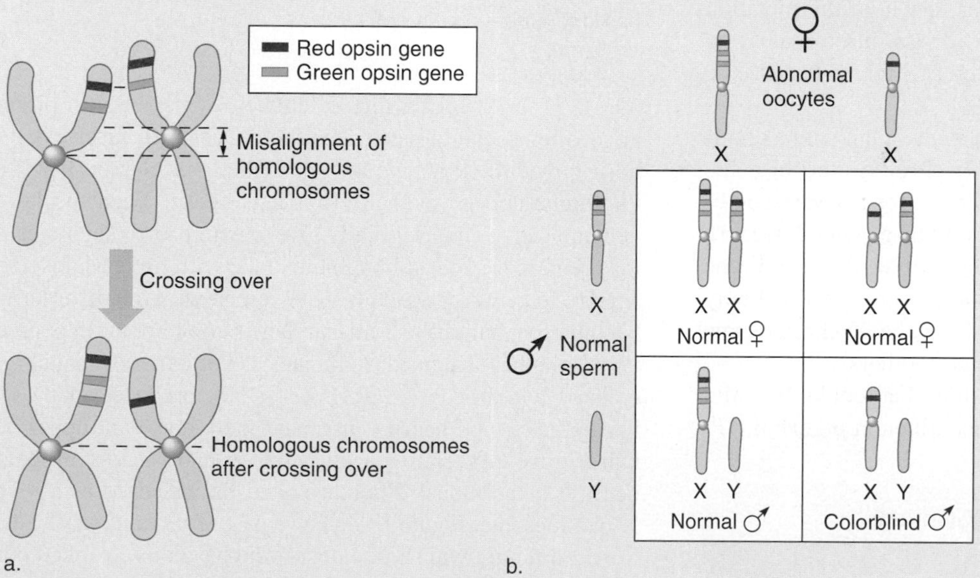

Questions for Discussion

1. How do the three types of colorblindness differ in genotype and phenotype?

2. How did Nathans discover the mutations that cause colorblindness?

3. Why is a female more likely to be a tetrachromat than a male?

Figure 1 How colorblindness arises. (a) The sequence similarities among the opsin genes responsible for color vision may cause chromosomes to misalign during meiosis in the female. Offspring may inherit too many, or too few, opsin genes. A son inheriting an X chromosome missing an opsin gene would be colorblind. A daughter would have to inherit two mutations to be colorblind. **(b)** A missing gene causes X-linked colorblindness.

The queen either had a new mutation or she inherited it. She passed it on through carrier daughters and one mildly affected son.

For many years, historians assumed that the royal families of Europe suffered from the more common form of hemophilia, type A (OMIM 306700). However, researchers in 2009 tested bits of bone from Tsarina Alexandra and Crown Prince Alexei for the wild type and mutant alleles for the genes that cause hemophilia A as well as B. Both genes are on the X chromosome. The result: Mother Alexandra was a carrier of the rarer hemophilia B, and son Crown Prince Alexei had the disease, which had been known from documents describing his severe bleeding episodes.

The transmission pattern of hemophilia B is consistent with the criteria for an X-linked recessive trait listed in table 6.2. A daughter can inherit an X-linked recessive disorder or trait if her father is affected and her mother is a carrier, because the daughter inherits one affected X chromosome from each parent. Without a biochemical test, though, an unaffected woman would not know she is a carrier for an X-linked recessive trait unless she has an affected son.

A woman whose brother has hemophilia B has a 1 in 2 risk of being a carrier. Both her parents are healthy, but her mother must be a carrier because her brother is affected. Her risk is the chance that she has inherited the X chromosome bearing the hemophilia allele from her mother. The chance of the woman conceiving a son is 1 in 2, and of that son inheriting hemophilia is 1 in 2. Using the product rule, the risk that she is a carrier and will have a son with hemophilia, out of all the possible children she can conceive, is $1/2 \times 1/2 \times 1/2$, or 1/8.

X-Linked Dominant Inheritance

Dominant X-linked conditions and traits are rare. Again, gene expression differs between the sexes (**table 6.3**). A female who inherits a dominant X-linked allele or in whom the mutation originates has the associated trait or illness, but a male who inherits the allele is usually more severely affected because he has no other allele to mask its effect.

The children of a normal man and a woman with a dominant, disease-causing allele on the X chromosome face the risks summarized in **figure 6.8**. However, for a severe condition, such as Rett syndrome (described in the chapter 2 opener), females may be too disabled to have children. Rett syndrome and similar X-linked dominant conditions therefore affect only girls, because sons would have to inherit the X chromosome bearing the mutation from their affected mothers.

An X-linked dominant condition that could be mild enough to be passed to children is incontinentia pigmenti (IP)

Table 6.3	Criteria for an X-Linked Dominant Trait

1. Expressed in females in one copy
2. Much more severe effects in males
3. High rates of miscarriage due to early lethality in males
4. Passed from male to all daughters but to no sons

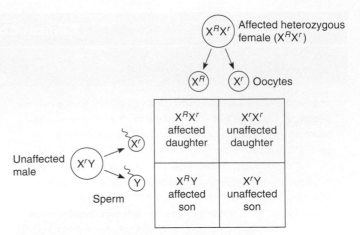

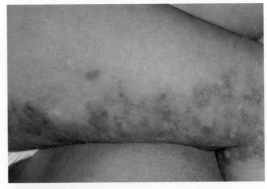

a.

b.

Figure 6.8 **X-linked dominant inheritance.** **(a)** A female who has an X-linked dominant trait has a 1 in 2 probability of passing it to her offspring, male or female. Males are generally more severely affected than females. **(b)** Note the characteristic patchy pigmentation on the leg of a girl who has incontinentia pigmenti.

(OMIM 308300). In affected females, swirls of skin pigment arise when melanin penetrates the deeper skin layers. A newborn girl with IP has yellow, pus-filled vesicles on her limbs that come and go over the first few weeks. Then the lesions become warty and eventually give way to brown splotches that may remain for life, although they fade with time. Males do not survive to be born, which is why women with the disorder who become pregnant have a miscarriage rate of about 25 percent.

Another X-linked dominant condition, congenital generalized hypertrichosis (OMIM 307150), produces many extra hair follicles, and hence denser and more abundant upper body hair (**figure 6.9**). Hair growth is milder and patchier in females because of hormonal differences and the presence of a second X chromosome. Figure 6.9*b* shows part of a pedigree of a large Mexican family with 19 members who have this X-linked dominant condition. The affected man in the pedigree passed the trait to all four daughters, but to none of his nine sons. Because sons inherit the X chromosome from their mother, and only the Y from their father, they could not have inherited the hairiness from their father. However, the mutation in this family is actually a piece of chromosome 4 that inserted into the X chromosome.

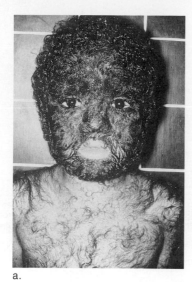

a.

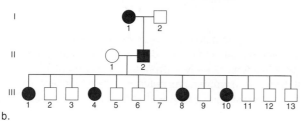

b.

Figure 6.9 An X-linked dominant condition. (a) This 6-year-old has congenital generalized hypertrichosis. **(b)** The affected male in the second generation passed the condition to all of his daughters and none of his sons. (The mutation in this family is actually a piece of chromosome 4 that inserted into the X chromosome.)

Solving a Problem of X-Linked Inheritance

Mendel's first law (segregation) applies to genes on the X chromosome. The same logic is used to solve problems as to trace traits transmitted on autosomes, with the added step of considering the X and Y chromosomes in Punnett squares. Follow these steps:

1. Look at the pattern of inheritance. Different frequencies of affected males and females in each generation may suggest X linkage. For an X-linked recessive trait:
   - An affected male has a carrier mother.
   - An unaffected female with an affected brother has a 50 percent (1 in 2) chance of being a carrier.
   - An affected female has a carrier or affected mother *and* an affected father.
   - A carrier (female) has a carrier mother or an affected father.

   For an X-linked dominant trait:
   - There may be no affected males, because they die early.
   - An affected female has an affected mother.
2. Draw the pedigree.
3. List all genotypes and phenotypes and their probabilities.
4. Assign genotypes and phenotypes to the parents. Consider clues in the phenotypes of relatives.
5. Determine how alleles separate into gametes for the genes of interest on the X and Y chromosomes.
6. Unite the gametes in a Punnett square.
7. Determine the phenotypic and genotypic ratios for the F_1 generation.
8. To predict further generations, use the genotypes of the F_1 and repeat steps 4 through 6.

Consider as an example Kallmann syndrome (OMIM 308700), which causes very poor or absent sense of smell and small testes or ovaries. It is X-linked recessive. Tanisha does not have Kallmann syndrome, but her brother Jamal and her maternal cousin Malcolm (her mother's sister's child) have it. Tanisha's and Malcolm's parents are unaffected, as is Tanisha's husband Sam. Tanisha and Sam wish to know the risk that a son would inherit the condition. Sam has no affected relatives.

Solution

Mode of inheritance: The trait is X-linked recessive because males are affected through carrier mothers.

K = wild type k = Kallmann syndrome

Genotypes	Phenotypes
$X^K X^K$, $X^K X^k$, $X^K Y$	normal
$X^k X^k$, $X^k Y$	affected

Individual	Genotype	Phenotype	Probability
Tanisha	$X^K X^k$ or $X^K X^K$	normal (carrier)	50% each
Jamal	$X^k Y$	affected	100%
Malcolm	$X^k Y$	affected	100%
Sam	$X^K Y$	normal	100%

Tanisha's gametes

if she is a carrier: X^K X^k
Sam's gametes: X^K Y

Punnett square

	X^K	X^k
X^K	$X^K X^K$	$X^K X^k$
Y	$X^K Y$	$X^k Y$

Interpretation: If Tanisha is a carrier, the probability that their son will have Kallmann syndrome is 50 percent, or 1 in 2. (This is a conditional probability. The chance that any son will have the condition is actually 1 in 4, because Tanisha also has a 50 percent chance of being genotype $X^K X^K$ and therefore not a carrier.)

Key Concepts Questions 6.2

1. How do males and females differ in their expression of X-linked traits?
2. Why are X-linked dominant conditions rare in males?
3. How does Mendel's first law apply to solving problems involving X-linked genes?

6.3 Sex-Limited and Sex-Influenced Traits

An X-linked recessive trait generally is more prevalent in males than females. Other situations, however, can affect gene expression in the sexes differently.

Sex-Limited Traits

A **sex-limited trait** affects a structure or function of the body that is present in only males or only females. The gene for such a trait may be X-linked or autosomal.

Understanding sex-limited inheritance is important in animal breeding. For example, a New Zealand cow named Marge, who has a mutation that makes her milk very low in saturated fat, is founding a commercial herd. Males play their part by transmitting the mutation, even though they do not make milk. In humans, beard growth is sex-limited. A woman does not grow a beard because she does not manufacture the hormones required for facial hair growth. She can, however, pass to her sons the genes specifying heavy beard growth.

Curiously, a father-to-be can affect the health of the mother-to-be. Although an inherited disease that causes symptoms associated with pregnancy is obviously sex-limited, the male genome may play a role by contributing to the development of supportive structures, such as the placenta. This is the case for preeclampsia, a sudden rise in blood pressure late in pregnancy. It kills 50,000 women worldwide each year. A study of 1.7 million pregnancies in Norway found that if a man's first wife had preeclampsia, his second wife had double the risk of developing it. Another study found that women whose mothers-in-law developed preeclampsia when pregnant with the women's husbands had twice the rate of developing the condition themselves. Perhaps a gene from the father affects the placenta in a way that elevates the pregnant woman's blood pressure.

Sex-Influenced Traits

In a **sex-influenced trait,** an allele is dominant in one sex but recessive in the other. Such a gene may be X-linked or autosomal. The difference in expression can be caused by hormonal differences between the sexes. For example, an autosomal gene for hair growth pattern has two alleles, one that produces hair all over the head and another that causes pattern baldness. The baldness allele is dominant in males but recessive in females, which is why more men than women are bald. A heterozygous male is bald, but a heterozygous female is not. A bald woman is homozygous recessive. Even a bald woman tends to have some wisps of hair, whereas an affected male may be completely hairless on the top of his head.

Key Concepts Questions 6.3

1. What is a sex-limited trait?
2. What is a sex-influenced trait?

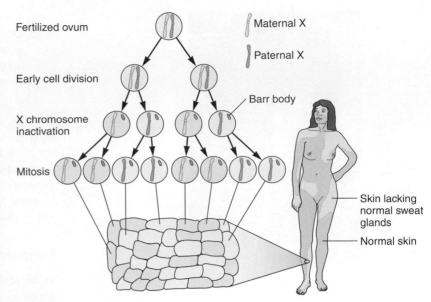

Figure 6.10 X inactivation. A female is a mosaic for expression of genes on the X chromosome because of the random inactivation of either the maternal or paternal X in each cell early in prenatal development. In anhidrotic ectodermal dysplasia, a woman has patches of skin that lack sweat glands and hair. (Colors distinguish cells with the inactivated X, not to depict skin color.)

6.4 X Inactivation

Females have two alleles for every gene on the X chromosomes, whereas males have only one. In mammals, a mechanism called **X inactivation** balances this apparent inequality in the expression of genes on the X chromosome.

Equaling Out the Sexes

By the time a female mammalian embryo consists of 8 cells, about 75 percent of the genes on one X chromosome in each cell are inactivated, and the remaining 25 percent are expressed to different degrees in different females. Which X chromosome is mostly turned off in each cell—the one inherited from the mother or the one from the father—is usually random. As a result, a female mammal expresses the X chromosome genes inherited from her father in some cells and those from her mother in others. She is, therefore, a mosaic for expression of most genes on the X chromosome (**figure 6.10**).

A gene called *XIST* in part of the X chromosome called the X inactivation center controls X inactivation. *XIST* encodes an RNA that binds to a specific site on the same (inactivated) X chromosome. From this point to the chromosome tip, most of the X chromosome is silenced. Figure 13.14 shows how *XIST* is used to turn off the extra chromosome that causes Down syndrome.

Once an X chromosome is inactivated in one cell, all its daughter cells have the same X chromosome shut off. Because the inactivation occurs early in development, the adult female has patches of tissue that differ in their expression of X-linked genes. With each cell in her body having only one active X chromosome, she is roughly equivalent to the male in gene expression.

X inactivation can alter the phenotype (gene expression), but not the genotype. The change is reversed in germline cells destined to become oocytes, and this is why a fertilized ovum does not have an inactivated X chromosome. X inactivation is an example of an **epigenetic** change—one that is passed from one cell generation to the next but that does not alter the DNA base sequence.

We can observe X inactivation at the cellular level because the turned-off chromosome absorbs a stain much faster than the active X. This differential staining occurs because inactivated DNA has chemical methyl (CH_3) groups bound to it that prevents it from being transcribed into RNA and also enables DNA to absorb stain.

X inactivation can be used to check the sex of an individual. The nucleus of a cell of a female, during interphase, has one dark-staining X chromosome called a Barr body. A cell from a male has no Barr body because his one X chromosome remains active.

Effect of X Inactivation on the Phenotype

The consequence of X inactivation on the phenotype can be interesting. For homozygous X-linked genotypes, X inactivation has no effect. Whichever X chromosome is turned off, the same allele is left to be expressed. For heterozygotes, however, X inactivation leads to expression of one allele or the other. This doesn't affect health if enough cells express the functional gene product. However, some traits reveal the X inactivation. The swirls of skin color in incontinentia pigmenti (IP) patients reflect patterns of X inactivation in skin cells (see figure 6.8b). Where the normal allele for melanin pigment is shut off, pale swirls develop. Where pigment is produced, brown swirls result.

A female who is heterozygous for an X-linked recessive gene can express the associated condition if the normal allele is inactivated in the tissues that the illness affects. Consider a carrier of hemophilia A. If the X chromosome carrying the normal allele for the clotting factor is turned off in the liver, then the woman's blood will clot slowly enough to cause mild hemophilia. (Luckily for her, slowed clotting time also greatly reduces her risk of cardiovascular disease caused by blood clots blocking circulation.) A carrier of an X-linked trait who expresses the phenotype is called a **manifesting heterozygote**.

Whether or not a manifesting heterozygote results from X inactivation depends upon how adept cells are at sharing. Consider two lysosomal storage disorders, which are deficiencies of specific enzymes that normally dismantle cellular debris in lysosomes. In carriers of Hunter syndrome (OMIM 309900, also called mucopolysaccharidosis II), cells that make the enzyme readily send it to neighboring cells that do not, essentially correcting the defect in cells that can't make the enzyme. Carriers of Hunter syndrome do not have symptoms because cells get enough enzyme. Boys with Hunter syndrome are deaf, intellectually disabled, have dwarfism and abnormal facial features, heart damage, and enlarged liver and spleen. In contrast, in Fabry disease (OMIM 301500), cells do not readily release the enzyme alpha-galactosidase A, so a female who is a heterozygote may have cells in the affected organs that lack the enzyme. She may develop mild symptoms of this disorder that causes skin lesions, abdominal pain, and kidney failure in boys.

A familiar example of X inactivation is the coat colors of tortoiseshell and calico cats. An X-linked gene confers brownish-black (dominant) or yellow-orange (recessive) color. A female cat heterozygous for this gene has patches of each color, forming a tortoiseshell pattern that reflects different cells expressing either of the two alleles (**figure 6.11**). The earlier the X inactivation, the larger the patches, because more cell divisions can occur after the event, producing more daughter cells. White patches may form due to epistasis by an autosomal gene that shuts off pigment synthesis. A cat with colored patches against such a white background is a calico. Tortoiseshell and calico cats are nearly always female. A male can have these coat patterns only if he inherits an extra X chromosome.

In humans, X inactivation can be used to identify carriers of some X-linked disorders. This is the case for Lesch-Nyhan syndrome (OMIM 300322), in which an affected boy has cerebral palsy; bites his fingers, toes, and lips to the point of mutilation; is intellectually disabled; and passes painful urinary stones. Mutation results in defective or absent HGPRT, an enzyme. A woman who carries Lesch-Nyhan syndrome can be detected when hairs from widely separated parts of her head are tested for HGPRT. (Hair is used for the test because it is accessible and produces the enzyme.) If some hairs contain

Figure 6.11 Visualizing X inactivation. X inactivation is obvious in a calico cat. X inactivation is rarely observable in humans because most cells do not remain together during development, as a cat's skin cells do.

HGPRT but others do not, she is a carrier. The hair cells that lack the enzyme have turned off the X chromosome that carries the normal allele; the hair cells that manufacture the normal enzyme have turned off the X chromosome that carries the disease-causing allele. The woman is healthy because her brain has enough HGPRT, but each son has a 50 percent chance of inheriting the disease.

X inactivation affects the severity of Rett syndrome, the X-linked dominant disorder discussed in the chapter 2 opener. Ninety-nine percent of cases arise anew, from mutations in X-bearing sperm cells. Rarely, Rett syndrome may be inherited from a woman who has a very mild case because, by chance, the X chromosomes bearing the mutation are silenced in her brain cells.

Theoretically, X inactivation evens out the sexes for expression of X-linked genes. In actuality, however, a female may *not* be equivalent, in gene expression, to a male because she has two cell populations, whereas a male has only one. One of a female's two cell populations has the X she inherited from her father active, and the other has the X chromosome she inherited from her mother active. For heterozygous X-linked genes, she would have some cells that manufacture the protein encoded by one allele, and some cells that produce the protein specified by the other allele. Although most heterozygous genes have the alleles about equally represented, sometimes X inactivation can be skewed. That is, most cells express the X inherited from the same parent. This unequal X inactivation pattern can happen if the two X chromosomes have different alleles for a gene that controls cell division rate, giving certain cells a survival advantage.

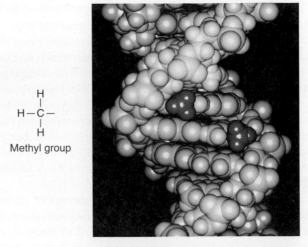

Figure 6.12 Methyl (CH_3) groups (red) "silence" certain genes.

Key Concepts Questions 6.4

1. How does X inactivation compensate for differences between males and females in the numbers of copies of genes on the X chromosome?

2. When does X inactivation begin?

3. Why and how are the effects of X inactivation noticeable in heterozygotes?

6.5 Parent-of-Origin Effects

In Mendel's pea experiments, it didn't matter whether a trait came from the male or female parent. For certain genes in mammals, however, parental origin does influence the phenotype such as age of onset or symptom severity. Two mechanisms of parent-of-origin effects are genomic imprinting and differences between the developmental timetables of sperm and oocytes.

Genomic Imprinting

In **genomic imprinting,** methyl (CH_3) groups cover a gene or several linked genes and prevent them from being accessed to synthesize protein (**figure 6.12**).

For a particular imprinted gene, the copy inherited from either the father or the mother is always covered with methyls, even in different individuals. That is, a particular gene might function if it came from the father, but not if it came from the mother, or vice versa. This parental effect on gene expression is seen as diseases that are always inherited from the mother or father. For example, central precocious puberty (OMIM 176400) is always inherited from the father. A gene called *MKRN3* from the father is covered in methyls, and therefore imprinted. In this condition, girls reach puberty before age 8 and boys before age 9.

Imprinting is an epigenetic alteration, which is a layer of meaning stamped upon a gene without changing its DNA sequence. The imprinting pattern is passed from cell to cell in mitosis, but not from individual to individual through meiosis. When silenced DNA is replicated during mitosis, the pattern of blocked genes is exactly placed, or imprinted, on the new DNA, covering the same genes as in the parental DNA (**figure 6.13**). In this way, the "imprint" of inactivation is perpetuated, as if each such gene "remembers" which parent it came from.

In meiosis, imprints are removed and reset. As oocyte and sperm form, the CH_3 groups shielding their imprinted genes are stripped away, and new patterns are set down, depending upon whether the fertilized ovum chromosomally is male (XY) or female (XX). In this way, women can have sons and men can have daughters without passing on sex-specific parental imprints.

The function of genomic imprinting isn't well understood. However, because many imprinted genes take part in early development, particularly of the brain, it may be a way to finely regulate the abundance of key proteins in the embryo. The fact that some genes lose their imprints after birth supports this idea of early importance. Also, imprinted genes are in clusters along a chromosome, and are controlled by other regions of DNA called imprinting centers. Perhaps one gene in a cluster is essential for early development, and the

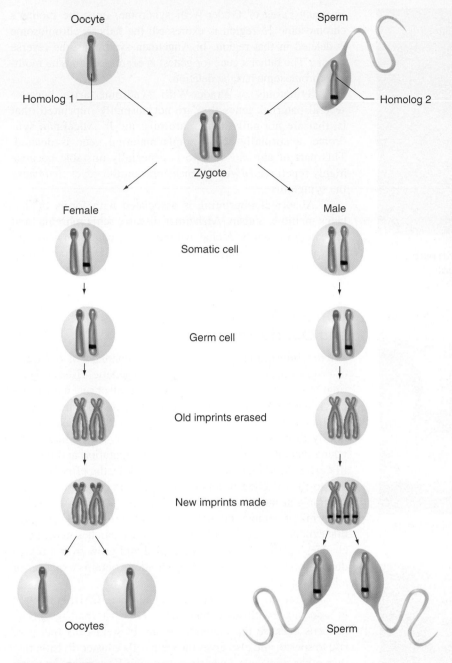

Oocyte

Homolog 1

Sperm

Homolog 2

Zygote

Female

Male

Somatic cell

Germ cell

Old imprints erased

New imprints made

Oocytes

Sperm

Figure 6.13 Genomic imprinting. Imprints are erased during meiosis, then reset according to the sex of the new individual.

others become imprinted simply because they are nearby—a bystander effect.

Genomic imprinting has implications for understanding early human development. It suggests that for mammals, two opposite-sex parents are necessary to produce a healthy embryo and placenta. This apparent requirement for opposite-sex parents was discovered in the early 1980s, through experiments on early mouse embryos and examination of certain rare pregnancy problems in humans.

Researchers created fertilized mouse ova that contained two male pronuclei or two female pronuclei, instead of one from each. Results were strange. When the fertilized ovum had two male genomes, a normal placenta developed, but the embryo was tiny and quickly stopped developing. A zygote with two female pronuclei developed into an embryo, but the placenta was very abnormal. Therefore, the male genome controls placenta development, and the female genome, embryo development.

The mouse results echoed abnormalities of human development. When two sperm fertilize an oocyte and the female pronucleus degenerates, an abnormal growth of placenta-like tissue called a hydatidiform mole forms. If a fertilized ovum contains two female genomes but no male genome, a mass of random differentiated tissue, called a teratoma, grows. A teratoma consists of a strange mix of tissues, such as skin and teeth (**figure 6.14**). When a hydatidiform mole or a teratoma develops, there is no embryo.

Genomic imprinting can explain incomplete penetrance, in which an individual is known to have inherited a genotype associated with a particular phenotype, but has no signs of the trait. This is the case for a person with normal fingers whose parent and child have polydactyly. An imprinted gene silences the dominant mutant allele.

Imprinting may be an important concern in assisted reproductive technologies that manipulate gametes to treat infertility. For example, the otherwise very rare Beckwith-Wiedemann syndrome (OMIM 130650) is more prevalent among the offspring of people who used *in vitro* fertilization and intracytoplasmic sperm injection (discussed in chapter 21) to become pregnant.

Imprinting Disorders in Humans

The number of imprinted genes in the human genome exceeds 156, and at least 60 of them affect health when abnormally expressed. The effects of genomic imprinting are revealed only when an individual has one copy of a normally imprinted allele and the other, active allele is inactivated or deleted.

Imprinting disorders can be dramatic, such as two syndromes that arise from small deletions in the same region of chromosome 15 (**figure 6.15**). A child with Prader-Willi syndrome (OMIM 176270) is small at birth and in infancy has difficulty gaining weight. Between ages 1 and 3, the child develops an obsession with eating and a very slow metabolism. Parents lock kitchen cabinets and refrigerators to keep their children from eating themselves to death by bursting digestive organs. The other condition resulting from the small deletion in chromosome 15, Angelman syndrome (OMIM 105830), causes autism and intellectual disability, an extended tongue, large jaw, poor muscle coordination, and convulsions that make the arms flap.

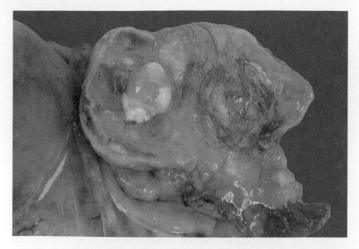

Figure 6.14 **A teratoma is an abnormal growth with an unorganized collection of structures.** Note the tooth, hair, and green fluid from a skin oil gland in this teratoma from an ovary.

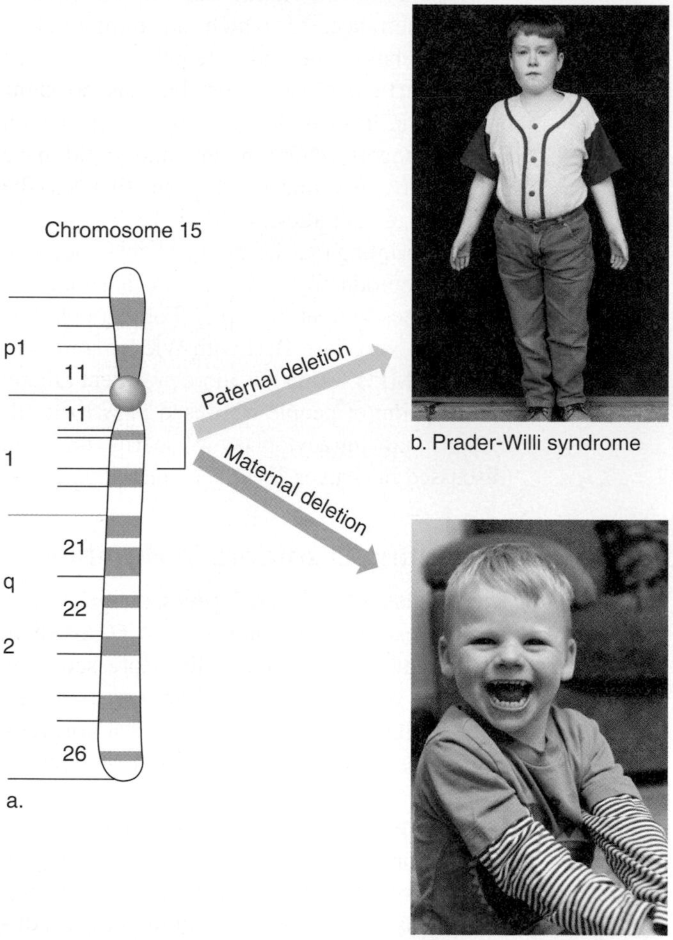

In many cases of Prader-Willi syndrome, only the mother's chromosome 15 region is expressed; the father's chromosome is deleted in that region. In Angelman syndrome, the reverse occurs: The father's gene (or genes) is expressed, and the mother's chromosome has the deletion.

Symptoms of Prader-Willi syndrome arise because several paternal genes that are not normally imprinted (that is, that are normally active) are missing. In Angelman syndrome, a normally active single maternal gene is deleted. This part of chromosome 15 is especially unstable because highly repetitive DNA sequences bracket the genes that cause the symptoms.

Abnormal imprinting is associated with forms of diabetes mellitus, autism, Alzheimer disease, schizophrenia, and male homosexuality. A clue that indicates a condition is associated with genomic imprinting is differing severity, depending upon whether it is inherited from the father or mother.

Different Timetables in Sperm and Oocyte Formation

Altered imprinting is not the only mechanism of a parent-of-origin effect on the phenotype of a genetic disease. Huntington disease (HD) has a younger age of onset, with faster progression and more severe symptoms, if a person inherits the mutation from his or her father. This is the case for the family discussed in the opener to chapter 4, in which three young daughters inherited their father's mutation and became ill during childhood or adolescence. In HD, the effect may be due to the different timetables of sperm and egg production. A female at puberty has about 400,000 eggs, each halted on the brink of completing meiosis, when the replication slippage that expands the gene could happen. But a man has many more chances for the gene to be miscopied and grow over a reproductive lifetime because each ejaculation contains a quarter of a billion sperm!

Increased risk of Noonan syndrome (OMIM 163950) is associated with an older father. This paternal age effect happens because the mutation occurs in stem cells that give rise to sperm, and also gives the stem cells a faster division rate than other stem cells. Over time, mutation-bearing sperm accumulate. Noonan syndrome causes a characteristic face, heart defects, intellectual disability, short stature, and higher risk of blood cancers.

Chromosome 15

p1

11

11

Paternal deletion

Maternal deletion

1

21

q

22

2

26

a.

b. Prader-Willi syndrome

c. Angelman syndrome

Figure 6.15 **Prader-Willi and Angelman syndromes result when the nonimprinted copy of a gene is deleted.** **(a)** Two distinct syndromes result from missing genetic material in the same region of chromosome 15. **(b)** Tyler has Prader-Willi syndrome, due to a deletion in the copy of the chromosome he inherited from his father. Note his small hands. **(c)** This child has Angelman syndrome, caused by a deletion in the chromosome 15 that he inherited from his mother.

Key Concepts Questions 6.5

1. What is genomic imprinting?
2. How does the binding of methyl groups imprint genes?
3. How can abnormal imprinting cause disease?
4. How can differences in the timetables of sperm and oocyte formation cause parent-of-origin effects?

Summary

6.1 Our Sexual Selves

1. Sexual identity includes sex chromosome makeup; gonadal specialization; phenotype (reproductive structures); and gender identity.

2. The human male is **heterogametic,** with an X and a Y chromosome. The female, with two X chromosomes, is **homogametic**.

3. The human Y chromosome has two pseudoautosomal regions and a large, male-specific region that does not recombine. Y-linked genes may correspond to X-linked genes, be similar to them, or be unique. Palindromic DNA sequences or inverted repeats promote gene loss.

4. If the *SRY* gene is expressed, undifferentiated gonads develop as testes. If *SRY* is not expressed, the gonads develop as ovaries, under the direction of other genes.

5. Starting 8 weeks after fertilization, the testes secrete a hormone that prevents development of female structures and secrete testosterone, which triggers development of male structures.

6. Testosterone converted to DHT controls male development. If *SRY* is not turned on, the Müllerian ducts develop into female structures as other genes are expressed.

7. Evidence supports a genetic inherited component to homosexuality.

8. **Sex ratio** is the number of males divided by the number of females multiplied by 1,000, for people of a particular age. Interfering with pregnancy outcomes can skew sex ratios.

6.2 Traits Inherited on Sex Chromosomes

9. **Y-linked** traits are passed from fathers to sons only.

10. Males are **hemizygous** for genes on the X chromosome and express phenotypes associated with these genes because they do not have another allele on a homolog. An **X-linked** trait passes from mother to son because he inherits his X from his mother and his Y from his father.

11. An X-linked allele may be dominant or recessive. X-linked dominant traits are more devastating to males.

6.3 Sex-Limited and Sex-Influenced Traits

12. **Sex-limited traits** may be autosomal or sex-linked, but they only affect one sex because of anatomical or hormonal gender differences.

13. A **sex-influenced trait** is dominant in one sex but recessive in the other.

6.4 X Inactivation

14. **X inactivation** shuts off one X chromosome in each cell in female mammals, making them mosaics for heterozygous X-linked genes. It evens out the dosages of genes on the sex chromosomes between the sexes. X inactivation is **epigenetic** because it does not alter the DNA sequence.

15. A female who expresses the phenotype corresponding to an X-linked gene she carries is a **manifesting heterozygote**.

6.5 Parent-of-Origin Effects

16. In **parent-of-origin effects,** the phenotype corresponding to a particular genotype differs depending upon whether the parent who passes the gene is female or male.

17. **Genomic imprinting** temporarily suppresses expression of some genes, leading to parent-of-origin effects.

18. Imprints are erased during meiosis and reassigned based upon the sex of a new individual.

19. Methyl groups that temporarily suppress gene expression are the physical basis of genomic imprinting.

20. Differences in development of sperm and oocytes explain some parent-of-origin effects.

www.mhhe.com/lewisgenetics11

Answers to all end-of-chapter questions can be found at **www.mhhe.com/lewisgenetics11**. You will also find additional practice quizzes, animations, videos, and vocabulary flashcards to help you master the material in this chapter.

Review Questions

1. How is sex expressed at the chromosomal, gonadal, phenotypic, and gender identity levels?

2. How do genes in the pseudoautosomal region of the Y chromosome differ from genes in the male-specific region (MSY)?

3. Describe the phenotypes of
   a. a person with a deletion of the *SRY* gene.
   b. a normal XX individual.
   c. an XY individual with a block in testosterone synthesis.

4. List the events required for a fetus to develop as a female.

5. Cite evidence that may point to a hereditary component to homosexuality.

6. Why is it unlikely one would see a woman who is homozygous for an X-linked dominant condition?

7. State two reasons why males with X-linked dominant conditions are very rare.

8. What is the basis of sex ratio at birth?

9. Traits that appear more frequently in one sex than the other may be caused by genes that are inherited in an X-linked, sex-limited, or sex-influenced fashion. How might you distinguish among these possibilities in a given individual?

10. Why are male calico cats very rare?

11. How might X inactivation cause patchy hairiness in women who have congenital generalized hypertrichosis, even though the disease-causing allele is dominant?

12. How does X inactivation even out the "doses" of X-linked genes between the sexes?

13. Cite evidence that genetic contributions from both parents are necessary for normal prenatal development.

14. Prader-Willi and Angelman syndromes are more common in children conceived with certain assisted reproductive technologies (*in vitro* fertilization and intracytoplasmic sperm injection) than among the general population. What process may these procedures disrupt?

Applied Questions

1. To answer the following questions, consider these population data on sex ratios:

Selected sex ratios at birth		Selected sex ratios after age 65	
Nation	Sex ratio	Nation	Sex ratio
Costa Rica	970	Rwanda	620
Tanzania	1,000	South Africa	630
Liechtenstein	1,010	France	700
South Africa	1,020	United States	720
United States	1,050	Qatar	990
Sweden	1,060	Montserrat	1,060
Italy	1,070	Bangladesh	1,160
China	1,130	Nigeria	990

   a. In Rwanda, South Africa, France, and the United States, males die, on average, significantly younger than females. What types of information might explain the difference?

   b. In Costa Rica, how many males at birth are there for every 100 females?

   c. In which country listed do males tend to live the longest?

2. In severe Hunter syndrome, lack of the enzyme iduronate sulfate sulfatase leads to buildup of certain carbohydrates swelling the liver, spleen, and heart. In mild cases, deafness may be the only symptom. Intellect is usually normal, and life span can be normal. Hunter syndrome is X-linked recessive. A man with mild Hunter syndrome has a child with a woman who is a carrier.

   a. What is the probability that a son inherits Hunter syndrome?

   b. What is the chance that a daughter inherits Hunter syndrome?

   c. What is the chance that a daughter is a carrier?

3. Amelogenesis imperfecta (OMIM 301200) is X-linked dominant. Affected males have extremely thin enamel on each tooth. Female carriers have grooved teeth from uneven deposition of enamel. Why might the phenotype differ between the sexes?

4. Explain how X inactivation might enable a woman to pass Rett syndrome to a son.

5. A drug used to treat cancer also shows promise in treating Angelman syndrome. The drug removes methyl groups from the paternal copy of a gene called *UBE3A* that is part of the imprinting region of chromosome 15. Explain how the drug might work.

Web Activities

1. Visit the National Center for Biotechnology Information (NCBI) website. Identify an X-linked disorder, then find it in OMIM and describe it.

2. At the Imprinted Gene Catalogue website, click on "search by species name" and then click on "complete list." Find two disorders that involve imprinting, one transmitted from the mother and one from the father, and use OMIM to describe them.

Case Studies and Research Results

1. For each case description, identify the principle at work from the list that follows. More than one answer per case may apply.

   A. Y-linked inheritance

   B. X-linked recessive

   C. X-linked dominant inheritance

   D. Sex-limited inheritance

   E. Sex-influenced inheritance

   F. X inactivation or manifesting heterozygote

   G. Uniparental disomy

   H. Imprinting abnormality

   a. A child with Russell-Silver syndrome (OMIM 180860) has very poor growth, a large head, a characteristic triangular face, and digestive problems. In some cases, a gene on chromosome 11 that is normally methylated in the father is not methylated.

   b. Six-year-old LeQuan inherited Fabry disease (OMIM 301500) from his mother, who is a heterozygote for the causative mutation. The gene, on the X chromosome, encodes a lysosomal enzyme. LeQuan would die before age 50 of heart failure, kidney failure, or a stroke, but fortunately he can be treated with twice-monthly infusions of the enzyme. His mother, Echinecea, recently began experiencing recurrent fevers, a burning pain in her hands and feet, a rash, and sensitivity to cold. She is experiencing mild Fabry disease.

c. The Chandler family has many male members who have a form of retinitis pigmentosa (RP) in which the cells that capture light energy in the retina degenerate, causing gradual visual loss. Several female members of the family presumed to be carriers because they have affected sons are tested for RP genes on chromosomes 1, 3, 6, and the X, but do not carry these RP genes. Many years ago, Rachel married her cousin Ross, who has the family's form of RP. They had six children. The three sons are all affected, but their daughters all have normal vision.

d. Simon's mother and her sister are breast cancer survivors, and their mother died of the disease. Simon's sister Maureen has a genetic test and learns that she, too, has inherited the *BRCA1* gene. Simon has two daughters, but doesn't want to be tested because he thinks a man cannot transmit a trait that affects a body part that is more developed in females.

e. Tribbles are extraterrestrial mammals that long ago invaded a starship on the television program *Star Trek*. A gene called *frizzled* causes kinky hair in female tribbles who inherit just one allele. However, two mutant alleles must be inherited for a male tribble to have kinky hair.

f. Perry died at age 16 of Lowe syndrome (OMIM 309000). He was slightly intellectually disabled, had visual problems (cataracts and glaucoma), seizures, poor muscle tone, and progressive kidney failure, which was ultimately fatal. His sister Lily is pregnant, and wonders whether she is a carrier of the disease that killed her brother. She remembers a doctor saying that her mother Yasmine was a carrier. Lily's physician determines that she is a carrier because she has cataracts, which is a clouding of the lenses. It has not yet affected her vision. When a prenatal test reveals that Lily's fetus is a female, her doctor tells her not to worry about Lowe syndrome.

g. Mating among Texas field crickets depends upon females responding to a male mating call. The sounds must arrive at a particular frequency to excite the females, who do not sing back in response. However, females can pass on a trait that confers frequency of singing.

h. When Winthrop was a baby, he was diagnosed with "failure to thrive." At 14 months of age, he suddenly took an interest in food, and his parents couldn't feed him fast enough. By age 4, Winthrop was obese, with disturbing behavior. He was so hungry that after he'd eaten his meal and everyone else's leftovers, he'd hunt through the garbage for more. Finally a psychiatrist who had a background in genetics diagnosed Prader-Willi syndrome. Testing showed that the allele for the Prader-Willi gene that Winthrop had inherited from his father was abnormally methylated.

i. Certain breeds of dogs have cryptorchidism, in which the testicles do not descend into the scrotum. The trait is passed through females.

2. Reginald has mild hemophilia A that he can control by taking a clotting factor. He marries Lydia, whom he met at the hospital where he and Lydia's brother, Marvin, receive their treatment. Lydia and Marvin's mother and father, Emma and Clyde, do not have hemophilia. What is the probability that Reginald and Lydia's son will inherit hemophilia A?

3. Harold works in a fish market, but the odor does not bother him because he has anosmia (OMIM 301700), an X-linked recessive lack of sense of smell. Harold's wife, Shirley, has a normal sense of smell. Harold's sister, Maude, also has a normal sense of smell, as does her husband, Phil, and daughter, Marsha, but their identical twin boys, Alvin and Simon, cannot detect odors. Harold and Maude's parents, Edgar and Florence, can smell normally. Draw a pedigree for this family, indicating people who must be carriers of the anosmia gene.

4. Caster Semenya is a South African sprinter who won in the 800-meter race at the World Championships in 2009. Because her time had dropped significantly from earlier races, the International Association of Athletics Federations asked her to take a gender test. The implication of the much-publicized request was that Semenya was really a male. For reasons of privacy, the results of her test were never released. Which gene would have been tested to determine Semenya's gender?

5. Maureen Seaberg is an author who has tetrachromia, among other sensory skills. Cosmetics companies ask her to consult on new products. Why is she of value to them?

Should parents submit their children's DNA for genetic testing to determine which sports they should play?

Multifactorial Traits

Learning Outcomes

7.1 Genes and the Environment Mold Traits

1. Distinguish between single-gene and polygenic traits.

2. Define multifactorial traits.

7.2 Polygenic Traits Are Continuously Varying

3. Explain how continuously varying traits reflect genes and the environment.

7.3 Methods to Investigate Multifactorial Traits

4. Explain how empiric risk differs from calculating a Mendelian frequency.

5. Define *heritability*.

6. Discuss what studies on adopted individuals and twins can reveal.

7. Explain what a genome-wide association study can reveal.

7.4 A Closer Look: Body Weight

8. Discuss tools and approaches used to study body weight.

The **BIG** Picture

Who we are and how we feel arises from an intricate interplay among our genes and environmental influences. Understanding genetic contributions to traits and illnesses can suggest how we can alter our environments to improve our lives.

The Complex Genetics of Athletics

A website offers a single-gene test that "gives parents and coaches early information on success in team or individual speed/power or endurance sports." The test is for variants of a gene that encodes a protein called actinin 3 that is expressed in skeletal muscle. One genotype is overrepresented among elite sprinters, another among world-class endurance runners. The test takes a simplistic view of a complex trait.

Athletic ability is multifactorial—many genes as well as environmental influences determine at which activities or sports a person may excel. Any single gene is unlikely to have a great influence. Environmental factors include exposure to pollution and toxins, as well as opportunities to participate in sports. The ability to work well on a team cannot be reduced to a simple string of DNA letters.

The idea to market athletic genes may have come from rare mutations in other genes that bestow great physical prowess. Members of a German family with a mutation in a "double muscle" gene are amazing weight lifters, while a Scandinavian family of Olympic skiers has a mutation that increases the number of red blood cells. Genes also influence metabolic rate, bone mineral density, fat storage, glucose use, and lung function.

Genetic testing to predict athletic success is genetic determinism—the idea that our genes solely determine who we are. Using such test results to choose a child's sport can stress a child with no interest in competing, or discourage a child who loves a particular sport.

7.1 Genes and the Environment Mold Traits

A woman who is a prolific writer has a daughter who becomes a successful novelist. An overweight man and woman have obese children. A man whose father suffers from alcoholism has the same problem. Are these characteristics—writing talent, obesity, and alcoholism—inherited or learned? These traits, and nearly all others, are not the result of an "either/or" mechanism, but reflect the input of many genes as well as environmental influences. Even single-gene disorders are modified by environmental factors and/or other genes. A child with cystic fibrosis, for example, has inherited a single-gene disorder, but her experiences reflect which variants of the gene she has, other genes that affect her immune system, the pathogens to which she is exposed, and the quality of the air she breathes (see Clinical Connection 4.2). This chapter considers characteristics that represent input from many genes, and the tools used to study them.

A trait can be described as either single-gene (Mendelian or monogenic) or **polygenic**. As its name implies, a polygenic trait reflects the activities of more than one gene. Both single-gene and polygenic traits can also be **multifactorial,** which means they are influenced by the environment. Lung cancer is a multifactorial trait (**figure 7.1**). Purely polygenic traits—those not influenced by the environment at all—are very rare. Eye color, discussed in chapter 4, is close to being purely polygenic.

Polygenic multifactorial traits include common ones, such as height, skin color, body weight, many illnesses, and behavioral conditions and tendencies. Behavioral traits are not inherently different from other types of traits; they involve the functioning of the brain, rather than another organ. Chapter 8 discusses them. A more popular term for "multifactorial" is complex, but multifactorial is more precise and is not confused with the general definition of "complex." The genes of a multifactorial trait are not more complicated than others. They follow Mendel's laws, but contribute only partly to a trait and are therefore more difficult to track.

Lung cancer caused by smoking illustrates the complexity of multifactorial traits. Variants of genes that increase the risk of becoming addicted to nicotine and of developing cancer come into play—but may not ever be expressed if a person never smokes and breathes only fresh air.

A polygenic multifactorial condition reflects additive contributions of several genes. Each gene confers a degree of susceptibility, but the input of these genes is not necessarily equal. While a rare allele may exert a large influence, several common alleles may each contribute only slightly to a trait. For example, three genes contribute significantly to the risk of developing type 2 diabetes mellitus, but other genes exert smaller effects.

Different genes may contribute different parts of a phenotype that was once thought to be due to the actions of a single gene. This is the case for migraine. A gene on chromosome 1 contributes sensitivity to sound; a gene on chromosome 5 produces the pulsating headache and sensitivity to light; and a gene on chromosome 8 is associated with nausea and vomiting. In addition, environmental influences trigger migraine in some people, such as eating certain foods. **Clinical Connection 7.1** takes a closer look at heart health, which reflects several multifactorial traits.

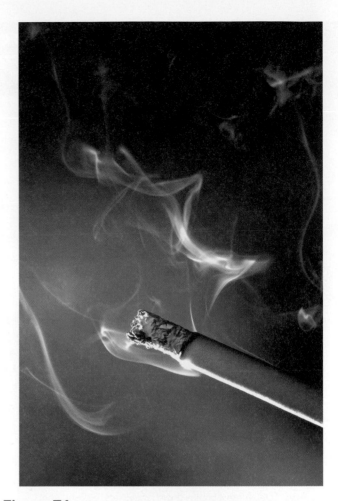

Figure 7.1 **Genetic and environmental factors contribute to lung cancer risk.** Mutations raise lung cancer risk in several ways: impairing DNA repair, promoting inflammation, blocking detoxification of carcinogens, keeping telomeres long, and promoting addiction. These genetic risk factors interact with each other and with environmental influences, such as smoking and breathing polluted air.

Key Concepts Questions 7.1

1. How do polygenic traits differ from Mendelian traits?

2. What are multifactorial traits?

3. Are the contributions of individual genes to a multifactorial, polygenic trait necessarily equal?

Many Genes Control Heart Health

Many types of cells and processes must interact for the heart and blood vessels (the cardiovascular system) to circulate blood, and many genes maintain the system. Effects of the environment are great, too, even on single-gene cardiovascular diseases. For example, intake of vitamin K, necessary for blood to clot, influences the severity of single-gene clotting disorders. Cardiovascular disease affects one in three individuals.

Genes control the heart and blood vessels in several ways: transporting lipids; blood clotting; blood pressure; and how well white blood cells stick to blood vessel walls. Lipids can only move in the circulation when bound to proteins to form large molecules called lipoproteins. Several genes encode the protein parts of lipoproteins, which are called apolipoproteins. Some types of lipoproteins carry lipids in the blood to tissues, where they are used, and other types of lipoproteins take lipids to the liver, where they are broken down into biochemicals that the body can excrete more easily. One allele of a gene that encodes apolipoprotein E, called *E4*, increases the risk of a heart attack threefold in people who smoke.

Maintaining a healthy heart and blood vessels requires a balance between enough lipids inside cells but not an excess outside cells. Several dozen genes control lipid levels in the blood and tissues by specifying enzymes that process lipids, proteins that transport them, or receptor proteins that admit lipids into cells.

An enzyme, lipoprotein lipase, lines the walls of the smallest blood vessels, where it breaks down fat packets released from the small intestine and liver. Lipoprotein lipase is activated by high-density lipoproteins (HDLs), and it breaks down low-density lipoproteins (LDLs). Because low LDL levels are associated with a healthy cardiovascular system in many families, low LDL is used as a biomarker of heart health. High HDL was once widely used as a biomarker for health, until cardiologists recognized what geneticists had long known—that in some families with no heart problems, HDL is low, usually a danger sign. People with elevated LDL that diet and exercise cannot control may take statin drugs, which block the enzyme the liver requires to synthesize cholesterol.

The fluidity of the blood is also critical to health. Overly active clotting factors or extra sticky white blood cells can cause clots to form that block blood flow, usually in blood vessels in the heart or in the legs. Poor clotting causes dangerous bleeding. Because clotting factors are proteins, clotting is genetically controlled.

Familial cholesterolemia (see figure 5.2) illustrates the multifactorial nature of cardiovascular disease. People who inherit one mutant allele of the LDL receptor gene typically suffer heart attacks in early adulthood, and the very rare individuals who inherit two mutations die much earlier. At least twelve other places in the genome include genes that also influence cholesterol deposition, as does diet. Many people with risky genotypes can forestall symptoms through diet and exercise.

Genetic test panels detect alleles of dozens of genes that each contributes risk to developing cardiovascular disease. More than 50 genes regulate blood pressure, and more than 95 contribute to inherited variation in blood cholesterol and triglyceride levels. Tests of gene expression can indicate which cholesterol-lowering drugs are most likely to be effective and tolerable for a particular individual. Computer analysis of multigene tests accounts for controllable environmental factors, such as exercising, not smoking, and maintaining a healthy weight (**table 1**). **Figure 1** shows an artery blocked by fatty plaque.

Table 1	Risk Factors for Cardiovascular Disease	
Uncontrollable	**Controllable**	
Age	Fatty diet	
Male sex	Hypertension	
Genes	Smoking	
Lipid metabolism	High serum cholesterol	
Apolipoproteins	Stress	
Lipoprotein lipase	Insufficient exercise	
Blood clotting	Obesity	
Fibrinogen	Diabetes	
Clotting factors		
Inflammation		
C-reactive protein		
Homocysteine metabolism		
Leukocyte adhesion		

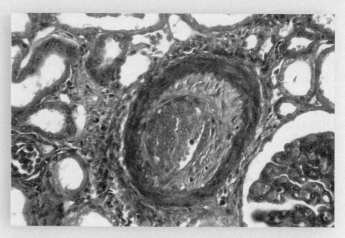

Figure 1 **Cardiovascular disease.** Genetic and dietary factors contribute to clogged arteries.

(Continued)

7.2 Polygenic Traits Are Continuously Varying

For a polygenic trait, the combined action of many genes often produces a "shades of grey" or "continuously varying" phenotype, also called a quantitative trait. DNA sequences that contribute to polygenic traits are called **quantitative trait loci,** or **QTLs.** A multifactorial trait is continuously varying if it is also polygenic. That is, it is the multi-gene component of the trait that contributes the continuing variation of the phenotype. The individual genes that confer a polygenic trait follow Mendel's laws, but together they do not produce single-gene phenotypic ratios. They all contribute to the phenotype, but without being dominant or recessive to each other. Single-gene traits are instead discrete or qualitative, often providing an "all-or-none" phenotype such as "normal" versus "affected."

A polygenic trait varies in populations, as our many nuances of hair color, body weight, and cholesterol levels demonstrate. Some genes contribute more to a polygenic trait than others. Within genes, alleles can have differing impacts depending upon exactly how they alter an encoded protein and how common they are in a population. For example, a mutation in the LDL receptor gene greatly raises blood serum cholesterol level. But because fewer than 1 percent of the individuals in most populations have this mutation, it contributes very little to the variation in cholesterol level at the population level. However, the mutation has a large impact on the person who has it.

Although the expression of a polygenic trait is continuous, we can categorize individuals into classes and calculate the frequencies of the classes. When we do this and plot the frequency for each phenotype class, a bell-shaped curve results. Even when different numbers of genes affect the trait, the curve takes the same shape, as the following examples show.

Fingerprint Patterns

The skin on the fingertips is folded into patterns of raised skin called dermal ridges that align to form loops, whorls, and arches. This pattern is a fingerprint. A technique called dermatoglyphics ("skin writing") compares the number of ridges that comprise these patterns to identify and distinguish individuals (**figure 7.2**). Dermatoglyphics is part of genetics because certain disorders (such as Down syndrome) include unusual ridge patterns. Forensic fingerprint analysis is also an application of dermatoglyphics.

Genes largely determine the number of ridges in a fingerprint, but the environment can affect them too. During weeks 6 through 13 of prenatal development, the ridge pattern can be

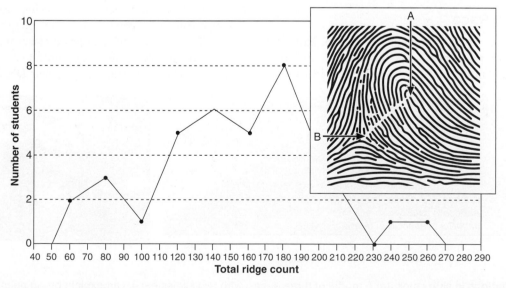

Figure 7.2 Anatomy of a fingerprint. Total ridge counts for a number of individuals, plotted on a bar graph, form an approximate bell-shaped curve. The number of ridges between landmark points A and B on this loop pattern is 12. Total ridge count includes the number of ridges on all fingers.

Figure 7.3 **The inheritance of height.** Eighty-three genetics students at the University of Notre Dame lined up by height in inches, revealing the continuously varying nature of height.

altered as the fetus touches the finger and toe pads to the wall of the amniotic sac. This early environmental effect explains why the fingerprints of identical twins, who share all genes, are in some cases not exactly alike.

We quantify a fingerprint with a measurement called a total ridge count, which tallies the numbers of ridges in whorls, loops, or arches. The average total ridge count in a male is 145, and in a female, 126. Plotting total ridge count reveals the bell curve of a continuously varying trait.

Height

The effect of the environment on height is obvious—people who do not eat enough do not reach their genetic potential for height. Students lined up according to height, but raised in two different decades and under different circumstances, illustrate the effects of genes and the environment on this continuously varying trait. Students from 1920 are on average considerably shorter than students from recent years. The tallest individuals from 1920 were 5′9″, whereas the tallest now are 6′5″. The difference is attributed to improved diet and better overall health and perhaps the fact that many men were killed in the first world war. At least 50 genes affect height. **Figure 7.3** shows the bell curve for height.

Skin Color and Race

More than 100 genes affect pigmentation in skin, hair, and the irises. Melanin pigments color the skin to different degrees in different individuals. In the skin, as in the iris

(see figure 4.6), melanocytes are cells that contain melanin in packets called melanosomes. Melanocytes extend between the tile-like skin cells, distributing pigment granules through the skin layers. Some melanin exits the melanocytes and enters the hardened cells in the skin's upper layers. Here the melanin breaks into pieces, and as the skin cells are pushed up toward the skin's surface as stem cells beneath them divide, the bits of melanin provide color. The pigment protects against DNA damage from ultraviolet radiation. Exposure to the sun increases melanin synthesis. **Figure 7.4a** shows a three-gene model for human skin color. This is an oversimplification, but it illustrates how several genes can contribute to a variable trait.

We all have about the same number of melanocytes per unit area of skin. People have different skin colors because they vary in melanosome number, size, and density of pigment distribution. Different skin colors arise from the number and distribution of melanin pieces in the cells in the uppermost skin layers.

Skin color is one physical trait that is used to distinguish race. The definition of race based largely on skin color is more a social construct than a biological concept, because skin color is only one of thousands of traits whose frequencies vary in different populations. From a genetic perspective, when referring to non-humans, races are groups within species that are distinguished by different allele frequencies. Humans are actually a lot less variable in appearance than other mammals. We may classify people by skin color because it is an obvious visible way to distinguish individuals, but this trait is *not* a reliable indicator of ancestry.

The concept of race based on skin color falls apart when considering many genes. That is, two people with very dark skin may be less alike than either is to another person with very light skin. For example, sub-Saharan Africans and Australian aborigines have dark skin, but are very dissimilar in other inherited characteristics. Their dark skins may reflect adaptation to life in a sunny, tropical climate rather than recent shared ancestry. Overall, 93 percent of varying inherited traits are no more common in people of one skin color than any other.

Testing DNA indicates that it makes more biological sense to classify people by ancestry rather than by skin color. In a sociology class, 100 students had their DNA tested for percent

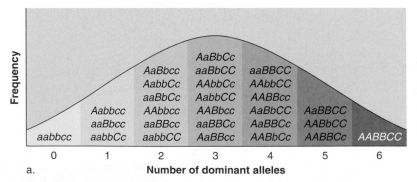

Figure 7.4 **Variations in skin color.** **(a)** A model of three genes, with two alleles each, can explain broad hues of human skin. In actuality, this trait likely involves many more than three genes. **(b)** Humans come in a great variety of skin colors. Skin color genes can assort in interesting ways. These beautiful young ladies, Marcia and Millie, are twins! Their father is Jamaican with dark skin and tight dark curls and their mother is European with fair skin and golden-brown hair.

contribution from "European white," "black African," "Asian," and "Native American" gene variants. Many students were surprised. A light-skinned black man learned that genetically he is approximately half black, half white. Another student who considered herself black was 58 percent white European. The U.S. census, in recognition of the complexity of classifying people into races based on skin color, began to allow "mixed race" as a category in 2000.

In a genetic sense the concept of race based on skin color has little meaning, but in a practical sense, racial groups *do* have different incidences of certain diseases, because of the tendency to choose partners within a group, which retains certain alleles. However, racial differences in disease prevalence may also result from social inequities, such as access to good nutrition or health care. Observations that populations of particular races have a higher incidence of certain illnesses have led to "race-based prescribing." For example, certain hypertension and heart disease drugs are marketed to African Americans, because this group has a higher incidence of these conditions than do people in other groups. But on the individual level a white person might be denied a drug that would work, or a black person given one that doesn't, if the treatment decision is based on a trait not directly related to how the body responds to a drug.

It is more accurate to prescribe drugs based on personal genotypes that determine whether or not a particular drug will work or have side effects, than by the color of a person's skin. For example, researchers cataloged 23 markers for genes that control drug metabolism in 354 people representing blacks (Bantu, Ethiopian, and Afro-Caribbean), whites (Norwegian, Armenian, and Ashkenazi Jews), and Asians (Chinese and New Guinean). The genetic markers fell into four very distinct groups that predict which of several blood thinners, chemotherapies, and painkillers will be effective— and these response groups did not at all match the groups based on skin color.

7.3 Methods to Investigate Multifactorial Traits

Predicting recurrence risks for polygenic traits is much more challenging than doing so for single-gene traits. Researchers use several strategies to investigate these traits.

Empiric Risk

Using Mendel's laws, it is possible to predict the risk that a single-gene trait will recur in a family from knowing the mode of inheritance—such as autosomal dominant or recessive. To predict the chance that a polygenic multifactorial trait will occur in a particular individual, geneticists use **empiric risk,** which is based on incidence in a specific population. **Incidence** is the rate at which a certain event occurs, such as the number of new cases of a disorder diagnosed per year in a population of known size. **Prevalence** is the proportion or number of individuals in a population who have a particular disorder at a specific time, such as during one year.

Empiric risk is not a calculation, but a population statistic based on observation. The population might be broad, such as an ethnic group or community, or genetically more well defined, such as families that have cystic fibrosis. Empiric risk increases with the severity of the disorder, the number of affected family members, and how closely related a person is to affected individuals. For example, empiric risk is used to predict the likelihood of a child being born with a neural tube defect (NTD). In the United States, the overall population risk of carrying a fetus with an NTD is about 1 in 1,000 (0.1 percent). For people of English, Irish, or Scottish ancestry, the risk is about 3 in 1,000. However, if a sibling has an NTD, for any ethnic group, the risk of recurrence increases to 3 percent, and if two siblings are affected, the risk to a third child is even greater.

If a trait has an inherited component, then it makes sense that the closer the relationship between two individuals, one of whom has the trait, the greater the probability that the second individual has the trait, too, because they share more genes. Studies of empiric risk support this logic. **Table 7.1** summarizes empiric risks for relatives of individuals with cleft lip (**figure 7.5**).

Because empiric risk is based solely on observation, it is useful to derive risks for disorders with poorly understood transmission patterns. For example, certain multifactorial disorders affect one sex more often than the other. Pyloric stenosis, an overgrowth of muscle at the juncture between the stomach and the small intestine, is five times more common

Table 7.1	Empiric Risk of Recurrence for Cleft Lip
Relationship to Affected Person	**Empiric Risk of Recurrence**
Identical twin	40.0%
Sibling	4.1%
Child	3.5%
Niece/nephew	0.8%
First cousin	0.3%
General population risk (no affected relatives)	0.1%

Figure 7.5 **Cleft lip.** Cleft lip is more likely in a person who has a relative with the condition. This child has had corrective surgery.

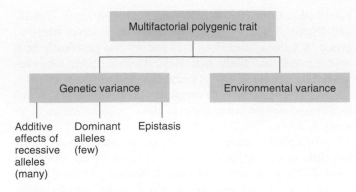

Figure 7.6 **Heritability estimates the genetic contribution to the variability of a trait.** Observed variance in a polygenic, multifactorial trait or illness reflects genetic and environmental contributions.

Table 7.2	Heritabilities for Some Human Traits
Trait	**Heritability**
Clubfoot	0.8
Height	0.8
Blood pressure	0.6
Body mass index	0.5
Verbal aptitude	0.7
Mathematical aptitude	0.3
Spelling aptitude	0.5
Total fingerprint ridge count	0.9
Intelligence	0.5–0.8
Total serum cholesterol	0.6

among males than females. The condition must be corrected surgically shortly after birth, or the newborn will be unable to digest foods. Empiric data show that the risk of recurrence for the brother of an affected brother is 3.8 percent, but the risk for the brother of an affected sister is 9.2 percent. An empiric risk, then, is based on real-world observations. The cause of the illness need not be known.

Heritability

Charles Darwin noted that some of the variation of a trait is due to inborn differences in populations, and some to differences in environmental influences. A measurement called **heritability,** designated H, estimates the proportion of the phenotypic variation for a trait that is due to genetic differences in a certain population at a certain time. The distinction between empiric risk and heritability is that empiric risk could result from non-genetic influences, whereas heritability focuses on the genetic component of the variation in a trait. Heritability refers to the degree of *variation* in a trait due to genetics, and not to the proportion of the trait itself attributed to genes.

Figure 7.6 outlines the factors that contribute to observed variation in a trait. Heritability equals 1.0 for a trait whose variability is completely the result of gene action, such as in a population of laboratory mice who share the same environment. Without environmental variability, genetic differences alone determine expression of the trait in the population. Variability of most traits, however, is due to differences among genes and environmental components. **Table 7.2** lists some traits and their heritabilities.

Heritability changes as the environment changes. For example, the heritability of skin color is higher in the winter months, when sun exposure is less likely to increase melanin synthesis. The same trait may be highly heritable in two populations, but certain variants much more common in one group due to long-term environmental differences. Populations in equatorial Africa, for example, have darker skin than sun-deprived Scandinavians.

Researchers use several statistical methods to estimate heritability. One way is to compare the actual proportion of pairs of people related in a certain manner who share a particular trait, to the expected proportion of pairs that would share it if it were inherited in a Mendelian fashion. The expected proportion is derived by knowing the blood relationships of the individuals and using a measurement called the **coefficient of relatedness,** which is the proportion of genes that two people related in a certain way share (**table 7.3**).

A parent and child share 50 percent of their genes, because of the mechanism of meiosis. Siblings share on average 50 percent of their genes, because they have a 50 percent chance of inheriting each allele for a gene from each parent. Genetic counselors use the designations of primary (1°), secondary (2°), and tertiary (3°) relatives when calculating risks (table 7.3 and **figure 7.7**). For extended or complicated pedigrees, the value of 1 in 2 or 50 percent between siblings and between parent-child pairs can be used to trace and calculate

Table 7.3	Coefficient of Relatedness for Pairs of Relatives		
Relationship	**Degree of Relationship**	**Percent Shared Genes (Coefficient of Relatedness)**	
Sibling to sibling	1°	50% (1/2)	
Parent to child	1°	50% (1/2)	
Uncle/aunt to niece/nephew	2°	25% (1/4)	
Grandparent to grandchild	2°	25% (1/4)	
First cousin to first cousin	3°	12 1/2% (1/8)	

the percentage of genes shared between people related in other ways.

If the heritability of a trait is very high, then of a group of 100 sibling pairs, nearly 50 would be expected to have the same phenotype, because siblings share on average 50 percent of their genes. Height is a trait for which heritability reflects the environmental influence of nutrition. Of 100 sibling pairs in a population, for example, 40 might be the same number of inches tall. Heritability for height among this group of sibling pairs is 0.40/0.50, or 80 percent, which is the observed phenotypic variation divided by the expected phenotypic variation if environment had no influence.

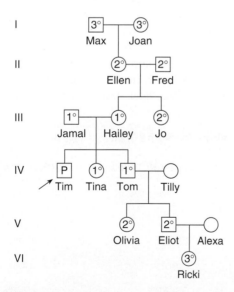

Figure 7.7 Tracing relatives. Tim has an inherited illness. A genetic counselor drew this pedigree to explain the approximate percentage of genes Tim shares with relatives. This information can be used to alert certain relatives to their risks.

("P" is the proband, or affected individual who initiated the study. See table 7.3 for definitions of 1°, 2°, and 3° relationships.)

Genetic variance for a polygenic trait is mostly due to the additive effects of recessive alleles of different genes. For some traits, a few dominant alleles can greatly influence the phenotype, but because they are rare, they do not contribute greatly to heritability. This is the case for heart disease caused by a faulty LDL receptor. Heritabilities for some traits or diseases may be underestimates if only mutations that change DNA base sequences are compared, and not also copy number variants (CNVs, which are differences in the numbers of copies of a DNA sequence).

Epistasis (interaction between alleles of different genes) can also influence heritability. To account for the fact that different genes affect a phenotype to differing degrees, geneticists calculate a "narrow" heritability that considers only additive recessive effects, and a "broad" heritability that also considers the effects of rare dominant alleles and epistasis. For LDL cholesterol level, for example, the narrow heritability is 0.36, but the broad heritability is 0.96, reflecting the fact that a rare dominant allele has a large impact.

Understanding multifactorial inheritance is important in agriculture. A breeder needs to know whether genetic or environmental influences contribute to variability in such traits as birth weight, milk yield, and egg hatchability. It is also valuable to know whether the genetic influences are additive or epistatic. The breeder can control the environment by adjusting the conditions under which animals are raised and crops grown, and control genetic effects by setting up crosses between particular individuals.

Studying multifactorial traits in humans is difficult, because information must be obtained from many families. Two special types of people, however, can help geneticists to tease apart the genetic and environmental components of the variability of multifactorial traits—adopted individuals and twins.

Adopted Individuals

An adopted person typically shares environmental influences, but not many gene variants, with the adoptive family. Conversely, adopted individuals share genes, but not the exact environment, with their biological parents. Therefore, biologists assume that similarities between adopted people and adoptive parents reflect mostly environmental influences, whereas similarities between adoptees and their biological parents reflect mostly genetic influences. Information on both sets of parents can reveal how heredity and the environment contribute to a trait.

Many early adoption studies used a database of all adopted children in Denmark and their families from 1924 to 1947. One study examined correlations between causes of death among biological and adoptive parents and adopted children. If a biological parent died of infection before age 50, the child he or she gave up for adoption was five times more likely to die of infection at a young age than a similar person in the general population. This may be because inherited variants in immune system genes increase susceptibility to certain infections. In support of this hypothesis, the risk that an adopted individual would die young from infection did not correlate with adoptive parents' death from infection before age 50. Researchers concluded that genetics mostly determines length of life, but they did find evidence of environmental influences. For example, if adoptive parents died before age 50 of cardiovascular disease, their adopted children were three times as likely to die of heart and blood vessel disease as a person in the general population. What environmental factor might explain this correlation?

Twins

Studies that use twins to separate the genetic from the environmental contribution to a phenotype provide more meaningful information than studying adopted individuals. Using twins to study genetic influence on traits dates to 1924, when German dermatologist Hermann Siemens reported that grades and teachers' comments were much more alike for identical twins than for fraternal twins. He proposed that genes contribute to intelligence based on this observation.

A trait that occurs more frequently in both members of identical (monozygotic or MZ) twin pairs than in both members of fraternal (dizygotic or DZ) twin pairs is at least partly controlled by heredity. Geneticists calculate the **concordance** of a trait as the percentage of pairs in which both twins express the trait among pairs of twins in whom at least one has the trait. Twins who differ in a trait are said to be discordant for it.

In one study, 142 MZ twin pairs and 142 DZ twin pairs took a "distorted tunes test," in which 26 familiar songs were played, each with at least one note altered. A person was considered "tune deaf" if he or she failed to detect the mistakes in three or more tunes. Concordance for "tune deafness" was 67 percent for MZ twins, but only 44 percent for DZ twins, indicating a considerable inherited component in the ability to accurately perceive musical pitch. **Table 7.4** compares twin types for a variety of hard-to-measure traits. (Figure 3.16 shows how DZ and MZ twins arise.)

Diseases caused by single genes that approach 100 percent penetrance, whether dominant or recessive, also approach 100 percent concordance in MZ twins. That is, if one identical twin has the disease, so does the other. However, among DZ twins, concordance generally is 50 percent for a dominant trait and 25 percent for a recessive trait. These are the Mendelian values that apply to any two non-twin siblings. For a polygenic trait with little environmental input, concordance values for MZ twins are significantly greater than for DZ twins. A trait molded mostly by the environment exhibits similar concordance values for both types of twins.

Comparing twin types assumes that both types of twins share similar experiences. In fact, MZ twins are often closer emotionally than DZ twins. This discrepancy between the closeness of the two types of twins can lead to misleading results. A study from the 1940s, for example, concluded that tuberculosis is inherited because concordance among MZ twins was higher than among DZ twins. Actually,

the infectious disease more readily passed between MZ twins because their parents kept them closer. However, we do inherit susceptibilities to some infectious diseases. MZ twins would share such genes, whereas DZ twins would only be as likely as any sibling pairs to do so.

A more informative way to assess the genetic component of a multifactorial trait is to study MZ twins who were separated at birth, then raised in very different environments. Much of the work using this "twins reared apart" approach has taken place at the University of Minnesota. Here, since 1987, thousands of sets of twins and triplets who were separated at birth have visited the laboratories of Thomas Bouchard. For a week or more, the twins and triplets are tested for physical and behavioral traits, including 24 blood types, handedness, direction of hair growth, fingerprint pattern, height, weight, functioning of all organ systems, intelligence, allergies, and dental patterns. The participants provide DNA samples. Researchers videotape facial expressions and body movements in different circumstances and probe participants' fears, interests, and superstitions.

Twins and triplets separated at birth provide natural experiments for distinguishing nature from nurture. Many of their common traits can be attributed to genetics, especially if their environments have been very different. Their differences tend to come from differences in upbringing, because their genes are identical (MZ twins and triplets) or similar (DZ twins and triplets).

Some MZ twins separated at birth and reunited later are remarkably similar, even when they grow up in very different adoptive families (**figure 7.8**). Idiosyncrasies are particularly striking. One pair of twins who met for the first time when they were in their thirties responded identically to questions; each paused for 30 seconds, rotated a gold necklace she was wearing three times, and then answered the question. Coincidence, or genetics?

The "twins reared apart" approach is not an ideal way to separate nature from nurture. MZ twins and other multiples share an environment in the uterus and possibly in early infancy that may affect later development. Siblings, whether

Table 7.4	Concordance Values for Some Traits in Twins	
Trait	MZ (Identical) Twins	DZ (Fraternal) Twins
Acne	14%	14%
Alzheimer disease	78%	39%
Anorexia nervosa	55%	7%
Autism	90%	4.5%
Bipolar disorder	33–80%	0–8%
Cleft lip with or without cleft palate	40%	3–6%
Hypertension	62%	48%
Schizophrenia	40–50%	10%

Figure 7.8 **Identical twins have much in common.** In addition to physical traits, MZ twins may share tastes, preferences, and behaviors. Studying MZ twins separated at birth and reunited is a way to assess which traits are inherited.

adoptive or biological, do not always share identical home environments. Differences in sex, general health, school and peer experiences, temperament, and personality affect each individual's perception of such environmental influences as parental affection and discipline.

Genome-Wide Association Studies

A newer tool to analyze multifactorial traits and diseases is a **genome-wide association study (GWAS),** which compares large sets of landmarks (genetic markers) across the genome between two large groups of people—one with a particular trait or disease and one without it. Identifying parts of the genome that are much more common among the people with the trait or illness can lead researchers to genes that contribute to the phenotype.

Genome-wide association studies use genetic markers (**table 7.5**). Single nucleotide polymorphisms (SNPs) and copy number variants (CNVs) describe the DNA base sequence. A SNP is a site in the genome that has a different DNA base in at least one percent of a population (**figure 7.9**). A CNV is a DNA sequence that repeats a different number of times in different individuals (**figure 7.10**). A CNV does not provide information in the way that a gene that encodes protein does, but it is another way to distinguish individuals. CNVs are very useful in forensic applications, discussed in chapter 14.

Gene expression patterns are also used in genome-wide association studies. These patterns represent which proteins

Table 7.5	Types of Information Used in Genome-Wide Association Studies
Marker Type	**Definition**
SNP	A single nucleotide polymorphism is a site in the genome that is a different DNA base in >1% of a population.
CNV	A copy number variant is a tandemly repeated DNA sequence, such as CGTA CGTA CGTA
Gene expression	The pattern of genes that are overexpressed and/or underexpressed in people with a particular trait or disease.
	Epigenetic signature of methyl groups binding

are overproduced or underproduced in people with the trait or illness, compared to unaffected controls. Another way to compare genomes is by the sites to which methyl (CH_3) groups bind, shutting off gene expression. This is an epigenetic change because it doesn't affect the DNA base sequence.

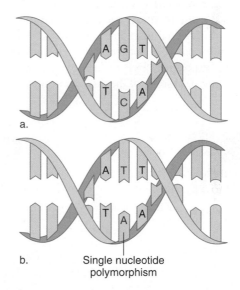

a.

b.   Single nucleotide polymorphism

Figure 7.9 **SNPs are sites of variability.** The DNA base pair in red is a SNP—a site that differs in more than 1 percent of a population. (The percentage may change as SNPs are identified in more individuals.)

GATTACA	Allele 1
GATTACAGATTACA	Allele 2
GATTACAGATTACAGATTACA	Allele 3
GATTACAGATTACAGATTACAGATTACA	Allele 4

Figure 7.10 For copy number variants, different numbers of repeats of a short DNA base sequence are considered to be different alleles.

To achieve statistical significance, a genome-wide association study must include at least 100,000 markers. It is the association of markers to a trait or disease that is informative (**figure 7.11**). Typically, genome-wide association studies use a million or more SNPs, grouped into half a million or so haplotypes. A specific "tag SNP" is used to identify a haplotype.

A GWAS is a stepwise focusing in on parts of the genome responsible to some degree for a trait (**figure 7.12**). In general, a group of people with the same condition or trait and a control group have their DNA isolated and genotyped for the 500,000 tag SNPs. Statistical algorithms identify the uniquely shared SNPs in the group with the trait or disorder. Repeating the process on additional populations narrows the SNPs and strengthens the association. It is important to validate a SNP association in different population groups, to be certain that it is the trait of interest that is being tracked, and not another part of the genome that members of one population share due to their common ancestry.

Several study designs are used in these investigations (**table 7.6**). In a **cohort study,** researchers follow a large group of individuals over time and measure many aspects of their health. The most famous is the Framingham Heart Study, which began tracking thousands of people and their descendants in Massachusetts in 1968. Nine thousand of them are participating in a genome-wide association study.

In a **case-control study,** each individual in one group is matched to an individual in another group who shares as many characteristics as possible, such as age, sex, activity level, and environmental exposures. SNP differences are then associated with the presence or absence of the disorder or trait. For example,

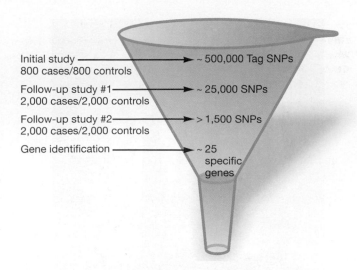

Figure 7.12 **A stepwise approach to gene discovery.** Genome-wide association study results must be validated in several different populations. Further research is necessary to go beyond association to demonstrate correlation and cause.

if 5,000 individuals with hypertension (high blood pressure) have particular DNA bases at six sites in the genome, and 5,000 matched individuals who do not have hypertension have different bases at only these six sites, then these genome regions may include genes whose protein products control blood pressure.

The **affected sibling pair** strategy follows the logic that because siblings share 50 percent of their genes, a trait or condition that many siblings share is likely to be inherited.

Researchers scan genomes for SNPs that most siblings who have the same condition share, but that siblings who do not both have the condition do not often share. Such genome regions may have genes that contribute to the condition.

A variation on the affected sibling pair strategy is **homozygosity mapping,** which is performed on families that are consanguineous—that is, the parents are related. The genomes of children whose parents share recent ancestors have more homozygous regions than do other children, and therefore greater likelihood that they have inherited two copies of a susceptibility or disease-causing mutation.

After a SNP association has been validated in diverse and large populations, the next step is gene identification. The human genome sequence near the SNPs might reveal "candidate" genes whose known functions explain the condition.

Common characteristics, such as height and body mass index, are also investigated with genome-wide association studies. One study examined facial features. Researchers measured eye and nose positions and dimensions, twenty lip descriptors, and the length of the space between the nose and the upper lip. The study identified variants of several genes already known to be mutant in specific syndromes that disrupt development of facial features.

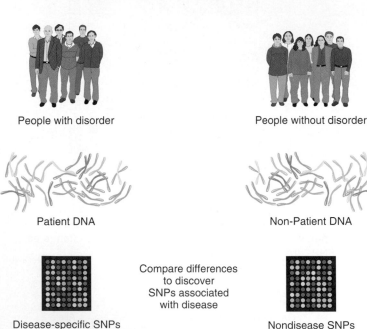

People with disorder   People without disorder

Patient DNA   Non-Patient DNA

Compare differences to discover SNPs associated with disease

Disease-specific SNPs   Nondisease SNPs

Figure 7.11 **Tracking genes in groups.** Genome-wide association studies seek DNA sequence variants that are shared with much greater frequency among individuals with the same illness or trait than among others. The squares are DNA microarrays, which display short, labeled DNA pieces (see figure 19.8). Different patterns indicate different alleles.

Table 7.6	Study Designs for Multifactorial Traits
Type of Study	**Definition**
Cohort	Researchers follow many people over time and measure several traits.
Case-control	People in two groups are individually matched for several characteristics and differences in SNP patterns identified.
Affected sibling pair	Identifies SNPs that siblings with the same condition share but siblings who do not share the condition do not have in common.
Homozygosity mapping	Disease-causing mutations identified in homozygous genome regions that children inherit from parents who are related to each other.

A genome-wide association study can explore the multifactorial aspects of traits and medical conditions. Consider a form of lung cancer (OMIM 211980) for which several chromosomes have susceptibility genes. An ongoing GWAS is seeking to identify smoking-related behaviors that may indicate an increased risk of inherited susceptibility leading to the clinical reality of lung cancer. These possibly informative behaviors include age at which smoking began, number of years smoking, number of years since quitting, and number of cigarettes smoked per day.

Genome-wide association studies have limitations, and are being replaced by direct analysis of the human genome sequence, as the functions of genes are being discovered. A conceptual limitation of genome-wide association studies is that they reveal associations between sets of information, and not causes. An *association* only means that one event or characteristic occurs when another occurs. A *correlation* is a directional association: If one measurement increases, so does the other, such as stress and blood pressure. In contrast, establishing a cause requires that a specific mechanism explains how one event makes another happen: *How* does stress elevate blood pressure? An association study does not provide information on a gene's function—it is more a discovery tool.

A practical limitation of genome-wide association studies is that they often identify parts of the genome that contribute only slightly to the risk of developing a disease. A genetic test that indicates a 1 percent increase in risk of developing cancer, for example, would not matter much to a smoker whose environmental risk is much higher. A gene that contributes so little risk of a cancer, for example, that it is not detected in a GWAS, may nonetheless cause that cancer in a particular family.

The way that a patient population is selected can introduce bias into a genome-wide association study. Samples drawn from clinics, for example, would not include the very mildly affected who are not ill enough to show up, or those who have died. Another source of error is that individuals in the control population might not actually be healthy. They might have problems other than the one being investigated.

The complicating factors discussed in chapter 5 also affect the accuracy of genome-wide association studies. Recall that a phenocopy is a trait or illness that resembles an inherited one, but has an environmental cause. Placing a person with anemia due to a drug reaction in a group with people who have an inherited anemia would be misleading. Genetic heterogeneity, in which different genes cause the same trait or condition, could also be a source of error. Epistasis, when one gene masks the effect of another, also confounds these studies, but as we learn more, such interactions are being taken into account.

Another limitation of genome-wide association studies is that people who share symptoms and a SNP pattern may share something *else* that accounts for the association, such as an environmental exposure that then can generate a false positive result. For example, mutations contribute to atherosclerosis risk, but so do infection, smoking, lack of exercise, and a fatty diet. These environmental factors are so common that if a GWAS isn't large enough, it might not correctly identify a genetic influence.

The success of a GWAS may depend on the quality of the question asked. The technique was very helpful, for example, in explaining why some people who live on the Solomon Islands have blond hair (**figure 7.13**). Most people living on

Figure 7.13 A striking phenotype. Blond hair among the residents of the Solomon Islands is due to a single base difference in a single gene.

these equatorial islands have dark hair and skin, similar to people who live in equatorial Africa. A case-control genome-wide association study on 43 blond Solomon Islanders and 42 dark-haired islanders clearly showed that the blonds were much more likely to have a particular SNP on chromosome 9. When researchers consulted the human genome sequence, they discovered in that interval a gene called tyrosine-related protein 1 (*TYRP1*). Its protein product controls melanin pigmentation in all vertebrate animals and the mutant gene causes a form of human albinism. A single DNA base change is responsible for the unusual blond hair of some Solomon Islanders.

Often, the old and the new techniques for dissecting multifactorial traits work well together. This is the case for stuttering. Concordance for MZ twins ranges from 20 to 83 percent, and for DZ twins, from 4 to 9 percent, suggesting a large inherited component. The risk of a first-degree relative of a person who stutters also stuttering is 15 percent based on empiric evidence, compared to the lifetime risk of stuttering in the general population of 5 percent, although part of that increase could be due to imitating an affected relative. A genome-wide association study on 100 families who have at least two members who stutter identified candidate genes on three chromosomes that contribute to the trait.

Table 7.7 reviews terms used to study multifactorial traits. The next section probes an example of such a trait—body weight.

Table 7.7	Terms Used in Evaluating Multifactorial Traits

Coefficient of relatedness   The proportion of genes shared by two people related in a particular way. Used to calculate heritability.

Concordance   The percentage of twin pairs in which both twins express a trait.

Empiric risk   The risk of recurrence of a trait or illness based on known incidence in a particular population.

Genome-wide association study   Detecting association between marker patterns and increased risk of developing a particular medical condition.

Heritability   The percentage of phenotypic variation for a trait that is attributable to genetic differences. It equals the ratio of the observed phenotypic variation to the expected phenotypic variation for a population of individuals.

Key Concepts Questions 7.3

1. What is empiric risk?

2. What is heritability?

3. How is the coefficient of relatedness used to calculate heritability?

4. How are adopted individuals and twins used to study environmental and inherited components of traits?

5. What type of information can genome-wide association studies provide?

7.4 A Closer Look: Body Weight

Weight is a multifactorial trait that illustrates the methods discussed in this chapter. Body weight reflects energy balance, which is the rate of food taken in versus the rate at which the body uses it for fuel. Excess food means, ultimately, excess weight. Being overweight or obese raises the risk of developing hypertension, diabetes, stroke, gallstones, sleep apnea, and some cancers.

Scientific studies of body weight use a measurement called body mass index (BMI), which is weight in proportion to height (**figure 7.14**). BMI makes sense—a person who weighs 170 pounds and is 6 feet tall is slim, whereas a person of the same weight who is 5 feet tall is obese. The tall person's BMI is 23; the short person's is 33.5.

Heritability for BMI is 0.55, which leaves room for environmental influences on our appetites and sizes. Dozens of genes affect how much we eat, how we use calories, and how fat is distributed in the body. The biochemical pathways and hormonal interactions that control weight may reveal points for drug intervention (**table 7.8**).

Genes That Affect Weight

Genetics became prominent in obesity research in 1994, when Jeffrey Friedman at Rockefeller University discovered a gene that encodes the protein hormone leptin in mice and in

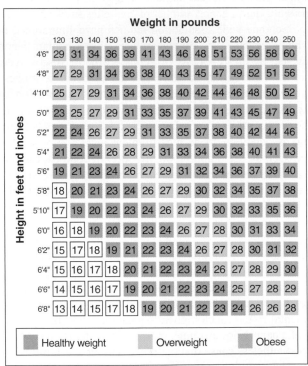

Developed by the National Center for Health Statistics in collaboration with the National Center for Chronic Disease Prevention and Health Promotion

Figure 7.14 Body mass index (BMI). BMI equals weight/height², with weight measured in kilograms and height measured in meters. This chart provides a shortcut—the calculations have been done and converted to the English system of measurement. Squares that are not filled in indicate underweight.

Table 7.8	Some Sites of Genetic Control of Body Weight		
Protein	**Function**	**OMIM**	**Effect on Appetite**
Leptin	Stimulates cells in hypothalamus to decrease appetite and metabolize nutrients	164160	↓
Leptin transporter	Enables leptin to cross from bloodstream into brain	601694	↓
Leptin receptor	Binds leptin on hypothalamus cell surfaces, triggering hormone's effects	601007	↓
Neuropeptide Y	Produced in hypothalamus when leptin levels are low and the individual loses weight	162640	↑
Melanocortin-4 receptor	Activated when leptin levels are high and the individual gains weight	155541	↓
Ghrelin	Signals hunger from stomach to brain in short term, stimulating neuropeptide Y	605353	↑
PYY	Signals satiety from stomach to brain	660781	↓
Stearoyl-CoA desaturase-1	Controls whether body stores or uses fat	604031	↑

humans. Normally, eating stimulates fat cells (adipocytes) to secrete leptin, which travels in the bloodstream to a region of the brain's hypothalamus, where it binds to receptors on nerve cells (neurons). Leptin binding signals the neurons to release another type of hormone that binds yet other types of receptors, which ultimately function as an appetite "brake," while speeding digestion of food already eaten. When a person hasn't eaten in several hours, leptin levels fall, which triggers the release of an appetite "accelerator."

The discovery of genes and proteins that affect appetite led to great interest in targeting them with drugs to either lose or gain weight. When Friedman gave mice extra leptin, they ate less and lost weight. Headlines soon proclaimed the new magic weight loss elixir, a biotech company paid $20 million for rights to the hormone, and clinical trials ensued. The idea was to give obese people leptin, assuming that they had a deficiency, to trick them into feeling full. Only about 15 percent of the people lost weight, but the other 85 percent didn't actually lack leptin. Instead, most of them had leptin resistance, which is a diminished ability to recognize the hormone due to defective leptin receptors. Giving these people leptin had no effect on their appetites. However, the discovery helped a few severely obese children with true leptin deficiency attain normal weights after years of daily leptin injections.

The stomach is another source of weight-related proteins. Ghrelin is a peptide (small protein) hormone produced in the stomach that responds to hunger, signaling the hypothalamus to produce more of the appetite accelerator. One of the ways that weight loss surgery may work is by decreasing ghrelin secretion by making the stomach smaller.

While leptin acts in the long term to maintain weight, the stomach's appetite control hormones function in the short term. All of these hormonal signals are integrated to finely control appetite in a way that maintains weight.

Identifying single genes that influence weight paved the way for considering the trait to be multifactorial. Researchers are investigating how combinations of genes control weight. One study looked at 21 genes in which mutations cause syndromes that include obesity, as well as 37 genes whose products participate in biochemical pathways related to weight. This approach identified many rare gene variants that could, in combinations, explain many people's tendency to gain weight.

Genome-wide association studies that compare gene expression patterns have also enhanced understanding of body weight. One study compared the sets of genes that are expressed in adipose (fat) tissue to other tissues. Samples from more than 1,600 people in Iceland revealed a set of genes whose products take part in inflammation and the immune response, but also contribute obesity-related traits.

Environmental Influences on Weight

Many studies on adopted individuals and twins suggest that obesity has a heritability of 75 percent. Because the heritability for BMI is lower than this, the discrepancy suggests that genes play a larger role in those who tend to gain weight easily. The role of genes in obesity is seen when populations that have an inherited tendency to easily gain weight experience a large and sudden plunge in the quality of the diet.

On the tiny island of Naura, in Western Samoa, the residents' lifestyles changed greatly when they found a market for the tons of bird droppings on their island as commercial fertilizer. The money led to inactivity and a high-calorie, high-fat diet, replacing an agricultural lifestyle and diet of fish and vegetables. Within a generation, two-thirds of the population had become obese, and a third had type 2 diabetes.

The Pima Indians offer another example of environmental effects on body weight. These people separated into two populations during the Middle Ages, one group settling in the Sierra Madre mountains of Mexico, the other in southern Arizona. By the 1970s, the Arizona Indians no longer farmed nor ate a low-calorie, low-fat diet, but instead consumed 40 percent of their calories from fat. With this extreme change in lifestyle, they developed the highest prevalence of obesity of any population on earth. Half of the Arizona group had diabetes by age 35, weighing, on average, 57 pounds (26 kilograms) more than their southern relatives, who still eat a low-fat diet and are very active.

The Pima Indians demonstrate that future obesity is not sealed in the genes at conception, but instead is much more likely to occur if the environment provides too many calories and too much fat. Geneticist James Neel expressed this idea as the "thrifty gene hypothesis" in 1962. He suggested that long ago, the hunter-gatherers who survived famine had genes that enabled them to store fat. Today, with food plentiful, the genetic tendency to retain fat is no longer healthful, but harmful. Unfortunately, for many of us, our genomes hold an energy-conserving legacy that works too well—it is much easier to gain weight than to lose it, for a sound evolutionary reason: survival.

The thrifty gene hypothesis also applies to people who were born after a full-term pregnancy, but were very low weight. To compensate for starvation conditions in the uterus, metabolism shifts, before birth, in a way that conserves calories—and the person later faces elevated risk of heart disease, stroke, obesity, osteoporosis, and type 2 diabetes. These are multifactorial conditions that, instead of arising from mutations, reflect epigenetic alterations of gene expression.

Another environmental influence on weight is the "gut microbiome," the types of bacteria that normally live in our digestive tracts (see section 2.5). Bacterial cells in our bodies actually outnumber our own cells. The actions of certain types of bacteria affect the number of calories that we extract from particular foods. An obese person has a different gut microbiome than a person who easily stays thin. The gut microbiome changes dramatically after weight-loss surgery, and finding a way to recreate this changed microbiome might one day provide an alternative to the surgery.

Perhaps nowhere are the complexities and challenges of gene-environment interactions more profound than in behavioral characteristics, nuances, quirks, and illnesses. The next chapter looks at a few of them.

Key Concepts Questions 7.4

1. How do leptin, ghrelin, and other proteins affect weight?
2. What is the significance of the difference in heritability for BMI and obesity?
3. What can populations that suddenly become sedentary and switch to a high-calorie diet reveal about environmental influences on weight?

Summary

7.1 Genes and the Environment Mold Traits

1. **Multifactorial traits** reflect influences of the environment and genes. A **polygenic trait** is determined by more than one gene and varies continuously in expression.

2. Single-gene traits are rare. For most traits, many genes contribute to a small, but not necessarily equal, degree.

7.2 Polygenic Traits Are Continuously Varying

3. Genes that contribute to polygenic traits are called **quantitative trait loci**. The frequency distribution of phenotypes for a polygenic trait forms a bell curve.

7.3 Methods to Investigate Multifactorial Traits

4. **Empiric risk** measures the likelihood that a multifactorial trait will recur based on **prevalence**. The risk rises with genetic closeness, severity, and number of affected relatives.

5. **Heritability** estimates the proportion of variation in a multifactorial trait due to genetics in a particular population at a particular time. The **coefficient of relatedness** is the proportion of genes that two people related in a certain way share.

6. Characteristics shared by adopted people and their biological parents are mostly inherited, whereas similarities between adopted people and their adoptive parents reflect environmental influences.

7. **Concordance** measures the frequency of expression of a trait in both members of MZ or DZ twin pairs. The more influence genes exert over a trait, the higher the differences in concordance between MZ and DZ twins.

8. **Genome-wide association studies** correlate genetic marker (SNP or CNV) patterns to increased disease risk. They may use a **cohort study** to follow a large group over time, or a **case-control study** on matched pairs.

9. The **affected sibling pair** strategy identifies homozygous regions that may include genes of interest. **Homozygosity mapping** identifies mutations in genome regions that are homozygous because the parents shared recent ancestors.

7.4 A Closer Look: Body Weight

10. Leptin and associated proteins affect appetite. Fat cells secrete leptin in response to eating, which decreases appetite. Populations that switch to a fatty, high-calorie diet and a less-active lifestyle reveal effects of the environment on weight.

Answers to all end-of-chapter questions can be found at
www.mhhe.com/lewisgenetics11. You will also find additional
practice quizzes, animations, videos, and vocabulary flashcards
to help you master the material in this chapter.

Review Questions

1. Explain how Mendel's laws apply to multifactorial traits.

2. Choose a single-gene disease and describe how environmental factors may affect the phenotype.

3. What is the difference between a Mendelian multifactorial trait and a polygenic multifactorial trait?

4. Do all genes that contribute to a polygenic trait do so to the same degree?

5. Explain why figures 7.2, 7.3, and 7.4 have the same bell shape, even though they represent different traits.

6. How can skin color have a different heritability at different times of the year?

7. Explain how the twins in figure 7.4 have such different skin colors.

8. In a large, diverse population, why are medium brown skin colors more common than very white or very black skin?

9. Which has a greater heritability—eye color or height? State a reason for your answer.

10. Describe the type of information in a(n)
    a. empiric risk calculation.
    b. twin study.
    c. adoption study.
    d. genome-wide association study.

11. Name three types of proteins that affect cardiovascular functioning and three that affect body weight.

12. What is a limitation of a genome-wide association study?

13. How will genome sequencing ultimately make a genome-wide association study unnecessary?

Applied Questions

1. "Heritability" is often used in the media to refer to the degree to which a trait is inherited. How is this definition different from the scientific one?

2. Would you take a drug that was prescribed to you based on your race? Cite a reason for your answer.

3. The incidence of obesity in the United States has doubled over the past two decades. Is this due more to genetic or environmental factors? Cite a reason for your answer.

4. One way to calculate heritability is to double the difference between the concordance values for MZ versus DZ twins. For multiple sclerosis, concordance for MZ twins is 30 percent, and for DZ twins, 3 percent. What is the heritability? What does the heritability suggest about the relative contributions of genes and the environment in causing MS?

5. In chickens, high body weight is a multifactorial trait. Several genes contribute small effects additively, and a few genes exert a great effect. Do the several genes provide broad heritability and the few genes narrow heritability, or vice versa?

6. Devise a genome-wide association study to assess whether restless legs syndrome is inherited, and if it is, where susceptibility or causative genes may be located.

7. The environmental epigenetics hypothesis states that early negative experiences, such as neglect, abuse, and extreme stress, increase the risk of developing depression, anxiety disorder, addictions, and obesity later in life, through effects on gene expression that persist. Suggest an experiment to test this hypothesis.

8. Guidelines from the American Academy of Ophthalmology support genetic tests for single-gene eye diseases such as the many types of retinitis pigmentosa, but do not advise use of genome-wide association study results to counsel patients who have age-related macular degeneration, which is multifactorial. What is the reasoning behind the recommendation?

9. Three large pharmaceutical companies are developing drugs based on a very few cases of people who have extremely low LDL levels, due to a homozygous recessive mutation in a gene called proprotein convertase subtilisin/kexin type 9 (*PCSK9*). The idea is that mimicking the effects of the mutation in people who don't have it will lower LDL level enough to combat elevated risk of heart disease. What other information would you like to have before enrolling in a clinical trial to test whether one of the new drugs can prevent heart disease?

10. Lung scarring due to idiopathic pulmonary fibrosis affects more than 50,000 people in the United States and is often

fatal within a few years of diagnosis. "Idiopathic" means "cause unknown," but the condition is set into motion by mutations in several genes (transforming growth factor B, surfactant protein C, mucin 5B, and human telomerase reverse transcriptase). Environmental influences are important too. These include certain viral infections and exposure to inhaled irritants including cigarette smoke. How do these contributing factors explain why idiopathic pulmonary fibrosis is considered to be a disease of aging?

11. Suggest a way to mimic the effects of weight-loss surgery.

Web Activities

1. Many genes contribute to lung cancer risk, especially among people who smoke. These genes include *p53*, *IL1A* and *IL1B, CYP1A1, EPHX1, TERT,* and *CRR*9. Search for one of these genes on the Internet and describe how mutations in it may contribute to causing lung cancer, or polymorphisms may be associated with increased risk.

2. Locate a website that deals with breeding show animals, farm animals, or crops to produce specific traits, such as litter size, degree of meat marbling, milk yield, or fruit ripening rate. Identify three traits with heritabilities that indicate a greater contribution from genes than from the environment.

3. Visit the Centers for Disease Control and Prevention (CDC) website. From the leading causes of death, list three that have high heritabilities, and three that do not. Base your decisions on common sense or data, and explain your selections.

4. Use OMIM to look up any of the following genes that encode proteins that affect cardiovascular health and explain what the proteins do: apolipoprotein E; LDL receptor; apolipoprotein A; angiotensinogen; beta-2 adrenergic receptor; toll-like receptor 4; C-reactive protein.

5. Go to the website for the Minnesota Center for Twin and Family Research (https://mctfr.psych.umn.edu). Click on "research." Select a trait or disorder and discuss what the investigators have discovered using twin studies.

Case Studies and Research Results

1. Researchers interviewed 49 MZ (identical) and 50 DZ (fraternal) pairs of twins in which one had Parkinson disease, asking about lifetime exposure to six chemical solvents that case reports had previously associated with the disease. Average age at diagnosis was 66, with symptoms starting 10 to 40 years after exposure began. The researchers found that long-term exposure to trichloroethylene increases Parkinson disease risk sixfold and to perchloroethylene ninefold, in both types of twins. Explain how the design and results of this study support either of two hypotheses: that Parkinson disease is inherited, or that it is caused by environmental exposure. Do you think that the results are good news or bad news, and why?

2. A team of criminologists not well versed in genetics used a genome-wide association study to examine the roots of criminal behavior. They investigated 3,000 adult sibling pairs with decreasing proportions of shared genomes: MZ twins (100 percent), DZ twins (50 percent), non-twin siblings (50 percent), half-siblings (25 percent), and first cousins (12.5 percent). A "delinquency scale" ranked participants' admission of such escalating behaviors as painting graffiti, lying to parents, running away, and stealing. The researchers also classified participants by whether they considered themselves to be "life-course persistents" (LCPs), who've been bad since childhood or "adolescence-limited" (AL) people, who've behaved badly only as teens.

The researchers discovered mostly what they expected: The closer the genetic relationship, the more alike the scores on the delinquency scale. They concluded that genetic factors explained 56–70 percent of the variance for lifelong criminals but only 35 percent for the teen-only type.

How is the media headline reporting the study— "Life of crime is in the genes, study claims"—an oversimplification? What do you think the results of the investigation indicate about inheriting criminality?

3. Researchers sampled pieces of the aorta (the largest blood vessel) in 96 patients. Each patient had a heart transplant, and the aorta sample came from the patients' blood vessel part that was stitched to the recipient heart. The researchers added oxidized lipids, the chemicals that cause atherosclerosis, to the aorta pieces and profiled which genes were turned on or off in response. The study revealed about 1,000 genes that are involved in inflammation of the aorta pieces. Suggest an application of these findings.

4. Marla and Anthony enjoy hiking and mountain climbing. They want to know whether their 2-year-old son, Spencer, will excel at these activities too. They send a sample of Spencer's cheek cells to a company that offers genetic tests for athletic ability and request a test on the angiotensin I–converting enzyme (*ACE*) gene. Marla and Anthony had read a study about 40 elite British

mountaineers, many of whom had a genotype that is rare among the general, non-mountain-climbing population. Look up what the gene does, and suggest how variants of it might affect athletic ability. Do you think that the parents should decide which sports Spencer tries based on the genetic test results? Explain your answer.

5. Concordance for the eating disorder anorexia nervosa for MZ twins is 55 percent, and for DZ twins, 7 percent. Ashley and Maggie are DZ twins. Maggie has anorexia nervosa. Should Ashley worry about an inherited tendency to develop the condition? Explain your answer.

6. A study looked at 200,000 SNPs throughout the genome for 1,820 people with premature hair graying and 1,820 without this trait. Those with the trait shared several SNPs on chromosome 9. What type of study is this?

7. An affected sibling pair study identified areas of chromosomes 1, 14, and 20 that are likely to harbor genes that predispose individuals toward or cause schizophrenia. Explain how such an investigation is conducted.

8. A study in England tested 20,430 people for alleles of 12 genes known to increase risk of becoming obese. (Each person had 24 alleles assessed, 12 from each parent.) Although the number of risk alleles correlated to BMI, the more a person exercised, the lower the BMI. Do these findings support a genetic deterministic view of the trait of body weight, or not? Cite a reason for your answer.

Child abuse and PTSD. *Inheriting a variant of a particular gene may influence how a person handles memories of child abuse throughout the rest of his or her life.*

Genetics of Behavior

Learning Outcomes

8.1 Genes and Behavior

1. Identify the physical basis of behavioral traits in the brain.

2. Explain how genes can affect behavior.

8.2 Sleep; 8.3 Intelligence and Intellectual Disability; 8.4 Drug Addiction; 8.5 Mood Disorders; 8.6 Schizophrenia; 8.7 Autism

3. Discuss the genetic and environmental influences on sleep, intelligence, addiction, mood, and risk of developing schizophrenia or autism.

The **BIG** Picture

Behavioral traits and disorders reflect effects and interactions of genes and environmental factors on the nervous system.

Genetic Predisposition to Posttraumatic Stress Disorder

A young woman who was physically and emotionally abused by an uncle during her childhood becomes a warm, caring, mother with a successful career. Another woman with a similar upbringing grows into an anxious adult who suffers from posttraumatic stress disorder (PTSD).

PTSD is a reaction to traumatic physical harm, or the perceived threat of harm, which persists long after the triggering event has passed. The person suffers from anxiety, flashbacks, and avoidance of memories of the event, resulting from inappropriate activation of the stress hormones that flooded the bloodstream during the original trauma. Often associated with combat experiences, PTSD also follows health scares, natural disasters, crimes, and other disturbing events. PTSD is rare.

A study of PTSD following child abuse indicates that although many factors contribute to the development of PTSD, a variant of one particular gene may play a significant role in why some people abused as children develop PTSD as adults and some do not. Researchers examined variants of a gene called *FKBP5* among 2,000 people who had been severely traumatized from abuse, either as children or adults. Individuals abused as children who developed PTSD were significantly more likely to have a specific variant of the gene than other individuals. The effect is both genetic and epigenetic. In people with the gene variant who were abused as children and developed PTSD, methyl groups on the gene are removed—an effect not seen in people who did not develop PTSD. As a result of the decreased methylation, receptors for stress hormones (glucocorticoids) malfunction, and the body cannot react and adapt to stress.

The study found the epigenetic change in white blood cells, but other experiments on cells growing in the laboratory showed that it also occurs and persists in progenitor cells that differentiate into brain nerve cells that will last a lifetime. In a person, the epigenetic change happens at a critical period in development, setting the stage in a child for an inability to fully use stress hormones later in life. The result can be PTSD.

8.1 Genes and Behavior

Behavior is a complex continuum of emotions, moods, intelligence, and personality that drives how we function on a daily basis. We are, to an extent, defined and judged by our many behaviors. They control how we communicate, cope with negative feelings, and react to stress. Behavioral disorders are common, with wide-ranging and sometimes overlapping symptoms. Many of our behaviors are in response to environmental factors, but *how* we respond may have genetic underpinnings, such as the chapter opener describes for reaction later in life to abuse during childhood. Understanding the biology behind behavior can help to develop treatments for behavior-based disorders.

Behavioral genetics considers nervous system function and variation, including the hard-to-define qualities of mood and mind. The human brain weighs about 3 pounds and resembles a giant gray walnut, but with the appearance and consistency of pudding. It consists of 100 billion nerve cells, or **neurons,** and at least a trillion other cells called **neuroglia,** which support and nurture the neurons.

Brain neurons connect and interact in complex networks that control all of the body. Branches from each of the 100 billion neurons in the brain form close associations, called synapses, with 1,000 to 10,000 other neurons. Neurons communicate across these tiny spaces using chemical signals called neurotransmitters. Networked neurons oversee broad functions such as sensation and perception, memory, reasoning, and muscular movements.

Genes affect behavior by controlling the production and distribution of neurotransmitters. **Figure 8.1** indicates the points where genes act in sending and receiving nervous system information. Enzymes oversee the synthesis of neurotransmitters and their transport from the sending (presynaptic) neuron across the synapse to receptors on the receiving (postsynaptic) neuron. Proteins called transporters ferry neurotransmitters from sending to receiving neurons, and proteins also form the subunits of receptors. Genes control the synthesis of myelin, a substance consisting of fats and proteins that coats neuron extensions called axons. Myelin insulates the neuron, which speeds neurotransmission.

Many behavioral traits and disorders, like other characteristics, probably reflect a major influence from a single gene, perhaps one whose protein product takes part directly in neurotransmission, but also small inputs from common gene variants. Researchers envision from 100 to 300 genes at play. Inheriting certain subsets of variants of these genes makes an individual susceptible to developing a certain disorder in the presence of a particular environmental stimulus. Some genes are implicated in more than one behavioral disorder.

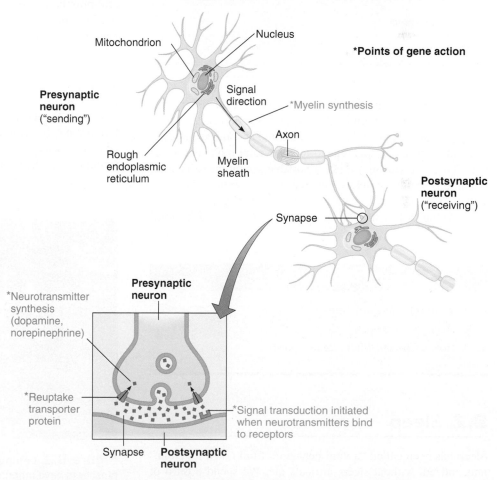

Figure 8.1 Neurotransmission. Many genes that affect behavior produce proteins that affect neurotransmission, which is sending a signal from one neuron to another across a synapse via a neurotransmitter molecule.

Table 8.1	Prevalence of Behavioral Disorders in the U.S. Population	
Condition		**Prevalence (%)**
Alzheimer disease		4.0
Anxiety		8.0
Phobias		2.5
Posttraumatic stress disorder		1.8
Generalized anxiety disorder		1.5
Obsessive compulsive disorder		1.2
Panic disorder		1.0
Attention deficit hyperactivity disorder		2.0
Autism spectrum disorders		0.1
Drug addiction		4.0
Eating disorders		3.0
Mood disorders		7.0
Major depressive disorder		6.0
Bipolar disorder		1.0
Schizophrenia		1.3

Deciphering the genetic components of behavioral traits and disorders uses traditional empiric risk, adoptee, and twin study data and, more and more, exome and genome sequencing to identify gene variants related to behavioral traits and disorders. This approach may make it possible to subtype behavioral conditions in a biologically meaningful way so that diagnoses will be personalized and effective treatments begun sooner.

Table 8.1 lists the prevalence of some behavioral disorders.

Key Concepts Questions 8.1

1. What is behavioral genetics?
2. What is the cellular basis of behavior?
3. How do genes influence behavior?

8.2 Sleep

Sleep has been called "a vital behavior of unknown function," and, indeed, without sleep, animals die. We spend a third of our lives in this mysterious state.

Genes influence sleep characteristics. When asked about sleep duration, schedule, quality, nap habits, and whether they are "night owls" or "morning people," MZ twins report significantly more in common than do DZ twins. This is true even for MZ twins separated at birth. Twin studies of brain wave patterns through four of the five stages of sleep confirm a hereditary influence. The fifth stage, REM sleep, is associated with dreaming and therefore may reflect the input of experience more than genes.

Narcolepsy

Researchers discovered the first gene related to sleep in 1999, for a condition called "narcolepsy with cataplexy" in dogs. Mutations in several genes cause narcolepsy in humans.

A person (or dog) with narcolepsy falls asleep suddenly several times a day. Extreme daytime sleepiness greatly disrupts daily activities. People with narcolepsy have a tenfold higher rate of car accidents. Another symptom is sleep paralysis, which is the inability to move for a few minutes after awakening. The most dramatic manifestation of narcolepsy is cataplexy. During these short and sudden episodes of muscle weakness, the jaw sags, the head drops, knees buckle, and the person falls to the ground. This often occurs during a bout of laughter or excitement. Narcolepsy with cataplexy affects only 0.02 to 0.06 percent of the general populations of North America and Europe, but the fact that it is much more common in certain families suggests a genetic component.

Studies on dogs led the way to discovery of one human narcolepsy gene. In 1999, researchers discovered mutations in a gene that encodes a receptor for a neuropeptide called hypocretin (OMIM 602358). In Doberman pinschers and Labrador retrievers, the receptor does not reach the cell surfaces of certain brain cells, preventing the cells from receiving signals to promote a state of awakeness. **Figure 8.2** shows a still frame of a film of narcoleptic dogs playing. Suddenly, they all collapse! A minute later, they get up and resume their antics. To induce

Figure 8.2 Letting sleeping dogs lie. These Doberman pinschers have inherited narcolepsy. They suddenly fall into a short but deep sleep while playing. Research on dogs with narcolepsy led to the discovery of a gene that affects sleep in humans.

a narcoleptic episode in puppies, researchers let them play with each other. Feeding older dogs meat excites them so much that they can take a while to finish a meal because they fall down in delight so often. Getting narcoleptic dogs to breed is difficult, too, for sex is even more exciting than play or food!

The hypocretin receptor gene turned out to be the same as a gene discovered a year earlier that controls eating, called orexin. Neurons that produce the protein in different parts of the brain account for the two effects. Researchers are studying hypocretin/orexin and the molecules with which they interact to develop drugs to treat insomnia. At least three other genes are associated with insomnia. The most severe condition is "fatal familial insomnia," which is a condition of infectious protein misfolding (see section 10.3).

Familial Advanced Sleep Phase Syndrome

Daily rhythms such as the sleep-wake cycle are set by cells that form a "circadian pacemaker" in two clusters of neurons in the brain called the suprachiasmatic nuclei. In these cells, certain "clock" genes are expressed in response to light or dark in the environment.

The function of clock genes is startling in families that have a mutation. For example, five generations of a family in Utah have familial advanced sleep phase syndrome (OMIM 604348). Affected people fall asleep at 7:30 each night and awaken suddenly at 4:30 A.M. In them a single DNA base mutation in a gene on chromosome 2 called *period 1* interferes with synchronization of the sleep-wake cycle with daily sunrise and sunset, which is an example of a circadian rhythm (**figure 8.3**).

Mutation at a different part of the *period 1* gene affects how early we tend to awaken. People with two adenines at this site (genotype *AA*) awaken an hour earlier than people with two guanines (genotype *GG*). Heterozygotes, genotype *AG,* awaken at times between those of the two homozygotes.

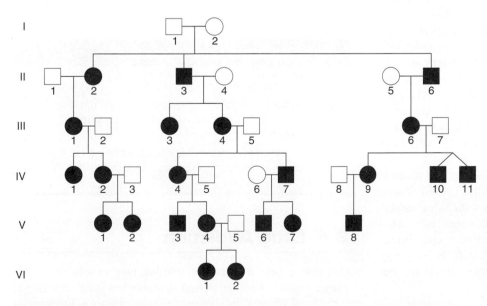

Figure 8.3 **Inheritance of a disrupted sleep-wake cycle.** This partial pedigree depicts affected members of a large family with an autosomal dominant form of familial advanced sleep phase syndrome.

Key Concepts Questions 8.2

1. What is the evidence that sleep habits are inherited?
2. Describe two mutations that alter sleep.

8.3 Intelligence and Intellectual Disability

Intelligence is a complex and variable trait that is subject to many genetic and environmental influences, and also to great subjectivity. Sir Francis Galton, a half first cousin of Charles Darwin, investigated genius, which he defined as "a man endowed with superior faculties." He identified successful and prominent people in Victorian-era English society, and then assessed success among their relatives. In his 1869 book, *Hereditary Genius,* Galton wrote that relatives of eminent people were more likely to also be successful than people in the general population. The closer the blood relationship, he concluded, the more likely the person was to succeed. This, he claimed, established a hereditary basis for intelligence.

Definitions of intelligence vary. In general, intelligence refers to the ability to reason, learn, remember, connect ideas, deduce, and create. The first intelligence tests, developed in the late nineteenth century, assessed sensory perception and reaction times to various stimuli. In 1904, Alfred Binet in France developed a test with questions based on language, numbers, and pictures to predict the success of developmentally disabled youngsters in school. The test was modified at Stanford University to assess white, middle-class Americans. An average score on this "intelligence quotient," or IQ test, is 100, with two-thirds of all people scoring between 85 and 115 in a bell curve or normal distribution (**figure 8.4**). An IQ between 50 and 70 is considered mild intellectual disability, and below 50, severe intellectual disability. IQ has been a fairly accurate predictor of success in school and work, but it does not solely represent biological influences on intelligence. Low IQ correlates with societal situations, such as poverty, a high divorce rate, failure to complete high school, incarceration (males), and having a child out of wedlock (females).

Psychologists use a general intelligence ability, called "*g,*" to represent the four basic skills that IQ encompasses. These are verbal fluency, mathematical reasoning, memory, and spatial visualization skills. The general intelligence measure eliminates the effect of unequal opportunities on assessment of intelligence.

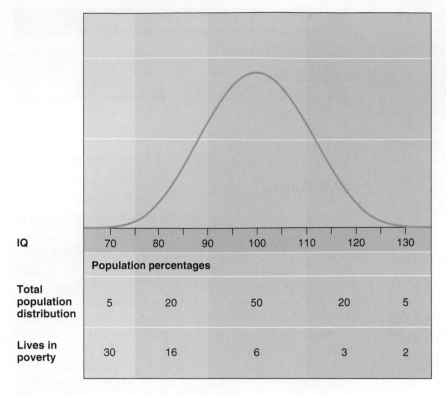

IQ	70	80	90	100	110	120	130
Population percentages							
Total population distribution	5	20		50		20	5
Lives in poverty	30	16		6		3	2

Figure 8.4 **Success and IQ.** IQ scores predict success in school and the workplace in U.S. society. The bell curve for IQ indicates that most people fall between 85 and 115. However, those living in poverty tend to have lower IQs.

Traditional methods such as adoption and twin studies provide a heritability of 0.5 to 0.8 for intelligence, indicating a high contribution from genetics. New types of studies confirm the heritability established with older techniques, and also identify specific genes that contribute to or affect intelligence.

Researchers used a variation of a genome-wide association study (see section 7.3) to evaluate change in intelligence over a lifetime. They used general intelligence measured with a standard test given to nearly 2,000 people in Scotland when they were 11 years old and again when they were older (65, 70, or 79). They asked the question: To what extent are differences in intelligence between childhood and old age due to genes or the environment? The researchers discovered a "substantial genetic correlation" of 0.62 for intelligence at both ends of the life span. That is, if someone was a smart 11-year-old, chances are excellent that he or she will be a smart senior citizen.

Genetics can explain some cases of intellectual disability, which affects 3 in 100 individuals in the United States. (Intellectual disability is a broad term that includes mental retardation.) Causes are many: abnormal genes and chromosomes, but also environmental exposure to toxins, infections, malnutrition, and noninherited birth defects. Probably at least 1,000 of our 20,325 or so genes affect the brain and can therefore impact intelligence.

Among the best-studied genetic causes of intellectual disability are Down syndrome (see *Bioethics: Choices for the Future* in Chapter 13 on page 250) and fragile X syndrome

(see Clinical Connection 12.1), and one of the most familiar environmental causes is fetal alcohol syndrome (see figure 3.21). Several ongoing projects are attempting to assign causative genes to children with intellectual disability that have not been diagnosed with these or other more common conditions.

Exome sequencing has been helpful in identifying causes of intellectual disability, some of which arise as new ("*de novo*") mutations. This is the strategy used to diagnose Nicholas Volker (see Clinical Connection 1.1) and Bea Rienhoff (see figure 4.18), children who have genetic conditions that do not include intellectual disability. Thousands of children with intellectual disability, but with no other affected family members and no diagnosis, are having their exomes sequenced and compared to those of their closest relatives to detect mutations. About 40 percent of such children find answers. In many of these cases, the child has an "atypical presentation" of a known inherited disorder. In other cases, the child may be the first person described with a newly recognized form of intellectual disability.

Identifying the causative gene behind intellectual disability in a child may not seem to be practical, but parents report great relief at finally, after years of seeking a diagnosis, getting an explanation. In cases in which the mutation is *de novo,* the parents have the reassurance that their other children will not be affected. The parents may also realize that the child's condition is not their fault. After researchers find new mutations that cause intellectual disability in more than one patient, they can use the discovery as a basis for developing new drugs, or realizing that an existing drug may be "repurposed" to treat the condition.

Key Concepts Questions 8.3

1. What is intelligence?
2. How does the general intelligence value improve upon using IQ score to assess the inheritance of intelligence?
3. How are newer genetic technologies being used to discover mutations that cause intellectual disability?

8.4 Drug Addiction

One person sees a loved one battling lung cancer and never smokes again. Another person actually has lung cancer, yet takes breaks from using her oxygen tank to smoke. Evidence is mounting that genes play a large role in making some individuals prone to addiction, and others not.

The Definition of Addiction

Drug addiction is compulsively seeking and taking a drug despite knowing its dangers. The two identifying characteristics are tolerance and dependence. Tolerance is the need to take more of the drug to achieve the same effects as time goes on. Dependence is the onset of withdrawal symptoms upon stopping use of the drug. Both tolerance and dependence contribute to the biological and psychological components of craving the drug. The behavior associated with drug addiction can be extremely difficult to break.

Drug addiction produces long-lasting brain changes. Craving and high risk of relapse remain even after a person has abstained for years. Heritability is 0.4 to 0.6, with a two- to threefold increase in risk among adopted individuals who have one affected biological parent. Twin studies also indicate an inherited component to drug addiction.

Brain imaging techniques localize the "seat" of drug addiction in the brain by highlighting the cell surface receptors that bind neurotransmitters when a person craves the drug. The brain changes that contribute to addiction occur in parts called the nucleus accumbens, the prefrontal cortex, and the ventral tegmental area, which are part of a larger set of brain structures called the limbic system (**figure 8.5**). The effects of cocaine seem to be largely confined to the nucleus accumbens, whereas alcohol affects the prefrontal cortex.

The specific genes and proteins that are implicated in addiction to different substances may vary, but several general routes of interference in brain function are at play. Proteins involved in drug addiction are those that

- produce neurotransmitters, such as enzymes;
- remove excess neurotransmitters from the synapse (called reuptake transporters);
- form receptors on the postsynaptic (receiving) neuron that are activated or inactivated when specific neurotransmitters bind; and
- convey chemical signals in the postsynaptic neuron.

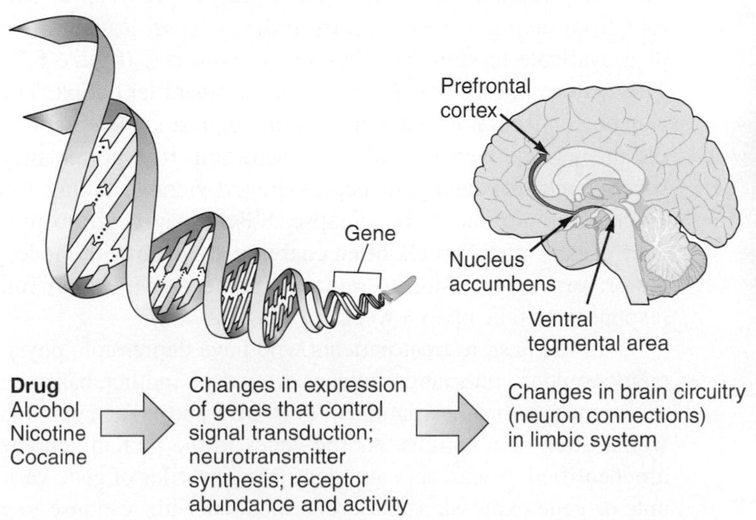

Drugs of Abuse

Our ancient ancestors must have discovered that ingesting certain natural substances, particularly from plants, provided a feeling of well-being. That tendency persists today.

Abused drugs are often plant-derived chemicals, such as cocaine, opium, and tetrahydrocannabinol (THC), the main active ingredient in marijuana. These substances bind to receptors on human neurons, which indicates that our bodies have versions of these substances. The human equivalents of the opiates are the endorphins and enkephalins, and the equivalent of THC is anandamide. The endorphins and enkephalins relieve pain. Anandamide modulates how brain cells respond to stimulation by binding to neurotransmitter receptors on presynaptic (sending) neurons. In contrast, neurotransmitters bind to receptors on postsynaptic neurons.

Amphetamines and LSD produce their effects by binding to receptors on neurons that normally bind neurotransmitters called trace amines, which are found throughout the brain at low levels. LSD causes effects similar to the symptoms of schizophrenia (see section 8.6), suggesting that the trace amine receptors, which are proteins, may be implicated in the illness.

People addicted to various drugs share certain gene variants that must be paired with environmental stimuli for addiction to occur. For example, people who are homozygous for the *A1* allele of the dopamine D(2) receptor gene variant are overrepresented among people with alcoholism and people with other addictions. Genome-wide association studies have found more than 50 chromosomal regions that may include genes that contribute to craving. Studies of gene expression flesh out this picture by providing a real-time view of biochemical changes that happen when a person craves a drug, and then takes it.

Discovering the genetic underpinnings of nicotine addiction is increasing our knowledge of addiction in general, and may have practical consequences. Genetics may explain how the nicotine in tobacco products causes lung cancer, and why nicotine is so highly addictive.

Each year, 35 million people try to quit smoking, yet only 7 percent succeed. It is easy to see on a whole-body level how this occurs: Nicotine levels peak 10 seconds after an inhalation and the resulting pleasurable release of dopamine in brain cells fades away in minutes. To keep the good feeling, smoking must continue.

On a molecular level, nicotine does damage through a five-part molecular assembly called a nicotinic receptor. The receptor normally binds the neurotransmitter acetylcholine, but it also binds the similarly shaped nicotine molecule (**figure 8.6**). Certain variants of the receptor bind nicotine very strongly,

Figure 8.5 **The events of addiction.** Addiction is manifest at several levels: at the molecular level, in neuron-neuron interactions in the brain, and in behavioral responses.

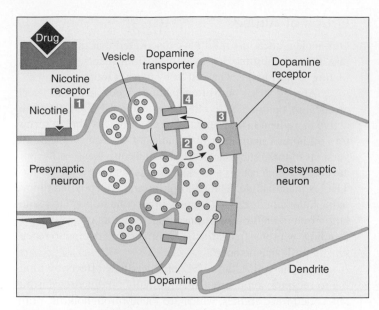

Figure 8.6 Nicotine's effects at the cellular level.
(1) Binding of nicotine to nicotinic receptors, which also bind the neurotransmitter acetylcholine, triggers release of dopamine **(2)** from vesicles into the synapse. Some dopamine binds receptors on the postsynaptic (receiving) neuron **(3)** and some dopamine re-enters the presynaptic neuron through a protein called the dopamine transporter **(4)**. Uptake of dopamine into the postsynaptic cell triggers the pleasurable feelings associated with smoking. The inset illustrates the mechanism of a smoking cessation drug, which blocks the nicotinic receptors.

which triggers a nerve impulse that, in turn, stimulates the pleasurable dopamine release. That may explain the addiction. These receptors are also located on several types of lung cells, where they bind carcinogens. So the nicotine in tobacco causes addiction and susceptibility to lung cancer, and it delivers the carcinogens right to the sensitive lung cells.

A different gene encodes each of the five parts of the nicotinic receptor. If two of the five parts are certain variants, then a person experiences desire to continue smoking after the first cigarette.

Key Concepts Questions 8.4

1. What is drug addiction?
2. How do genes control addiction?
3. How can a human become addicted to a chemical from plants?
4. Explain the basis of nicotine addiction.

8.5 Mood Disorders

We all have moods, but mood disorders, which affect millions of people, impair the ability to function on a day-to-day basis. Context is important in evaluating extreme moods. Symptoms that may lead to a diagnosis of depression if they occur for no apparent reason are normal in the context of experiencing profound grief. The two most prevalent mood disorders are **major depressive disorder** and **bipolar disorder** (also called manic-depression).

Major Depressive Disorder

Major depressive disorder is more than being "down in the dumps," nor can a person simply "snap out of it." Clinical depression is a disabling collection of symptoms that go beyond sadness. A person with depression no longer enjoys favorite activities. He or she is tired, and has trouble making decisions, concentrating, and recalling details. Motivation lags, anxiety and irritability rise, and crying may start unexpectedly, often several times a day. People with depression have difficulty falling and staying asleep, which intensifies the daytime fatigue. Routine, simple tasks may become overwhelming, and the ability to work or study may become very difficult. Weight may drop as interest in food, as in most everything else, diminishes. Fifteen percent of people hospitalized for severe, recurrent depression ultimately end their lives.

Genes and the environment contribute about equally to depression. The general population risk is 10 percent, but risk to someone with an affected first-degree relative (parent or sibling) is 20 to 30 percent. Many of the genes that contribute to depression are probably not yet identified, but researchers found a few of them by focusing on the disturbed sleep. They looked for genes known to control circadian rhythms (sleep, hormone levels, and body temperature), and discovered eight that show altered expression in the brains of people who had depression compared to brains from people who hadn't been depressed.

A likely cause of depression is deficiency of the neurotransmitter serotonin in synapses. It affects mood, emotion, appetite, and sleep. Antidepressant drugs called selective serotonin reuptake inhibitors (SSRIs) prevent presynaptic neurons from taking up serotonin from the synapse, leaving more of it available to stimulate the postsynaptic cell (**figure 8.7**). This action apparently offsets the neurotransmitter deficit. The genetic connection is that a protein in the plasma membrane of the presynaptic neuron, called the serotonin transporter, may function too efficiently in depressed individuals, taking too much serotonin out of the synapse. SSRIs may begin to produce effects after 1 week, often enabling a person with moderate or severe depression to return to some activities, but full response can take up to 8 weeks.

In the past, to treat patients who have depression, physicians would try one antidepressant drug after another, based on experience with other patients. This trial-and-error approach would often take months. As the genes in the human genome are identified, researchers are identifying profiles of gene variants or gene expression that are associated with response to a particular drug. In the future, physicians will be able to select the most effective drug with the fewest adverse effects to help a patient overcome depression.

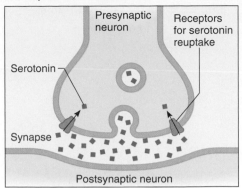
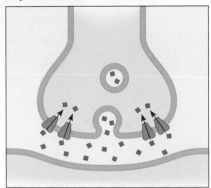
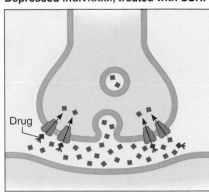

Figure 8.7 **How an antidepressant is thought to work.** Selective serotonin reuptake inhibitors (SSRIs) block the reuptake of serotonin, leaving more of the neurotransmitter in synapses. This corrects a neurotransmitter deficit that presumably causes the symptoms. Overactive or overabundant reuptake receptors can cause the deficit. The precise mechanism of SSRIs is not well understood, and the different drugs may work in slightly different ways.

Bipolar Disorder

Bipolar disorder is much rarer than depression. It affects 1 percent of the population and has a general population lifetime risk of 0.5 to 1.0 percent. Weeks or months of depression alternate with periods of mania, when the person is hyperactive and restless, and may experience a rush of ideas and excitement. A person who is normally frugal might, when manic, spend lavishly. In one subtype, the "up" times are termed hypomania, and they seem more a temporary reprieve from depression than the bizarre behavior of full mania. Bipolar disorder with hypomania may appear to be depression. This is an important clinical distinction because different drugs are used to treat depression and bipolar disorder.

Many gene variants contribute to bipolar disorder. Early genetic studies looked at large Amish families, in whom the manic phase of the disorder was very obvious amid their restrained lifestyle. But studies in other families implicated different genes. Studies disagree because many gene variant combinations cause or contribute to the phenotype of bipolar disorder, but only a few such variants are seen in any one family. A systematic scrutiny of all studies of the genes behind bipolar disorder suggests as much as 10 percent of the genome—that is, hundreds of genes—are part of the clinical picture.

Key Concepts Questions 8.5

1. What are the symptoms of major depressive disorder?
2. To what extent is depression inherited?
3. What are some genes that can cause or raise the risk of developing depression?
4. What are the symptoms of bipolar disorder?
5. How is the role of genes in bipolar disorder complex?

8.6 Schizophrenia

Schizophrenia is a debilitating loss of the ability to organize thoughts and perceptions, which leads to a withdrawal from reality. It is a form of psychosis, which is a disorder of thought and sense of self. In contrast, the mood disorders are emotional, and the dementias are cognitive. Various forms of schizophrenia together affect 1 percent of the world's population. Ten percent of affected individuals commit suicide.

Identifying genetic contributions to schizophrenia illustrates the difficulties in analyzing a behavioral condition. Some of the symptoms are also associated with other illnesses; many genes cause or contribute to schizophrenia; and several environmental factors mimic the condition.

Signs and Symptoms

The first signs of schizophrenia often affect thinking. In late childhood or early adolescence, a person might develop trouble paying attention in school, and learning may become difficult as memory falters and information-processing skills diminish. Symptoms of psychosis begin between ages 17 and 27 for males and 20 and 37 for females. These include delusions and hallucinations, sometimes heard, sometimes seen. A person with schizophrenia may hear a voice giving instructions. What others recognize as irrational fears, such as being followed by monsters, are very real to the person with schizophrenia. Speech reflects the garbled thought process, skipping from topic to topic with no obvious thread of logic. Responses may be inappropriate, such as laughing at sad news. Artwork by a person with schizophrenia can display the characteristic fragmentation of the mind (**figure 8.8**). (Schizophrenia means "split mind," but it does not cause a split or multiple personality.) In many patients the course of schizophrenia plateaus (evens out) or becomes episodic. It is not a continuous decline, as is the case for dementia.

Figure 8.8 Schizophrenia alters thinking. People with schizophrenia communicate the disarray of their thoughts in characteristically disjointed drawings.

Table 8.2	Empiric Risks for Schizophrenia
Relationship	**Risk**
MZ twin	48%
DZ twin	17%
Child	13%
Sibling	9%
Parent	6%
Half sibling	6%
Grandchild	5%
Niece/nephew	4%
Aunt/uncle	2%
First cousin	2%
General population	1%

Genetic Associations

A heritability of 0.8 and empiric risk values indicate a strong role for genes in causing schizophrenia (**table 8.2**). However, it is possible to develop some of the symptoms, such as disordered thinking, from living with and imitating people who have schizophrenia. Although concordance is high, a person who has an identical twin with schizophrenia has a 52 percent chance of *not* developing it. Therefore, the condition has a significant environmental component, too.

Dozens of genes may interact with an environmental trigger or triggers to cause schizophrenia. Genome-wide screens of families with schizophrenia reveal at least twenty-four sites where affected siblings share alleles much more often than the 50 percent of the time that Mendel's first law predicts. These regions may harbor genes that contribute to development of schizophrenia. People with schizophrenia are particularly likely to have rare duplications or deletions of DNA that arise *de novo*—in them—rather than being inherited.

Environmental Influences

Several environmental factors may increase the risk of developing schizophrenia. These include birth complications, fetal oxygen deprivation, herpesvirus infection at birth, and malnutrition or traumatic brain injury in the mother. Infection during pregnancy is a particularly well-studied environmental factor in elevating risk of schizophrenia. When a pregnant woman is infected, her immune system bathes the brain of the embryo or fetus with cytokines, which cause inflammation and can subtly alter brain development. It is likely the mother's immune response that increases the risk because many pathogens cannot cross the placenta and reach the fetus.

The idea that maternal infection can contribute to schizophrenia came from observations that many people with schizophrenia were born in the winter, especially in the years of flu pandemics. Research supports this observation. Blood stored from 12,000 pregnant women during the 1950s and 1960s shows an association between high levels of the cytokine interleukin-8 and having a child who developed schizophrenia.

Environmental influences after birth can also elevate the risk of developing schizophrenia. A study in Europe of 7,500 people with schizophrenia identified four such risk factors: living in cities; migration; regular marijuana use; and victimization during childhood, such as abuse, neglect, and bullying. The opening essay to chapter 11 discusses another environmental factor that can greatly increase the risk of schizophrenia: starvation.

An Epigenetic Explanation

One class of mutant genes that cause schizophrenia suggests how the condition might arise. These genes encode enzymes called methyltransferases, which transfer methyl groups (CH_3; see figure 6.12) to shield specific genes at specific times, silencing their expression. This control of gene expression is epigenetic, because the DNA sequences are not altered. Researchers measured methyltransferase activity in neurons from the prefrontal cortex region of the brains of individuals who had died at different ages, ranging from embryos to 80-year-olds, from nonpsychiatric causes, to construct a "human prefrontal cortex epigenome across the life span." The epigenome revealed that as we age, more genes are silenced. The prefrontal cortex is involved in addiction and also controls executive function, which is the ability to plan, organize, learn, remember, and relate past events to current ones.

In schizophrenia, methyltransferases in the prefrontal cortex silence genes to too great an extent. Researchers hypothesize that in some people who develop schizophrenia, methyltransferases worked too well during a critical period in development—when the fetus becomes the newborn. This is the time when a person first reacts to the outside environment. It then takes years for the symptoms of disordered thinking to arise. Methyltransferase overactivity is also seen in some people who have bipolar disorder.

Key Concepts Questions 8.6

1. What are symptoms of schizophrenia?
2. How do genes and environmental factors influence susceptibility to schizophrenia?

8.7 Autism

Autism spectrum disorder (OMIM 290850) impairs socialization and communication skills, with symptoms appearing before age 3 even if this is recognized in retrospect. One in 68 children in the United States has autism, and boys outnumber girls four to one. About 25 percent of affected children have seizures as they grow older. Although 70 percent of people with autism have intellectual disability, others may be very intelligent.

In severe autism, a toddler may lose vocabulary words and cease starting conversations. He or she stops playing with other children, preferring to be alone. Rocking back and forth or clutching a treasured object and using repetitive motions are soothing behaviors for some children. The child may refer to herself by her name, rather than "me," not make eye contact, and appear oblivious to nonverbal cues such as facial expressions or tone of voice. Special education programs that provide a rigid routine can help an individual with autism to connect with the world. People who have autism may have the socialization effects but not the language deficits. (Older classification called this Asperger syndrome.)

In autism, many genes contribute to different degrees against a backdrop of environmental influences. Environmental triggers include prenatal exposures to rubella (German measles) and to the drug valproate. Evidence is accumulating that exposure to folic acid supplements late in gestation can cause epigenetic changes that increase the risk of autism (see Case Studies and Research Results #5). (Scientific evidence does not support a link to the mercury compound once used in vaccines.) A better way to identify environmental risk factors may be to consider different genetic subtypes, so that studies compare individuals with the same underlying problem.

Heritability of autism is high: about 90 percent. Although siblings of affected children are at a 15 percent risk of being affected, compared to the less than 1 percent for the general population, in some MZ (identical) twin pairs, one has autism and one does not. More than 30 susceptibility or causative genes have been identified. In about 15 percent of people with autism, the condition is part of a syndrome, including single-gene disorders such as Rett syndrome (see the chapter 2 opener), and chromosomal conditions such as fragile X syndrome (see Clinical Connection 12.1) and Down syndrome (see Bioethics: Choices for the Future in chapter 13). A diagnostic workup for autism includes tests for these conditions, as well as a test called chromosomal microarray analysis, which detects copy number variants (tiny deletions and duplications of DNA).

Our understanding of the genetics of autism has paralleled development of genetic technology. At first, the rare individuals who had autism as well as an abnormal chromosome pointed the way to a place in the genome—the abnormal chromosome—that likely harbors a predisposing or causative mutation. Then genome-wide association studies added mutations in more genes to the list.

Today exome and whole-genome sequencing projects are enabling researchers to identify many more genes behind autism. Experiments focus on candidate genes within the huge datasets that these technologies generate. **Figure 8.9** illustrates three general strategies for isolating autism genes (or other genes for a shared phenotype) from exome or genome data.

The first approach, "multiple unrelated affected subjects," compares the exome or genome sequences of many unrelated individuals with autism to "reference" genomes from healthy people compiled in a population database, searching for genes in which the affected children share unusual variants that are not in the healthy people. In one study, researchers began by comparing genome sequences from affected members of 55 families that each had several people with autism to reference genomes. They identified 153 parts of the genome. Then the researchers made probes—copies of those identified DNA sequences that incorporate a fluorescent label—and applied them to the DNA of 2,175 children with autism and 5,801 children without autism. In those 153 genome regions, the investigators narrowed down 15 sites, with a total of 24 variants, seen in the affected children and associated with a twofold increase in the risk of autism.

The second approach, "trio analysis," compares the DNA in parent-child trios. These studies either find that parents are carriers of autosomal recessive or X-linked conditions, or that the autism in the child arose *de novo*. A study in Canada using whole-genome sequencing identified genes for half of 32 families with autism. About 60 percent of the mutations were inherited and 40 percent had arisen *de novo*.

The third approach, "recessive analysis," compares DNA sequences in families in which more than one child is affected but the parents are not, seeking genes that are mutant in two copies in the children with autism. A study that used this approach looked at three very large families from the Middle East that have many cousin-cousin marriages, which are at higher risk for autosomal recessive conditions due to inheritance of mutations from shared ancestors. When the researchers discovered mutations in three genes already known to cause other diseases, they hypothesized that some cases of autism are really mild or atypical versions of recognized syndromes.

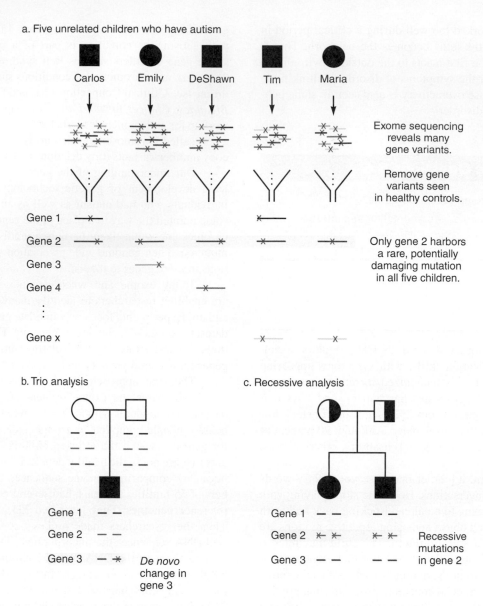

a. Five unrelated children who have autism

Carlos   Emily   DeShawn   Tim   Maria

Exome sequencing reveals many gene variants.

Remove gene variants seen in healthy controls.

Gene 1

Gene 2   Only gene 2 harbors a rare, potentially damaging mutation in all five children.

Gene 3

Gene 4

Gene x

b. Trio analysis

Gene 1 — —

Gene 2 — —

Gene 3 — *   *De novo* change in gene 3

c. Recessive analysis

Gene 1 — —   — —

Gene 2 *—* *—*   *—* *—*   Recessive mutations in gene 2

Gene 3 — —   — —

Figure 8.9 **Three approaches to use exome sequencing to discover genes that contribute to autism.** **(a)** Comparison to a reference genome identifies genes that unrelated people with autism share. **(b)** Trio analysis detects *de novo* mutations in a child. **(c)** Recessive analysis detects mutations in families with more than one affected child.

The researchers found seven other known autosomal recessive conditions that also cause autism. These disorders are pleiotropic—they have several manifestations, and in some children, autism is the only symptom.

After identifying mutations that might cause or contribute to autism based on their occurrence in people with the condition, the next step is to consider what these genes do. How might malfunction of the gene impair socialization and communication skills? For example, two types of proteins found at synapses, called **neurexins** and **neuroligins,** are likely part of the autism picture because their functions make sense (**figure 8.10**).

Three types of neurexins emanate from the "sending" presynaptic neuron, and a neuroligin extends toward the

neurexin from the "receiving" postsynaptic neuron. When the two protein types meet, they draw the neurons closer together so that one can signal the other, using the neurotransmitter glutamate. Mutations misfold the two proteins so that they cannot interact as they normally do, which disrupts the sending of signals from one nerve cell to another. If this interference happens in early childhood, when synapses naturally form in response to experiences in brain parts that oversee learning and memory, the behaviors of autism could be the result.

Identifying the genes behind autism and the other behavioral (also known as neuropsychiatric) disorders discussed in this chapter can have clinical benefits. Genetic information can add precision to a traditional diagnosis based on symptoms,

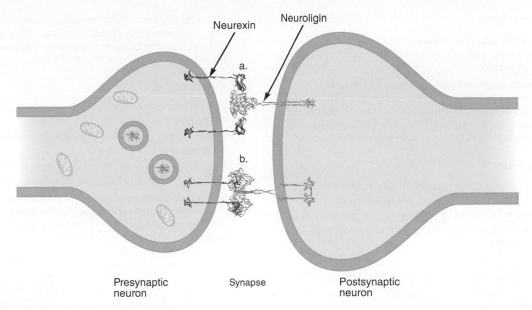

Neurexin   Neuroligin

a.

b.

Presynaptic   Synapse   Postsynaptic
neuron   neuron

Figure 8.10 Understanding autism. An emerging view of a cause of autism points to changes in the nervous system that occur as a child begins to explore the environment. One such system consists of two types of proteins that strengthen synaptic connections among neurons associated with learning and memory **(a).** The proteins are called neurexins (in presynaptic neurons) and neuroligins (in postsynaptic neurons). In a child with autism, the neurexin and neuroligin connections that would form following early experiences may not be strong enough **(b).**

define clinical subtypes detectable with older diagnostic tests, and provide a way to predict risk that relatives will develop the condition, which is greater than the general population risk. Finally, identifying genes behind behavioral disorders is important in drug discovery research. One day we may have genetically based personalized treatments for subtypes of the multifactorial conditions that we have traditionally considered together based on their phenotypes.

Key Concepts Questions 8.7

1. What is autism?
2. What is the evidence that genes contribute to autism risk?
3. What are some environmental factors that contribute to the development of autism?
4. How has changing genetic technology led to the discovery of genes implicated in autism?

Summary

8.1 Genes and Behavior

1. Genes affect how the brain responds to environmental stimuli. **Neurons** and **neuroglia** are the two types of brain cells.
2. Candidate genes for behavioral traits and disorders affect neurotransmission.
3. Old and new genetic tools are being used to describe the biological causes of various behavioral disorders.
4. Variants of hundreds of genes contribute to different degrees to behavioral disorders.

8.2 Sleep

5. Twin studies and single-gene disorders that affect the sleep-wake cycle reveal a large inherited component to sleep behavior.
6. The *period* genes enable a person to respond to day and night environmental cues.

8.3 Intelligence and Intellectual Disability

7. The general intelligence (*g*) value measures the inherited portion of IQ that may underlie population variance in IQ test performance.
8. Studies indicate that general intelligence is maintained throughout life. The fact that many chromosomal disorders affect intelligence suggests high heritability.
9. Exome sequencing is identifying mutations that cause or contribute to intellectual disability.

8.4 Drug Addiction

10. Drug addiction arises from tolerance and dependence. Addiction produces stable changes in certain brain parts.
11. Candidate genes for drug addiction include the dopamine D(2) receptor and variants in nicotinic receptor parts.

8.5 Mood Disorders

12. **Major depressive disorder** is common and associated with mutations in genes that regulate circadian rhythms and deficits of serotonin.

13. **Bipolar disorder** is depressive periods and periods of mania or hypomania. Hundreds of genes may raise the risk of developing this disorder. Different families have different combinations of these gene variants, some of which, under certain environmental conditions, can lead to the disorder.

8.6 Schizophrenia

14. Schizophrenia greatly disrupts the ability to think and perceive the world. Onset is typically in early adulthood, and the course is episodic or steady.

15. Empiric risk estimates and heritability indicate a large genetic component.

16. Many genes and environmental influences are associated with schizophrenia. Overactive methyltransferases acting around the time of birth provide an epigenetic explanation for the origin of schizophrenia.

8.7 Autism

17. Autism is a loss of communication and social skills beginning in early childhood.

18. Exposure to certain pathogens and chemicals can increase risk of autism.

19. Sequencing exomes and genomes is revealing mutations that cause autism. Mutations may be inherited or new.

20. **Neuroligins** and **neurexins** are proteins embedded in the plasma membranes that join neurons in response to environmental stimuli and may be abnormal in some cases of autism.

www.mhhe.com/lewisgenetics11

Answers to all end-of-chapter questions can be found at **www.mhhe.com/lewisgenetics11**. You will also find additional practice quizzes, animations, videos, and vocabulary flashcards to help you master the material in this chapter.

Review Questions

1. What are the two major types of cells in the brain, and what do they do?

2. Why are behavioral traits nearly always multifactorial?

3. List the pathways or mechanisms that include proteins that, when absent or atypical, cause variations in behavior.

4. How do epigenetic changes connect nurture (the environment) to nature (DNA)?

5. What is the evidence that the Utah family with familial advanced sleep phase syndrome has a genetic condition rather than them all just becoming used to keeping weird hours?

6. Choose a behavior discussed in the chapter and identify the region of the brain and/or a specific molecule that is affected.

7. Choose a study discussed in the chapter, or find another one, that demonstrates something about the role of a gene in a behavioral condition, and explain how the gene affects behavior.

8. Why is investigating the role of genetics in a behavioral trait or condition more complicated than determining the genetic contribution to a Mendelian trait or disorder?

9. What were some of the prejudices that were part of studying the inheritance of intelligence?

10. Why is it so difficult to identify factors that cause intellectual disability?

11. What are the two defining characteristics of addiction?

12. Select a drug mentioned in the chapter and explain what it does to the nervous system.

13. Describe how exome sequencing is being used to better understand the inherited component of a trait, behavior, or disorder discussed in the chapter.

14. What is the evidence that our bodies have their own uses for cocaine, THC, and opium?

15. What is an environmental factor that may influence the development of schizophrenia?

16. Select a behavioral disorder mentioned in this text, or another. Compare and contrast the type of information derived from classical studies on families versus genome-wide association studies or exome or genome sequencing.

17. Distinguish among the three approaches to autism gene discovery that figure 8.9 shows.

18. How might abnormal interactions between neurexins and neuroligins at brain synapses cause autism?

Applied Questions

1. Consult chapter 7 and suggest two genes in which mutations might raise the risk of developing an eating disorder.

2. A company is attempting to market a test for variants of genes that control the neurotransmitters vasopressin and oxytocin. The test is based on experiments, many in rodents, which associate these genes with monogamy and parental ability. What might be a problem with such a test?

3. Many older individuals experience advanced sleep phase syndrome. Even though this condition is probably a normal part of aging, how might research on the Utah family with an inherited form of the condition help researchers develop a drug to help the elderly sleep through the night and awaken later in the morning?

4. How does the subunit structure of the nicotinic receptor provide a mechanism for epistasis? (Epistasis is one gene affecting expression of another; see chapter 5.)

5. A television and film star went into rehab for "sex addiction," much to the embarrassment of his wife. Describe how you would evaluate whether this diagnosis is valid, either in an individual or in a large study.

6. What might be the advantages and disadvantages of a genetic test done at birth that indicates whether a person is at high risk for developing a drug addiction?

7. In some studies of depression and bipolar disorder, correlations to specific gene variants are only evident when participants are considered in subgroups based on symptoms. What might be a biological basis for this finding?

8. A study found that the risk of schizophrenia among spouses of people with schizophrenia who have no affected blood relatives is 2 percent. What might this indicate about the causes of schizophrenia?

9. In the United States, the incidence of autism has dramatically increased since 2012.

   a. Does this finding better support a genetic cause or an environmental cause for autism?

   b. What is a nongenetic factor that might explain the increased incidence of autism?

10. A "markers for addiction" gene panel scans people's DNA for variants in genes that are associated with addiction tendency. The eleven genes encode proteins that are ion channels, cell surface molecules, receptors, enzymes, or cell adhesion molecules. Look up one of them in *Mendelian Inheritance in Man* (or elsewhere) and explain its role in increasing the risk of nicotine addiction. The markers are: *CHRNA3, CHRNA5, CHRNB3, CLCA1, CTNNA3, GABRA4, KCNJ6, NRXN1,* and *TRPC7.*

Web Activities

1. Use OMIM to learn about one of the eight genes that disrupt circadian rhythms and are also implicated in major depressive disorder. Explain how the gene, when mutant, might have this effect. The genes are: *BMAL1, PER1, PER2, PER3, NR1D1, DBP, BHLHE40,* and *BHLHE42.*

2. Choose one of the following genes, read about it in OMIM or elsewhere, and discuss how mutations in the gene might contribute to intellectual disability: *FMR1, ARID1B, EFTUD2, SCN2A* and *SMAD4.*

3. Choose one of the following genes, read about it in OMIM or elsewhere, and discuss how mutations might contribute to autism: *AFF2, AMT,CAPRIN1, CHD7, FOXP1, GRIN2B, KCNQZ, LAMC3, MECP2, NLGN1, NRXN1, PEX7, POMGNT1, SCN1A, SCN2A, SYNE1, VIP, VPS13B.*

4. Look at the website for a project that is using exome sequencing to find *de novo* mutations that cause intellectual disability, and describe either what the researchers are doing, or how a family reacts to the news that their child's mutation has been discovered. Projects include: FORGE (Finding of Rare Disease Genes in Canada), Deciphering Developmental Disorders (United Kingdom), the German Mental Retardation Network, and the ClinVar program from the National Human Genome Research Institute (United States).

5. Go to the website for a drug used to treat any of the conditions mentioned in the chapter and explain how the drug is thought to work. (The smoking cessation drug varenicline [Chantix] has an interesting mechanism.)

Forensics Focus

1. When 43-year-old F.F. discovered his soon-to-be-ex wife in bed with her boyfriend, he shot and killed them both. The defense ordered a pre-trial forensic psychiatric work-up that included genotyping for the enzyme monoamine oxidase *A* (*MAOA,* OMIM 309850), which breaks down the neurotransmitters serotonin, dopamine, and noradrenalin, and for the serotonin transporter (*SLC6A4,* OMIM 182138). A "high-activity" allele for *MAOA* is associated with violence in people who also suffered child abuse. Inheriting one or two "short" alleles of *SLC6A4* is associated with depression and suicidal thoughts, in people who have experienced great stress. F.F. had the high-activity MAOA genotype, two short *SLC6A4* alleles, and a lifetime of abuse and stress. However, the judge

ruled that the science was not far enough along to admit the genotyping results as evidence.

a. Under what conditions or situations do you think it is valid to include genotyping results in cases like this?

b. In a Dutch family, a mutation disables *MAOA*, causing "a syndrome of borderline mental retardation and abnormal behavior," according to one report. Family members had committed arson, attempted rape, and shown exhibitionism. How can both high and low levels of an enzyme each cause behavioral problems?

c. What is a limitation to use of behavioral genotyping in a criminal trial?

Case Studies and Research Results

1. Starting in 1998, the US government mandated addition of folic acid to grains and cereals to lower the risk of certain birth defects (neural tube defects or NTDs) that can arise from a deficiency of this B vitamin. In addition, for many years women have taken high doses of folic acid throughout pregnancy to prevent NTDs, even though they happen on day 28.

   Ted Brown, Mohammed Junaid and their colleagues at the New York State Institute for Basic Research in Developmental Disabilities hypothesized that fetal exposure to excess folic acid may explain the increase in diagnoses of autism since supplementation of the vitamin began. Folic acid affects gene expression by methylating genes (see Figure 6.12). They reported in 2011 that in white blood cell precursors (a cell type easy to study), exposure to high levels of folic acid altered methylation of areas that are rich in cytosine and guanine-rich at the starts of certain genes. These regions, called promoters, control how much of a gene's product is made. Folic acid altered expression of more than 1,000 genes in the cells, in a dose-dependent manner.

   In 2014, the researchers reported that key developmental genes, including those implicated in autism in humans, showed different methylation patterns in the brains of mouse pups if their mothers received folic acid supplementation while pregnant.

   a. Is the relationship between folic acid exposure and autism an association, a correlation, or a cause? (You might have to look these up.)

   b. Suggest further experiments to investigate the connection between prenatal folic acid exposure and autism.

   c. What information is needed to determine exactly when during pregnancy folic acid supplementation is most beneficial?

   d. A report from another research group in 2013, using data from children, concluded that folic acid supplementation started 4 weeks before pregnancy does not increase the risk of autism. The data on folic acid exposure go up to the 8th week of pregnancy. Do you think the conclusions are valid? Cite a reason for your answer.

   e. Look up information on the idea that vaccines cause autism and compare the evidence for it to the evidence that suggests that folic acid exposure plays a role in autism.

   f. If the connection between folic acid exposure and autism is confirmed, suggest how the information might be used to help children with autism.

2. Until the 1990s, bipolar disorder was thought not to affect children under age 18. Psychiatrists today maintain that fewer than 1 percent of children have bipolar disorder. However, the percentage of children being diagnosed with bipolar disorder has soared since 2000, along with the publication of many books written by parents of affected children, and appearances on TV talk shows of affected children and their parents.

   a. Does this pattern of increasing disease incidence suggest a genetic or an environmental cause?

   b. Suggest another explanation for the recent apparent increase in incidence of bipolar disorder in children.

   c. Most children are diagnosed with bipolar disorder based on their answers to questions. What might a genetic diagnosis entail?

3. Variants of a gene called *CYP2A6* are associated with the speed at which a person metabolizes nicotine. A "fast-metabolizer" smokes more. Design a clinical trial using this information to test whether the *CYP2A6* genotype can be used to predict which individuals will respond to specific smoking cessation medications.

4. In a short e-book by Karen Russell called "Sleep Donation," overuse of lit screens leads to a pandemic of an infectious, deadly insomnia due to abnormal expression of the gene encoding a protein mentioned in the chapter. Name the implicated gene, and state whether it would have to be overexpressed or underexpressed to cause insomnia. (The book is fiction.)

DNA is the genetic material. *DNA bursts from this treated bacterial cell. The DNA in a human cell would unravel to nearly 6 feet, yet it is wound into the nucleus of a microscopic cell.*

DNA Structure and Replication

Learning Outcomes

9.1 Experiments Identify and Describe the Genetic Material

1. Describe the experiments that showed that DNA is the genetic material and protein is not.

2. Explain how Watson and Crick deduced the structure of DNA.

9.2 DNA Structure

3. List the components of a DNA nucleotide building block.

4. Explain how nucleotides are joined into two chains to form the strands of a DNA molecule.

9.3 DNA Replication—Maintaining Genetic Information

5. Explain the semiconservative mechanism of DNA replication.

6. List the steps of DNA replication.

7. Explain how the polymerase chain reaction amplifies DNA outside cells.

9.4 Sequencing DNA

8. Explain the basic strategy used to determine the base sequence of a DNA molecule.

9. Explain how next-generation sequencing improves upon Sanger sequencing.

The **BIG** Picture

DNA is the basis of life because of three qualities: It holds information, it copies itself, and it changes.

On the Meaning of Gene

To a biologist, *gene* has a specific definition—a sequence of DNA that tells a cell how to assemble amino acids into a particular protein. To others, "gene" has different meanings:

To folksinger Arlo Guthrie, *gene* means aging without signs of the Huntington disease that claimed his father, legendary folksinger Woody Guthrie.

To rare cats in New England, *gene* means extra toes.

To Adolph Hitler and others who have dehumanized those not like themselves, the concept of *gene* was abused to justify genocide.

To a smoker, a *gene* may mean lung cancer develops.

To a redhead in a family of brunettes, *gene* means an attractive variant.

To a woman whose mother and sisters had breast cancer, *gene* means escape from their fate—and survivor guilt.

To a lucky few, *gene* means a mutation that locks HIV out of their cells.

To people with diabetes, *gene* means safer insulin.

To a forensic entomologist, *gene* means a clue in the guts of maggots devouring a corpse.

To scientists-turned-entrepreneurs, *gene* means money.

Collectively, our genes mean that we are very much more alike than different from one another.

9.1 Experiments Identify and Describe the Genetic Material

"A genetic material must carry out two jobs: duplicate itself and control the development of the rest of the cell in a specific way," wrote Francis Crick, codiscoverer with James Watson of the three-dimensional structure of DNA in 1953. Only DNA fulfills these requirements.

DNA was first described in the mid-nineteenth century, when Swiss physician and biochemist Friedrich Miescher isolated nuclei from white blood cells in pus on soiled bandages. He discovered in the nuclei, an unusual acidic substance containing nitrogen and phosphorus. He and others found it in cells from a variety of sources. Because the material resided in cell nuclei, Miescher called it *nuclein* in an 1871 paper; subsequently, it was called a nucleic acid. Few people appreciated the importance of Miescher's discovery at the time, when inherited disease was widely blamed on protein.

In 1902, English physician Archibald Garrod was the first to provide evidence linking inherited disease and protein. He noted that people who had certain inborn errors of metabolism lacked certain enzymes. Other researchers added evidence of a link between heredity and enzymes from other species, such as fruit flies with unusual eye colors and bread molds with nutritional deficiencies. Both organisms had absent or malfunctioning specific enzymes. As researchers wondered about the connection between enzymes and heredity, they returned to Miescher's discovery of nucleic acids.

DNA *Is* the Hereditary Molecule

In 1928, English microbiologist Frederick Griffith took the first step in identifying DNA as the genetic material. He was studying pneumonia in the years after the 1918 flu pandemic. Griffith noticed that mice with a certain form of pneumonia harbored one of two types of *Streptococcus pneumoniae* bacteria. Type R bacteria were rough in texture. Type S bacteria were smooth because they were enclosed in a polysaccharide (a type of carbohydrate) capsule. Mice injected with type R bacteria did not develop pneumonia (**figure 9.1a**), but mice injected with type S did (figure 9.1b). The polysaccharide coat shielded the bacteria from the mouse immune system, enabling them to cause severe (virulent) infection. Injecting mice with unaltered type R or type S bacteria served as control experiments, which represent the situation without the experimental intervention.

When type S bacteria were heated, which killed them, they no longer could cause pneumonia in mice. However, when Griffith injected mice with a mixture of type R bacteria plus heat-killed type S bacteria—neither of which, alone, was deadly to the mice—the mice died of pneumonia (figure 9.1d). Their bodies contained live type S bacteria, encased in polysaccharide.

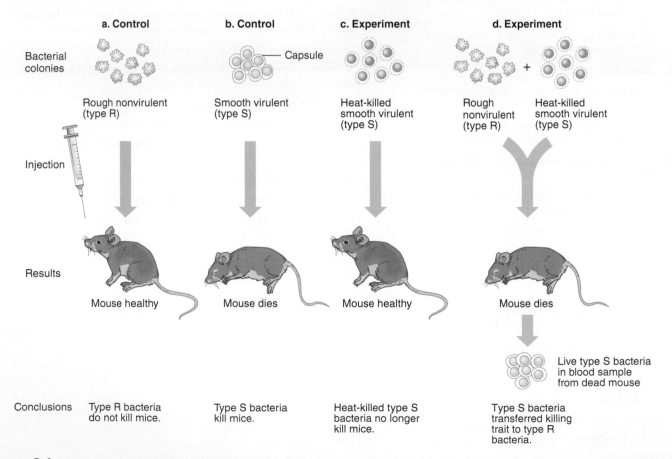

Figure 9.1 **Discovery of bacterial transformation.** Griffith's experiments showed that a molecule in a lethal type of bacteria can transform nonkilling (nonvirulent) bacteria into killers (virulent).

Griffith termed the apparent conversion of one bacterial type into another "transformation." How did it happen? What component of the dead, smooth bacteria transformed type R to type S?

U.S. physicians Oswald Avery, Colin MacLeod, and Maclyn McCarty hypothesized that a nucleic acid might be Griffith's "transforming principle." They observed that treating broken-open type S bacteria with a protease—an enzyme that dismantles protein—did not prevent the transformation of a nonvirulent to a virulent strain, but treating it with deoxyribonuclease (or DNase), an enzyme that dismantles DNA only, did disrupt transformation. In 1944, they confirmed that DNA transformed the bacteria. They isolated DNA from heat-killed type S bacteria and injected it with type R bacteria into mice (**figure 9.2**). The mice died, and their bodies contained active type S bacteria. The conclusion: DNA passed from type S bacteria into type R, enabling the type R to manufacture the smooth coat necessary for infection. Once type R bacteria encase themselves in smooth coats, they are no longer type R.

Protein *Is Not* the Hereditary Molecule

Science seeks answers by eliminating explanations. It provides evidence in support of a hypothesis, not proof, because conclusions can change when new data become available. To identify the genetic material, researchers also had to show that protein does *not* transmit genetic information. To do this, in 1953, U.S. microbiologists Alfred Hershey and Martha Chase used *Escherichia coli* bacteria infected with a virus that consisted of a protein "head" surrounding DNA. Viruses infect bacterial cells by injecting their DNA (or RNA) into them. Infected bacteria may then produce many more viruses. The viral protein coats remain outside the bacterial cells.

Researchers can analyze viruses by growing them on culture medium that contains a radioactive chemical that the viruses take up. The "labeled" viral nucleic acid then emits radiation, which can be detected in several ways. Hershey and Chase knew that protein contains sulfur but not phosphorus, and that nucleic acids contain phosphorus but not sulfur. Both elements also come in radioactive forms. When Hershey and Chase grew

viruses in the presence of radioactive sulfur, the viral *protein coats* took up and emitted radioactivity, but when they ran the experiment using radioactive phosphorus, the viral *DNA* emitted radioactivity. If protein is the genetic material, then the infected bacteria would have radioactive sulfur. But if DNA is the genetic material, then the bacteria would have radioactive phosphorus.

Hershey and Chase grew one batch of virus in a medium containing radioactive sulfur (designated ^{35}S) and another in a medium containing radioactive phosphorus (designated ^{32}P). The viruses grown on sulfur had their protein marked, but not their DNA, because protein incorporates sulfur but DNA does not. Conversely, the viruses grown on labeled phosphorus had their DNA marked, but not their protein, because this element is found in DNA but not protein.

After allowing several minutes for the virus particles to bind to the bacteria and inject their DNA into them, Hershey and Chase agitated each mixture in a blender, shaking free the empty virus protein coats. The contents of each blender were collected in test tubes, then centrifuged (spun at high speed). This settled the bacteria at the bottom of each tube because the lighter virus coats drift down more slowly than bacteria.

At the end of the procedure, Hershey and Chase examined fractions containing the virus coats from the top of each test tube and the infected bacteria that had settled to the bottom (**figure 9.3**). In the tube containing viruses labeled with sulfur, the virus coats were radioactive, but the virus-infected bacteria, containing viral DNA, were not. In the other tube, where the virus had incorporated radioactive phosphorus, the virus coats carried no radioactive label, but the infected bacteria were radioactive. Therefore, the part of the virus that could enter bacteria and direct them to mass produce more virus was the part that had incorporated phosphorus—the DNA. The genetic material is DNA, and not protein.

Discovering the Structure of DNA

In 1909, Russian-American biochemist Phoebus Levene identified the 5-carbon sugar **ribose** as part of some nucleic acids, and in 1929, he discovered a similar sugar—**deoxyribose**—in other nucleic acids. He had revealed a major chemical distinction between RNA and DNA: RNA has ribose, and DNA has deoxyribose. (Recall that RNA serves as a carrier of the information in a DNA molecule that instructs the cell to manufacture a particular protein.)

Levene then discovered that the three parts of a nucleic acid—a sugar, a nitrogen-containing base, and a phosphorus-containing component—are present in equal proportions. He deduced that a nucleic acid building block must contain one of each component. Furthermore, although the sugar and phosphate portions of nucleic acids were always the same, the nitrogen-containing bases were of four types. Scientists at first thought that the bases were present in equal amounts, but if this were so, DNA could not encode as much information as it could if the number of each base type varied. Imagine how much less useful a written language would be if it had to use all the letters with equal frequency.

In the early 1950s, two lines of experimental evidence converged to provide the direct clues that finally revealed

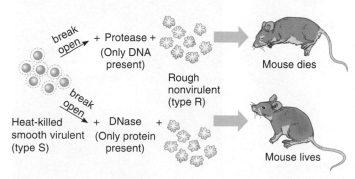

Figure 9.2 **DNA is the "transforming principle."** Avery, MacLeod, and McCarty identified DNA as Griffith's transforming principle. By adding enzymes that either destroy proteins (protease) or DNA (deoxyribonuclease or DNase) to bacteria that were broken apart to release their contents, the researchers demonstrated that DNA transforms bacteria—and that protein does not.

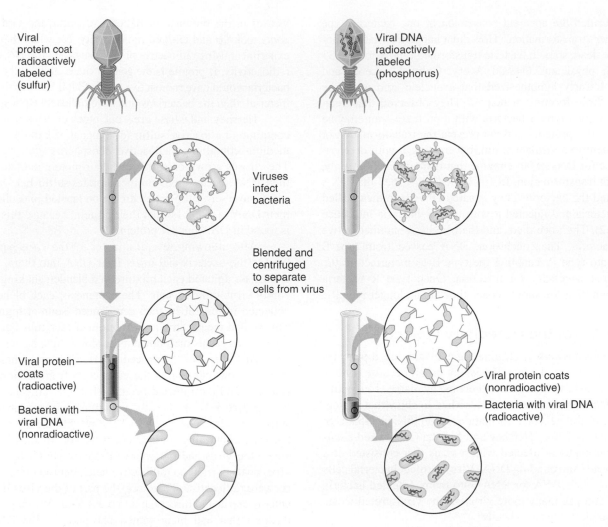

Viral protein coat radioactively labeled (sulfur)

Viral DNA radioactively labeled (phosphorus)

Viruses infect bacteria

Blended and centrifuged to separate cells from virus

Viral protein coats (radioactive)

Bacteria with viral DNA (nonradioactive)

Viral protein coats (nonradioactive)

Bacteria with viral DNA (radioactive)

Figure 9.3 **DNA is the hereditary material; protein is not.** Hershey and Chase used different radioactive molecules to distinguish the viral protein coat from the genetic material (DNA). These "blender experiments" showed that the virus transfers DNA, and not protein, to the bacterium. Therefore, DNA is the genetic material. The blender experiments used particular types of sulfur and phosphorus atoms that emit detectable radiation.

DNA's structure. Austrian-American biochemist Erwin Chargaff showed that DNA in several species contains equal amounts of the bases **adenine** (A) and **thymine** (T) and equal amounts of the bases **guanine** (G) and **cytosine** (C). Next, English physicist Maurice Wilkins and English chemist Rosalind Franklin bombarded DNA with X rays using a technique called X-ray diffraction, then deduced the overall structure of the molecule from the patterns in which the X rays were deflected.

Rosalind Franklin provided a pivotal clue to deducing the three-dimensional structure of DNA. She distinguished two forms of DNA—a dry, crystalline "A" form, and the wetter type seen in cells, the "B" form. It took her 100 hours to obtain "photo 51" of the B form in May 1952 (**figure 9.4**). When a graduate student showed photo 51 to Wilkins, who

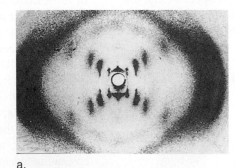

a.

Figure 9.4 **Deciphering DNA structure. (a)** Rosalind Franklin's "photo 51" of B DNA was critical to Watson and Crick's deduction of the three-dimensional structure of the molecule. The "X" in the center indicates a helix, and the darkened regions reveal symmetrically organized subunits. **(b)** Franklin died very young, of cancer.

b. Rosalind Franklin 1920–1958

showed it to Watson at the end of January in 1953, the men realized that the remarkable symmetry of the molecule, indicated by the positions of the phosphates in the photograph, best fit the shape of a sleek helix.

The race was on. During February, famed biochemist Linus Pauling suggested a triple helix structure for DNA. This was incorrect. Meanwhile, Watson and Crick, certain of the sugar-phosphate backbone largely from photo 51, turned their attention to the bases. They found the answer not through sophisticated chemistry or crystallography, but using cardboard cutouts of the DNA components.

On Saturday morning, February 28, Watson arrived early for a meeting with Crick. While he was waiting, he played with cardboard cutouts of the four DNA bases, pairing A with A, then A with G. When he assembled A next to T, and G next to C, he noted the similar shapes, and suddenly all the pieces fit. He had been modeling the chemical attractions (hydrogen bonds) between the DNA bases that create the "steps" of the double helix. When Crick arrived 40 minutes later, the two quickly realized they had solved the puzzle of DNA's structure (**figure 9.5**). They published their findings in the April 25, 1953 issue of *Nature* magazine, without ever having done an experiment.

Watson, Crick, and Wilkins eventually received the Nobel Prize. In 1958, Franklin died at the age of 37 from ovarian cancer, and the Nobel can only be awarded to a living person. However, she has become a heroine for her long-unappreciated role in deciphering the structure of DNA. In 2010, researchers found six boxes of lost correspondence from Crick, mostly to Wilkins. The letters, photos, and postcards reveal how upset the elder Wilkins was to be in a race with the young upstarts Watson and Crick.

Table 9.1 summarizes some of the experiments that led to the discovery of the structure of the genetic material. However, researchers have yet to visualize an actual DNA molecule. Transmission electron microscopy can image down

Figure 9.5 **Watson and Crick.** Prints of this famed, if posed, photo fetched a high price when signed and sold at celebrations of DNA's fiftieth anniversary in 2003. Crick, told to point to the model, picked up a slide rule.

to seven copies of crystallized DNA double helices. What we know about DNA structure and function has been deduced and inferred from larger-scale experiments and observations.

Key Concepts Questions 9.1

1. What are the two requirements for a genetic material?
2. Describe the experiments that revealed the nature of the genetic material.
3. Describe the experiments that revealed the structure of the genetic material.

Table 9.1	The Road to the Double Helix	
Investigator	**Contribution**	**Timeline**
Friedrich Miescher	Isolated nuclein in white blood cell nuclei	1869
Frederick Griffith	Transferred killing ability between types of bacteria	1928
Oswald Avery, Colin MacLeod, and Maclyn McCarty	Discovered that DNA transmits killing ability in bacteria	1940s
Alfred Hershey and Martha Chase	Determined that the part of a virus that infects and replicates is its nucleic acid and not its protein	1950
Phoebus Levene, Erwin Chargaff, Maurice Wilkins, and Rosalind Franklin	Discovered DNA components, proportions, and positions	1909–early 1950s
James Watson and Francis Crick	Elucidated DNA's three-dimensional structure	1953
James Watson	Had his genome sequenced	2008

9.2 DNA Structure

A **gene** is a section of a DNA molecule whose sequence of building blocks specifies the sequence of amino acids in a particular protein. The activity of the protein imparts the phenotype. The fact that different building blocks combine to form nucleic acids enables them to carry information, as the letters of an alphabet combine to form words.

Inherited traits are diverse because proteins have diverse functions (see table 10.1). Malfunctioning or inactive proteins can devastate health. Most of the amino acids that are assembled into proteins come from the diet or from breaking down proteins in the cell. The body synthesizes the others.

The structure of DNA is easiest to understand if we begin with the smallest components. A single building block of DNA is a **nucleotide** (**figure 9.6**). It consists of one deoxyribose sugar, one phosphate group (a phosphorus atom bonded to four oxygen atoms), and one nitrogenous base. The bases adenine (A) and guanine (G) are **purines,** which have a two-ring structure. Cytosine (C) and thymine (T) are **pyrimidines,** which have a single-ring structure (**figure 9.7**).

The nitrogenous bases are the information-containing parts of DNA because they form sequences. DNA sequences are measured in numbers of base pairs. The terms kilobase (kb) and megabase (mb) are used to abbreviate a thousand and a million DNA bases, respectively. A particular gene, for example, may be 1,400 bases, or 1.4 kilobases (kb) long.

Nucleotides are joined into long chains when strong attachments called phosphodiester bonds form between the deoxyribose sugars and the phosphates, creating a continuous **sugar-phosphate backbone** (**figure 9.8**). Two such chains of nucleotides align head-to-toe, as **figure 9.9** depicts.

The opposing orientation of the two nucleotide chains in a DNA molecule is called **antiparallelism**. The head-to-toe configuration derives from the structure of the sugar-phosphate backbone, which we can follow by assigning numbers to the five carbons of the sugars based on their relative positions in the molecule (**figure 9.10**). The carbons are numbered from 1 to 5, starting with the first carbon moving clockwise from the oxygen in each sugar in figure 9.10. In **figure 9.11**, one chain runs from the #5 carbon (top of the figure) to the #3 carbon, but the chain aligned with it runs from the #3 to the #5 carbon. These ends are depicted as "5'" and "3'", pronounced "5 prime" and "3 prime."

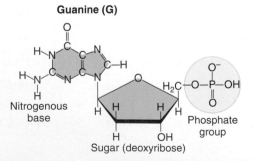

Figure 9.6 **Nucleotides.** A nucleotide of a nucleic acid consists of a 5-carbon sugar, a phosphate group, and an organic, nitrogenous base (G, A, C, or T).

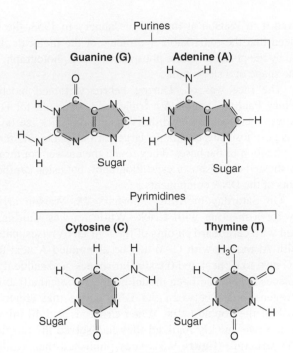

Figure 9.7 **DNA bases are the informational parts of nucleotides.** Adenine and guanine are purines, each composed of a six-membered organic ring plus a five-membered ring. Cytosine and thymine are pyrimidines, each with a single six-membered ring. (Within the molecules, C, H, N and O, are atoms of carbon, hydrogen, nitrogen and oxygen, respectively.)

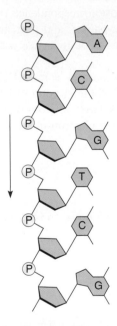

Figure 9.8 **A chain of nucleotides.** A single DNA strand consists of a chain of nucleotides that forms when the deoxyribose sugars (green) and phosphates (yellow) bond to create a sugar-phosphate backbone. The bases A, C, G, and T are blue.

The symmetrical DNA double helix forms when nucleotides containing A pair with those containing T, and nucleotides containing G pair with those carrying C. Because purines

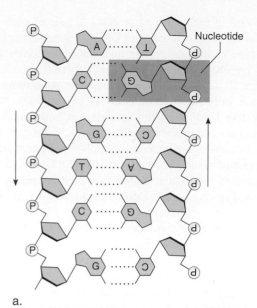

Nucleotide

a.

b.

Figure 9.9 **DNA consists of two chains of nucleotides in an antiparallel configuration.** **(a)** Hydrogen bonds hold the nitrogenous bases of one strand to the nitrogenous bases of the second strand (dotted lines). Note that the sugars point in opposite directions—that is, the strands are antiparallel. **(b)** These zebras are assuming an antiparallel stance.

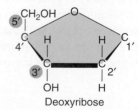

Deoxyribose

Figure 9.10 **Numbering the five carbons of deoxyribose.** The antiparallel nature of the DNA double helix becomes apparent when the carbons in the sugar are numbered. The 5′ (prime) and 3′ carbons establish the directionality of each DNA strand.

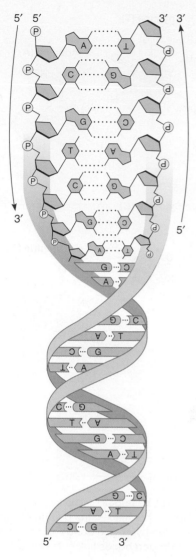

Figure 9.11 **DNA is directional.** Antiparallelism in a DNA molecule arises from the orientation of the deoxyribose sugars. One-half of the double helix runs in a 5′ to 3′ direction, and the other half runs in a 3′ to 5′ direction.

have two rings and pyrimidines one, the consistent pairing of a purine with a pyrimidine ensures that the double helix has the same width throughout, as Watson discovered using cardboard cutouts. These specific purine-pyrimidine couples are called **complementary base pairs**.

Chemical attractions called hydrogen bonds hold the base pairs together. In a hydrogen bond, a hydrogen atom on one molecule is attracted to an oxygen atom or nitrogen atom on another molecule, as **figure 9.12** depicts. The locations of the hydrogen bonds that form between A and T and between G and C make the bases complementary to each other and generate the highly symmetrical DNA double helix. Hydrogen bonds are weak individually, but over the many bases of a DNA molecule impart great strength. Two hydrogen bonds join A and T, and three hydrogen bonds join G and C. Finally, DNA forms a double helix when the antiparallel, base-paired strands twist about one another in a regular fashion. The double-stranded,

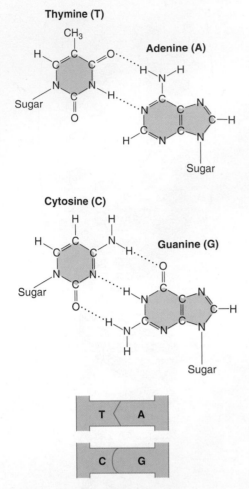

Figure 9.12 **DNA base pairs.** The key to the constant width of the DNA double helix is the pairing of purines with pyrimidines. Two hydrogen bonds join adenine and thymine; three hydrogen bonds link cytosine and guanine.

is composed of eight histone proteins (a pair of each of four types plus the 147 nucleotides of DNA entwined around them). A fifth type of histone protein anchors nucleosomes to short "linker" regions of DNA. The linkers tighten the nucleosomes into fibers 30 nanometers (nm) in diameter. As a result, at any given time, only small sections of the DNA double helix are exposed, like holding wound up string in both hands with a section in between that is a single, outstretched strand. Chemical modification of the histones controls when particular DNA sequences unwind and become accessible for the cell to use the information in the base sequence to synthesize protein. (This is discussed in chapter 11.) DNA unwinds further in localized regions when it is replicated.

Altogether, the chromosome substance is called **chromatin,** which means "colored material." Chromatin is not just DNA; it is about 30 percent histone proteins, 30 percent DNA scaffold and other proteins that bind DNA, 30 percent DNA, and 10 percent RNA. Specific points along the chromatin attach it, in great loops, to the inner face of the nuclear membrane, placing particular chromosome parts in particular locations in the nucleus. Chromatin attachment sites are not random, and evidence is emerging that this placement determines or reflects which genes a cell is using to make proteins. Progeria, the inherited disease that resembles rapid aging described in the chapter 3 opener, disrupts chromatin binding to the nuclear envelope.

Key Concepts Questions 9.2

1. What are the components of the DNA double helix?
2. Explain the basis of antiparallelism.
3. Explain the basis of complementary base pairing.
4. How are very long DNA molecules wound so that they fit inside cell nuclei?

9.3 DNA Replication— Maintaining Genetic Information

DNA must be replicated so that the information it holds can be maintained and passed to future cell generations, even as that information is accessed to guide the manufacture of proteins. As soon as Watson and Crick deciphered the structure of DNA, its mechanism for replication became obvious. They ended their report on the structure of DNA with the statement, *"It has not escaped our notice that the specific pairing we have postulated immediately suggests a possible copying mechanism for the genetic material."*

DNA Replication Is Semiconservative

Watson and Crick envisioned the two halves of the DNA double helix unwinding and separating, exposing unpaired bases that would attract their complements from free, unattached nucleotides available in the cell. In this way, two identical double helices

helical structure of DNA gives it 50 times the strength of single-stranded DNA. A single strand of DNA would not form a helix.

DNA molecules are very long. The DNA of the smallest human chromosome, if stretched out, would be 14 millimeters (a millimeter is a thousandth of a meter) long, but it is packed into a chromosome that, during cell division, is only 2 micrometers (a micrometer is a millionth of a meter) long. To fit inside the nucleus, the DNA molecule must fold so tightly that its compacted length shrinks by a factor of 7,000.

Several types of proteins compress DNA without damaging or tangling it. Scaffold proteins form frameworks that guide DNA strands. On a smaller scale, the DNA coils around proteins called **histones,** forming structures that resemble beads on a string. A DNA "bead" is called a **nucleosome.** The compaction of a molecule of DNA is a little like wrapping a very long, thin piece of thread around your fingers, to keep it from unraveling and tangling.

DNA wraps at several levels, until it is compacted into a chromatid (a chromosome consisting of one double helix, in the unreplicated form) (**figure 9.13**). Specifically, a nucleosome

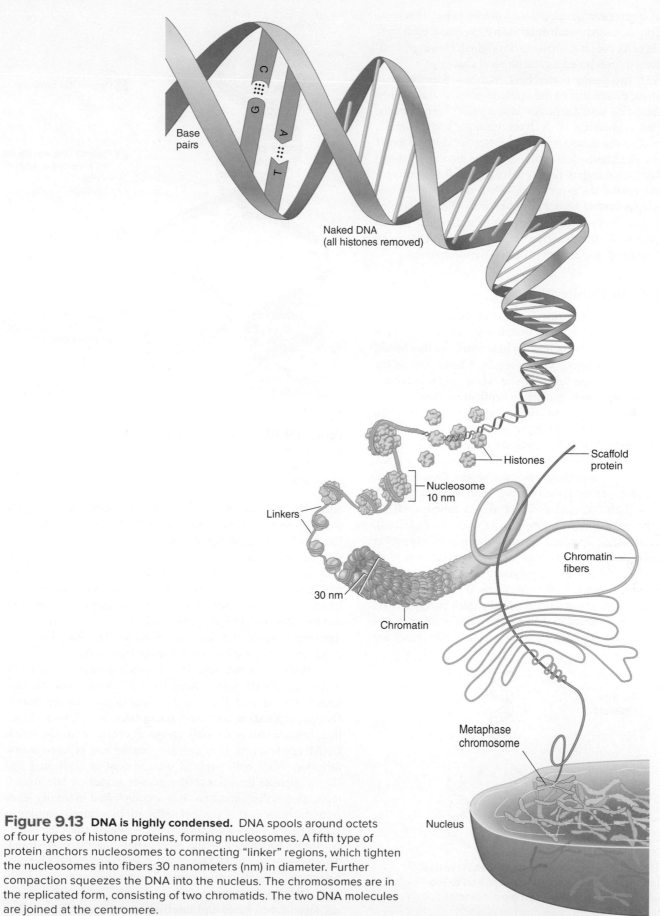

Base
pairs

Naked DNA
(all histones removed)

Histones

Scaffold
protein

Nucleosome
10 nm

Linkers

Chromatin
fibers

30 nm

Chromatin

Metaphase
chromosome

Nucleus

Figure 9.13 DNA is highly condensed. DNA spools around octets of four types of histone proteins, forming nucleosomes. A fifth type of protein anchors nucleosomes to connecting "linker" regions, which tighten the nucleosomes into fibers 30 nanometers (nm) in diameter. Further compaction squeezes the DNA into the nucleus. The chromosomes are in the replicated form, consisting of two chromatids. The two DNA molecules are joined at the centromere.

would form from one original, parental double helix. This route to replication is called **semiconservative,** because each new DNA double helix conserves half of the original. However, separating the long strands posed a great physical challenge.

In 1957, two young researchers, Matthew Meselson and Franklin Stahl, demonstrated the semiconservative mechanism of DNA replication with a series of "density shift" experiments. They labeled replicating DNA from bacteria with a dense, heavy form of nitrogen, and traced the pattern of distribution of the nitrogen. The higher-density nitrogen was incorporated into one strand of each daughter double helix (**figure 9.14**). Alternative mechanisms that the experiments ruled out were replicating a daughter DNA double helix built of entirely "heavy" labeled nucleotides (a conservative mechanism) or a daughter double helix in which both strands were composed of joined pieces of "light" and "heavy" nucleotides (a dispersive mechanism).

Steps of DNA Replication

DNA replication occurs during S phase of the cell cycle (see figure 2.14). When DNA is replicated, it unwinds, locally separates, the hydrogen bonds holding the base pairs together break, and two identical nucleotide chains are built from one, as the bases form pairs (**figure 9.15**). A site where DNA is locally opened, resembling a fork, is called a **replication fork**.

DNA replication begins when an unwinding protein called a helicase breaks the hydrogen bonds that connect a base pair (**figure 9.16**). The helicase opens up a localized area, enabling other enzymes to guide assembly of a new DNA strand. Binding proteins hold the two single strands apart. Another enzyme, primase, then attracts complementary RNA nucleotides to build a short piece of RNA, called an RNA primer, at the start of each segment of DNA to be replicated. The RNA primer is required because the major replication enzyme, **DNA polymerase** (DNAP), can only add bases to an existing nucleic acid strand. (A polymerase is an enzyme that builds a polymer, which is a chain of chemical building blocks.)

Next, the RNA primer attracts DNAP, which brings in DNA nucleotides complementary to the exposed bases on the parental strand; this strand serves as a mold, or template. New bases are

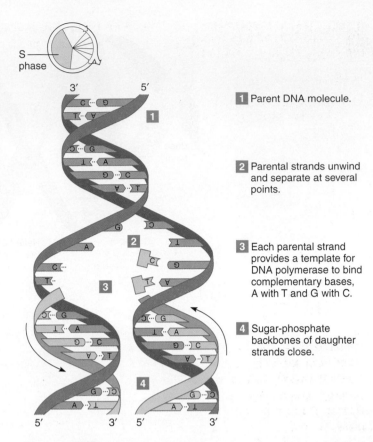

Figure 9.15 Overview of DNA replication. After experiments demonstrated the semiconservative nature of DNA replication, the next challenge was to decipher the steps of the process.

1. Parent DNA molecule.
2. Parental strands unwind and separate at several points.
3. Each parental strand provides a template for DNA polymerase to bind complementary bases, A with T and G with C.
4. Sugar-phosphate backbones of daughter strands close.

added one at a time, starting at the RNA primer. The new DNA strand grows as hydrogen bonds form between the complementary bases and the sugar-phosphate backbone links the newly incorporated nucleotides into a strong chain. DNA polymerase carries out both of these activities, and also removes the RNA primer once replication is under way and replaces it with the correct DNA bases. Nucleotides are abundant in cells, and are synthesized from dietary nutrients. The base-pairing rules ensure that each base on the parental strand pulls in its complements. The many hydrogen bonds connecting the bases stabilize the huge molecule.

Because of the antiparallel configuration of the DNA molecule, DNAP works directionally, adding new nucleotides to the exposed 3′ end of the sugar in the growing strand. Overall, replication adds new nucleotides in a 5′ to 3′ direction, because this is the only chemical configuration in which DNAP can function. How can the growing fork proceed in one direction, when both parental strands must be replicated and run in opposite directions? The answer is that on one strand, replication is discontinuous. It is accomplished in small pieces from the inner part of the fork outward, in a pattern similar to backstitching. Next, an enzyme called a **ligase** seals the sugar-phosphate backbones of the pieces, building the new strand. These pieces, up to 150 nucleotides long, are called Okazaki fragments, after their discoverer (figure 9.16).

DNA polymerase also "proofreads" as it goes, excising mismatched bases and inserting correct ones. Yet another

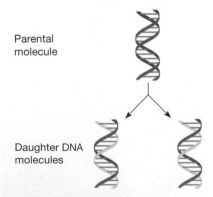

Figure 9.14 DNA replication is semiconservative. The two daughter double helices are identical to the original parental double helix. Light blue indicates newly-replicated DNA.

Parental molecule

Daughter DNA molecules

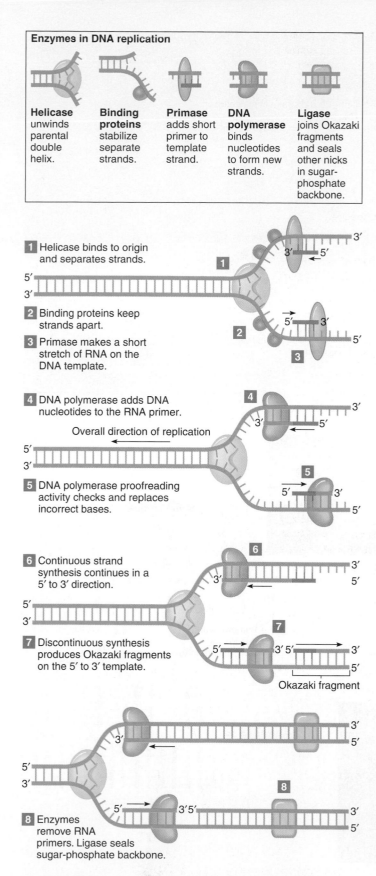

1 Helicase binds to origin and separates strands.

2 Binding proteins keep strands apart.

3 Primase makes a short stretch of RNA on the DNA template.

4 DNA polymerase adds DNA nucleotides to the RNA primer.

Overall direction of replication

5 DNA polymerase proofreading activity checks and replaces incorrect bases.

6 Continuous strand synthesis continues in a 5′ to 3′ direction.

7 Discontinuous synthesis produces Okazaki fragments on the 5′ to 3′ template.

Okazaki fragment

8 Enzymes remove RNA primers. Ligase seals sugar-phosphate backbone.

Figure 9.16 **Activities at the replication fork.** DNA replication takes many steps. (The continuous strand is also called the leading strand, and the discontinuous strand is the lagging strand.)

Enzymes in DNA replication

Helicase unwinds parental double helix.

Binding proteins stabilize separate strands.

Primase adds short primer to template strand.

DNA polymerase binds nucleotides to form new strands.

Ligase joins Okazaki fragments and seals other nicks in sugar-phosphate backbone.

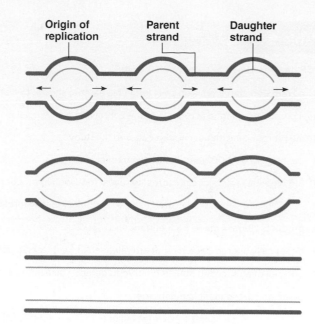

Figure 9.17 **DNA replication bubbles.** DNA replication occurs simultaneously along the 150 million nucleotides of the average human chromosome. The sites of replication resemble bubbles that coalesce as the daughter double helices form.

enzyme, called an annealing helicase, rewinds any sections of the DNA molecule that remain unwound. Finally, ligases seal the entire sugar-phosphate backbone. Ligase comes from a Latin word meaning "to tie."

Human DNA replicates about 50 bases per second. To accomplish the job in about an hour rather than the month it would take if starting from only one point, a human chromosome replicates simultaneously at hundreds of points, called replication bubbles, and the pieces are joined (**figure 9.17**). As a human body grows to 37 trillion or so cells, DNA replication occurs about 100 quadrillion times. However, telomeres (chromosome tips) do not replicate, and the chromosomes shrink with each cell division.

Polymerase Chain Reaction

DNA replication is necessary in a cell to perpetuate genetic information. Researchers use DNA replication conducted outside cells in a biotechnology called DNA amplification. It has diverse applications (**table 9.2**)

The first and best-known DNA amplification technique is the **polymerase chain reaction** (**PCR**). PCR uses DNA polymerase to rapidly replicate a specific DNA sequence in a test tube. **Figure 9.18** presents the steps in amplifying DNA using PCR to identify a specific DNA sequence in a virus causing respiratory illness. The (numbers) in the next paragraph correspond to the numbers in the figure.

PCR requires the following: (1) knowing a target DNA sequence from the suspected pathogen; (2) two types of lab-made, single-stranded, short pieces of DNA called primers— these are complementary in sequence to opposite ends of the target sequence; (3) many copies of the four types of DNA

Table 9.2	Uses of PCR

PCR has been used to amplify DNA from:

- a cremated man, from skin cells left in his electric shaver, to diagnose an inherited disease in his children.

- a preserved quagga (a relative of the zebra) and a marsupial wolf, both extinct.

- microorganisms that cannot be cultured for study.

- the brain of a 7,000-year-old human mummy.

- the digestive tracts of carnivores, to reveal food web interactions.

- roadkills and carcasses washed ashore, to identify locally threatened species.

- products illegally made from endangered species.

- genetically altered bacteria that are released in field tests, to follow their dispersion.

- one cell of an 8-celled human embryo to detect a disease-related genotype.

- poached moose meat in hamburger.

- remains in Jesse James's grave, to make a positive identification.

- the guts of genital crab lice on a rape victim, which matched the DNA of the suspect.

- fur from Snowball, a cat that linked a murder suspect to a crime.

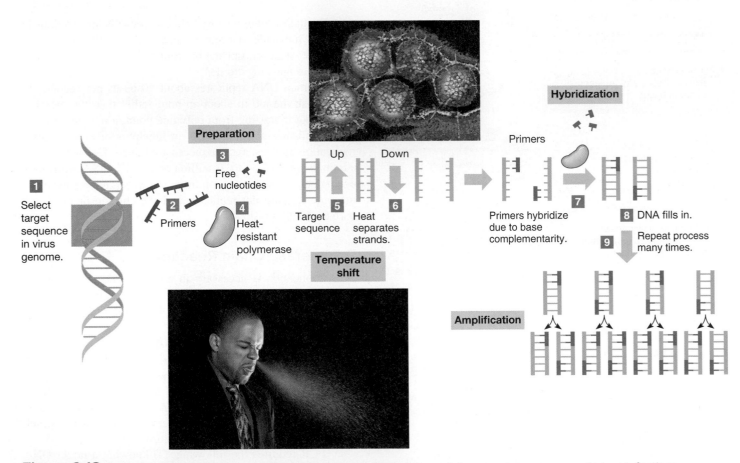

Figure 9.18 PCR amplifies a specific nucleic acid sequence. Many types of viruses cause upper respiratory infections. PCR is not typically used to diagnose such a common and mild illness, but this figure illustrates how it would be used on secretions from the sneezing person to identify an adenovirus. Many viruses have RNA as the genetic material, and a different form of PCR, called reverse PCR, is used to analyze them.

Infidelity Testing

Afraid your significant other is cheating? Send us a DNA sample, and we'll find the proof.

Bridgette came home a day early from a business trip to find her husband Roy drinking coffee in their kitchen with Tiffany, his business associate. They were laughing so hard that it took a few moments for them to notice Bridgette standing there. When they did, Tiffany blushed and Roy knocked over her coffee mug, then they both stammered that they were discussing an acquisition.

Bridgette went upstairs to unpack. Flinging her purse on the bed, she noticed several strands of brown hair on her pillow. Bridgette's hair was brown, too, but she never left it on her pillow like that. She also noticed a crumpled tissue on the floor, partway under the bed.

Bridgette had read an article on the plane about companies that test "abandoned DNA." She went back downstairs for some plastic bags, and picked up Tiffany's coffee mug, carrying it all back upstairs. In the bedroom, she quickly collected evidence—the telltale hairs, the discarded used tissue, and a cotton swab rubbed along the inside rim of the lipstick-stained mug. Bridgette e-mailed gotchaDNA.com and received a cheek swab collection kit a few days later, which she used to send in her own DNA for comparison, plus the $600 fee. Then she waited.

The technicians at gotchaDNA.com extracted the DNA from the samples and amplified it. First they checked for Y chromosome markers, which they found on the crumpled tissue. Then they looked for several short tandem repeats (STRs), which are short DNA sequences that are found in certain places in the genome but in different numbers of repeats in different individuals. STRs are used as markers to identify individuals in forensic investigations (see figures 14.10 and 14.11). The STRs confirmed Bridgette's suspicion—the DNA on the mug that Tiffany had used and in the hair cells matched each other, and not Bridgette's DNA. Tiffany, or at least her hair, had somehow found its way onto Bridgette's pillow.

Cells use DNA to manufacture protein. People use DNA to identify people. Dozens of companies offer "infidelity DNA testing." Although a few websites provide documents for attesting that the samples are given willingly, many do not, and even list suggested sources of DNA for "adultery tracing." These sources include underwear, toothbrushes, dental floss, nail clippings, gum, cigarette butts, and razor clippings.

Questions for Discussion

1. In the United Kingdom, a law was enacted to prohibit sampling of a celebrity's DNA after someone tried to steal hair from Prince Harry to determine whether or not Prince Charles is his biological father. The United States has no such law. Do you think that one is warranted? (A half waffle whose other half was consumed by Barack Obama was auctioned on eBay, with claims that it contained the presidential DNA.)

2. Do you think that DNA data obtained without consent should be admissible in a court of law? State a reason for your answer.

3. Discuss one reason in support of infidelity testing of DNA and one reason against it.

4. Identify the individuals in the scenario whom you believe behaved unethically.

---

nucleotides; (4) Taq1, which is a DNA polymerase from a bacterium that lives in hot springs. The enzyme eases PCR because it withstands heat, which is necessary to separate the DNA strands.

In the first step of PCR, the temperature is raised (5) and then lowered (6), which separates the two strands of the target DNA. Next (7), many copies of the two short DNA primers and Taq1 DNA polymerase are added. Primers bind by complementary base pairing to the separated target strands. The polymerase then fills in the bases opposite their complements, creating two daughter strands from each separated target double helix (8). The four double helices resulting from the first round of amplification then serve as the target sequences for the next round (9), and the process continues by again raising the temperature.

Pieces of identical DNA accumulate exponentially. The number of amplified pieces of DNA equals 2^n, where n is the number of temperature cycles. After 30 cycles, PCR yields more than 10 billion copies of the target DNA sequence. PCR is useful in forensic investigations to amplify small DNA samples (see *Bioethics: Choices for the Future*).

A GLIMPSE OF HISTORY

PCR was born in the mind of Kary Mullis on a moonlit night in northern California in 1983. As he drove the hills, Mullis was thinking about the precision of DNA replication, and a way to tap into it popped into his mind. He excitedly explained his idea to his girlfriend and then went home to think it through. "It was difficult for me to sleep with deoxyribonuclear bombs exploding in my brain," he wrote much later.

The idea behind PCR was so simple that Mullis had trouble convincing his superiors at Cetus Corporation that he was onto something. Over the next year, he used the technique to amplify a well-studied gene. Mullis published a landmark paper in 1985 and filed patent applications, launching the field of DNA amplification. He received a $10,000 bonus for his invention, which the company sold to another for $300 million. Mullis did, however, win a Nobel Prize, in 1993.

Key Concepts Questions 9.3

1. Why must DNA be replicated?
2. What does semiconservative replication mean?
3. Explain the steps of DNA replication.
4. How can DNA be replicated fast enough to sustain the cell?
5. How is the polymerase chain reaction based on DNA replication?

9.4 Sequencing DNA

In 1977, Frederick Sanger, a British biochemist and two-time Nobel Prize winner, invented a way to determine the base sequence of a small piece of DNA. "Sanger sequencing" remains the conceptual basis for techniques that today can sequence an entire human genome in a day. Sanger's method is still used to sequence individual genes, or to check the accuracy of a selected sequenced area of a genome. The ability to sequence millions of small pieces at once is the basis of newer methods, called "next-generation sequencing," that can handle much larger DNA molecules much faster.

Sanger sequencing generates a series of DNA fragments of identical sequence that are complementary to the DNA sequence of interest, which serves as a template strand. The fragments differ in length from each other by one end base, as follows:

Sequence of interest:	T A C G C A G T A C
Complementary sequence:	**A** T G C G T C A T G
Series of fragments:	**T** G C G T C A T G
	G C G T C A T G
	C G T C A T G
	G T C A T G
	T C A T G
	C A T G
	A T G
	T G
	G

DNA sequencing technologies are based on clever applications of chemistry. Sanger sequencing uses an approach called "chain termination." In today's version, PCR replicates a sequence of DNA many times, incorporating chemically altered bases that bear a fluorescent tag, a different color for A, T, C and G. When a tagged base is incorporated into a growing DNA chain, the DNA replication halts because another base cannot be added. It is a little like someone breaking into a line of people waiting to get into an event who turns around and

prevents others from extending the line. The result, in PCR, is a partial DNA molecule.

When chain termination happens many times, at different points in the replication of many copies of the same sequence, a collection of partial molecules results. The end bases reveal the sequence, as indicated previously The pieces are separated by size. A flash of laser light excites the four fluorescent tags and produces signals captured in a readout (**figures 9.19** and **9.20**). An algorithm deduces and displays the DNA sequence.

In 2007, new DNA sequencing "platforms" began to become available to researchers. They follow the basic approach

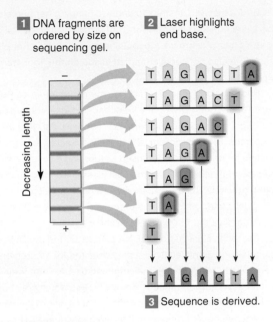

Figure 9.19 Reading a DNA sequence. A computer algorithm detects and records the end base from a series of size-ordered DNA fragments.

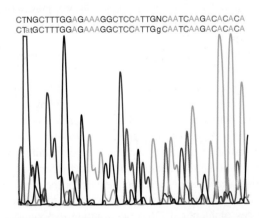

Figure 9.20 DNA sequence data. In automated first-generation DNA sequencing, a readout of sequenced DNA is a series of wavelengths that represent the terminal DNA base labeled with a fluorescent molecule.

of Sanger sequencing, but use different chemistries and materials as well as different ways to display the DNA pieces. The new techniques still fragment the human genome into millions of DNA strands, but then capture the pieces onto a solid surface such as glass or beads, and then make identical copies of each fragment on its unique bead or position on the glass. A reiterative process called "sequencing by synthesis" then reads one base at a time on the tens of millions of clusters or beads, yielding 40 billion to 600 billion bases of sequence information per experiment.

Several companies provide variations of next-generation sequencing (NGS). One product attaches short DNA pieces in a flow cell, which resembles a microscope slide. Another device uses tiny DNA-fringed beads in a water-oil mixture. A laser reads off the bases corresponding to the fluorescently tagged bases that are added and stream past the strands, binding their complements.

Another approach uses nanomaterials to distinguish how each of the four nucleotide bases disrupts an electrical field as a DNA molecule passes through nanopores at about 1,000 bases per second. One such material is a one-atom-thick sheet of carbon called grapheme that is strong, very thin, and conducts electricity. DNA is threaded through nanopores in it. Each of the four DNA base types disrupts an electrical field in a slightly different way. An algorithm converts the voltage changes into the DNA sequence that they represent. The voltage approach is also used on semiconductor chips.

The new ways to sequence DNA, some of which are company secrets, build on Sanger's invention, but differ in scale. Next-generation sequencing reads millions of sequences at once, and so the general approach is also called "massively parallel DNA sequencing." The sequenced short pieces are called "reads." A sequencing experiment will yield many copies of each and every subset of contiguous bases that makes up a target sequence or genome. The more times any particular site in the target is represented among the pieces, the more accurate the deduced sequence—like learning more from reading this book over and over, from different starting points. For example, thirtyfold coverage of a genome sequence is more accurate than twentyfold coverage. Thanks to NGS, a human genome can be sequenced in under a day—the first sequencing took 10 years!

Chapter 22 continues the discussion of how to overlap short DNA sequences to derive a genome sequence. The chapters in between address what is really important—what our genes do, and what all those sequences of A, T, C, and G can tell us about who we are, where we came from, and even where we may be headed.

Key Concepts Questions 9.4

1. Explain how Sanger sequencing works.
2. How are many copies of DNA pieces generated to obtain the sequence?
3. Why must several copies of a genome be cut up to sequence it?
4. What are next-generation sequencing technologies?

Summary

9.1 Experiments Identify and Describe the Genetic Material

1. DNA encodes information that the cell uses to synthesize protein. DNA can also be replicated, passing on its information during cell division.

2. Many scientists contributed to the discovery of DNA as the hereditary material. Miescher identified DNA in white blood cell nuclei. Garrod connected heredity to enzyme abnormalities. Griffith identified a "transforming principle" that transmitted infectiousness in pneumonia-causing bacteria; Avery, MacLeod, and McCarty discovered that the transforming principle is DNA; and Hershey and Chase confirmed that the genetic material is DNA and not protein.

3. Levene described the three components of a DNA building block (**nucleotide**) and found that they appear in DNA in equal amounts, and that nucleic acids include the sugar **ribose** or **deoxyribose**. Chargaff discovered that the amount of **adenine** (A) equals the amount of **thymine** (T), and the amount of **guanine** (G) equals that of **cytosine** (C). Rosalind Franklin showed that the molecule is a certain type of helix. Watson and Crick deduced DNA's structure.

9.2 DNA Structure

4. A **gene** encodes a protein. A nucleotide building block of a gene consists of a deoxyribose, a phosphate, and a nitrogenous base. A and G are **purines;** C and T are **pyrimidines.**

5. The rungs of the DNA double helix consist of hydrogen-bonded **complementary base pairs** (A with T, and C with G). The rails are chains of alternating sugars and phosphates (the **sugar-phosphate backbone)** that run **antiparallel** to each other. DNA is highly coiled, and complexed with protein to form **chromatin.** DNA winds around **histone** proteins and forms beadlike structures called **nucleosomes**.

9.3 DNA Replication—Maintaining Genetic Information

6. Meselson and Stahl demonstrated the **semiconservative** nature of DNA replication with density shift experiments.

7. During replication, the DNA unwinds locally at several sites. **Replication forks** form as hydrogen bonds break between base pairs. Primase builds short RNA primers, which DNA sequences eventually replace.

Next, **DNA polymerase** fills in DNA bases, and **ligase** seals remaining gaps, forming the sugar-phosphate backbone.

8. Replication proceeds in a 5′ to 3′ direction, so the process must be discontinuous in short stretches on one strand.

9. Nucleic acid amplification (**polymerase chain reaction; PCR**) uses the power and precision of DNA replication enzymes to selectively mass produce selected DNA sequences. In PCR, primers corresponding to a DNA sequence of interest direct polymerization of supplied nucleotides to make many copies.

9.4 Sequencing DNA

10. Sanger sequencing deduces a DNA sequence by aligning pieces that differ from each other by the end base. Variations on this theme label, cut, and/or immobilize the DNA pieces in different ways, greatly speeding the process.

11. Next-generation sequencing uses a massively-parallel approach, and different types of materials on which to immobilize DNA pieces, to read and overlap millions of pieces at once. It greatly speeds DNA sequencing.

www.mhhe.com/lewisgenetics11

Answers to all end-of-chapter questions can be found at **www.mhhe.com/lewisgenetics11**. You will also find additional practice quizzes, animations, videos, and vocabulary flashcards to help you master the material in this chapter.

Review Questions

1. List the components of a nucleotide.

2. How does a purine differ from a pyrimidine?

3. Distinguish between the locations of the phosphodiester bonds and the hydrogen bonds in a DNA molecule.

4. Why must DNA be replicated?

5. Why would a DNA structure in which each base type could form hydrogen bonds with any of the other three base types not produce a molecule that is easily replicated?

6. What part of the DNA molecule encodes information?

7. Explain how DNA is a directional molecule in a chemical sense.

8. What characteristic of the DNA structure does the following statement refer to? "New DNA forms in the 5′ to 3′ direction, but the template strand is read in the 3′ to 5′ direction."

9. Match the experiment described in #s 1–5 to a concept it illustrates from a-e.
   1. Density shift experiments
   2. Discovery of an acidic substance that includes nitrogen and phosphorus on dirty bandages
   3. "Blender experiments" that showed that the part of a virus that infects bacteria contains phosphorus, but not sulfur
   4. Determination that DNA contains equal amounts of guanine and cytosine, and of adenine and thymine
   5. Discovery that bacteria can transfer a "factor" that transforms a harmless strain into a lethal one

   a. First experiments in identifying hereditary material
   b. Complementary base pairing is part of DNA structure and maintains a symmetrical double helix
   c. Identification of nuclein
   d. DNA, not protein, is the hereditary material
   e. DNA replication is semiconservative

10. Place the following enzymes in the order in which they function in DNA replication: ligase, primase, helicase, and DNA polymerase.

11. How can very long DNA molecules fit into a cell's nucleus?

12. Place in increasing size order: nucleosome, histone protein, and chromatin.

13. How are very long strands of DNA replicated without twisting into a huge tangle?

14. List the steps in DNA replication.

15. Why must DNA be replicated continuously as well as discontinuously?

16. How does RNA participate in DNA replication?

17. Is downloading a document from the Internet analogous to replicating DNA? Cite a reason for your answer.

Applied Questions

1. Bloom syndrome (OMIM 210900) causes short stature, sun sensitivity, and susceptibility to certain types of cancer. An autosomal recessive mutation in the gene that encodes ligase causes the condition. Which step in DNA replication is affected?

2. DNA contains the information that a cell uses to synthesize a particular protein. How do proteins assist in DNA replication?

3. A person with deficient or abnormal ligase may have an increased cancer risk and chromosome breaks that cannot heal. The person is, nevertheless, alive. Why are there no people who lack DNA polymerase?

4. Write the sequence of a strand of DNA replicated using each of the following base sequences as a template:

   a. T C G A G A A T C T C G A T T
   b. C C G T A T A G C C G G T A C
   c. A T C G G A T C G C T A C T G

5. Which do you think was the more far-reaching accomplishment, determining the structure of DNA, or sequencing the human genome? State a reason for your answer.

6. Cite an example of how knowing a DNA sequence could be abused, and an example of how knowing a DNA sequence could be helpful.

7. People often use the phrase "the gene for" to describe traits that do not necessarily or directly arise from a protein's actions, such as "a gene for jealousy" or "a gene for acting." How would you explain to them what a gene actually is?

8. The first several DNA bases in the gene that encodes serum albumin, which is an abundant blood protein as well as abundant in egg white, are as follows:

   A G C T T T T C T C T T C T G T C A A C

   This is the strand that holds the information the cell uses to construct the protein. Write the sequence of the complementary DNA strand.

9. To diagnose a rare form of encephalitis (brain inflammation), a researcher needs a million copies of a viral gene. She decides to use PCR on a sample of the patient's cerebrospinal fluid. If one cycle takes 2 minutes, how long will it take to obtain a millionfold amplification?

Web Activities

1. The Frozen Ark project is an international consortium of zoos, laboratories, and museums that is preserving DNA samples from endangered animal species. Consult http://www.frozenark.org

   a. Describe an endangered species. What do you think is the value of this project?
   b. Do you think the project should be extended to include organisms other than animals? Cite a reason for your answer.
   c. What would be the difficulties encountered in attempting to increase population sizes of endangered species using stored DNA?

2. Visit the Cystic Fibrosis Mutation Database website (http://genet.sickkids.on.ca/app) or another mutation database. Select twenty contiguous bases of the sequence for a mutant gene and write the complementary sequence.

3. Find further information on one of the inventors or researchers mentioned in the chapter, and describe their inspiration or thought process.

4. Consult a website for a company that supplies next-generation sequencing, such as Illumina, and explain how their product works.

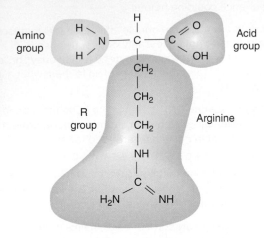

Amino group

Acid group

R group

Arginine

In argininosuccinic aciduria, the body cannot manufacture an enzyme required to make the amino acid arginine. Several symptoms result, but a heart disease drug can treat this inborn error of metabolism.

Gene Action: From DNA to Protein

Learning Outcomes

10.1 Transcription Copies the Information in DNA

1. List the major types of RNA molecules and their functions.

2. Explain the importance of transcription factors.

3. List the steps of transcription.

10.2 Translation of a Protein

4. Discuss how researchers deduced the genetic code.

5. List the steps of protein synthesis.

10.3 Processing a Protein

6. Define the four components of a protein's shape.

7. Explain the importance of protein folding.

The **BIG** Picture

DNA sequences are the blueprints of life. Cells must maintain this information, yet also access it to manufacture proteins. RNA acts as the go-between, linking DNA sequences to the amino acid sequences of proteins.

An Inborn Error of Arginine Production

Genes instruct cells to build proteins from 20 types of amino acids. An amino acid has four parts: a central carbon atom bonds to a hydrogen atom, an amino group (NH_2), an acid group (COOH), and an "R" group that distinguishes the 20 types. The body synthesizes 10 of the 20 amino acids, and must obtain the rest, termed "essential," from food. Some amino acids, such as arginine, are essential only during childhood.

To be healthy, the body must make or obtain all 20 types of amino acids. In argininosuccinic aciduria (ASA; OMIM 207900), the body cannot produce an enzyme required to make arginine. ASA is autosomal recessive and affects 1 in 70,000 newborns.

Lack of arginine causes seizures and coma in children. The brain damage happens because without arginine, nitrogen atoms released from broken-down proteins, instead of being excreted in the urine, bond with hydrogen atoms to form ammonia (NH_3). The ammonia harms brain neurons. A different symptom, however, led researchers to a drug already in use for a different disease.

Children with ASA may also have developmental and cognitive delays, very high blood pressure that does not respond to standard medications, and enlarged hearts. Researchers at Baylor College of Medicine working with a teenager with ASA who had all of these symptoms discovered how to help him. They knew that arginine was also necessary to manufacture the tiny but essential molecule nitric oxide (NO). So the researchers tried a drug that heart disease patients use to treat chest pain caused by lack of NO. The drug enables the body to produce the needed NO another way.

Within days of receiving the heart drug, the young man's blood pressure dropped to normal. Over the following weeks, his heart returned to a normal size, and he began doing much better in school. He started to socialize, and test scores for verbal memory and problem solving increased dramatically.

10.1 Transcription Copies the Information in DNA

Only about 1.5 percent of the DNA of the human genome encodes protein. This part is the **exome**. Much of the rest of the genome controls how, where, and when proteins are made.

Our genes encode 20,325 types of proteins. A protein consists of one or more long chains of amino acids, called polypeptides. A short sequence of amino acids is a peptide, and the bonds that join amino acids are peptide bonds. Proteins have a great variety of functions; **table 10.1** lists some of them.

A cell uses two processes to manufacture proteins using genetic instructions. **Transcription** first synthesizes an RNA molecule that is complementary to one strand of the DNA double helix for a particular gene. The RNA copy is taken out of the nucleus and into the cytoplasm. There, the process of **translation** uses the information in the RNA to manufacture a protein by aligning

and joining specified amino acids. Finally, the protein folds into a specific three-dimensional form necessary for its function.

Accessing the genome is a huge, ongoing task. Cells replicate their DNA only during S phase of the cell cycle. In contrast, transcription and translation occur continuously, except during M phase. Transcription and translation supply the proteins essential for life, as well as those that give a cell its specialized characteristics.

Shortly after Watson and Crick published their structure of DNA in 1953, they described the relationship between nucleic acids and proteins as a directional flow of information called the "central dogma" (**figure 10.1**). As Francis Crick explained in 1957, *"The specificity of a piece of nucleic acid is expressed solely by the sequence of its bases, and this sequence is a code for the amino acid sequence of a particular protein."* This statement inspired more than a decade of intense research to discover exactly how cells make proteins. The process centers around RNA.

RNA Structure and Types

RNA is the bridge between gene and protein. RNA and DNA share an intimate relationship, as **figure 10.2** depicts. The bases of an RNA sequence are complementary to those of one strand of the double helix, which is called the **template strand**. An enzyme, **RNA polymerase,** builds an RNA molecule. The other, nontemplate strand of the DNA double helix is called the **coding strand**.

Table 10.1	Protein Diversity in the Human Body
Protein	**Function**
Actin, myosin, dystrophin	Muscle contraction
Antibodies, antigens, cytokines	Immunity
Carbohydrases, lipases, proteases, nucleases	Digestion (digestive enzymes)
Casein	Milk protein
Collagen, elastin, fibrillin	Connective tissue
Colony-stimulating factors, erythropoietin	Blood cell formation
DNA and RNA polymerase	DNA replication, gene expression
Ferritin	Iron transport in blood
Fibrin, thrombin	Blood clotting
Growth factors, kinases, cyclins	Cell division
Hemoglobin, myoglobin	Oxygen transport
Insulin, glucagon	Control of blood glucose level
Keratin	Hair structure
Tubulin, actin	Cell movements
Tumor suppressors	Cell cycle regulation

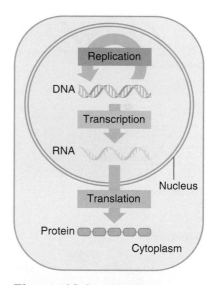

Figure 10.1 DNA to RNA to protein. Some of the information stored in DNA is copied to RNA (transcription), some of which is used to assemble amino acids into proteins (translation). DNA replication perpetuates genetic information. This figure repeats throughout the chapter, with the part under discussion highlighted.

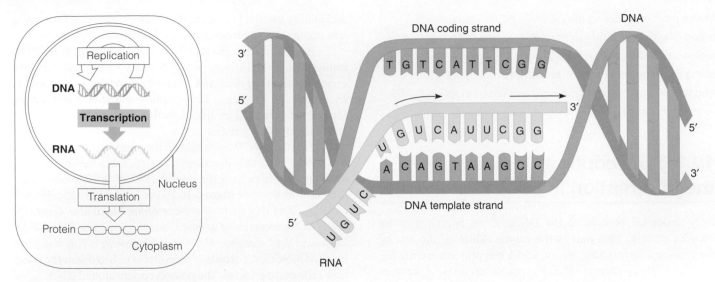

Figure 10.2 **The relationship among RNA, the DNA template strand, and the DNA coding strand.** The RNA sequence is complementary to the DNA template strand and is the same sequence as the DNA coding strand, with uracil (U) in place of thymine (T).

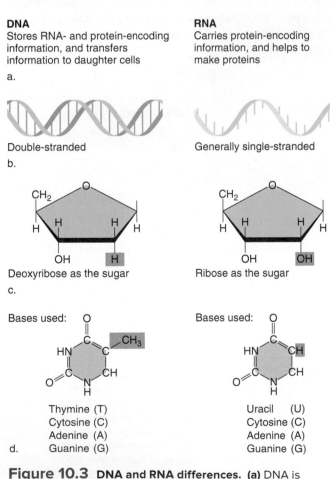

Figure 10.3 **DNA and RNA differences. (a)** DNA is double-stranded; RNA is usually single-stranded **(b).** DNA nucleotides include deoxyribose; RNA nucleotides have ribose **(c).** Finally, DNA nucleotides include the pyrimidine thymine, whereas RNA has uracil **(d).**

RNA and DNA have similarities and differences (**figure 10.3** and **table 10.2**). Both are nucleic acids, consisting of sequences of nitrogen-containing bases joined by sugar-phosphate backbones. However, RNA is usually single-stranded, whereas DNA is double-stranded. Also, RNA has the pyrimidine base **uracil** where DNA has thymine. RNA (*ribo*nucleic acid) nucleotides include the sugar ribose. DNA (*deoxyribo*nucleic acid) nucleotides include the sugar deoxyribose. Functionally, DNA stores genetic information, whereas RNA controls how that information is used. The presence of the —OH at the 5′ position of ribose makes RNA much less stable than DNA, which is critical in its function as a short-lived carrier of genetic information.

As RNA is synthesized along DNA, it folds into a three-dimensional shape, or **conformation,** that arises from complementary base pairing within the same RNA molecule. For example, a sequence of AAUUUCC might hydrogen bond to

Table 10.2	How DNA and RNA Differ
DNA	**RNA**
1. Usually double-stranded	1. Usually single-stranded
2. Thymine as a base	2. Uracil as a base
3. Deoxyribose as the sugar	3. Ribose as the sugar
4. Maintains protein-encoding information	4. Carries protein-encoding information and controls how information is used
5. Cannot function as an enzyme	5. Can function as an enzyme
6. Persists	6. Transient

Table 10.3	Major Types of RNA	
Type of RNA	Size (number of nucleotides)	Function
Messenger RNA (mRNA)	500–4,500+	Encodes amino acid sequence
Ribosomal RNA (rRNA)	100–3,000	Associates with proteins to form ribosomes, which structurally support and catalyze protein synthesis
Transfer RNA (tRNA)	75–80	Transports specific amino acids to the ribosome for protein synthesis

a sequence of UUAAAGG—its complement—elsewhere in the same molecule, a little like touching elbows to knees. Conformation is very important for RNA's functioning. The three major types of RNA are messenger RNA, ribosomal RNA, and transfer RNA (**table 10.3**). Other classes of RNA control which genes are expressed (transcribed and translated) under specific circumstances. Table 11.1 describes them.

Messenger RNA (mRNA) carries the information that specifies a particular protein. Each three mRNA bases in a row form a genetic code word, or **codon,** that specifies a certain amino acid. Because genes vary in length, so do mature mRNA molecules. Most mRNAs are 500 to 4,500 bases long. Differentiated cells can carry out specialized functions because they express certain subsets of genes—that is, they produce certain mRNA molecules, which are also called transcripts. The information in the transcripts is then used to manufacture the encoded proteins. A muscle cell, for example, has many mRNAs that specify the contractile proteins actin and myosin, whereas a skin cell contains many mRNAs that specify scaly keratin proteins.

To use the information in an mRNA sequence, a cell requires the two other major classes of RNA. **Ribosomal RNA (rRNA)** molecules range from 100 to nearly 3,000 nucleotides long. Ribosomal RNAs associate with certain proteins to form a ribosome. Recall from chapter 2 that a ribosome is an organelle made up of many different protein and RNA subunits. Overall, a ribosome functions as a machine to assemble and link amino acids to form proteins (**figure 10.4**).

A ribosome has two subunits that are separate in the cytoplasm but join at the site of initiation of protein synthesis. The larger ribosomal subunit has three types of rRNA

molecules, and the small subunit has one. Ribosomal RNA, however, is more than a structural support. Certain rRNAs catalyze the formation of the peptide bonds between amino acids. Such an RNA with enzymatic function is called a ribozyme. Other rRNAs help to align the ribosome and mRNA.

The third major type of RNA molecule, **transfer RNA (tRNA),** binds an mRNA codon at one end and a specific amino acid at the other. A tRNA molecule is only 75 to 80 nucleotides long. Some of its bases form weak chemical bonds with each other, folding the tRNA into loops in a characteristic cloverleaf shape (**figure 10.5**). One loop of the tRNA has three bases in a row that form the **anticodon,** which is complementary to an mRNA codon. The end of the tRNA opposite the anticodon strongly bonds to a specific amino acid. A tRNA with a particular anticodon sequence always carries the same amino acid. (Organisms have 20 types of amino acids.) For example, a tRNA with the anticodon sequence GAA always picks up the amino acid phenylalanine. Enzymes attach amino acids to tRNAs that bear the appropriate anticodons, where they form chemical bonds (**figure 10.6**).

Transcription Factors

If all of the genes in the human genome were being transcribed and translated at the same time, chaos would result. It would be like trying to open all of the files and programs on a computer at once. Instead, accessing genetic information is economical.

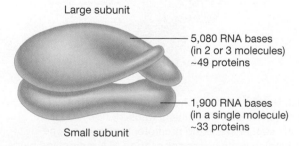

Figure 10.4 **The ribosome.** A ribosome from a eukaryotic cell has two subunits; together, they consist of 82 proteins and four rRNA molecules.

Large subunit

5,080 RNA bases (in 2 or 3 molecules) ~49 proteins

1,900 RNA bases (in a single molecule) ~33 proteins

Small subunit

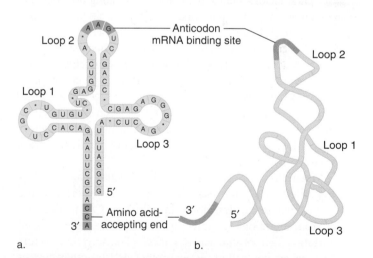

Figure 10.5 **Transfer RNA.** **(a)** Certain nucleotide bases within a tRNA hydrogen bond with each other, giving the molecule a "cloverleaf" conformation that can be represented in two dimensions. The darker bases at the top form the anticodon, the sequence that binds a complementary mRNA codon. Each tRNA terminates with the sequence CCA, where a particular amino acid covalently bonds. **(b)** A three-dimensional representation of a tRNA depicts the loops that interact with the ribosome.

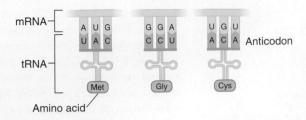

Figure 10.6 A tRNA with a particular anticodon sequence always binds the same type of amino acid.

In 1961, French biologists François Jacob and Jacques Monod described the remarkable ability *of E. coli* bacteria to produce the enzymes needed to break down the sugar lactose only when the sugar is actually present. They named a set of coordinately controlled genes that function in the same process an operon, writing in 1961, "The genome contains not only a series of blueprints, but a coordinated program of protein synthesis and means of controlling its execution." Jacob and Monod were ahead of their time in describing the complexity of the genome.

Different cell types express different subsets of genes. To manage this, proteins called **transcription factors** come together and interact, forming an apparatus that binds DNA at certain sequences and initiates transcription at specific sites on chromosomes. The transcription factors respond to signals from outside the cell, such as hormones and growth factors, and form a pocket for RNA polymerase to bind and begin building an RNA chain. Transcription factors include regions called binding domains that guide them to the genes they control. The DNA binding domains have very colorful names, such as "helix-turn-helix," "zinc fingers," and "leucine zippers," that reflect their distinctive shapes.

The few types of transcription factors work in combinations, providing great specificity in controlling gene expression. Overall, transcription factors link the genome to the environment. For example, lack of oxygen, such as from choking or smoking, sends signals that activate transcription factors to turn on dozens of genes that enable cells to handle the stress of low-oxygen conditions.

Mutations in transcription factor genes can be devastating. Rett syndrome (see the chapter 2 opener) and the homeotic mutations described in Clinical Connection 3.1 result from mutations in transcription factors. Overexpressed transcription factors can cause cancer. Transcription factors are themselves controlled by each other and by other classes of molecules.

Steps of Transcription

Transcription is described in three steps: initiation, elongation, and termination. The process is called transcription because it copies the information, but stays in the genetic language of nucleotide bases.

Transcription factors and RNA polymerase (RNAP) "know" where to bind to DNA to begin transcribing a specific gene. In transcription initiation, transcription factors and RNA polymerase are attracted to a **promoter,** which is a special sequence that signals the start of the gene, like a capital letter at the start of a sentence.

Figure 10.7 illustrates transcription factor binding, which sets up a site to receive RNA polymerase. The first transcription factor to bind, called a TATA binding protein, is chemically attracted to a DNA sequence called a TATA box—the base sequence TATA surrounded by long stretches of G and C. Once the first transcription factor binds, it attracts others in groups. Finally RNA polymerase joins the complex, binding just in front of the start of the gene sequence. The assembly of these components is transcription initiation.

In the next stage, transcription elongation, enzymes unwind the DNA double helix locally, and free RNA nucleotides bond with exposed complementary bases on the DNA template strand (see figure 10.2). RNA polymerase adds the RNA nucleotides in the sequence the DNA specifies, moving along the DNA strand in a 3′ to 5′ direction, synthesizing the RNA molecule in a 5′ to 3′ direction. A terminator sequence in the DNA indicates where the gene's RNA-encoding region ends, like the period at the end of a sentence. When this spot is reached, the third stage, transcription termination, occurs

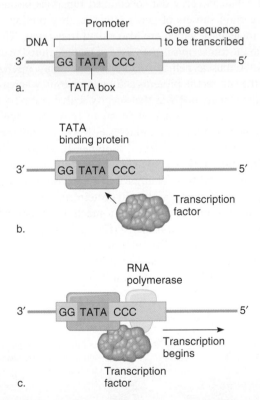

Figure 10.7 Setting the stage for transcription to begin. **(a)** Proteins that initiate transcription recognize specific sequences in the promoter region of a gene. **(b)** A binding protein recognizes the TATA region and binds to the DNA. This allows other transcription factors to bind. **(c)** The bound transcription factors form a pocket that allows RNA polymerase to bind and begin making RNA.

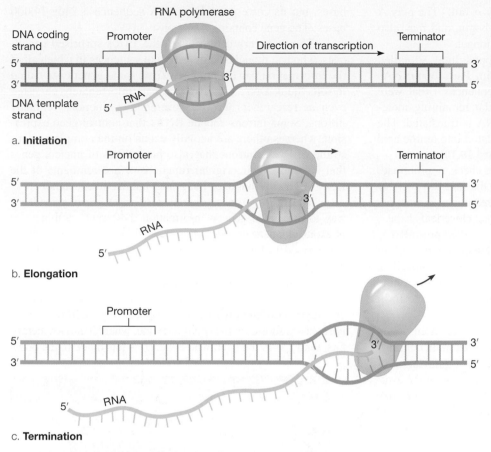

a. **Initiation**

b. **Elongation**

c. **Termination**

Figure 10.8 **Transcription of RNA from DNA.** Transcription occurs in three stages: initiation, elongation, and termination. Initiation is the control point that determines which genes are transcribed. RNA nucleotides are added during elongation. A terminator sequence signals the end of transcription.

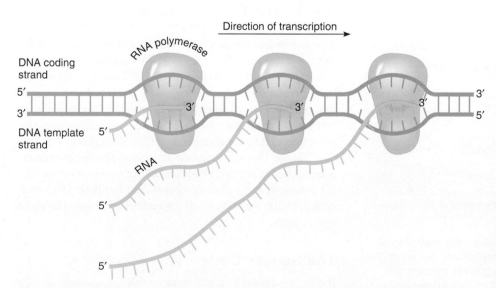

Figure 10.9 **Many identical copies of RNA are transcribed simultaneously.** Usually 100 or more DNA bases lie between RNA polymerases.

(**figure 10.8**). A typical rate of transcription in humans is 20 bases per second.

RNA is transcribed using only a gene's template strand. However, different genes on the same chromosome may be transcribed from different strands of the double helix. The coding strand of the DNA is so-called because its sequence is identical to that of the RNA, except with thymine (T) in place of uracil (U). Several RNAs may be transcribed from the same DNA template strand simultaneously (**figure 10.9**). Because mRNA is short-lived, with about half of it degraded every 10 minutes, a cell must constantly transcribe certain genes to maintain supplies of essential proteins.

To determine the sequence of RNA bases transcribed from a gene, write the RNA bases that are complementary to the template DNA strand, using uracil opposite adenine. For example, a DNA template strand that has the sequence

3′ C C T A G C T A C 5′

is transcribed into RNA with the sequence

5′ G G A U C G A U G 3′

and the coding DNA sequence is

5′ G G A T C G A T G 3′

RNA Processing

In bacteria, RNA is translated into protein as soon as it is transcribed from DNA because a nucleus does not physically separate the two processes. In eukaryotic cells like ours, mRNA must first exit the nucleus to enter the cytoplasm, where ribosomes are located. Messenger RNA is altered in several steps before it is translated in these more complex cells.

First, after mRNA is transcribed, a short sequence of modified nucleotides, called a cap, is added to the 5′ end of the molecule. The cap consists of a backwardly inserted guanine (G), which attracts an enzyme that adds methyl groups (CH_3) to the G and one or two adjacent nucleotides. This methylated cap is a recognition site for protein synthesis. At the 3′ end, a special polymerase adds about 200 adenines, forming a "poly A tail." The poly A tail is necessary for protein synthesis to begin, and may also stabilize the mRNA so that it stays intact longer.

Further changes occur to the capped, poly A tailed mRNA before it is translated into protein. Parts of mRNAs called **introns** (short for "intervening sequences") that were transcribed are removed. The ends of the remaining molecule are spliced together before the mRNA is translated. The parts of mRNA that remain and are translated into amino acid sequences (protein) are called **exons (figure 10.10)**.

Once introns are spliced out, enzymes check, or proofread, the remaining mRNA. Messenger RNAs that are too short or too long may be held in the nucleus. Proofreading also monitors tRNAs, ensuring that they assume the correct cloverleaf shape.

Prior to intron removal, the mRNA is called pre-mRNA. Introns control their own removal. They associate with certain proteins to form small nuclear ribonucleoproteins (snRNPs), or "snurps." Four snurps form a structure called a spliceosome that cuts introns out and attaches exons to each other to form the mature mRNA that exits the nucleus.

Introns range in size from 65 to 10,000 or more bases; the average intron is 3,365 bases. The average exon, in contrast, is only 145 bases long. The number, size, and organization of introns vary from gene to gene. The coding portion of the average human gene is 1,340 bases, whereas the average total size of a gene is 27,000 bases. The dystrophin gene is 2,500,000 bases, but its corresponding mRNA sequence is only 14,000 bases! The gene contains 80 introns.

The discovery of introns in the 1970s surprised geneticists, who had thought genes were like sentences in which all of the information has meaning. At first, some geneticists called introns "junk DNA"—a term that has unfortunately persisted even as researchers have discovered the functions of many introns. Some introns encode RNAs that control gene expression, whereas others are actually exons on the complementary strand of DNA. Introns may also be vestiges of ancient genes that have lost their original function, or are remnants of the DNA of viruses that once integrated into a chromosome.

The intron/exon organization of most genes provides a way to maximize genetic information. Different combinations of exons of a gene encode different versions of the protein product, termed isoforms. From 40 to 60 percent of human genes encode isoforms, and the mechanism of combining exons of a gene in different ways is called **alternate splicing**. In this way, cell types can use versions of the same protein in slightly different ways in different tissues. For example, a protein that transports fats is shorter in the small intestine, where it carries dietary fats, than it is in the liver, where it carries fats made in the body.

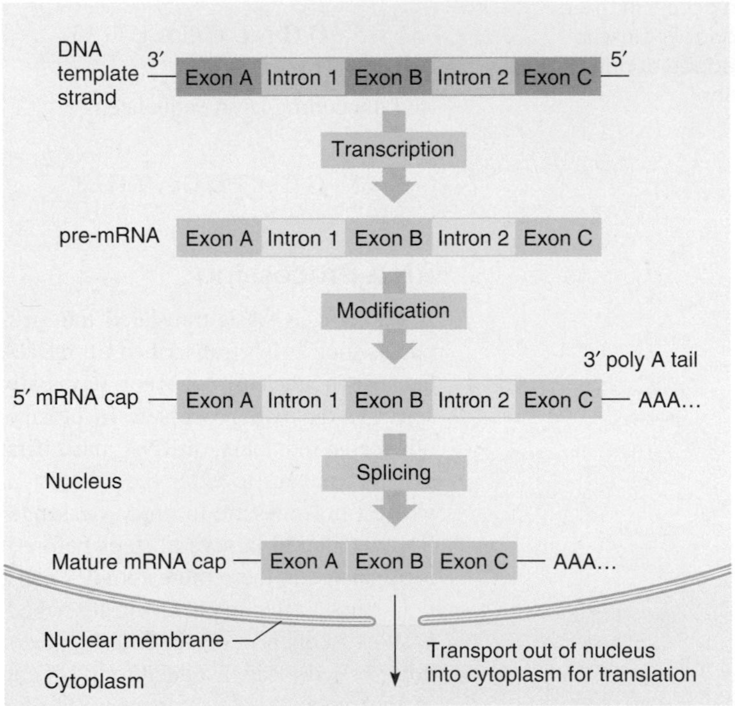

Key Concepts Questions 10.1

1. Explain how the structures and functions of DNA and RNA differ.

2. Explain how messenger RNA transmits instructions to build proteins.

3. What is the function of transfer RNAs in synthesizing proteins?

4. List the steps of transcription.

5. What is alternate splicing?

Figure 10.10 Messenger RNA processing—the maturing of the message. Several steps process pre-mRNA into mature mRNA. First, a large region of DNA containing the gene is transcribed. Then a modified nucleotide cap and poly A tail are added and introns are spliced out. Finally, the intact, mature mRNA is sent out of the nucleus.

10.2 Translation of a Protein

Translation assembles a protein using the information in the mRNA sequence. Particular mRNA codons correspond to particular amino acids (**figure 10.11**). This correspondence between the chemical languages of mRNA and protein is the **genetic code** (**table 10.4**). Translation takes place on free ribosomes in the cytoplasm as well as on ribosomes that are embedded in the endoplasmic reticulum (ER).

In the 1960s, researchers described and deciphered the genetic code. They used logic and clever experiments on simple genetic systems such as viruses, and synthetic DNA molecules, to discover which mRNA codons correspond to which amino acids.

The Genetic Code

Before researchers could match mRNA codons to the amino acids they encode, they had to establish certain facts about the code.

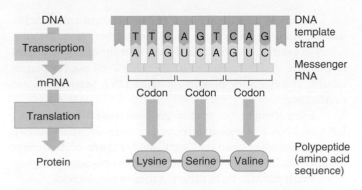

DNA

Transcription

mRNA

Translation

Protein

DNA template strand

Messenger RNA

T T C A G T C A G C
A A G U C A G U C

Codon Codon Codon

Lysine — Serine — Valine

Polypeptide (amino acid sequence)

Figure 10.11 From DNA to RNA to protein. Messenger RNA is transcribed from a locally unwound portion of DNA. In translation, transfer RNA matches mRNA codons with amino acids.

1. The Code Is Triplet.

The number of different protein building blocks (20) exceeds the number of different mRNA building blocks (4). Therefore, each codon must include more than one mRNA base. If a codon consisted of only one mRNA base, then codons could specify only four different amino acids, one corresponding to each of the four bases: A, C, G, and U. If each codon consisted of two bases, then only 16 (4^2) different amino acids could be specified, one corresponding to each of the 16 possible combinations of two RNA bases. If a codon consisted of three bases, then the genetic code could specify as many as 64 (4^3) different amino acids, sufficient to encode the 20 different amino acids that make up proteins. Therefore, the minimum number of bases in a codon is three.

Francis Crick and his coworkers showed that the code is triplet by adding or removing one, two, or three bases to or from a viral gene with a well-known sequence and protein product. Altering the DNA sequence by one or two bases produced a different amino acid sequence. This happened because the change disrupted the **reading frame,** which is the sequence of amino acids encoded from a certain starting point in a DNA sequence. However, adding or deleting three contiguous bases added or deleted only one amino acid in the protein without disrupting the reading frame. The rest of the amino acid sequence was retained. The code, the researchers deduced, is triplet (**figure 10.12**).

Further experiments confirmed the triplet nature of the genetic code. Adding a base at one point in the gene and deleting a base at another point disrupted the reading frame only between these sites. The result was a protein with a stretch of the wrong amino acids, like a sentence with a few misspelled words in the middle.

Table 10.4	The Genetic Code*

		Second Letter							
		U		**C**		**A**		**G**	
U	UUU	Phenylalanine (Phe)	UCU	Serine (Ser)	UAU	Tyrosine (Tyr)	UGU	Cysteine (Cys)	**U**
	UUC		UCC		UAC		UGC		**C**
	UUA	Leucine (Leu)	UCA		UAA	"stop"	UGA	"stop"	**A**
	UUG		UCG		UAG	"stop"	UGG	Tryptophan (Trp)	**G**
C	CUU	Leucine (Leu)	CCU	Proline (Pro)	CAU	Histidine (His)	CGU	Arginine (Arg)	**U**
	CUC		CCC		CAC		CGC		**C**
	CUA		CCA		CAA	Glutamine (Gln)	CGA		**A**
	CUG		CCG		CAG		CGG		**G**
A	AUU	Isoleucine (Ile)	ACU	Threonine (Thr)	AAU	Asparagine (Asn)	AGU	Serine (Ser)	**U**
	AUC		ACC		AAC		AGC		**C**
	AUA		ACA		AAA	Lysine (Lys)	AGA	Arginine (Arg)	**A**
	AUG	Methionine (Met) and "start"	ACG		AAG		AGG		**G**
G	GUU	Valine (Val)	GCU	Alanine (Ala)	GAU	Aspartic acid (Asp)	GGU	Glycine (Gly)	**U**
	GUC		GCC		GAC		GGC		**C**
	GUA		GCA		GAA	Glutamic acid (Glu)	GGA		**A**
	GUG		GCG		GAG		GGG		**G**

The genetic code consists of mRNA triplets and the amino acids that they specify.

First Letter (vertical label) *Third Letter* (vertical label)

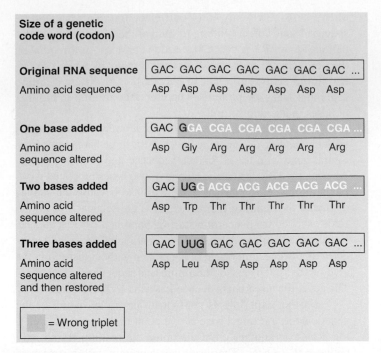

Size of a genetic code word (codon)							
Original RNA sequence	GAC	GAC	GAC	GAC	GAC	GAC	GAC ...
Amino acid sequence	Asp	Asp	Asp	Asp	Asp	Asp	Asp
One base added	GAC	G GA	CGA	CGA	CGA	CGA	CGA ...
Amino acid sequence altered	Asp	Gly	Arg	Arg	Arg	Arg	Arg
Two bases added	GAC	UG G	ACG	ACG	ACG	ACG	ACG ...
Amino acid sequence altered	Asp	Trp	Thr	Thr	Thr	Thr	Thr
Three bases added	GAC	UUG	GAC	GAC	GAC	GAC	GAC ...
Amino acid sequence altered and then restored	Asp	Leu	Asp	Asp	Asp	Asp	Asp

☐ = Wrong triplet

Figure 10.12 **Three at a time.** Adding or deleting one or two nucleotides in a DNA sequence results in a frameshift that disrupts the encoded amino acid sequence. Adding or deleting three bases does not disrupt the reading frame because the code is triplet. This is a simplified representation of the Crick experiment.

2. The Code Does Not Overlap.

Consider a hypothetical mRNA sequence:

AUGCCCAAG

If the genetic code is triplet and a DNA sequence is "read" in a nonoverlapping manner, then this sequence has only three codons and specifies three amino acids:

AUGCCCAAG

AUG (methionine)

CCC (proline)

AAG (lysine)

If the DNA sequence is overlapping, however, the sequence specifies seven codons:

AUGCCCAAG

AUG (methionine)

UGC (cysteine)

GCC (alanine)

CCC (proline)

CCA (proline)

CAA (glutamine)

AAG (lysine)

In an overlapping DNA sequence, certain amino acids would always follow certain others, constraining protein structure. For example, AUG would always be followed by an amino acid whose codon begins with UG. This does not happen in nature. Therefore, the protein-encoding DNA sequence is not overlapping.

Even though the genetic code is nonoverlapping, any DNA or RNA sequence can be read in three different reading frames, depending upon the "start" base. **Figure 10.13** depicts the three reading frames for the sequence just discussed, slightly extended. It encodes three different trios of amino acids.

3. The Code Includes Controls.

The genetic code includes directions for starting and stopping translation. The codon AUG signals "start," and the codons UGA, UAA, and UAG signify "stop." A sequence of DNA that does not include a stop codon is called an **open reading frame,** and is a sign of a possible protein-encoding gene. If a DNA sequence was just random, assuming equal numbers of each base, a stop codon would arise about every 21 codons.

Another form of "punctuation" in the genetic code is a short sequence of bases at the start of each mRNA that enables the mRNA to form hydrogen bonds with rRNA in a ribosome. It is called a leader sequence.

4. The Code Is The Same in All Species.

All species use the same mRNA codons to specify the same amino acids, and therefore the same genetic code. References to the "human genetic code" usually mean a human genome sequence. The simplest explanation for the "universality" of the genetic code is that all life evolved from a common ancestor. No other mechanism as efficient at directing cellular activities has emerged and persisted. The only known exceptions to the universality of the genetic code are a few codons in mitochondria and in certain single-celled eukaryotes (ciliated protozoa).

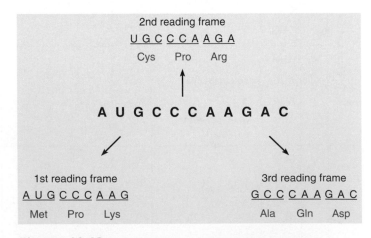

Figure 10.13 **Reading frames—where the sequence begins.** A sequence of DNA has three reading frames.

Once researchers knew that a codon consists of three bases and the code is nonoverlapping, the next step was to match codons to amino acids. A Glimpse of History on this page explains the clever experiments that revealed the genetic code. Sixty of the possible 64 codons specify amino acids, three codons indicate "stop," and one encodes both the amino acid methionine and "start." Some amino acids are specified by more than one codon. For example, UUU and UUC encode phenylalanine.

Different codons that specify the same amino acid are termed **synonymous codons,** just as synonyms are words with the same meaning. The genetic code is termed "degenerate" because most amino acids are not uniquely specified. Several synonymous codons differ by the base in the third position. The corresponding base of a tRNA's anticodon is called the "wobble" position because it can bind to more than one type of base in synonymous codons. The degeneracy of the genetic code protects against mutation, because changes in the DNA that substitute a synonymous codon do not alter the protein's amino acid sequence. **Nonsynonymous codons** encode different amino acids.

The human genome project picked up where the genetic code experiments of the 1960s left off by identifying the DNA sequences that are transcribed into tRNAs. Sixty-one different tRNAs could exist, one for each codon that specifies an amino acid (the 64 triplets minus 3 stop codons). However, only 49 different genes encode tRNAs because the same type of tRNA can detect synonymous codons that differ only in whether the wobble (third) position is U or C. The same type of tRNA, for example, binds to both UUU and UUC codons, which specify phenylalanine.

Building a Protein

Protein synthesis requires several participants: mRNA, tRNA molecules carrying amino acids, ribosomes, energy-storing molecules such as adenosine triphosphate (ATP) and guanosine triphosphate (GTP), and protein factors. These pieces meet during translation initiation (**figure 10.14**). Chemical bonds hold the components together.

First, the mRNA leader sequence forms hydrogen bonds with a short sequence of rRNA in a small ribosomal subunit. The first mRNA codon to specify an amino acid is always AUG, which attracts an initiator tRNA that carries the amino acid methionine (abbreviated Met). This methionine signifies the start of a polypeptide. The small ribosomal subunit, the mRNA bonded to it, and the initiator tRNA with its attached methionine form the initiation complex at the appropriate AUG codon of the mRNA.

To begin elongation, a large ribosomal subunit bonds to the initiation complex. The codon adjacent to the initiation codon (AUG), which is GGA in **figure 10.15**, then bonds to its complementary anticodon, which is part of a free tRNA that carries the amino acid glycine. The two amino acids (Met and Gly in the example), still attached to their tRNAs, align.

The part of the ribosome that holds the mRNA and tRNAs together can be described as having two sites. The positions of the sites on the ribosome remain the same with respect to each other as translation proceeds, but they cover different parts of the mRNA as the ribosome moves. The P ("peptide") site holds the growing amino acid chain, and the A ("acceptor") site next to it holds the next amino acid to be added to the chain. In figure 10.15a, when the forming protein consists of only the first two amino acids, Met occupies the P site and Gly the A site.

The amino acids link by a type of chemical bond called a peptide bond, with the help of rRNA that functions as a ribozyme (an RNA with enzymatic activity). Then the first tRNA is released to pick up another amino acid of the same type and be used again. Special enzymes ensure that tRNAs always bind the correct amino acids. The ribosome moves down the mRNA by one codon, so that the former A site is now the P site (figure 10.15b). The attached mRNA is now bound to a single tRNA, with two amino acids extending from it at the P site. This is the start of a polypeptide.

Next, the ribosome moves down the mRNA by another codon, and a third tRNA brings in its amino acid (Cys in figure 10.15b). This third amino acid aligns with the other two and forms a peptide bond to the second amino acid in the

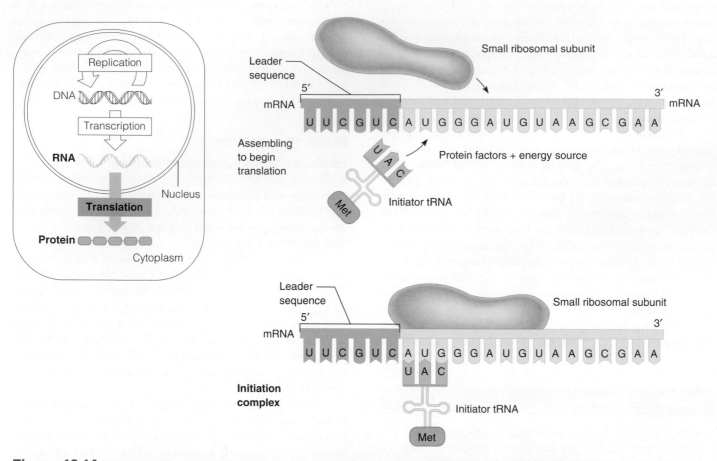

Figure 10.14 **Translation begins as the initiation complex forms.** Initiation of translation brings together a small ribosomal subunit, mRNA, and an initiator tRNA, and aligns them in the proper orientation to begin translation.

growing chain. The tRNA attached to the second amino acid is released and recycled. The polypeptide builds one amino acid at a time. Each piece is brought in by a tRNA whose anticodon corresponds to a consecutive mRNA codon as the ribosome moves down the mRNA (figure 10.15c).

Elongation halts when the A site of the ribosome has a "stop" codon (UGA, UAG, or UAA), because no tRNA molecules correspond to it. A protein release factor starts to free the polypeptide. The last tRNA leaves the ribosome, the ribosomal subunits separate and are recycled, and the new polypeptide is released (**figure 10.16**).

Protein synthesis is economical. A cell can produce large numbers of a particular protein molecule from just one or two copies of a gene. A plasma cell in the immune system, for example, manufactures 2,000 identical antibody molecules per second. To mass produce proteins at this rate, RNA, ribosomes, enzymes, and other proteins are continually recycled. In addition, transcription produces many copies of a particular mRNA, and each mRNA may bind dozens of ribosomes, as **figure 10.17** shows. As soon as one ribosome has moved far

enough along the mRNA to leave space, another ribosome attaches. In this way, many copies of the encoded protein are made from the same mRNA.

Some proteins undergo further alterations, called posttranslational modifications, before they can function. For example, insulin, which is 51 amino acids long, is initially translated as the 80-amino-acid-long polypeptide proinsulin. Enzymes cut it to 51. Some proteins must have sugars attached or polypeptides must aggregate for them to become active.

Key Concepts Questions 10.2

1. What are the general characteristics of the genetic code?

2. What are the steps of translation?

3. How is translation economical?

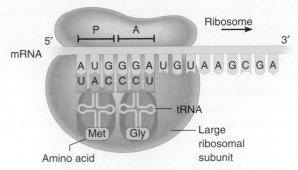

a. **Second amino acid joins initiation complex.**

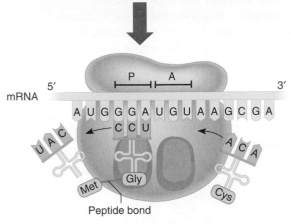

b. **First peptide bond forms as new amino acid arrives.**

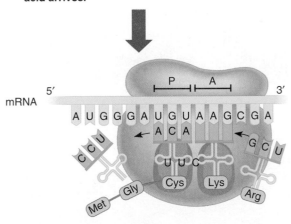

c. **Amino acid chain extends.**

Figure 10.15 **Building a polypeptide. (a)** A large ribosomal subunit binds to the initiation complex, and a tRNA bearing a second amino acid (glycine, in this example) forms hydrogen bonds between its anticodon and the mRNA's second codon at the A site. The first amino acid, methionine, occupies the P site. **(b)** The methionine brought in by the first tRNA forms a peptide bond with the amino acid brought in by the second tRNA, and a third tRNA arrives, in this example carrying the amino acid cysteine, at the temporarily vacated A site. **(c)** A fourth and then fifth amino acid are linked to the growing polypeptide chain. The process continues until a stop codon is reached.

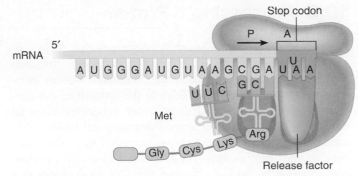

a. Ribosome reaches stop codon.

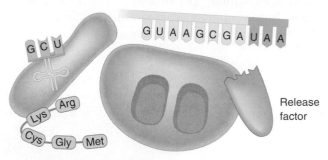

b. Components disassemble.

Figure 10.16 **Terminating translation. (a)** A protein release factor binds to the stop codon, releasing the completed polypeptide from the tRNA and **(b)** freeing all of the components of the translation complex.

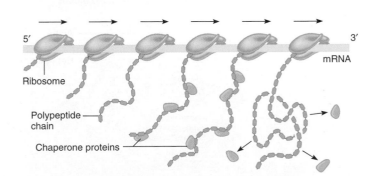

Figure 10.17 **Making many copies of a protein.** Several ribosomes can simultaneously translate a protein from a single mRNA. These ribosomes hold different-sized polypeptides—the closer to the end of a gene, the longer the polypeptide. Proteins called chaperones help fold the polypeptide.

10.3 Processing a Protein

Proteins fold into one or more three-dimensional shapes, or **conformations**. This folding is based on chemistry: attraction and repulsion between atoms of the proteins as well as interactions of proteins with chemicals in the immediate environment. For example, thousands of water molecules surround a growing chain of amino acids. Because some amino acids are attracted to water and some are repelled by it, the water contorts the protein's shape. Sulfur atoms also affect protein conformation by bridging the two types of amino acids that contain them.

The conformation of a protein is described at several levels (**figure 10.18**). The amino acid sequence of a polypeptide chain is its **primary (1°) structure**. The illustration for the chapter opener, of arginine, shows the parts of an amino acid. Chemical attractions between amino acids that are close

together in the 1° structure fold the polypeptide chain into its **secondary (2°) structure,** which may form loops, coils, barrels, helices, or sheets. Two common secondary structures are an alpha helix and a beta-pleated sheet. Secondary structures wind into larger **tertiary (3°) structures** as more widely separated amino acids attract or repel in response to water molecules. Finally, proteins consisting of more than one polypeptide form a **quaternary (4°) structure**. Hemoglobin, the blood protein that carries oxygen, has four polypeptide chains (see figure 11.1). The liver protein ferritin has 20 identical polypeptides of 200 amino acids each. In contrast, the muscle protein myoglobin is a single polypeptide chain.

Mutations may alter the primary structure of a protein if the genetic change is nonsynonymous, which means that it changes the amino acid. In contrast, more than one tertiary or quaternary structure may be possible for a protein if an amino acid chain can fold in different ways.

H_2N— Ala — Thr — Cys — Tyr — Glu — Gly —COOH

a. **Primary structure**—the sequence of amino acids in a polypeptide chain

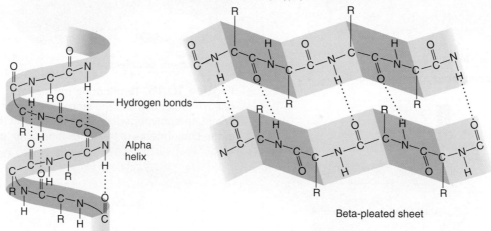

b. **Secondary structure**—loops, coils, sheets, or other shapes formed by hydrogen bonds between neighboring carboxyl and amino groups

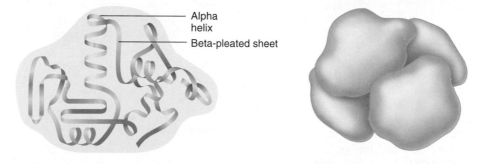

c. **Tertiary structure**—three-dimensional forms shaped by bonds between R groups, interactions between R groups and water

d. **Quaternary structure**—protein complexes formed by bonds between separate polypeptides

Figure 10.18 **Four levels of protein structure. (a)** The amino acid sequence of a polypeptide forms the primary structure. Each amino acid has an amino end (H_2N) and a carboxyl end (COOH), and each of the 20 types of amino acids is distinguished by an R group. **(b)** Hydrogen bonds between non–R groups create secondary structures such as helices and sheets. The tertiary structure **(c)** arises when R groups interact, folding the polypeptide in three dimensions and forming a unique shape. **(d)** If different polypeptide units must interact to be functional, the protein has a quaternary structure.

Protein Folding and Misfolding

Proteins begin to fold within a minute after the amino acid chain winds away from the ribosome. A small protein might contort into its final, functional form in one step, taking microseconds. Larger proteins may fold into a series of short-lived intermediates before assuming their final, functional forms.

Proteins start to move toward their destinations as they are being synthesized. In some proteins, part of the start of the amino acid chain forms a tag that helps direct the protein in the cell. The first few amino acids in a protein that will be secreted or lodge in a membrane form a "signal sequence" that leads it and the ribosome to which it binds into a pore in the ER membrane. Once in the ER, the protein enters the secretory network (see figure 2.6). Proteins destined for the mitochondria bear a different signal sequence. (Mitochondria manufacture their own proteins but also use many proteins that are encoded in DNA sequences in the nucleus.)

Signal sequences are not found on proteins synthesized on free ribosomes in the cytoplasm. These proteins may function right where they are made, such as the protein tubules and filaments of the cytoskeleton (see figure 2.11) or enzymes that take part in metabolism. Some proteins travel to and function in the nucleus, such as transcription factors. Proteins destined for the nucleus are synthesized on free ribosomes.

Various proteins assist in this precise folding, whatever the destination. **Chaperone proteins** stabilize partially folded regions in their correct form, and prevent a protein from getting "stuck" in an intermediate form, which would affect its function. Chaperone proteins are being developed as drugs to treat diseases that result from misfolded proteins. Other proteins help new chemical bonds to form as the final shape arises, and yet others monitor the accuracy of folding. If a protein misfolds, an "unfolded protein response" occurs in which protein synthesis slows or even stops, and transcription of genes that encode chaperone proteins and the other folding proteins accelerates, quickly restoring proper protein folding.

If a protein misfolds despite these protections, cells have ways to either refold the protein correctly, or get rid of it. Misfolded proteins are sent out of the ER back into the cytoplasm, where they are "tagged" with yet another protein, called ubiquitin. A misfolded protein bearing just one ubiquitin tag may straighten and refold correctly, but a protein with more than one tag is taken to another cellular machine called a **proteasome** (**figure 10.19**). A proteasome is a tunnel-like multi-protein structure. As a protein moves through the opening, it is stretched out, chopped up, and its peptide pieces degraded into amino acids, a little like a wood chipper. The amino acids may be recycled to build new proteins.

Proteasomes also destroy properly folded proteins that are in excess or no longer needed. For example, a cell must dismantle excess transcription factors, or the genes that they control may remain activated or repressed for too long. Proteasomes also destroy proteins from pathogens, such as viruses.

Proteins misfold in two ways: from a mutation, or by having more than one conformation. A mutation may change the amino acid sequence in a way that alters the attractions and repulsions between parts of the protein, contorting it. In many cases of cystic fibrosis, CFTR protein misfolds in the endoplasmic reticulum, attracts too many chaperones, and gets sent to the cytoplasm, never reaching the plasma membrane where normally it would form a chloride channel (see Clinical Connection 4.2).

Some proteins have more than one conformation, and one of them may become "infectious," converting the others to more copies of itself. The two forms of the same protein have identical amino acid sequences, but fold differently. It is a little like a person with a casual, friendly posture, versus the person hunched over and raising the arms in a threatening stance. Such an infectious protein is called a "prion" (pronounced "pree-on").

In several disorders that affect the brain, misfolded proteins aggregate, forming masses that clog the proteasomes and block them from processing any malformed proteins. Different proteins are affected in different disorders. In Huntington disease, for example, extra glutamines in the protein huntingtin cause it to obstruct proteasomes. Misfolded proteins that clog proteasomes also form in the disorders listed in **table 10.5**, but it isn't always clear whether the accumulated proteins cause the disease or are a response to it. Chapter 12 discusses some of these disorders further.

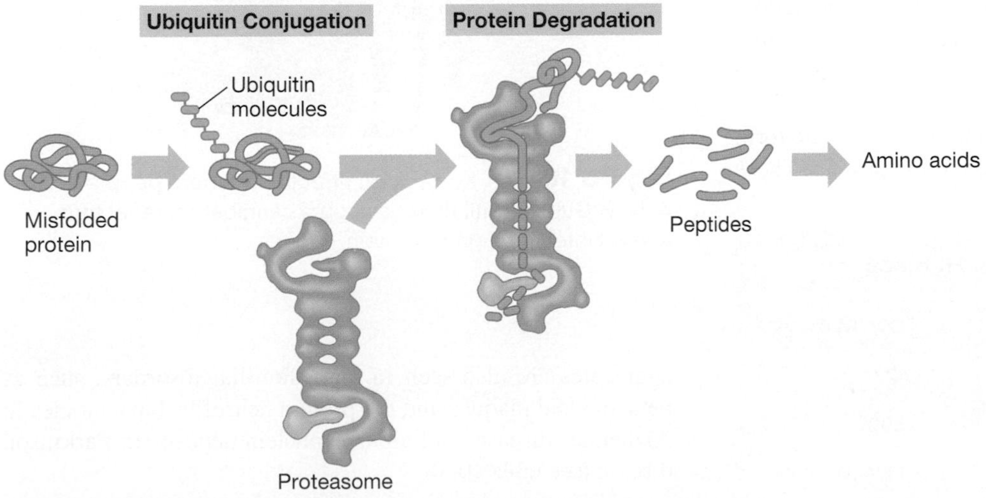

Prion Diseases

Prion diseases were first described in sheep, which develop a disease called scrapie when they eat prion-infected brains of other sheep. The name denotes the way the sick sheep

Figure 10.19 **Proteasomes provide quality control.** Ubiquitin binds to a misfolded protein and escorts it to a proteasome. The proteasome, which is composed of several proteins, encases the misfolded protein, straightening and dismantling it.

Table 10.5	Disorders Associated with Protein Misfolding	
Disease	**Misfolded Protein**	**OMIM (protein)**
Alzheimer disease	Amyloid beta precursor protein	104760
	Tau proteins	157140
Familial amyotrophic lateral sclerosis	Superoxide dismutase, TDP-43	147450, 605078
Frontotemporal dementia	Tau proteins, TDP-43	600274, 605078
Huntington disease	Huntingtin	143100
Parkinson disease	Alpha synuclein	163890
Lewy body dementia	Alpha synuclein	163890
PKU	Phenylalanine hydroxylase	261600
Prion disorders	Prion protein	176640

(All but Huntington disease are genetically heterogeneic; that is, abnormalities in different proteins cause similar syndromes.)

rub against things to scratch themselves. Their brains became riddled with holes. More than 85 animal species develop similar disorders. **Table 10.6** lists prion disorders of humans.

The first prion disease recognized in humans was kuru, which affected the native Fore people who lived in the remote mountains of Papua New Guinea (**figure 10.20**). *Kuru* means "to shake." The disease began with wobbly legs, then trembling and whole-body shaking. Uncontrollable laughter led to the name "laughing disease." Speech slurred, thinking slowed, and the person became unable to walk or eat. Death came within a year. The disease, which affected mostly women and children, was traced to a ritual in which the people ate their war heroes. When the women and children prepared the brains, prions entered cuts and they became infected.

Not many people knew about the plight of the Fore in the 1950s, but in the mid-1990s, prion diseases dominated the headlines when "mad cow disease" in the United Kingdom led to a human version, called variant Creutzfeldt-Jakob disease. More than 120 people ate infectious prions in beef and became ill.

Prions cause disease both by spreading the alternate form (infectious or mutant), and by aggregation of the protein. These

Figure 10.20 Kuru. Kuru affected the Fore people of New Guinea until they gave up a cannibalism ritual that spread infectious prion protein.

Table 10.6	Prion Disorders of Humans
Disorder	**OMIM #**
Creutzfeldt-Jakob disease	123400
Fatal familial insomnia	600072
Gerstmann-Sträussler-Scheinker disease*	137440

* See *Case Studies and Research Results* question 4 in chapter 21.

aggregates are also seen in more familiar disorders, such as beta amyloid plaques and tau protein neurofibrillary tangles in Alzheimer disease, and alpha synuclein deposits in Parkinson disease (see table 10.5).

The rare prion diseases like kuru, scrapie, and mad cow disease, as well as the more common ones, are all disorders

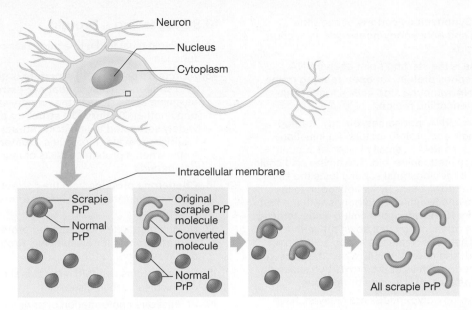

Figure 10.21 Prions change shape. Scrapie is a disease of sheep that was the first recognized prion disease. A single scrapie prion protein (PrP) contacts a normal PrP and changes it into the scrapie conformation. As the change spreads, disease results. Accumulated scrapie prion proteins clog brain tissue, eventually causing symptoms.

of protein folding. The infectious forms have beta-pleated sheets in places where their normal counterparts have alpha helices. **Figure 10.21** depicts schematically how normal proteins become prions, which can directly destroy brain areas and aggregate. In rare cases of these conditions, a mutation misfolds the protein. Drug discovery efforts are being directed toward refolding misfolded proteins.

> ## Key Concepts Questions 10.3
>
> 1. How are proteins folded?
> 2. How does a cell handle misfolded proteins?
> 3. How can a protein that has more than one conformation cause a disease?

Summary

10.1 Transcription Copies the Information in DNA

1. The exome is the part of the genome that encodes protein. Much of the rest controls protein synthesis.

2. **Transcription** makes a copy of DNA's information into a molecule of RNA. **Translation** uses the information in RNA to connect amino acids into proteins.

3. RNA is transcribed from the **template strand** of DNA. The other DNA strand is called the **coding strand**.

4. RNA is a single-stranded nucleic acid similar to DNA but with uracil and ribose rather than thymine and deoxyribose.

5. **Messenger RNA (mRNA)** carries a protein-encoding gene's information in the form of three-base units called **codons. Ribosomal RNA (rRNA)** associates with certain proteins to form ribosomes, which are where proteins are synthesized. **Transfer RNA (tRNA)** is cloverleaf-shaped, with a three-base **anticodon** that is complementary to mRNA on one end and bonds to a particular amino acid on the other end.

6. **Transcription factors** regulate which genes are transcribed in a particular cell type under particular conditions.

7. Transcription begins when transcription factors help **RNA polymerase (RNAP)** bind to a gene's **promoter**. RNAP then adds RNA nucleotides to a growing chain, in a sequence complementary to the DNA template strand.

8. After a gene is transcribed, the mRNA receives a "cap" of modified nucleotides at the 5′ end and a poly A tail at the 3′ end.

9. Many genes do not encode information in a continuous manner. After transcription, **introns** are removed and **exons** are translated into protein. Introns may outnumber and outsize exons. **Alternate splicing** increases protein diversity.

10.2 Translation of a Protein

10. Each three consecutive mRNA bases form a codon that specifies a particular amino acid. The **genetic code** is the correspondence between each codon and the amino acid it specifies. Of the 64 different possible codons, 60 specify amino acids, one specifies the amino acid methionine and "start," and three signal "stop." Because, theoretically, 61 codons specify the 20 amino acids, more than one type of codon may encode a single amino acid. The genetic code is nonoverlapping, triplet, universal,

and degenerate. **Synonymous codons** encode the same amino acid, and **nonsynonymous codons** encode different amino acids.

11. The **reading frame** is the starting point of an mRNA sequence that encodes protein. An **open reading frame** is a stretch of mRNA without a stop codon that might indicate a protein-encoding region.

12. Translation requires tRNA, ribosomes, energy-storage molecules, enzymes, and protein factors. An initiation complex forms when mRNA, a small ribosomal subunit, and a tRNA carrying methionine join. The amino acid chain elongates when a large ribosomal subunit joins the small one. Next, a second tRNA binds by its anticodon to the next mRNA codon, and its amino acid bonds with the first amino acid. Transfer RNAs add more amino acids, forming a polypeptide. The ribosome moves down the mRNA as the chain grows. The P site bears the amino acid chain, and the A site holds the newest tRNA. When the ribosome reaches a "stop" codon, it falls apart into its two subunits and is released. The new polypeptide breaks free.

13. After translation, some polypeptides are cleaved, have sugars added, or aggregate. The cell uses or secretes the protein.

10.3 Processing a Protein

14. A protein must fold into a particular **conformation** to function.

15. A protein's **primary (1°) structure** is its amino acid sequence. Its **secondary (2°) structure** forms as amino acids close in the primary structure attract one another. **Tertiary (3°) structure** appears as more widely separated amino acids attract or repel in response to water molecules. **Quaternary (4°) structure** forms when a protein consists of more than one polypeptide.

16. **Chaperone proteins** help the correct conformation arise. Other proteins help new bonds form and oversee folding accuracy.

17. Ubiquitin attaches to misfolded proteins and escorts them to **proteasomes** for dismantling. Protein misfolding causes diseases.

18. Some proteins can fold into several conformations, some of which can cause disease.

19. At least one conformation of prion protein is infectious, causing disease.

www.mhhe.com/lewisgenetics11

Answers to all end-of-chapter questions can be found at **www.mhhe.com/lewisgenetics11**. You will also find additional practice quizzes, animations, videos, and vocabulary flashcards to help you master the material in this chapter.

Review Questions

1. Explain how complementary base pairing is responsible for
   a. the structure of the DNA double helix.
   b. DNA replication.
   c. transcription of RNA from DNA.
   d. the attachment of mRNA to a ribosome.
   e. codon/anticodon pairing.
   f. tRNA conformation.

2. A retrovirus has RNA as its genetic material. When it infects a cell, it uses enzymes to copy its RNA into DNA, which then integrates into the host cell's chromosome. Is this flow of genetic information consistent with the central dogma? Why or why not?

3. What are the functions of these proteins?
   a. RNA polymerase
   b. ubiquitin
   c. a chaperone protein
   d. a transcription factor

4. Explain where a hydrogen bond forms and where a peptide bond forms in the transmission of genetic information.

5. List the differences between RNA and DNA.

6. Where in a cell do DNA replication, transcription, and translation occur?

7. How does transcription control cell specialization?

8. How can the same mRNA codon be at an A site on a ribosome at one time, but at a P site at another time?

9. Describe the events of transcription initiation.

10. List the three major types of RNA and their functions.

11. Describe three ways RNA is altered after it is transcribed.

12. What are the components of a ribosome?

13. Why would an overlapping genetic code be restrictive?

14. Why would two-nucleotide codons be insufficient to encode the number of amino acids in biological proteins?

15. How are the processes of transcription and translation economical?

16. What factors determine how a protein folds into its characteristic conformation?

17. How do a protein's primary, secondary, and tertiary structures affect conformation? Which is the most important determinant of conformation?

18. Explain how a protein can be infectious.

Applied Questions

1. List the RNA sequence transcribed from the DNA template sequence TTACACTTGCTTGAGAGTC.

2. Reconstruct the corresponding DNA template sequence from the partial mRNA sequence GCUAUCUGUCAUAAAAGAGGA.

3. List three different mRNA sequences that could encode the amino acid sequence

   histidine-alanine-arginine-serine-leucine-valine-cysteine.

4. Write a DNA sequence that would encode the amino acid sequence valine-tryptophan-lysine-proline-phenylalanine-threonine.

5. In the film *Jurassic Park,* which is about cloned dinosaurs, a cartoon character named Mr. DNA talks about the billions of genetic codes in DNA. Why is this statement incorrect?

6. Titin is a muscle protein named for its size—its gene has the largest known coding sequence of 80,781 DNA bases. How many amino acids long is it?

7. An extraterrestrial life form has a triplet genetic code with five different bases. How many different amino acids can this code specify, assuming no degeneracy?

8. In malignant hyperthermia, a person develops a life-threateningly high fever after taking certain types of anesthetic drugs. In one family, mutation deletes three contiguous bases in exon 44. How many amino acids are missing from the protein?

9. The protein that serves as a receptor that allows insulin to enter cells has a different number of amino acids in a fetus and in an adult. Explain how this may happen.

10. Frontotemporal dementia causes social difficulties, poor executive function (decision making), insomnia, depression, addictive behaviors, and suicide. Athletes and people who sustain severe head injuries, such as from football or combat, are at much higher risk of developing frontotemporal dementia. Autopsies reveal that this condition is due to either of two types of misfolded proteins. How might the mechanism underlying prion disorders explain these findings?

11. A person with SHORT syndrome has a very unusual face that is somewhat triangular and flat, with a small nose, deep-set eyes, low ears, and a downturned mouth. The person is thin, short, and has hyperextensible joints. A child with SHORT syndrome loses baby teeth very late. Part of the coding strand sequence for the gene *PIK3R1* (OMIM 269880), which causes the syndrome, is GAAAACCTT.

    a. What is the corresponding mRNA sequence?
    b. Which three amino acids does the mRNA sequence encode?

12. In Kallmann syndrome, sexual organs are underdeveloped, the person is infertile, and has no sense of smell. Three mutations are known in one of the six genes that cause the syndrome. For each mutation, indicate which DNA bases might be changed to alter a:

    a. lysine (Lys) to threonine (Thr)
    b. serine (Ser) to leucine (Leu)
    c. alanine (Ala) to valine (Val)

13. In "hypomyelination with atrophy of the basal ganglia and cerebellum" (H-ABC; OMIM 612438), a mutation in a gene that encodes a form of the cytoskeletal protein tubulin, *TUBB4A,* changes an aspartic acid (Asp) to an asparagine (Asn). The resulting brain shrinkage causes seizures, developmental delay, and a spastic gait. Only 22 cases are known.

    a. Is the mutation synonymous or nonsynonymous?
    b. Consult the genetic code table (table 10.4) and determine two mutations that could account for the amino acid changes.
    c. What might be the transcription pattern in the body for this gene?

Web Activities

1. Use the Web to find out how the ubiquitin-proteasome system is overtaxed or disabled in a neurodegenerative disease such as Alzheimer disease, Parkinson disease, Huntington disease, amyotrophic lateral sclerosis, or Lewy body dementia. (Find websites for these disorders and discuss how the mechanism involves proteasomes.)

Case Studies and Research Results

1. Five patients meet at a clinic for families with early-onset Parkinson disease. This condition causes rigidity, tremors, and other motor symptoms. Only 2 percent of cases of Parkinson disease are inherited. The five patients all have mutations in a gene that encodes the protein parkin, which has 12 exons. For each patient, indicate whether the mutation shortens, lengthens, or does not change the size of the protein, or if it isn't possible to tell what the effect might be.

    a. Manny Filipo's *parkin* gene is missing exon 3.
    b. Frank Myer's *parkin* gene has a duplication in intron 4.
    c. Theresa Ruzi's *parkin* gene lacks six contiguous nucleotides in exon 1.
    d. Elyse Fitzsimmon's *parkin* gene has an altered splice site between exon 8 and intron 8.
    e. Scott Shapiro's *parkin* gene is deleted.

2. The human genome sequence encodes many more mRNA transcripts than there are genes. Why isn't the number the same?

3. Francis Crick envisioned *"20 different kinds of adaptor molecule, one for each amino acid, and 20 different enzymes to join the amino acid to their adaptors."* What type of molecule was Dr. Crick describing?

4. In the 1990s, several people with Parkinson disease received implants of fetal cells into their brains. Years later, after the patients died, autopsies showed that the areas of the fetal implants in their brains had the transmissible form of alpha synuclein, aggregated into deposits called Lewy bodies. Explain how the fetal cells changed since they were implanted.

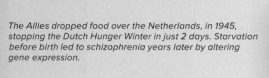

The Allies dropped food over the Netherlands, in 1945, stopping the Dutch Hunger Winter in just 2 days. Starvation before birth led to schizophrenia years later by altering gene expression.

Gene Expression and Epigenetics

Learning Outcomes

11.1 Gene Expression Through Time and Tissue

1. Define *epigenetics*.

2. Explain how globin chain switching, development of organs, and the types of proteins cells make over time illustrate gene expression.

11.2 Control of Gene Expression

3. Explain how small molecules binding to histone proteins control gene expression by remodeling chromatin.

4. Explain how microRNAs control transcription.

11.3 Maximizing Genetic Information

5. Explain how division of genes into exons and introns maximizes the number of encoded proteins.

11.4 Most of the Human Genome Does *Not* Encode Protein

6. Discuss how viral DNA, noncoding RNAs, and repeated sequences account for large proportions of the human genome.

The **BIG** Picture

Discovering the nature of the genetic material, determining the structure of DNA, deciphering the genetic code, and sequencing the human genome led to today's challenge: learning how genes are expressed through tissue and time.

The Dutch Hunger Winter

"Nature versus nurture" implies that genes and the environment work separately, but environmental conditions can greatly affect gene expression, which can affect health. For example, starvation before birth can alter gene expression in a way that may manifest as schizophrenia years later.

From February through April 1945, the "Dutch Hunger Winter," the Nazis blocked all food supplies from entering six large cities in western Holland. As malnutrition weakened and killed people, a cruel experiment took place. Children who had been starved before birth were much more likely to develop schizophrenia years later than their siblings born in better times. The key factor in setting the stage for future poor health was not birth weight, as had been thought, but exposure to dangerous conditions during the first weeks of pregnancy.

In the ongoing Dutch Famine Study, researchers at Columbia University and Leiden University in the Netherlands discovered the link between prenatal malnutrition and schizophrenia because they knew the exact time of the starvation, and the exact calorie intake, from food ration records. They obtained the schizophrenia diagnoses from psychiatric registries and military induction records.

Prenatal nutrition affects an adult phenotype because starvation alters the pattern in which methyl groups (CH_3) bind DNA, selectively silencing genes. The DNA of people born in the months after the famine, when studied 60 years later, had different methylation patterns in the gene that encodes insulin-like growth factor 2 (IGF-2), than siblings born following healthier gestations. Because IGF-2 controls expression of genes that affect thinking, researchers hypothesize that schizophrenia may develop in some people born into famine when IGF-2 has too few methyl groups and is overexpressed in the brain.

11.1 Gene Expression Through Time and Tissue

A genome is like an orchestra. Just as not all of the musical instruments play with the same intensity at every moment, not all genes are expressed continually at the same levels. Some genes are always transcribed and translated, in all cells. Because they keep cells running, these genes are sometimes called "housekeeping genes." Other genes have more specialized roles, and become active as a cell differentiates.

Before the field of genomics began in the 1990s, the study of genetics proceeded one gene at a time, like hearing the separate contributions of a violin, a viola, and a flute. Many genetic investigations today, in contrast, track the crescendos of gene activity that parallel events in an organism's life. This new view introduced the element of time to genetic analysis. Unlike the gene maps of old, which ordered genes linearly on chromosomes, new types of maps are more like networks that depict the timing of gene expression in unfolding programs of development and response to the environment.

The discoveries of the 1950s and 1960s on DNA structure and function answered some questions about the control of gene expression while raising many more. How does a bone cell "know" to transcribe the genes that control the synthesis of collagen and not to transcribe genes that specify muscle proteins? What causes the proportions of blood cell types to shift into leukemia? How do chemical groups "know" to shield DNA from transcription in one circumstance, yet expose it in others?

Changes to the chemical groups that associate with DNA greatly affect which parts of the genome are accessible to transcription factors and under which conditions. Such changes to the molecules that bind to DNA that are transmitted to daughter cells when the cell divides are termed **epigenetic,** which means "outside the gene." Figure 6.12 shows how methyl groups bind to DNA, causing epigenetic changes.

Epigenetic changes do not alter the DNA base sequence and are reversible. They are passed from one cell generation to the next. These changes may affect the next generation of individuals if the conditions to which a fetus is exposed become dangerous. This is what happened to the survivors of the Dutch Hunger Winter described in the chapter opener. For a few sites in the genome, an epigenetic change may persist through meiosis to a third generation, but this appears to be rare. Specific classes of proteins and RNA molecules carry out epigenetic changes. Much of the genome encodes these modifiers of gene expression. The human genome, then, is a little like a device that comes with a long, detailed instruction manual.

This chapter looks at how cells access the information in DNA. We begin with two examples of gene expression at the molecular and organ levels: (1) hemoglobin switching during development and (2) specialization of the two major parts of the pancreas.

Globin Chain Switching

The globin proteins transport oxygen in the blood. They vividly illustrate control of gene expression because, in a process called globin chain switching, they assemble into different hemoglobin molecules depending upon stage of development (**figures 11.1** and **11.2**). A hemoglobin molecule in the blood of an adult consists of four polypeptide chains, each wound into a globular conformation. Two of the chains are 146 amino acids long and are called "beta" (β). The other two chains are 141 amino acids long and are termed "alpha" (α). Genes on different chromosomes encode the alpha and beta globin polypeptide chains.

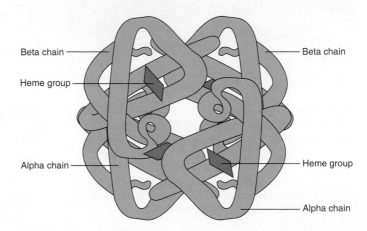

Figure 11.1 **The structure of hemoglobin.** A hemoglobin molecule is made up of two globular protein chains from the beta (β) globin group and two from the alpha (α) globin group. Each globin surrounds an iron-containing chemical group called a heme.

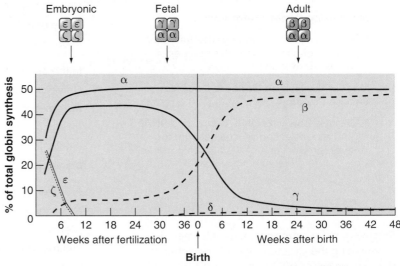

Figure 11.2 **Globin chain switching.** The subunit composition of human hemoglobin changes as the concentration of oxygen in the environment changes. Each globin quartet has two polypeptide chains encoded by genes in the alpha (α) globin cluster (chromosome 16) and two polypeptide chains from the beta (β) globin cluster (chromosome 11). With the switch from the placenta to the newborn's lungs to obtain oxygen, beta globin begins to replace gamma (γ) globin.

As a human develops, different globin polypeptide chains are used to make molecules of hemoglobin. The different forms of hemoglobin are necessary because of changes in blood oxygen levels that happen when a newborn begins breathing and no longer receives oxygen through the placenta. The promoters of the globin genes include binding sites for transcription factors, which orchestrate the changing hemoglobin molecules through development. Other DNA sequences in the globin gene clusters turn off expression of genes no longer needed.

The chemical basis for globin chain switching is that different globin polypeptide chains attract oxygen molecules to different degrees. In the embryo, as the placenta forms, hemoglobin consists first of two epsilon (ε) chains, which are in the beta globin group, and two zeta (ζ) chains, which are in the alpha globin group. About 4 percent of the hemoglobin in the embryo includes beta chains. This percentage gradually increases. Globin chains are manufactured first in the yolk sac in the embryo, then in the liver and spleen in the fetus, and finally primarily in the bone marrow after birth.

As the embryo develops into a fetus, the epsilon and zeta globin polypeptide chains decrease in number, as gamma (γ) and alpha chains accumulate. Hemoglobin consisting of two gamma and two alpha chains is called fetal hemoglobin. Because the gamma globin subunits bind very strongly to oxygen released from maternal red blood cells into the placenta, fetal blood carries 20 to 30 percent more oxygen than an adult's blood. As the fetus matures, beta chains gradually replace the gamma chains. At birth, however, the hemoglobin is not fully of the adult type—fetal hemoglobin (two gamma and two alpha chains) comprises from 50 to 85 percent of the blood. By 4 months of age, the proportion drops to 10 to 15 percent, and by 4 years, less than 1 percent of the child's hemoglobin is the fetal form.

Building Tissues and Organs

Blood is a structurally simple tissue that is easy to obtain and study because its components are easily separated. A solid gland or organ, with a distinctive three-dimensional form compared to the fluid blood, and constructed from specialized cells and in many cases more than one type of tissue, is much more complex. Its specific, solid organization must be maintained throughout a lifetime of growth, repair, and changing external conditions. Cells must maintain their specializations.

In all tissues and organs, genes are turned on and off during development, as stem cells self-renew and yield more specialized daughter cells. Researchers isolate individual stem cells and then see which combinations of growth factors, hormones, and other biochemicals must be added to steer development toward a particular cell type.

The pancreas has an interesting organization. It is a dual gland, with two types of cell clusters. The exocrine part releases digestive enzymes into ducts, whereas the endocrine part secretes polypeptide hormones that control nutrient use directly into the bloodstream. The endocrine cell clusters are called pancreatic islets.

The complexity of the pancreas unfolds in the embryo, when ducts form. Within duct walls reside rare stem cells and progenitor cells (see figure 2.20). A transcription factor is activated and controls expression of other genes in a way that stimulates some progenitor cells to divide. Certain daughter cells follow an exocrine pathway and will produce digestive enzymes. Other progenitor cells respond to different signals and divide to yield daughter cells that follow the endocrine pathway.

Figure 11.3 shows the differentiated cell types that form from the two cell lineages in the pancreas. The most familiar

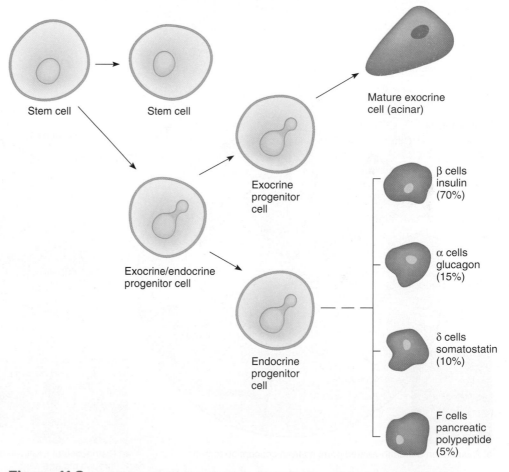

Figure 11.3 **Building a pancreas.** A single type of stem cell theoretically divides to give rise to an exocrine/endocrine progenitor cell that in turn divides to yield more restricted progenitor cells that give rise to both mature exocrine and endocrine cells. The endocrine progenitor cell in turn divides, yielding specialized daughter cells that produce specific hormones.

pancreatic hormone is insulin, which the beta cells of the pancreas secrete. The absence of insulin, or the inability of cells to recognize it, causes diabetes mellitus. If pancreatic stem cells can be isolated and cultured, it might be possible to coax a person with diabetes to produce new and functional pancreatic beta cells. Many people with diabetes who cannot make insulin take it as a drug.

Proteomics

A more complete portrait of gene expression emerges through **proteomics,** which is an area of genetics that identifies and analyzes all the proteins made in a cell, tissue, gland, organ, or body. **Figure 11.4** depicts a global way of comparing the relative contributions of major categories of proteins from conception through birth and from conception through old age. The differences in the levels of proteins made at these times make sense. Transcription factors are more abundant before birth because of the extensive cell differentiation of this period, as organs form. During the prenatal period, enzymes are less abundant, perhaps because the fetus receives some enzymes through the placenta.

Immunoglobulins appear after birth, when the immune system begins to function.

Another way to look at the proteome is by specific functions, which has led to the creation of various "ome" words. Genes whose encoded proteins control lipid synthesis, for example, constitute the "lipidome," and proteins that monitor carbohydrates form the "glycome." "Omics" designations sort the thousands of proteins a human cell can manufacture. However, identifying proteins is only a first step. The next hurdle is to determine how proteins with related functions interact. Researchers call these relationships among genes "interactomes" or "connectomes." Figure 1.9 shows a "diseaseome" of disease pairs linked by shared gene expression.

Gene expression profiles for different cell types under various conditions are the basis for many new medical tests that assess risk, diagnose disease, or monitor response to treatment. For example, 55 genes are overexpressed and 480 underexpressed in cells of a prostate cancer that has a very high likelihood of spreading, but not in a prostate cancer that will probably not spread. A test based on such findings can assist physicians in deciding which patients can safely delay or avoid invasive and risky treatment.

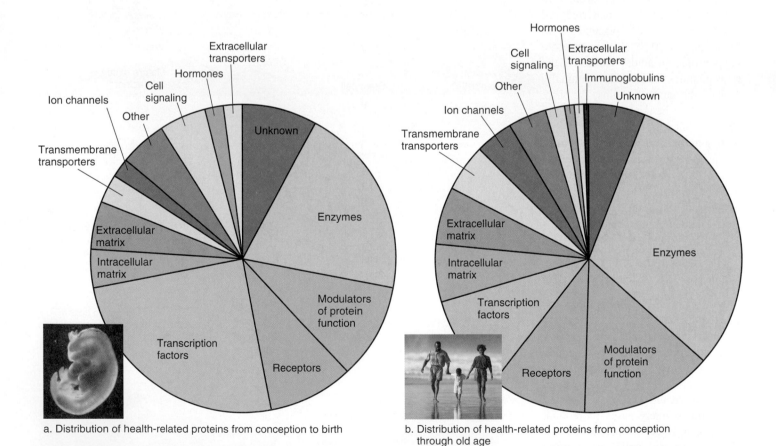

a. Distribution of health-related proteins from conception to birth

b. Distribution of health-related proteins from conception through old age

Figure 11.4 Our proteomes change over time. We can categorize genes by their protein products, and chart the relative abundance of each protein class at different stages of development. The pie chart in **(a)** considers 13 categories of proteins that when abnormal or missing cause disease, and their relative abundance, before birth. The pie chart in **(b)** displays the same protein categories from conception to old age, plus one activated after birth, the immunoglobulins.

11.2 Control of Gene Expression

The human genome includes the blueprints for building proteins as well as instructions for when and where to do so. A protein-encoding gene contains controls of its own expression. One is the promoter sequence. Recall from chapter 10 that the promoter is the part of the gene where RNA polymerase and transcription factors bind, marking the start point of transcription. Variations in the promoter sequence of a gene can affect how quickly the encoded protein is synthesized. For example, a form of early-onset Alzheimer disease is caused by a mutation in the promoter for the gene that encodes amyloid precursor protein (see Clinical Connection 5.1). In people who have the mutation, the sticky protein accumulates in the brain twice as fast as normal because it isn't cleared quickly enough. A second way that expression of a gene can exceed normal pace is if a person has more than one copy of it.

Cells specialize as different combinations of transcription factors bind the genes that provide the cell's characteristics. Transcription factors can bind at the promoter or to a DNA sequence away from the gene that the transcription factor controls, providing long-distance regulation.

Control of gene expression happens in several steps. In **chromatin remodeling,** the histone proteins associated with DNA interact with other chemical groups in ways that expose some sections of DNA to transcription factors and shield other sections, blocking their expression. Later in the protein production process, small RNAs called, appropriately, **microRNAs,** bind certain mRNAs, preventing their translation into protein. Overall, these two processes—chromatin remodeling and microRNAs—determine the ebb and flow of different proteins, enabling cells to adapt to changing conditions.

Chromatin Remodeling

Recall from figure 9.13 that DNA associates with proteins and RNA to form chromatin. For many years, biologists thought that the histone proteins that wind long DNA molecules into nucleosomes were little more than tiny spools. However, histones do much more. Enzymes add or delete small organic chemical groups to histones, affecting expression of the protein-encoding genes that histone proteins bind.

The three major types of small molecules that bind to histones are acetyl groups, methyl groups, and phosphate groups (**figure 11.5**). The key to the role histones play in controlling gene expression lies in acetyl groups (CH_3CO_2). They

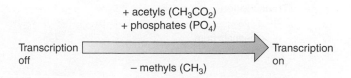

Figure 11.5 **Chromatin remodeling.** Chromatin remodeling adds or removes certain organic chemical groups to or from histones. The pattern of binding controls whether the DNA wrapped around the histones is transcribed or not.

bind to very specific sites on certain histones, particularly to the amino acid lysine.

Figure 11.6 shows how acetyl binding can shift histone interactions in a way that allows transcription to begin. A series of proteins moves the histone complex away from the TATA box, exposing it enough for RNA polymerase to bind and transcription to begin (see figure 10.7). First, a group of proteins called an enhanceosome attracts the enzyme (acetylase) that adds acetyl groups to specific lysines on specific histones, which neutralizes the histones' positive charge. Because DNA carries a negative charge, this change moves the histones away from the DNA, making room for transcription factors to bind and begin transcription. Enzymes called deacetylases remove acetyl groups, which shuts off gene expression. In most cell types, about 2 percent of the chromatin is "open." Researchers can separate these regions from the rest of the DNA and deduce which genes are being transcribed in a particular cell type under particular conditions.

Methyl groups (CH_3) are also added to or taken away from histones. When CH_3 binds to a specific amino acid in a specific histone type, a protein is attracted that prevents transcription. As CH_3 groups are added, methylation spreads from the tail of one histone to the adjacent histone, propagating the gene silencing. Methyl groups also control gene expression by binding to DNA directly, to cytosines at about 16,000 places in the genome, turning off transcription. The "methylome" is the collection of all the methylated sites in the genome. Different cell types have different genes that are methylated and turned off.

The modified state of chromatin can be passed on when DNA replicates. These changes in gene expression are heritable from cell generation to cell generation, but they do not alter the DNA base sequence—that is, they are epigenetic. Addition and removal of acetyl groups, methyl groups, or phosphate groups are examples of epigenetic changes. Effects of methylation can sometimes be seen when MZ (identical) twins inherit the same disease-causing genotype, but only one twin is sick. The reason for the discordance may be different patterns of methylation of one or more genes.

Enzymes that add or delete acetyl, methyl, and phosphate groups must be in a balance that controls which genes are expressed and which are silenced. Upset this balance, and disease can result. In a blood cancer called mixed lineage leukemia (see figure 18.19), for example, a single abnormal protein binds to more than 150 genes and alters their associated chromatin. Among these genes are several that normally stimulate frequent cell division in the stem cells that give rise to blood cells. In the leukemia, these

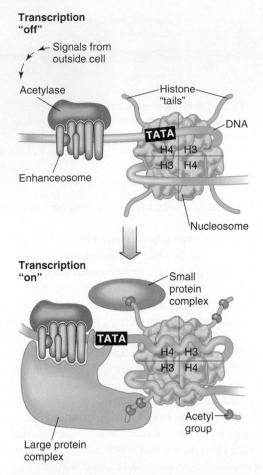

Transcription "off"

Signals from outside cell

Acetylase

Enhanceosome

Histone "tails"

DNA

TATA

H4 H3
H3 H4

Nucleosome

Transcription "on"

Small protein complex

TATA

H4 H3
H3 H4

Acetyl group

Large protein complex

Figure 11.6 **Acetylated histones allow transcription to begin.** Once acetyl groups are added to particular amino acids in the tails of certain histones, the TATA box becomes accessible to transcription factors. H3 and H4 are histone types.

overexpressed genes send the affected white blood cells back in developmental time, to a state in which the rapid cell division causes cancer. One limitation to altering chromatin remodeling to treat inherited disease is that this action could affect the expression of many genes—not just the one implicated in the disease.

MicroRNAs

Chromatin remodeling determines which genes are transcribed. MicroRNAs act later in gene expression, preventing the translation of mRNA transcripts into protein. If chromatin remodeling is considered as an on/off switch to transcription, then the control of microRNAs is more like that of a dimmer switch—fine-tuning gene expression at a later stage. **Figure 11.7** schematically compares chromatin

modeling and the actions of microRNAs, and **figure 11.8** places them in the overall flow of genetic information.

MicroRNAs are so-named because they are small—just 21 or 22 bases long. For many years, researchers accidentally threw them out when doing experiments searching for longer molecules. In the cell, microRNAs are cut from precursor molecules. They are a type of noncoding RNA, which means that they do not encode an amino acid sequence.

The human genome has at least 2,555 distinct sequences of microRNAs that regulate at least one-third of the protein-encoding genes. The DNA sequences that encode microRNAs are found in protein-encoding parts of the genome as well as in the vast regions that do not encode protein. A typical human cell has from 1,000 to 200,000 microRNAs.

Each type of microRNA binds to complementary sequences within a particular set of mRNAs. When a microRNA

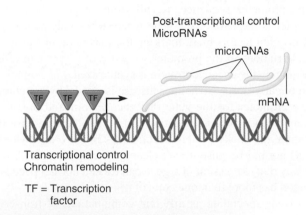

Post-transcriptional control MicroRNAs

microRNAs

mRNA

TF TF TF

Transcriptional control Chromatin remodeling

TF = Transcription factor

Figure 11.7 **Control of gene expression.** Chromatin opens to allow transcription factors to bind, whereas microRNAs bind to specific mRNAs, blocking their translation into protein.

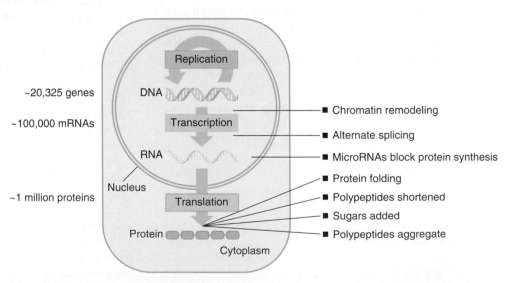

~20,325 genes

~100,000 mRNAs

~1 million proteins

Replication

DNA

Transcription

RNA

Nucleus

Translation

Protein

Cytoplasm

- Chromatin remodeling
- Alternate splicing
- MicroRNAs block protein synthesis
- Protein folding
- Polypeptides shortened
- Sugars added
- Polypeptides aggregate

Figure 11.8 **A summary of the events of gene expression.** Chromatin remodeling determines which genes are transcribed. Alternate splicing creates different forms of a protein. MicroRNAs bind to mRNAs by complementary base pairing, blocking translation. After translation, a protein must fold a certain way. Certain polypeptides are shortened, attached to sugars, and/or aggregated.

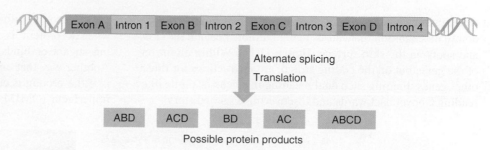

Figure 11.9 **Exons provide flexibility in gene structure that expands gene number.** Alternate splicing enables a cell to manufacture different versions of a protein by adding or deleting parts. Introns are removed and exons retained.

binds a "target" mRNA, it prevents translation. Because a single type of microRNA has many targets, it controls the expression of sets of genes. In turn, a single type of mRNA can bind several different microRNAs. Researchers use computational tools (bioinformatics) to analyze these complex interactions.

Within the patterns of microRNA function may lie clues to new ways to fight disease, because these controls of gene expression have stood the test of evolutionary time. That is, discovering the mRNAs that a certain microRNA targets may reveal genes that function together. Discovering these relationships can suggest new uses for and ways to combine existing drugs.

The first practical applications of identifying specific microRNA activities are in cancer. Certain microRNAs are either more or less abundant in cancer cells than in healthy cells of the same type from which the cancer cells formed. Restoring the levels of microRNAs that normally suppress the too-rapid cell cycling of cancer, or blocking production of microRNAs that are too abundant in cancer, could help to return cells to a normal cell division rate. In a related technology called RNA interference (RNAi), small, synthetic, double-stranded RNA molecules are introduced into selected cell types, where they block gene expression in the same manner as naturally occurring microRNAs (see figure 19.10).

Key Concepts Questions 11.2

1. What is the role of histone proteins in controlling gene expression?
2. How do acetyl, phosphate, and methyl groups control transcription?
3. How do microRNAs control gene expression?

11.3 Maximizing Genetic Information

The human genome maximizes information in the 20,325 genes that encode about 100,000 mRNAs, which in turn specify more than a million proteins. Figure 11.8 depicts this increase in information from gene to RNA to protein on the left, and the mechanisms that maximize the information on the right.

Several events account for the fact that proteins outnumber genes. The "genes in pieces" pattern of exons and introns and alternate splicing make it possible for one store of information—the gene—to encode different versions of a protein, much as a few items of clothing can be assembled into many outfits by combining them in different ways (**figure 11.9**).

The different proteins that result from different uses of the information in a gene are called isoforms. The driving force behind which version of a particular protein a cell makes is circumstance. For example, when an infection begins, an immune system cell first secretes a short version of an antibody molecule that is displayed on the cell's surface, where it alerts other cells. As the infection progresses, the cell transcribes an additional exon that extends the antibody in a way that enables it to be secreted into the bloodstream, where it attacks the pathogen. Exons from different genes can also be combined into a single mRNA. These "chimeric RNA transcripts" are unusual; some cause cancer.

Alternate splicing explains how a long sequence of DNA can specify more mRNAs than genes. On a part of chromosome 22, for example, 245 genes yield 642 mRNA transcripts. About 90 percent of all human genes are alternately spliced.

Another way that the genome holds more information than it may appear to is that a DNA sequence that is an intron in one context may encode protein (function as an exon) in another. Consider prostate specific antigen (PSA), which is a protein on certain cell surfaces that is overproduced in some cases of an enlarged prostate gland and in some prostate cancers. The gene for PSA has five exons and four introns. It is alternately spliced to encode seven isoforms. One of them, called PSA-linked molecule (PSA-LM), consists of the first exon and the fourth intron (**figure 11.10**). The two proteins (PSA and PSA-LM) work against one another. When the level of one is high, the other is low. Blood tests that measure levels of both proteins may more accurately assess the risk of developing prostate cancer than testing PSA alone.

Yet another way that introns may increase the number of proteins compared to genes is that a DNA sequence that is an intron in one gene's template strand may encode protein on the coding strand. This is so for the gene for neurofibromin, which,

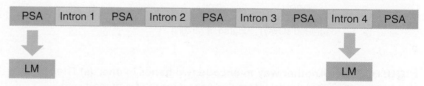

Figure 11.10 **Two genes from one.** Embedded in the PSA gene are two protein-encoding sequences. The PSA portion consists of five exons, and the PSA-LM part consists of two exons, one of which lies within an intron of PSA. (Not drawn to scale; introns are much larger than exons.)

when mutant, causes neurofibromatosis type 1, an autosomal dominant condition that causes benign tumors beneath the skin and spots on the skin surface (**figure 11.11**). Within an intron of the gene, but on the coding strand, are instructions for three other genes. Finding such dual meaning in a gene is a little like reading a novel backwards and discovering a second story!

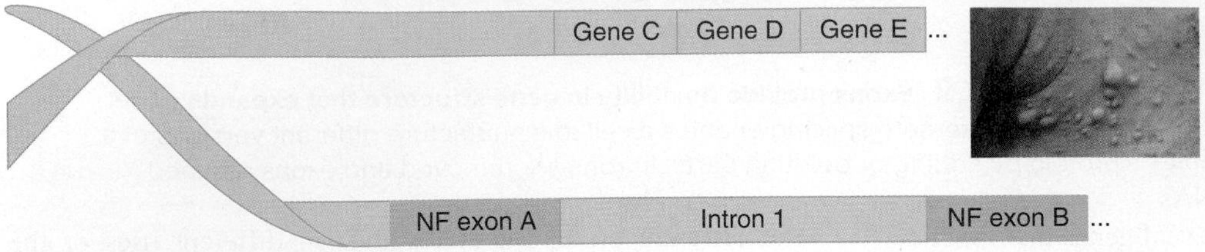

Figure 11.11 **Genes in introns.** An intron of the neurofibromin gene harbors three genes on the opposite strand. The inset shows neurofibromas.

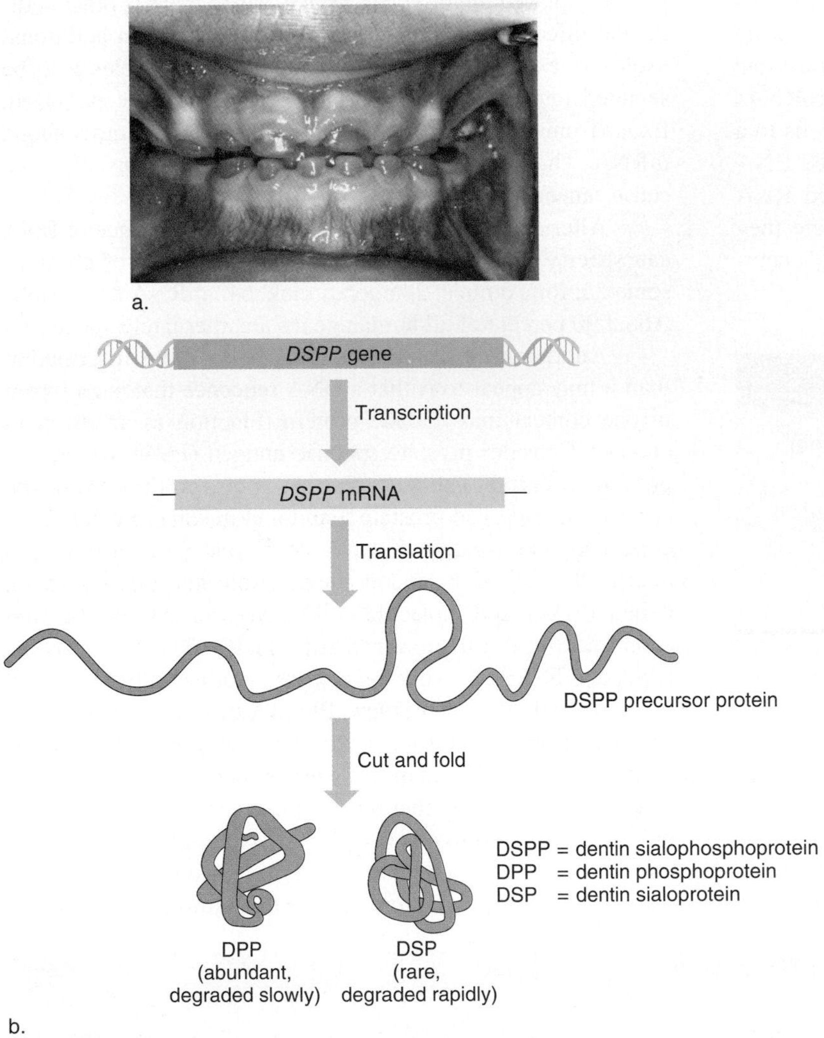

Figure 11.12 **Another way to encode two genes in one.** **(a)** The misshapen, discolored, and enamel-stripped teeth of a person with dentinogenesis imperfecta were at first associated with deficiency of the protein DPP. Then researchers discovered that DSP is deficient, too, but is very scant. **(b)** Both DPP and DSP are cut from the same larger protein, then separated, but DSP is degraded faster.

Genome information is maximized after translation when a protein is modified into different forms by adding sugars or lipids to create glycoproteins and glycolipids. Another way that one gene can encode more than one protein is if the protein is cut in two. This happens in dentinogenesis imperfecta (OMIM 125490), which is an autosomal dominant condition that causes discolored, misshapen teeth with peeling enamel (**figure 11.12**). The dentin, which is the bonelike substance beneath the enamel that forms the bulk of the tooth, is abnormal. Dentin is a complex mixture of extracellular matrix proteins, 90 percent of which are common collagen molecules. However, two proteins are unique to dentin: the abundant dentin phosphoprotein (DPP) and the rare dentin sialoprotein (DSP). DSP regulates mineral deposition into dentin, and DPP controls maturation of the mineralized dentin. A single gene encodes both. DPP and DSP are translated from a single mRNA molecule as the precursor protein dentin sialophosphoprotein (DSPP). Then they are separated. The two proteins are present in differing amounts because one (DSP) is degraded much faster than the other (DPP). The disease results from abnormal comparative levels of the two final proteins, DSP and DPP.

Key Concepts Questions 11.3

1. Why do the number of different proteins encoded in the genome differ from the number of genes?

2. Explain how alternate splicing, introns that encode protein, protein modifications, and cutting a precursor protein maximize the number of proteins that DNA encodes.

11.4 Most of the Human Genome Does *Not* Encode Protein

Only about 1.5 percent of human DNA encodes protein. The rest includes viral sequences, sequences that encode RNAs other than mRNA (called noncoding or ncRNAs), introns, promoters and other control sequences, and repeated sequences (**table 11.1**). In fact, most of the genome is transcribed—a DNA sequence is not "junk" if it does not encode protein.

Table 11.1   Some Non-protein-Encoding Parts of the Human Genome

Type of Sequence	Function or Characteristic
Viral DNA	Evidence of past infection
Noncoding RNA genes	
tRNA genes	Connect mRNA codon to amino acid
rRNA genes	Parts of ribosomes
Long noncoding RNAs	Control of gene expression
Pseudogenes	DNA sequences very similar to known genes that are not translated
Piwi-interacting RNA (piRNA)	Keeps transposons out of germline
Large intergenic noncoding RNAs	Between genes
Small nucleolar RNAs (snoRNAs)	Process rRNA in nucleolus
Small nuclear RNAs (snRNAs)	Parts of spliceosomes
Telomerase RNA	Adds bases to chromosome tips
Xist RNA	Inactivates one X chromosome in cells of females
Introns	Parts of genes that are cut out of mRNA
Promoters and other control sequences	Guide enzymes that carry out DNA replication, transcription, or translation
Small interfering RNAs (siRNAs)	Control translation
MicroRNAs (miRNAs)	Control translation of many genes
Repeats	
Transposons	Repeats that move around the genome
Telomeres	Protect chromosome tips
Centromeres	Largest constriction in a chromosome, providing attachment points for spindle fibers
Duplications of 10 to 300 kilobases	Unknown
Simple short repeats	Unknown

Viral DNA

Our genomes include DNA sequences that represent viruses. Viruses are nonliving infectious particles that consist of a nucleic acid (DNA or RNA) encased in a protein coat (see Clinical Connection 17.1). A virus replicates using a cell's transcriptional and translational machinery to mass-produce itself. New viruses may exit the cell, or the viral nucleic acid may remain in a host cell. A DNA virus may take over directly, inserting into a chromosome or remaining outside the nucleus in a circle called an episome. An RNA virus first uses an enzyme (reverse transcriptase) to copy its genetic material into DNA, which then inserts into a host chromosome.

About 100,000 sequences in our DNA, of varying lengths and comprising about 8 percent of the genome, were once a type of RNA virus called a retrovirus. The name refers to a retrovirus' direction of genetic information transfer, which is opposite DNA to RNA to protein. Retroviral sequences in our chromosomes are termed "endogenous" because they are carried from generation to generation of the host, rather than acquired as an acute infection. The retroviruses whose genetic material is in our chromosomes are called human endogenous retroviruses, or HERVs.

By comparing HERV sequences to similar viruses in other primates, researchers traced HERVs to a sequence representing a virus that infected our ancestors' genomes about 5 million years ago. Since then, HERV sequences have exchanged parts (recombined) and mutated to the extent that they no longer make us sick. Harmless HERVs silently pass from human generation to generation as parts of our chromosomes. They increase in number with time, as **figure 11.13** shows.

Noncoding RNAs

Much more of the human genome is transcribed than would be predicted based on the number and diversity of proteins that a human body can produce. The two general classes of RNAs are coding (the mRNAs) and noncoding (ncRNAs), which include everything else. The best-studied noncoding RNAs are the tRNAs and rRNAs.

The rate of transcription of a cell's tRNA genes is attuned to cell specialization. The proteins of a skeletal muscle cell, for example, require different amounts of certain amino acids than the proteins of a white blood cell, and therefore different amounts of the corresponding tRNAs. Human tRNA genes are dispersed among the chromosomes in clusters. Altogether, our 500 or so types of tRNA genes account for 0.1 percent of the genome.

The 243 types of rRNA genes are grouped on six chromosomes, each cluster harboring 150 to 200 copies of a 44,000-base repeat sequence. Once transcribed from these clustered genes, the rRNAs go to the nucleolus, where another type of ncRNA called small nucleolar RNA (snoRNA) cuts them into their final forms.

Hundreds of thousands of noncoding RNAs are neither tRNA nor rRNA, nor snoRNAs, nor microRNAs, nor the other less abundant types described in table 11.1. Some noncoding RNAs correspond to DNA sequences called **pseudogenes**. A pseudogene is very similar in sequence to a protein-encoding

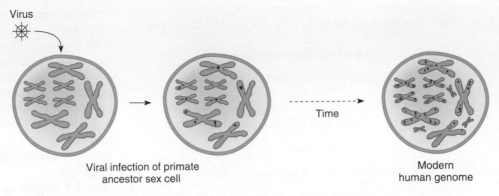

Virus

Viral infection of primate
ancestor sex cell

Time

Modern
human genome

Figure 11.13 **The human genome includes viral DNA sequences.** Most, if not all, of them do not harm us.

gene that may be transcribed, but it is not translated into protein. A pseudogene is altered in sequence from an ancestral gene in a way that may impair its translation or folding. Pseudogenes may be remnants of genes past, variants that diverged from the normal sequence too greatly to encode a working protein.

Since the sequencing of the human genome, a similarly huge project called ENCODE has discovered the functions of non-protein-encoding parts of the genome, including about 12,000 **long noncoding RNAs**. Many of these sequences likely control gene expression, because they reside in the nucleus, where they are physically associated with chromatin. Long noncoding RNAs represent exons, introns, and regions between genes. A third of them are found only in the genomes of primates, and many are transcribed only in the brain. These findings suggest that the long noncoding RNAs may hold clues to what makes us human—a topic discussed in chapter 16.

Repeats

The human genome is riddled with highly repetitive sequences that may be a different type of information than a protein's amino acid sequence. Perhaps repeat size or number constitute another type of molecular language. Or, perhaps some types of repeats help to hold a chromosome together.

The most abundant type of repeat is a sequence of DNA that can move about the genome. It is called a transposable element, or **transposon** for short. Geneticist Barbara McClintock originally identified transposons in corn in the 1940s, and they were rediscovered in bacteria in the 1960s. Transposons comprise about 45 percent of the human genome sequence, and typically are present in many copies. Some transposons include parts that encode enzymes that cut them out of one chromosomal site and integrate them into another. Unstable transposons may lie behind inherited diseases that have several symptoms, because they insert into

different genes. This is the case for Rett syndrome (see the chapter 2 opener).

An example of a specific type of repeat is an Alu sequence. Each Alu repeat is about 300 bases long, and a human genome may contain 300,000 to 500,000 of them. Alu repeats comprise 2 to 3 percent of the genome, and they have been increasing in number over time because they can copy themselves. We don't know exactly what these common repeats do, if anything. They may serve as attachment points for proteins that bind newly replicated DNA to parental strands before anaphase of mitosis, when replicated chromosomes pull apart.

Rarer classes of repeats comprise telomeres, centromeres, duplications of 10,000 to 300,000 bases, copies of pseudogenes, and simple repeats of one, two, or three bases. In fact, the entire human genome may have duplicated once or even twice.

Repeats may make sense in light of evolution, past and future. Pseudogenes are likely vestiges of genes that functioned in our nonhuman ancestors. Perhaps the repeats that seem to have no obvious function today will serve as raw material from which future genes may arise by mutation.

Discovery of the intricate controls of gene expression has led to a new definition of a gene, greatly expanded from the one-gene, one-protein idea of years past. A gene is a DNA sequence that contributes to a phenotype or function, plus the sequences, both in the gene and outside it, that control its expression. Chapter 12 looks at different types of changes in the DNA sequence—mutations—and their consequences.

Key Concepts Questions 11.4

1. What can RNA do in addition to encoding protein?
2. What are some types of noncoding RNAs?
3. What type of noncoding RNA might reflect our past?

Summary

11.1 Gene Expression Through Time and Tissue

1. Changes in gene expression occur over time at the molecular and organ levels. **Epigenetic** changes to DNA alter gene expression, but do not change the DNA sequence.

2. **Proteomics** catalogs the types of proteins in particular cells, tissues, organs, or entire organisms under specified conditions.

11.2 Control of Gene Expression

3. Acetylation of certain histone proteins enables the transcription of associated genes, whereas phosphorylation and methylation prevent transcription. The effect of these three molecules is called **chromatin remodeling.**

4. **MicroRNAs** bind to certain mRNAs, blocking translation.

11.3 Maximizing Genetic Information

5. A small part of the genome encodes protein, but the number of proteins is much greater than the number of genes.

6. Alternate splicing, use of introns, protein modification, and cutting proteins translated from a single gene contribute to protein diversity.

11.4 Most of the Human Genome Does *Not* Encode Protein

7. The non-protein-encoding part of the genome includes viral sequences, noncoding RNAs, **pseudogenes,** introns, **transposons,** promoters and other controls, and repeats.

8. **Long noncoding RNAs** control gene expression.

www.mhhe.com/lewisgenetics11

Answers to all end-of-chapter questions can be found at **www.mhhe.com/lewisgenetics11.** You will also find additional practice quizzes, animations, videos, and vocabulary flashcards to help you master the material in this chapter.

Review Questions

1. Why is control of gene expression necessary?

2. Define *epigenetics*.

3. Distinguish between the type of information that epigenetics provides and the information in the DNA sequence of a protein-encoding gene.

4. Describe three types of cells and how they differ in gene expression from each other.

5. What is the environmental signal that stimulates globin switching?

6. How does development of the pancreas illustrate differential gene expression?

7. Explain how a mutation in a promoter can affect gene expression.

8. How do histones control gene expression, yet genes also control histones?

9. What controls whether histones enable DNA wrapped around them to be transcribed?

10. State two ways that methyl groups control gene expression.

11. Name a mechanism that silences transcription of a gene and a mechanism that blocks translation of an mRNA.

12. Why might a computational algorithm be necessary to evaluate microRNA function in the human genome?

13. Describe three ways that the number of proteins exceeds the number of protein-encoding genes in the human genome.

14. How can alternate splicing generate more than one type of protein from the information in a gene?

15. In the 1960s, a gene was defined as a continuous sequence of DNA, located permanently at one place on a chromosome, that specifies a sequence of amino acids from one strand. List three ways this definition has changed.

16. Give an example of a discovery mentioned in the chapter that changed the way we think about the genome.

17. What is the evidence that some long noncoding RNAs may hold clues to human evolution?

Applied Questions

1. The World Anti-Doping Agency warns against gene doping, which it defines as "the non-therapeutic use of cells, genes, genetic elements, or of the modulation of gene expression, having the capacity to improve athletic performance." The organization lists the following genes as candidates for gene doping when overexpressed:

    Insulin-like growth factor (*IGF-1*)

    Growth hormone (*GH*)

Erythropoietin (*EPO*)

Vascular endothelial growth factor (*VEGF*)

Fibroblast growth factor (*FGF*)

Select one of these genes and explain how its overexpression might improve athletic performance.

2. What might be the effect of a mutation in the part of the gamma globin gene that normally binds a transcription factor?

3. Seema Sethi and her co-workers at Wayne State University compared the pattern of microRNAs in breast cancer tumors from women whose disease did not spread to the brain to the pattern in breast tumors that did spread to the brain. Suggest a clinical application of this finding.

4. Several new drugs inhibit the enzymes that either put acetyl groups on histones or take them off. Would a drug that combats a cancer caused by too little expression of a gene that normally suppresses cell division add or remove acetyl groups?

5. Chromosome 7 has 863 protein-encoding genes, but encodes many more proteins. The average gene is 69,877 bases, but the average mRNA is 2,639 bases. Explain both of these observations.

6. How many different proteins encompassing two exons can be produced from a gene that has three exons?

7. Researchers took pieces of the aorta (the largest blood vessel) from patients undergoing a heart transplant that were to be discarded as medical waste. They grew the pieces in lab dishes and added chemicals that stimulate inflammation, inducing atherosclerosis. The researchers identified the genes that were transcribed when the cells became inflamed, and concluded that about 1,000 genes are expressed in atherosclerosis. Explain how expression of a thousand genes differs from inheriting thousands of mutations.

Web Activities

1. Gene expression profiling tests began to be marketed several years ago. Search for "Oncotype DX," "MammaPrint," or "gene expression profiling in cancer" and describe how classifying a cancer this way can improve diagnosis and/or treatment. (Or apply this question to a different type of disease.)

2. The government's Genotype-Tissue Expression (GTEx; https://commonfund.nih.gov/GTEx/) project is a database of gene expression profiles of 24 tissues (parts of organs) from 190 people who died while healthy.
   a. What type of data are compared?
   b. Suggest a way that a researcher can use this type of information.

3. Look up each of the following conditions using OMIM or another source, and describe how they arise from altered chromatin: alpha-thalassemia, ICF syndrome, Rett syndrome, Rubinstein-Taybi syndrome.

Forensics Focus

1. Establishing time of death is critical information in a murder investigation. Forensic entomologists can estimate the "postmortem interval" (PMI), or the time at which insects began to deposit eggs on the corpse, by sampling larvae of specific insect species and consulting developmental charts to determine the stage. The investigators then count the hours backwards to estimate the PMI. Blowflies are often used for this purpose, but their three larval stages look remarkably alike in shape and color, and development rate varies with environmental conditions. With luck, researchers can count back 6 hours from the developmental time for the largest larvae to estimate the time of death.

   In many cases, a window of 6 hours is not precise enough to narrow down suspects when the victim visited several places and interacted with many people in the hours before death. Suggest a way that gene expression profiling might be used to more precisely define the PMI and extrapolate a probable time of death.

Case Studies and Research Results

1. To make a "reprogrammed" induced pluripotent stem (iPS) cell (see figure 2.22), researchers expose fibroblasts taken from skin to "cocktails" that include transcription factors. The fibroblasts divide and give rise to iPS cells, which, when exposed to other transcription factors, divide and yield daughter cells that specialize in distinctive ways that make them different from the original fibroblasts. How do transcription factors orchestrate these changes in cell type?

2. A study investigated "genomic signatures of global fitness" to identify gene expression patterns that indicate that a course of exercise is beneficial. In the study, sixty sedentary women representing different ethnic groups

underwent a 3-month exercise program. Blood samples were taken before and after the training period, and mRNAs identified. The study sought to identify specific genes whose expression changes in those women who experienced the greatest and least improvements in fitness, as measured by strength, lipids in the blood, and respiratory capacity. The study identified 39 genes that were expressed more in the women who became fit than in those who didn't.

a. Look up one of these genes and describe how it may contribute to better fitness: *TIGD7, UQCRH, PSMA6, WDR12, TFB2M,* and *USP15*.

b. What is a limitation of this study?

3. Using an enzyme called DNAse 1, researchers can determine which parts of the genome are in the "open chromatin" configuration in a particular cell. How could this technique be used to develop a new cancer treatment?

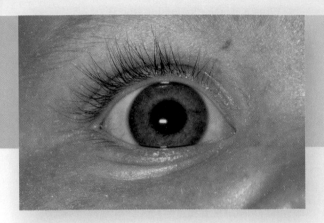

The whites of the eyes are bluish in a person with the "brittle bone disease" osteogenesis imperfecta type I. The color fades with time.

Gene Mutation

One Mutation, Multiple Effects: Osteogenesis Imperfecta

Shirley Banks, 73, became aware of her family's unusual condition as a young child. "My oldest brother had many fractures, and the doctors told him to eat high-calcium foods. We lived on a farm! All the dairy made no difference because of the mutation, but nobody knew." Several cousins easily broke bones too, and Shirley had the family's "brittle bone disease," but didn't yet realize it. Years later, her son Todd would inherit the family legacy: osteogenesis imperfecta (OI) type I (OMIM 166200).

"Todd had his first fracture at 2, when he tripped and broke his leg. I became suspicious because my brother's child broke bones, too. Finally, a doctor who had seen the disease as an intern explained it, and we were diagnosed," Shirley says. She realized her grandmother and a great uncle, who were two of seven in that generation, had the autosomal dominant disease, too.

The Banks family's mutation is in a gene that encodes the connective tissue protein collagen. Their type causes up to 100 fractures in a lifetime, which readily heal, but other forms may break bones before birth, proving lethal in infancy. When genetic testing became available, it was at first for only one type of OI, and some parents who had different types were falsely accused of abusing their children. Eight forms of OI are now recognized.

Only a few cases of OI are known from history. An Egyptian mummy from 1000 B.C. had it, as did ninth-century Viking "Ivan the Boneless," who was reportedly carried into battle aboard a shield and whose remains were exhumed and burnt by King William I, forever obscuring the true diagnosis.

Shirley's only broken bones are in her toes, but the disease causes other problems. She wears digital hearing aids and

Learning Outcomes

12.1 The Nature of Mutations

1. Distinguish between mutation and mutant.

2. Distinguish between mutation and polymorphism.

12.2 A Closer Look at Two Mutations

3. Describe mutations in the genes that encode beta globin and collagen.

12.3 Allelic Disorders

4. Provide examples of how mutations in a single gene can cause more than one illness.

12.4 Causes of Mutation

5. Explain the chemical basis of a spontaneous mutation.

6. Describe ways that researchers induce mutations.

12.5 Types of Mutations

7. Describe the two types of single-base mutations.

8. Explain the consequences of a splice-site mutation.

9. Discuss mutations that add, remove, or move DNA nucleotides.

12.6 The Importance of Position

10. Give examples of how the location of a mutation in a gene affects the phenotype.

11. Describe a conditional mutation.

12.7 DNA Repair

12. What types of damage do DNA repair mechanisms counter?

13. Describe the types of DNA repair.

The **BIG** Picture

On a species level, mutations provide the variation necessary for life to continue. On an individual level, mutations cause many illnesses, although a few mutations are helpful. DNA repair mechanisms protect against DNA damage.

the whites of her eyes (sclerae) have a bluish cast. Her tissues are fragile, and she bled profusely during surgery. Shirley and Todd have high pressure in their eyeballs (glaucoma) because their corneas are abnormally thin, which makes the pressure read lower than it actually is. They need higher doses of medication than other people with glaucoma. OI vividly illustrates the fact that one mutation can have multiple effects.

12.1 The Nature of Mutations

A **mutation** is a change in a DNA sequence that is rare in a population and typically affects the phenotype. "Mutate" refers to the process of altering a DNA sequence. Mutations range from substitution of a single DNA base; to deletion or duplication of tens, hundreds, thousands, or even millions of bases; to missing or extra entire chromosomes. This chapter discusses smaller-scale mutations, and chapter 13 considers mutation at the chromosomal level. However, the extent of mutation is a continuum.

Mutation can affect any part of the genome: sequences that encode proteins or control transcription; introns; repeats; and sites critical to intron removal and exon splicing. Not all DNA sequences are equally likely to mutate.

The effects of mutation vary. Mutations may impair a function, have no effect, or even be beneficial. A deleterious (harmful) mutation can stop or slow production of a protein, overproduce it, or impair the protein's function—such as altering its secretion, location, or interaction with another protein. The effect of a mutation is called a "loss-of-function" when the gene's product is reduced or absent, or a "gain-of-function" when the gene's activity changes. Most mutations are recessive and cause a loss-of-function (see figure 4.8). Gain-of-function mutations tend to be dominant and are also called "toxic."

The terms *mutation* and *polymorphism* each denote a genetic change. Recall from chapter 7 that a single nucleotide polymorphism, or SNP, is a single base change. So are many mutations. The distinction between mutation and polymorphism is largely artificial, reflecting frequency in a particular population, in which a mutation is much rarer than a polymorphism. If a genetic change greatly impairs health, individuals with it are unlikely to reproduce, and the mutant allele remains uncommon. A polymorphism that does not harm health, elevates risk of illness only slightly, or is even beneficial, will remain prevalent in a population or even increase in frequency. A genetic change that is a mutation in one population may be a harmless polymorphism in another. This is why considering a patient's ancestry is important in interpreting genetic test results.

Not all mutations are harmful, in contrast to their depiction in science fiction. For example, a mutation protects against HIV infection. About 1 percent of the general population is homozygous for a recessive allele that encodes a cell surface protein called CCR5 (see figure 17.11). To infect an immune system cell, HIV must bind CCR5 and another protein. Because the mutation prevents CCR5 from moving to the cell surface from inside the cell, HIV cannot bind. Heterozygotes for this mutation are partially protected against HIV infection. The opener to chapter 17 describes how mimicking *CCR5* mutation treats HIV infection.

The term *mutation* refers to genotype—that is, a change at the DNA or chromosome level. The familiar term **mutant** refers to phenotype. The nature of a mutant phenotype depends upon how the mutation affects the gene's product or activity, and usually connotes an abnormal or unusual characteristic. However, a mutant phenotype may also be an uncommon variant that is nevertheless "normal," such as red hair.

In an evolutionary sense, mutation has been essential to life, because it produces individuals with variant phenotypes who are better able to survive specific environmental challenges, including illnesses. Our evolutionary relatedness to other species enables us to learn from mutations in nonhuman species (**figure 12.1**).

A mutation may be present in all the cells of an individual or just in some cells. In a **germline mutation,** the change occurs during the DNA replication that precedes *meiosis*. The resulting gamete and all the cells that descend from it after fertilization have the mutation—that is, every cell in the body. Germline mutations are transmitted to the next generation of individuals. In contrast, a **somatic mutation** happens during DNA replication before a *mitotic* cell division, and is passed to the next generation of cells, not individuals. All the cells that descend from the original changed cell are altered, but they might only comprise a small part of the body. Somatic mutations are more likely to occur in cells that divide often, such as skin and blood cells, because there are more opportunities for replication errors. Such errors occur spontaneously or in response to exposure to toxins.

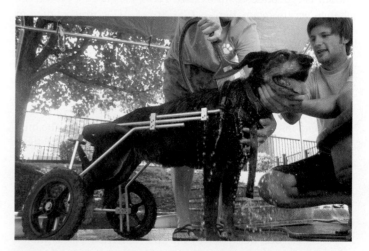

Figure 12.1 Animal models of human disease. This dog has amyotrophic lateral sclerosis (Lou Gehrig's disease), which also affects humans. Mutation in the same gene—superoxide dismutase 1—causes about 2 percent of human cases, as well as the canine cases.

New tools that enable researchers to detect mutations and polymorphisms throughout the genome, and in single cells from different organs of the same individual, reveal that the body has more genomically different cell populations than had been thought. A person may have cells from a twin that died before birth, or that descend from a cell that underwent a somatic mutation. Most women who have been pregnant retain cells from their fetuses. Most of us are, in some way, genomic mosaics.

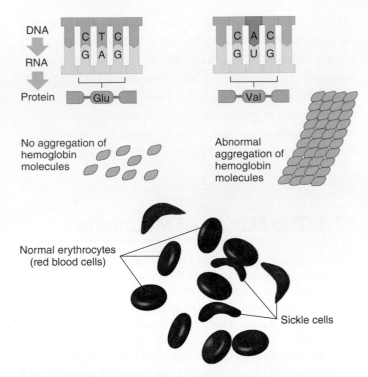

Figure 12.2 **Sickle cell disease results from a single DNA base change that substitutes one amino acid in the protein (valine replaces glutamic acid).** This changes the surfaces of the molecules, and they aggregate into long, curved rods that deform the red blood cell. The illustration shows the appearance of sickled cells.

Key Concepts Questions 12.1

1. Where do mutations occur in the genome?
2. How are mutations and polymorphisms alike and how do they differ?
3. Why is the ability of DNA to mutate important in evolution?
4. Distinguish between the consequences of a germline versus a somatic mutation.

12.2 A Closer Look at Two Mutations

Identifying how a mutation causes symptoms has clinical applications, and also reveals the workings of biology. Following are two examples of well-studied mutations that cause disease.

The Beta Globin Gene Revisited

The first genetic illness understood at the molecular level was sickle cell disease. The tiny mutation responsible for sickle cell disease is a substitution of the amino acid valine for the glutamic acid that is normally the sixth amino acid in the beta globin polypeptide chain (**figure 12.2**). At the DNA level, the change was even smaller—a CTC to a CAC, corresponding to RNA codons GAG and GUG. Valine at this position changes the surfaces of hemoglobin molecules so that in low-oxygen conditions they attach at many more points than they would if the wild type glutamic acid were at the site. The aggregated hemoglobin molecules form ropelike cables that at first make red blood cells sticky and able to deform. Then the red blood cells bend into rigid, fragile, sickle-shaped structures. The misshapen cells lodge in narrow blood vessels, cutting off local blood supplies. Once a blockage occurs, sickling speeds up and spreads, as the oxygen level falls. The result is great pain in the blocked body parts, particularly the hands, feet, and intestines. The bones ache, and depletion of normal red blood cells causes the great fatigue of anemia.

Sickle cell disease was the first inherited illness linked to a molecular abnormality, but it wasn't the first known condition that results from a mutation in the beta globin gene. In 1925, Thomas Cooley and Pearl Lee described severe anemia in Italian children, and in the decade following, others described a milder version of "Cooley's anemia," also in

Italian children. The disease was named thalassemia, from the Greek for "sea," in light of its high prevalence in the Mediterranean area. The two disorders turned out to be the same. The severe form, sometimes called thalassemia major, results from a homozygous mutation in the beta globin gene at a site other than the one that causes sickle cell disease. The milder form, called thalassemia minor, affects some individuals who are heterozygous for the mutation.

Once researchers had worked out the structure of hemoglobin, and learned that different globins function in the embryo and fetus (see figure 11.2), the molecular basis of thalassemia became clear. The disorder that is common in the Mediterranean is more accurately called beta thalassemia (OMIM 141900), because the symptoms result from too few beta globin chains. Without them, not enough hemoglobin molecules are assembled to effectively deliver oxygen to tissues. Fatigue and bone pain arise during the first year of life as the child depletes fetal hemoglobin, and the "adult" beta globin genes are not transcribed and translated on schedule.

As severe beta thalassemia progresses, red blood cells die because the excess of alpha globin chains prevents formation of hemoglobin molecules. Liberated iron slowly destroys the heart, liver, and endocrine glands. Periodic blood transfusions can control the anemia, but they hasten iron buildup and organ damage. Drugs called chelators that entrap the iron can extend life past early adulthood, but they are very costly and not available in some nations.

In 1904, young medical intern Ernest Irons noted "many pear-shaped and elongated forms" in a blood sample from a dental student in Chicago who had anemia. Irons sketched this first view of sickle cell disease at the cellular level, and reported his findings to his supervisor, physician James Herrick. Alas, Herrick published the work without including Irons and has been credited with the discovery ever since.

In 1949, Linus Pauling discovered that hemoglobin from healthy people and from people with the anemia, when placed in a solution in an electrically charged field, moved to different positions. Hemoglobin molecules from the parents of people with the anemia, who were carriers, moved to both positions.

The difference between the two types of hemoglobin lay in beta globin. Recall from figure 11.1 that adult hemoglobin consists of two alpha polypeptide subunits and two beta subunits. Protein chemist V. M. Ingram took a shortcut to localize the mutation in the 146-amino-acid-long beta subunit. He cut normal and sickle hemoglobin with a protein-digesting enzyme, separated the pieces, stained them, and displayed them on filter paper. The patterns of fragments—known as peptide fingerprints—were different for the two types of hemoglobin. This meant, Ingram deduced, that the two molecules differ in amino acid sequence. Then he discovered the difference. One piece of the molecule in the fingerprint, fragment four, occupied a different position for each of the two types of hemoglobin. Because this peptide was only 8 amino acids long, Ingram needed to decipher only that short sequence to find the site of the mutation. It was a little like finding which sentence on a page contains a miskey.

Disorders of Orderly Collagen

Much of the human body consists of the protein collagen, which is a major component of connective tissue. Collagen accounts for more than 60 percent of the protein in bone and cartilage and provides 50 to 90 percent of the dry weight of skin, ligaments, tendons, and the dentin of teeth. Collagen is in parts of the eyes and the blood vessel linings, and it separates cell types in tissues.

Genetic control of collagen synthesis and distribution is complex; more than thirty-five collagen genes encode more than twenty types of collagen molecules. Other genes affect collagen, too. Mutations in the genes that encode collagen, not surprisingly, lead to a variety of medical conditions (**table 12.1**). These disorders are particularly devastating, not only because collagen is nearly everywhere, but because collagen has an extremely precise conformation that is easily disrupted, even by slight alterations that might have little effect in proteins with other shapes (**figure 12.3**).

Collagen is trimmed from a longer precursor molecule called procollagen, which consists of many repeats of the amino acid sequence *glycine-proline-modified proline*. Three procollagen chains entwine. Two of the chains are identical and are encoded by one gene, and the other is encoded by a second gene and has a different amino acid sequence. The electrical charges and interactions of the amino acids with water coil the procollagen chains into a very regular triple helix, with space in the middle only for tiny glycine. Enzymes snip off the ragged ends of the polypeptides, forming mature collagen. The collagen fibrils continue to associate with each other outside the cell, building the fibrils and networks that hold the body together. **Figure 12.4** shows the characteristic stretchy skin that results from unassembled collagen molecules.

Table 12.1	Some Collagen Disorders		
Disorder	**OMIM**	**Genetic Defect (Genotype)**	**Signs and Symptoms (Phenotype)**
Alport syndrome	203780	Mutation in type IV collagen interferes with tissue boundaries	Deafness and inflamed kidneys
Aortic aneurysm	100070	Missense mutation substitutes Arg for Gly in α1 gene	Aorta bursts
Chondrodysplasia	302950	Deletion, insertion, or missense mutation replaces Gly with bulky amino acids	Stunted growth, deformed joints
Dystrophic epidermolysis bullosa	226600	Collagen fibrils that attach epidermis to dermis break down	Skin blisters on any touch
Ehlers-Danlos syndrome	130050	Missense mutations replace Gly with bulky amino acids; deletions or missense mutations disrupt intron/exon splicing	Stretchy, easily scarred skin, lax joints
Osteoarthritis	165720	Missense mutation substitutes Cys for Arg in α1 gene	Painful joints
Osteogenesis imperfecta type I	166200	Inactivation of α allele reduces collagen triple helices by 50%	Easily broken bones; blue eye whites; deafness
Stickler syndrome	108300	Nonsense mutation in procollagen	Joint pain, degeneration of vitreous gel and retina

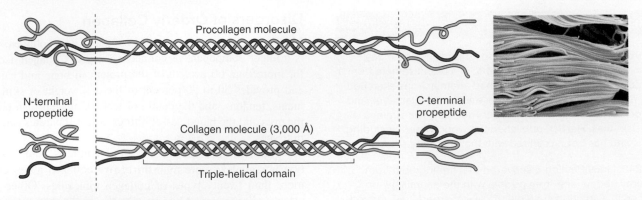

Figure 12.3 **Collagen has a very precise conformation.** The α1 collagen gene encodes the two blue polypeptide chains, and the α2 procollagen gene encodes the third (red) chain. The procollagen triple helix is shortened before it becomes functional, forming the fibrils and networks that comprise much of the human body. The inset shows aligned collagen fibrils.

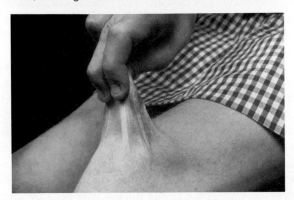

Figure 12.4 **Collagen mutations can affect the skin.** Ehlers-Danlos syndrome type I causes very extensible joints and stretchy skin, due to inability of collagen molecules to assemble properly.

So important is the precision of collagen formation that a mutation that controls placement of a single hydroxyl chemical group ($-OH^-$) on collagen causes a severe form of osteogenesis imperfecta ("brittle bone disease"). The chapter opener describes a milder form of osteogenesis imperfecta. Other collagen mutations cause missing procollagen chains, kinks in the triple helix, and defects in aggregation outside the cell.

Aortic aneurysm is a serious connective tissue abnormality that can occur by itself or as part of Marfan syndrome (see figure 5.7). Detection of mutations that cause Marfan syndrome before symptoms arise can be lifesaving, because frequent ultrasound exams can detect aortic weakening early enough to patch the vessel before it bursts. **Table 12.2** describes how a few other mutations impair health.

Table 12.2	How Mutations Cause Disease			
Disorder	**OMIM**	**Protein**	**Genetic Defect (Genotype)**	**Signs and Symptoms (Phenotype)**
Cystic fibrosis	602421	Cystic fibrosis transmembrane regulator (CFTR)	Missing amino acid or other defect alters conformation of chloride channels in certain epithelial cell plasma membranes. Water enters cells, drying out secretions.	Frequent lung infection, pancreatic insufficiency
Duchenne muscular dystrophy	310200	Dystrophin	Deletion eliminates dystrophin, which normally binds to inner face of muscle cell plasma membranes, maintaining cellular integrity. Cells and muscles weaken.	Gradual loss of muscle function
Familial hypercholesterolemia	143890	LDL receptor	Deficient LDL receptors cause cholesterol to accumulate in blood.	High blood cholesterol, early heart disease
Hemophilia A	306700	Factor VIII	Absent or deficient clotting factor causes hard-to-control bleeding.	Slow or absent blood clotting
Huntington disease	143100	Huntingtin	Extra bases in the gene add amino acids to the protein product, which impairs certain transcription factors and proteasomes.	Uncontrollable movements, personality changes
Marfan syndrome	154700	Fibrillin or transforming growth factor β receptor	Deficient connective tissue protein in lens and aorta.	Long limbs, weakened aorta, spindly fingers, sunken chest, lens dislocation
Neurofibromatosis type 1	162200	Neurofibromin	Defect in protein that normally suppresses activity of a gene that causes cell division.	Benign tumors of nervous tissue beneath skin

Key Concepts Questions 12.2

1. Describe how mutations affect the beta globin gene.
2. Explain why collagen genes are very prone to mutation.

12.3 Allelic Disorders

Annotation of the human genome is changing the way we describe single-gene disorders. In the past, geneticists were inconsistent when assigning disease names to mutations. For one gene, different mutations cause differing degrees or different subsets of symptoms of one named syndrome. Yet for another gene, different mutations cause different named disorders. For example, all mutations in the *CFTR* gene cause cystic fibrosis, which may include the full spectrum of impaired breathing and digestion, or just male infertility or frequent bronchitis. CF can affect different tissues in different individuals. Yet different mutations in the beta globin gene cause sickle cell disease and beta thalassemia. They have been considered two different disorders, although they both affect the same molecule in the blood.

Adding to the inconsistency in distinguishing diseases is a gene such as *lamin A*. Mutations in *lamin A* cause different disorders that affect very different tissues. They include the rapid aging disorder Hutchinson-Gilford progeria syndrome (see the chapter 3 opener), muscular dystrophies, and a heart disorder. Lamin A proteins form a network beneath the inner nuclear membrane that interacts with chromatin. Different mutations affect lamin A's interactions with chromatin in ways that cause the diverse associated disorders.

As researchers discovered more cases of different diseases arising from mutations in the same gene, it became clear that the phenomenon is not unusual, and is not merely a matter of how we name diseases. Different disease phenotypes caused by mutations in the same gene are termed **allelic disorders**. A compelling example is that of young Nicholas Volker, whose severe and mysterious intestinal disease was an "atypical presentation" of a known immune system condition. Clinical Connection 1.1 tells his famous story.

The same gene can underlie different diseases in different ways (**table 12.3**). A pair of allelic disorders may result from mutations in different parts of the gene; be localized (a single base change) or catastrophic (a missing gene); or alter the protein in ways that affect its interactions with other proteins.

Allelic disorders may arise from a mutation that affects a protein that is used in different tissues. For example, autosomal recessive mutations in a gene called *B3GALT6* (OMIM 615291) cause two conditions: spondyloepimetaphyseal dysplasia with joint laxity, type 1 (SEMD-JL1) and Ehlers-Danlos syndrome progeroid type. People with SEMD-JLI have prominent eyes with blue sclerae, long upper lips, small jaws with cleft palate, poor muscle tone, bowing of the back and limbs, and splayed fingers with short nails. Enlarged joints cause hip dislocation, clubfeet, and a bent spine that ultimately makes breathing impossible (**figure 12.5a** and **b**). Different mutations in

Table 12.3	Allelic Diseases	
Gene	**Function**	**Associated Diseases**
ATP7A	Copper transport	Menkes ("kinky hair") disease; peripheral neuropathy
DMD	Dystrophin muscle protein	Duchenne and Becker muscular dystrophy
FBN1	Encodes fibrillin-1, which forms tiny fibrils outside cells; a connective tissue protein	Marfan syndrome; stiff skin syndrome (scleroderma)
FGFR3	Fibroblast growth factor	2 types of dwarfs
GBA	Glucocerebrosidase	Gaucher disease; Parkinson disease
PSEN1	Presenilin 1 (enzyme part that trims membrane proteins)	Acne inversa; Alzheimer disease
RET	Oncogene (causes cancer)	Multiple endocrine neoplasia; Hirschsprung disease
TRPV4	Calcium channel	Peripheral neuropathy; spinal muscular atrophy

B3GALT6 cause the connective tissue disorder Ehlers-Danlos syndrome progeroid type, which has different symptoms: hyperextensible joints and loose elastic skin that scars easily. *B3GALT6* normally encodes an enzyme that attaches sugars to proteins, forming a component of the extracellular matrix called a proteoglycan. Researchers have not yet determined how a single type of enzyme deficiency affects skin, bones,

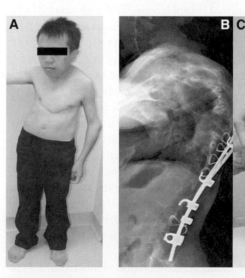

Figure 12.5 Mutations in *B3GALT6* cause allelic disorders. Panel A shows a 34-year-old man with the skeletal symptoms of SEMD-JL1. Note the curvature of his spine in panel B. Panel C shows a 5-year-old girl with Ehlers-Danlos syndrome progeroid type, in which the mutations primarily affect connective tissue of joints and her skin. Mutations in this gene cause other conditions, too.

cartilage, ligaments, and tendons in ways that cause different clinical presentations.

Allelic disorders challenge the concept of diagnosis. A clinician diagnoses a condition based on symptoms, obvious abnormalities at the whole-body level, and perhaps results of biochemical tests. A geneticist would define allelic disorders by sharing mutations in the same gene. Allelic disorders, in which different phenotypes correspond to mutations in the same gene, are in a way opposite genetic heterogeneity (discussed in chapter 5), in which mutations in different genes produce the same or very similar phenotypes.

Key Concepts Questions 12.3

1. How are the ways that cystic fibrosis and sickle cell disease named inconsistent?

2. Explain how mutations in *lamin A* cause wide-ranging effects on the phenotype.

3. Name a pair of allelic diseases and the gene from which they arise.

12.4 Causes of Mutation

A mutation can occur spontaneously or be induced by exposure to a chemical or radiation. An agent that causes mutation is called a **mutagen**.

Spontaneous Mutation

A spontaneous mutation can be a surprise. For example, two healthy people of normal height have a child with achondroplasia, an autosomal dominant form of dwarfism (see figure 5.1*a*). How could this happen when no other family members are affected? If the mutation is dominant, why are the parents of normal height? The child has a genetic condition, but he did not inherit it. Instead, he originated it. His siblings have no higher risk of inheriting the condition than anyone in the general population, but each of his children will face a 50 percent chance of inheriting it. The boy's achondroplasia arose from a *de novo,* or new, mutation in a parent's gamete. This is a spontaneous mutation—that is, it is not caused by a mutagen. A spontaneous mutation usually originates as an error in DNA replication. Chapter 4 explained how exome sequencing on children with symptoms that resemble those of an inherited syndrome, but whose parents do not have the associated mutations, may have *de novo* mutations.

One cause of spontaneous mutation stems from the tendency of free DNA bases to exist in two slightly different chemical structures, called tautomers. For extremely brief times, each base is in an unstable tautomeric form. If, by chance, such an unstable base is inserted into newly forming DNA, an error will be generated and perpetuated when that strand replicates. **Figure 12.6** shows how this can happen.

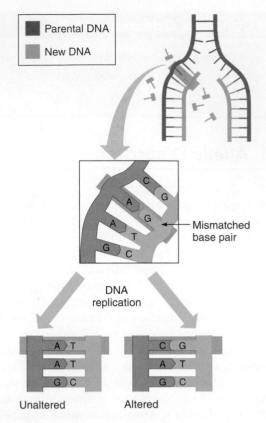

Figure 12.6 Spontaneous mutation. DNA bases exist fleetingly in alternate forms. A replication fork encountering a base in its unstable form can insert a noncomplementary base. After another round of replication, one of the daughter cells has a different base pair than the one in the corresponding position in the original DNA. (This figure depicts two rounds of DNA replication.)

Spontaneous Mutation Rate

Different genes have different spontaneous mutation rates. The gene that, when mutant, causes neurofibromatosis type 1 (NF1; see figure 11.11), for example, has a very high mutation rate, arising in 40 to 100 of every million gametes. NF1 affects 1 in 3,000 births, about half in families with no prior cases. The gene's large size may contribute to its high mutability—there are more ways for its sequence to change, just as there are more opportunities for a misspelling to occur in a long sentence than in a short one. In contrast, the mutation that causes the clotting disease hemophilia B happens in only one to ten of every million gametes formed.

A mutation may appear to be spontaneous and new (*de novo*) if it is in a child with a syndrome but not in genetic tests conducted on blood samples from either parent, and therefore should not affect siblings. But then a second child has the syndrome. An unusual circumstance called **gonadal mosaicism** may explain how this happens. A parent is a gonadal mosaic if only some sperm or oocytes have the mutation, because a spontaneous mutation occurred in the developing testis or ovary, and was transmitted only to the cells descended from the original cell bearing the mutation. In a gonadal mosaic the mutation is not in the blood cells used in genetic testing of the parents.

Sequencing of human genomes has revealed that spontaneous mutations are more common than had been thought. For example, at least 10 percent of heart defects in newborns are due to mutations not found in the parents. Each of us has about 175 spontaneously mutated alleles. Mitochondrial genes mutate at a higher rate than genes in the nucleus because they cannot repair DNA (see section 12.7).

Estimates of the spontaneous mutation rate for a particular gene are usually derived from observations of new, dominant conditions, such as achondroplasia. This is possible because a new dominant mutation is detectable simply by observing the phenotype. In contrast, a new recessive mutation would not be obvious until two heterozygotes produced a homozygous recessive offspring with a noticeable phenotype.

The spontaneous mutation rate for autosomal genes can be estimated using the formula: number of *de novo* cases/2X, where X is the number of individuals examined. The denominator has a factor of 2 to account for the nonmutated homologous chromosome.

Spontaneous mutation rates in human genes are difficult to assess because our generation time is long—usually 20 to 30 years. In bacteria, a new generation arises every half hour or so, and mutation is therefore much more frequent. The genetic material of viruses also spontaneously mutates rapidly, because they reproduce quickly and do not repair DNA errors.

Mutational Hot Spots

In some genes, mutations are more likely to occur in regions called hot spots, where sequences are repetitive. It is as if the molecules that guide and carry out replication become "confused" by short repeated sequences, much as an editor scanning a manuscript might miss the spelling errors in the words "hipppopotamus" and "bananana" (**figure 12.7**).

The increased incidence of mutations in repeats has a physical basis. Within a gene, when DNA strands locally unwind to replicate in symmetrical or repeated sequences, bases located on the same strand may pair. A stretch of ATATAT might pair with TATATA elsewhere on the same strand, creating a loop that interferes with replication and repair enzymes. Errors may result. For example, mutations in the gene for clotting factor IX, which causes hemophilia B, occur 10 to 100 times as often at any of 11 sites in the gene that have extensive direct repeats of CG than they do elsewhere in the gene.

Small additions and deletions of DNA bases are more likely to occur near sequences called palindromes (figure 12.7). These sequences read the same, in a 5′ to 3′ direction, on complementary strands. Put another way, the sequence on one strand is the reverse of the sequence on the complementary strand. Palindromes probably increase the spontaneous mutation rate by disturbing replication.

The blood disorder alpha thalassemia (OMIM 141800) illustrates the confusing effect of direct (as opposed to inverted) repeats of an entire gene. A person who does not have the disorder has four genes that specify alpha globin chains, two next to each other on each chromosome 16. Homologs with repeated genes can misalign during meiosis

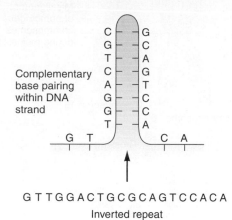

Figure 12.7 **DNA symmetry may increase the likelihood of mutation.** These examples show repetitive and symmetrical DNA sequences that may "confuse" replication enzymes, causing errors.

when the first sequence on one chromosome lies opposite the second sequence on the homolog. Crossing over can result in a sperm or oocyte that has one or three alpha globin genes instead of the normal two (**figure 12.8**). Fertilization with a normal gamete then results in a zygote with one extra or one missing alpha globin gene. At least three dozen conditions result from this unequal crossing over, including colorblindness (see Clinical Connection 6.1).

The number of alpha globin genes affects health. A person with only three alpha globin genes produces enough hemoglobin, and is a healthy carrier. Individuals with only two copies of the gene are mildly anemic and tire easily, and a person with a single alpha globin gene is severely anemic. A fetus lacking all alpha globin genes does not survive.

Induced Mutation

Researchers can infer a gene's normal function by observing what happens when mutation alters it. Because the spontaneous mutation rate is much too low to be a practical source of genetic variants for experiments, researchers make mutants, using mutagens on model organisms. These types of experiments can yield insights into human health.

Intentional Use of Mutagens

Chemicals or radiation are used to induce mutation. Alkylating agents, for example, are chemicals that remove a DNA base, which is replaced with any of the four bases—three of which

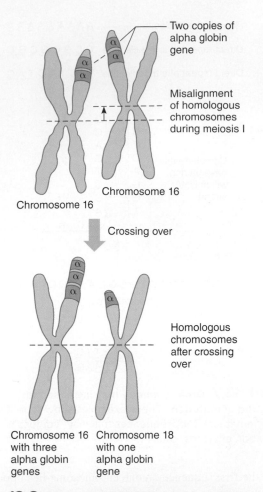

Two copies of alpha globin gene

Misalignment of homologous chromosomes during meiosis I

Chromosome 16

Chromosome 16

Crossing over

Homologous chromosomes after crossing over

Chromosome 16 with three alpha globin genes

Chromosome 18 with one alpha globin gene

Figure 12.8 **Gene duplication and deletion.** The repeated alpha globin genes are prone to mutation by mispairing during meiosis.

are a mismatch against the complementary strand. Dyes called acridines add or remove a single DNA base. Because the DNA sequence is read three bases in a row, adding or deleting a single base can destroy a gene's information, altering the amino acid sequence of the encoded protein. Several other mutagenic chemicals alter base pairs, so that an A-T replaces a G-C, or vice versa. X rays and other forms of radiation delete a few bases or break chromosomes.

Researchers have developed several ways to test the mutagenicity of a substance. The best known, the Ames test, assesses how likely a substance is to harm the DNA of rapidly reproducing bacteria. One version of the test uses a strain of *Salmonella* that cannot grow when the amino acid histidine is absent from its medium. If exposure to a substance enables bacteria to grow on the deficient medium, then a gene has mutated that allows it to do so.

In a variation of the Ames test, researchers exposed human connective tissue cells growing in culture to liquefied cigarette smoke. The chemicals from the smoke cut chromosomes through both DNA strands. Broken chromosomes can join with each other in different ways that can activate cancer-causing genes. Hence, the experiment may have modeled one way that cigarettes cause cancer. Because many mutagens are

also carcinogens (cancer-causing agents), the substances that the Ames test identifies as mutagens may also cause cancer. Common products that contain mutagens are hair dye, smoked meats, certain flame retardants used in children's sleepwear, and food additives.

A limitation of using a mutagen is that it cannot cause a specific mutation. In contrast, a technique called site-directed mutagenesis changes a gene in a desired way. A gene is mass-produced, but it includes an intentionally substituted base, just as an error in a manuscript is printed in every copy of a book. Site-directed mutagenesis is faster and more precise than waiting for nature or a mutagen to produce a useful variant.

Accidental Exposures to Mutagens

Some mutagen exposure is unintentional. This occurs from workplace contact before the danger is known; from industrial accidents; from medical treatments such as chemotherapy and radiation; from exposure to weapons that emit radiation; and from natural disasters that damage radiation-emitting equipment. For example, on April 25, 1986, between 1:23 and 1:24 A.M., Reactor 4 at the Chernobyl Nuclear Power Station in Ukraine exploded, sending a great plume of radioactive isotopes into the air that spread for thousands of miles. The reactor had been undergoing a test, its safety systems temporarily disabled, when it overloaded and rapidly flared out of control. Twenty-eight people died of acute radiation exposure in the days following the explosion.

Acute radiation poisoning is not genetic. Evidence of a mutagenic effect is the ten-fold increased rate of thyroid cancer among children who were living in nearby Belarus. The thyroid glands of young people soaked up iodine, which in a radioactive form bathed Belarus in the days after the explosion.

One way that researchers tracked mutation rates after the Chernobyl explosion was to compare the lengths of short DNA repeats, called minisatellite sequences, in children born in 1994 and in their parents, who lived in the exposed district at the time of the accident and have remained there. Minisatellites are the same length in all cells of an individual. A minisatellite size in a child that does not match the size of either parent indicates that a mutation occurred in a parent's gamete. Such a mutation was twice as likely to occur in exposed families as in families living elsewhere. Because mutation rates of non-repeated DNA sequences are too low to provide useful information on the effects of radiation exposure, investigators track minisatellites as a sensitive test of change.

Natural Exposure to Mutagens

Simply being alive exposes us to radiation that can cause mutation. Natural environmental sources of radiation account for 81% of our exposure, including cosmic rays, sunlight, and radioactive minerals in the earth's crust, such as radon. Medical X rays and occupational radiation hazards add risk. Job sites with increased radiation exposure include weapons facilities, research laboratories, health care facilities, nuclear power plants, and certain manufacturing plants.

Most of the potentially mutagenic radiation we are exposed to is ionizing, which means that it has sufficient energy to remove electrons from atoms. Unstable atoms that emit ionizing radiation exist naturally, and we make them. Ionizing radiation breaks the DNA sugar-phosphate backbone.

Ionizing radiation is of three major types. Alpha radiation is the least energetic and most short-lived, and the skin absorbs most of it. Uranium and radium emit alpha radiation. Beta radiation can penetrate the body farther, and emitters include tritium (a form of hydrogen), carbon-14, and strontium-70. Both alpha and beta rays tend not to harm health, although they can do damage if inhaled or eaten. In contrast is the third type of ionizing radiation, gamma rays. These can penetrate the body, damaging tissues. Plutonium and cesium isotopes used in weapons emit gamma rays, and this form of radiation is used to kill cancer cells.

X rays are the major source of exposure to human-made radiation. They have less energy and do not penetrate the body to the extent that gamma rays do.

The effects of radiation damage to DNA depend upon the functions of the mutated genes. Mutations in oncogenes or tumor suppressor genes, discussed in chapter 18, can cause cancer. Radiation damage can be widespread, too. Exposing cells to radiation and then culturing them causes a genome-wide destabilization, so that mutations may occur even after the cell has divided a few times. Cell culture studies have also identified a "bystander effect," when radiation harms cells not directly exposed.

Chemical mutagens are in the environment, too. Evaluating the risk that a specific chemical exposure will cause a mutation is very difficult, largely because people vary greatly in inherited susceptibilities, and are exposed to many chemicals. The risk that exposure to a certain chemical will cause a mutation is often less than the natural variability in susceptibility within a population, making it nearly impossible to track the true source and mechanism of any mutational event.

Key Concepts Questions 12.4

1. How do spontaneous mutations occur?
2. How does the DNA sequence affect the likelihood of mutations?
3. What are some mutagens?
4. Why are the effects of mutagens encountered in the environment difficult to assess?

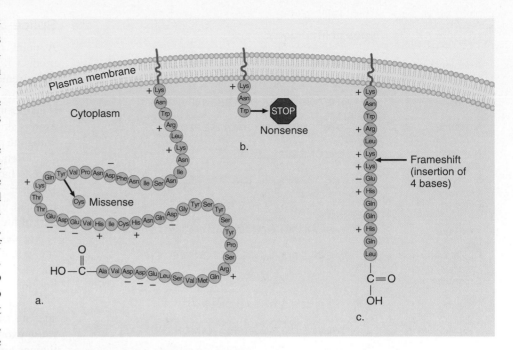

Figure 12.9 **Different mutations in a gene can cause the same disorder.** In familial hypercholesterolemia, several types of mutations alter the LDL receptor normally anchored in the plasma membrane. LDL receptor **(a)** bears a missense mutation—a cysteine substitutes for a tyrosine, bending the receptor enough to impair its function. The short LDL receptor in **(b)** results from a nonsense mutation, in which a stop codon replaces a tryptophan codon. In **(c)**, a 4-base insertion adds an amino acid and then alters the reading frame; + and − indicate electrical charge.

12.5 Types of Mutations

Mutations are classified by whether they remove, alter, or add a function, or by how they structurally alter DNA. The same single-gene disorder can result from different types of mutations, as **figure 12.9** shows for familial hypercholesterolemia. **Table 12.4** summarizes types of mutations using an analogy to an English sentence.

Point Mutations

A **point mutation** is a change in a single DNA base. It is a **transition** if a purine replaces a purine (A to G or G to A) or a pyrimidine replaces a pyrimidine (C to T or T to C). It is a **transversion** if a purine replaces a pyrimidine or vice versa (A or G to T or C). Addition or deletion of a single DNA base is also considered to be a point mutation. A point mutation can have any of several consequences—or it may have no obvious effect at all on the phenotype, acting as a silent mutation.

Missense and Nonsense Mutations

A point mutation that changes a codon that normally specifies a particular amino acid into one that codes for a different amino acid is called a **missense mutation**. If the substituted amino acid alters the protein's conformation significantly or occurs at a site critical to its function, signs or symptoms of disease or an

Table 12.4 Types of Mutations

A sentence comprised of three-letter words is analogous to the effect of mutations on a gene's DNA sequence:

Normal	THE ONE BIG FLY HAD ONE RED EYE
Missense	TH Q ONE BIG FLY HAD ONE RED EYE
Nonsense	THE ONE BIG ▭
Frameshift	THE ONE QBI GFL YHA DON ERE DEY
Deletion	THE ONE BIG ▭ HAD ONE RED EYE
Insertion	THE ONE BIG WET FLY HAD ONE RED EYE
Duplication	THE ONE BIG FLY FLY HAD ONE RED EYE
Expanding mutation	
Generation 1	THE ONE BIG FLY HAD ONE RED EYE
Generation 2	THE ONE BIG FLY FLY FLY HAD ONE RED EYE
Generation 3	THE ONE BIG FLY FLY FLY FLY FLY HAD ONE RED EYE

observable variant of a trait may result. About a third of missense mutations harm health.

The point mutation that causes sickle cell disease (see figure 12.2) is a missense mutation. The DNA sequence CTC encodes the mRNA codon GAG, which specifies glutamic acid. In sickle cell disease, the mutation changes the DNA sequence to CAC, which encodes GUG in the mRNA, specifying valine. This mutation changes the protein's shape, altering its function.

A point mutation that changes a codon specifying an amino acid into a "stop" codon—UAA, UAG, or UGA in mRNA—is a **nonsense mutation**. A premature stop codon is one that occurs before the natural end of the gene. It shortens the protein product, which can greatly influence the phenotype. For example, in factor XI deficiency (OMIM 264900), which is a blood clotting disorder, a GAA codon specifying glutamic acid is changed to UAA, signifying "stop." The shortened clotting factor cannot halt the profuse bleeding that occurs during surgery or from injury. Nonsense mutations are predictable by considering which codons can mutate to a "stop" codon.

In the opposite of a nonsense mutation, a normal stop codon mutates into a codon that specifies an amino acid. The resulting protein is longer than normal, because translation continues through what is normally a stop codon.

Point mutations can control transcription, affecting the quantity rather than the quality of a protein. For example, in 15 percent of people who have Becker muscular dystrophy (OMIM 310200)—a milder adult-onset form of the condition—the muscle protein dystrophin is normal, but its levels are reduced. The mutation is in the promoter for the dystrophin gene. This slows transcription, and dystrophin protein is scarce. The other 85 percent of individuals who have Becker muscular dystrophy have shortened proteins, not too few normal-length proteins.

Splice-Site Mutations

A point mutation can greatly affect a gene's product if it alters a site where introns are normally removed from the mRNA. This is called a **splice-site mutation**. It can affect the phenotype if an intron is translated into amino acids, or if an exon is skipped instead of being translated, shortening the protein.

Retaining an intron is unusual because most introns have stop codons in all reading frames. However, if a stop codon is not encountered, a retained intron adds bases to the protein-coding part of an mRNA. For example, in a family with severe cystic fibrosis, mutation alters an intron site so that it is not removed from the mRNA. The encoded protein is too bulky to move to its normal position in the plasma membrane.

A missense mutation that creates an intron splicing site where there should not be one can cause **exon skipping,** which removes a few contiguous amino acids. An entire exon is "skipped" when the mRNA is translated into protein, as if it were an intron. It is a little like leaving out a word when cutting and pasting a sentence in a document. An exon-skipping mutation is a deletion at the mRNA level, but it is a point (single-base) mutation at the DNA level. For example, a disorder called familial dysautonomia (OMIM 223900) (FD) can result from exon skipping in the gene encoding an enzyme necessary for the survival of certain neurons that control sensation and involuntary responses. The *In Their Own Words* box on page 223 describes life with FD.

A peculiarity of some disorders caused by exon skipping is that some cells ignore the mutation and manufacture a normal protein from the affected gene—after all, the amino acid sequence information is still there. Depending upon which cells actually make the full encoded protein, the phenotype may be less severe than in individuals with the same disorder but with a different type of mutation in an exon.

Studies on cells from individuals who have or have died from FD reveal that the cells in which the exon is skipped are the cells that contribute to symptoms. That is, many cells from the brain and spinal cord skip the exon, but cells from muscle, lung, liver, white blood cells, and glands produce normal-length proteins. Coaxing nervous system cells in affected children to produce the full protein could provide a treatment. Clinical trials are examining the ability of several natural compounds to restore normal processing of the FD gene's information.

In the case of familial dysautonomia, a skipped exon causes the disease. Harnessing exon skipping can also be used to *treat* genetic disease (**figure 12.10**). This is the case for Duchenne muscular dystrophy (DMD; see figure 2.1), a severe form of the disease. In most cases, a mutation in exon 50 of this 79-exon gene disrupts the reading frame (see figure 10.13), which generates a nonsense mutation. As a result, no full dystrophin protein is produced. However, in the milder Becker muscular dystrophy (BMD), a different mutation yields an

Familial Dysautonomia: Rebekah's Story

Familial dysautonomia (FD) is a rare genetic disorder that affects the autonomic and peripheral nervous systems. It causes pneumonia, vomiting and retching, extremely high fevers, chills, rapid heartbeat, rashes, and seizures. FD also reduces sensation of pain, heat, and cold. Problems with balance and coordination include motor difficulties that affect eating, swallowing, and breathing. Most people with FD have a feeding tube, and some have learning disabilities. Most individuals develop scoliosis, usually requiring corrective spine surgery. In short, FD affects every organ and system in the body.

FD is difficult to diagnose because so few physicians have ever seen it. Lynn Lieberman describes how doctors finally figured out why her daughter Rebekah was so sick. "After more than twelve local hospitalizations and a variety of tests, we traveled to a major children's teaching hospital, hoping that a fresh team of doctors would identify Rebekah's condition. One doctor knew immediately that she had FD. He recognized the pattern of 'dysautonomic crises.' Two more symptoms, which we hadn't even noticed, were diagnostic indicators. Individuals with FD do not cry tears, and they lack papillae (bumps) on the tip of the tongue. Our Eastern European, Jewish heritage was also a clue, because FD is one of a number of diseases primarily affecting this population."

Rebekah has done remarkably well, and describes her experience with FD.:

> Being a girl with FD has many challenges. There are medical problems and so much more. Like most people with FD, I have had scoliosis. That's when your spine becomes all twisted. When I was in middle school, I had a very bad curve in my back and it affected my breathing and health. I have had surgery to fix my spine. I had to stay in the hospital for a while and it took a long time to recover. But, now, my spine is as straight as an arrow. Well, almost, anyway.
>
> A lot of people with FD have problems with breathing. When I was younger, I had problems getting enough oxygen while I slept, which is called sleep apnea. I used oxygen while I was sleeping, but since I had my tonsils

out, I haven't had to use the oxygen that much. Some of my friends with FD use oxygen at night, or even all the time. The first of them died when I was in the seventh grade. That is very hard.

> I am not ashamed of having FD. I take my medication and do everything else I need to do so I can stay as healthy as possible. My friends with FD and I have a lot of things in common, so we understand what each other are going through. We are not treated like everyone else except by our close friends and family. Sometimes, when people don't really know us, they treat us like we are "special." We do not like that. We would just like to be treated like everyone else.
>
> Like any other teenager, I like to go shopping, spend time with friends, and go to movies. I also like it when I go to camp so I am able to see all of my friends. I have trouble working in school, but I do my best. I am a senior this year, and I'm excited about graduating. I love working with children and someday I would love to own my own daycare. I want to go to college to study Early Childhood Education. So, I take one step at a time so I am able to make my goal."

© Rebekah Lieberman

Rebekah Lieberman loves children and writing.

---

abnormal form of dystrophin that has partial activity, and symptoms are much milder and life span much longer. An experimental drug, consisting of a complementary, or "antisense" DNA-like molecule, binds to exon 50 and its immediate neighbors, mimicking exon skipping. The silencing of part of the gene is designed to restore the reading frame, so that the DMD patient can manufacture the altered form of the protein made in the milder form of the disease. So far the drug enables muscles of affected boys to produce functional dystrophin and

restores some mobility. The challenge in this type of treatment will be to deliver the drug to enough cells to make a difference in quality of life and survival.

Deletions and Insertions

In genes, the number 3 is very important, because triplets of DNA bases specify amino acids. Adding or deleting a number of bases that is not a multiple of three devastates a gene's function

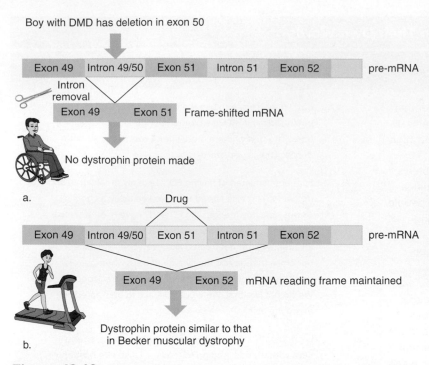

Boy with DMD has deletion in exon 50

| Exon 49 | Intron 49/50 | Exon 51 | Intron 51 | Exon 52 | pre-mRNA

Intron removal

| Exon 49 | Exon 51 | Frame-shifted mRNA

No dystrophin protein made

a.

Drug

| Exon 49 | Intron 49/50 | Exon 51 | Intron 51 | Exon 52 | pre-mRNA

| Exon 49 | Exon 52 | mRNA reading frame maintained

Dystrophin protein similar to that in Becker muscular dystrophy

b.

Figure 12.10 Inducing exon-skipping to treat Duchenne muscular dystrophy. (a) In DMD a mutation in exon 50 disrupts the reading frame, preventing synthesis of dystrophin protein. **(b)** An experimental drug that binds to the part of the dystrophin gene implicated in DMD and surrounding regions restores the reading frame, resulting in some protein production.

because it disrupts the gene's reading frame, which refers to the nucleotide position where the DNA begins to encode protein. Most exons are "readable" (have no stop codons) in only one of the three possible reading frames. A reading frame that is readable in that it is translatable into protein is called an open reading frame. A change that alters the reading frame is called a **frameshift mutation**.

A **deletion mutation** removes DNA. A deletion that removes three or a multiple of three bases will not cause a frameshift, but can still alter the phenotype. Deletions range from a single DNA nucleotide to thousands of bases to large parts of chromosomes. Many common inherited disorders result from deletions, including male infertility caused by tiny deletions in the Y chromosome.

An **insertion mutation** adds DNA, which can offset the reading frame. In one form of Gaucher disease, for example, an inserted single DNA base prevents production of an enzyme that normally breaks down glycolipids in lysosomes. The resulting buildup of glycolipid enlarges the liver and spleen and causes easily fractured bones and neurological impairment.

One type of insertion mutation repeats part of a gene's sequence. The insertion is usually adjacent or close to the original sequence, like a typographical error repeating a word word. Two copies of a gene next to each other is a type of mutation called a **tandem duplication**. A form of Charcot-Marie-Tooth disease (OMIM 118200), which causes numb hands and feet, results from a 1.5-million-base-long tandem duplication.

Figure 12.9 compares the effects on protein sequence of missense, nonsense, and frameshift mutations in the gene that

encodes the LDL receptor, causing familial hypercholesterolemia (see figure 5.2). These three mutations exert very different effects on the protein. A missense mutation replaces one amino acid with another, bending the protein in a way that impairs its function. A nonsense mutation greatly shortens the protein. A frameshift mutation introduces a section of amino acids that is not in the wild type protein.

Pseudogenes and Transposons Revisited

Recall from chapter 11 that a pseudogene is a DNA sequence that is very similar to the sequence of a protein-encoding gene. A pseudogene is not translated into protein, although it may be transcribed. The pseudogene may have descended from the original gene sequence, which was duplicated when DNA strands misaligned during meiosis, similar to the situation depicted in figure 12.8 for the alpha globin gene. When this happens, a gene and its copy end up right next to each other on the chromosome. The original gene or the copy then mutates to such an extent that it is no longer functional and becomes a pseudogene. Its duplicate lives on as the functional gene.

A pseudogene is not translated, but its presence can interfere with the expression of the functional gene and cause a mutation. For example, some cases of Gaucher disease result from a crossover between the working gene and its pseudogene, which has 96 percent of the same sequence and is located 16,000 bases away. The result is a fusion gene, which is part functional gene and part of the pseudogene. The fusion gene does not retain enough of the normal gene sequence to enable the cell to synthesize the encoded enzyme, and Gaucher disease results. Gaucher disease is a lysosomal storage disease that causes fatigue, bruising, anemia, and weak bones. The phenotype is very variable, and for many patients, supplying the enzyme eliminates symptoms.

Chapter 11 also discussed transposons. These "jumping genes" can alter gene function in several ways. They can disrupt the site they jump from, shut off transcription of the gene they jump into, or alter the reading frame of their destination if they are not a multiple of three bases. For example, a boy with X-linked hemophilia A had a transposon in his factor VIII gene—a sequence that was also in his carrier mother's genome, but on her chromosome 22. Apparently, in the oocyte, the transposon jumped into the factor VIII gene on the X chromosome, causing the boy's hemophilia.

Expanding Repeats

In a type of mutation called an **expanding repeat,** a gene actually grows as a small part of the DNA sequence repeats. The Mervar family described in the opener to chapter 4 offers an extreme example of the devastation that this type of mutation can cause.

One of the first expanding repeat diseases studied was myotonic dystrophy, which begins earlier and causes more severe symptoms from one generation to the next. A grandfather might experience mild weakness in his forearms, but his daughter may have more noticeable arm and leg weakness and a flat facial expression. Her children might have severe muscle weakness.

For many years, clinicians thought that the "anticipation"—the worsening of symptoms over generations—was psychological. Then, with the ability to sequence genes, researchers found that myotonic dystrophy indeed worsens with each generation because the gene expands! The gene that causes the first recognized type of myotonic dystrophy, on chromosome 19, has an area rich in repeats of the DNA triplet CTG. The wild type repeat number is 5 to 37; a person with the disorder has from 50 to thousands of copies (**figure 12.11**).

Expanding triplet repeats have been discovered in more than fifteen human inherited disorders. Usually, a repeat number of fewer than 40 copies is stably transmitted to the next generation and doesn't produce symptoms. Larger repeats are unstable, growing with each generation and causing symptoms that are more severe and begin sooner. **Clinical Connection 12.1** describes the triplet repeat disorder fragile X syndrome.

The mechanism behind triplet repeat disorders lies in the DNA sequence. The bases of the repeated triplets implicated in the expansion diseases, unlike others, bond to each other in ways that bend the DNA strand into shapes, such as hairpins. These shapes then interfere with replication, which causes the expansion. Once these repeats are translated, the extra-long proteins harm cells in several ways. They

■ bind to parts of transcription factors that have stretches of amino acid repeats similar to or matching the expanded repeat;

■ block proteasomes, enabling misfolded proteins to persist; and

■ trigger apoptosis.

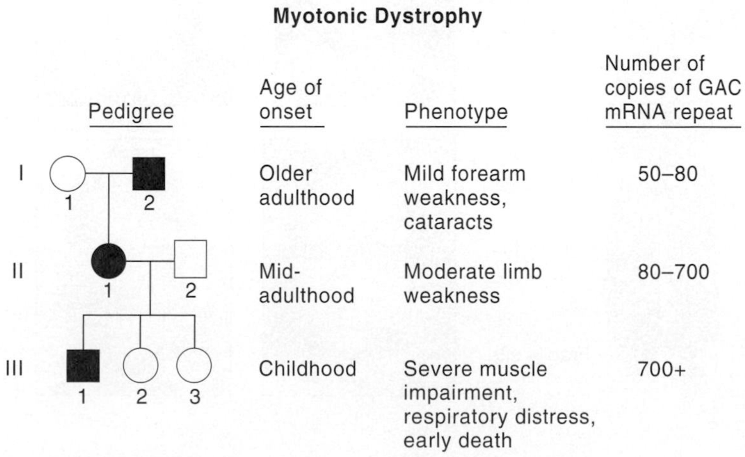

Triplet repeat proteins may also enter the nucleus, even though their wild type versions function only in the cytoplasm, or vice versa.

The triplet repeat disorders cause a "dominant toxic gain-of-function," illustrated in figure 4.8. They cause something novel to happen, rather than removing a function, such as a recessive enzyme deficiency. The idea of a gain-of-function arose from the observation that deletions of these genes do not cause symptoms. Several triplet repeat disorders are "polyglutamine diseases" that have repeats of the mRNA codon CAG, which encodes the amino acid glutamine.

For some triplet repeat disorders, the mutation blocks gene expression before a protein is even made. In myotonic dystrophy type 1, the expansion is in the initial untranslated region of the gene on chromosome 19, resulting in a huge mRNA. When genetic testing became available for the disorder, researchers discovered a second form of the illness in patients who had wild type alleles for the chromosome 19 gene. They have myotonic dystrophy type 2, caused by an expanding *quadruple* repeat of CCTG in a gene on chromosome 3. Affected individuals have more than 100 copies of the repeat, compared to the normal maximum of 10 copies.

When researchers realized that this second repeat mutation for myotonic dystrophy was in a non-protein-encoding part of the gene—an intron—a mechanism of disease became apparent: The mRNA is not processed normally and as a result cannot exit the nucleus. In myotonic dystrophy type 1, the excess material is added to the start of the gene; in type 2, it appears in an intron that is not excised. The bulky mRNAs bind to a protein that, in turn, alters intron splicing in several other genes. Deficiency of the proteins encoded by these final affected genes causes the symptoms.

A lesson learned from the expanding repeat disorders is that a DNA sequence is more than just one language that can be translated into another. Whether a sequence is random—CGT CGT ATG CAT CAG, for example—or highly repetitive—such as CAG CAG CAG CAG and on and on—can affect transcription, translation, or the ways that proteins interact.

Copy Number Variants

In addition to differing slightly in our DNA sequences, we differ in the numbers of copies of particular DNA sequences. These sequences that vary in number from person to person are called **copy number variants (CNVs)**. Our genomes have hundreds to thousands of them, and they account for about a quarter of the genome. Sequencing the human genome missed CNVs because the technology used at that time detected any DNA sequence only once. It was a little like searching this book for the word "variant," and not the number of times it is used.

Myotonic Dystrophy

Pedigree	Age of onset	Phenotype	Number of copies of GAC mRNA repeat
I 1 2	Older adulthood	Mild forearm weakness, cataracts	50–80
II 1 2	Mid-adulthood	Moderate limb weakness	80–700
III 1 2 3	Childhood	Severe muscle impairment, respiratory distress, early death	700+

Figure 12.11 **Expanding genes explain anticipation.** In some disorders, symptoms that worsen from one generation to the next—termed *anticipation*—have a physical basis: The gene is expanding as the number of repeats grows.

Fragile X Mutations Affect Boys and Their Grandfathers

Fragile X syndrome is the most common inherited form of intellectual disability and also accounts for 3 percent of all cases of autism. In the 1940s, geneticists thought that a gene on the X chromosome caused intellectual disability, which was then termed mental retardation, because more affected individuals were male. In 1969, a clue emerged to the genetic basis of X-linked intellectual disability. Two brothers with the condition and their mother had an unusual X chromosome. The tips at one chromosome end dangled by a thin thread (**figure 1a**). When grown in culture medium lacking folic acid, this part of the X chromosome was very prone to breaking—hence, the name fragile X syndrome. Worldwide, it affects 1 in 2,000 males, accounting for 4 to 8 percent of all males with intellectual disability. One in 4,000 females is affected. They usually have milder cases because their cells have normal X chromosomes too.

Youngsters with fragile X syndrome look normal, but MRI scans show brain differences as early as age 2. By young adulthood, their faces are very long and narrow, with long jaws and protruding ears. The testicles are very large. They may have learning disabilities, repetitive speech, hyperactivity, shyness, social anxiety, a short attention span, language delays, and temper outbursts.

Fragile X syndrome is inherited in an unusual pattern. The syndrome should be transmitted like any X-linked trait, from carrier mother to affected son. However, penetrance is incomplete. One-fifth of males who inherit the chromosomal abnormality have no symptoms. Because they transmit the affected chromosome to all their daughters—half of whom have some mental impairment—they are called "transmitting males." A transmitting male's grandchildren may inherit fragile X syndrome.

A triplet repeat mutation causes fragile X syndrome. In unaffected individuals, the fragile X area contains 29 or 30 repeats of the sequence CGG, in a gene called the fragile X mental retardation gene (*FMR1*). In people who have the fragile chromosome and show its effects, this region is expanded to 200 to 2,000 CGG repeats. Transmitting males, as well as females with mild symptoms, or who have affected sons, may have a "premutation" of 55 to 200 repeats. People with the premutation may develop mild neurological problems, such as tremors and poor balance. About a fifth of women with the premutation have infertility due to ovarian failure. People with 40 to 54 repeats may develop the mild symptoms associated with premutation, too.

The *FMR1* gene encodes fragile X mental retardation protein (FMRP). This protein, when abnormal, binds to and disables several mRNA molecules whose encoded proteins are crucial for brain neuron function.

Mysteries remain about fragile X syndrome. A distinct type of disorder is seen in the maternal grandfathers of boys who have fragile X syndrome. Clinicians noticed that mothers of boys with fragile X syndrome very often reported the same symptoms in their fathers—tremors, balance problems, and then cognitive or psychiatric difficulties. The grandfathers were sometimes misdiagnosed with Parkinson disease due to the tremors. However, Parkinson's patients can walk a straight line, while the grandfathers cannot. The grandfathers' symptoms worsen with time and can lead to premature death (**table 1**).

Table 1	Prevalence of FXTAS in Grandfathers of Fragile X Syndrome Grandsons	
Age		**Prevalence**
50s		17%
60s		38%
80 +		75%

Further investigation led to the description of the new condition, called fragile X-associated tremor/ataxia syndrome (FXTAS, OMIM 300623). (Ataxia is poor balance and coordination.) The disorder has been studied in brains obtained after the grandfathers died and in mice. Like the granddads, the mice are fine until middle age, when they develop tremors, balance problems, as well as nervousness and memory impairment. Perhaps the symptoms of FXTAS arise from excess FMR1 mRNA, which attracts and disables other mRNAs.

The discovery of FXTAS has genetic counseling implications. As neurologists learn to distinguish this disorder from others, such

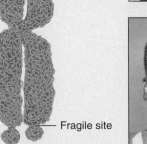
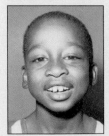

— Fragile site

a.

b.

Figure 1 **Fragile X syndrome.** A fragile site on the tip of the long arm of the X chromosome **(a)** is associated with intellectual disability and a characteristic long face that becomes pronounced with age **(b)**.

Copy number is a different form of information than DNA sequence differences. A language metaphor is useful to distinguish point mutations and single nucleotide polymorphisms (SNPs) from CNVs. If a wild type short sequence and a variant with two SNPs are written as:

The fat rat sat on a red cat   (wild type)

The fat rat sat in a red hat   (two SNPs)

then the sequence with two CNVs might be:

The fat fat rat sat on a red red red cat

CNVs may contribute significantly to the differences among us. A CNV can range in size from a few DNA bases to millions, and copies may lie next to each other on a chromosome ("tandem") or might be far away—even parts of other chromosomes. A duplication is a type of copy number variant.

CNVs may have no effect on the phenotype, or they can disrupt a gene's function and harm health. A CNV may have a direct effect by inserting into a protein-encoding gene and offsetting its reading frame, or have an indirect effect by destabilizing surrounding sequences. CNVs are particularly common among people who have behavioral disorders, such as attention deficit hyperactivity disorder (ADHD), autism, and schizophrenia.

Whole Exome Sequencing Explains Atypical Cases

Increasingly, researchers are using whole exome sequencing to identify rare mutations in patients who have unusual presentations of known genetic disorders. For example, exome sequencing revealed that an adolescent who has autism and mild intellectual disability actually has Prader-Willi syndrome (PWS), the imprinting disorder shown in figure 6.15. PWS is usually inherited as a deletion, including up to five genes, from a region on chromosome 15 from the father, because the mother's region of the chromosome is normally silenced. The young man did not have the classic symptoms of compulsive overeating and obesity, nor did one chromosome 15 harbor a detectable deletion. He had not received a diagnosis of PWS, and so his parents had gone from doctor to doctor, seeking a diagnosis, for 12 years.

A program at the Baylor College of Medicine to apply exome sequencing to find the genetic causes of intellectual disability, autism, and developmental delay in patients who have not found a diagnosis revealed that one chromosome 15 in the young man, which originated in him and was therefore a spontaneous (*de novo*) mutation, deletes a single base in a gene called *MAGEL2*. The tiny mutation has a big effect: It offsets the reading frame, generating a stop codon, and so the encoded protein is too short to function. Once the researchers knew to look at small changes in this gene, they diagnosed several other youngsters who had tested negative for a large deletion in the area on chromosome 15. These patients had different types of point mutations.

Key Concepts Questions 12.5

1. What is a point mutation?
2. Distinguish between a transversion and a transition.
3. Distinguish between a missense mutation and a nonsense mutation.
4. Explain how mutations outside the coding region of a gene can affect its function.
5. How do insertions and deletions affect gene function?
6. How do pseudogenes and transposons affect gene function?
7. Describe expanded repeat mutations and their effects.
8. What are copy number variants?
9. How is exome sequencing useful in identifying mutations?

12.6 The Importance of Position

The degree to which a mutation alters a phenotype depends upon where in the gene the change occurs, and how the mutation affects the folding, conformation, activity, or abundance of an encoded protein. A mutation that replaces an amino acid with a very similar one would probably not affect the phenotype greatly, because it wouldn't substantially change the conformation of the protein. Even substituting a very different amino acid would not have much effect if the change is in part of the protein not crucial to its function. The effects of specific mutations are well-studied in hemoglobin.

Globin Variants

Hundreds of globin gene mutations have been known for years. Mutations in these genes can cause anemia with or without sickling, or cause cyanosis (a blue pallor due to poor oxygen binding). Rarely, a mutation boosts the molecule's affinity for oxygen. Some globin gene variants exert no effect at all and are considered "clinically silent" (**table 12.5**).

Different mutations at the same site in a gene can have different effects. For example, hemoglobin S and hemoglobin C result from mutations that change the sixth amino acid in the beta globin polypeptide, but in different ways. Homozygotes for hemoglobin S have sickle cell disease, yet homozygotes for hemoglobin C are healthy. Both types of homozygotes are resistant to malaria because the unusual hemoglobin alters the shapes and surfaces of red blood cells in ways that keep out the parasite that causes the illness (see figure 15.15).

An interesting consequence of certain mutations in either the alpha or beta globin chains is hemoglobin M. Normally, the iron in hemoglobin is in the ferrous form, which means that it has two positive charges. In hemoglobin M, the mutation stabilizes the ferric form, which has three positive charges and cannot bind oxygen. Fortunately, an enzyme converts the abnormal ferric iron to the normal ferrous form, so that the only symptom is usually cyanosis (**figure 12.12**). The condition has been known for more than 200 years in a small town in Japan, where many people have autosomal dominant "blackmouth."

Even more noticeable than people with blackmouth are the "blue people of Troublesome Creek." Seven generations ago, in 1820, a French orphan named Martin Fugate who settled in this area of Kentucky brought in a recessive gene that causes a condition called recessive congenital methemoglobinemia. (Exposure to certain drugs causes a

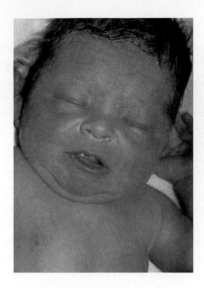

Figure 12.12 Cyanosis. Some mutations that alter hemoglobin cause a bluish cast to the skin called cyanosis.

noninherited form.) Martin Fugate was missing an enzyme (cytochrome b_5 methemoglobin reductase) that normally catalyzes a reaction that converts a type of hemoglobin with poor oxygen affinity, called methemoglobin, back into normal hemoglobin by adding an electron. Martin's skin was blue. His wife, Elizabeth Smith, was a carrier for this very rare disease. Four of their seven children were blue. After extensive inbreeding in the isolated community—their son married his aunt, for example—a large pedigree of "blue people" of both sexes arose.

In "blue person disease," excess oxygen-poor hemoglobin causes a dark blue complexion. Carriers may have bluish

Table 12.5	Globin Mutations	
Associated Phenotype	**Name**	**Mutation**
Clinically silent	Hb Wayne	Single-base deletion in alpha gene causes frameshift, changing amino acids 139–141 and adding amino acids
	Hb Grady	Nine extra bases add three amino acids between amino acids 118 and 119 of alpha chain
Oxygen binding	Hb Chesapeake	Change from arginine to leucine at amino acid 92 of beta chain
	Hb McKees Rocks	Change from tyrosine to STOP codon at amino acid 145 in beta chain
Anemia	Hb Constant Spring	Change from STOP codon to glutamine elongates alpha chain
	Hb S	Change from glutamic acid to valine at amino acid 6 in beta chain causes sickling
	Hb Leiden	Amino acid 6 deleted from beta chain
Protection against malaria	Hb C	Change from glutamic acid to lysine at amino acid 6 in beta chain causes sickling

lips and fingernails at birth, which usually improve. Treatment is simple: A tablet of methylene blue, a commonly used dye, adds the electron back to methemoglobin, converting it to normal hemoglobin.

In most members of the Fugate family, blueness was the only symptom. Normally, less than 1 percent of hemoglobin molecules are the methemoglobin form, which binds less oxygen. The Fugates had 10 to 20 percent in this form. People with the inherited condition who have more than 20 percent methemoglobin may suffer seizures, heart failure, and even death. The disease remains very rare and is still seen in the Kentucky family and among certain families in Alaska and Algeria, and among Navajo Indians.

Factors That Lessen the Effects of Mutation

Mutation is a natural consequence of DNA's ability to change. This flexibility is essential for evolution because it generates new variants, some of which may resist environmental change and enable a population or even a species to survive. However, many factors minimize the negative effects of mutations on phenotypes.

The genetic code protects against mutation to an extent. Recall from chapter 10 that synonymous codons specify the same amino acid. Mutation in the third codon position is called "silent" because the two codons are synonymous. For example, a change from RNA codon CAA to CAG does not change the amino acid, glutamine, so a protein whose gene contains the change would not change. However, synonymous codons can affect splicing out of introns and mRNA stability differently, so expression of genes that have mutations that result in synonymous codons can differ.

Other genetic code nuances prevent synthesis of very altered proteins. For example, mutations in the second codon position sometimes replace one amino acid with another that has a similar conformation, minimizing disruption of the protein's shape. GCC mutated to GGC, for instance, replaces alanine with equally small glycine.

A **conditional mutation** affects the phenotype only under certain circumstances. This can be protective if an individual avoids the exposures that trigger symptoms. Consider a common variant of the X-linked gene that encodes glucose 6-phosphate dehydrogenase (G6PD), an enzyme that immature red blood cells use to extract energy from glucose. One hundred million people worldwide have G6PD deficiency (OMIM 305900). It can cause life-threatening hemolytic anemia, but only under rather unusual conditions—eating fava beans or taking certain antimalarial drugs (**figure 12.13**).

In the fifth century B.C., the Greek mathematician Pythagoras wouldn't allow his followers to consume fava beans, because he had discovered that it would sicken some of them. During the Second World War, several soldiers taking the antimalarial drug primaquine developed hemolytic anemia. What these, and other triggering substances have in common is

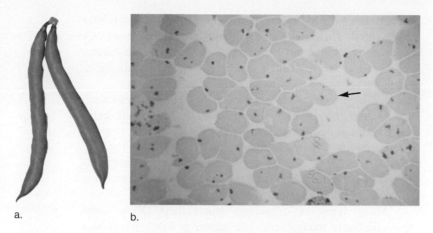

a.   b.

Figure 12.13 **Sickness and circumstance.** A conditional mutation causes some cases of G6PD deficiency hemolytic anemia. Exposure to two biochemicals in fava beans **(a)** unfolds hemoglobin molecules, causing spots called Heinz bodies **(b)** and bending red blood cells out of shape. They then burst.

that they "stress" red blood cells by exposing them to oxidants, chemicals that strip electrons from other compounds. Without the enzyme, the red blood cells burst.

Another protection against mutation occurs in stem cells. When a stem cell divides to yield another stem cell and a progenitor or differentiated cell, the oldest DNA strands segregate with the stem cell, and the most recently replicated DNA strands go to the more specialized daughter cells (see figure 2.20) This makes sense in organs where stem cells very actively yield specialized daughter cells, such as the skin and small intestine. Because mutations occur when DNA replicates, this skewed distribution of chromosomes sends the DNA most likely to harbor mutations into cells that will soon be shed (from a towel rubbed on skin or in a bowel movement) while keeping mutations away from the stem cells that must continually regenerate the tissues.

Key Concepts Questions 12.6

1. What characteristics of a mutation determine its effects?
2. Describe some mutations in globin genes.
3. Name three phenomena that lessen the effects of mutations.

12.7 DNA Repair

Any manufacturing facility tests a product in several ways to see whether it has been assembled correctly. Mistakes in production are rectified before the item goes on the market—most of the time. The same is true for a cell's manufacture of DNA.

Damage to DNA becomes important when the genetic material is replicated, because the error is passed on to daughter cells. In response to damage, the cell may die by apoptosis or it may repair the error. If the cell doesn't die or the error is not repaired, cancer may result. Fortunately, DNA replication is very accurate—only 1 in 100 million or so bases is incorrectly incorporated. This is quite an accomplishment, because DNA replicates approximately 10^{16} times during an average human lifetime. However, most such mutations occur in somatic cells, and do not affect the phenotype.

DNA polymerase as well as "DNA damage response" genes oversee the accuracy of replication. In DNA repair, a cell detects damage and then signaling systems in the cell respond by repairing the damage or signaling apoptosis to kill the cell. More than 50 DNA damage response genes have been identified. Mitochondrial DNA cannot repair itself, which accounts for its higher mutation rate.

Many types of organisms repair their DNA, some more efficiently than others. The master at DNA repair is a large, reddish microbe, *Deinococcus radiodurans*. It tolerates 1,000 times the radiation level that a person can, and it can even live amidst the intense radiation of a nuclear reactor. The bacterium realigns its radiation-shattered pieces of DNA. Then enzymes bring in new nucleotides and assemble the pieces.

Types of DNA Repair

Exposure to radiation is a fact of life. The Earth, since its beginning, has been periodically bathed in UV radiation. Volcanoes, comets, meteorites, and supernovas all depleted ozone in the atmosphere, which allowed ultraviolet wavelengths of light to reach organisms. The shorter wavelengths—UVA—are not dangerous, but the longer UVB wavelengths damage DNA by forming an extra covalent bond between adjacent (same-strand) pyrimidines, particularly thymines (**figure 12.14**). The linked thymines are called thymine dimers. Their extra bonds kink the double helix enough to disrupt replication and permit insertion of a noncomplementary base. For example, an A might be inserted opposite a G or C, instead of opposite a T. Thymine dimers also disrupt transcription.

Early in the evolution of life, organisms that could survive UV damage had an advantage. Enzymes enabled them to do this, and because enzymes are gene-encoded, DNA repair came to persist.

In many modern species, three types of DNA repair check the genetic material for mismatched base pairs. In the first type, enzymes called photolyases absorb energy from visible light and use it to detect and bind to pyrimidine dimers, then break the extra bonds. This type of repair, called photoreactivation, enables UV-damaged fungi to recover from exposure to sunlight. Humans do not have this type of DNA repair.

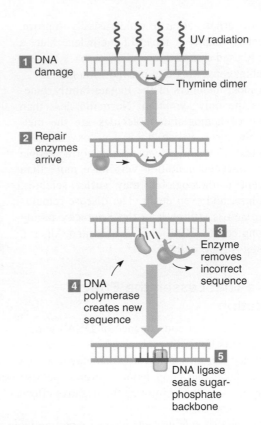

Figure 12.14 Excision repair. Human DNA damaged by UV light is repaired by excision repair, which removes and replaces the pyrimidine dimer and a few surrounding bases.

In the early 1960s, researchers discovered a second type of DNA self-mending, called **excision repair,** in mutant *E. coli* that were unable to repair UV-induced DNA damage. Enzymes cut the bond between the DNA sugar and base and excise the pyrimidine dimer and surrounding bases (see figure 12.14). Then, a DNA polymerase fills in the correct nucleotides, using the exposed template as a guide. DNA polymerase also detects and corrects mismatched bases in newly replicated DNA.

Humans have two types of excision repair. **Nucleotide excision repair** replaces up to 30 nucleotides and removes errors that result from several types of insults, including exposure to chemical carcinogens, UVB in sunlight, and oxidative damage. Thirty different proteins carry out nucleotide excision repair.

The second type of excision repair, **base excision repair,** replaces one to five nucleotides at a time, but specifically corrects errors that result from oxidative damage. Oxygen free radicals are highly reactive forms of oxygen that arise during chemical reactions such as those of metabolism and transcription. Free radicals damage DNA. Genes that are very actively transcribed face greater oxidative damage from free radicals; base excision repair targets this type of damage.

A third mechanism of DNA sequence correction is **mismatch repair**. Enzymes "proofread" newly replicated DNA for small loops that emerge from the double helix. The enzymes excise the mismatched base so that it can be replaced (**figure 12.15**). These loops emerge from where the two strands do not precisely align, but instead slip and misalign. This happens where very short DNA sequences repeat. These sequences, called microsatellites, are scattered throughout the genome. Like minisatellites, microsatellite lengths can vary from person to person, but within an individual, they are usually the same length. Excision and mismatch repair differ in the cause of the error—UV-induced pyrimidine dimers versus replication errors—and in the types of enzymes involved.

Excision repair and mismatch repair in human cells relieve the strain on thymine dimers or replace incorrectly inserted bases. Another form of repair can heal a broken sugar-phosphate backbone in both strands, which can result from exposure to ionizing radiation or oxidative damage. Such a double-stranded break is especially damaging because it breaks a chromosome, which can cause cancer. At least two types of multiprotein complexes reseal the sugar-phosphate backbone, either by rejoining the broken ends or recombining with DNA on the unaffected homolog.

In yet another type of DNA repair called damage tolerance, a "wrong" DNA base is left in place, but replication and transcription proceed. "Sloppy" DNA polymerases, with looser adherence to the base-pairing rules, read past the error, randomly inserting any other base. It is a little like retaining a misspelled word in a sentence—usually the meaning remains clear.

Figure 12.16 summarizes DNA repair mechanisms.

DNA Repair Disorders

The ability to repair DNA is crucial to health. If both copies of a repair gene are mutant, a disorder can result. Heterozygotes who have one mutant repair gene may be more sensitive to damage from environmental toxins.

A well-studied DNA repair gene encodes a protein called p53. It controls whether DNA is repaired and the cell salvaged, or the cell dies by apoptosis (see figure 18.3). Signals from outside the cell activate p53 protein to aggregate into complexes of four proteins. These quartets bind certain genes that slow the cell cycle, enabling repair to take place. If the damage is too severe, the p53 protein quartets instead increase the rate of transcription of genes that promote apoptosis, and the cell dies.

In DNA repair disorders, chromosome breakage caused by factors such as radiation cannot be repaired. Mutations in repair genes greatly increase susceptibility to certain types of cancer following exposure to ionizing radiation or chemicals that affect cell division. These conditions develop because errors in the DNA sequence accumulate and are perpetuated to a much greater extent than they are in people with functioning repair systems. We conclude this chapter with a closer look at repair disorders. Chapter 18 discusses other cancer susceptibility genes that disrupt DNA repair.

Trichothiodystrophy (OMIM 601675)

At least five genes can cause trichothiodystrophy. At its worst, this condition causes dwarfism, intellectual disability, and failure to develop, in addition to brittle hair and scaly skin, both

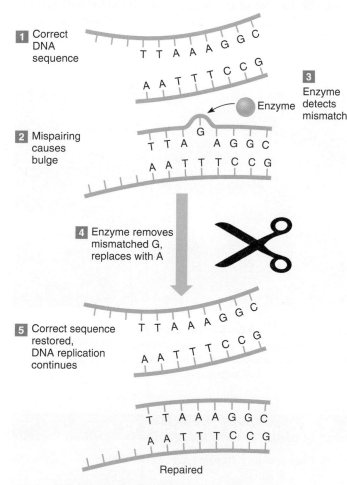

Figure 12.15 Mismatch repair. In this form of DNA repair, enzymes detect loops and bulges in newly replicated DNA that indicate mispairing. The enzymes correct the error. Highly repeated sequences are more prone to this type of error.

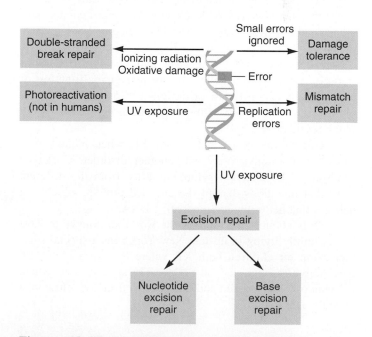

Figure 12.16 DNA repair mechanisms.

with low sulfur content. Although the child may appear to be normal for a year or two, growth soon slows dramatically, signs of premature aging begin, and life ends early. Hearing and vision may fail. Symptoms reflect accumulating oxidative damage. Individuals have faulty nucleotide excision repair, base excision repair, or both. This repair disorder is unusual in that it does not increase cancer risk.

Inherited Colon Cancer

Hereditary nonpolyposis colon cancer (HNPCC) (OMIM 120435 and others, also known as Lynch syndrome) was linked to a DNA repair defect when researchers discovered different-length short repeated sequences of DNA (microsatellites) within an individual. People with this type of colon cancer have a breakdown of mismatch repair, which normally keeps a person's microsatellites all the same length. HNPCC affects 1 in 200 people, and mutations in any of at least 7 genes can increase the risk of developing it.

HNPCC accounts for 3 percent of newly diagnosed colorectal cancers. Genetic testing for this condition in all people newly diagnosed with colon cancer is advised because if they have a mutation, their relatives can be tested. If healthy relatives test positive, frequent colonoscopies can detect disease early, at a more treatable stage. Penetrance of HNPCC is about 45 percent by age 70—considered a high cancer risk.

Xeroderma Pigmentosum (XP) (OMIM 278700)

A child with XP must stay indoors in artificial light, because even the briefest exposure to sunlight causes painful blisters. Failing to cover up and use sunblock can result in skin cancer (**figure 12.17**). More than half of all children with XP develop the cancer before they reach their teens. People with XP have a 1,000-fold increased risk of developing skin cancer compared to others, and a 10-fold increased risk of developing tumors inside the body.

XP is autosomal recessive, and results from mutations in any of seven genes. It can reflect malfunction of nucleotide excision repair or deficient "sloppy" DNA polymerase, both of which allow thymine dimers to stay and block replication.

One of the genes that causes XP, when mutant, also causes trichothiodystrophy and another disorder, Cockayne syndrome. The different symptoms arise from the different ways that mutations disrupt the encoded protein, which is a helicase that helps unwind replicating DNA.

Only about 250 people in the world are known to have XP. A family living in upstate New York runs a special summer camp for children with XP, where they turn night into day. Activities take place at night, or in special areas where the windows are covered and light comes from low-ultraviolet incandescent lightbulbs.

Ataxia Telangiectasis (AT) (OMIM 208900)

This multisymptom disorder is the result of a defect in a kinase that functions as a cell cycle checkpoint (see figure 2.14). In AT, cells proceed through the cell cycle without pausing just after replication to inspect the new DNA and to repair any mispaired bases. Some cells die through apoptosis if the damage is too great to repair. Because of the malfunctioning cell cycle, individuals who have this autosomal recessive disorder have 50 times the risk of developing cancer, particularly of the blood. About 40 percent of individuals with ataxia telangiectasis have cancer by age 30. Additional symptoms include poor balance and coordination (ataxia), red marks on the face (telangiectasia), delayed sexual maturation, and high risk of infection and diabetes mellitus.

AT is rare, but heterozygotes are not. They make up from 0.5 to 1.4 percent of various populations. Carriers may have mild radiation sensitivity, which causes a two- to sixfold increase in cancer risk over that of the general population. For people who are AT carriers, dental or medical X rays may cause cancer.

DNA's changeability, so vital for evolution of a species, comes at the cost of occasional harm to individuals. Each of us harbors about 175 new mutations, many old ones, and many polymorphisms, although most are hidden in the recessive state.

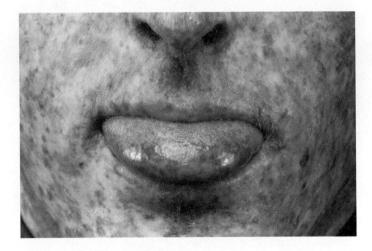

Figure 12.17 A DNA repair disorder. The marks on this person's face result from sun exposure. Xeroderma pigmentosum, an impairment of excision repair, makes the skin highly sensitive to the sun.

Key Concepts Questions 12.7

1. Describe the different types of DNA repair mechanisms.
2. Describe disorders that result from faulty DNA repair.

Summary

12.1 The Nature of Mutations

1. A **mutation** is a change in a gene's nucleotide base sequence that is rare in a population and can cause a **mutant** phenotype. A polymorphism is a more common and typically less harmful genetic change.

2. A **germline mutation** originates in meiosis, affects all cells of an individual, and can be transmitted to the next generation in gametes. A **somatic mutation** originates in mitosis and affects a subset of cells.

3. A mutation disrupts the function or abundance of a protein or introduces a new function. Most loss-of-function mutations are recessive, and most gain-of-function mutations are dominant.

12.2 A Closer Look at Two Mutations

4. Mutations in the beta globin and collagen genes cause a variety of disorders.

12.3 Allelic Disorders

5. Whether different mutations in a gene cause the same or distinct illnesses varies; nomenclature is inconsistent.

6. **Allelic disorders** have different phenotypes but result from mutations in the same gene.

12.4 Causes of Mutation

7. A spontaneous mutation arises due to chemical phenomena or to an error in DNA replication. Spontaneous mutation rate is characteristic of a gene and is more likely in repeats. In **gonadal mosaicism,** only some gametes have a spontaneous mutation.

8. **Mutagens** are chemicals or radiation that delete, substitute, or add bases. An organism may be exposed to a mutagen intentionally, accidentally, or naturally.

12.5 Types of Mutations

9. A **point mutation** alters a single DNA base. It may be a **transition** (purine to purine or pyrimidine to pyrimidine) or a **transversion** (purine to pyrimidine or vice versa). A **missense mutation** substitutes one amino acid for another, while a **nonsense mutation** substitutes a "stop" codon for a codon that specifies an amino acid, shortening the protein product. Splice-site mutations add or delete amino acids.

10. A **deletion mutation** removes genetic material and an **insertion mutation** adds it. A **frameshift mutation** alters the sequence of amino acids (**reading frame**). A **tandem duplication** is a copy of a gene next to the original.

11. A pseudogene results when a duplicate of a gene mutates. It may disrupt chromosome pairing, causing mutation.

12. Transposons may disrupt the functions of genes they jump into.

13. **Expanding repeat** mutations add stretches of the same amino acid to a protein. They expand because they attract each other, which affects replication.

14. **Copy number variants** are DNA sequences that are repeated a different number of times among individuals. They may have no effect on phenotype or may directly or indirectly cause disease.

12.6 The Importance of Position

15. Mutations in the globin genes may affect the ability of the blood to transport oxygen, or they may have no effect.

16. Synonymous codons limit the effects of mutation. Changes in the second codon position may substitute a similarly shaped amino acid.

17. **Conditional mutations** are expressed only in response to certain environmental triggers.

18. Sending the most recently replicated DNA into cells headed for differentiation, while sending older strands into stem cells, protects against mutation.

12.7 DNA Repair

19. DNA polymerase proofreads DNA, but repair enzymes correct errors in other ways.

20. Photoreactivation repair uses light energy to split pyrimidine dimers.

21. In **excision repair,** pyrimidine dimers are removed and the area filled in correctly. **Nucleotide excision repair** replaces up to 30 nucleotides from various sources of mutation. **Base excision repair** fixes up to five bases that paired incorrectly due to oxidative damage.

22. **Mismatch repair** proofreads newly replicated DNA for loops that indicate noncomplementary base pairing.

23. DNA repair fixes the sugar-phosphate backbone. Damage tolerance enables replication to continue beyond a mismatch.

24. Mutations in repair genes break chromosomes and increase cancer risk.

www.mhhe.com/lewisgenetics11

Answers to all end-of-chapter questions can be found at **www.mhhe.com/lewisgenetics11.** You will also find additional practice quizzes, animations, videos, and vocabulary flashcards to help you master the material in this chapter.

Review Questions

1. How do a mutation and a polymorphism differ?

2. Distinguish between a germline and a somatic mutation.

3. How is a human body a genomic mosaic?

4. Explain how a mutation causes sickle cell disease.

5. Why is the collagen molecule especially likely to be altered by mutation?

6. Describe a pair of allelic disorders.

7. How can DNA spontaneously mutate?

8. Distinguish between
   a. a transition and transversion.
   b. a missense mutation and nonsense mutation.

9. List two types of mutations that can alter the reading frame.

10. List four ways that DNA can mutate without affecting the phenotype.

11. Cite two ways a jumping gene can disrupt gene function.

12. What is a molecular explanation for the worsening of an inherited illness over generations?

13. How can short repeats within a gene, long triplet repeats within a gene, and repeated genes cause disease?

14. How does a copy number variant differ from a missense mutation?

15. How can a mutation that retains an intron's sequence and a triplet repeat mutation have a similar effect on a gene's encoded protein?

16. Explain how a single-base mutation can encode a protein that is missing many amino acids.

17. Cite three ways in which the genetic code protects against the effects of mutation.

18. What is a conditional mutation?

19. How do excision and mismatch repair differ?

20. Explain how semiconservative DNA replication makes it possible for stem cells to receive the DNA least likely to bear mutations.

Applied Questions

1. Which of the following DNA sequences is most likely to mutate? Cite a reason for your answer.
   a. CTGAGTCGTGACTCGTACGTCAT
   b. ATGCTCAGCTACTCAGCTACGGA
   c. ATCCTAGTCGTCGTCGTCGTCATC

2. Consider the following sequence of part of an mRNA molecule:

   A U G U U G U C A A A A G C A U G G C G G C C A

   Introduce the following changes to the sequence, and indicate the effect, if any, on the encoded amino acid sequence:
   a. a missense mutation
   b. a nonsense mutation
   c. a frameshift mutation
   d. a silent mutation
   e. a transversion
   f. a transition
   g. a tandem duplication
   h. a deletion

3. Retinitis pigmentosa causes night blindness and loss of peripheral vision before age 20. A form of X-linked retinitis pigmentosa (OMIM 300455) is caused by a frameshift mutation that deletes 199 amino acids. How can a simple mutation have such a great effect?

4. A mutation that changes a C to a T causes a form of Ehlers-Danlos syndrome, forming a "stop" codon and shortened procollagen. Consult the genetic code (see table 10.4) and suggest one way that this can happen.

5. Part of the mRNA sequence of an exon of a gene that encodes a blood protein is:

   A U G A C U C A U C G C U G U A G U U U A C G A

   Consult the genetic code to answer the following questions:
   a. What is the sequence of amino acids that this mRNA encodes?
   b. What is the sequence if a point mutation changes the tenth base from a C to an A?
   c. What is the effect of a point mutation that changes the fifteenth base from a U to an A?
   d. How does the encoded amino acid sequence change if a C is inserted between the fourth and fifth bases?
   e. Which would be more devastating to the encoded amino acid sequence, insertion of three bases in a row, or insertion of two bases in a row?

6. Susceptibility to developing prion diseases arises from a mutation from aspartic acid (Asp) to asparagine (Asn). Which nucleotide base changes make this happen?

7. Two teenage boys meet at a clinic to treat muscular dystrophy. The boy who is more severely affected has a two-base insertion at the start of his dystrophin gene. The other boy has the same two-base insertion but also has a third base inserted a few bases away. Why is the second boy's illness milder?

8. Two missense mutations in the gene that encodes an enzyme called superoxide dismutase cause amyotrophic lateral sclerosis (ALS, or Lou Gehrig's disease). This disorder causes loss of neurological function over

a 5-year period. One mutation alters the amino acid asparagine (Asn) to lysine (Lys). The other changes an isoleucine (Ile) to a threonine (Thr). List the codons involved and describe how single-base mutations alter the amino acids they specify.

9. A man develops Huntington disease at age 48. He is tested and his mutant gene has 54 repeats. His 20-year-old daughter is tested, and she has a gene with 68 repeats. Explain how the gene likely expanded.

10. Certain antibiotic drugs suppress nonsense mutations by inserting a random amino acid into the protein corresponding to the site of the mutation in the gene. Explain how this happens, and how this finding might be applied to treat a genetic disease.

11. In one family, Tay-Sachs disease stems from a four-base insertion, which changes an amino-acid-encoding codon into a "stop" codon. What type of mutation is this?

12. Epidermolytic hyperkeratosis (OMIM 607602) is an autosomal dominant condition that produces scaly skin. It can be caused by a missense mutation that substitutes a histidine (His) amino acid for an arginine (Arg). Write the mRNA codons that could account for this change.

13. A biotechnology company has encapsulated DNA repair enzymes in fatty bubbles called liposomes. Why would this be a valuable addition to a suntanning lotion?

Web Activities

1. Look at one of the following websites and select a disease. Describe the mutation in a particular family.

   Center for Jewish Genetics
   www.jewishgenetics.org
   CheckOrphan
   www.checkorphan.org
   Genetic and Rare Disorders Information Center
   www.genome.gov/10000409
   Global Genes Project
   Globalgenes.org
   National Organization for Rare Diseases
   www.rarediseases.org

2. Children with Hutchinson-Gilford progeria syndrome (see the opener to chapter 3) age extremely rapidly. In 18 of 20 children, a single base change in the *lamin A* gene alters a C to a T, but this mutation removes 50 amino

acids from the encoded protein. In all 20 children, the parents do not have the mutation.

   a. Is the mutation in the 18 children *de novo* or induced? What is the evidence for this distinction?

   b. How can a change in a single base remove 50 amino acids?

   c. Using OMIM, list and describe six other disorders caused by mutation in the *lamin A* gene.

3. Select an image at positiveexposure.org and describe the symptoms of the disorder that affects the person in the photograph.

4. Look up three of the following triplet repeat disorders at www.omim.org (or elsewhere) and compare the actual repeat, the normal copy number, and the disease copy number: fragile X syndrome, Friedreich ataxia, Haw River syndrome, Jacobsen syndrome, spinal and bulbar muscular atrophy, and spinocerebellar ataxia (any type).

Forensics Focus

1. Late one night, a man broke into Colleen's apartment and raped her. He wore a mask and it was dark, so she couldn't see his face, but she yanked out some of his long, greasy hair. The forensic investigator, after examining the hair, asked Colleen if her boyfriend had been in the bed or with her earlier in the evening, because the hairs were of two genotypes for one of the repeated sequences that was analyzed. What is another explanation for finding two genotypes?

Case Studies and Research Results

1. Tom Staniford is the national paracycling champion in the United Kingdom. He and only a few other people in the world have MDP syndrome, so newly recognized that it doesn't have an OMIM listing yet. MDP stands for the symptoms of mandibular hypoplasia (underdeveloped jaw), deafness, and progeroid features. The syndrome causes extreme lack of fat and hearing loss. Genome sequencing of Staniford and three other patients found mutations in a gene called *POLD1*. The parents of all four people with the syndrome had

their genomes sequenced too, but none of them have mutations in *POLD1*. Suggest two ways that the disease could have arisen in the four skinny people.

2. How has the ability to sequence genomes revealed that somatic mutations and spontaneous mutations are more common than thought?

3. A 3-year-old boy has mild autism, intellectual disability, and developmental delay—all common symptoms. But he also has small hands, wide eyes, and he is very short. He

has violent outbursts and long tantrums. The collection of signs and symptoms alert his pediatrician that he may have a genetic syndrome, although he is the only person in his family to have these symptoms. The boy's parents take him to a several specialists. Genetic tests eliminate several possibilities, but he remains undiagnosed, and does not appear to be outgrowing the symptoms, as doctors had initially said he might.

a. What technology might help to diagnose the child?

b. What information would sequencing the parents provide?

c. If a mutation is found in the boy that accounts for the symptoms, but is not in either parent, then what is the risk of the condition affecting a future sibling (assuming the parents are not gonadal mosaics)?

4. Latika and Keyshauna meet at a clinic for college students who have cystic fibrosis. Latika's mutation results in exon skipping. Keyshauna's mutation is a nonsense mutation. Which young woman probably has more severe symptoms? Cite a reason for your answer.

5. Marshall and Angela have skin cancer resulting from xeroderma pigmentosum. They meet at an event for teenagers with cancer. However, their mutations affect different genes. They decide to marry but not to have children because they believe that each child would have a 25 percent chance of inheriting XP because it is autosomal recessive. Are they correct? Why or why not?

6. Two girls and a boy in a Pakistani family have a form of deafness caused by a mutation in the gene that encodes a protein called tricellulin (OMIM 610153). The normal protein attaches epithelial (lining) cells in groups of three in the inner ear in a way that is crucial to hearing. Consider a pedigree for the family:

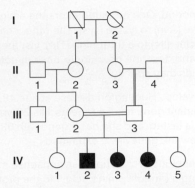

a. What is the mode of inheritance for this form of deafness, and how do you know this?

b. This form of deafness is rare worldwide, but more common among Pakistani families, many of whose pedigrees have double horizontal lines. What does the double line mean, and how does it account for the increased prevalence of this form of deafness in the population?

c. The affected children have a partial sequence for the tricellulin gene of C T G C A A T G T. Unaffected family members have the sequence C T G C A G T G T. What are the amino acid differences encoded in these sequences?

7. Presenilin 1 is one of the genes that, when mutant, causes familial Alzheimer disease. It is also expressed in the heart. Certain mutations cause a condition that leads to heart failure. In the Esposito family, all of the relatives who have or had heart failure have the following partial sequence for the presenilin 1 gene: G A T G A T G G C G G G. Family members with healthy hearts have the sequence G A T G G T G G C G G G. How do the encoded amino acid sequences differ between the healthy and sick family members for this part of the gene?

A duplication of 1.5 million DNA bases at the tip of a tiny chromosome causes Ambras syndrome. The condition is a form of hypertrichosis ("extra hairs") and may have been the inspiration for the werewolf legend.

Chromosomes

Learning Outcomes

13.1 Portrait of a Chromosome

1. List the major parts of a chromosome.
2. List the types of chromosomes based on centromere position.

13.2 Detecting Chromosomes

3. Describe ways that chromosomes are obtained, prepared, and detected.

13.3 Atypical Chromosome Number

4. Explain how atypical chromosome numbers arise.
5. Describe syndromes associated with incorrect chromosome number.

13.4 Atypical Chromosome Structure

6. Explain how atypical chromosome structures arise.
7. Describe syndromes associated with specific variants of chromosome structure.

13.5 Uniparental Disomy—A Double Dose from One Parent

8. Explain how a person could inherit both copies of a DNA sequence from one parent.
9. Describe how inheriting both copies of DNA from one parent can affect health.

The BIG Picture

A human genome has 20,000-plus protein-encoding genes dispersed among 24 chromosome types. Abnormalities in chromosome number or structure can have sweeping effects, but mutation is a continuum. Chromosome-level illnesses reflect disruption of individual genes or their regulation.

The Curious Chromosomes of Werewolves

People who have historically inspired tales of werewolves may actually have had Ambras syndrome (OMIM 145701), in which silky straight hair streams down the face and curls from the ears, cloaking the shoulders. Only 50 cases have been described since the Middle Ages. The name comes from a slave from the Canary Islands who served in the court of France in the late sixteenth century. He and his hairy children and grandchildren sat for a portrait at a castle called Ambras. Others with Ambras syndrome ended up in circus sideshows, billed as bearded ladies, wolf girls, monkey men, and "Jo-Jo the dog-faced man."

Much more recently, geneticists at Columbia University analyzed single nucleotide polymorphisms (SNPs) across the genomes of a father and son with Ambras syndrome. The researchers discovered that a duplication of 1.5 million bases near one tip of chromosome 17 sets into motion the mixed signals that prevent hair follicles from becoming spaced apart during prenatal development. The result is hypertrichosis—"extra hairs."

Although the father and son trace their condition to chromosome 17, the story actually begins with a mutation in a gene called *Trps1* on chromosome 8. Usually in the syndrome the gene is deleted or that part of the chromosome is inverted. *Trps1* is a transcription factor (see chapter 11) that regulates several genes involved in hair and bone development, which explains the additional symptoms of coarse facial features and long nose. Symptoms can also result from disruption of any of the genes that this transcription factor controls. One gene, *Sox9*, regulates stem cells deep in developing hair follicles. The father and son's duplicated DNA sequence places *Sox9* farther out on chromosome 17 and the protein that *Trps1* encodes cannot find it. So even if *Trps1* is wild type, if Trps1 protein cannot locate a target gene, the same phenotype arises as if *Trps1* itself were mutant.

The genetic explanation is complicated, but it improves on older ideas: that people become werewolves by sleeping outdoors under a summer full moon, drinking from the pawprint of a wolf, or from a devil's curse.

13.1 Portrait of a Chromosome

Mutations range from single-base changes to entire extra sets of chromosomes. A mutation is considered a chromosomal aberration if it is large enough to be seen with a light microscope using stains and/or fluorescent probes to highlight missing, extra, or moved genetic material.

In general, too little genetic material has more severe effects on health than too much. Most extensive chromosome abnormalities present in all cells of an embryo or fetus disrupt or halt prenatal development. As a result, only 0.65 percent of all newborns have chromosomal abnormalities that produce symptoms. An additional 0.20 percent of newborns have chromosomal rearrangements in which chromosome parts have flipped or swapped, but they do not produce symptoms unless they disrupt the structures or functions of genes that affect health.

Cytogenetics is the area of genetics that links chromosome variations to specific traits, including illnesses. This chapter explores several ways that chromosomes can be atypical (used synonymously with abnormal) and affect health. Actual cases introduce some of them.

Required Parts: Telomeres and Centromeres

A chromosome consists primarily of DNA and proteins with a small amount of RNA. It is duplicated and transmitted, via mitosis or meiosis, to the next cell generation. Chromosomes have long been described and distinguished by size and shape, using stains and dyes to contrast dark **heterochromatin** with the lighter **euchromatin** (**figure 13.1**). Heterochromatin consists mostly of highly repetitive DNA sequences, whereas euchromatin has more protein-encoding sequences.

A chromosome must include structures that enable it to replicate and remain intact. Everything else is informational cargo (protein-encoding genes and their controls). The essential parts of a chromosome are as follows:

- telomeres;
- origin of replication sites, where replication forks begin to form; and
- the centromere.

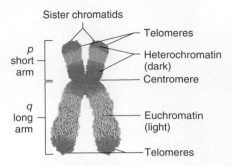

Figure 13.1 **Portrait of a chromosome.** Tightly wound, highly repetitive heterochromatin forms the centromere (the largest constriction) and the telomeres (the tips) of chromosomes. Elsewhere, lighter-staining euchromatin includes many protein-encoding genes. The centromere divides this chromosome into a short arm (*p*) and a long arm (*q*). This chromosome is in the replicated form.

Recall from figure 2.18 that **telomeres** are chromosome tips. In humans, each telomere repeats the sequence TTAGGG. In most cell types, telomeres shorten with each mitotic cell division.

The **centromere** is the largest constriction of a chromosome. It is where spindle fibers attach when the cell divides. A chromosome without a centromere is no longer a chromosome. It vanishes from the cell as soon as division begins because there is no way to attach to the spindle.

Centromeres, like chromosomes, consist mostly of DNA and protein. Many of the hundreds of thousands of DNA bases that form the centromere are copies of a specific 171-base DNA sequence. The size and number of repeats are similar in many species, although the sequence differs. The similarity among species suggests that these repeats have a structural role in maintaining chromosomes rather than an informational one from their sequence. Certain centromere-associated proteins are synthesized only when mitosis is imminent, forming a structure called a kinetochore that contacts the spindle fibers, enabling the cell to divide.

Centromeres replicate toward the end of S phase. A protein that may control the process is centromere protein A, or CENP-A. Molecules of CENP-A stay with centromeres as chromosomes replicate, covering about half a million DNA base pairs. When the replicated (sister) chromatids separate at anaphase, each member of the pair retains some CENP-A. The protein therefore passes to the next cell generation, but it is *not* DNA. This is an epigenetic change.

Centromeres lie within vast stretches of heterochromatin. The arms of the chromosome extend outward from the centromere. Gradually, the DNA includes more protein-encoding sequences with distance from the centromere. Gene density varies greatly among chromosomes. Chromosome 21 is a gene "desert," harboring a million-base stretch with no protein-encoding genes at all. Chromosome 22, in contrast, is a gene "jungle." These two tiniest chromosomes are remarkably similar in size, which is why they are out of place in karyotypes. Chromosome 22 contains 545 genes to chromosome 21's 225!

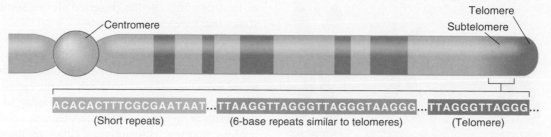

Figure 13.2 **Subtelomeres.** The repetitive sequence of a telomere gradually diversifies toward the centromere. The centromere is depicted as a buttonlike structure to more easily distinguish it, but it is composed of DNA like the rest of the chromosome.

The chromosome parts that lie between protein-rich areas and the telomeres are termed subtelomeres (**figure 13.2**). These areas extend from 8,000 to 300,000 bases inward toward the centromere from the telomeres. Subtelomeres include some protein-encoding genes and therefore bridge the gene-rich regions and the telomere repeats. The transition is gradual. Areas of 50 to 250 bases, right next to the telomeres, consist of 6-base repeats, many of them very similar to the TTAGGG of the telomeres. Then, moving inward from the 6-base zone are many shorter repeats. Their function isn't known. Finally the sequence diversifies and protein-encoding genes appear.

At least 500 protein-encoding genes lie in the total subtelomere regions. About half are members of multigene families (groups of genes of very similar sequence next to each other) that include pseudogenes. These multigene families may reflect recent evolution: Apes and chimps have only one or two genes for many of the large gene families in humans. Such gene organization is one explanation for why our genome sequence is so very similar to that of our primate cousins—but we are clearly different animals. Our genomes differ more in gene copy number and chromosomal organization than in DNA base sequence.

Karyotypes Chart Chromosomes

Even in this age of genomics, the standard chromosome chart, or **karyotype,** remains a major clinical tool. A karyotype displays chromosomes in pairs by size and by physical landmarks that appear during mitotic metaphase, when DNA coils tightly. **Figure 13.3** shows a karyotype with an extra chromosome.

The 24 human chromosome types are numbered from largest to smallest—1 to 22. The other two chromosomes are the X and the Y. Early attempts to size-order chromosomes resulted in generalized groupings because many of the chromosomes are of similar size. Use of dyes and stains made it easier to distinguish chromosomes because they form patterns of bands.

Centromere position is one physical feature of chromosomes. A chromosome is **metacentric** if the centromere divides it into two arms of approximately equal length. It is **submetacentric** if the centromere establishes one long arm and one short arm, and **acrocentric** if it pinches off only a small amount of material toward one end (**figure 13.4**). Some species have telocentric chromosomes that have only one arm, but humans do not. The long arm of a chromosome is designated *q,* and the short arm *p* (*p* stands for "petite").

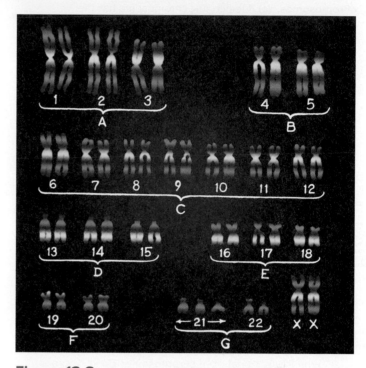

Figure 13.3 **A karyotype displays chromosome pairs in size order.** Note the extra chromosome 21 that causes trisomy 21 Down syndrome. (Color-enhanced; A-G denote generalized groups.)

Five human chromosomes (13, 14, 15, 21, and 22) have bloblike ends, called satellites, that extend from a thinner, stalklike bridge from the rest of the chromosome. The stalk regions do not bind stains well. The stalks carry many copies of genes encoding ribosomal RNA and ribosomal proteins. These areas coalesce to form the nucleolus, a structure in the nucleus where ribosomal building blocks are produced and assembled (see figure 2.4).

Karyotypes are useful at several levels. When a baby is born with the distinctive facial features of Down syndrome, a karyotype confirms the clinical diagnosis. Within families, karyotypes are used to identify relatives with a particular chromosome aberration that can affect health. In one family, several adults died from a rare form of kidney cancer. Karyotypes revealed that the affected individuals all had an exchange, called a **translocation,** between chromosomes 3 and 8. When karyotypes showed that two healthy young family members

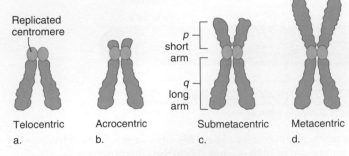

Figure 13.4 **Centromere position distinguishes chromosomes.** **(a)** A telocentric chromosome has the centromere near one end, although telomere DNA sequences are at the tip. Humans do not have telocentric chromosomes. **(b)** An acrocentric chromosome has the centromere near an end. **(c)** A submetacentric chromosome's centromere creates a long arm (*q*) and a short arm (*p*). **(d)** A metacentric chromosome's centromere establishes more equal-sized arms.

had the translocation, physicians examined and monitored their kidneys. Cancer was found very early and successfully treated.

Karyotypes of individuals from different populations can reveal the effects of environmental toxins, if abnormalities appear only in a group exposed to a contaminant. Because chemicals and radiation that can cause cancer and birth defects often break chromosomes into fragments or rings, detecting this genetic damage can alert physicians to the possibility that certain cancers may appear in the population. Chapter 16 explores clues to evolution seen among chromosomes.

Key Concepts Questions 13.1

1. What are the basic parts of a chromosome?
2. What characteristics of chromosomes do karyotypes display?

13.2 Detecting Chromosomes

Laboratory tests to detect or analyze chromosomes can be done on cells sampled at any stage of prenatal development or at any age, but many such tests are conducted on cells from the fetus or the structures that support it. Direct methods collect cells and separate the chromosomes, then add a stain or a DNA probe (a labeled piece of DNA that binds its complement) to distinguish the different chromosomes, or collect DNA from the maternal bloodstream and sequence it. Indirect methods detect changes in levels of biochemicals or clinical findings (such as symptoms or ultrasound scans) associated with a particular chromosomal condition.

Direct Visualization of Chromosomes

Two technologies provide images of chromosomes from a fetus: **amniocentesis** and **chorionic villus sampling** (**CVS**). The images are organized into karyotypes that show the normal 46 chromosomes or abnormal conditions, such as extra

or missing chromosomes, inverted chromosomes, or chromosomes that have exchanged parts.

Any cell other than a mature red blood cell (which lacks a nucleus) can be used to examine chromosomes, but some cells are easier to obtain and culture than others. Skinlike cells collected from the inside of the cheek are the easiest to obtain for a chromosome test. The cells are collected using a brush swirled in the mouth or by spitting into a tube. Genetics classes sometimes analyze students' chromosomes either of these ways. White blood cells are commonly used as the source of DNA in a clinical setting. A person might require a chromosome test if he or she has a family history of a chromosome abnormality or seeks medical help because of infertility.

Chromosome tests are commonly performed on cells from fetuses. Couples who receive a prenatal diagnosis of a chromosome abnormality can arrange for treatment of the newborn, if possible; learn more about the condition and contact support groups and plan care; or terminate the pregnancy. These choices are best made after a genetic counselor or physician provides information on the medical condition and treatment options.

Amniocentesis

The first fetal karyotype was constructed in 1966 using amniocentesis. In this procedure, a doctor removes a small sample of amniotic fluid from the uterus with a needle passed through the woman's abdominal wall (**figure 13.5a**). The fluid contains a few shed fetal cells, which are cultured for a week to 10 days, and then 20 cells are karyotyped. The sampled amniotic fluid may also be examined for deficient, excess, or abnormal biochemicals that could indicate an inborn error of metabolism. Tests for specific single-gene disorders may be done on cells in the amniotic fluid and are tailored to family history. Ultrasound is used to follow the needle's movement and to visualize the fetus or its parts (figure 13.5c).

Amniocentesis can detect approximately 1,000 of the more than 5,000 known chromosomal and biochemical problems. The most common chromosomal abnormality detected is one extra chromosome, called a **trisomy**. Amniocentesis is usually performed between 14 and 16 weeks gestation, when the fetus isn't yet very large but amniotic fluid is plentiful. Amniocentesis can be carried out anytime after this point.

Doctors recommend amniocentesis if the risk that the fetus has a detectable condition exceeds the risk that the procedure will cause a miscarriage. Until recently, this risk cutoff was age 35, when the risk to the fetus of a detectable chromosome problem about equals the risk of amniocentesis causing pregnancy loss—1 in 350. While it is still true that the risk of a chromosomal problem increases steeply after maternal age 35, amniocentesis has become much safer since the statistics were obtained that have been used for most risk estimates (**figure 13.6**). A more recent risk estimate for amniocentesis causing miscarriage of about 1 in 1,600 pregnancies has led some physicians to offer amniocentesis to women under 35. The procedure is also warranted if a couple has had several spontaneous abortions or children with birth defects or a known chromosome abnormality, irrespective of maternal age.

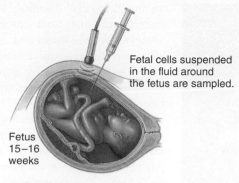

Fetal cells suspended in the fluid around the fetus are sampled.

Fetus 15–16 weeks

a. Amniocentesis

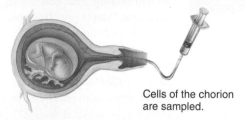

Cells of the chorion are sampled.

b. Chorionic villus sampling

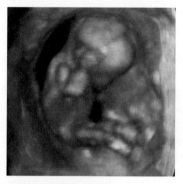

c. Ultrasound image

Figure 13.5 **Checking fetal chromosomes, directly and indirectly.** **(a)** Amniocentesis shows chromosomes from fetal cells in amniotic fluid. **(b)** Chorionic villus sampling examines chromosomes from cells of structures that will develop into the placenta. **(c)** Ultrasound images the fetus and may reveal an anomaly that is associated with a specific chromosome abnormality.

Chorionic Villus Sampling

During the 10th through 12th week of pregnancy, chorionic villus sampling obtains cells from the chorionic villi, which are fingerlike structures that develop into the placenta (see figure 13.5*b*). A karyotype is prepared directly from the collected cells, rather than first culturing them, as in amniocentesis. Results are ready in days.

Because chorionic villus cells descend from the fertilized ovum, their chromosomes should be identical to those of the embryo and fetus. Occasionally, a chromosomal aberration occurs only in a cell of the embryo, or only in a chorionic villus

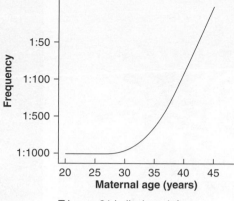

Trisomy 21 in liveborn infants

Figure 13.6 **The maternal age effect.** The risk of conceiving an offspring with trisomy 21 rises dramatically with maternal age.

cell. This results in chromosomal mosaicism—the karyotype of a villus cell differs from that of an embryo cell. Chromosomal mosaicism has great clinical consequences. If CVS indicates an abnormality in villus cells that is not also in the fetus, then a couple may elect to terminate the pregnancy when the fetus is actually chromosomally normal. In the opposite situation, the results of the CVS may be normal, but the fetus has atypical chromosomes.

CVS is slightly less accurate than amniocentesis, and in about 1 in 1,000 to 3,000 procedures, it halts development of the feet and/or hands and may be lethal. Also, CVS does not sample amniotic fluid, so tests for inborn errors of metabolism are not possible. The advantage of CVS is earlier results, but the disadvantage is a greater risk of spontaneous abortion. However, CVS has become much safer in recent years.

Preparing Cells and Chromosomes

The difficulty in distinguishing chromosomes is physical—it is challenging to prepare a cell in which chromosomes do not overlap (**figure 13.7**). To count chromosomes, scientists had to find a way to capture them when they are most condensed—during cell division—and also spread them apart. Since the 1950s, cytogeneticists have used colchicine, an extract of the crocus plant, to arrest cells during division.

How to untangle chromosomes was solved by accident in 1951. A technician mistakenly washed white blood cells being prepared for chromosome analysis in a salt solution that was less concentrated than the interiors of the cells. Water rushed into the cells, swelling them and separating the chromosomes. Then cell biologists found that drawing cell-rich fluid into a pipette and dropping it onto a microscope slide prepared with stain burst the cells and freed the mass of chromosomes. Adding a glass coverslip spread the chromosomes so they could be counted. Researchers finally could see that the number of chromosomes in a diploid human cell is 46, and that the number in gametes is 23.

Karyotypes were once constructed using a microscope to locate a cell where the chromosomes were not touching, photographing the cell, developing a print, cutting out the

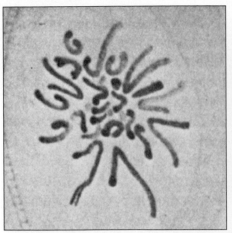

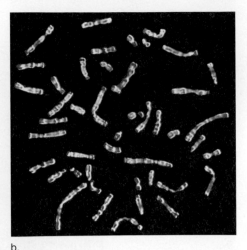

a.   b.

Figure 13.7 **Viewing chromosomes, then and now.** **(a)** The earliest drawings of chromosomes, by German biologist Walter Flemming, date from 1882. His depiction captures the random distribution of chromosomes as they splash down on a slide. **(b)** Highlighted bands distinguish the chromosomes in this micrograph.

individual chromosomes, and arranging them into a size-ordered chart. Today, a computer scans ruptured cells in a drop of stain and selects one in which the chromosomes are the most discernable. Image analysis software recognizes the band patterns of each stained chromosome pair, sorts the structures into a size-ordered chart, and prints the karyotype. If the software recognizes an atypical band pattern, a database pulls out identical or similar karyotypes from records of other patients.

The first karyotypes used dyes to stain chromosomes a uniform color. Chromosomes were grouped into size classes, designated A through G, in decreasing size order. In 1959, scientists described the first chromosomal abnormalities—Down syndrome (an extra chromosome 21), Turner syndrome (also called XO syndrome, a female with only one X chromosome), and Klinefelter syndrome (also called XXY syndrome, a male with an extra X chromosome). These first chromosome stains could highlight only large deletions and duplications. In 1967,

an intellectually disabled child with material missing from chromosome 4 would have been diagnosed as having a "B-group chromosome" disorder.

Describing smaller chromosomal aberrations required better ways to distinguish chromosomes. By the 1970s, new stains created banding patterns unique to each chromosome. These stains are specific for AT-rich or GC-rich stretches of DNA, or for heterochromatin, which is dark-staining. A band represents at least 5 to 10 million DNA bases. Synchronizing the cell cycle revealed even more bands per chromosome.

A technique called fluorescence *in situ* hybridization (FISH) introduced the ability to highlight individual genes. FISH is more precise and targeted than conventional chromosome staining because it uses DNA probes. The probes are attached to molecules that fluoresce when illuminated, producing a flash of color precisely where the probe binds to a complementary DNA sequence among the chromosomes in a patient's sample. Using a FISH probe is a little like a search engine finding the word "hippopotamus" in a book compared to pulling out all words that have the letters *h, p,* and *o.*

FISH can "paint" entire karyotypes by probing each chromosome with several different fluorescent molecules. A computer integrates the images and creates a unique false color for each chromosome. Many laboratories that perform amniocentesis or CVS use FISH probes for chromosomes 13, 18, 21, and the X and Y to quickly identify the most common problems. In **figure 13.8**, FISH reveals the extra

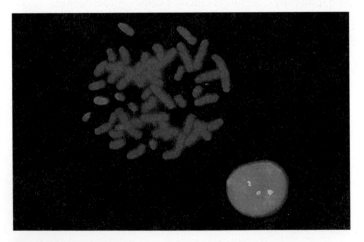

Figure 13.8 **FISHing for genes and chromosomes.** FISH shows three fluorescent dots that correspond to three copies of chromosome 21. Each dot represents a specific DNA sequence to which the fluorescently labeled probe complementary base pairs.

Table 13.1   Chromosome Shorthand

Abbreviation	What It Means
46,XY	Normal male
46,XX	Normal female
45,X	Turner syndrome (female)
47,XXY	Klinefelter syndrome (male)
47,XYY	Jacobs syndrome (male)
46,XY, del (7q)	A male missing part of the long arm of chromosome 7
47,XX, + 21	A female with trisomy 21 Down syndrome
46,XY, t(7;9)(p21.1; q34.1)	A male with a translocation between the short arm of chromosome 7 at band 21.1 and the long arm of chromosome 9 at band 34.1
48, XXYY	A male with an extra X and an extra Y

Figure 13.9 Ideogram. An ideogram is a schematic chromosome map. It indicates chromosome arm (*p* or *q*) and major regions delineated by banding patterns. This figure is repeated in the context of more specific ways to depict chromosomes in figure 22.1.

chromosome 21 in cells from a fetus with trisomy 21 Down syndrome.

Once amniocentesis or CVS produces a karyotype, the pertinent information is abbreviated by listing chromosome number, sex chromosome constitution, and atypical autosomes. Symbols describe the type of aberration, such as a deletion or translocation; numbers correspond to specific bands. A normal male is 46,XY; a normal female is 46,XX. Band notations describe gene locations. For example, the gene encoding the β-globin subunit of hemoglobin is located at 11p15.5. **Table 13.1** gives examples of chromosomal shorthand.

A graphical representation called an ideogram displays chromosome information (**figure 13.9**). The chromosome is divided into arms and numbered regions and subregions. Specific gene loci are sometimes listed on the right side. The sequencing and annotation (description and localization) of specific genes is making chromosome depictions, such as ideograms, outdated.

Indirect Detection of Extra Chromosomes

Amniocentesis and chorionic villus sampling provide a diagnosis, because they directly detect chromosome abnormalities known to cause specific clinical conditions. Indirect methods that measure or detect a characteristic that is associated with having a chromosomal abnormality are used as screening tests to identify individuals at elevated risk—not to provide a diagnosis. Their use may decline in coming years as more tests become based on cell-free fetal DNA in the maternal bloodstream, including complete sequencing of fetal genomes.

Maternal serum markers (**table 13.2**) are biochemicals whose levels in the blood are within a certain range in a pregnant woman carrying a fetus with the normal number of chromosomes, but may lie outside that range in fetuses whose cells have an extra copy of a certain chromosome. The more markers tested, the more predictive the information.

Table 13.2   Maternal Serum Markers

Marker	Normal Function
High Level	
Beta human chorionic gonadotropin (hCG)	Implantation of the embryo
Inhibin A	Lowers levels of follicle-stimulating hormone
Low Level	
Estradiol (UE3)	Female hormone that binds AFP
Alpha fetoprotein (AFP)	Binds estradiol
Pregnancy-associated plasma protein (PAPP-A)	Cell division

What these biomarkers actually do is not as important for testing purposes as their concentrations.

Testing maternal serum markers didn't begin with chromosomal conditions, but for neural tube defects (NTDs). Recall from chapter 3 that in an NTD, part of the brain or spinal cord protrudes. In the 1980s, a researcher who had a child with an NTD developed a test based on the finding that the level of alpha fetoprotein (AFP) is higher in fetuses with an open neural tube defect. AFP is made in the yolk sac and leaves the fetal circulation and enters the maternal bloodstream at a certain rate; too much indicates an opening.

As data accumulated to detect elevated AFP, researchers noted that the level of AFP is lower in fetuses that have trisomy 21 Down syndrome. After that, the markers listed in table 13.2 were incorporated into testing, too. Altogether, combined with considering maternal age and ultrasound findings, maternal serum markers offer a greater than 90 percent probability of detecting trisomies 13, 18, or 21 if they are present. (This does not mean that the chance of a condition being present exceeds 90 percent.)

For trisomy 21 Down syndrome, ultrasound findings include short limb lengths, flattened noses, and excess fluid at the back of the neck (called nuchal translucency). However, these signs can occur in fetuses who do not have Down syndrome, too.

Maternal serum marker tests are offered to all pregnant women because they are safe and inexpensive. Cut-off levels for the results based on personal statistics (such as race, other medical conditions like diabetes, and a twin pregnancy) are used to identify fetuses at elevated risk. The result may be returned as a value called "MoM," which stands for "multiples of the median." The MoM value indicates how far individual results deviate from the normal range of concentration for each marker. A MoM above 2.0 means that the level of the biomarker is twice as high or low than the average in a normal pregnancy, and this is considered elevated risk. Parents of possibly affected fetuses are then counseled about undergoing the more invasive amniocentesis or chorionic villus sampling procedures to confirm or rule out the trisomy.

The accuracy of multiple maternal serum marker screening rises to 99 percent when cell-free fetal DNA is also considered. Tests for this DNA are a form of "noninvasive prenatal diagnosis," or NIPD (ultrasound is noninvasive, too). Small pieces of DNA are normally in a pregnant woman's bloodstream, and up to 20 percent of those pieces actually come from the placenta, and therefore represent the fetus (although mosaicism is possible, as in CVS). The fetal pieces can be separated because they are shorter (**figure 13.10**). The earliest versions of these tests detect trisomies 13, 18, and 21 (discussed in the next section) by finding proportionately more of the sequences from these chromosomes than in an unaffected fetus. These fetal DNA tests are done at 10 weeks or later. They may prove most useful in identifying women who are over 35 or have family histories of the chromosome problems but are actually carrying an unaffected fetus, enabling them to avoid more invasive and therefore riskier diagnostic procedures.

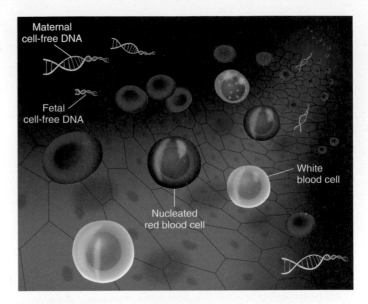

Figure 13.10 **Testing fetal DNA.** The fact that pieces of cell-free fetal DNA in the maternal circulation are much shorter than pieces of maternal DNA provides a way to collect the fetal material and sequence the genome by overlapping pieces. Comparison of a fetal genome sequence to that of the parents can reveal inherited diseases and new mutations.

Key Concepts Questions 13.2

1. From which cells are chromosomes typically visualized?

2. Describe how amniocentesis and chorionic villus sampling directly reveal fetal chromosomes.

3. Describe indirect ways to detect abnormal chromosome numbers.

4. Explain how presence of an extra chromosome is deduced from testing cell-free fetal DNA.

13.3 Atypical Chromosome Number

A human karyotype is atypical (abnormal) if the number of chromosomes in a somatic cell is not 46, or if individual chromosomes have extra, missing, or rearranged genetic material. More discriminating technologies can detect very small numbers of extra or missing nucleotides (microduplications and microdeletions). As a result, more people are being diagnosed with chromosomal abnormalities than in the days when stains made all chromosomes look alike.

Atypical chromosomes account for at least 50 percent of spontaneous abortions, yet only 0.65 percent of newborns have them. Therefore, most embryos and fetuses with atypical chromosomes stop developing before birth. **Table 13.3** summarizes the types of chromosome variants in the order in which they are discussed.

Table 13.3	Chromosome Abnormalities
Type of Abnormality	**Definition**
Polyploidy	Extra chromosome sets
Aneuploidy	An extra or missing chromosome
Monosomy	One chromosome absent
Trisomy	One chromosome extra
Deletion	Part of a chromosome missing
Duplication	Part of a chromosome present twice
Translocation	Two chromosomes join long arms or exchange parts
Inversion	Segment of chromosome reversed
Isochromosome	A chromosome with identical arms
Ring chromosome	A chromosome that forms a ring due to deletions in telomeres, which cause ends to adhere

Table 13.4	Organizations for Families with Chromosome Abnormalities
Hope for Trisomy 13 + 18	www.hopefortrisomy13and18.org/
National Down Syndrome Society	www.ndss.org
Rainbows Down Under	http://www.trisomyonline.org/rock.htm
Support Organization for Trisomy 18, 13 and Related Disorders (SOFT)	www.trisomy.org
The XXYY Project	XXYYsyndrome.org
Tracking Rare Incidence Syndromes (TRIS)	http://web.coehs.siu.edu/grants/tris/
Trisomy 18 Foundation	trisomy18.org
UNIQUE—the Rare Chromosome Disorder Support Group	http://www.rarechromo.org/html/home.asp

Often families with the same chromosome disorder form organizations, such as the "International 22q11.2 Deletion Syndrome Foundation," a group of families with members missing DNA on the short arm of chromosome 22. **Table 13.4** lists some of these organizations.

Polyploidy

The most extreme upset in chromosome number is an entire extra set. A cell with extra sets of chromosomes is **polyploid**. An individual whose cells have three copies of each chromosome is a triploid (designated 3N, for three sets of chromosomes).

Two-thirds of all triploids result from fertilization of an oocyte by two sperm. The other cases arise from formation of a diploid gamete, such as when a normal haploid sperm fertilizes a diploid oocyte. Triploids account for 17 percent of spontaneous abortions (**figure 13.11**). Very rarely, an infant survives as long as a few days, with defects in nearly all organs. However, certain human cells may be polyploid. The liver, for example, has some tetraploid (4N) and even octaploid (8N) cells.

Polyploids are very common among flowering plants, including roses, cotton, barley, and wheat, and in some insects. Fish farmers raise triploid salmon, which cannot breed.

Aneuploidy

Cells missing a single chromosome or having an extra chromosome are **aneuploid,** which means "not good set." Rarely, aneuploids can have more than one missing or extra chromosome, indicating defective meiosis in a parent. A normal chromosome number is **euploid,** which means "good set."

Most autosomal aneuploids (with a missing or extra non-sex chromosome) are spontaneously aborted. Those that survive have specific syndromes, with symptoms depending upon which chromosomes are missing or extra. Intellectual disability is common in aneuploidy because development of the brain is so complex and of such long duration that nearly any chromosome-scale disruption affects genes whose protein products affect the brain. Sex chromosome aneuploidy usually produces milder symptoms.

Most children born with the wrong number of chromosomes have an extra chromosome (a **trisomy**) rather than a missing one (a **monosomy**). Most monosomies are so severe that an affected embryo ceases developing. Trisomies and monosomies are named for the chromosomes involved, and in

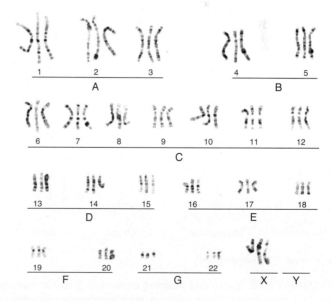

Figure 13.11 Polyploids in humans are nearly always lethal. Individuals with three copies of each chromosome (triploids) in every cell account for 17 percent of all spontaneous abortions and 3 percent of stillbirths and newborn deaths.

the past the associated syndromes were named for the discoverers. Today, cytogenetic terminology is used because it is more precise. For example, Down syndrome can result from a trisomy or a translocation. The distinction is important in genetic counseling. Translocation Down syndrome, although accounting for only 4 percent of cases, has a much higher recurrence risk within a family than the trisomy form, a point we return to later in the chapter.

The meiotic error that causes aneuploidy is called **nondisjunction**. Recall that in normal meiosis, homologs separate and each of the resulting gametes receives only one member of each chromosome pair. In nondisjunction, a chromosome pair fails to separate at anaphase of either the first or second meiotic division. This produces a sperm or oocyte that has two copies of a particular chromosome, or none, rather than the normal one copy (**figure 13.12**). When such a gamete meets its partner at fertilization, the resulting zygote has either 45 or 47 chromosomes, instead of the normal 46. Different trisomies tend to be caused by nondisjunction in the male or female, at meiosis I or II.

A cell can have a missing or extra chromosome in 49 ways—an extra or missing copy of each of the 22 autosomes, plus the five abnormal types of sex chromosome combinations—Y, X, XXX, XXY, and XYY. (Some individuals have four or even five sex chromosomes.) However, only nine types of aneuploids are recognized in newborns. Others are seen in spontaneous abortions or fertilized ova intended for *in vitro* fertilization.

Most of the 50 percent of spontaneous abortions that result from extra or missing chromosomes are 45, X individuals (missing an X chromosome), triploids, or trisomy 16. About

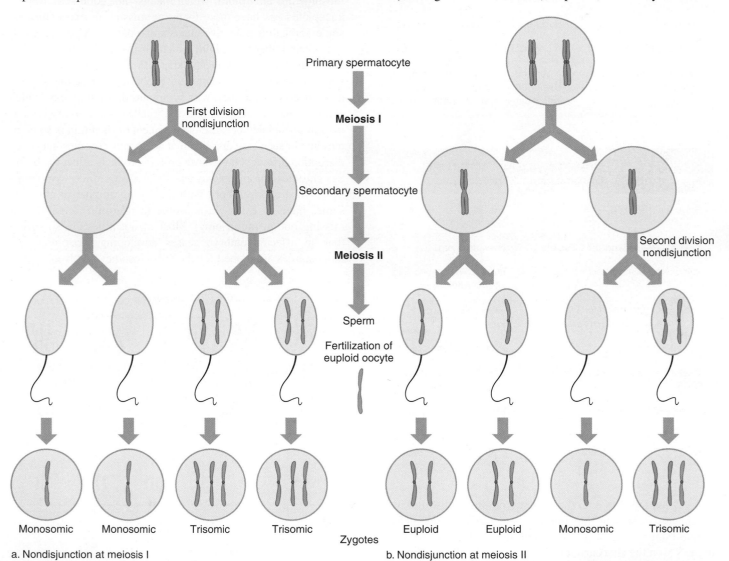

a. Nondisjunction at meiosis I

b. Nondisjunction at meiosis II

Figure 13.12 **Extra and missing chromosomes—aneuploidy.** Unequal division of chromosome pairs can occur at either the first or second meiotic division. **(a)** A single pair of chromosomes is unevenly partitioned into the two cells arising from meiosis I in a male. The result: Two sperm cells have two copies of the chromosome, and two sperm cells have no copies. When a sperm cell with two copies of the chromosome fertilizes a normal oocyte, the zygote is trisomic; when a sperm cell lacking the chromosome fertilizes a normal oocyte, the zygote is monosomic. **(b)** This nondisjunction occurs at meiosis II. Because the two products of the first division are unaffected, two of the mature sperm are normal and two are aneuploid. Oocytes can undergo nondisjunction as well, leading to zygotes with extra or missing chromosomes when normal sperm cells fertilize them.

9 percent of spontaneous abortions are trisomy 13, 18, or 21. More than 95 percent of newborns with atypical chromosome numbers have an extra 13, 18, or 21, or an extra or missing X or Y chromosome. These conditions are all rare at birth—together they affect only 0.1 percent of all children. However, nondisjunction occurs in 5 percent of recognized pregnancies.

Types of chromosome abnormalities differ between the sexes. Atypical oocytes mostly have extra or missing chromosomes, whereas atypical sperm more often have structural variants, such as inversions or translocations, discussed later in the chapter.

Aneuploidy and polyploidy also arise during mitosis, producing groups of somatic cells with the extra or missing chromosome. An individual with two chromosomally distinct cell populations is a mosaic. If only a few cells are altered, health may not be affected. However, a mitotic abnormality that occurs early in development, so that many cells descend from the unusual one, can affect health. A chromosomal mosaic for a trisomy may have a mild version of the associated condition. This is usually the case for the 1 to 2 percent of people with Down syndrome who are mosaic. The phenotype depends upon which cells have the extra chromosome. Unfortunately, prenatal testing cannot reveal which cells are affected.

Autosomal Aneuploids

Most autosomal aneuploids cease developing long before birth. Following are cases and descriptions of the most common autosomal aneuploids among liveborns. The most frequently seen extra autosomes in newborns are chromosomes 21, 18, and 13 because these chromosomes carry many fewer protein-encoding genes than the other autosomes, compared to their total amount of DNA. Therefore, extra copies of these chromosomes are tolerated well enough for some fetuses with them to survive to be born (**table 13.5**).

Trisomy 21—David's Story

When David G. was born in 1994, doctors told his 19-year-old mother, Toni, to put him into an institution. "They said he wouldn't walk, talk, or do anything. Today, I want to bring him back and say look, he walks and talks and has graduated high school," says Toni.

Like other teens, David has held part-time jobs, gone to dances, and uses a computer. But he is unlike most other teens in that his cells have an extra chromosome 21, which limits his

intellectual abilities. "Maybe he's not book smart, but when you look around at what he can do, he's smart," Toni says. His speech is difficult to understand, and he has facial features characteristic of Down syndrome, but he has a winning personality and close friends.

Sometimes David gets into unusual situations because he thinks literally. He once dialed 911 when he stubbed his toe, because he'd been told to do just that when he was hurt.

Today David lives in a group home and attends community college.

The most common autosomal aneuploid among liveborns is trisomy 21, because this chromosome has the fewest genes. The extra folds in the eyelids, called epicanthal folds, and flat face prompted Sir John Langdon Haydon Down to term the condition *mongoloid* when he described it in 1866. As the medical superintendent of a facility for the profoundly intellectually disabled, Down noted that about 10 percent of his patients resembled people of Mongolian heritage. The resemblance is superficial. People of all ethnic groups are affected.

A person with Down syndrome is usually short and has straight, sparse hair and a tongue protruding through thick lips. The hands have an atypical pattern of creases, the joints are loose, and poor reflexes and muscle tone give a "floppy" appearance. Developmental milestones (such as sitting, standing, and walking) come slowly, and toilet training may take several years. Intelligence varies greatly. Parents of a child with Down syndrome can help their child reach maximal potential by providing a stimulating environment (**figure 13.13**).

Many people with Down syndrome have physical problems, including heart and kidney defects and hearing and visual loss. A suppressed immune system can make influenza deadly. Digestive system blockages are common and may require surgical correction. A child with Down syndrome is 15 times more likely to develop leukemia than a child who does not have the syndrome, but this is still only a 1 percent risk. However, people with Down syndrome are somewhat protected against developing solid tumors. Many of the medical problems associated with Down syndrome are treatable, so that average life expectancy is now sixty. In 1910, life expectancy was only 9 years.

Some people with Down syndrome older than 40 develop the black fibers and tangles of amyloid beta protein in their brains characteristic of Alzheimer disease, although they usually do not become severely demented (see Clinical Connection 5.1). The chance of a person with trisomy 21 developing Alzheimer disease is 25 percent, compared to 6 percent for the general population. A gene on chromosome 21 causes one inherited form of Alzheimer disease. Perhaps the extra copy of the gene in trisomy 21 has a similar effect to a mutation in the gene that causes Alzheimer disease, such as causing amyloid beta buildup.

Before the sequencing of the human genome, researchers studied people who have a third copy of part of chromosome 21 to see which genes cause which symptoms. This approach led to the discovery that genes near the tip of the long arm of the chromosome contribute most of the abnormalities. Two specific genes control many aspects of Down syndrome by controlling a third gene, which encodes a transcription factor and therefore affects many other genes.

Table 13.5	Comparing and Contrasting Trisomies 13, 18, and 21	
Type of Trisomy	**Incidence at Birth**	**Percent of Conceptions That Survive 1 Year After Birth**
13 (Patau)	1/12,500–1/21,700	<5%
18 (Edward)	1/6,000–1/10,000	<5%
21 (Down)	1/800–1/826	85%

Figure 13.13 **Trisomy 21.** Many years ago, people with Down syndrome were institutionalized. Today, thanks to tremendous strides in both medical care and special education, people with the condition can hold jobs and attend college. This young lady enjoys painting.

A newer approach, **genome editing,** is being used to study Down syndrome in cells. Section 19.4 discusses how genome editing cuts double-stranded DNA and allows insertion of a specific sequence. For trisomy 21 Down syndrome, the technique applies the mechanism that shuts off one X chromosome in the cells of females. Researchers inserted the DNA that encodes the long noncoding RNA sequence called XIST, which normally shuts off one X chromosome (see figure 6.10), into one chromosome 21 of induced pluripotent stem cells (see figure 2.22) made from skin cells of a boy with trisomy 21. The shut-off chromosome 21 formed a Barr body, which normally happens when XIST silences one X chromosome in a female cell. (Male cells do not normally have Barr bodies because they have only one X chromosome.) The treated male cells had Barr bodies (**figure 13.14**), meaning one of the three chromosome 21s had been turned off. Observing how the cells change may reveal characteristics of trisomy 21 cells that might suggest new targets for intervention.

The likelihood of giving birth to a child with trisomy 21 Down syndrome increases dramatically with the age of the mother (see figure 13.6). However, 80 percent of children with trisomy 21 are born to women under age 35, because younger women are more fertile and, until noninvasive prenatal testing on cell-free fetal DNA became available in 2011, less likely to have prenatal tests than older women. About 90 percent of trisomy 21 conceptions are due to nondisjunction during meiosis I in the female. The 10 percent of cases due to the male result from nondisjunction during meiosis I or II. The chance that trisomy 21 will recur in a family, based on empirical data (how often it actually does recur in families), is 1 percent.

The age factor in trisomy 21 Down syndrome and other trisomies may reflect the fact that the older a woman is, the longer her oocytes have been arrested on the brink of completing meiosis. During this 15 to 45 years, oocytes may have been exposed to toxins, viruses, and radiation. A second explanation for the maternal age effect is that females have a pool of immature aneuploid oocytes resulting from spindle abnormalities that cause nondisjunction. As a woman ages, selectively releasing normal oocytes each month, the atypical ones remain, much as black jelly beans accumulate as people preferentially eat the colored ones.

The association between maternal age and Down syndrome has been recognized for a long time, because affected individuals were often the youngest children in large families. Before the chromosome connection was made in 1959, the syndrome was attributed to syphilis, tuberculosis, thyroid malfunction, alcoholism, emotional trauma, or "maternal reproductive exhaustion." The increased risk of Down syndrome correlates to maternal age, not to the number of children in the family. The *Bioethics: Choices for the Future* essay on page 250 examines the effect of prenatal diagnosis on the prevalence of people with trisomy 21 Down syndrome.

Trisomy 18—Anthony's Story

When an ultrasound scan early in pregnancy revealed a small fetus with low-set ears, a small jaw, a pocket of fluid in the brain, and a clenched fist, the parents-to-be, Elisa and Brendan, were offered amniocentesis to view the fetus's chromosomes. The signs on the scan suggested an extra chromosome 18, and amniocentesis confirmed it.

Although Elisa and Brendan were upset to learn what lay ahead, they continued the pregnancy. The fetus remained small, as Elisa swelled hugely with three times the normal volume of amniotic fluid. Further ultrasound scans revealed that only one of the kidneys worked, the heart had holes between the chambers, and part of the intestine lay outside the stomach in a sac. The child would be severely developmentally delayed and intellectually disabled. Anthony was delivered at 36 weeks gestation, after his heart rate became erratic during a routine prenatal visit. He lived only 22 days.

Trisomy 18 (**figure 13.15**) (Edwards syndrome) and trisomy 13 (Patau syndrome) were described in a research report in 1960. Usually affected individuals do not survive to be born.

Most children who have trisomy 18 have great physical and intellectual disabilities, with developmental skills stalled at the

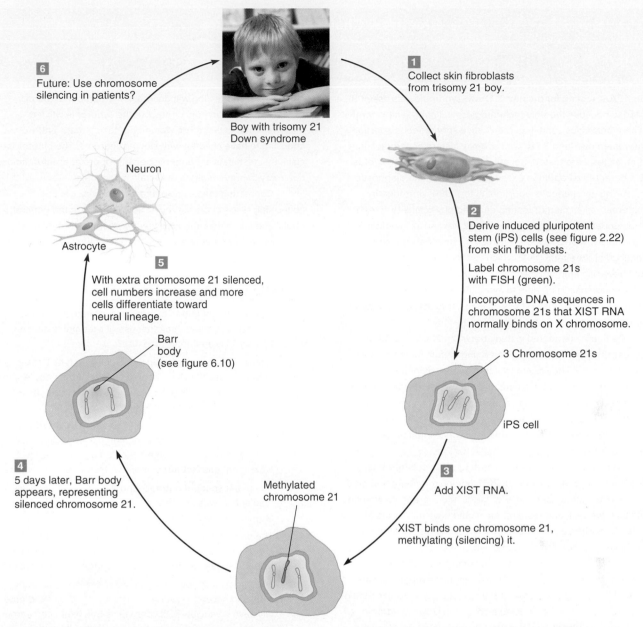

Figure 13.14 Turning off the extra chromosome 21 of Down syndrome. One application of genome editing uses the mechanism of X inactivation to "turn off" the extra chromosome of trisomy 21, silencing the extra genetic information—in stem cells from an affected individual. Observing what happens as the stem cells are stimulated to differentiate as neurons may reveal new ways to treat the syndrome.

The labels within the figure read:

6 Future: Use chromosome silencing in patients?

Boy with trisomy 21 Down syndrome

1 Collect skin fibroblasts from trisomy 21 boy.

Neuron

Astrocyte

5 With extra chromosome 21 silenced, cell numbers increase and more cells differentiate toward neural lineage.

Barr body (see figure 6.10)

2 Derive induced pluripotent stem (iPS) cells (see figure 2.22) from skin fibroblasts.

Label chromosome 21s with FISH (green).

Incorporate DNA sequences in chromosome 21s that XIST RNA normally binds on X chromosome.

3 Chromosome 21s

iPS cell

4 5 days later, Barr body appears, representing silenced chromosome 21.

Methylated chromosome 21

3 Add XIST RNA.

XIST binds one chromosome 21, methylating (silencing) it.

6-month level. Major abnormalities include heart defects, a displaced liver, growth retardation, and oddly clenched fists. Milder signs include overlapping fingers, a narrow and flat skull, low-set ears, a small mouth, unusual fingerprints, and "rocker-bottom" feet. Most cases of trisomy 18 arise from nondisjunction in meiosis II of the oocyte. **In Their Own Words** on page 251 describes one of the oldest people known to have trisomy 18.

Trisomy 13—Tykesia's Story

At 15 months of age, Tykesia is a "long-term survivor" of trisomy 13. She is small for her age, at the 5th percentile for weight, but she is happy, curious and playful. Her physical skills, however, lag. She can finally, with great effort, sit up, but cannot yet crawl. She has about 20 minor seizures a day, which look like startles, and has difficulty eating because of persistent acid reflux. She is also missing a rib. Early surgeries corrected a cleft lip and palate, removed an extra finger and toe, and corrected a hernia. Blood vessels leading from the heart to the lungs that did not close as they normally should before birth did so by the time Tykesia was 6 months old. She is intellectually disabled, but her parents hope she will live long enough to attend preschool. Despite these challenges, Tykesia's case is mild—she has her sight and hearing, unlike many others with trisomy 13.

Will Trisomy 21 Down Syndrome Disappear?

In the 1970s, testing for trisomy 21 Down syndrome was riskier than it is today because the only technologies available—amniocentesis and chorionic villus sampling—can cause spontaneous abortion. Only women over age 35 or with a family history of chromosome abnormalities were tested, because their risk of carrying a fetus with a detectable condition exceeded the risk of the diagnostic test causing spontaneous abortion. In the 1980s, introduction of maternal serum marker testing enabled physicians to screen women under 35, identifying those at higher risk as candidates for the more invasive tests. Then in 2011 came tests based on detecting cell-free fetal DNA, which can identify women who would have had amniocentesis or CVS based on age but who are at low enough risk to be able to avoid these more invasive tests. All of this history means that it is now easier to test for trisomy 21 Down syndrome.

Denmark conducted a study between 2000 and 2006 to assess the effect of offering screening (serum markers) to all pregnant women. They found that the number of infants born with trisomy 21 fell by 50 percent, the number diagnosed before birth increased by 30 percent, and the number of amniocentesis and CVS procedures done annually decreased by 50 percent. Countering this trend are the medical advances that have greatly extended life span for people with trisomy 21.

When noninvasive prenatal testing using cell-free fetal DNA became widely available in 2011, finding out about trisomy 21 became safer, earlier, and more accurate. More pregnant women could take the test, and researchers would have more data to track trends. Soon after, a prominent bioethicist and his student, from New Zealand, published a paper in which they argue that to do the most good, ". . . we accept that in some cases, the perceived disadvantages resulting from a Down syndrome pregnancy (to child and family) may outweigh the perceived good from the child's life. . . . For some families, raising a child with Down syndrome will be immensely difficult, and it is for this reason that we allow the option of termination."

Increased testing will decrease the number of trisomy 21 births, the researchers wrote, but the syndrome will not disappear due to test error, women not taking tests, and women choosing to have children with Down syndrome. The bioethicists claim that testing is not eugenic because it provides information, not a way to prevent births of certain individuals.

Countering the claims of the bioethicists, genetic counseling researchers surveyed 1,800 people in the general public and found that attitudes toward people with Down syndrome and their families have not been negatively affected by the new testing.

Questions for Discussion

1. Look up the definition of eugenics, or read about it in chapter 15. Do you think that use of prenatal testing for trisomy 21 is a eugenic measure?

2. A researcher who has pioneered cell-free fetal DNA testing to detect trisomies has received threats from people accusing her of killing babies. She counters that the test will actually save lives. Explain how the test can do this.

3. The accuracy statistics on cell-free fetal DNA testing are based on studies of women at high risk of having a fetus with an extra chromosome based on age or family history. How might this fact affect more widespread use of the test?

4. The paper sparked outrage, and the professor who was the lead author was forced to resign as bioethics director at a large university. Should he have been punished?

5. The paper mentions only women and their choices. To what extent do you think fathers' opinions are important in the decision to take a prenatal test and then act on the result?

6. What impact do you think the wider availability of early, safe, and accurate testing for trisomy 21 Down syndrome will have on society? Discuss views that might come from a pregnant woman, a parent of an adult child with Down syndrome, and a physician who offers prenatal testing.

Trisomy 13 has a different set of signs and symptoms than trisomy 18. Most striking is fusion of the developing eyes, so that a fetus has one large eyelike structure in the center of the face. More common is a small or absent eye. Major abnormalities affect the heart, kidneys, brain, face, and limbs. The nose is often malformed, and cleft lip and/or palate is present in a small head. There may be extra fingers and toes. Ultrasound examination of an affected newborn may reveal an extra spleen, atypical liver, rotated intestines, and an abnormal pancreas. A few individuals have survived until adulthood, but they do not progress developmentally beyond the 6-month level.

Sex Chromosome Aneuploids: Female

People with sex chromosome aneuploidy have extra or missing sex chromosomes. **Table 13.6** indicates how these aneuploids can arise. Some conditions can result from nondisjunction in meiosis in the male *or* female. Sex chromosome aneuploids are generally much less serious than autosomal aneuploids.

XO Syndrome—Miranda's Story

At age 17, Miranda still looked about 12. She was short, her breasts had never developed, and she had never menstruated.

Some Individuals With Trisomies Survive Childhood

Babies born with trisomy 18 rarely survive to reach their first birthdays, but Greg D. (not his real name) has reached his twenty-first. No one knows why. He has the complete extra chromosome in all of his cells. His mother devotes her life to caring for him. Here, in her own words:

I was 42 when I became pregnant. I didn't have amniocentesis, because I didn't think there would be anything wrong. I thought he'd be a healthy bouncing baby boy, until he was born.

They called him a vegetable at birth. His godmother is an RN and she was in the delivery room. She knew there was something wrong. The first doctor to see him knew what was wrong. He said, "Take him home and love him and he'll die in your arms." They said he'd be blind and hearing impaired, and he's not. And he has full-blown trisomy 18; he's not a mosaic.

Greg is low functioning. He's had pneumonia three times this year. He lived through a *C. difficile* infection. He has fevers, heart and bowel problems, and urinary tract infections, and he breaks his fragile bones. He doesn't walk or talk. He weighs 55 pounds, and he won't get any bigger. But he keeps on going.

He's definitely not a vegetable. He's a little person. He reacts appropriately. I use a suction machine and when I put the part near him, he opens his mouth. He knows. He recognizes people. His face lights up when he sees people he knows. I can't quite comprehend that he's that rare, but according to the geneticist, Greg is believed to be one of the oldest in the world with trisomy 18.

He loves his special needs school, and I'm heartsick because he's aging out of everything. He loves music, and being on the school bus with other kids. I'll have to stop calling him a kid!

He vocalizes. I know when he's wet his diaper. He makes happy noises. We like watching game shows on TV. He likes the different-colored lights, the sound of the clapping. He picks up on the excitement. He lies in a big La-Z-Boy chair, and I sit on the couch right beside him. We like *Wheel of Fortune, The Price Is Right,* and *Let's Make a Deal.*

There's a light inside those eyes; I can see it. He amazes me.

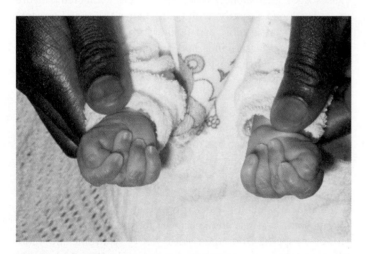

Figure 13.15 **Trisomy 18.** An infant with trisomy 18 clenches the fists in a characteristic way, with fingers overlapping.

Her sister Charlotte, 2 years younger, looked older. When Miranda turned 16, her physician suggested that she have her chromosomes checked. Miranda's karyotype revealed absence of an X chromosome, called Turner or XO syndrome. The diagnosis explained other problems, such as poor hearing, high blood pressure, low thyroid function, and the "beauty marks" that dotted her skin.

In 1938, at a medical conference, a U.S. endocrinologist named Henry Turner described seven young women, aged 15 to 23, who were sexually undeveloped, short, had folds of skin on the back of their necks, and had malformed elbows. (Eight years earlier, an English physician had described the syndrome in young girls, so it is called Ullrich syndrome in the United Kingdom.) Alerted to what would become known as Turner syndrome in the United States, other physicians soon began identifying such patients. Physicians assumed that a hormonal insufficiency caused the symptoms. They were right, but there was more to the story—a chromosomal imbalance caused the hormone deficit.

In 1954, at a London hospital, a physician discovered that cells from Turner patients do not have a Barr body, the dark spot that indicates the silenced X chromosome. By 1959, karyotyping confirmed the presence of only one X chromosome. Later, researchers learned that only 50 percent of affected individuals are XO. The rest are missing only part of an X chromosome or are mosaics, with only some cells missing an X.

Like the autosomal aneuploids, Turner syndrome, now called XO syndrome, is found more frequently among spontaneously aborted fetuses than among newborns—99 percent of XO fetuses are not born. The syndrome affects 1 in 2,500 female births. However, if amniocentesis or CVS was not done, a person with XO syndrome would likely not know she has a chromosome abnormality until she lags in sexual development. Two X chromosomes are necessary for normal sexual development in females.

At birth, a girl with XO syndrome looks normal, except for puffy hands and feet caused by impaired lymph flow. In childhood, signs of XO syndrome include wide-set nipples, soft nails that turn up at the tips, slight webbing at the back of the

Table 13.6 — How Nondisjunction Leads to Sex Chromosome Aneuploids

Situation	Oocyte	Sperm	Consequence
Normal	X	Y	46,XY normal male
	X	X	46,XX normal female
Female nondisjunction	XX	Y	47,XXY Klinefelter syndrome
	XX	X	47,XXX triplo-X
		Y	45,Y nonviable
		X	45,X Turner syndrome
Male nondisjunction	X		45,X Turner syndrome
(meiosis I)	X	XY	47,XXY Klinefelter syndrome
Male nondisjunction	X	XX	47,XXX triplo-X
(meiosis II)	X	YY	47,XYY Jacobs syndrome
	X		45,X Turner syndrome
Male and female nondisjunction	XX	YY	48, XXYY syndrome

neck, short stature, coarse facial features, and a low hairline at the back of the head. About half of people with XO syndrome have impaired hearing and frequent ear infections due to a small defect in the shape of the coiled part of the inner ear. They cannot hear certain frequencies of sound.

At sexual maturity, sparse body hair develops, but the girls do not ovulate or menstruate, and their breasts do not develop. The uterus is very small, but the vagina and cervix are normal size. In the ovaries, oocytes mature too fast, depleting the supply during infancy. Intelligence is normal. XO syndrome may impair the ability to solve math problems that entail envisioning objects in three-dimensional space, and may cause memory deficits. Low doses of hormones (estrogen and progesterone) are given to stimulate development of secondary sexual structures for girls diagnosed before puberty, and prompt use of growth hormone can maximize height.

Individuals who are mosaics (only some cells lack the second X chromosome) may have children, but their offspring are at high risk of having too many or too few chromosomes. XO syndrome is unrelated to the age of the mother. The effects of XO syndrome continue past the reproductive years. Life span is shortened slightly. Adults with XO syndrome are more likely to develop certain disorders than the general population, including osteoporosis, types 1 and 2 diabetes, and colon cancer. The many signs and symptoms of XO syndrome result from loss of specific genes.

Triplo-X

About 1 in every 1,000 females has an extra X chromosome in each of her cells, a condition called triplo-X. The only symptoms are tall stature and menstrual irregularities. Although triplo-X females are rarely intellectually disabled, they tend to be less intelligent than their siblings. The lack of symptoms reflects the protective effect of X inactivation—all but one of the X chromosomes is inactivated.

Sex Chromosome Aneuploids: Male

Any individual with a Y chromosome is a male. A man can have one or several extra X or Y chromosomes.

XXY Syndrome—Stefan's Story

Looking back, Stefan Schwarz's only indication of XXY syndrome was small testes. When his extra X chromosome was detected when he was 25, suddenly his personality quirks made sense. "I was very shy, reserved, and had trouble making friends. I would fly into rages for no apparent reason. My parents knew when I was very young that there was something about me that wasn't right," he recalls.

Psychologists, psychiatrists, and therapists diagnosed learning disabilities, and one told Stefan he "was stupid and lazy, and would never amount to anything." But Stefan earned two bachelor's degrees and is a successful software engineer. He heads a support group for men with XXY syndrome.

About 1 in 500 males has the extra X chromosome that causes XXY (Klinefelter) syndrome. Severely affected men are underdeveloped sexually, with rudimentary testes and prostate glands and sparse pubic and facial hair. They have very long arms and legs, large hands and feet, and may develop breast tissue. XXY syndrome is the most common genetic or chromosomal cause of male infertility.

Testosterone injections during adolescence can limit limb lengthening and stimulate development of secondary sexual characteristics. Boys and men with XXY syndrome may be slow to learn, but they are usually not intellectually disabled unless they have more than two X chromosomes, which is rare.

Men with XXY syndrome have fathered children, with medical help. Doctors select sperm that contain only one sex chromosome and use the sperm to fertilize oocytes *in vitro*. However, sperm from men with XXY syndrome are more likely to have extra chromosomes—usually X or Y, but also autosomes—than sperm from men who do not have XXY syndrome.

XXYY Syndrome—Devon's Story

Devon's parents suspected early on that he was different. His problems were so common that it was years before a chromosome check revealed an extra X and an extra Y.

Devon was late to sit, crawl, walk, and talk. In preschool, he had frequent outbursts and made inappropriate comments. He was tall and clumsy, and drooled and choked easily. Devon would run about flapping his arms, then hide under a chair. Severe ulcers formed on his legs.

By the second grade, Devon's difficulties alarmed his special education teacher, who suggested to Devon's parents that they and their son have their chromosomes checked. The parents' chromosomes were normal. Devon must have been conceived from a very unusual oocyte meeting a very unusual sperm, both arising from nondisjunction. The extra sex chromosomes explained many of the boy's problems, and even a few that hadn't been recognized, such as curved pinkies, flat feet, and scoliosis. He began receiving testosterone injections so that his teen years would be more normal than his difficult childhood had been.

About 1 in 17,000 newborn boys have an extra X chromosome *and* an extra Y chromosome, and, until recently, were diagnosed with Klinefelter syndrome. These XXYY males have more severe behavioral problems than males with XXY syndrome and tend to develop foot and leg ulcers, resulting from poor venous circulation. Attention deficit disorder, obsessive compulsive disorder, autism, and learning disabilities typically develop by adolescence. In the teen years, testosterone level is low, development of secondary sexual characteristics is delayed, and the testes are undescended. A man with XXYY syndrome is infertile.

XYY Syndrome

In 1961, a tall, healthy man, known for his boisterous behavior, had a chromosome check after fathering a child with Down syndrome. The man had an extra Y chromosome. A few other cases were detected over the next several years.

In 1965, researcher Patricia Jacobs published results of a survey among 197 inmates at Carstairs, a high-security prison in Scotland. Of twelve men with unusual chromosomes, seven had an extra Y. Jacobs's findings were repeated for mental institutions, and soon after, Newsweek magazine ran a cover story on "congenital criminals," blaming such behavior on the extra Y chromosome. Having an extra Y, also known as Jacobs syndrome, became a legal defense for committing a violent crime.

In the early 1970s, newborn screens began in hospital nurseries in England, Canada, Denmark, and Boston. Social workers and psychologists visited XYY boys to offer "anticipatory guidance" to the parents on how to deal with their toddling future criminals. By 1974, geneticists and others halted the program, pointing out that singling out these boys on the basis of a few statistical studies was inviting self-fulfilling prophecy.

One male in 1,000 has an extra Y chromosome. Today, we know that 96 percent of XYY males are not destined to become criminals, but instead may be very tall and have acne. Problems with speech and understanding language are subtle. The mother of one young man who has an extra Y chromosome and is in middle school, but is intellectually at a third-grade level, explains his difficulties: *I have two older children and he was different from the beginning. He did not talk until speech therapy at age 5. He cannot tell a joke—he does not have the ability to relate information and put it together on his own, so a lot of his language is adopted from what he sees and hears. He is emotionally age 5 to 7, and he is almost 12 and just started to dress himself fully.*

The continued prevalence of XYY among mental-penal institution populations may be more psychological than biological. Large body size may lead teachers, employers, parents, and others to expect more of these people, and a few XYY individuals may deal with this stress aggressively.

XYY syndrome can arise from nondisjunction in the male, producing a sperm with two Y chromosomes that fertilizes an X-bearing oocyte. Geneticists have never observed a sex chromosome constitution of one Y and no X. Because the Y chromosome carries little genetic material, and the gene-packed X chromosome would not be present, the absence of so many genes makes development beyond a few cell divisions in a YO embryo impossible.

Key Concepts Questions 13.3

1. How many chromosomes are in a normal human karyotype?
2. What is a polyploid?
3. Explain how nondisjunction generates aneuploids.
4. Compare the severities of monosomies and trisomies, and of sex chromosome and autosomal aneuploidy.
5. Describe the most common aneuploid conditions.
6. Explain how a chromosomal mosaic arises.

13.4 Atypical Chromosome Structure

A chromosome can be structurally atypical in several ways. It may have too much genetic material, too little, or a stretch of DNA that is inverted or moved and inserted into a different type of chromosome (**figure 13.16**). Atypical chromosomes are balanced if they have the normal amount of genetic material or unbalanced if excess or deficient DNA results.

Deletions and Duplications

Ashley's Story

Ashley Naylor had $5p^-$ syndrome, in which part of the short arm of one copy of chromosome 5 is missing (*figure 13.17*). It is also called cri-du-chat syndrome because of the peculiar sound that children make.

Ashley defied her doctors' expectations. She learned to walk with the aid of a walker and express herself using sign language and a communication device. With early intervention and education, Ashley found the resources and additional encouragement she needed to succeed. She died shortly before her sixth birthday from pneumonia.

Deletions and **duplications** are missing and extra DNA sequences, respectively. They are types of copy number variants (CNVs), introduced in chapter 12. The more genes involved, the more severe the associated syndrome. **Figure 13.18** depicts a common duplication, of part of chromosome 15. Deletions and

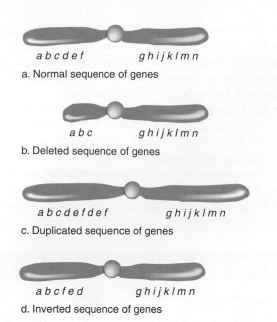

a b c d e f   g h i j k l m n

a. Normal sequence of genes

a b c   g h i j k l m n

b. Deleted sequence of genes

a b c d e f d e f   g h i j k l m n

c. Duplicated sequence of genes

a b c f e d   g h i j k l m n

d. Inverted sequence of genes

Figure 13.16 **Chromosome abnormalities.** If a hypothetical normal gene sequence appears as shown in **(a),** then **(b)** represents a deletion, **(c)** a duplication, and **(d)** an inversion.

Figure 13.17 **Ashley Naylor brought great joy to her family and community during her short life.** She had 5p⁻ syndrome.

Courtesy of Kathy Naylor.

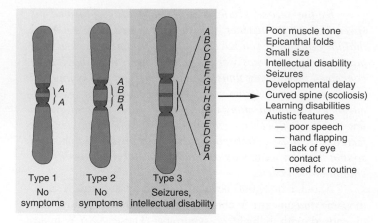

Type 1
No
symptoms

Type 2
No
symptoms

Type 3
Seizures,
intellectual disability

Poor muscle tone
Epicanthal folds
Small size
Intellectual disability
Seizures
Developmental delay
Curved spine (scoliosis)
Learning disabilities
Autistic features
  — poor speech
  — hand flapping
  — lack of eye
     contact
  — need for routine

Figure 13.18 **A duplication.** A study of duplications of parts of chromosome 15 revealed that small duplications do not affect the phenotype, but larger ones may. The letters indicate specific DNA sequences, which serve as markers to compare chromosome regions. Note that the duplication is also inverted.

duplications often arise "*de novo*," which means that neither parent has the abnormality, and it is therefore new.

A technique called **comparative genomic hybridization** is used to detect very small CNVs, which are also termed microdeletions and microduplications. The technique compares the abundance of copies of a particular CNV in the same amount of DNA from two people—one with a medical condition, one healthy. Comparative genomic hybridization is being used increasingly to help narrow down diagnoses for children with autism, intellectual disability, learning disabilities, or unusual behavior. For example, the technique showed that a young boy who had difficulty concentrating and sleeping and would often scream for no apparent reason had a small duplication in chromosome 7. A young girl plagued with head-banging behavior, digestive difficulties, severe constipation, and great sensitivity to sound had a microdeletion in chromosome 16. Other microdeletions cause male infertility.

Deletions and duplications can arise from chromosome rearrangements. These include translocations, inversions, and ring chromosomes.

Translocation Down Syndrome

Rhiannon's Story

When Rhiannon P. was born, while her parents marveled at her beauty, the obstetrician was disturbed by her facial features: the broad, tilted eyes and sunken nose looked like the face of a child with Down syndrome. The doctor might not have noticed, except that he knew that the mother, Felicia, had had two spontaneous abortions, a family history suggesting a chromosome problem. So the doctor looked for the telltale single crease in the palms of people with Down syndrome, and found it. Gently, he told Felicia and her husband Matt that he'd like to do a chromosome check.

Two days later, the new parents learned that their daughter has an unusual form of Down syndrome that she inherited from one of them, rather than the more common "extra chromosome" form. Since Matt's mother and sister had also had several miscarriages, the exchanged chromosomes likely came from his side. Karyotypes of Matt and Felicia confirmed this: Matt was a translocation carrier. One of his chromosome 14s had attached to one of his chromosome 21s, and distribution of the unusual chromosome in meiosis had led to various imbalances, depicted in *figure 13.19*.

Rhiannon has very mild Down syndrome. She has none of the physical problems associated with the condition, and she does well in school with the help of a special education teacher. Matt and Felicia chose to see the bright side—each conception will have a one-in-three chance of having balanced chromosomes. Some day they hope to give Rhiannon a brother or sister.

In a translocation, different (nonhomologous) chromosomes exchange or combine parts. Translocations can be inherited because they can be present in carriers, who have the normal amount of genetic material, but it is rearranged. A translocation can affect the phenotype if it breaks a gene or leads to duplications or deletions in the chromosomes of offspring.

There are two major types of translocations, and rarer types. In a **Robertsonian translocation,** the short arms of two different acrocentric chromosomes break, leaving sticky ends on the two long arms that join, forming a single, large chromosome with two long arms (chromosome 14/21 in figure 13.19). The tiny short arms are lost, but their DNA sequences are repeated elsewhere in the genome, so the loss does not cause symptoms. The person with the large, translocated chromosome, called a **translocation carrier,** has 45 chromosomes, but may not have symptoms if no crucial genes have been deleted or damaged. Even so, he or she may produce unbalanced gametes—sperm or oocytes with too many or too few genes. This can lead to spontaneous abortion or birth defects.

In 1 in 20 cases of Down syndrome, a parent has a Robertsonian translocation between chromosome 21 and another, usually chromosome 14. That parent produces some gametes that lack either of the involved chromosomes and some gametes that have extra material from one of the translocated chromosomes. In such a case, each fertilized ovum has a 1 in 2 chance

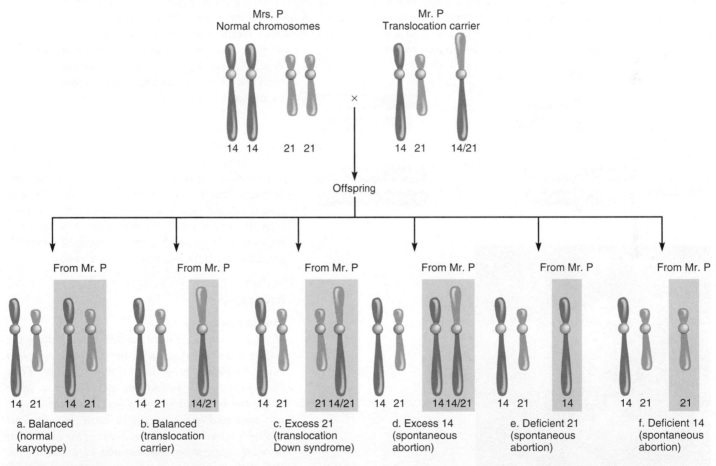

Figure 13.19 **A Robertsonian translocation.** Mr. P. has only 45 chromosomes because the long arm of one chromosome 14 has joined the long arm of one chromosome 21. He has no symptoms. Mr. P. makes six types of sperm cells, and they determine the fates of offspring. **(a)** A sperm with one normal chromosome 14 and one normal 21 yields a normal child. **(b)** A sperm carrying the translocated chromosome produces a child who is a translocation carrier, like Mr. P. **(c)** If a sperm contains Mr. P.'s normal 21 and his translocated chromosome, the child receives too much chromosome 21 material and has Down syndrome. **(d)** A sperm containing the translocated chromosome and a normal 14 leads to excess chromosome 14 material, which is lethal in the embryo or fetus. If a sperm lacks either chromosome 21 **(e)** or 14 **(f)**, it leads to monosomies, which are lethal prenatally. (Chromosome arm lengths are not precisely accurate.)

of ending in spontaneous abortion, and a 1 in 6 chance of developing into an individual with Down syndrome. The risk of giving birth to a child with Down syndrome is theoretically 1 in 3, because the spontaneous abortions are not births. However, because some Down syndrome fetuses spontaneously abort, the actual risk of a couple in this situation having a child with Down syndrome is about 15 percent. The other two outcomes—a fetus with normal chromosomes or a translocation carrier like the parent—have normal phenotypes. Either a male or a female can be a translocation carrier, and the condition is not related to age.

In a **reciprocal translocation,** the second major type of translocation, two different chromosomes exchange parts (**figure 13.20**). About 1 in 600 people is a carrier for a reciprocal translocation. FISH can be used to highlight the involved chromosomes. If the chromosome exchange does not break any genes, then a person who has both translocated chromosomes is healthy and a translocation carrier. He or she has the normal amount of genetic material, but it is rearranged. A reciprocal translocation carrier can have symptoms if one of the two breakpoints lies in a gene, disrupting its function. For example, a translocation between chromosomes 11 and 22 causes infertility in males and recurrent pregnancy loss in females. Sometimes, a *de novo* reciprocal translocation arises in a gamete. This can lead to a new individual with a disorder, if fertilization occurs, as opposed to inheriting a translocated chromosome from a parent who is a carrier.

Reciprocal translocations usually occur in specific chromosomes that have unstable parts. Vulnerable parts of chromosomes arise where the DNA is so symmetrical in sequence that complementary base pairing occurs within the same DNA strand, folding it into loops during DNA replication. These contortions can break both DNA strands, which can enable parts of two different chromosomes to switch places.

Unbalanced gametes can result from inheriting or originating a reciprocal translocation, just as they can from a Robertsonian translocation. The four possibilities are (1) transmitting two normal copies of the two involved chromosomes; (2) two abnormal copies, with no effect on the phenotype of an offspring unless a vital gene is disrupted; or (3 and 4) transmitting either translocated chromosome, which introduces extra or missing genetic material and likely would affect the phenotype.

A rare type of translocation is an insertional translocation, in which part of one chromosome inserts into a nonhomologous chromosome. Symptoms may result if the inserted DNA disrupts a vital gene or if crucial DNA sequences are lost or present in excess.

A carrier of any type of translocation can produce unbalanced gametes—sperm or oocytes that have deletions or duplications of some of the genes in the translocated chromosomes. The phenotype depends upon the genes that the rearrangement disrupts and whether genes are extra or missing. A translocation and a deletion can cause the same syndrome if they affect the same part of a chromosome.

A genetic counselor suspects a translocation when a family has a history of birth defects, pregnancy loss, and/or stillbirth. Prenatal testing may also reveal a translocation in a fetus, which can then be traced back to a parent who is a translocation carrier. If neither parent has the translocation in cells typically used to check chromosomes, then the translocation is *de novo* (new), or only some of a parent's gametes have it (that is, he or she is a mosaic).

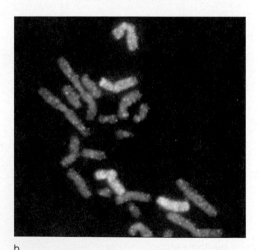

a.

b.

Figure 13.20 **A reciprocal translocation.** In a reciprocal translocation, two nonhomologous chromosomes exchange parts. In **(a)**, genes *C, D,* and *E* on the blue chromosome exchange positions with genes *M* and *N* on the red chromosome. Part **(b)** highlights a reciprocal translocation using FISH. The pink chromosome with the dab of blue, and the blue chromosome with a small section of pink, are the translocated chromosomes.

Inversions

Madison's Story

Madison and Grant were excited about getting the results of the amniocentesis because Madison had never carried a pregnancy this far before. Then the doctor's office called and asked them to come in for the results. Expecting bad news, the couple was surprised and confused to learn that the fetus had an inverted chromosome. Some of the bands that normally appear on chromosome 11 were flipped. Before the genetic counselor would describe which genes might be affected, she advised that the parents-to-be have their chromosomes checked. Karyotyping revealed that Madison had the same inversion as the fetus. Because Madison was healthy, the unusual chromosome would likely not harm their daughter. When she was older, however, she might, like her mother, experience pregnancy loss.

An inverted sequence of chromosome bands disrupts important genes and harms health in only 5 to 10 percent of cases. If neither parent has the inversion, then it arose in a gamete. This event might have occurred either *de novo* or because the ovary or testis is mosaic. The specific effects of an inverted chromosome may depend upon which genes the flip disrupts. Consulting the human genome sequence can help to identify genes that might be implicated in a particular inversion.

Like a translocation carrier, an adult who is heterozygous for an inversion can be healthy, but have reproductive problems. One woman had an inversion in the long arm of chromosome 15 and had two spontaneous abortions, two stillbirths, and two children who died within days of birth. She did eventually give birth to a healthy child. How did the inversion cause these problems?

Inversions with such devastating effects can be traced to meiosis, when a crossover occurs between the inverted chromosome segment and the noninverted homolog. To allow the genes to align, the inverted chromosome forms a loop. When crossovers occur within the loop, some areas are duplicated and some deleted in the resulting recombinant chromosomes. In inversions, the atypical chromosomes result from the chromatids that crossed over.

Two types of inversions are distinguished by the position of the centromere relative to the inverted section. **Figure 13.21** shows a **paracentric inversion**, which does not include the centromere. A single crossover within the inverted segment gives rise to one normal, one inversion, and two very atypical chromatids. One abnormal chromatid retains both centromeres and is termed dicentric. When the cell divides, the two centromeres are pulled to opposite sides of the cell, and the chromatid breaks, leaving pieces with extra or missing segments. The second type of atypical chromatid resulting from a crossover within an inversion loop is a small piece that lacks a centromere, called an acentric fragment. When the cell divides, the fragment is lost because a centromere is required for cell division.

A **pericentric inversion** includes the centromere within the loop. A crossover in it produces two chromatids that have duplications and deletions, but one centromere each, plus one normal and one inversion chromatid. (**figure 13.22**).

Isochromosomes and Ring Chromosomes

An **isochromosome** is the result of another meiotic error that leads to unbalanced genetic material. It is a chromosome that has identical arms. An isochromosome forms when, during division, the centromeres part in the wrong plane (**figure 13.23**). Isochromosomes are known for chromosomes 12 and 21 and for the long arms of the X and the Y. Some women with Turner syndrome are not the more common XO, but have an isochromosome with the long arm of the X chromosome duplicated but the short arm absent.

Chromosomes shaped like rings form in 1 out of 25,000 conceptions. Ring chromosomes

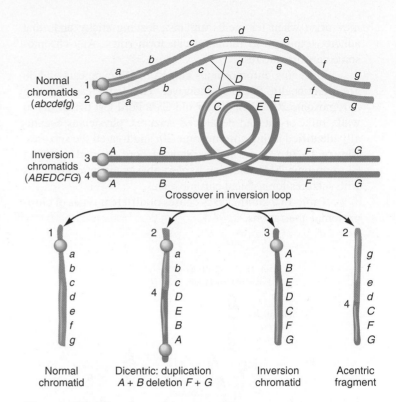

Figure 13.21 Paracentric inversion. A paracentric inversion in one chromosome leads to one normal chromatid, one inverted chromatid, one with two centromeres (dicentric), and one with no centromere (an acentric fragment) if a crossover occurs with the normal homolog. The letters *a* through *g* denote genes.

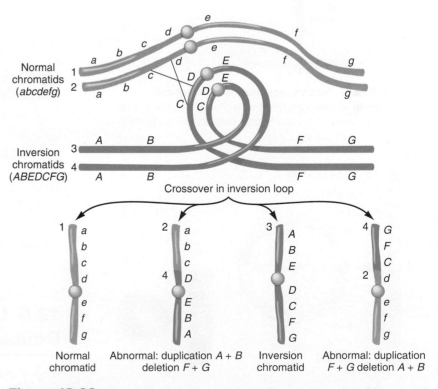

Figure 13.22 Pericentric inversion. A pericentric inversion in one chromosome leads to two chromatids with duplications and deletions, one normal chromatid, and one inverted chromatid if a crossover occurs with the normal homolog.

may arise when telomeres are lost, leaving sticky ends that adhere. Exposure to radiation can form rings. Any chromosome can circularize.

Most ring chromosomes consist of DNA repeats and do not affect health. Some do, however. This is the case for ring chromosome 20. When 6-year-old Cara Ford lost the ability to walk, talk, or eat and developed seizures, physicians eventually identified a ring chromosome 20, and treated the seizures, (**figure 13.24**). Cara's father, Stewart Ford, started the Ring Chromosome 20 Foundation to alert physicians to test patients with seizures for atypical chromosomes.

Table 13.7 summarizes causes of different types of chromosomal aberrations.

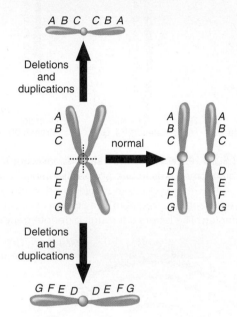

Figure 13.23 Isochromosomes have identical arms. They form when chromatids divide along the wrong plane (in this depiction, horizontally rather than vertically).

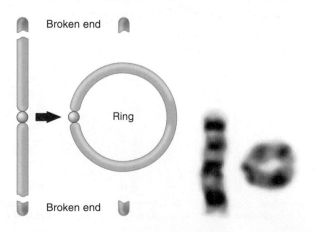

Figure 13.24 A ring chromosome. A ring chromosome may form if the chromosome's tips (telomeres) break, forming sticky ends. Genes can be lost or disrupted, possibly causing symptoms. Ring chromosome 20, for example, causes seizures.

Table 13.7	Causes of Chromosome Aberrations
Abnormalities	**Causes**
Numerical Abnormalities	
Polyploidy	Error in cell division (meiosis or mitosis) in which not all chromatid pairs separate in anaphase
	Multiple fertilization
Aneuploidy	Nondisjunction (in meiosis or mitosis) leading to lost or extra chromosomes
Structural Abnormalities	
Deletions and duplications	Translocation
	Crossover between a chromosome that has a pericentric inversion and its noninverted homolog
Translocation	Exchange between nonhomologous chromosomes
Inversion	Breakage and reunion of fragment in same chromosome, but with wrong orientation
Dicentric and acentric	Crossover between a chromosome with a paracentric inversion and its noninverted homolog
Ring chromosome	A chromosome loses telomeres and the ends fuse, forming a circle

Key Concepts Questions 13.4

1. Which chromosome rearrangements can cause deletions and duplications?
2. Distinguish between the two types of translocations.
3. What must occur for a translocation or inversion to cause symptoms?
4. How do isochromosomes and ring chromosomes arise?

13.5 Uniparental Disomy—A Double Dose from One Parent

If nondisjunction occurs in both a sperm and oocyte that join, a pair of chromosomes (or their parts) can come solely from one parent, rather than one from each parent, as Mendel's law of segregation predicts. For example, if a sperm that does not have a chromosome 14 fertilizes an ovum that has two chromosome 14s,

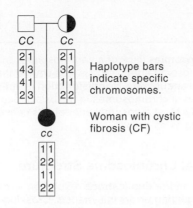

CC   Cc

2 1	2 1
4 3	3 2
4 1	1 1
2 3	2 2

Haplotype bars indicate specific chromosomes.

Woman with cystic fibrosis (CF)

cc

| 1 1 |
| 2 2 |
| 1 1 |
| 2 2 |

Figure 13.25 Uniparental disomy. Uniparental disomy doubles part of one parent's genetic contribution. In this family, the woman with CF inherited two copies of her mother's chromosome 7, and neither of her father's. Unfortunately, it was the chromosome with the disease-causing allele that she inherited in a double dose.

an individual with the normal 46 chromosomes results, but the two chromosome 14s come only from the female.

Inheriting two chromosomes or chromosome segments from one parent is called **uniparental disomy (UPD)** ("two bodies from one parent"). UPD can also arise from a trisomic embryo in which some cells lose the extra chromosome, leaving two homologs from one parent. For example, an embryo may have trisomy 21, with the extra chromosome 21 coming from the father. If in some cells the chromosome 21 from the mother is lost, then both remaining copies of the chromosome are from the father.

Because UPD requires the simultaneous occurrence of two very rare events—either nondisjunction of the same chromosome in sperm and oocyte, or trisomy followed by chromosome loss—it is very rare. In addition, many cases of UPD are probably never seen, because bringing together identical homologs inherited from one parent could give the fertilized ovum homozygous lethal alleles. Development would halt. Other cases of UPD may go undetected if they cause known recessive conditions and both parents are assumed to be carriers, when only one parent contributed to the offspring's illness. This was how UPD was discovered.

In 1988, Arthur Beaudet of the Baylor College of Medicine saw an unusual patient with cystic fibrosis (see Clinical Connection 4.2). In comparing *CFTR* alleles of the patient to those of her parents, Beaudet found that only the mother was a carrier—the father had two normal alleles. Beaudet constructed haplotypes for each parent's chromosome 7, which includes the *CFTR* gene, and he found that the daughter had two copies from her mother, and none from her father (**figure 13.25**). How did this happen?

Apparently, in the patient's mother, nondisjunction of chromosome 7 in meiosis II led to formation of an oocyte

bearing two identical copies of the chromosome, instead of the usual one copy. A sperm that had also undergone nondisjunction and did not have a chromosome 7 then fertilized the abnormal oocyte. The mother's extra genetic material compensated for the father's deficit, but unfortunately, the child inherited a double dose of the mother's chromosome that carried the mutant *CFTR* allele. In effect, inheriting two of the same chromosome from one parent shatters the protection that combining genetic material from two individuals offers. This protection is the defining characteristic of sexual reproduction.

UPD may also cause disease if it removes the contribution of the important parent for an imprinted gene. Recall from chapter 6 that an imprinted gene is expressed if it comes from one parent, but silenced if it comes from the other (see figure 6.13). If UPD removes the parental genetic material that must be present for a critical gene to be expressed, a mutant phenotype results. The classic example of UPD disrupting imprinting is 20 to 30 percent of Prader-Willi syndrome and Angelman syndrome cases (see figure 6.15). These disorders arise from mutations in different genes that are closely linked in a region of the long arm of chromosome 15, where imprinting occurs. They both cause intellectual disability and a variety of other symptoms, but are quite distinct.

Some children with Prader-Willi syndrome have two parts of the long arm of chromosome 15 from their mothers. The disease results because the father's Prader-Willi gene must be expressed for the child to avoid the associated illness. For Angelman syndrome, the situation is reversed. Children have a double dose of their father's DNA in the same chromosomal region implicated in Prader-Willi syndrome, with no maternal contribution. The mother's gene must be present for health.

People usually learn their chromosomal makeup only when something goes wrong—when they have a family history of reproductive problems, exposure to a toxin, cancer, or symptoms of a known chromosomal disorder. While researchers analyze human genome sequences, chromosome studies will continue to be part of medical care—beginning before birth.

Key Concepts Questions 13.5

1. What is uniparental disomy?
2. What are two ways that uniparental disomy can arise?
3. What are two ways that uniparental disomy can cause disease?

Summary

13.1 Portrait of a Chromosome

1. **Cytogenetics** is the study of chromosome aberrations and their effects on phenotypes.

2. **Heterochromatin** stains darkly and harbors many DNA repeats. **Euchromatin** is light staining and contains many protein-encoding genes.

3. A chromosome consists of DNA and proteins. Essential parts are the **telomeres, centromeres,** and origin of replication sites.

4. Centromeres include DNA repeats and proteins that enable the cell to divide.

5. Subtelomeres have telomere-like repeats that gradually change inward toward the centromere, as protein-encoding genes predominate.

6. Chromosomes are distinguishable by size, centromere position, satellites, and staining patterns. They are displayed in **karyotypes**.

7. A **metacentric** chromosome has two fairly equal arms. A **submetacentric** chromosome has a large arm and a short arm. An **acrocentric** chromosome's centromere is near a tip, so that it has one long arm and one very short arm.

13.2 Detecting Chromosomes

8. **Amniocentesis** and **chorionic villus sampling (CVS)** directly obtain and analyze DNA from fetal cells or their supporting structures.

9. Fluorescence *in situ* hybridization (FISH) provides more specific chromosome bands than dyes. Ideograms display chromosome bands.

10. Chromosomal shorthand indicates chromosome number, sex chromosome constitution, and type of abnormality.

11. Maternal serum markers and ultrasound findings are used to screen for elevated risk of a **trisomy** in a fetus.

12. Cell-free fetal DNA is used in noninvasive prenatal diagnosis of trisomies.

13.3 Atypical Chromosome Number

13. A **euploid** somatic human cell has 22 pairs of autosomes and one pair of sex chromosomes.

14. **Polyploid** cells have extra chromosome sets.

15. **Aneuploids** have extra or missing chromosomes. **Trisomies** (an extra chromosome) are less harmful than **monosomies** (lack of a chromosome), and sex chromosome aneuploidy is less severe than autosomal aneuploidy. **Nondisjunction** is uneven distribution of chromosomes in meiosis. It causes aneuploidy. Most autosomal aneuploids cease developing as embryos.

13.4 Atypical Chromosome Structure

16. **Deletions** and/or **duplications** result from crossing over after pairing errors in synapsis. Crossing over in an inversion heterozygote can also generate deletions and duplications. Microdeletions and microduplications explain many disorders.

17. In a **Robertsonian translocation,** the short arms of two acrocentric chromosomes break, leaving sticky ends on the long arms that join to form an unusual, large chromosome.

18. In a **reciprocal translocation,** two nonhomologous chromosomes exchange parts.

19. An insertional translocation places a DNA sequence from one chromosome into a nonhomologous chromosome.

20. A **translocation carrier** may have an associated phenotype and produces some unbalanced gametes.

21. A heterozygote for an inversion may have reproductive problems if a crossover occurs between the inverted region and the noninverted homolog, generating deletions and duplications. A **paracentric inversion** does not include the centromere; a **pericentric inversion** does.

22. **Isochromosomes** repeat one chromosome arm but delete the other. They form when the centromere divides in the wrong plane in meiosis. Ring chromosomes form when telomeres are removed, leaving sticky ends that adhere.

13.5 Uniparental Disomy—A Double Dose from One Parent

23. In **uniparental disomy,** a chromosome or part of one doubly represents one parent. It can result from nondisjunction in both gametes, or from a trisomic cell that loses a chromosome.

24. Uniparental disomy causes symptoms if it creates a homozygous recessive state associated with an illness, or if it affects an imprinted gene.

www.mhhe.com/lewisgenetics11

Answers to all end-of-chapter questions can be found at **www.mhhe.com/lewisgenetics11**. You will also find additional practice quizzes, animations, videos, and vocabulary flashcards to help you master the material in this chapter.

Review Questions

1. Identify the structures and/or DNA sequences that must be present for a chromosome to carry information and withstand the forces of cell division.

2. How does the DNA sequence change with distance from the telomere?

3. Distinguish among a euploid, aneuploid, and polyploid.

4. Explain how analysis of fetal DNA in the maternal circulation can provide information on a trisomy.

5. How are amniocentesis and CVS alike yet different?

6. Distinguish between a genetic screen and a diagnostic test for a trisomy in a fetus.

7. What happens during meiosis to produce
   a. an aneuploid?
   b. a polyploid?
   c. increased risk of trisomy 21 in the offspring of a woman over age 40 at the time of conception?
   d. recurrent spontaneous abortions to a couple in which the man has a pericentric inversion?
   e. several children with Down syndrome in a family where one parent is a translocation carrier?

8. A human liver has patches of octaploid cells—they have eight sets of chromosomes. Explain how this might arise.

9. Describe an individual with each of the following chromosomes. List gender and possible phenotype.
   a. 47,XXX
   b. 45,X
   c. 47,XX, trisomy 21
   d. XXYY

10. Which chromosomal anomaly might you expect to find more frequently among members of the National Basketball Association than in the general public? Cite a reason.

11. Explain the difference between a fetus found to have a trisomy and one found to be triploid.

12. About 80 percent of cases of Edward syndrome are caused by trisomy 18; 10 percent are caused by mosaic trisomy 18, and 10 percent are attributed to translocation. Distinguish among these three chromosome aberrations.

13. List three examples illustrating the idea that the amount of genetic material involved in a chromosomal aberration affects the severity of the associated phenotype.

14. List three types of chromosomal aberrations that can cause duplications and/or deletions, and explain how they do so.

15. Distinguish among three types of translocations.

16. Why would having the same inversion on both members of a homologous chromosome pair *not* lead to unbalanced gametes, as having only one inverted chromosome would?

17. Define or describe the following technologies:
   a. FISH
   b. amniocentesis
   c. chorionic villus sampling
   d. cell-free fetal DNA analysis

18. How many chromosomes would a person have who has Klinefelter syndrome and also trisomy 21?

19. Explain why a female cannot have XXY syndrome and a male cannot have XO syndrome.

20. List three causes of Turner syndrome.

21. Explain how the sequence of genes on an isochromosome differs from normal.

Applied Questions

1. Describe the evidence that a fetus has trisomy 21 that each of the following procedures or technologies provides:
   a. amniocentesis
   b. chorionic villus sampling
   c. maternal serum markers
   d. cell-free fetal DNA

2. Amniocentesis indicates that a fetus has the chromosomal constitution 46, XX,del(5)(p15). What does this mean? What might the child's phenotype be?

3. In a college genetics laboratory course, a healthy student constructs a karyotype from a cell from the inside of her cheek. She finds only one chromosome 3 and one chromosome 21, plus two unusual chromosomes that do not seem to have matching partners.
   a. What type of chromosomal abnormality does she have?
   b. Why doesn't she have any symptoms?
   c. Would you expect any of her relatives to have any particular medical problems?

4. A fetus ceases developing in the uterus. Several of its cells are karyotyped. Approximately 75 percent of the cells are diploid, and 25 percent are tetraploid (four copies of each chromosome). What happened, and when in development did it probably occur?

5. Distinguish among Down syndrome caused by aneuploidy, mosaicism, and translocation.

6. A couple has a son diagnosed with XXY syndrome. Explain how the son's chromosome constitution could have arisen from either parent.

7. DiGeorge syndrome (OMIM 188400) causes atypical parathyroid glands, a heart defect, and an

underdeveloped thymus gland. About 85 percent of patients have a microdeletion of part of chromosome 22. A girl, her mother, and a maternal aunt have very mild DiGeorge syndrome. They all have a reciprocal translocation of chromosomes 22 and 2.

   a. How can a microdeletion and a translocation cause the same symptoms?

   b. Why were the people with the translocation less severely affected than the people with the microdeletion?

   c. What other problems might arise in the family with the translocation?

8. From 2 to 6 percent of people with autism have an extra chromosome that consists of two long arms of chromosome 15. It includes two copies of the chromosome 15 centromere. Two normal copies of the chromosome are also present. What type of chromosome abnormality in a gamete can lead to this karyotype, which is called isodicentric 15?

9. Consider a science fiction plot. How could Robertsonian translocations, which occur in 1 in 1,000 people, lead to formation of a new species?

Web Activities

1. Go to one of the websites listed in table 13.4, or find a similar disease organization, and learn about daily life with a particular chromosomal abnormality. Identify a challenge or problem common to several chromosomal syndromes, and describe how families cope with the problem.

2. Go to the website for the Genetic Science Learning Center at the Eccles Institute of Human Genetics at the University of Utah. Follow the instructions to create a karyotype.

3. Visit the website for the Human Genome Landmarks poster. Select a chromosome, and use Mendelian Inheritance in Man (OMIM) to describe four traits or disorders associated with it. Or, consult the website for the Human Chromosome Launchpad for information on four genes carried on a specific chromosome.

Case Studies and Research Results

1. An ultrasound of a pregnant woman detects a fetus and a similarly sized and shaped structure that has disorganized remnants of facial features at one end. Amniocentesis on both structures reveals that the fetus is 46,XX, but cells of the other structure are 47,XX,trisomy 2. No cases of trisomy 2 infants have ever been reported. However, individuals who are mosaics for trisomy 2 have a collection of defects, including a rotated and underdeveloped small intestine, a small head, a hole in the diaphragm, and seizures.

   a. How do the chromosomes of cells from the fetus and the other structure differ?

   b. What is the process that occurred during meiosis to yield the bizarre structure?

   c. Use this information to explain why children with a complete extra chromosome 2 are not seen, even though people with an extra chromosome 21 can live many years.

   d. List two factors that determine the type and severity of abnormalities in an individual who is an aneuploid mosaic.

2. Two sets of parents who have children with Down syndrome meet at a clinic. The Phelps know that their son has trisomy 21. The Watkins have two affected children, and Mrs. Watkins has had two spontaneous abortions. Why should the Watkins be more concerned about future reproductive problems than the Phelps? How are the offspring of the two families different, even though they have the same symptoms?

3. The genomes of four of 291 people with intellectual disability have a microdeletion in chromosome 17q21.3. The children have large noses, delayed speech, and mild intellectual disability. Each had a parent with an inversion in the same part of chromosome 17.

   a. Which arm of chromosome 17 is implicated in this syndrome?

   b. How can an inversion in a parent's chromosome cause a deletion in a child's chromosome?

   c. What other type of chromosome abnormality might occur in these children's siblings?

4. A 38-year-old woman, Dasheen, has amniocentesis. She learns that the fetus she is carrying has an inversion in chromosome 9 and a duplication in chromosome 18. She and her husband Franco have their chromosomes tested, and they learn that she has the duplication and Franco has the inversion. Both of the parents are healthy. Should they be concerned about the health of the fetus? Cite a reason for your answer.

5. What is a difficulty of applying the research technique to silence the extra chromosome 21 that causes trisomy 21 Down syndrome for use in treating the condition?

PART **4** Population Genetics

A forensic scientist consults a DNA profile. The black bars represent short tandem repeats that form patterns used to exclude suspects in a crime.

CHAPTER

14

Constant Allele Frequencies

Learning Outcomes

14.1 Population Genetics Underlies Evolution

1. State the unit of information of genetics at the population level.

2. Define *gene pool.*

3. List the five processes that cause microevolutionary change.

4. State the consequence of macroevolutionary change.

14.2 Constant Allele Frequencies

5. State the genotypes represented in each part of the Hardy-Weinberg equation.

6. Explain the conditions necessary for Hardy-Weinberg equilibrium.

14.3 Applying Hardy-Weinberg Equilibrium

7. Explain how the Hardy-Weinberg equilibrium uses population incidence statistics to predict the probability of a particular phenotype.

14.4 DNA Profiling Uses Hardy-Weinberg Assumptions

8. Explain how parts of the genome that are in Hardy-Weinberg equilibrium can be used to identify individuals.

The **BIG** Picture

Human genetics at the population level considers allele frequencies. Some parts of the genome that have changed over time enable us to trace our origins, migrations, and relationships. Allele frequencies that do not change in response to environmental factors provide a way to distinguish individuals.

Postconviction DNA Testing

Josiah Sutton had served 4 1/2 years of a 25-year sentence for rape when he was exonerated, thanks to the Innocence Project. The nonprofit legal clinic and public policy organization created in 1992 has used DNA retesting to free more than 312 wrongfully convicted prisoners, most of whom were "poor, forgotten, and have used up all legal avenues for relief," according to the website (www .innocenceproject.com). Sutton became a suspect after a woman in Houston identified him and a friend 5 days after she had been raped, threatened with a gun, and left in a field. The two young men supplied saliva and blood samples, from which DNA profiles were done and compared to DNA profiles from semen found in the victim and in her car. At the trial, a crime lab employee testified that the probability that Sutton's DNA matched that of the evidence by chance was 1 in 694,000, leading to a conviction. Jurors ignored the fact that Sutton's physical description did not match the victim's description of her assailant.

The DNA evidence came from more than one individual, yielded different results when the testing was repeated, and most importantly, looked at only seven of the parts of the genome that are typically compared in a DNA profile, or fingerprint. Doing the test correctly and considering more forensic "alleles" revised the statistics dramatically: Sutton's pattern was shared not with 1 in 694,000 black men, as had originally been claimed, but with 1 in 16.

While in jail, Sutton read about DNA profiling and requested independent testing, but was refused. Then journalists investigating the Houston crime laboratory learned of his case and alerted the

14.1 Population Genetics Underlies Evolution

The language of genetics at the family and individual levels is written in DNA sequences. At the population level, the language of genetics is allele (gene variant) frequencies. It is at the population level that genetics goes beyond science, embracing information from history, anthropology, human behavior, and sociology. Population genetics enables us to trace our beginnings, understand our diversity today, and imagine the future.

A biological **population** is any group of members of the same species in a given geographical area that can mate and produce fertile offspring (**figure 14.1**). A population in a sociological sense may be more restrictive, such as ethnic groups or economic strata. **Population genetics** is a branch of genetics that considers all the alleles of all the genes in a biological population, which constitute the **gene pool**. The "pool" refers to the collection of gametes in the population; an offspring represents two gametes from the pool. Alleles can move between populations when individuals migrate and mate. This movement, termed gene flow, underlies evolution, which is explored in the next two chapters.

Thinking about genes at the population level begins by considering frequencies—that is, how often a particular gene variant occurs in a particular population. Such frequencies can be calculated for alleles, genotypes, or phenotypes, and may include single base mutations or numbers of short, repeated DNA sequences. For example, an allele frequency for the cystic fibrosis gene (*CFTR*) might be the number of $\Delta F508$ alleles among the residents of San Francisco. $\Delta F508$ is the most common allele that, when homozygous, causes the disorder. The allele frequency derives from the two $\Delta F508$ alleles in each person with CF, plus alleles carried in heterozygotes, considered as a proportion of all alleles for that gene in the gene pool of San Francisco. The genotype frequencies are the proportions of heterozygotes and the two types of homozygotes in the population. Finally, a phenotypic frequency is simply the percentage of people in the population who have CF (or who do not). With multiple alleles for a single gene, the situation becomes more complex because there are many more phenotypes and genotypes to consider.

Phenotypic frequencies are determined empirically—that is, by observing how common a condition or trait is in a population. Genetic counselors use phenotype frequency to estimate the risk that a particular inherited disorder will occur in an individual when there is no family history of the illness. **Table 14.1** shows disease incidence for phenylketonuria (PKU), an inborn error of metabolism that causes intellectual disability unless the person follows a special, low-protein diet from birth. Note how the frequency differs in different populations. A person from Turkey without a family history of PKU would have a higher risk of having an affected child than a person from Japan.

On a broader level, shifting allele frequencies in populations reflect small steps of genetic change, called **microevolution**. These small, step-by-step changes alter genotype frequencies and underlie evolution. Genotype frequencies rarely stay constant. They can change under any of the following conditions:

1. Individuals of one genotype are more likely to choose to reproduce with each other than with individuals of other genotypes (*nonrandom mating*).
2. Individuals move between populations (*migration*).
3. Random sampling of gametes alters allele frequencies (*genetic drift*).

Figure 14.1 **A biological population is a group of interbreeding organisms living in the same place.** Populations of sexually reproducing organisms include many genetic variants. This genetic diversity gives the group a flexibility that enhances species survival. To us, these hippos look alike, but they can undoubtedly recognize phenotypic differences in each other.

Table 14.1	Frequency of PKU in Various Populations	
Population		**Frequency of PKU**
Chinese		1/16,000
Irish, Scottish, Yemenite Jews		1/5,000
Japanese		1/119,000
Swedes		1/30,000
Turks		1/2,600
U.S. Caucasians		1/10,000

4. New alleles arise (*mutation*).
5. People with a particular genotype are more likely to produce viable, fertile offspring under a specific environmental condition than individuals with other genotypes (*natural selection*).

In today's world, all of these conditions, except mutation, are quite common. Therefore, genetic equilibrium—when allele frequencies are *not* changing—is rare. Put another way, given our tendency to pick our own partners and move about, microevolution is not only possible, but also nearly unavoidable. (Chapter 15 considers these factors in depth.)

When sufficient microevolutionary changes accumulate to keep two fertile organisms of opposite sex from producing fertile offspring together, a new species forms. Changes that are great enough to result in speciation are termed **macroevolution**. Speciation can occur through many small changes over time, and/or a few changes that greatly affect the phenotype.

Key Concepts Questions 14.1

1. What is a biological population?
2. What is a gene pool?
3. What are microevolution and macroevolution?
4. What are the five factors that can change genotype frequencies?

14.2 Constant Allele Frequencies

Before we consider the pervasive genetic evidence for evolution, this chapter discusses the interesting, but unusual, situation in which frequencies for certain alleles stay constant. This is a condition called **Hardy-Weinberg equilibrium**.

Hardy-Weinberg Equilibrium

In 1908, Cambridge University mathematician Godfrey Harold Hardy (1877–1947) and Wilhelm Weinberg (1862–1937), a German physician interested in genetics, independently used algebra to explain how allele frequencies can be used to predict phenotypic and genotypic frequencies in populations of diploid, sexually reproducing organisms.

Hardy unintentionally cofounded the field of population genetics with a simple letter published in the journal *Science*. The letter began with a curious mix of modesty and condescension:

> *I am reluctant to intrude in a discussion concerning matters of which I have no expert knowledge, and I should have expected the very simple point which I wish to make to have been familiar to biologists.*

Hardy continued to explain how mathematically inept biologists had incorrectly deduced from Mendel's work that dominant traits would increase in populations while recessive traits would become rarer. At first glance, this seems logical. However, it is untrue because recessive alleles enter populations by mutation or migration and are maintained in heterozygotes. Recessive alleles also become more common when they confer a reproductive advantage, thanks to natural selection.

Hardy and Weinberg disproved the assumption that dominant traits increase while recessive traits decrease using the language of algebra. The expression of population genetics in algebraic terms begins with the simple equation

$$p + q = 1.0$$

where p represents the frequency of all dominant alleles for a gene and q represents the frequency of all recessive alleles. The expression "$p + q = 1.0$" means that all the dominant alleles and all the recessive alleles comprise all the alleles for that gene in a population.

Next, Hardy and Weinberg described the possible genotypes for a gene with two alleles using the binomial expansion

$$p^2 + 2pq + q^2 = 1.0$$

In this equation, p^2 represents the proportion of homozygous dominant individuals, q^2 represents the proportion of homozygous recessive individuals, and $2pq$ represents the proportion of heterozygotes (**figure 14.2**). The letter p designates the frequency of a dominant allele, and q is the frequency of a recessive allele. **Figure 14.3** shows how the binomial expansion is derived from allele frequencies. Note that the derivation is conceptually the same as tracing alleles in a monohybrid cross, in which the heterozygote forms in two ways: a from the mother and A from the father, or vice versa (see figure 4.4).

The binomial expansion used to describe genes in populations became known as the Hardy-Weinberg equation. It can reveal the changes in allele frequency that underlie evolution.

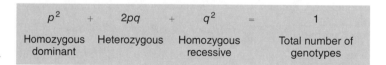

p^2	+	$2pq$	+	q^2	=	1
Homozygous dominant		Heterozygous		Homozygous recessive		Total number of genotypes

Figure 14.2 **The Hardy-Weinberg equation in English.**

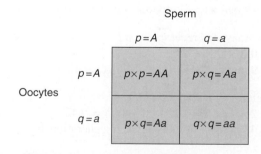

Figure 14.3 **Source of the Hardy-Weinberg equation.** A variation on a Punnett square reveals how random mating in a population in which gene *A* has two alleles—*A* and *a*—generates genotypes *aa*, *AA*, and *Aa*, in the relationship $p^2 + 2pq + q^2$.

If the proportion of genotypes remains the same from generation to generation, as the equation indicates, then that gene is not evolving (changing). This situation, Hardy-Weinberg equilibrium, is theoretical. It happens only if the population is large and undisturbed. That is, its members mate at random, do not migrate, and there is no genetic drift, mutation, or natural selection.

Hardy-Weinberg equilibrium is rare for protein-encoding genes that affect the phenotype, because an organism's appearance and health affect its ability to reproduce. Harmful allele combinations are weeded out of the population while helpful ones are passed on. Hardy-Weinberg equilibrium *is* seen in many DNA sequences that do not affect the phenotype. However, if a gene in Hardy-Weinberg equilibrium is closely linked on its chromosome to a gene that is subject to natural selection, it may be pulled from equilibrium by being inherited along with the selected gene. This bystander effect is called **linkage disequilibrium** (see figure 16.13).

Solving a Problem Using the Hardy-Weinberg Equation

To understand Hardy-Weinberg equilibrium, it helps to follow the frequency of two alleles of a gene from one generation to the next. Mendel's laws underlie such calculations.

Consider an autosomal recessive trait: a middle finger shorter than the second and fourth fingers. If we know the frequencies of the dominant and recessive alleles, then we can calculate the frequencies of the genotypes and phenotypes and trace the trait through the next generation. The dominant allele D confers normal-length fingers; the recessive allele d confers a short middle finger (**figure 14.4**). We can deduce the frequencies of the dominant and recessive alleles by observing the frequency of homozygous recessives, because this phenotype—short finger—reflects only one genotype. If 9 out of 100 individuals in a population have short fingers—genotype dd—the frequency is 9/100 or 0.09. Since dd equals q^2, then q equals 0.3. Since $p + q = 1.0$, knowing that q is 0.3 tells us that p is 0.7.

Next, we can calculate the proportions of the three genotypes that arise when gametes combine at random:

Homozygous dominant = DD
$= 0.7 \times 0.7 = 0.49$
$= 49$ percent of individuals in generation 1

Homozygous recessive = dd
$= 0.3 \times 0.3 = 0.09$
$= 9$ percent of individuals in generation 1

Heterozygous = $Dd + dD$
$= 2pq = (0.7)(0.3) + (0.3)(0.7) = 0.42$
$= 42$ percent of individuals in generation 1

The proportion of homozygous individuals is calculated by multiplying the allele frequency for the recessive or dominant allele by itself. The heterozygous calculation is $2pq$ because D can combine with d in two ways—a D sperm with a d egg, and a d sperm with a D egg.

In this population, 9 percent of the individuals have a short middle finger. Now jump ahead a few generations, and assume that people choose mates irrespective of finger length. This means that each genotype of a female (DD, Dd, or dd) is equally likely to mate with each of the three types of males (DD, Dd, or dd), and vice versa. **Table 14.2** multiplies the genotype frequencies for each possible mating, which leads to offspring in the familiar proportions of 49 percent DD, 42 percent Dd, and 9 percent dd. This gene, therefore, is in Hardy-Weinberg equilibrium—the allele and genotype frequencies do not change from one generation to the next.

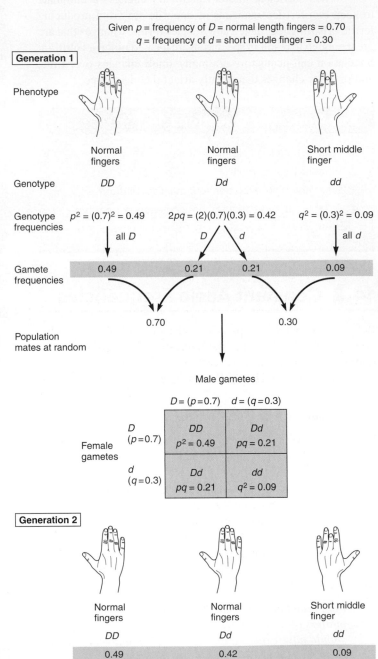

Figure 14.4 Hardy-Weinberg equilibrium. In Hardy-Weinberg equilibrium, allele frequencies remain constant from one generation to the next.

Table 14.2	Hardy-Weinberg Equilibrium—When Allele Frequencies Stay Constant				
POSSIBLE MATINGS			**FREQUENCY OF OFFSPRING GENOTYPES**		
Male	Female	Proportion in Population	*DD*	*Dd*	*dd*
0.49 *DD*	0.49 *DD*	0.2401 (*DD* × *DD*)	0.2401		
0.49 *DD*	0.42 *Dd*	0.2058 (*DD* × *Dd*)	0.1029	0.1029	
0.49 *DD*	0.09 *dd*	0.0441 (*DD* × *dd*)		0.0441	
0.42 *Dd*	0.49 *DD*	0.2058 (*Dd* × *DD*)	0.1029	0.1029	
0.42 *Dd*	0.42 *Dd*	0.1764 (*Dd* × *Dd*)	0.0441	0.0882	0.0441
0.42 *Dd*	0.09 *dd*	0.0378 (*Dd* × *dd*)		0.0189	0.0189
0.09 *dd*	0.49 *DD*	0.0441 (*dd* × *DD*)		0.0441	
0.09 *dd*	0.42 *Dd*	0.0378 (*dd* × *Dd*)		0.0189	0.0189
0.09 *dd*	0.09 *dd*	0.0081 (*dd* × *dd*)			0.0081
		Resulting offspring frequencies:	0.49	0.42	0.09
			DD	*Dd*	*dd*

Key Concepts Questions 14.2

1. What does the Hardy-Weinberg equation describe?
2. Explain the components of the Hardy-Weinberg equation.
3. Why aren't all genes under Hardy-Weinberg equilibrium?

14.3 Applying Hardy-Weinberg Equilibrium

A young woman pregnant for the first time learns that her health care providers offer all patients a blood test to see if they are carriers for cystic fibrosis. The test detects several dozen of the most common CF mutations. The woman had never thought about CF. She and her partner are from the same population (Caucasians of European descent) and neither has a family history of the disease. What is her risk of being a carrier? If it is not high, perhaps she will not take the test. The Hardy-Weinberg equation can answer the patient's question by determining the probability that she and her partner are carriers for CF. If they are carriers, then Mendel's first law predicts that the risk that an offspring inherits the disease is 1 in 4, or 25%. The couple might then opt for prenatal testing.

To derive carrier risks, the Hardy-Weinberg equation is applied to population statistics on genetic disease incidence. To determine allele frequencies for autosomal recessively

inherited characteristics, we need to know the frequency of one genotype in the population. This is typically the homozygous recessive class, because its phenotype indicates its genotype.

The incidence (frequency) of an autosomal recessive disorder in a population is used to help calculate the risk that a particular person is a heterozygote. Returning to the example of CF, the incidence of the disease, and therefore also of carriers, may vary greatly in different populations (**table 14.3**).

CF affects 1 in 2,000 Caucasian newborns. Therefore, the homozygous recessive frequency—*cc* if *c* represents the disease-causing allele—is 1/2,000, or 0.0005 in the population. This equals q^2. The square root of q^2 is about 0.022, which equals the frequency of the *c* allele. If *q* equals 0.022, then *p*, or $1 - q$, equals 0.978. Carrier frequency is equal to $2pq$, which equals (2)(0.978)(0.022), or 0.043—about 1 in 23. **Figure 14.5** summarizes these calculations.

Because there is no CF in the woman's family, her risk of having an affected child, based on population statistics, is low.

Table 14.3	Carrier Frequency for Cystic Fibrosis
Population Group	**Carrier Frequency**
African Americans	1 in 66
Asian Americans	1 in 150
Caucasians of European descent	1 in 23
Hispanic Americans	1 in 46

Cystic Fibrosis

incidence (autosomal recessive class) = 1/2,000 = 0.0005

$\therefore q^2 = 0.0005$

$\therefore q = \sqrt{0.0005} = 0.022$

$\therefore p = 1 - q = 1 - 0.022 = 0.978$

$\therefore$ carrier frequency = $2pq$ = (2) (0.978) (0.022) = 0.043 = 1/23

Figure 14.5 Calculating the carrier frequency given population incidence: Autosomal recessive.

The chance of *each* potential parent being a carrier is about 4.3 percent, or 1 in 23. The chance that *both* are carriers is 1/23 multiplied by 1/23—or 1 in 529—because the probability that two independent events will occur equals the product of the probability that each event will happen alone (the product rule). However, if they *are* both carriers, each of their children would face a 1 in 4 chance of inheriting the illness, based on Mendel's first law of gene segregation. Therefore, the risk that these two unrelated Caucasian individuals with no family history of CF will have an affected child is $1/4 \times 1/23 \times 1/23$, or 1 in 2,116. The woman takes the test, but is much less worried than she was when she first learned it was an option.

For X-linked traits, different predictions of allele frequencies apply to males and females. For a female, who can be homozygous recessive, homozygous dominant, or a heterozygote, the standard Hardy-Weinberg equation of $p^2 + 2pq + q^2$ applies. However, in males, the allele frequency is the phenotypic frequency, because a male who inherits an X-linked recessive allele exhibits it in his phenotype.

The incidence of X-linked hemophilia A, for example, is 1 in 10,000 male (X^hY) births. Therefore, q (the frequency of the h allele) equals 0.0001. Using the formula $p + q = 1$, the frequency of the wild type allele is 0.9999. The incidence of carriers (X^HX^h), who are all female, equals $2pq$, or (2)(0.0001)(0.9999), which equals 0.00019; this is 0.0002, or 0.02 percent, which equals about 1 in 5,000. The incidence of a female having hemophilia A (X^hX^h) is q^2, or $(0.0001)^2$. This is about 1 in 100 million. **Figure 14.6** summarizes these calculations.

Neat allele frequencies such as 0.6 and 0.4, or 0.7 and 0.3, are unusual. In actuality, single-gene disorders are very rare, and so the q component of the Hardy-Weinberg equation contributes little. Because this means that the value of p approaches 1, the carrier frequency, $2pq$, is very close to $2q$. Thus, the carrier frequency is approximately twice the frequency of the rare, disease-causing allele.

Consider Tay-Sachs disease, which occurs in 1 in 3,600 Ashkenazim (Jewish people of Eastern European descent). This means that q^2 equals 1/3,600, or about 0.0003. The square root, q, equals 0.017. The frequency of the dominant allele (p) is then $1 - 0.017$, or 0.983. What is the likelihood that an Ashkenazi carries Tay-Sachs disease? It is $2pq$, or (2)(0.983)(0.017), or 0.033. This is very close to double the frequency of the mutant allele (q), 0.017. Modifications of the Hardy-Weinberg equation are used to analyze genes that have more than two alleles.

Hemophilia A

incidence = 1/10,000 male births = 0.0001

$\therefore q = 0.0001$

$\therefore p = 1 - q \quad 1 - 0.0001 = 0.9999$

$\therefore$ carrier frequency (females) = $2pq$ = (2) (0.9999) (0.0001) = 0.00019 = about 1/5,000

$\therefore$ affected females = q^2 = (0.0001) (0.0001) = 1/100 million

Figure 14.6 Calculating the carrier frequency given population incidence: X-linked recessive.

Key Concepts Questions 14.3

1. How are allele frequencies in populations inferred?
2. How does the Hardy-Weinberg equation apply to X-linked traits?
3. What is the approximate carrier frequency for very rare disorders?

14.4 DNA Profiling Uses Hardy-Weinberg Assumptions

Hardy-Weinberg equilibrium is useful in understanding the conditions necessary for evolution to occur, but in a very practical sense it provides the foundation for **DNA profiling,** described in the chapter opener. The connection between the theoretical and the practical is that many parts of the genome that do not affect the phenotype, such as some short repeated sequences that do not encode amino acids and are not closely linked to genes that do, are in Hardy-Weinberg equilibrium. Variability in these sequences can be used to identify individuals if the frequencies are known in particular populations, and several sites in the genome are considered at the same time and the product rule applied.

In DNA profiling, the number of copies of a repeated sequence is considered to be an allele in that it is information—just not protein-encoding information. A person is classified as a heterozygote or a homozygote based on the number of copies of the same repeat at the same chromosomal locus on the two homologs. A homozygote has the same number of copies of a repeat on both homologs, such as individual 2 in **figure 14.7**. A heterozygote has two different numbers of copies of a repeat, such as the other two individuals in the figure. The copy numbers are distributed in the next generation according to Mendel's law of segregation. A child of individual 1 and individual 2 in figure 14.7, for example, could have any of the two possible combinations of the parental copy numbers, one from each parent: 2 repeats and 3 repeats, or 4 repeats and 3 repeats.

DNA profiling was pioneered on detecting copy number variants of very short repeats and using them to identify or

Figure 14.7 **DNA profiling detects differing numbers of repeats at specific chromosomal loci.** Individuals 1 and 3 are heterozygotes for the number of copies of a 5-base sequence at a particular chromosomal locus. Individual 2 is a homozygote, with the same number of repeats on the two copies of the chromosome. (Repeat number is considered an allele.)

distinguish individuals. In general, the technique calculates the probability that certain combinations of repeat numbers will be in two DNA sources by chance. For example, if a DNA profile of skin cells taken from under the fingernails of an assault victim matches the profile from a suspect's hair, and the likelihood is very low that those two samples would match by chance, that is strong evidence of guilt rather than a coincidental similarity. DNA evidence is more often valuable in excluding a suspect, and should be considered along with other types of evidence.

An Evolving Technology

Obtaining a DNA profile is a molecular technique, but interpretation requires statistical analysis of population genetic data. The methods for generating a DNA profile grew out of genetic marker testing that identified DNA sequences closely linked to disease-causing genes. Genetic markers for disease were used in the 1980s, before knowing the sequences of human disease-causing genes made more

direct tests possible. Sir Alec Jeffreys at Leicester University in the United Kingdom was the first investigator to use DNA typing in forensic cases (**figure 14.8**). He followed DNA sequences called variable number of tandem repeats—VNTRs—that are each 10 to 80 bases long (**table 14.4**). **Clinical Connection 14.1** describes one of the earliest legal cases using DNA profiling.

DNA for profiling can come from any cell with a nucleus, including hair, skin, secretions, white blood cells, and cells scraped from the inside of the cheek. Sir Jeffreys cut DNA with naturally occurring, scissorlike enzymes, called restriction enzymes, that cut at specific sequences. He then separated the DNA pieces using a technique called agarose gel electrophoresis. The DNA fragments carry a negative electrical charge because of the phosphate groups. When placed in an electrical field on a gel strip, the DNA pieces move toward the positively charged end by size. The shorter the piece, the faster it travels.

Figure 14.8 **The inventor of DNA profiling.** Sir Alec Jeffreys introduced DNA profiling using variable number of tandem repeats (VNTRs) in the 1980s. Today the technique uses shorter repeats that are more likely to remain intact in harsh environments.

Table 14.4	Characteristics of Repeats Used in DNA Profiling			
Type	**Repeat Length**	**Distribution**	**Example**	**Fragment Sizes**
VNTRs (minisatellites)	10–80 bases	Not uniform	TTCGGGTTGT	50–1,500 bases
STRs (microsatellites)	2–10 bases	More uniform	ACTT	50–500 bases

DNA Profiling: Molecular Genetics Meets Population Genetics

DNA profiling is a powerful tool in forensic investigations, agriculture, paternity testing, and historical research. Until 1986, it was unheard of outside of scientific circles. A dramatic rape case changed that.

Tommie Lee Andrews watched his victims months before he attacked so that he knew when they would be home alone. On a balmy Sunday night in May 1986, Andrews awaited Nancy Hodge, a young computer operator at Disney World in Orlando, Florida. The burly man surprised her when she was in her bathroom removing her contact lenses. He covered her face, then raped and brutalized her repeatedly.

Andrews was very careful not to leave fingerprints, threads, hairs, or any other indication that he had ever been in Hodge's home. But he left DNA. Thanks to a clear-thinking crime victim and scientifically savvy lawyers, Andrews was soon at the center of a trial that would judge the technology that helped to convict him.

After the attack, Hodge went to the hospital, where she provided a vaginal secretion sample containing sperm. Two district attorneys who had read about DNA testing sent the sperm to a biotechnology company that extracted DNA and cut it with restriction enzymes. The sperm's DNA pieces were then mixed with labeled DNA probes that bound to complementary sequences.

The same extracting, cutting, and probing of DNA was done on white blood cells from Hodge and Andrews, who had been held as a suspect in several assaults. When the radioactive DNA pieces from each sample, which were the sequences where the probes had bound, were separated and displayed by size, the resulting pattern of bands—the DNA profile—matched exactly for the sperm sample and Andrews' blood, differing from Hodge's DNA (**figure 1**). (This differs from the STR approach used today.)

Andrews' allele frequencies were compared to those for a representative African American population.

At the first trial the judge, fearful too much technical information would overwhelm the jury, did not allow the prosecution to cite population-based statistics. Without the appropriate allele frequencies, DNA profiling was reduced to just a comparison of smeary lines on test papers to see whether the patterns of DNA pieces in the forensic sperm sample looked like those for Andrews' white blood cells. Although population statistics indicated that the possibility that Andrews' DNA would match the evidence by chance was 1 in 10 billion, the prosecution was not allowed to mention this critical interpretation.

After a mistrial was declared, the prosecution cited the precedent of using population statistics to derive databases on standard blood types. When Andrews stood trial just 3 months later for raping a different woman, the judge permitted population analysis. This time, Andrews was convicted.

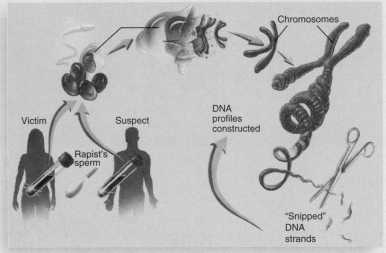

Figure 1 **DNA profiling.** The 1986 Andrews rape case used radioactive DNA probes on sperm from the victim's body, her white blood cells, and the suspect's white blood cells.

The pattern that forms when the different-sized fragments stop moving creates a pattern of smears. A person heterozygous for a particular VNTR—meaning a different number of copies of the repeat on each of the two homologs—would have a DNA profile with two bands at that site. A homozygote would have just one band, because his or her DNA would have only one size piece corresponding to that part of the genome. **Figure 14.9** shows one of these older DNA profiles, done not to identify a criminal but to test whether Dolly the cloned sheep actually came from the claimed breast cell from a ewe. She did.

Researchers began to use a different type of repeat, called a **short tandem repeat** (**STR**), because shorter DNA molecules were more likely to persist in a violent situation, such as an explosion, fire, or natural disaster. Today the Federal

Bureau of Investigation (FBI) in the United States tracks 13 STRs for DNA profiling, which generates 26 data points, because a person has two copies of each STR. The FBI system, called CODIS for Combined DNA Index System, shares DNA profiles among local, state, and federal crime laboratories. The power of the technique lies in the fact that the number of repeats of each of the 13 STRs occur with certain probabilities in any particular population. The probability that any two unrelated individuals have the same 26 CODIS markers (13 pairs) by chance is 1 in 250 trillion.

STRs range in size from two to ten bases, but most used for forensic applications are four bases long. **Figure 14.10** shows the 13 STRs used in DNA profiling. The technique also includes the genotype for a gene called amelogenin to

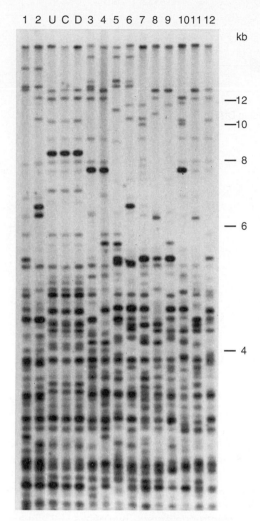

Figure 14.9 Comparing DNA profiles. These DNA profiles compare VNTRs from the DNA of Dolly the cloned sheep (lane D), fresh donor udder tissue (U), and cultured donor udder tissue (C). The other twelve lanes represent other sheep. The match between Dolly and the two versions of her nucleus donor is obvious. Dolly was born in 1996 and died in 2003—young for a ewe, perhaps due to her first cell's earlier life.

determine the gender of the person who left the DNA sample. Amelogenin is present on both the X and Y chromosomes, but is six bases shorter on the X chromosome.

Today a technique called capillary electrophoresis is used to separate DNA pieces (**figure 14.11**). Fluorescently labeled short pieces of DNA that correspond to parts of the genome including the CODIS STRs are applied to sample DNA, where they bind to their complements. Then the polymerase chain reaction (PCR; see figure 9.18) copies the selected sequences, amplifying all 13 CODIS markers simultaneously. Instead of the slab of gel used when the field began, the DNA pieces travel through hair-thin tubes, called capillaries. A laser shone through a window excites the fluorescent dyes that tag each DNA piece, and the strength of the fluorescence is measured and displayed as a peak on a read-out. A single tall peak for a particular STR indicates that the person is a homozygote—for example, six copies of the repeat on each homolog. Two peaks of lesser intensity indicate that the person is a heterozygote, such as having six copies on one homolog but eight on its mate.

The FBI maintains a National DNA Index (NDIS). It has DNA from more than 10 million convicted offenders, 1.5 million arrested individuals, and half a million samples collected from crime scenes but never associated with an individual. The FBI claims that NDIS has aided more than 213,000 investigations.

If DNA is so degraded that even STRs are destroyed, mitochondrial DNA (mtDNA) is often used instead, particularly two regions of repeats that are highly variable in populations. Because a single cell can yield hundreds or thousands of copies of the mitochondrial genome, even extremely small forensic samples can yield this DNA. MtDNA analysis was critical in analyzing evidence from the September 11 terrorist attacks, most of which was extremely degraded.

Using Population Statistics to Interpret DNA Profiles

In forensics in general, the more clues, the better. The power of DNA profiling comes from tracking repeats on several chromosomes. The numbers of copies of a repeat are assigned

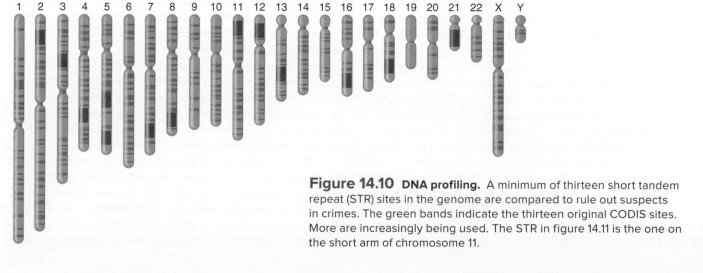

Figure 14.10 DNA profiling. A minimum of thirteen short tandem repeat (STR) sites in the genome are compared to rule out suspects in crimes. The green bands indicate the thirteen original CODIS sites. More are increasingly being used. The STR in figure 14.11 is the one on the short arm of chromosome 11.

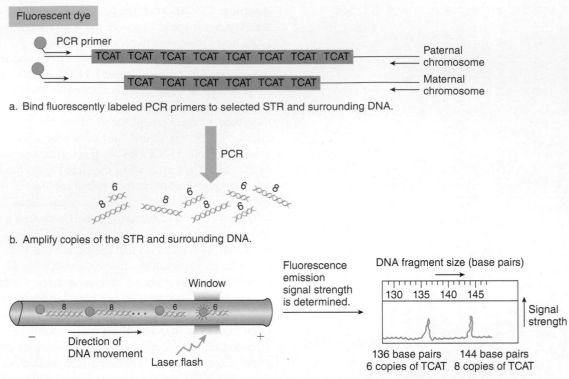

a. Bind fluorescently labeled PCR primers to selected STR and surrounding DNA.

PCR

b. Amplify copies of the STR and surrounding DNA.

Fluorescence emission signal strength is determined.

DNA fragment size (base pairs)

Window

Direction of DNA movement

Laser flash

Signal strength

136 base pairs   144 base pairs
6 copies of TCAT   8 copies of TCAT

c. Run amplified copies of labeled STR-containing DNA pieces through capillary electrophoresis device. Electropherogram shows two peaks for heterozygotes (two fragment sizes) and one larger peak for homozygotes (same fragment size from both homologs). This person is heterozygous for the TCAT STR on chromosome 11.

Figure 14.11 Several steps identify STRs. In this example, the STR is TH01, located on chromosome 11. **(a)** Short pieces of DNA called primers that are complementary to regions near the STR of interest are added to the DNA sample. Each primer is bound to a fluorescent dye. **(b)** The primers direct the polymerase chain reaction to make many copies of the specific STR. **(c)** The amplified DNA pieces—in this case either six or eight copies of the 4-base repeat TCAT—enter a gel electrophoresis device, where the smaller pieces move toward the positive charge faster than the larger pieces. Laser excitation of the dye bound to the primers emits a signal that is converted into an image called an electropherogram. Two peaks indicate a heterozygote and a single, larger peak a homozygote—for this STR.

probabilities (likelihood of being present) based on their observed frequencies in the population from which the source of the DNA comes. Considering repeats on different chromosomes makes it possible to use the product rule to calculate the probabilities of particular combinations of repeat numbers occurring in a population, based on Mendel's law of independent assortment.

The Hardy-Weinberg equation and the product rule are used to derive the statistics that back up a DNA profile. First, the peak patterns of the electropherograms indicate whether an individual is a homozygote or a heterozygote for each repeat. Genotype frequencies are then calculated using parts of the Hardy-Weinberg equation. That is, p^2 and q^2 denote each of the two homozygotes for a two-allele repeat, and $2pq$ represents the heterozygote. Then the frequencies are multiplied to reveal the probability that this particular combination of repeat copy numbers would occur in a particular population. Logic then enters the equation. If the combination is very rare in the population the suspect comes from, and if it is found both in the suspect's DNA and in crime scene evidence, such as a rape victim's body or stolen property, the suspect's guilt appears highly likely. **Figure 14.12** summarizes the procedure for considering multiple STRs.

For the sequences used in DNA profiling, Hardy-Weinberg equilibrium is assumed. When it doesn't apply, problems can arise. For example, the requirement of nonrandom mating for Hardy-Weinberg equilibrium wouldn't be met in a community with a few very large families where distant relatives might inadvertently marry each other—a situation in many small towns. A particular DNA profile for one person might be shared by his or her cousins. In one case, a young man was convicted of rape based on a DNA profile—which he shared with his father, the actual rapist. Considering a larger number of repeat sites can minimize such complications. If more repeat sites had been considered in the rape case, chances are that they would have revealed a polymorphism that the son had inherited from his mother, but that the guilty father lacked. Presence of the polymorphism would have indicated that the son was not guilty, but a close male relative might be.

The accuracy and meaning of a DNA profile depend upon the population that is the source of the allele frequencies. If populations are too broadly defined, then allele frequencies are typically low, leading to very low estimates of the likelihood that a suspect matches evidence based on chance. In one oft-quoted trial, the prosecutor concluded, *The chance of the DNA fingerprint of the*

Figure 14.12 **To solve a crime.** A man was found brutally murdered, with bits of skin and blood beneath a fingernail. The evidence was sent to a forensics lab, where the patterns of five STRs in the blood from beneath the fingernail was compared to patterns for the same STRs in blood from the victim and blood from a suspect. The pattern for the crime scene evidence matched that for the suspect. Allele frequencies from the man's ethnic group used in the Hardy-Weinberg equation yielded the probability that his DNA matched that of the skin and blood under the murdered man's fingernail by chance. This is a hypothetical case. Using the full set of 13 STRs yields much more significant probabilities. The numbers under the peaks are the numbers of copies of the STR. The decimals are allele frequencies.

DNA collected
from evidence
at crime scene
(blood, skin under
victim's fingernail)

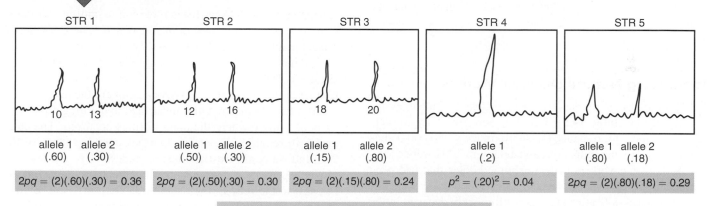

STR 1	STR 2	STR 3	STR 4	STR 5
10 13	12 16	18 20		
allele 1 allele 2 (.60) (.30)	allele 1 allele 2 (.50) (.30)	allele 1 allele 2 (.15) (.80)	allele 1 (.2)	allele 1 allele 2 (.80) (.18)
$2pq = (2)(.60)(.30) = 0.36$	$2pq = (2)(.50)(.30) = 0.30$	$2pq = (2)(.15)(.80) = 0.24$	$p^2 = (.20)^2 = 0.04$	$2pq = (2)(.80)(.18) = 0.29$

$$0.36 \times 0.30 \times 0.24 \times 0.04 \times 0.29 \approx 0.00031 \approx 1/3{,}226$$

Conclusion: The probability that another person in the suspect's population group has the same pattern of these alleles is approximately 1 in 3,226.

cells in the evidence matching blood of the defendant by chance is 1 in 738 trillion. The numbers were accurate, but did they really reflect the gene pool compositions of actual populations?

The first DNA profiling databases relied on populations that weren't realistic enough to yield valid statistics. They placed different groups into just three categories: Caucasian, black, or Hispanic. People from Poland, Greece, or Sweden were all considered Caucasian, and a dark-skinned person from Jamaica and one from Somalia would be considered blacks. Perhaps the most incongruous of all were the Hispanics. Cubans and Puerto Ricans are part African, whereas people from Mexico and Guatemala have mostly Native American gene variants. Spanish and Argentinians have neither black African nor Native American genetic backgrounds. Native Americans and Asians were left out altogether. Analysis of just three databases—Caucasian, black, and Hispanic—revealed significantly more homozygous recessives for certain polymorphic

genes than the Hardy-Weinberg equation would predict, confirming that allele frequencies were not in equilibrium.

More restrictive ethnic databases are required to apply allele frequencies to interpret DNA profiles. A frequency of 1 in 1,000 for an allele in all Caucasians may actually be much higher or lower in a subgroup that marry among themselves. However, even narrowly defined ethnic databases may be insufficient to interpret DNA profiles from people of mixed heritage, such as someone whose mother was Scottish/French and whose father was Greek/German.

Using DNA Profiling to Identify Victims

DNA profiling was first used in criminal cases and to identify human remains from plane crashes. Then terrorist attacks and natural disasters took the scope of DNA profiling to a new level.

Identifying World Trade Center Victims

In late September 2001, a company that provides breast cancer tests received three unusual types of DNA samples:

- evidence from the World Trade Center in New York City;
- cheek brush scrapings from relatives of people missing from the site; and
- "reference samples" from the victims' toothbrushes, razors, and hairbrushes.

Technologists analyzed the DNA for copy numbers of the thirteen standard STRs and the sex chromosomes (the amelogenin gene). STR analysis worked on pieces of soft tissue, but bone bits that persisted despite the ongoing fire at the site required hardier mtDNA analysis. If the DNA pattern in crime scene evidence matched DNA from a victim's toothbrush, identification was fairly certain. Forensic investigators used the DNA results to match family members to victims. DNA profiling provides much more reliable information on identity than traditional forensic identifiers such as dental patterns, scars, and fingerprints, and objects found with the evidence, such as jewelry.

Identifying Natural Disaster Victims

Different types of disasters present different challenges for DNA profiling (**table 14.5**). Whereas New York City workers searched rubble for remains, the approximately 250,000 bodies strewn about by the Indian Ocean tsunami in 2004 were everywhere. Disaster workers had to exhume bodies that had been buried quickly to stem the spread of infectious disease. Remains that were accessible after the waves hit quickly decayed in the hot, wet climate. These conditions, combined with the lack of roads and labs, led to 75 percent of the bodies being identified by standard dental record analysis, and 10 percent from fingerprints. Fewer than half of 1 percent of the victims were identified by their DNA.

Forensic scientists had learned from 9/11 the importance of matching victim DNA to that of relatives, to avoid errors when two people matched at several genome sites by chance. In New York City, many of those relatives were from nearby neighborhoods; in the 2004 tsunami, 12 countries were directly affected and victims came from 30 countries. Entire families were washed away, leaving few and many times no relatives to provide DNA, even if everyday evidence such as toothbrushes had remained.

To compensate for the barriers to implementing DNA profiling in mass disasters, Sir Alec Jeffreys advises assessing 15 to 20 STRs, and some investigators recommend using 50. Tragic as these disasters were, they have spurred forensic scientists to develop ways to better integrate many types of evidence, including that found in DNA sequences.

Reuniting Holocaust Survivors

A happier use of DNA profiling is to reunite families who were torn apart in the Holocaust of World War II. The DNA Shoah project maintains a DNA database of many of the 300,000 survivors, including some of the 10,000-plus Holocaust orphans. (*Shoah* is Hebrew for *holocaust*.) Michael Hammer and colleagues at the University of Arizona compare the data to DNA profiles from human remains unearthed in building projects in parts of Europe where the mass killings occurred. The challenges in reuniting Holocaust families combine those of the 9/11 and tsunami investigations: degraded DNA and few surviving relatives and descendants. The Shoah project is linking the past to the present by matching DNA profiles. The goal: "Using science and technology to reunite families, identify the missing and educate future generations," according to the program's website.

Table 14.5	Challenges to DNA Profiling in Mass Disasters

- Climate that hastens decay
- Inability to reach remains
- No laboratory facilities
- Number of casualties
- Lack of relatives
- Destruction of personal item evidence
- Poor DNA quality (too fragmented, scarce, degraded)
- Lack of availability of DNA probes and statistics for population

Genetic Privacy

Before the information age, population genetics was an academic discipline that was more theoretical than practical. Today, with the combination of information technology, genome-wide association studies, genome sequencing, and shortcuts to identify people by SNP or copy number patterns, population genetics presents a powerful way to identify individuals. The *Bioethics: Choices for the Future* on page 275 discusses how the Supreme Court dealt with a matter of genetic information use.

The human genome is 3.2 billion bits of information, each of which can be one of four possibilities: A, C, T, or G. Because of this huge capacity for diversity, our genomes can vary many more ways than there are people—about 10 billion worldwide. Only 30 to 80 genome sites need be considered to uniquely describe each person. This is why

Should DNA Collected Today Be Used to Solve a Past Crime?

In 2003, a masked man broke into a woman's home in Salisbury, Maryland, and raped her. She couldn't describe her assailant, but the man had left behind his DNA, which forensic investigators collected and stored. The crime remained unsolved—until 2009.

Alonzo Jay King was arrested for assault in 2009 in Wicomico County, Maryland, for "menacing a group of people with a shotgun." State law required that he provide a DNA sample collected on a cheek swab, because the crime was considered violent. Maryland law also holds that taking a DNA sample and entering the information in a database is legal as long as the suspect has been arraigned, but that the sample be destroyed if the suspect is not convicted.

When Federal Bureau of Investigation forensic scientists ran King's DNA profile against a database of half a million unidentified DNA samples taken from crime scenes (part of CODIS), the sample matched DNA on the underwear of the victim from the 2003 unsolved rape case. When a second sample from King matched, too, he was brought before a grand jury, indicted, then tried and convicted of the older crime and sentenced to life in prison without parole. The conviction was based on the DNA from the scene of the later crime. But a court of appeals ruled that taking of the DNA in 2009 had been an unlawful seizure and unreasonable search, and set the conviction aside.

The case went to the Supreme Court, which ruled in 2013 that taking DNA from King was like taking fingerprints or a photograph—a noninvasive way to establish or confirm identity. The court deemed DNA profiling a "legitimate police booking procedure." The five judges in the assenting opinion argued for the accuracy of DNA profiling compared to other forensic methods, and the fact that the parts of the genome that are compared have nothing to do with either traits or health, so privacy was not at issue. The dissenting opinion of four judges focused more on common sense—DNA wasn't needed to identify King in 2009. Everyone knew who he was. What wasn't known, and the DNA revealed, was that he'd committed the *earlier* crime. Whether or not DNA collected at one time for one crime can be used to convict an individual for a *different* crime is the bioethical issue.

Questions for Discussion

1. What should be criteria for requiring a suspect to provide a DNA sample to law enforcement officials?

2. Why would the DNA profile not reveal anything about the arrested individual's health?

3. Do you agree with the majority opinion or the dissenting opinion? Cite a reason for your answer.

4. How is a DNA profile like a fingerprint and unlike one?

5. Why is destroying DNA evidence futile?

6. How do you think the rape victim might have felt about the use of DNA testing in this case?

---

forensic tests can compare only 13 STRs to rule out or establish identity.

The ease of assigning highly individualized genetic name tags may be helpful in forensics, but it poses privacy issues. Consider a "DNA dragnet," which is a forensic approach that compiles DNA profiles of all residents of a town where a violent crime is unsolved. Sir Jeffreys conducted some of the first DNA dragnets in the late 1980s. The largest to date occurred in 1998 in Germany, where more than 16,000 men had their DNA profiled in a search for the man who raped and murdered an 11-year-old. The dragnet indeed caught the killer.

A controversial application of DNA profiling is a familial DNA search, which is based on the fact that close relatives share large portions of their genomes. DNA from a crime scene is compared to DNA in databases from convicted felons. If nearly half of the CODIS sites match, then a first-degree relative of the convict becomes a suspect. This might be a son, brother, or the father of a man in prison, for example.

In the case of California's "Grim Sleeper," who killed many young women over more than two decades, the suspect was finally arrested when DNA found on a victim closely matched that of a young man in prison for trafficking weapons. The young man did not have brothers, was too young to have a son old enough to be the Grim Sleeper, but did have a father who could have committed the crimes. After an intense investigation, Los Angeles police collected DNA from the father, Lonnie David Franklin Jr., from a discarded pizza slice. A search of his home turned up 180 photographs of potential victims, which were posted on the police department's website so that the public could help in identifying them. Yet for every criminal that familial DNA searches identify, some innocent people are accused, based on sharing CODIS markers with convicted felons.

DNA profiling is based on population genetics, but it requires logic to avoid false accusations. Investigators must consider how DNA came to be at a crime scene, which is called DNA transfer, or "touch DNA." Primary transfer occurs when the suspect's DNA is on an object, such as a glove or a weapon. Secondary transfer occurs when the person who touches the object had earlier contacted another person's DNA, such as a rapist having first shaken hands with someone. In one complicated case, a professor at an Ivy League school was accused of his wife's murder. It turned out that when he and his wife shared a towel earlier that day, his skin cells had been transferred to her face. Later, the murderer, wearing gloves, touched her face, picked up the husband's cells, and unknowingly transferred them to his weapon. Reconstructing this scenario helped to exonerate the husband.

The Hardy-Weinberg equilibrium that makes DNA profiling possible is extremely rare in the real world, for most genes. The next chapter considers the familiar circumstances that change allele frequencies.

Key Concepts Questions 14.4

1. How does the concept of Hardy-Weinberg equilibrium enable interpretation of DNA profiles?

2. What type of DNA sequence is used to construct DNA profiles?

3. How do the terms heterozygote and homozygote apply to repeated DNA sequences?

4. How is probability used to interpret DNA profiles?

5. Describe how capillary electrophoresis is used to detect STRs to provide a DNA profile.

6. What is the source of the power behind DNA profiling?

7. How can researchers improve the accuracy of the interpretation of DNA profiles?

Summary

14.1 Population Genetics Underlies Evolution

1. A **population** is a group of interbreeding members of the same species in a particular area. Their genes and all the variants constitute the **gene pool**.

2. **Population genetics** considers allele, genotype, and phenotype frequencies to reveal **microevolution**. Phenotypic frequencies can be determined empirically, then used in algebraic expressions to derive other frequencies. Changes great enough to cause speciation are termed **macroevolution**.

3. Genotype frequencies change if migration, nonrandom mating, genetic drift, mutations, or natural selection operate.

14.2 Constant Allele Frequencies

4. In **Hardy-Weinberg equilibrium**, genotype frequencies are not changing. Hardy and Weinberg proposed an algebraic equation to explain the constancy of allele frequencies. This would show why dominant traits do not increase and recessive traits do not decrease in populations. If a gene in Hardy-Weinberg equilibrium is tightly linked to a gene subject to natural selection, a condition called **linkage disequilibrium,** the equilibrium may be broken.

5. The Hardy-Weinberg equation is a binomial expansion used to represent genotypes in a population. In Hardy-Weinberg equilibrium, gamete frequencies do not change as they recombine in the next generation. Evolution is not occurring. When the equation $p^2 + 2pq + q^2$ represents a gene with one dominant and one recessive allele, p^2 corresponds to the frequency of homozygous dominant individuals; $2pq$ stands for heterozygotes; and q^2 represents the frequency of the homozygous recessive class. The frequency of the dominant allele is p, and of the recessive allele, q.

14.3 Applying Hardy-Weinberg Equilibrium

6. If we know either p or q, we can calculate genotype frequencies, such as carrier risks. Often such information comes from knowing the q^2 class, which corresponds to the frequency of homozygous recessive individuals in a population.

7. For X-linked recessive traits, the mutant allele frequency for males equals the trait frequency. For very rare disorders or traits, the value of p approaches 1, so the carrier frequency ($2pq$) is approximately twice the frequency of the rare trait (q).

14.4 DNA Profiling Uses Hardy-Weinberg Assumptions

8. The numbers of copies of repeated DNA sequences that do not encode protein are presumably in Hardy-Weinberg equilibrium and can be compared to establish individual DNA profiles.

9. To obtain a **DNA profile,** determine numbers of **short tandem repeats (STRs)** and multiply population-based allele frequencies to derive the probability that profiles from two sources match by chance.

10. Forensic DNA profiles consider 13 STR loci and the amelogenin gene size to determine sex.

www.mhhe.com/lewisgenetics11

Answers to all end-of-chapter questions can be found at **www.mhhe.com/lewisgenetics11**. You will also find additional practice quizzes, animations, videos, and vocabulary flashcards to help you master the material in this chapter.

Review Questions

1. Define *gene pool*.

2. Why are Hardy-Weinberg calculations more complicated if a gene has many alleles that affect the phenotype?

3. How can evolution occur at a microscopic and macroscopic level?

4. Explain the differences among an allele frequency, a phenotypic frequency, and a genotypic frequency.

5. What are the conditions under which Hardy-Weinberg equilibrium cannot be met?

6. Why is knowing the incidence of a homozygous recessive condition in a population important in deriving allele frequencies?

7. How is the Hardy-Weinberg equation used to predict the recurrence of X-linked recessive traits?

8. Why were VNTR sequences replaced with STRs for obtaining a DNA profile?

9. How would a heterozygote and homozygote for the same STR be represented differently on an electropherogram?

10. Why are specific population databases needed to interpret DNA profiles?

11. What is the basis of assigning a probability value to a particular copy number variant? Where do the probabilities come from?

12. Under what circumstances is analysis of repeats in mtDNA valuable?

13. Explain and provide an example of how a familial DNA search can lead to a false accusation.

Applied Questions

1. "We like him, he seems to have a terrific gene pool," say the parents upon meeting their daughter's boyfriend. Why doesn't their statement make sense?

2. Two couples want to know their risk of conceiving a child with cystic fibrosis. In one couple, neither partner has a family history of the disease; in the other, one partner knows he is a carrier. How do their risks differ?

3. How does calculation of allele frequencies differ for an X-linked trait or disorder compared to one that is autosomal recessive?

4. Profiling of Y chromosome DNA implicated Thomas Jefferson in fathering a child of his slave, discussed in chapter 1. What might have been a problem with the conclusion?

5. Torsion dystonia (OMIM 128100) is a movement disorder that affects 1 in 1,000 Jewish people of Eastern European descent (Ashkenazim). What is the carrier frequency in this population?

6. Maple syrup urine disease (MSUD) (see Clinical Connection 2.1) is autosomal recessive and causes intellectual and physical disability, difficulty feeding, and a sweet odor to urine. In Costa Rica, 1 in 8,000 newborns inherits the condition. What is the carrier frequency of MSUD in this population?

7. Ability to taste phenylthiocarbamide (PTC) (OMIM 607751) is mostly determined by the gene *PTC*. The letters *T* and *t* are used here to simplify analysis. *TT* individuals taste a strong, bitter taste; *Tt* people experience a slightly bitter taste; *tt* individuals taste nothing.

   A fifth-grade class of 20 students tastes PTC that has been applied to small pieces of paper, rating the experience as "very yucky" (*TT*), "I can taste it" (*Tt*), and "I can't taste it" (*tt*). For homework, the students test their parents, with these results:

   Of 6 *TT* students, 4 have 2 *TT* parents; and two have one parent who is *TT* and one parent who is *Tt*.

   Of 4 students who are *Tt*, 2 have 2 parents who are *Tt*, and 2 have one parent who is *TT* and one parent who is *tt*.

   Of the 10 students who can't taste PTC, 4 have 2 parents who also are *tt*, but 4 students have one parent who is *Tt* and one who is *tt*. The remaining 2 students have 2 *Tt* parents.

   Calculate the frequencies of the *T* and *t* alleles in the two generations. Is Hardy-Weinberg equilibrium maintained, or is this gene evolving?

8. The examples of DNA profiling in the chapter concern criminals or natural disasters. Explain how the 13 STR sites can be used to rule out paternity.

Web Activities

1. Consult the website for the Innocence Project or a legal site such as http://www.denverda.org/dna/Familial_DNA_Database_Searches.htm and discuss how DNA evidence was used to help convict or exonerate someone.

2. Go to the FBI's Biometric Analysis page (www.fbi.gov/about-us/lab/biometric-analysis/codis/ndis-statistics) and click on the symbol for your state. List the numbers of DNA samples on file for different categories. Do you think that it is useful or ethical to store DNA samples from people who have been arrested but not yet convicted of a crime?'

3. The website www.cstl.nist.gov/strbase/ has information on STRs for dogs, cats, cattle, and horses, with entertaining names such as the meowPLEX test for cats. Look up the number of STRs for one of these animals, and describe a scenario in which such a test might provide useful information.

Forensics Focus

1. A DNA profiling method that is simplified so that it can be used in the field uses only three STRs. How should the three be selected to provide the best distinguishing of individuals?

2. The Indian Ocean tsunami of 2004 gave forensic investigators the idea to form and implement a Global DNA Response Team to collect samples after a natural disaster. What rules should such a program have?

3. Irene is 80 years old and lives alone with her black cat Moe. One day when she is taking out the garbage, a man jumps out from behind a garbage can and shoves her. As she struggles to get up, he runs inside, pushes open her door, enters, and grabs her purse, which is next to a slumbering Moe on the kitchen table. Moe, sensing something is wrong or perhaps just upset that his nap has been interrupted, scratches the man, who yelps and forcefully flings the cat against a wall, yanking out one of the animal's claws in the process.

   Meanwhile, outside, Irene is back on her feet, trying to get her bearings when the fleeing thief knocks her down again. This time she blacks out, and the man escapes. A few minutes later a neighbor finds her and calls the police. At the crime scene, they collect a drop of blood on the table, a curly blond hair, a few straight black hairs, several gray hairs, and a cat's claw with human skin under it. The police send Irene to the hospital, Moe to the veterinary clinic, and the claw, blood, and hairs to the state forensics lab. The police then thoroughly search the immediate area and neighborhood, but do not find any suspects. When Irene regains consciousness, she remembers nothing about her attackers.

   a. List DNA tests that could help identify the perpetrator.

   b. A familial DNA search closely matches two of the forensic samples to a 58-year-old man in prison for murder. One sample matches at 12 CODIS sites and the other at 10. The investigators find that the convict has four sons and four nephews. What should the police do with this information?

4. DNA dragnets have been so successful that some people have suggested storing DNA samples of everyone at birth, so that a DNA profile could be obtained from anyone at any time. Do you think that this is a good idea or not? Cite reasons for your answer.

5. Rufus the cat was discovered in a trash can by his owners, his body covered in cuts and bite marks and bits of gray fur clinging to his claws—gray fur that looked a lot like the coat of Killer, the huge hound next door. Fearful that Killer might attack their other felines, Rufus' distraught owners brought his body to a vet, demanding forensic analysis. The vet suggested that the hair might have come from a squirrel, but agreed to send samples to a genetic testing lab. Identify the samples that the vet might have sent, and what information each could contribute to the case.

6. In a crime in Israel, a man knocked a woman unconscious and raped her. He didn't leave any hairs at the crime scene, but he left eyeglasses with unusual frames, and an optician helped police locate him. The man also left a half-eaten lollipop at the scene. DNA from blood taken from the suspect matched DNA from cheek-lining cells collected from the base of the telltale lollipop at four repeat loci on different chromosomes. Allele frequencies from the man's ethnic group in Israel are listed beside the profile pattern shown here:

STR1	Homozygote	Allele frequency .2
STR2	Allele 1 = .3	
	Allele 2 = .7	
STR3	Homozygote	Allele frequency = .1
STR4	Allele 1 = .4	
	Allele 2 = .2	

   a. How would the electropherogram look different for STRs 1 and 3 compared to STRs 2 and 4?

   b. What is the probability that the suspect's DNA matches that of the lollipop rapist by chance? (Do the calculation.)

   c. The man's population group is highly inbred. How does this affect the accuracy or reliability of the DNA profile? (P.S.—He was so frightened by the DNA analysis that he confessed!)

Case Studies and Research Results

1. An extra row of eyelashes is an autosomal recessive trait seen in 900 of the 10,000 residents of an island in the South Pacific. Greta knows that she is a heterozygote for this gene, because her eyelashes are normal, but she has an affected parent. She wants to have children with a homozygous dominant man so that the trait will not affect her offspring. What is the probability that a person with normal eyelashes in this population is a homozygote for this gene?

2. Glutaric aciduria type 1 (OMIM 231680) was the first disease investigated at the Clinic For Special Children, which was founded in 1989 in the heart of the Old Order Amish and Mennonite community in Lancaster, Pennsylvania (see Clinical Connection 15.1). The disease causes severe movement problems that lead to paralysis, brain damage, and early death. In this population, 0.25 percent of newborns have the disorder. Researchers at the clinic developed a special formula with the levels of amino acids tailored to counter the metabolic abnormality. Children who use the formula can avoid the symptoms.

   a. What percentage of this population are carriers for glutaric aciduria type 1?

   b. What effect might the treatment have on the mutant allele frequency in this population in the future?

The ability to digest lactose (milk sugar) became more prevalent in populations after agriculture introduced dairy foods—thanks to evolution.

Changing Allele Frequencies

Learning Outcomes

15.1 Nonrandom Mating

1. Explain how nonrandom mating changes allele frequencies in populations.

15.2 Migration

2. Explain how migration changes allele frequencies in populations.

15.3 Genetic Drift

3. Explain how the random fluctuations of genetic drift affect genetic diversity.

4. Discuss how founder effects and population bottlenecks amplify genetic drift.

15.4 Mutation

5. Discuss how mutation affects population genetic structure.

15.5 Natural Selection

6. Provide examples of negative, positive, and artificial selection.

7. Explain how balanced polymorphism maintains diseases in populations.

15.6 Eugenics

8. Explain how eugenics attempts to alter allele frequencies.

The BIG Picture

Several forces mold populations, which over time drive evolution. They are nonrandom mating, migration, genetic drift, mutation, and natural selection.

The Evolution of Lactose Tolerance

For millions of people who have lactose (milk sugar) intolerance, dairy food causes cramps, bloating, gas, and diarrhea. They no longer produce lactase, which is an enzyme made in early childhood that breaks down the milk sugar lactose into more easily digested sugars. But people who have lactose intolerance may represent the "normal," or wild type condition. Only 35 percent of people in the world can digest lactose into adulthood. People who *can* digest dairy foods have lactase persistence (OMIM 223000). One gene controls the ability to digest milk sugar.

Clues in DNA suggest that agriculture drove the differences in our abilities to digest lactose. As dairy farming spread around the world, from 5,000 to 10,000 years ago, people who had gene variants enabling them to digest milk into adulthood had an advantage. They could eat a greater variety of the now more plentiful foods, were healthier, and had more children. Over time, populations that consumed dairy foods had more people with lactase persistence. In contrast, in populations with few or no dairy foods, lactose intolerance was not a problem, and so those gene variants persisted.

The link between lactose intolerance and agriculture is why today, the European American population only has 10 percent lactose intolerance. Among Asian Americans, who eat far less dairy, 90 percent have lactose intolerance. That is, the inability to digest lactose doesn't bother them. Seventy-five percent of African Americans and Native Americans have lactose intolerance.

15.1 Nonrandom Mating

Many aspects of modern life alter allele frequencies, and so many genes are not in Hardy-Weinberg equilibrium, defined as unchanging allele frequencies from generation to generation. Religious restrictions and personal preferences guide our choices of mates. Wars and persecution kill certain populations. Economic and political systems enable some groups to have more children. We travel, shuttling genes in and out of populations. Natural disasters and new diseases reduce populations to a few individuals, who then rebuild their numbers, at the expense of genetic diversity. These factors, plus mutation and a reshuffling of genes at each generation, make a gene pool very fluid.

The ever-present and interacting forces of nonrandom mating, migration, genetic drift, mutation, and natural selection shape populations at the allele level. Changing allele frequencies can change genotype frequencies, which in turn can change phenotype frequencies. In a series of illustrations throughout this chapter, colored shapes represent alleles. Figure 15.16 combines the illustrations to summarize the chapter. We begin our look at the forces that change allele frequencies in populations with nonrandom mating (**figure 15.1**).

In the theoretical state of Hardy-Weinberg equilibrium, individuals of all genotypes are equally likely to successfully mate and to choose partners at random. In reality we choose partners for many reasons: physical appearance, ethnic background, intelligence, and shared interests, to name a few (**figure 15.2**). This nonrandom mating is a major factor in changing allele frequencies in human populations.

Nonrandom mating occurs when certain individuals contribute more to the next generation than others. This is common in agriculture when semen from one prize bull is used to inseminate thousands of cows, and a similar situation has happened when many families used the same sperm donor to conceive children. One such man fathered 150 children. High prevalence of an otherwise rare inherited condition can be due to nonrandom mating. For example, a form of albinism is uncommon in the general U.S. population, but it affects 1 in 200 Hopi Indians who live in Arizona. The reason for the trait's prevalence is cultural—men with albinism often stay back and help the women, rather than risk severe sunburn in the fields with the other men. They contribute more children to the population because they have more contact with the women.

The events of history can lead to nonrandom mating patterns. When a group of people is subservient to another, genes tend to "flow" from one group to the other as the males of the ruling class have children with females of the underclass—often forcibly. Historical records and DNA sequences show this directional gene flow phenomenon. For example, Y chromosome analysis suggests that Genghis Khan, a Mongolian warrior who lived from 1162 to 1227, had sex with so many women that today, 1 in every 200 males living between Afghanistan

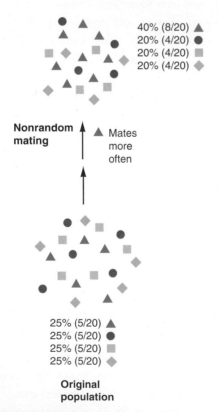

40% (8/20) ▲
20% (4/20) ●
20% (4/20) ■
20% (4/20) ◆

Nonrandom mating ↑ ▲ Mates more often

25% (5/20) ▲
25% (5/20) ●
25% (5/20) ■
25% (5/20) ◆

Original population

Figure 15.1 Nonrandom mating alters allele frequencies. More successful mating among individuals with the blue triangle allele will skew allele frequencies in the next generation.

Figure 15.2 Nonrandom mating. We marry people similar to ourselves about 80 percent of the time. In the 1990s, worldwide, about one-third of all marriages were between people who were born fewer than 10 miles apart. The Internet may change that statistic!

and northeast China shares his Y—that's 16 million men. The number is so high because his many male descendants also passed on the distinctive Y.

Traits may mix randomly in the next generation if we are unaware of them or do not consider them in choosing partners. In populations where AIDS is extremely rare or nonexistent, for example, the two mutations that render a person resistant to HIV infection are in Hardy-Weinberg equilibrium. This would change, over time, if HIV arrives, because the people with these mutations would become more likely to survive to produce offspring—and some of them would perpetuate the protective mutation. Natural selection would intervene, ultimately altering allele frequencies.

Many blood types are in Hardy-Weinberg equilibrium because we do not choose partners by blood type. Yet sometimes the opposite situation occurs. People with mutations in the same gene meet when their families participate in programs for people with the associated disorder. For example, more than two-thirds of relatives visiting a camp for children with cystic fibrosis are likely to be carriers, compared to the 1 in 23 or fewer in large population groups.

People can avoid genetic disease with controlled mate choice and reproduction. In a program that began in New York City called Dor Yeshorim, for example, young people take tests for more than a dozen genetic disorders that are much more common among Jewish people of Eastern European descent (Ashkenazim). Results are stored in a confidential database. Two people wishing to have children together can find out if they are carriers for the same disorder. If so, they may elect not to have children. Thousands of people have been tested, and the program is partly responsible for the near-disappearance of Tay-Sachs disease among Ashkenazi Jews. The very few cases each year are usually in non-Jews, because they have not been tested.

A population that practices consanguinity has very nonrandom mating. Recall from chapter 4 that in a consanguineous relationship, "blood" relatives have children together. On the family level, this practice increases the likelihood that harmful recessive alleles from shared ancestors will be combined and passed to offspring, causing disease. The birth defect rate in offspring is 2.5 times the normal rate of about 3 percent. On a population level, consanguinity decreases genetic diversity. The proportion of homozygotes rises as the proportion of heterozygotes falls.

Some populations encourage marriage between cousins, which increases the incidence of certain recessive disorders. In certain parts of the middle east, Africa, and India, 20 to 50 percent of marriages are between cousins, or uncles and nieces. The tools of molecular genetics can reveal these relationships. Researchers traced DNA sequences on the Y chromosome and in mitochondria among residents of an ancient, geographically isolated "micropopulation" on the island of Sardinia, near Italy. They consulted archival records dating from the village's founding by 200 settlers around 1000 A.D. to determine familial relationships. Between 1640 and 1870, the population doubled, reaching 1,200 by 1990. Fifty percent of the present population descends from just two paternal and four maternal lines, and 86 percent of the people have the same X chromosome. Researchers are analyzing disorders that are especially prevalent in this population, which include hypertension and a kidney disorder.

Worldwide, about 960 million married couples are related, and know of their relationship. Also contributing to nonrandom mating is endogamy, which is marriage within a community. In an endogamous society, spouses may be distantly related and be unaware of the connection.

15.2 Migration

Large cities, with their pockets of ethnicity, defy Hardy-Weinberg equilibrium by their very existence. Waves of immigrants formed the population of New York City, for example. The original Dutch settlers of the 1600s had different alleles than those in today's metropolis of English, Irish, Slavics, Africans, Hispanics, Italians, Asians, and many others. **Figure 15.3** depicts the effect on allele and genotype frequencies when individuals join a migrating population. Clues to past migrations lie in historical documents as well as in differing allele frequencies in regions defined by geographical or language barriers.

The frequency of the allele that causes galactokinase deficiency (OMIM 230200) in several European populations reveals how people with this autosomal recessive disorder migrated (**figure 15.4**). Galactokinase deficiency causes cataracts (clouding of the lens) in infants. It is very common among a population

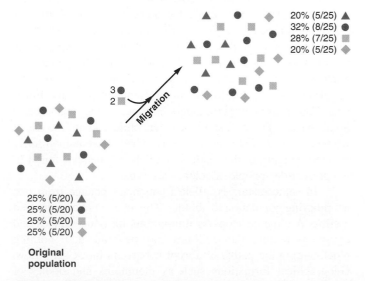

Figure 15.3 **Migration alters allele frequencies.** If the population travels and picks up new individuals, allele frequencies can change.

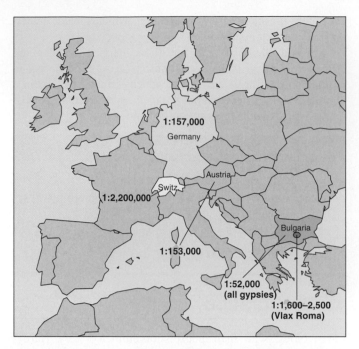

Figure 15.4 Galactokinase deficiency in Europe illustrates a cline. This autosomal recessive disorder that causes blindness varies in prevalence across Europe. It is most common among the Vlax Roma gypsies in Bulgaria. The condition becomes much rarer to the west, as the shading from dark to light green indicates.

of 800,000 gypsies, called the Vlax Roma, who live in Bulgaria. It affects 1 in 1,600 to 2,500 people among them, and 5 percent of the people are carriers. But among all gypsies in Bulgaria as a whole, the incidence drops to 1 in 52,000. As the map in figure 15.4 shows, the disease becomes rarer to the west. This pattern may have arisen when people with the allele settled in Bulgaria, with only a few individuals or families moving westward.

Allele frequencies often reflect who ruled whom. For example, the frequency of ABO blood types in certain parts of the world today mirrors past Arab rule. The distribution of ABO blood types is very similar in northern Africa, the Near East, and southern Spain. These are precisely the regions where Arabs ruled until 1492. The uneven distribution of allele frequencies can also reveal when and where nomadic peoples stopped. For example, in the eighteenth century, European Caucasians called trekboers migrated to the Cape area of South Africa. The men stayed and had children with the native women of the Nama tribe. The mixed society remained fairly isolated, leading to the distinctive allele frequencies found in the present-day people of color of the area.

In some instances, allele frequencies change from one neighboring population to another. This phenomenon is termed a **cline**. A cline develops as immigrants introduce alleles and emigrants remove them. Clines may be gradual, if nothing blocks migration paths, or abrupt if barriers block gene flow. Geographical formations such as mountains and bodies of water may block migration, maintaining population differences in allele frequencies on either side of the barrier. Language differences may also isolate alleles, if people who cannot communicate tend not to have children together. In Italy, for example, certain blood types are more common among people who speak the same dialect than among people who live in the same area but do not share language.

Allele frequencies up and down the lush strip of fertile land that hugs the Nile River illustrate the concept of clines. Researchers found a gradual change in mitochondrial DNA sequences in 224 people who live on either side of the Nile, an area settled 15,000 years ago. The farther apart two individuals live along the Nile, the less alike their mtDNA. This is consistent with evidence from mummies and historical records that indicate the area was once kingdoms separated by wars and language differences. If the region had been one large interacting settlement, then the DNA sequences would have been more evenly distributed.

Key Concepts Questions 15.2

1. How does migration alter allele frequencies?
2. What are clines?
3. Name two factors that can create great differences in allele frequencies.

15.3 Genetic Drift

Genetic drift is a characteristic of all populations. It refers to fluctuations in allele frequencies from generation to generation that happen by chance, to gametes. The allele frequency changes that cause genetic drift occur at random and are unpredictable.

The effects of genetic drift are accelerated when the population becomes very small, and sampling changes allele frequencies. Populations may shrink, amplifying genetic drift, in several circumstances: migration, a natural disaster or geographical barrier that isolates small groups, or the consequences of human behavior (**figure 15.5**). Members of a small ethnic community within a larger population might have children only among themselves, keeping certain alleles more prevalent within the smaller group. For example, the skin-lightening condition vitiligo (OMIM 193200) is much more common in a small community isolated in the mountains of northern Romania than elsewhere in the nation. **Clinical Connection 15.1** discusses genetic diseases among the Old Order Amish and Mennonite populations of North America, in which genetic drift has isolated and amplified disease-causing alleles brought from Europe more than 300 years ago.

Two situations that can accelerate genetic drift are a **founder effect** and a **population bottleneck,** which both decrease population size. A founder effect results when some individuals leave a larger group or become reproductively isolated from them. In contrast, a population bottleneck is a large decrease in the size of an original population.

The Founder Effect

In a founder effect, a small group leaves a population to found a new settlement, and the new colony has different allele frequencies than the original population (**table 15.1**). Geneticists recognize a founder effect in a community known from local history to have descended from a few founders who have inherited traits and illnesses that are rare elsewhere.

A founder effect is easiest to trace when historical or genealogical records are available. This is the case for the 2.5 million Afrikaners of South Africa, who descend from a small group of Dutch, French, and German immigrants who had very large families. In the nineteenth century, some Afrikaners migrated northeast to the Transvaal Province, where they lived in isolation until the Boer War in 1902 introduced better transportation. Today, 30,000 Afrikaners who have porphyria varigata (see figures 5.5 and 5.6) descended from one couple who came from the Netherlands in 1688! Their many children also had large families, passing on and amplifying the dominant mutation.

Another type of evidence for a founder effect is when all individuals in a population with a certain illness have the same mutation, which present-day patients inherited from shared ancestors. The Bulgarian gypsies (see figure 15.4) who have galactokinase deficiency, for example, all have a mutation that is extremely rare elsewhere. In contrast, a population with several mutations that cause the same disorder is more likely to have picked up those variants from unrelated people joining the group.

Very often when a disease-associated allele is identical among people in the same population, so is the DNA surrounding the gene. This pattern indicates that a portion of a chromosome, rather than just the disease-causing gene, has been passed among the members of the population from its founders. For this reason, many studies that trace founder effects examine haplotypes that include tightly linked genes (see figure 16.13).

Founder effects are also evident in more common illnesses, where populations have different mutations in the same gene. *BRCA1* breast cancer, for example, is most prevalent among Ashkenazi Jewish people. Nearly all affected individuals have the same 3-base-pair deletion. In contrast, *BRCA1* breast cancer is rare in blacks, but it affects families from the Ivory Coast in Africa, the Bahamas, and the southeastern United States. They share a 10-base-pair deletion, probably inherited from West Africans ancestral to all three modern groups. Slaves brought the disease to the United States and the Bahamas between 1619 and 1808, but some of their relatives who stayed in Africa have perpetuated the mutant allele there.

Population Bottlenecks

A population bottleneck occurs when many members of a group die, and only a few are left to replenish the numbers. The new population has only those alleles in the small group that survived

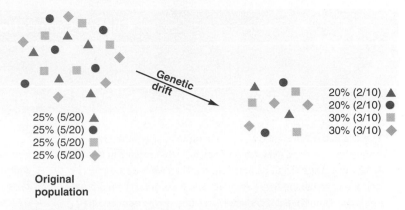

Figure 15.5 Genetic drift alters allele frequencies. If members of a population leave or do not reproduce, allele frequencies can change by chance sampling of a small population. When half of this population does not contribute to the next generation, two genotypes increase in frequency and two decrease.

Table 15.1	Founder Populations		
Population	Number of Founders	Number of Generations	Population Size Today
Costa Rica	4,000	12	2,500,000
Finland	500	80–100	5,000,000
Hutterites	80	14	36,000
Japan	1,000	80–100	120,000,000
Iceland	25,000	40	300,000
Newfoundland	25,000	16	500,000
Quebec	2,500	12–16	6,000,000
Sardinia	500	400	1,660,000

the catastrophe. An allele in the remnant population might become more common in the replenished population than it was in the original larger group. Therefore, the new population has a much more restricted gene pool than the larger ancestral population, with some variants amplified, others diminished.

Population bottlenecks can occur when people (or other animals) colonize islands. An extreme example is seen among the Pingelapese people of the eastern Caroline Islands in Micronesia. Four to 10 percent are born with "Pingelapese blindness," an autosomal recessive combination of colorblindness, nearsightedness, and cataracts also called achromatopsia (OMIM 603096). Elsewhere, only 1 in 20,000 to 50,000 people inherits the condition. Nearly 30 percent of the Pingelapese are carriers. The prevalence of the blindness among the Pingelapese stems from a typhoon in 1780 that killed all but nine males and ten females who founded the present population. This severe population bottleneck, plus geographic and cultural isolation, increased the frequency of the blindness gene as the population resurged.

The Clinic for Special Children: The Founder Effect and "Plain" Populations

The Old Order Amish and Mennonite people, called the Plain Populations, arrived in North America from Switzerland in the early 1700s, to escape religious persecution. They were descendants of a group called the Swiss Anabaptists. The earliest immigrants settled in Pennsylvania, and additional migrations from Europe established small farming communities in Ohio. The Plain people spread across the Midwest and into Canada, isolating themselves and maintaining simpler ways of living. Today more than a million of their descendants live in hundreds of communities in South, Central, and North America. The genetic traits that are overrepresented in these populations originated in Europe and are still there today.

Genetic drift plus a powerful founder effect, nonrandom mating, and migration contributed to the higher prevalence of certain inherited disorders among these people, who lacked access to health care services. Some families, however, would go to Children's Hospital of Philadelphia for diagnosis and treatment. Geneticists took an interest, in helping the families and discovering causes of inherited disease, and the Plain People have thus contributed greatly to our modern knowledge of genetic disease. Now several nonprofit health care centers are bringing genetic testing, exome sequencing, and new therapies to the Plain people. Earlier and more accurate diagnoses are making more treatments possible.

Statistics reveal founder effects in the Plain populations. In Lancaster County, Pennsylvania, for example, maple syrup urine disease (MSUD; OMIM 248600; see Clinical Connnection 2.1) affects 1 in 400 newborns, but affects only 1 in 225,000 newborns in the general population. Some conditions are rare variants of more common disorders, such as "Amish cerebral palsy." A physician from Philadelphia, Holmes Morton, discovered that the disease was an inborn error of metabolism called glutaric acidemia type 1 (OMIM 231670), and not due to lack of oxygen at birth, as others had thought. He went from farm to farm, tracking cases against genealogical records and drawing pedigrees, finding that nearly every family with MSUD traced its roots back to a couple who passed on the recessive alleles. They had come to the settlement in 1730.

Dr. Morton founded the Clinic for Special Children in 1989, constructed at the site of a cornfield in Strasburg, Pennsylvania. The goal was to use genetic tools to diagnose children early, when they were still healthy enough to treat.

Since the late 1990s the Clinic for Special Children has embraced genomic technology. For example, in the case of an Amish infant born with very thick skin, no hair, and a quickly developing life-threatening bacterial infection, comparing genetic markers to those of her seven healthy siblings revealed where her genome differed. A gene in the region, *RAG1*, caused severe combined immune deficiency (OMIM 179615). A stem cell transplant from a sister saved her life.

Research on the Plain people helps the wider community, too. For example, several related Amish children who have autism and seizures led researchers to mutations in a gene, *CNTNAP2* (OMIM 604569), that causes some cases of unexplained autism, seizures, schizophrenia, and language problems in the broader population (**figure 1**). Depression, bipolar disorder, and attention deficit disorder are other conditions that affect many populations, but may stand out among the Plain people.

The Clinic for Special Children's approach of catching genetic disease early and treating symptoms as they arise has

Figure 1 **The Clinic for Special Children.** Genetic drift amplifies mutations from Europe among Plain populations. The child seated with Dr. Kevin Strauss, Medical Director at the Clinic for Special Children, inherited a homozygous single-base deletion in a gene called *CNTNAP2*, which causes seizures and autism (cortical dysplasia-focal epilepsy syndrome, OMIM 610042). The family is Beachy Amish, a sect.

(Continued)

worked. They can now treat nearly half of the 110 genetic disorders that they detect. The clinic can also recognize conditions that are lethal, saving children from pointless and painful treatments that physicians less familiar with the diseases unique to the Plain communities might provide.

The Amish and Mennonites not only have at least 100 rare diseases, but have their own variants of the more common single-gene disorders, including cystic fibrosis, phenylketonuria, fragile X syndrome, and clotting and immune disorders. **Table 1** lists some of these conditions.

Questions for Discussion

1. Public perception is that diseases of the Amish are unique to them. Explain why this is not true.

2. If natural selection removes deleterious alleles from a population, how have certain single-gene disorders become much more prevalent among the Plain people than in other groups?

3. Because the Amish do not permit prenatal diagnosis, inborn errors of metabolism are detected with tests on cord blood that midwives collect. Explain how such analysis can lead to diagnosis of a single-gene disorder, even if the child does not have symptoms.

4. Research a condition listed in Table 1 and describe treatments that may alter the course of a child's life if detected early.

Table 1	Selected Genetic Disorders More Common in Plain Populations	
Disease	**Gene (OMIM#)**	**Symptoms**
Achromatopsia 2	CNGA3 (216900)	Colorblindness
Crigler-Najjar 1 syndrome	UGT1A1 (218800)	Jaundice in newborn
Ellis-van Creveld syndrome	LBN, EVC (225500)	Dwarfism, heart disease, polydactyly, fused wrist bones
Liebenberg syndrome	LBNBG (186550)	Hands develop as feet (see Clinical Connection 3.1)
Malignant hyperthermia	APOA4 (145600)	Death on exposure to certain anesthetics
Periodic fever, familial	TNFRSF1A (142680)	Fever, abdominal pain, skin lesions, muscle pain
Pierson syndrome	LAMB2 (150325)	Scarred kidneys, weak muscles, narrow pupils
Salla disease	SLC17A5 (604369)	Poor muscle tone, uncoordinated movement, intellectual disability, coarse facial features

A more widespread population bottleneck occurred as a consequence of the early human expansion from Africa, discussed in chapter 16. As numbers dwindled during the journeys and then were replenished as people settled down, mating among relatives led, over time, to an increase in homozygous recessive genotypes compared to ancestral populations that maintained their genetic diversity in Africa. These bottlenecks are reflected today in the persistence of genetic diversity among African populations. The lack of genetic diversity in some modern human populations is evident as "runs of homozygosity," which are chromosome regions where alleles of individual genes are identical. Runs of homozygosity represent regions that are inherited from shared ancestors. They are common, for example, in highly purebred dogs.

Figure 15.6 illustrates schematically the dwindling genetic diversity that results from a population bottleneck. Today's cheetahs live in just two isolated populations of a few thousand animals in South and East Africa. Their numbers once exceeded 10,000. The South African cheetahs are so alike genetically that even unrelated animals can accept skin grafts from each other. Researchers attribute the cheetahs' genetic uniformity to two bottlenecks—at the end of the most recent ice age, when habitats changed, and another following mass slaughter by humans in the nineteenth century.

However, the good health of the animals today indicates that the alleles that have persisted enable the cheetahs to thrive in their environment.

Human-wrought disasters that kill many people can cause population bottlenecks that greatly alter gene pools because aggression is typically directed against particular groups, while a typhoon indiscriminately kills anyone in its path. For example, after the many waves of killings, called pogroms, of Jewish people, only a few thousand remained in Eastern Europe by the end of the eighteenth century. Then their numbers grew again, and from 1800 to 1939, the Jewish population in eastern Europe swelled to several million, only to be decimated again by the Holocaust.

Until recently, Jewish people tended to have children only with each other. Both of these factors—nonrandom mating and repeated population bottlenecks—changed allele frequencies and contributed to the incidence of certain inherited diseases seen among the Ashkenazi Jewish people that is ten times higher than in other populations. Several genetic testing companies offer "Jewish genetic disease" panels that are not meant to discriminate or stereotype, but are based on a genetic fact of life—some illnesses are more common in certain populations, due to human behavior. However, DNA itself does not discriminate. The "Jewish"

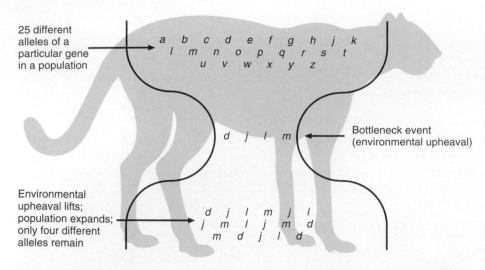

25 different alleles of a particular gene in a population

a b c d e f g h j k
l m n o p q r s t
u v w x y z

d j l m ← Bottleneck event (environmental upheaval)

Environmental upheaval lifts; population expands; only four different alleles remain

d j l m j l
j m l j m d
m d j l d

Figure 15.6 Population bottlenecks. A population bottleneck occurs when the size of a genetically diverse population falls, remains at this level for a time, and then expands again. The new population loses some genetic diversity if alleles are lost.

mutations can arise anew in anyone. **Table 15.2** describes some inherited diseases that are more common among Ashkenazi Jewish populations.

Key Concepts Questions 15.3

1. What is genetic drift?
2. Explain how founder effects and population bottlenecks amplify the effects of genetic drift.

15.4 Mutation

A continual source of genetic variation in populations is mutation, which changes one allele into another (**figure 15.7**). Genetic variability also arises from crossing over and independent assortment during meiosis, but these events recombine existing traits rather than introduce new ones.

If a DNA base change occurs in a part of a gene that encodes part of a protein necessary for its function, then an altered trait may result. Another way that genetic change can occur from generation to generation is in the numbers of repeats of copy number variants (CNVs), which function as alleles.

Natural selection, discussed in the next section, eliminates alleles that adversely affect reproduction. Yet harmful recessive alleles are maintained in heterozygotes and are reintroduced by new mutation. Therefore, all populations have some alleles that would be harmful if homozygous. The collection of such deleterious alleles in a population is called its **genetic load**.

Human behavior and the events of history can influence the diversity of mutations in a population. Consider phenylketonuria (PKU), an autosomal recessive condition that causes intellectual disability unless a special diet restricts one amino acid type. PKU mutations worldwide are very diverse, indicating that the disease has arisen anew more than once. Conversely, mutations that are found in many groups of people are probably more ancient, having occurred before those groups separated.

Table 15.2	Autosomal Recessive Genetic Diseases Prevalent Among Ashkenazi Jewish Populations		
Disorder	**Gene (OMIM#)**	**Signs and Symptoms (Phenotype)**	**Carrier Frequency**
Bloom syndrome	*RECQL3* 210900	Sun sensitivity, short stature, poor immunity, impaired fertility, increased cancer risk	1/110
Breast cancer	*BRCA1 BRCA2* 113705, 600185	Malignant breast tumor	3/100
Canavan disease	*ASPA* 271900	Brain degeneration, seizures, developmental delay, early death	1/40
Familial dysautonomia	*IKAKAP* 223900	No tears, cold hands and feet, skin blotching, drooling, difficulty swallowing, excess sweating	1/32
Gaucher disease	*GBA* 231000	Enlarged liver and spleen, bone degeneration, nervous system impairment	1/12
Niemann-Pick disease type A	*SMPD1* 257200	Lipid accumulation in cells, particularly in the brain; intellectual and physical disability, death by age 3	1/90
Tay-Sachs disease	*HEXA* 272800	Brain degeneration causing intellectual disability, paralysis, blindness, death by age 4	1/26
Fanconi anemia type C	*FANCA* 227650	Deficiencies of all blood cell types, poor growth, increased cancer risk	1/89

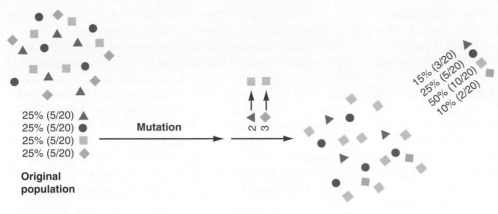

Figure 15.7 **Mutation alters allele frequencies.** If one allele changes into another from one generation to the next, genotype frequencies can change.

Mutations found only in a small geographical region are more likely to be of recent origin, perhaps set apart due to genetic drift. These mutations have had less time to spread. For example, people from Turkey, Norway, Jewish people from Yemen living in Israel, and French-Canadians have their "own" PKU alleles. The Yemeni mutation is so distinctive—a large deletion, compared to point mutations in other populations—that researchers used court and religious records to trace the mutation from families leaving the capital of Yemen north to three towns, and then to Israel.

The contribution that mutation makes to counter Hardy-Weinberg equilibrium is small compared to the influence of migration and nonrandom mating, because mutation is rare. Natural selection has the greatest influence. The spontaneous mutation rate is about 170 bases per haploid genome in each gamete. Each of us probably has at least five "lethal equivalents"—alleles or allele combinations that if homozygous would kill us or make us too sick to have children.

15.5 Natural Selection

Environmental change can alter allele frequencies when individuals with certain phenotypes are more likely to survive and reproduce than others. This differential survival to reproduce guided by environmental change is **natural selection (figure 15.8)**. The chapter opener chronicles natural selection acting on gene variants that enables people to digest the sugar lactose.

Inability to digest lactose is actually the wild type condition, because it predominated before people began domesticating mammals and drinking their milk. Most of the mutations that introduced lactase persistence are point mutations. Another dietary illustration of natural selection involves copy number variants (CNVs). Members of populations that follow high-starch diets tend to have more copies of the gene that encodes salivary amylase, the digestive enzyme that begins to break down starch in the mouth. Members of populations that follow low-starch diets have fewer copies of the gene, and presumably less of the enzyme.

In natural selection, reproductive success is all-important, because this is what transmits favorable alleles and weeds out the unfavorable ones, ultimately impacting population structure and therefore driving microevolution. In the common phrase used synonymously with natural selection—"survival of the fittest"—"fit" actually refers to reproductive success, not to physical prowess or intelligence (unless those traits lead directly to reproductive success). In a Darwinian sense, an unattractive and out-of-shape parent of ten is more "fit" than a gorgeous triathlete with one child.

Negative and Positive Selection

Natural selection acts negatively if a trait diminishes in a population because it adversely affects reproduction. Natural selection acts positively if a trait becomes more common in a population because it enhances the chances of reproductive success. However, "negative" and "positive" selection are not

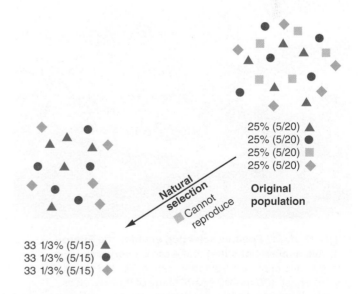

Figure 15.8 **Natural selection alters allele frequencies.** If health conditions impair the ability of individuals who have certain alleles to reproduce, allele frequencies can change.

value judgments—they simply refer to a trait's decreasing in a population or increasing. In Darwin's time, natural selection was thought to be primarily negative. Today, the ability to sequence DNA has revealed "signatures" of positive selection in the human genome.

One sign of positive selection is a gene in humans that has a counterpart in other primates, but in humans has at least one distinctive difference in the amino acid sequence, not just the DNA sequence. A change in the DNA sequence that does *not* substitute an amino acid does not change the protein, and therefore has no effect on the phenotype. Such a change isn't subject to natural selection, which acts on phenotypes.

The allele for lactase persistence illustrates positive selection. The allele is part of a large haplotype (genes linked on a chromosome) that shows very little sequence diversity among individuals. The persistence of the DNA sequence of an allele that has greatly increased in frequency in the population over a very short time—since agriculture began—indicates positive selection. Chapter 16 offers several examples of traits that were positively selected in human evolution.

A dramatic example of positive selection in humans is seen in native people who live more than two miles above sea level on the Tibetan plateau (**figure 15.9**). The Tibetans thrive in the oxygen-poor air that makes others very ill and not likely to reproduce. Even the Han Chinese, who are a related population that lives in the lowlands, cannot survive up in the mountains for long without developing the headache, dizziness, ringing in the ears, heart palpitations, breathlessness, fatigue, sleep disturbance, lack of appetite, and confusion of chronic mountain sickness. If they ascend to great heights, their bloodstreams become crammed with extra red blood cells that extract maximal oxygen from the thin air, raising the risk of a heart attack and stroke.

In people other than the native Tibetans exposed to low oxygen conditions, a protein called hypoxia inducible factor 2 signals the kidneys to release the hormone EPO (erythropoietin), which stimulates the bone marrow to make more red blood cells, increasing hemoglobin levels in the blood. Native Tibetans have eight unique SNPs in the gene *EPAS1* that encodes hypoxia inducible factor 2. Their apparently unique version of the protein enables them to thrive in the thin air. Whole-genome sequencing has revealed two other genes that affect adaptation to thin air. Because the Tibetans have resided in the area only a few thousand years, the adaptation to low oxygen levels is an example of very rapid and recent evolution.

Artificial Selection

Natural selection acts on preexisting genetic variants. It is uncontrolled and largely unpredictable. In contrast, artificial selection is controlled breeding to perpetuate individuals with a particular phenotype, such as a crop plant or purebred dog. Darwin's idea of natural selection grew from his observations of artificially selected pigeons.

We have created our pets by controlling their evolution (**figure 15.10**). Pets arose from initial domestication from wild species, followed by artificial selection. Cats were domesticated in the Near East when agriculture began, about 10,000 years ago. They descended from one of five subspecies of wildcats. Dogs were domesticated in Southeast Asia about 40,000 years ago, from gray wolves. Most of today's dog breeds were artificially selected in the 1800s, but some, such as the Australian bulldog and the silken windhound, were bred in the 1990s. The intense inbreeding required to fashion breeds cuts their genetic diversity, resulting in more than 300 inherited diseases, from bladder stones in Dalmatians to hip problems in St. Bernards. Yet the differences among dog breeds for some traits stem from

Figure 15.9 Positive selection enables the Sherpa to thrive at elevations that make others very sick. The Sherpa have migrated from their native Tibet to Nepal over the past 300 to 400 years. Many of them work as mountaineering guides, especially for climbers ascending Mt. Everest. They are short, strong, and hardy. A variant of the hypoxia inducible factor 2 (*EPAS1*) gene is responsible for their astonishing adaptation to low oxygen conditions.

Figure 15.10 Dogs small and large. It is hard to believe that the diminutive Chihuahua and the Great Dane are members of the same species. Thanks to artificial selection and intense inbreeding, body size in domesticated dogs varies more than it does in any other terrestrial mammal.

only a few genes. For example, variants in only three genes account for 95 percent of the variability in the texture and length of canine coats.

The recent history of dog domestication explains why only a few genes define breeds. We select oddities and quirks in dogs, and these are often the consequence of a mutation in a single gene, rather than the collection of variants in many genes that may underlie a multifactorial trait that hasn't been artificially selected. In nature, that mutant may not have survived; we intentionally breed to select it. For example, ear floppiness and dwarfism, both seen in basset hounds, are each the result of a mutation in a single gene.

Analysis of the genomes of domesticated dogs and their wild relatives confirmed what we know from history—that our pets are no longer very much like wolves. Specifically, linkage disequilibrium (blocks of SNPs next to each other on chromosomes) tends to distinguish modern breeds. The fact that breeds share few SNP blocks indicates that we have selected away many ancestral DNA sequences. In addition, runs of homozygosity—long DNA sequences identical on both chromosomes—reflect intense inbreeding. The fact that domesticated breeds differ from each other less than wild dogs differ from each other indicates population bottlenecks that accompanied dog breeding, narrowing their gene pools.

Three Examples of Natural Selection in Action

Microevolution is compellingly illustrated in three health care challenges: tuberculosis, HIV infection, and antibiotic resistance.

Tuberculosis and Natural Selection

Natural selection drives the appearance, spread, and return of infectious diseases. If infection kills before the host's reproductive age or impairs fertility, its spread will ultimately remove from the population individuals susceptible to infection. Disease incidence falls as only resistant survivors remain. If conditions change, the disease may resurge. This has happened with tuberculosis (TB).

Humans have had an interesting evolutionary relationship with the bacterium that causes tuberculosis for at least 70,000 years. The effects of natural selection are evident in recent history. When TB first appeared in the Plains Indians of the Qu'Appelle Valley Reservation in Saskatchewan, Canada in the mid-1880s, it struck swiftly, infecting many organs. Ten percent of the population died. But by 1921, TB tended to affect only the lungs, and only 7 percent of the population died annually from it. By 1950, mortality was down to 0.2 percent. Some people were symptomless carriers.

Outbreaks of TB ran similar courses in other human populations. The disease appeared in crowded settlements where the bacteria easily spread in exhaled droplets. In the 1700s, TB raged through the cities of Europe. Immigrants brought it to U.S. cities. But TB incidence and virulence fell dramatically in the cities of the industrialized world in the first half of the

twentieth century—before antibiotic drugs were discovered. What tamed tuberculosis?

Natural selection, operating on both the bacterial and human populations, lessened the virulence of the infection. Some people inherited resistance and passed this beneficial trait on. At the same time, the most virulent bacteria killed their hosts so quickly that the victims had no time to spread the infection and they were too sick to have many children. As the deadliest bacteria were selected out of the population (negative selection), and as people who inherited resistance mutations contributed more to the next generation (positive selection), TB gradually evolved from a severe, acute, systemic infection to a rare chronic lung infection. Then in the late 1980s, conditions ideal for the infection's return arose.

At first, complacency led to the resurgence of TB. Then the U.S. government decreased funding for TB research because the infection was considered "cured." Patients thought themselves cured when antibiotics helped in a few months, abandoning the drugs yet unknowingly continuing to spread the bacteria. With increased air travel, people began spreading different strains of the bacteria around the world. Treatment in the 1950s—isolating patients for 18 months or longer in facilities called sanitoria—was actually more effective than antibiotic drugs because it quarantined infectious individuals (**figure 15.11**). Then AIDS happened, providing millions of vulnerable human lungs. The bacterial populations soared, mutations accumulated in 39 genes, and variants resistant to antibiotic drugs arose and had a selective advantage. Today, a third of all HIV-infected people also have TB. Someone with HIV faces a fifty-fold increased risk of contracting TB, and can pass it to anyone, not only people with HIV infection, in just a sneeze or cough. In some parts of the world, such as Russia, tuberculosis is now resistant to so many drugs that health officials have changed the terminology from "multidrug resistant" to "extensively drug resistant."

The observations that bacteria that cause TB are becoming resistant to many types of antibiotic drugs and are increasing in genetic diversity provide powerful evidence for evolution.

Figure 15.11 **Treating tuberculosis the old-fashioned way.** In the 1950s, people with tuberculosis were sent away from the crowds—where the disease easily spread— to isolated "sanitoria," such as this one.

However, new genetic tests that detect infection in under 2 hours may counter these changes. In the past, diagnosis took up to 2 months, and during that time many patients spread the infection.

Evolving HIV

Mutations in viruses accumulate rapidly because their genetic material replicates often and errors are not repaired. Like bacteria, the viruses in a human body form a population, including naturally occurring genetic variants. In HIV infection, natural selection controls the diversity of viral variants in a human body as the disease progresses. The human immune system and drugs to slow the infection are the environmental factors that select (favor) resistant viral variants.

HIV infection proceeds in three stages, both from the human and the viral perspective (**figure 15.12**). Initially, a person infected with HIV may experience an acute phase, with fever, night sweats, rash, and swollen glands. In a second period, lasting from 2 to 15 years, health usually returns. In a third stage, immunity collapses, the virus replicates explosively, and opportunistic infections and cancer may eventually cause death.

The HIV population changes and expands throughout the course of infection, even when the patient seems to stay the same for a long time. New mutants continuously arise, and they alter such traits as speed of replication and the patterns of molecules on the viral surface.

In the first stage of HIV infection, as the person battles acute symptoms, viral variants that replicate swiftly predominate. In the second stage, the immune system starts to fight back and symptoms abate, as viral replication slows and many viruses are destroyed. Now natural selection acts. Certain viral variants reproduce and mutate, giving rise to a diverse viral population. Ironically, drugs used to treat AIDS may further select against the weakest HIV variants. Gradually, the HIV population overtakes the immune system cells, but years may pass before immunity begins to noticeably decline. The third stage, full-blown AIDS, occurs when the virus overwhelms the immune system. With the selective pressure off, viral diversity again diminishes, and the fastest-replicating HIV variants predominate. The timecourse of HIV infection reflects the value of genetic diversity, which enables survival of a population encountering an environmental threat. When that threat—an immune system attack or drugs—wipes out sensitive variants, one genotype may ultimately prevail.

The fact that HIV diversifies early in the course of infection is why the most effective treatment strategy is to take combinations of drugs right after diagnosis. The drugs target several viral variants simultaneously in different ways, slowing the course of the infection. For many people, thanks to declining viral genetic diversity, HIV infection has become a chronic illness rather than the swift killer that it once was.

Antibiotic Resistance

Many antibiotic drugs are no longer effective in treating bacterial infections. Two million antibiotic-resistant infections occur in the United States each year, causing 23,000 deaths.

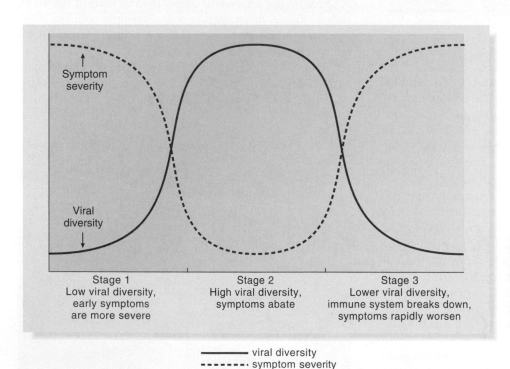

Stage 1
Low viral diversity, early symptoms are more severe

Stage 2
High viral diversity, symptoms abate

Stage 3
Lower viral diversity, immune system breaks down, symptoms rapidly worsen

——— viral diversity
----- symptom severity

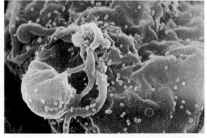

Figure 15.12 **Natural selection controls the genetic diversity of an HIV population in a person's body.** Before the immune system gathers strength, and after it breaks down, HIV diversity is low. A rapidly reproducing viral strain predominates, although new mutations continually arise. During the 2- to 15-year latency period, viral variants that can evade the immune system gradually accumulate. The inset shows HIV (green dots) on a human white blood cell (pink).

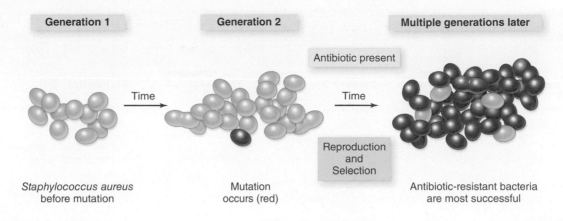

Generation 1   **Generation 2**   **Multiple generations later**

Antibiotic present

Time   Time

Reproduction
and
Selection

Staphylococcus aureus   Mutation   Antibiotic-resistant bacteria
before mutation   occurs (red)   are most successful

Figure 15.13 **Antibiotic resistance.** Resistance to an antibiotic drug develops when a mutation in a bacterial cell gives the cell the ability to survive and reproduce in the presence of the drug. Over time, bacteria harboring the mutation come to take over the population. To the person, this means return or resurgence of the infectious disease.

Antibiotic resistance arises from the interplay between mutation and natural selection.

Genetic variants arise in populations of pathogenic bacteria in our bodies, some of which enable the microorganisms to survive in the presence of a particular antibiotic drug. When an infected person takes the drug, symptoms abate as sensitive bacteria die. However, resistant mutants reproduce, taking over the niche the antibiotic-sensitive bacteria vacated (**figure 15.13**). Soon, enough antibiotic-resistant bacteria accumulate to once again cause symptoms as the immune system responds. Usually antibiotic resistance genes already exist in the bacterial populations, and exposure to the drug selects the resistant bacteria. However, some antibiotics actually induce mutation.

Resistant bacteria circumvent antibiotic actions in several ways. Penicillin kills bacteria by tearing apart their cell walls. Resistant microbes produce enzyme variants that dismantle penicillin, or have altered cell walls that the drug cannot bind. Erythromycin, streptomycin, tetracycline, and gentamicin are antibiotic drugs that kill bacteria by attacking their ribosomes, which are different from ribosomes in a human. Drugs cannot bind the ribosomes of resistant bacteria.

Antibiotic resistance can begin in either of two ways. Bacterial DNA can mutate, passing the resistance from one bacterial generation to the next by cell division. Or, groups of resistance genes are transmitted on mobile pieces of DNA called transposons, which move from cell to cell as part of DNA circles called plasmids. Bacteria usually pass transposons to similar bacteria, but in the unnatural environment of a hospital, genes may flit to any bacterium, quickly passing drug resistance. This is what has happened with infection by the bacterium *Staphylococcus aureus*.

S. aureus is normally present in low numbers in the nose and on the skin, but in high numbers it causes boils, food poisoning, toxic shock syndrome, pneumonia, and wound infections. *S. aureus* became resistant to penicillin soon after the drug was introduced in the 1940s. A related penicillin, methicillin, worked for a time, but resistant bacterial strains appeared in 2000, called MRSA, for methicillin-resistant *S. aureus*. By 2014, MRSA had become a global concern, accounting for

90% of *S. aureus* infections in some nations. Another antibiotic, vancomycin, works against some cases of MRSA infection, but *S. aureus* is becoming resistant to it, too. Some plasmids harbor resistance genes to both drugs.

In the rise of MRSA infection, natural selection benefits the pathogen, not us. That is, bacteria that can resist the drugs that we use to fight them will survive and reproduce, ensuring that *S. aureus* infection continues.

Balanced Polymorphism

If natural selection eliminates individuals with detrimental phenotypes from a population, then how do harmful mutant alleles remain in a gene pool? Harmful recessive alleles are replaced in two ways: by new mutation, and by persistence in heterozygotes.

A recessive condition can remain prevalent if the heterozygote enjoys a health advantage that affects reproduction, such as resisting an infectious disease or surviving an environmental threat. This "heterozygous advantage" that maintains a recessive, disease-causing allele in a population is called **balanced polymorphism**. Recall that *polymorphism* means variant; the effect is *balanced* because the protective effect of the noninherited condition counters the negative effect of the deleterious allele in two copies, maintaining its frequency in the population. Balanced polymorphism is a type of balancing selection, which more generally refers to maintaining heterozygotes in a population. A few examples follow, and these and others are summarized in **table 15.3**.

Sickle Cell Disease and Malaria

Sickle cell disease is an autosomal recessive disorder that causes anemia, joint pain, a swollen spleen, and frequent, severe infections. It is the classic example of balanced polymorphism: carriers are resistant to malaria, or develop very mild cases.

Malaria is an infection by the parasite *Plasmodium falciparum* and related species that causes cycles of chills and fever. The parasite spends the first stage of its life cycle in the salivary glands of the mosquito *Anopheles gambiae*. When an

Table 15.3	Balanced Polymorphism			
Disease 1 (inherited, carrier)	**Protects against** →	**Disease 2**	**Because** →	**Mechanism**
Sickle cell disease		Malaria		Atypical red blood cells cannot retain parasites
G6PD deficiency		Malaria		Parasite cannot reproduce in atypical red blood cells
Phenylketonuria		Fungal infection in fetuses		Elevated phenylalanine inactivates fungal toxin
Prion protein mutation		Transmissible spongiform encephalopathy		Prion protein cannot misfold in presence of infectious prion protein
Cystic fibrosis		Diarrheal disease (cholera, typhoid fever)		Fewer chloride channels in intestinal cells prevent water loss
Smith-Lemli-Opitz syndrome		Cardiovascular disease		Lowered serum cholesterol

infected female mosquito draws blood from a human, malaria parasites enter red blood cells, which transport the parasites to the liver (**figure 15.14a**). The red blood cells burst, releasing parasites throughout the body.

In sickle cell disease, the blood becomes unwelcoming to the malaria parasite in several ways. The blood is thicker than normal, which may hamper the parasite's ability to infect. The bent shape of the cells has a dual effect (**figure 15.14b**). It prevents parasites from growing, and also blocks them from producing a protein required to go to the red blood cell surface and enable the cell to bind to other types of host cells, spreading the infection. In addition, many red blood cells burst too soon, expelling the parasites. A sickle cell disease carrier's blood has enough atypical cells to hamper the actions of the parasite, but usually not enough sickled cells to block circulation. Two other

globin abnormalities that block malaria parasites are alpha thalassemia and hemoglobin C.

A clue to the protective effect of being a carrier for sickle cell disease came from striking differences in the incidence of the two diseases in different parts of the world (**figure 15.15**). In the United States, 8 percent of African Americans are sickle cell carriers, whereas in parts of Africa, up to 45 to 50 percent are carriers. Although Africans had known about a painful disease that shortened life, the sickled cells weren't recognized until 1910 (see section 12.2). In 1949, British geneticist Anthony Allison found that the frequency of sickle cell carriers in tropical Africa was higher in regions where malaria rages all year long. Blood tests from children hospitalized with malaria showed that nearly all were homozygous for the wild type sickle cell allele. The few sickle cell carriers among them had the mildest cases of

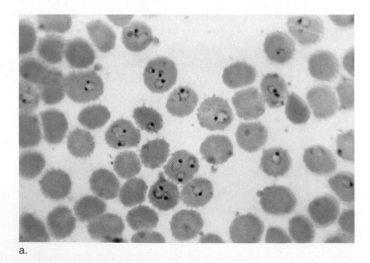
a.

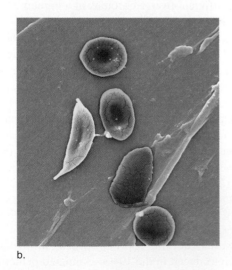
b.

Figure 15.14 **Balanced polymorphism.** **(a)** For part of the life cycle, one parasite that causes malaria, *Plasmodium falciparum,* lives inside red blood cells (dark areas). **(b)** Carriers of sickle cell disease do not contract malaria, or have very mild cases, because the misshapen red blood cells are inhospitable to the parasite.

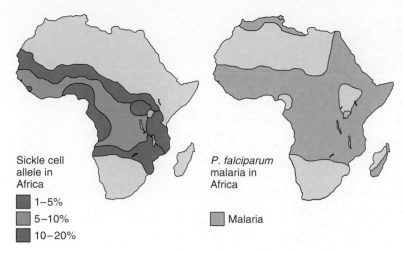

Sickle cell
allele in
Africa

■ 1–5%
■ 5–10%
■ 10–20%

P. falciparum
malaria in
Africa

■ Malaria

Figure 15.15 **Balanced polymorphism.** Comparing the distribution of people with malaria and people with sickle cell disease in Africa reveals balanced polymorphism. Carriers for sickle cell disease are resistant to malaria because changes in the blood caused by the sickle cell allele are not severe enough to impair health, but they do inhibit the malaria parasite.

malaria. Was malaria enabling the sickle cell allele to persist by felling people who did not inherit it? The fact that sickle cell disease is rarer where malaria is rare supports the idea that sickle cell heterozygosity protects against the infection.

Further evidence of a sickle cell carrier's advantage in a malaria-ridden environment is the fact that the rise of sickle cell disease parallels the cultivation of crops that provide breeding grounds for *Anopheles* mosquitoes. About 1000 B.C., Malayo-Polynesian sailors from Southeast Asia traveled in canoes to East Africa, bringing new crops of bananas, yams, taros, and coconuts. When the jungle was cleared to grow these crops, the open space provided breeding grounds for the mosquitoes. The insects, in turn, offered a habitat for part of the life cycle of the malaria parasite.

The sickle cell allele may have been brought to Africa by people migrating from Southern Arabia and India, or it may have arisen directly by mutation in East Africa. However it happened, people who inherited one copy of the sickle cell allele survived or never contracted malaria—the essence of natural selection. These healthy carriers had more children and passed the protective allele to approximately half of them. Gradually, the frequency of the sickle cell allele in East Africa rose from 0.1 percent to 45 percent in 35 generations. Carriers paid the price for this genetic protection, however, whenever two of them produced a child with sickle cell disease.

A cycle set in. Settlements with large numbers of sickle cell carriers escaped malaria. They were strong enough to clear even more land to grow food, and support the disease-bearing mosquitoes. Today, however, in African nations that control malaria well, the frequency of the sickle cell allele has decreased. The selective pressure is off. A Glimpse of History discusses a time when malaria was common in parts of the United States.

A GLIMPSE OF HISTORY

For three centuries, malaria plagued the United States, as human activities opened niches for the mosquitoes that carried the parasites that cause the disease. Starting with Christopher Columbus, European explorers brought the disease, but in mild forms that did not spread. It wasn't until slaves brought the parasite in from western Africa that malaria became a deadly infectious disease in the United States.

By 1776, swarms of mosquitoes carried malaria from Georgia up through Pennsylvania, as pioneers took the disease westward. Clearing forests and grasslands and digging canals opened up vast new environments for the mosquitoes. Some Africans were protected because they were heterozygous (carriers) for sickle cell disease, but Caucasians were vulnerable, and Native Americans the most vulnerable of all. Settlers from France and Spain brought the disease to Louisiana and Texas, respectively. By 1850, summertime epidemics of malaria stretched from Florida up through the middle of New England, and even into the valleys of California. By the 1920s, public health workers began to use dynamite to dig drainage ditches to remove the mosquitoes' habitat (**figure 1**).

By the beginning of the Second World War, malaria in the United States was largely confined to the South, but watery conditions around military training grounds again spread the disease. In 1942, the Office of Malaria Control in War Areas began a huge effort to eradicate the disease. Nearly 5 million homes were sprayed with DDT, and residents used mosquito netting at night, as is common today in parts of Africa. By 1951, the disease was gone. The Office of Malaria Control in War Areas eventually became the Centers for Disease Control and Prevention. Today only a thousand or so cases of malaria are reported in the country, nearly all from travelers to areas where malaria is endemic. Had malaria not been stopped in the United States, natural selection might have favored carriers of sickle cell disease.

Figure 1 **Stopping the spread of malaria.** In the 1920s, public health workers blew up tree stumps to drain the standing water in which mosquito larvae thrive.

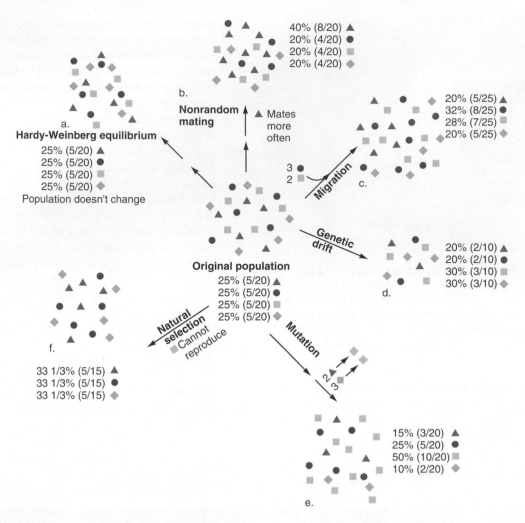

Figure 15.16 Forces that change allele frequencies.

Cystic Fibrosis and Diarrheal Disease

Balanced polymorphism may explain why CF is so common—its cellular defect protects against certain diarrheal illnesses. Diarrheal disease epidemics have left their mark on many human populations, and continue to be a major killer in the developing world.

Severe diarrhea rapidly dehydrates the body and leads to shock, kidney and heart failure, and death in days. In cholera, bacteria produce a toxin that opens chloride channels in cells of the small intestine. As salt (NaCl) leaves the intestinal cells, water rushes out, producing diarrhea. The CFTR protein does just the opposite, closing chloride channels and trapping salt and water in cells, which dries out mucus and other secretions. A person with CF is very unlikely to contract cholera, because the toxin cannot open the chloride channels in the small intestine cells.

CF carriers enjoy the mixed blessing of balanced polymorphism. They do not have enough abnormal chloride channels to cause the labored breathing and clogged pancreas of CF, but they have enough of a defect to block the cholera toxin.

During cholera epidemics throughout history, individuals carrying mutant CF alleles had a selective advantage, and they disproportionately transmitted those alleles to future generations.

Because CF arose in western Europe and cholera originated in Africa, an initial increase in CF heterozygosity may have been a response to a different diarrheal infection—typhoid fever. The causative bacterium, *Salmonella typhi,* rather than producing a toxin, enters cells lining the small intestine—but only if CFTR channels are present. The cells of people with severe CF manufacture CFTR proteins that never reach the cell surface, and therefore bacteria cannot get in. Cells of CF carriers admit some bacteria. Protection against infections that produce diarrhea may therefore have kept CF in populations.

Human societies are highly complex, and so the forces of evolutionary change—nonrandom mating, migration, genetic drift, mutation, and natural selection—act and interact all the time, to different degrees for different traits and illnesses. **Figure 15.16** summarizes the forces contributing alone, and **table 15.4** lists examples in the chapter with the mechanisms that they illustrate.

Table 15.4	Forces that Change Allele Frequencies	

Mechanism of Allele Frequency Change	Examples
Nonrandom mating	Agriculture Hopi Indians with albinism Genghis Khan's Y chromosome
Migration	Galactokinase deficiency in Europe ABO blood type distribution Clines along the Nile
Genetic drift	
Founder effect	Disorders among Old Order Amish and Mennonites Afrikaners and porphyria variegata
Population bottleneck	Pingelapese blindness Cheetahs Pogroms against Ashkenazi Jews
Mutation	Chapters 12 and 13
Natural selection	Lactose intolerance Tibetan adaptation to high altitude TB incidence and virulence HIV infection Antibiotic resistance in bacteria Sickle cell disease and malaria CF and diarrheal disease

Key Concepts Questions 15.5

1. How do cycles of infectious disease prevalence and virulence reflect natural selection?

2. Why don't disease-causing alleles vanish from populations?

3. Explain how balanced polymorphism maintains an inherited disease.

15.6 Eugenics

We usually think of artificial selection in the context of Darwin's pet pigeons or purebred cats, dogs, and horses. We also practice artificial selection through control of our own reproduction, through individual choices as well as at the societal level.

Some people attempt to control the genes in their offspring. They do this by seeking mates with certain characteristics, by choosing egg or sperm donors with particular traits for use in assisted reproductive technologies (see Chapter 21), or by ending pregnancies after a test reveals or predicts a devastating

disorder. On a societal level, **eugenics** refers to programs or policies that control human reproduction with the intent of changing the genetic structure of the population. Eugenics works in two directions. Positive eugenics creates incentives for reproduction among those considered superior; negative eugenics interferes with reproduction of those judged inferior. Obviously, eugenic measures are highly subjective. **Table 15.5** lists some famous examples of eugenics, and *Bioethics: Choices for the Future* on page 297 considers a personal viewpoint.

The word "eugenics" was coined in 1883 by Sir Francis Galton to mean "good in birth." He defined eugenics as "the science of improvement of the human race germplasm through better breeding." One vocal supporter of the eugenics movement was Sir Ronald Aylmer Fisher. In 1930, he tried to apply the principles of population genetics to human society, writing that those at the top of a society tend to be "genetically infertile," producing fewer children than the less-affluent classes. This difference in family size, he claimed, was the reason why civilizations ultimately topple. He offered several practical suggestions to remedy this situation, including state monetary gifts to high-income families for each child born to them.

Early in the twentieth century, eugenics focused on maintaining purity. One prominent geneticist, Luther Burbank, realized the value of genetic diversity at the beginning of a eugenic effort. Known for selecting interesting plants and crossing them to breed plants with useful characteristics, Burbank in 1906 applied his ideas to people. In a book called *The Training of the Human Plant,* he encouraged immigration to the United States so that advantageous combinations of traits would appear as the new Americans interbred. Burbank's plan ran into problems, however, at the selection stage, which allowed only those with "desirable" traits to reproduce.

On the East Coast of the United States, Charles Davenport led the eugenics movement. In 1910, he established the Eugenics Record Office at Cold Spring Harbor, New York. There he headed a massive effort to compile data from institutions, prisons, circuses, and general society. He attributed nearly every trait to a single gene. "Feeblemindedness," for example, was an autosomal recessive trait. It was a catch-all phrase for a person with low intelligence (as measured on an IQ test) and such characteristics as "criminality," "promiscuity," and "social dependency." In one famous case, a young woman named Carrie Buck was ordered to be sterilized when she, her mother, and her illegitimate infant daughter Vivian were declared feebleminded. Carrie had been raped by a relative of her foster parents, and was actually an average student. A crude pedigree drawn at the time showed Carrie Buck and her "inherited trait" of feeblemindedness. **Figure 15.17** shows the unveiling of a roadside plaque honoring Carrie Buck near her hometown.

Other nations practiced eugenics. From 1934 until 1976, the Swedish government forced certain individuals to be sterilized as part of a "scientific and modern way of changing society for the better," according to one historian. At first, only mentally ill people were sterilized, but poor, single mothers were later included. The women's movement in the 1970s pushed for an end to forced sterilizations.

Table 15.5	A Chronology of Eugenics-Related Events
1883	Sir Francis Galton coins the term *eugenics*.
1889	Sir Francis Galton's writings are published in the book *Natural Inheritance*.
1896	Connecticut enacts law forbidding sex with a person who has epilepsy or is "feebleminded" or an "imbecile."
1904	Galton establishes the Eugenics Record Office at the University of London to keep family records.
1907	First eugenic law in the United States orders sterilization of institutionalized intellectually disabled males and criminal males when experts recommend it.
1910	Eugenics Record Office founded in Cold Spring Harbor, New York, to collect family and institutional data.
1924	Immigration Act limits entry into the United States of "idiots, imbeciles, feebleminded, epileptics, insane persons," and restricts immigration to 7 percent of the U.S. population from a particular country according to the 1890 census—keeping out those from southern and eastern Europe.
1927	Supreme Court (*Buck vs. Bell*) upholds compulsory sterilization of the intellectually disabled by a vote of 8 to 1, leading to many state laws.
1934	Eugenic sterilization law of Nazi Germany orders sterilization of individuals with conditions thought to be inherited, including epilepsy, schizophrenia, and blindness, depending upon rulings in Genetic Health Courts.
1939	Nazis begin killing 5,000 children with birth defects or intellectual disability, then 70,000 "unfit" adults.
1956	U.S. state eugenic sterilization laws are repealed, but 58,000 people have already been sterilized.
1965	U.S. immigration laws reformed, lifting many restrictions.
1980s	California's Center for Germinal Choice established, where Nobel Prize winners can deposit sperm to inseminate selected women.
1990s	In the U.S., state laws passed to prevent health insurance or employment discrimination based on genotype.
2003	Many governments recommend certain genetic tests, and enact legislation to prevent genetic discrimination.
2004	Genocide of black Africans in Sudan.
2009	U.S. Genetic Information Nondiscrimination Act enacted, but is limited in scope.

Figure 15.17 Eugenics aimed to abolish "feeblemindedness." In 1927, 17-year-old Carrie Buck, of Charlottesville, stood trial for the crime of having a mother who lived in an asylum for the feebleminded, and having a daughter out-of-wedlock (following rape) also deemed feebleminded, as was Carrie herself, though she was a B student in school. Ruled Sir Oliver Wendell Holmes, Jr., "three generations of imbeciles are enough." Carrie Buck was the first person sterilized to prevent having another "socially inadequate offspring." This roadside plaque is to remember her and eugenics laws in the United States.

Two Views of Neural Tube Defects

Genetic tests enable people to make reproductive choices that can alter allele frequencies in populations. Identifying carriers of a recessive illness, who then may decide not to have children together, is one way to remove disease-causing alleles from a population. Screening pregnant women for fetal anomalies, then terminating affected pregnancies, also alters disease prevalence and, if the disorder has a genetic component, allele frequencies. This is the case for neural tube defects (NTDs), which are multifactorial.

An NTD forms at the end of the first month, when the embryo's neural tube does not completely close. An opening in the head (anencephaly) usually ends in miscarriage, stillbirth, or a newborn who dies within days. An opening in the spinal cord (spina bifida) causes paralysis but the person can live into adulthood and have normal intelligence. Surgery can help to preserve functions.

Many people informed that their fetus has a neural tube defect end the pregnancy. Blaine Deatherage-Newsom has a different view of population screening for NTDs because he has one. Blaine was born in 1979 with spina bifida. Paralyzed from the armpits down, he has endured much physical pain, but he has also achieved a great deal. While in high school, he put the question, "If we had the technology to eliminate disabilities from the population, would that be good public policy?" on the Internet—initiating a global discussion. He wrote:

> I was born with spina bifida and hydrocephalus. I hear that when parents have a test and find out that their unborn child has spina bifida, in more than 95 percent of the cases they choose to have an abortion. I also went to an exhibit at the Oregon Museum of Science and Industry several years ago where the exhibit described a child born with spina bifida and hydrocephalus, and . . . asked people to vote on whether the child should live or die. I voted that the child should live, but when I voted, the child was losing by quite a few votes.

> When these things happen, I get worried. I wonder if people are saying that they think the world would be a better place without me. I wonder if people just think the lives of people with disabilities are so full of misery and suffering that they think we would be better off dead. It's true that my life has suffering (especially when I'm having one of my 11 surgeries so far), but most of the time I am very happy and I like my life very much. My mom says she can't imagine the world without me, and she is convinced that everyone who has a chance to know me thinks that the world is a far better place because I'm in it.

Today Blaine works for a not-for-profit organization that refurbishes computer equipment for community service organizations.

Questions for Discussion

1. Is the decision to end a pregnancy that would otherwise lead to the birth of a child with a neural tube defect a eugenic measure or not? State a reason for your answer.

2. People with certain medical conditions or limitations, such as those with hearing loss, object to genetic tests that would ultimately decrease their numbers in the population. How would you feel if you had such a condition?

3. Do you think that eugenics will resurge as personal genome sequencing and genetic testing become more widespread?

Excerpt by Blaine Deatherage-Newsom, "If we could eliminate disabilities from the population, should we? Results of a survey on the Internet." Reprinted by permission.

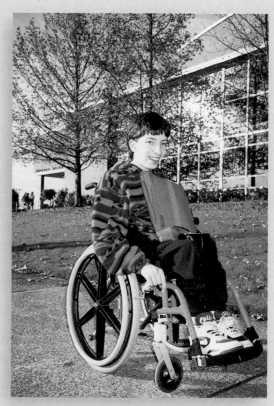

Blaine Deatherage-Newsom as a teen.

In 1994, China passed the Maternal and Infant Health Care Law. It proposed "ensuring the quality of the newborn population" and forbade procreation between two people if physical exams showed "genetic disease of a serious nature" that included intellectual disability, mental illness, and seizures, conditions that are ill-defined in the law and, if inherited, typically multifactorial.

Another guise of eugenics is war, if the fighting groups differ genetically. Throughout history, war and conflict have altered gene pools, sometimes dramatically. These effects are eugenic when they take the form of rape of women of one group by men from another, with the intent of "diluting" the genes of the rape victims. In recent years in Rwanda, Congo, and Darfur, the conquerors claimed that their intent was to diminish the genetic contributions of their victims and spread their own genes. That statement is a definition of eugenics.

Modern genetics is sometimes compared to eugenics because genetic technologies may affect reproductive choices and can influence which alleles are passed to the next generation. However, medical genetics and eugenics differ in their intent. Eugenics aims to allow only people with certain "valuable" genotypes to reproduce, for the supposed benefit of the population as a whole. The goal of medical genetics, in contrast, is to prevent and alleviate suffering.

Humanity arose in Africa some 200,000 years ago. As pockets of peoples spread across the globe, our behaviors and intelligence introduced culture. The next chapter explores some of our journeys, through clues to the past in the sequences of our DNA.

Key Concepts Questions 15.6

1. What is eugenics?
2. Distinguish between positive and negative eugenics.
3. How do eugenics and medical genetics differ?

Summary

15.1 Nonrandom Mating

1. Hardy-Weinberg equilibrium assumes all individuals mate with the same frequency and choose mates without regard to phenotype. This rarely happens. We choose mates based on certain characteristics, and some people have many more children than others.

2. DNA sequences that do not cause a phenotype important in mate selection or reproduction may be in Hardy-Weinberg equilibrium.

3. Consanguinity increases the proportion of homozygotes in a population, which may lead to increased incidence of recessive illnesses or traits.

15.2 Migration

4. **Clines** are changes in allele frequencies from one area to another. They reflect geographical barriers or linguistic differences and may be abrupt or gradual.

5. Human migration patterns through history explain many cline boundaries. Forces behind migration include escape from persecution and a nomadic lifestyle.

15.3 Genetic Drift

6. **Genetic drift** is the random fluctuation of allele frequencies from generation to generation. New allele frequencies may result from chance sampling.

7. A **founder effect** occurs when a few individuals found a settlement and their alleles form a new gene pool, amplifying their alleles and eliminating others.

8. A **population bottleneck** is a narrowing of genetic diversity that occurs after many members of a population die and the few survivors rebuild the gene pool.

9. Founder effects and population bottlenecks can amplify the impact of genetic drift.

15.4 Mutation

10. Mutation continually introduces new alleles into populations. It can occur as a consequence of DNA replication errors.

11. The **genetic load** is the collection of deleterious alleles in a population.

15.5 Natural Selection

12. Environmental conditions influence allele frequencies via **natural selection**. Alleles that do not enable an individual to reproduce in a particular environment are selected against and diminish in the population, unless conditions change. Beneficial alleles are retained.

13. In **balanced polymorphism,** the frequencies of some deleterious alleles are maintained when heterozygotes have a reproductive advantage under certain conditions.

15.6 Eugenics

14. **Eugenics** is the control of individual reproduction to serve a societal goal.

15. Positive eugenics encourages those deemed superior to reproduce. Negative eugenics restricts reproduction of those considered inferior.

16. Some aspects of genetic technology affect reproductive choices and allele frequencies, but the goal is to alleviate or prevent suffering, not to change society.

www.mhhe.com/lewisgenetics11

Answers to all end-of-chapter questions can be found at **www.mhhe.com/lewisgenetics11.** You will also find additional practice quizzes, animations, videos, and vocabulary flashcards to help you master the material in this chapter.

Review Questions

1. Give examples of how each of the following can alter allele frequencies from Hardy-Weinberg equilibrium:
   a. nonrandom mating
   b. migration
   c. a population bottleneck
   d. mutation

2. How might a mutant allele that causes an inherited illness in homozygotes persist in a population?

3. Why can increasing homozygosity in a population be detrimental?

4. Describe two scenarios in human populations, one of which accounts for a gradual cline, and one for an abrupt cline.

5. How does a knowledge of history, sociology, and anthropology help geneticists to interpret allele frequency data?

6. Define *genetic drift*.

7. How does a founder effect or a population bottleneck amplify the effect of genetic drift?

8. Explain the influence of natural selection on
   a. the virulence of tuberculosis.
   b. the changing degree of genetic diversity in an HIV population during infection.
   c. bacterial resistance to antibiotics.

9. Explain how negative and positive selection can operate on a population at the same time.

10. What do artificial selection and natural selection have in common and how do they differ?

11. Give an example of an inherited disease allele that protects against an infectious illness.

12. How do genetic drift, nonrandom mating, and natural selection interact?

13. How have human activities contributed to balanced polymorphism between sickle cell disease and malaria? What effect, if any, might climate change have on the carrier frequency for sickle cell disease?

14. Cite three examples of eugenic actions or policies.

15. Distinguish between positive and negative selection, and between positive and negative eugenics. How do selection and eugenics differ?

Applied Questions

1. Begin with the original population represented at the center of figure 15.16, and deduce the overall, final effect of the following changes:

   ■ Two yellow square individuals join the population when they stop by on a trip and stay awhile.

   ■ Four red circle individuals are asked to leave as punishment for criminal behavior.

   ■ A blue triangle man has sex with many females, adding five blue triangles to the next generation.

   ■ A green diamond female produces an oocyte with a mutation that adds a yellow square to the next generation.

   ■ A new infectious disease affects only blue triangles and yellow squares, removing two of each from the next generation.

2. Before 1500 A.D., medieval Gaelic society in Ireland isolated itself from the rest of Europe, physically and culturally. Men in the group are called "descendants of Niall," and they have Y chromosomes inherited from a single shared ancestor. In the society, men took several partners, and sons born out of wedlock were fully accepted. Today, in a corner of northwest Ireland, one in five men has the "descendant of Niall" Y chromosome. In all of Ireland, the percentage of Y chromosomes with the Niall signature is 8.2 percent. In western Scotland, where the Celtic language is similar to Gaelic, 7.3 percent of the males have the telltale Niall Y. In the United States, among those of European descent, it is 2 percent. Worldwide, the Niall Y chromosome makes up only 0.13 percent of the total. What concept from the chapter do these findings illustrate?

3. Ten castaways are shipwrecked on an island. The first mate has blue eyes, the others brown. A coconut falls on the first mate, killing him. The coconut accident is a chance event affecting a small population. What chapter concept does this example illustrate?.

4. The Old Order Amish of Lancaster, Pennsylvania have more cases of polydactyly (extra fingers and toes) than the rest of the world combined. All of the affected individuals descend from the same person, in whom the dominant mutation originated. Does this illustrate a population bottleneck, a founder effect, or natural selection? Give a reason for your answer.

5. Population bottlenecks are evident today in Arab communities, Israel, India, Thailand, Scandinavia, some African nations, and especially among indigenous peoples such as Native Americans. Research one of these populations and describe a genetic condition that its members have that is rare among other groups of people, and how a population bottleneck occurred.

6. The ability to taste bitter substances is advantageous in avoiding poisons, but might keep people from eating bitter vegetables that contain chemicals that protect against cancer. Devise an experiment, perhaps based on population data, to test either hypothesis—that the ability to taste bitter substances is either protective or harmful.

7. The *CCR5* mutation discussed in the Chapter 17 opener that keeps HIV out of human cells also blocks infection by the bacterium that causes plague. Seven centuries ago, in Europe, the "Black Death" plague epidemic increased the protective allele in the population. Today the mutation

makes 3 million people in the United States and the United Kingdom resistant to HIV infection. Is the increase in incidence of this allele due to nonrandom mating or natural selection?

8. Use the information in chapters 14 and 15 to explain why

   a. porphyria variegata is more prevalent among Afrikaners than other South African populations.

   b. cystic fibrosis and sickle cell disease remain common.

   c. the Sherpa from Tibet can tolerate thin air.

   d. the Amish in Lancaster County have a high incidence of genetic diseases that are very rare elsewhere.

   e. the frequency of the allele that causes galactokinase deficiency varies across Europe.

9. Which principles discussed in this chapter do the following science fiction plots illustrate?

   a. In the novel *The Passage*, researchers infect a dozen criminals bound for death row with viruses that turn them into vampires. The infected criminals, called "virals," escape and ravage the Earth, killing 90 percent of the population and turning most of the remaining 10 percent into vampires. A century later, the numbers of infected humans have greatly increased, but not their genetic diversity. Each person has one of the twelve original virally infected human genomes.

   b. In *When Worlds Collide*, 100 people leave a ruined Earth to colonize a new planet.

   c. In *The Time Machine*, set in the distant future on Earth, one group of people is forced to live on the surface while another group is forced to live in caves. Over many generations, the groups diverge. The Morlocks that live below ground have dark skin, dark hair, and are very aggressive, whereas the Eloi that live above ground are blond, fair-skinned, and meek.

   d. In *Children of the Damned*, all of the women in a small town are suddenly impregnated by genetically identical beings from another planet.

   e. In Dean Koontz's novel *The Taking*, giant mutant fungi kill nearly everyone on Earth, sparing only young children and the few adults who protect them. The human race must re-establish itself from the survivors.

10. In Chester's Mill, Maine, a huge, transparent dome suddenly descends and isolates everything within from the rest of the world. If the novel and the TV show that tell this tale ("Under the Dome") were to go on forever, what might happen to allele frequencies within the trapped human population?

11. Syndrome X consists of obesity, type 2 diabetes, hypertension, and heart disease. Researchers sampled blood from nearly all of the 2,188 residents of the Pacific Island of Kosrae, and found that 1,709 of them are part

of the same pedigree. The incidence of syndrome X is much higher in this population than for other populations. Suggest a reason for this finding, and indicate why it would be difficult to study these particular traits, even in an isolated population.

12. By which mechanisms discussed in this chapter do the following situations alter Hardy-Weinberg equilibrium?

   a. In ovalocytosis (OMIM 166910), a protein that anchors the red blood cell plasma membrane to the cytoplasm is abnormal, making the membrane so rigid that parasites that cause malaria cannot enter.

   b. In the mid-1700s, a multitoed male cat from England crossed the sea and settled in Boston, where he left behind many kittens, about half of whom also had extra digits. People loved the odd felines and bred them. Today, in Boston and nearby regions, multitoed cats are much more common than in other parts of the United States.

   c. Many slaves in the United States arrived in groups from Nigeria, which is an area in Africa with many ethnic subgroups. They landed at a few sites and settled on widely dispersed plantations. Once emancipated, former slaves in the South were free to travel and disperse.

   d. About 300,000 people in the United States have Alzheimer disease caused by a mutation in the presenilin-2 gene. They all belong to five families that came from two small villages in Germany that migrated to Russia in the 1760s and then to the United States from 1870 through 1920.

13. Explain how medical treatments can counter natural selection.

14. A challenging environment can either kill individuals whose genetic susceptibilities and characteristics make it difficult to survive or reproduce, or encourage such individuals to migrate to more comfortable surroundings. Describe the different effects of these alternatives on the genetic structure of the population.

15. African Americans develop a form of end-stage kidney disease associated with elevated blood pressure that Europeans do not. Two variants in a gene on chromosome 22, called *ApoL1*, cause the condition. The encoded protein is secreted into the blood, but only the forms in the African Americans who have the kidney disease also kill the parasites that cause African sleeping sickness. The mutations persist because they protect against African sleeping sickness. What phenomenon described in the chapter does this situation illustrate?

16. Describe an event in history that likely led to a population bottleneck.

Web Activities

1. Go to the Centers for Disease Control and Prevention website and access the journal *Emerging Infectious Diseases*. Using this resource, describe an infectious disease that is evolving, and cite the evidence for this.

2. Do a Google search for a pair of disorders listed in table 15.3 (balanced polymorphism) and discuss how the carrier status of the inherited disease protects against the second condition.

3. Go to the Image Archive on the American Eugenics Movement website. Look at several images, and either find one that presents a genetic disorder and describe it, or find an image that presents biologically incorrect information, and explain the error.

Forensics Focus

1. In the 1870s, prison inspector and self-described sociologist Richard Dugdale noticed that many inmates at his facility in Ulster County, New York, were related. He began studying them, calling the family "Jukes," although he kept records of their real names. Dugdale traced the family back seven generations to a son of Dutch settlers, named Max, who was a pioneer and lived off the land. Margaret, "the mother of criminals," as Dugdale wrote in his 1877 book *The Jukes: A Study in Crime, Pauperism, Disease and Heredity,* married one of Max's sons, and the couple supposedly gave rise to 540 of the 709 criminals on Dugdale's watch. Dugdale attributed the Jukes' less desirable traits to heredity.

   The Jukes study influenced social scientists to probe other families with misfits who were all Caucasian, descended from colonial settlers, and poor. Poverty was not seen as an economic problem, but due to inborn degeneracy that if left unchecked would cost society greatly.

   Dugdale's book fed the fledgling eugenics movement. In 1911, researchers at the Eugenics Record Office in Cold Spring Harbor described the Jukes' phenotype as "feeblemindedness, indolence, licentiousness, and dishonesty." The Jukes story and others were used to support compulsory sterilization of those deemed unfit. But the original research on the Jukes family was flawed, and its accuracy never questioned. Less notorious Jukes family members served in respected professions, some even holding public office. The Jukes were vindicated in 2003, when archives at the State University of New York at Albany revealed the original names of the people in Dugdale's account; most were not even related. The Jukes family curse was more legend than fact.

   a. Was the original jailing of the people called Jukes eugenic or not? Cite a reason for your answer.

   b. How could studies on one family harm others?

   c. Cite an example of an idea based on eugenics today or in the recent past.

   d. If you were a contemporary of Dugdale's, what type of evidence would you have sought to counter his ideas?

Case Studies and Research Results

1. Lana S. seemed to be a healthy newborn. In fact, she seemed to be hitting developmental milestones ahead of schedule, trying to lift her head up at only 3 weeks. But then she rapidly lost skills. Her head flopped, she stopped trying to turn over, and her arms and legs became spastic. When she no longer made eye contact, her anxious parents took her to the pediatrician. She referred the family to a pediatric neurologist, who was puzzled. "She has all the symptoms of Canavan disease, but that can't be. She's not Jewish." Explain how the neurologist was incorrect.

2. The population of India is divided into many castes, and the people follow strict rules governing who can marry whom. Researchers compared several genes among 265 Indians of different castes and 750 people from Africa, Europe, and Asia. The study found that the genes of higher Indian castes most closely resembled those of Europeans, and that the genes of the lowest castes most closely resembled those of Asians. In addition, maternally inherited genes (mitochondrial DNA) more closely resembled Asian versions of those genes, but paternally inherited genes (on the Y chromosome) more closely resembled European DNA sequences. Construct an historical scenario to account for these observations.

3. A magazine article featured parents who filed a "wrongful birth" lawsuit against their doctor for failing to offer prenatal testing for spina bifida, which their daughter was born with in 2003—even though they love the child dearly. They will not say whether they would have ended the pregnancy had they known about the birth defect. If they had ended it, would that have been a eugenic act? Explain your answer.

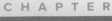

Who were the little people of Flores?

Human Ancestry and Evolution

Learning Outcomes

16.1 Human Origins

1. How can DNA sequences provide information about our ancestry?

2. Describe our ancestors.

3. What can we learn from indigenous peoples about our origins?

16.2 Methods to Study Molecular Evolution

4. How do chromosome banding patterns and protein sequences reveal evolution?

5. What is a "molecular clock"?

6. How are mitochondrial DNA and Y chromosome sequences used to track human ancestry?

7. Explain how haplotypes provide clues to ancient migrations.

16.3 The Peopling of the Planet

8. What does mitochondrial Eve represent?

9. How did people expand out of Africa?

16.4 What Makes Us Human?

10. How does the human genome differ from the genomes of other primates?

11. What traits are unique to humans?

12. List genes that distinguish us from our closest relatives.

 The **BIG** Picture

Our genes and genomes hold clues to our deep past and our present diversity. How will our species continue to evolve?

The Little Lady of Flores

The Nage people, who live on the island of Flores in Indonesia, speak of the Ebu Gogo, short hairy people thought to be mythical—until a team of Australian and Indonesian archaeologists arrived in 2003. They discovered, 17 feet beneath a cave floor, the near-complete skeleton of a female who fit the legendary description, plus pieces of seven other individuals. The ancient remains represent a people named *Homo floresiensis*.

The little people of Flores were half our height, with a brain about half the size of ours but with well-developed frontal lobes, suggesting that they were smart enough to use tools and fire and to hunt. They must have arrived on the island by raft, so some investigators suggest that the people had a language to coordinate the journey. *Homo floresiensis* had large teeth and feet, no chin, and a receding forehead. The little lady weighed about 55 pounds.

The people may have exhibited "island dwarfism," which is an effect of natural selection on small, isolated island populations. With limited resources, individuals who need less food are more likely to survive to reproduce. Over time under these conditions, average body size decreases. The little people hunted local little elephants.

Evidence indicates that the Flores people lived on the island from 95,000 to as recently as 12,000 years ago, but Portuguese traders report having seen the people as recently as the seventeenth century. Some researchers suggest that they may still exist.

Who were these little people? The wrist bones are like those of a chimp, yet the skull is more like that of *Homo erectus,* a more robust human ancestor and contemporary. Were the Flores people modern humans with a medical condition, such as microcephaly (a small head), Laron syndrome (see section 3.6), or thyroid hormone deficiency? Or were they a distinct species? The Flores people's bones are not fossilized, and disintegrate easily. So far researchers have been unable to obtain enough DNA from the bones to determine who they really were—or are.

16.1 Human Origins

Who are we and where did we come from? We have sparse evidence of our beginnings—pieces of a puzzle in time, some out of sequence, many missing. Traditionally, paleontologists (scientists who study evidence of ancient life) have consulted the record in the earth's rocks—fossils—to glimpse the ancestors of *Homo sapiens,* our own species. Researchers assign approximate ages to fossils by observing which rock layers fossils are in, and by extrapolating the passage of time from the ratios of certain radioactive chemicals in surrounding rock.

Fossils aren't the only way to peek into species' origins and relationships. Modern organisms also provide intriguing clues to the past in their DNA. Sequences of DNA change over time due to mutation. Frequencies of gene variants (alleles) change in populations by the forces of nonrandom mating, migration, genetic drift, and natural selection, the topics of Chapter 15.

The premise behind DNA sequence comparisons is that the more closely related two species, the more similar their sequences. Similar DNA sequences are more likely to come from individuals or species sharing ancestors than from the exact same set of spontaneous mutations occurring by chance. By analogy, it is more likely that two women wearing the same combination of clothes and accessories bought them at the same store than that each happened to assemble the same collection of items from different sources. Rarely, DNA from ancient specimens can add to what we know from DNA sequences of modern organisms.

Treelike diagrams are used to depict evolutionary relationships, based on fossil evidence and/or inferred from DNA sequence similarities. Branchpoints on the diagrams represent divergence from shared ancestors. Overall, evolution is shown as a series of branches as species diverged, driven by allele frequencies changing in response to the forces discussed in chapter 15: nonrandom mating, genetic drift, migration, mutation, and natural selection. Evolution is *not* a linear morphing of one type of organism into another—a common misunderstanding. Humans and chimps diverged from a shared ancestor; humans didn't form directly from chimps. Similarly, two

second cousins share great-grandparents, but one cousin did not descend from the other.

This chapter explores human origins and considers how genetic and genomic evidence adds to our view of our evolution. It concludes with a look at more recent events in our ancestry and a consideration of which characteristics distinguish our species.

Our Place in the Primate Family Tree

A species includes organisms that can successfully produce healthy offspring only among themselves. *Homo sapiens* ("the wise human") probably first appeared during the Pleistocene epoch, about 200,000 years ago.

We and animals ancestral only to us are members of a taxonomic (biological classification) "tribe" called the Hominini that diverged from an ancestor we shared with other African apes (gorillas, chimpanzees, and bonobos) about 6 million years ago, in Africa. **Figure 16.1** shows broader taxonomic groups that indicate our place among the primates as well as those of our ancestors mentioned in the chapter.

We know of at least three types of Hominini (also called hominins) who lived shortly (in geological time) after the split from our ancestor shared with chimpanzees: *Ardipithecus kadabba* from Ethiopia, *Sahelanthropus tchadensis* from Chad, and *Orrorin tugenensis* from Kenya. Fossil evidence for these ancestors is very incomplete. We know a little bit more about another hominin, *Ardipithecus ramidus,* who lived more recently, about 4.4 million years ago (**figure 16.2**). "Ardi" was discovered in the Afar rift valley of Ethiopia in 1994, an area whose rocks hold fossilized remains of human ancestors going back nearly 6 million years. From skeletons, paleontologists deduced that Ardi was taller than more ancient hominins and was partly able to walk on two legs (bipedal), perhaps while on the forest floor, but could also use all four limbs to climb trees. Ardi likely ate plants.

Fossil evidence is more complete for our ancestors who lived 2 to 4 million years ago. They walked fully upright and conquered vast new habitats on the plains. Several species of a hominin called *Australopithecus* lived at this time, probably following a hunter-gatherer lifestyle. More than one species could coexist because the animals lived in small, widely separated groups that probably never came into contact. The australopithecines were gradually replaced with members of our own genus, *Homo.* The following sections introduce a few of these ancestors known from rare fossil remains and what our computer modeling and imaginations can fill in. DNA evidence becomes important when we reach the archaic humans.

Australopithecus

The australopithecines had a mix of apelike and humanlike characteristics. They had flat skull bases, as do all modern primates except humans. Australopithecine legs were shorter than ours but their arms were longer. The angle of preserved bones from the pelvis, and discovery of *Australopithecus* fossils with those of grazing animals, suggest that this ape-human had left

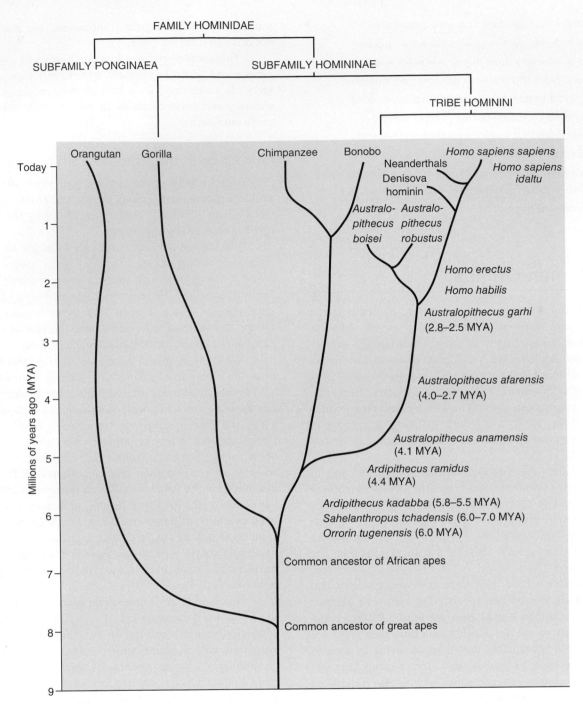

Figure 16.1 Our place on the human family tree. Our closest ancestors, the archaic humans, are just below us at the upper right of the tree.

the forest to venture onto the grasslands. Paleontologists have pieced together what little we know about the australopithecines from fossil finds. These hominins lived too long ago to have left DNA evidence.

The oldest species of australopithecine known, *Australopithecus anamensis,* lived about 4.1 million years ago. A partial skeleton, named Lucy, represents an individual who lived about 3.6 million years ago in the Afar river basin of Ethiopia (**figure 16.3**), where evidence of our ancestors goes

back 6 million years. Lucy was a member of *Australopithecus afarensis.* She was about 4 feet tall, but partial skeletons from slightly older males of her species stood about 5 feet 3 inches tall, and were more humanlike than chimplike. Stone tools found with *A. afarensis* from 3.4 million years ago show distinctive cut marks that indicate these hominins sliced meat from bones and removed, and presumably ate, the marrow. The jaws were strong enough to crack nuts. Lucy died, with arthritis, at about age 20.

Figure 16.2 *Ardipithecus ramidus* **lived about 4.4 million years ago, in Ethiopia.** We do not have very much evidence from this ancestor.

Figure 16.3 *Australopithecus afarensis.* About 3.6 million years ago, Lucy walked upright in the grasses along a lake in the Afar region of Ethiopia, about 6 miles from where the Dikika infant would live 300,000 years later. Lucy likely skimmed the shores for crabs, turtles, and crocodile eggs to eat. The Afar region is the only place known to have evidence of our ancestors that spans 6 million years.

Other fossils offer additional clues to australopithecine life. Two parallel paths of humanlike footprints, preserved in volcanic ash in the Laetoli area of Tanzania, are contemporary with Lucy. A family may have left the prints, which are from a large and small individual walking close together, with a third following in the steps of the larger animal in front. A fossilized 3-year-old girl from 3.3 million years ago, the "Dikika infant," apparently died in a flash flood, perhaps while trying to catch fish in rivulets of the Awash river delta near where Lucy lived 300,000 years earlier.

Toward the end of the australopithecine reign, *Australopithecus garhi* may have coexisted with the earliest members of *Homo*. *A. garhi* fossils from the Afar region date from about 2.5 million years ago. Remains of an antelope found near the australopithecine fossils suggest butchering. The ends of the long bones had been cleanly cut with tools, the marrow removed, meat stripped, and the tongue cleanly sliced off. *A. garhi* stood about 4.5 feet tall, and like the Dikika infant and Lucy, the long legs were like those of a human, but the long arms were more like those of an ape. The small cranium and large teeth hinted at apelike ancestors.

Early Homo

We do not know how *Homo* replaced *Australopithecus*, or even whether there was just one or several species of early *Homo*. Some australopithecines were "dead ends" that died off. Clues suggest that by 2.3 million years ago, *Australopithecus* coexisted with hominins of genus *Homo*. Based on fossil evidence paleontologists have described several *Homo* species, but more recent fossil finds that reveal variability in bone structure may lead to revision in the classification.

The earliest *Homo* species recognized before the new fossil descriptions was *Homo habilis,* who was a more humanlike cave dweller that cared intensively for its young. *Habilis* means handy, and this primate was the first to use tools for tasks more challenging than stripping meat from bones. *H. habilis* may have descended from hominins who ate a more varied diet than other ape-humans, enabling them to live in a wider range of habitats.

H. habilis coexisted with and was followed by *Homo erectus* during the Paleolithic Age (**table 16.1**). One of the first *H. erectus* individuals described, Nariokotome Boy, influenced

Table 16.1	Cultural Ages	
Age	**Time (years ago)**	**Defining Skills**
Paleolithic	750,000 to 15,000	Earliest chipped tools
Mesolithic	15,000 to 10,000	Cutting tools, bows and arrows
Neolithic	10,000 to present	Complex tools, agriculture

descriptions of the species as tall and thin for two decades, until discovery of more specimens revealed that *H. erectus* matured faster than we do, but their heights and builds varied. Another famed *H. erectus* fossil, named "Daka" for the place where he was found in the Afar region, is from an individual who lived about a million years ago. He had a shallow forehead, massive brow ridges, a brain about a third smaller than ours, and strong, thick legs. Daka lived on a grassland, with elephants, wilde-beests, hippos, antelopes, many types of pigs, and giant hyenas.

H. erectus left fossil evidence of cooperation, social organization, tools, and use of fire. Fossilized teeth and jaws suggest that they ate meat. The distribution of fossils indicates that they lived in families of male-female pairs (most primates have harems). The male hunted, and the female nurtured the young. They were the first to have an angled skull base that enabled them to produce a greater range of sounds, making speech possible. *H. erectus* fossils have been found in China, Java, Africa, Europe, and Southeast Asia, indicating that these animals could migrate farther than earlier primates.

In 2013, publication of discovery of a complete hom-inin skull from Dmanisi, Georgia, dated to about 1.8 million years ago, has challenged the idea that several species of *Homo* diverged in Africa before some of them migrated out. The jaw to the Georgian skull had already been discovered, and together jaw and skull form the framework for what must have been a massive head, suggesting it was from a male. It has a long face and large teeth, with a small cranium, and was found near ani-mal bones bearing cutmarks from butchery, and plant remains.

The reconstruction, called Skull 5, is puzzling because it has features thought to have come from different species of *Homo*. The finding of remains from several other individu-als nearby confirmed that these early people were physically as diverse as any five people are today (**figure 16.4**). Researchers used computer-aided methods to intricately measure and com-pare parts of the preserved faces. Some researchers are now considering whether a single species of *Homo*, perhaps most like *Homo erectus*, dwelled in and left Africa, with variations no more distinct than those seen among modern people. Had the back of the skull and facial bones been discovered separately, the fossil might have been classified as different species! DNA evidence can flesh out fossil evidence for more recent members of *Homo*.

Archaic Humans

Homo erectus, or whatever we come to call the more ancient of our genus, may have persisted until as recently as 35,000 years ago. Meanwhile, the branching of evolution led to several types

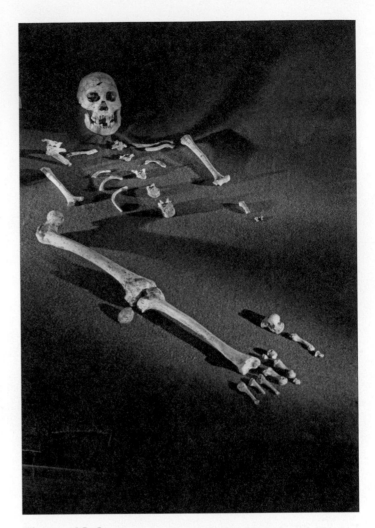

Figure 16.4 **Reconstructing *Homo erectus*.**
Paleontologists assembled fossilized bones from several *Homo erectus* individuals from the Dmanisi site in the country of Georgia to try to understand what this hominin looked like. It lived about 1.8 million years ago.

of hominins that were more like us and that we call the archaic humans. We call ourselves "anatomically modern humans." **Figure 16.5** depicts how we might all be related.

As some of our ancestors left Africa and expanded beyond that continent, eventually populations would have begun to overlap geographically. If populations interbred, then gene flow would appear, leaving evidence in preserved ancient DNA as well as in the genomes of modern peoples. Both have happened. While the older technique of comparing fossils tended to focus on how species and populations differed, clues in DNA are showing us that groups recently thought to be sepa-rate may actually have shared genetic material.

We probably do not yet know anything about some of the peoples who migrated out of Africa over the past tens of thou-sands of years. The following sections describe the three types of recent hominin ancestors about which we know the most: *Homo sapiens idaltu*, the Neanderthals, and the Denisovans. They may have branched from a shared ancestor as far back as 700,000 years ago or even earlier. We do not know the extent

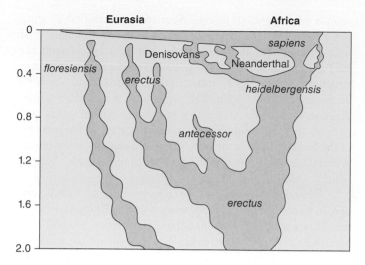

Figure 16.5 Archaic humans. Several types of archaic humans likely co-existed.

to which they coexisted, and we do not know all of the types of archaic humans that might have lived.

Homo Sapiens idaltu

Fossils of *Homo sapiens idaltu* from Ethiopia reveal that by 156,000 years ago, our ancestors did not look very different from ourselves. *H. sapiens idaltu* probably resembled an Australian aborigine, with a large and powerful build and dark skin.

In 1997, paleoanthropologist Tim White was driving by the village of Herto, along a bend of the Awash River, once home to australopithecines. Seasonal rains had driven the nomadic people and their livestock away, and in the cleared ground a hippopotamus skull protruded near tools made of obsidian, a glasslike rock. Three humanlike skulls were slightly farther away—one, from a child, shattered and scattered into more than 200 pieces, including baby teeth. The researchers named the hominin *Homo sapiens idaltu*, which in the local language means "elder."

The most intact skull was slightly longer, and the brain slightly larger, than those of modern humans. Fine, parallel lines had been etched along the base of the skull. The dome of the skull had not been damaged, as it would have been had cannibalism been practiced. The skull was very smooth, and there were no other bones nearby. Might the skulls have been gently separated from the bodies, saved, and touched, as modern cultures do to honor the dead?

Other fossils filled in the story. Evidence of catfish and hippos indicate that the Awash River had flooded, forming a freshwater lake. The hippo and buffalo bones bore marks made with tools that had probably sliced off meat. Some bones were broken in ways that suggested that the people ate the marrow. The tools were of a sophisticated design compared to the flaked tools from a million years ago. Overall, the scene evoked an image of ancestors who not only understood the concept of death, but who practiced mortuary rituals.

By 70,000 years ago, humans, still mostly confined to Africa, used more intricately carved tools made of bones, and red rock that bore highly symmetrical hatchmarks, which may indicate early counting. Groups of hominins may have been very isolated on the vast continent. The first *Homo* may have left Africa around 100,000 years ago, as a founder group of about a thousand individuals, most of them male, who died out by 70,000 years ago. Because of the isolation, it's possible that even as *H. sapiens idaltu* and perhaps others yet to be discovered were far along the road to modern humanity, pockets of *H. erectus* may have persisted, perhaps until as recently as 35,000 years ago.

The Neanderthals

Quarry workers blasted out the first bones of *Homo neanderthalensis,* better known as the Neanderthals (or Neandertals) in Neander Valley, Germany, in 1856. These hominins had skeletons wider than ours with shorter arms and legs, and distinctive faces with prominent brow ridges, sloping foreheads, and jutting faces. Some individuals had the large teeth and jaws of Neanderthals, but the small chins of modern humans. Neanderthals used tools, and their bones buried with flowers suggests that they were capable of spirituality. Starch granules discovered within dental plaque on Neanderthal teeth indicate that these early people cooked the grains that they ate.

About 400,000 years ago, the forebears of the Neanderthals began to leave Africa, heading toward Europe and west Asia but not establishing themselves there until about 230,000 years ago. The Neanderthals left Africa in separate migrations from our forerunners, which is why modern Africans do not have Neanderthal DNA sequence variants in their genomes, but Europeans and Asians do. We have DNA evidence of Neanderthals because they lived in cold areas, which preserved the genetic material. In contrast, the DNA of earlier *Homo* and *Australopithecus* was not only much older, but would have degraded rapidly in the heat of Africa.

About 50,000 years ago, our ancestors left Africa and encountered Neanderthals in the Middle East en route to Eurasia. The two types of hominins interbred—a discovery deduced from shared DNA sequence variants. When researchers first began to sequence the DNA from Neanderthal bones, they concluded that interbreeding was highly unlikely, but sequencing the entire genome changed the long-held story of our separateness. Our early views of the Neanderthals, based on a few mitochondrial DNA sequences, was like trying to guess the end of a novel by reading only a small part of the story.

The researchers who sequenced the Neanderthal genome used dental drills to delicately remove pulverized bone "dust" from three bones from females found in a cave in Croatia, which kept the bones intact for future studies. DNA sequencing in a "clean room" overcame the problem of the investigators' DNA contaminating the samples, and other techniques avoided sequencing bacterial or fungal DNA, another source of error in ancient DNA studies. Then researchers compared the

Neanderthal genome sequence to those of modern humans from Asia, Europe, and Africa. By comparing SNPs—sites where the DNA base differs—it became clear that the genomes of modern Europeans and Asians have about 2 percent of their DNA in common with Neanderthals. DNA from a 50,000-year-old Neanderthal baby's rib found in a cave in Russia is more similar to modern human genomes than is Neanderthal DNA from Europe, suggesting at least some interbreeding occurred in Russia. However, the evidence cannot reveal whether the interbreeding was intense over a short period of time, or once-in-a-while mating that happened over many years. Discoveries of individual genes and their variants have added to our knowledge of these people. Variants of the *FOXP2* gene might have enabled them to vocalize and possibly speak, and some specimens had a mutation in the *MC1R* gene that gave them pale skin and red hair (**figure 16.6**).

More interesting than what our genomes share with those of the Neanderthals is how our genomes differ, because this reveals what makes us human. Sequencing identified 73 genes that encode proteins that are different in modern humans from their counterparts in Neanderthals, and these affect such varied functions as sperm motility, wound healing, immunity, hair structure, bone shapes, transcription control, and cognitive development. In addition, 212 regions of the modern human genome show signs of positive selection—linked genes with variants that are uniquely human and not seen in the genomes of the Neanderthals or chimps. These 212 regions of the human genome are said to have undergone "selective sweeps" because natural selection favored them.

Neanderthals may have lived as recently as 30,000 years ago, in warm caves in Gibraltar when northern Europe was under ice. They might have inhabited these caves since 100,000 years ago. Then, the fossil record indicates, the Neanderthals vanished. According to their DNA, however, they may have been assimilated as different groups interbred. Without a time machine, we may never know exactly what happened.

The Denisovans

In 2010, we met a new ancestor, the Denisovans, originally described from DNA in a pinky finger bone found in Denisova Cave in the Altai Mountains of southern Siberia (**figure 16.7**). Researchers thought the bone was from a Neanderthal, but the mitochondrial DNA sequence was unlike that of either Neanderthals or modern humans. Within 2 years, researchers led by Svante Pääbo, at the Max Planck Institute for Evolutionary Anthropology in Leipzig, had sequenced the genome from two preserved molars from other individuals, and we learned more about these ancestors. The teeth were found in different layers of rock in the cave wall, and so were left there at different times. Their genomes were quite different, suggesting that the Denisovan population may have lived long enough to generate significant genetic diversity.

The Denisovan girl or young woman represented by the pinky bone, nicknamed Denise, lived between 32,000 to 50,000 years ago, according to both fossil and DNA evidence. The genome sequence reveals that Denise had dark skin and

Figure 16.6 A Neanderthal phenotype. Mutations found in Neanderthal DNA suggest that some of them may have had pale skin and red hair. Others had darker hair and skin. About 2 percent of the genomes of modern people of European or Asian ancestry arose from long-ago breeding with Neanderthals. The genomes of modern Africans do not have Neanderthal DNA.

Figure 16.7 The Denisovan cave. A Siberian cave holds abundant evidence of archaic humans.

brown eyes and hair. About 3.5 percent of her genome persists today in people from Papau New Guinea, Melanesia, Oceania, Polynesia, and in some aboriginal Australians.

Anthropologists are now trying to decipher the relationships among the archaic humans, and how these peoples might have converged to eventually give rise to anatomically modern humans. An intriguing addition to the fossil collection is the Altai Neanderthals, who lived in the same cave from which the Denisovans take their name. The Neanderthal genomes from the cave have many regions of extensive homozygosity, which suggests consanguinity—brothers and sisters having children together. Comparing genomes of the archaic humans to our own indicates that the Denisovans and Neanderthals were more closely related to each other than either was to us, and that Neanderthal genes flowed into older Denisovan genomes. As complicated as this emerging picture is, we do not yet have all of the puzzle pieces—the Denisovan genome clearly indicates admixture with at least one other, yet unknown, type of archaic human. Several small populations of archaic humans probably existed and eventually mixed for at least 100,000 years before we modern humans, retaining some archaic genes, emerged and persisted.

Modern Humans

Cave art from about 14,000 years ago indicates that by that time, our ancestors had developed fine-hand coordination and could use symbols. These were milestones in cultural evolution. By 10,000 years ago, people had expanded from the Middle East across Europe, bringing agricultural practices.

Early Farmers

DNA evidence has glimpsed one scene from the spread of agriculture. Researchers sequenced mitochondrial DNA haplotypes from 21 bodies found in a graveyard in Germany, about 100 miles south of Berlin, and compared them to DNA from 36 modern Eurasian populations. The bodies were from about 7,100 years ago. The comparisons yielded a clear cline, with genetic similarities indicating a long-ago migration of early farmers from Turkey, Syria, Iraq, and other Near Eastern cultures westward from the Balkans, north along the Danube into central Europe—not just the spread of their agricultural techniques by word of mouth. The migration took centuries. According to the DNA, the farmers arrived in Europe and encountered hunter-gatherers descended from the original population from 40,000 B.C., and the two groups of people interbred. **Figure 16.8** shows what the ancient farmers from 7,100 years ago might have looked like.

Ötzi the Ice Man

In 1991, hikers in the Ötztaler Alps of northern Italy discovered an ancient man frozen in the ice (**figure 16.9**). Named Ötzi, the Ice Man was on a mountain more than 10,000 feet high 5,200 years ago when he perished. He wore furry leggings, leather suspenders, a loincloth, fanny pack, bearskin cap and cape, and sandal-like snowshoes. He had stained his skin to fashion

Figure 16.8 **Farmers from 7,100 years ago probably looked like modern people.** *Karol Schauer, State Museum of Prehistory in Halle (Saale), Saxony-Anhalt, Germany.*

Figure 16.9 **A 5,200-year-old man.** Hikers discovered Ötzi the Ice Man in the Ötztaler Alps of northern Italy in 1991.

tattoos, and indentations in his ears suggest that he might have worn earrings. He carried mushrooms that had antibiotic properties. Berries found with him place the season as late summer or early fall. His last meal was ibex and venison.

Ötzi died following a fight. He had a knife in one hand, cuts and bruises, and an arrowhead embedded in his left shoulder that nicked a vital artery. The wound bore blood from two other individuals, and his cape had the blood of a third person. Mosses found on his body may have been wound dressings. When X rays indicated swelling on the right side of Ötzi's brain, researchers delicately removed two tiny samples of the tissue and analyzed the proteins present. Several of the proteins were related to blood clotting. Whether Ötzi fell and hit his head or was attacked remains a mystery. However he perished, Ötzi landed in a ditch, where snow covered him and preserved his body, including his genetic material. DNA profiling indicates that he belonged to the same gene pool as modern people living in the area, which is near the Italian-Austrian border.

Indigenous Peoples

An indigenous group of people is one that can trace its ancestry back farther in a particular geographical region than any other group, and has retained its uniqueness in cultural practices, social organization, and/or language. The group has remained physically or culturally isolated among colonists, and has therefore kept its gene pool separate, too. In those gene pools lie clues to adaptations to past ways of life, and by comparison to other modern genomes, clues to how we are continuing to evolve.

Today less than 6 percent of the world population is indigenous, accounting for about 380 million individuals in 5,000 groups, living in 72 nations. They range from just a few dozen people to sizeable portions of a country's population. Some live in distinct tribes yet go to school and work and dress just like anyone else. But some indigenous tribes are not very different in lifestyle from their hunter-gatherer ancestors.

Indigenous peoples living today include the Khoisan (bushmen) and Pygmies of Africa, the Etas of Japan, the Hill People of New Guinea, and a Brazilian tribe, the Arawete, who number only 130 individuals. A fascinating look into the past comes from the genomes of the Khoisan, the modern people whose roots go back the farthest. These hunter-gatherers live today, as they have for millennia, in the Kalahari Desert in southern Africa (**figure 16.10**). They are also known as "San" or "Bushmen," and they speak with a language that uses several "click" sounds.

Researchers compared the complete genome sequences of a Khoisan man named !Gubi to that of Archbishop Desmond Tutu, a well-known South African civil rights activist and a member of the majority Bantu group, as well as partial genome sequences of three other Khoisan who live near each other. The results indicate that the great genetic diversity from which humanity sprung in Africa persists in the Khoisan today, whose genomes are as different from each other as a modern European genome is from that of a modern Asian. The four Khoisan genomes and Desmond Tutu's differ from each other at more than a million places. Comparing the Khoisan genomes to those of later arriving groups may indicate adaptations to an agricultural way of life among more recent Africans.

The Khoisan have gene variants that reflect their lifestyle in the desert: a variant of the actinin-3 muscle gene that promotes sprinting over distance running; a gene variant that encodes a chloride channel that conserves water; and the "bitter taste" gene, something that would enable a hunter-gatherer to avoid poison and perhaps locate medicine. They lack a gene variant that in other populations protects against malaria. Selection would have ignored it in the dry climate where malaria-bearing mosquitoes cannot live.

a.

b.

Figure 16.10 **The most ancient modern humans.** The Khoisan are the modern people whose genetic roots go back the farthest. Their genomes are very diverse, with gene variants that reflect their long adaptation to a hunter-gatherer lifestyle **(a)**. They live in the Kalahari Desert in Botswana and Namibia **(b)**.

16.2 Methods to Study Molecular Evolution

Fossils paint an incomplete picture of the past because they are scarce and provide evidence of only certain parts of certain organisms. We can also glimpse the past through the informational molecules of life in rare preserved specimens such as the Neanderthals and Denisovans, and by inferring past relationships from the DNA, chromosomes, and proteins of organisms living today. Molecules of DNA and protein change in sequence over time as mutations occur and are perpetuated. The more alike a gene or protein sequence is in two species, the more closely related the two are presumed to be—that is, the more recently they shared an ancestor. An alternate hypothesis, that two unrelated species have exactly the same sequence of DNA nucleotides by chance, seems less likely.

Comparing genome, DNA or protein sequences, and chromosome banding patterns constitutes the field of **molecular evolution.** Knowing the mutation rates for specific genes provides a way to measure the passage of time using a sequence-based "molecular clock."

Comparing Genes and Genomes

We can assess similarities in DNA sequences between two species for a piece of DNA, a single gene, a chromosome segment, a chromosome, mitochondrial DNA, or an entire exome or genome.

For some gene variants, similarities among species can be startling. People with Waardenburg syndrome (OMIM 148820), for example, have a white forelock of hair; wide-spaced, light-colored eyes; and hearing impairment. The mutant gene is very similar in sequence in cats, horses, mice, and minks, who have light-colored coats and eyes and are deaf (**figure 16.11**). The phenotype stems from abnormal movements of pigment cells in the embryo's outermost layer.

In general, DNA sequences that encode protein are often very similar among closely related species. The related species presumably inherited the gene from a shared ancestor, and a change in that gene would not persist in a population unless it provided a selective advantage. At the same time, natural selection weeded out proteins that did not promote survival to reproduce.

Similar DNA or amino acid sequences in different species are said to be "highly conserved." Sequences that are similar in closely related species but that do not encode protein often control transcription or translation, and so are also vital and therefore subject to natural selection. In contrast, some genome regions that vary widely among species do not affect the phenotype, and are therefore not acted upon by natural selection. Within a protein-encoding gene, the exons tend to be highly conserved, but the introns, which are removed from the corresponding RNA, are not.

Comparing Chromosomes

Before gene and genome sequencing became possible, researchers considered similarities in chromosome banding patterns to assess evolutionary relatedness. Human chromosome banding patterns most closely match those of chimpanzees, then gorillas, and then orangutans (**table 16.2**). The karyotypes of humans, chimpanzees, and gorillas differ from each other by nine inversions, one translocation, and one chromosome fusion.

Chromosome banding patterns are like puzzle pieces. If both copies of human chromosome 2 were broken in half, we

a. b. c.

Figure 16.11 **The same mutation can cause similar effects in different species.** A mutation in mice **(a)**, cats **(b)**, and horses **(c)**, and other mammals causes light eye color, hearing or other neurological impairment, and a fair forelock.

Table 16.2	Percent of Common Chromosome Bands Between Humans and Other Species
Chimpanzees	99+%
Gorillas	99+%
Orangutans	99+%
African green monkeys	95%
Domestic cats	35%
Mice	7%

Table 16.3	Cytochrome *c* Evolution	
	Organism	Number of amino acid differences from humans
	Chimpanzee	0
	Rhesus monkey	1
	Rabbit	9
	Cow	10
	Pigeon	12
	Bullfrog	20
	Fruit fly	24
	Wheat germ	37
	Yeast	42

would have 48 chromosomes, as the three species of apes do, instead of 46. Human chromosome 2 therefore arose from a fusion event. The banding pattern of chromosome 1 in humans, chimps, gorillas, and orangutans matches that of two small chromosomes in the African green monkey, suggesting that this monkey was ancestral to the other primates. Most of the karyotype differences among these three primates and more primitive primates are translocations.

We can also compare chromosome patterns between species that are not as closely related as we are to other primates. All mammals, for example, have identically banded X chromosomes. One section of human chromosome 1 that is alike in humans, apes, and monkeys is remarkably similar to parts of chromosomes in cats and mice. A human even shares chromosomal segments with a horse, but our karyotype is much less like that of the aardvark, the most primitive placental mammal.

The information in comparing chromosome bands from stains is broad and imprecise compared to information from DNA sequences, because a band can contain many genes that differ from those within a band at a corresponding locus in another species' genome. In contrast, DNA probes used in a FISH analysis highlight specific genes (see figure 13.8). FISH can indicate direct correspondence of gene order, or **synteny**, between species, which is solid evidence of close evolutionary relationships. For example, versions of 11 genes are closely linked on human chromosome 21, mouse chromosome 16, and chromosome U10 in cows.

Comparing Proteins

Many different types of organisms use the same proteins, with only slight variations in amino acid sequence. The similarity of protein sequences is compelling evidence for descent from shared ancestors—that is, evolution. Many proteins in humans and chimps are alike in 99 percent of their amino acids, and several are identical. When analyzing a gene's function, researchers routinely consult databases of known genes in many other organisms.

Cytochrome *c* is one of the most ancient and well-studied proteins. It helps to extract energy from nutrients in the mitochondria. The more closely related two species are, the more alike their cytochrome *c* amino acid sequence is (**table 16.3**). Human cytochrome *c*, for example, differs from the horse version by 12 amino acids, and from kangaroo cytochrome *c* by 8 amino acids.

The human protein is identical to chimpanzee cytochrome *c*. The homeotic genes, discussed in Clinical Connection 3.1, are another class of genes that has changed little across evolutionary time.

Molecular Clocks

A clock measures the passage of time as its hands move through a certain degree of a circle in a specific and constant interval of time—a second, a minute, or an hour. In the same way, an informational molecule can be used as a "molecular clock" if its building blocks are replaced at a known and constant rate.

The similarity of nuclear DNA sequences in different species can be used to estimate the time when the organisms diverged from a common ancestor, if the rate of base substitution mutation is known. For example, many nuclear genes studied in humans and chimpanzees differ in 5 percent of their bases, and substitutions occur at a rate of 1 percent per 1 million years. Therefore, 5 million years have presumably passed since the two species diverged. Mitochondrial DNA (mtDNA) sequences may also be tracked in molecular clock studies, as we will soon see.

Timescales based on fossil evidence and molecular clocks can be superimposed on evolutionary tree diagrams constructed from DNA or protein sequence data. However, evolutionary trees can become complex when data can be arranged into different tree configurations. A tree for seventeen mammalian species, for example, can be constructed in 10,395 different ways! The sequence in which the data are entered into tree-building computer programs influences the tree's shape, which is vital to interpreting species relationships. With new sequence information, the tree possibilities change.

Researchers use statistical methods to identify an evolutionary tree out of many possible that is likely to represent what really happened. An algorithm connects all evolutionary tree sequence data using the fewest possible number of mutational events to account for observed DNA base sequence differences. Because mutations are rare events, the tree that requires the fewest mutations is more likely to reflect reality. Statistical methods can account for different mutation rates for different genes and for different positions within genes, and for incomplete data.

Using Mitochondrial and Y Chromosome DNA Sequences

To track ancient human migration patterns, researchers use the types of genetic markers that are used to track traits in modern families, an approach called genetic genealogy or genetic ancestry. These markers include single nucleotide polymorphisms (SNPs), short tandem repeats (STRs or microsatellites), and other copy number variants (CNVs). Markers of mitochondrial DNA (mtDNA) are used to trace the female lineage, and markers of Y chromosome sequences to trace the male lineage. Markers also follow DNA sequences that are part of autosomes, called "ancestry informative markers." Sequencing mtDNA and Y chromosome DNA sequences provides information on only some of a person's ancestors, as the pedigree in **figure 16.12** illustrates. It is easy to see that the contribution of a particular ancestral DNA sequence decreases as the number of generations increases.

MtDNA is ideal for monitoring recent events because it mutates faster than DNA in the nucleus. Its sequences change by 2 to 3 percent per million years. Mutations accumulate faster because mtDNA has no DNA repair. Another advantage of typing mtDNA is that it is more abundant than nuclear DNA because mitochondria have several copies of it, and a cell has many mitochondria. When researchers are lucky enough to find fossils or ancient humans, mtDNA is the most likely DNA to be recovered.

For many years, researchers compared certain rapidly-mutating DNA sequences in mitochondria, but today they compare complete mitochondrial genomes, or "mitogenomes." A study of mitogenomes of the Ashkenazi Jews, for example, showed that many maternal lineages came, not from the Near East, as had long been thought from the more limited mtDNA analysis, but from many places in Europe. A picture is emerging of Ashkenazi men traveling in Europe, where they married local women, who converted to Judaism.

Most of the Y chromosome DNA sequence offers the advantage of not recombining. Crossing over, which it could only do with an X chromosome because there is no second Y, would break the linkage from the past generation and therefore make tracing relationships impossible.

Sets of SNPs along mitochondrial and Y chromosome DNA define long DNA stretches termed **haplogroups.** (The haplotypes used to describe linkage in chapter 5 refer to shorter DNA sequences.) Haplogroups track with geography, rather than with racial or ethnic groups, which are social designations and not biological ones.

Y haplogroups are classified from "A" through "T," with several subgroups, called subclades, indicated by alternating letters and numbers. Haplogroups and their subgroups also describe mtDNA. Populations can be classified into both mtDNA and Y chromosome groups, indicating the sources of female and male lineages, respectively. Sub-Saharan Africans, for example, have Y haplogroups E1, E2, and E3a and mtDNA haplogroups L3. Europeans, however, have Y haplogroups R, I, E3b, and J, and their mtDNA haplogroup is R, which includes three subgroups.

Admixture

The first molecular evolution studies used Y chromosome and mitochondrial DNA sequences because their gender-specificity made them easiest to track. Because we now know of millions of SNPs across the genome, and can sequence genomes, we can fill in more of the blanks in the history of human migrations, by tracing transmission of haplogroups and haplotypes. New combinations of gene variants (alleles) that arise when individuals from two previously distinct populations have children together is termed **genetic admixture**. It often follows migrations. All modern human populations are admixed.

We can trace admixture by following how the pattern of haplotypes changes from one generation to the next if a crossover

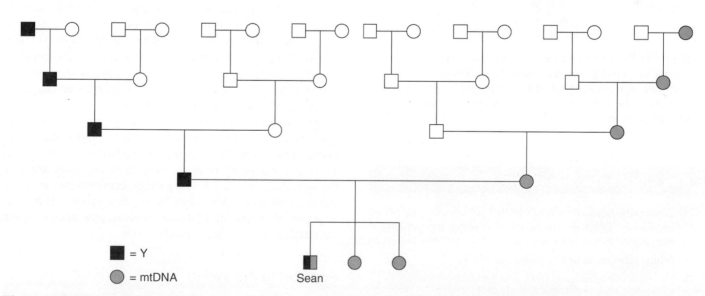

Figure 16.12 Genetic genealogy. Y chromosome and mtDNA sequences represent only some of a person's ancestors. Considering autosomal sequences can capture the contributions of other relatives, represented as the unfilled symbols in the interior of this pedigree. Sean's symbol has two shades because his Y and mtDNA were tested.

occurs. If we know the crossover rate for a particular gene that is involved in the exchange of genetic material, then we can extrapolate back to when admixture occurred, assuming a generation time, such as 25 years. **Figure 16.13** illustrates schematically what this chromosomal exchange over time looks like.

Computational tools use changing haplotype data to reconstruct what one researcher calls "snapshots in time" as far back as 2 million years ago, by following admixture. One study, for example, looked at 2,400 35,000-base parts of the genome, with a generation time of 25 years, to estimate that an ethnic group called the Yoruba left their native southwestern Nigeria in 1550 and in 1790, times already known from history to correspond to the start of the slave trade and its peak. Another investigation traced introduction of west Eurasian gene variants into southern Africa 40 to 60 generations ago.

For recent times, admixture estimates are more precise. For example, researchers traced a slight increase in the height of certain pygmy groups in Africa to admixture with DNA from their taller neighbors, the Bantu, about 325 years ago, equal to 13 generations. Recent admixture from outside Africa also explains why some Ethiopians have light skin and eyes.

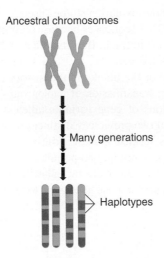

Ancestral chromosomes

Many generations

Haplotypes

Figure 16.13 **In admixture, crossing over recombines segments of linked genes representing ancestral populations.** Linkage can be so strong that detecting one SNP can represent the entire block. This tight linkage is called linkage disequilibrium.

Key Concepts Questions 16.2

1. What is the basis of comparing DNA sequences, protein sequences, chromosome banding patterns, and genome sequences to learn about origins and evolution?

2. What is the basis of the molecular clock?

3. How are mitochondrial DNA and Y chromosome sequences used to trace lineages?

4. How does admixture provide information about the past?

16.3 The Peopling of the Planet

Fossil evidence and extrapolating and inferring relationships from DNA sequence data provide clues to the major movements that peopled the planet (**figure 16.14**). The evidence so far has been like reading chapters from different parts of a novel and trying to understand the whole story. Three such chapters in the saga of modern human origins stand out: our beginnings 200,000 years ago; expansion from Africa; and populating the New World.

Mitochondrial Eve

Theoretically, if a particular sequence of mtDNA could have mutated to yield the mtDNA sequences in modern humans, then that ancestral sequence may represent a very early human or humanlike female—a mitochondrial "Eve," or metaphorical first woman. **Figure 16.15** shows how one maternal line may have persisted.

When might this theoretical "first" woman, the most recent female ancestor common to us all, have lived? Researchers in the mid 1980s compared mtDNA sequences for protein-encoding as well as noncoding DNA regions in a variety of people, including Africans, African Americans, Europeans, New Guineans, and Australians. They deduced that the hypothesized ancestral woman lived about 200,000 years ago, in Africa. More recent analysis of mtDNA from 600 living East Africans estimated 170,000 years ago for the beginning of the modern human line, which is remarkably close to the date of the *H. sapiens idaltu* fossils. One way to reach this time estimate is by comparing how much the mtDNA sequence differs among modern humans to how much it differs between humans and chimps. The differences in mtDNA sequences among contemporary humans are 1/25 the difference between humans and chimps. The two species diverged about 5 million years ago, according to extrapolation from fossil and molecular evidence. Multiplying 1/25 by 5 million gives a value of 200,000 years ago, assuming that the mtDNA mutation rate is constant over time.

Where did Eve live? The locations of fossil evidence, such as *H. sapiens idaltu* skulls, support an African origin, and Charles Darwin suggested it, too. In addition, studies comparing mitochondrial and nuclear DNA sequences among modern populations consistently find that Africans have the most numerous and diverse mutations. For this to be so, African populations must have existed in place for longer than other populations, because it takes time for mutations to accumulate. In many evolutionary trees constructed by computer analysis, the individuals whose DNA sequences form the bases are from Africa. That is, other modern human populations all have at least part of an ancestral African genome, plus mutations that occurred after their ancestors left Africa.

Populating the World

Data from mtDNA, Y chromosome DNA, and markers on the autosomes indicate that the peopling of the world was a series of founder effects as groups left Africa, perhaps when the

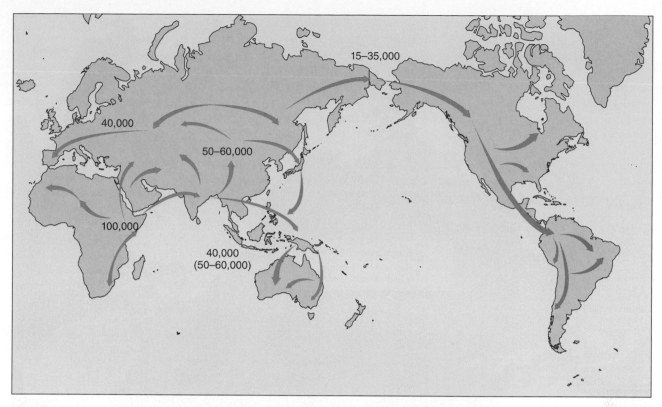

Figure 16.14 **The peopling of the planet.** This map depicts major migratory paths based on DNA haplogroup information, mostly from mtDNA and the Y chromosome. Fossil and DNA evidence suggest that humanity arose in East Africa, but a southern African origin is possible too.

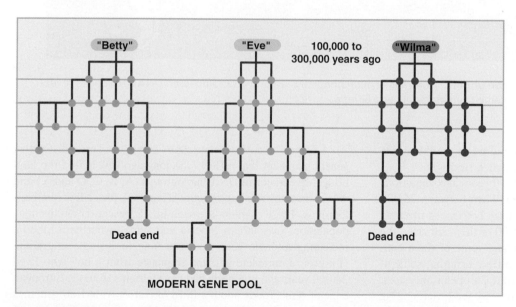

Figure 16.15 **Mitochondrial Eve.** According to the mitochondrial Eve hypothesis, modern mtDNA retains some sequences from a figurative first woman, "Eve," who lived in Africa 300,000 to 100,000 years ago. In this schematic illustration, the lines represent generations, and the circles, females. Lineages cease whenever a woman does not have a daughter to pass on the mtDNA.

and genes flowed from one region to another. Geographical and climatic barriers periodically shrank human populations, while natural selection and genetic drift narrowed the African gene pool. At the same time, new mutations among the groups that left Africa established the haplogroups that are consulted to trace non-African ancestry today. Human evolution continued in Africa as groups left and populated other parts of the world. Global climate change, such as ice ages, forced people into habitable areas, called refugia. Today gene flow continues, both out of and back into Africa.

People spread across Eurasia by 40,000 years ago, as well as elsewhere, and lastly through southern Siberia and Mongolia. By 20,000 years ago, humanity was everywhere except the Americas and Antarctica (**figure 16.16**). From here

Sahara desert periodically grew wetter. A large expansion out of Africa occurred about 50,000 years ago, but estimates range very widely. These migrations yielded "chains of colonies" that may have overlapped and merged when neighbors met,

people could cross the Bering Land Bridge, which emerged between Siberia and Alaska during times when the glaciers had retreated. The land bridge stretched for about 1,000 miles from north to south, appearing as winds from the southwest blew

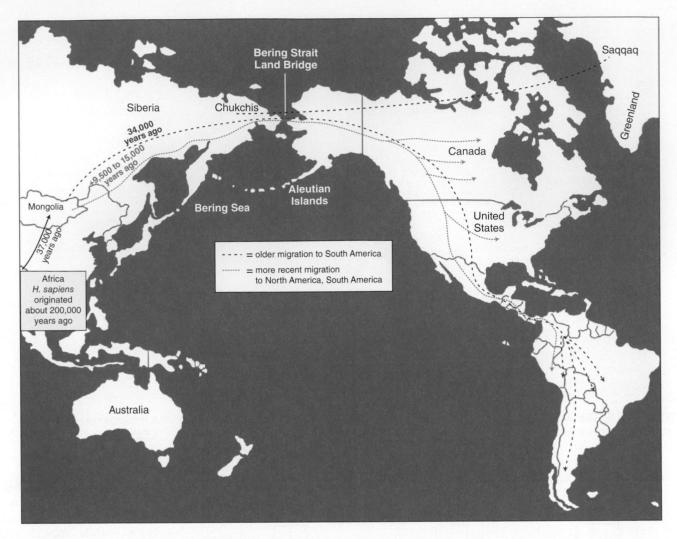

Figure 16.16 **Tracing Native American origins.** Analyses of mitochondrial DNA and Y chromosome DNA sequences reveal that the ancestors of Native Americans came from Mongolia and Siberia.

snow away. The areas for several hundred miles on either side of the bridge, and the bridge itself, are called Beringia.

From 40,000 years ago until 23,000 years ago, mutations occurred in the population of Beringia. Then between 23,000 and 19,000 years ago, a severe population bottleneck affected the people in Beringia. Only about 1,000 of them survived the journey over the bridge from Siberia, and some of them continued southward along the Pacific coastline bringing some of the mutations that distinguish them from European ancestors. As the ice age ended about 18,000 years ago, the tiny founding population in the Americas began a period of rapid expansion that lasted 3,000 years. This amplified alleles that had survived, as alleles unique to people who had perished vanished from the population. The people spread south and east through the Americas as the first Native Americans, a process that may have taken about 2,000 years. The mitochondrial DNA evidence is particularly valuable because the coastal migratory path is now under water, hiding archeological clues.

Today, Native Americans carry a very distinctive genetic nametag that reflects the long-ago trek across the land bridge: five major mtDNA haplogroups (A, B, C, D and X) and two Y chromosome haplogroups (C and Q), plus a few rare haplogroups. These markers are seen in all present-day indigenous populations in southern Siberia and some in northern Siberia, indicating a single gene pool traveling in a single migration. However, a complete genome sequence from a boy who lived 24,000 years ago in Siberia had gene variants from the European edge of Asia. Today about a third of Native American genomes come from western Eurasia and two-thirds from eastern Asia.

DNA sequence information extrapolated back from present-day Native American populations is consistent with molecular clock data from ancient DNA. In addition, Native American populations have an STR marker and some mtDNA haplogroups that are not seen in eastern Siberian peoples, indicating mutations that happened after crossing the Bering Strait.

A comparison of 678 autosomal STR markers from 29 Native American populations and 49 other indigenous groups worldwide found that Native Americans are very different from other populations, yet are very much like each other—as might be expected from a multigenerational journey southward by a small but hardy group, along the coastline of the Americas.

By 14,000 years ago, Native Americans had arrived inland, indicated in fossilized excrement. DNA from the skeleton of an adolescent girl discovered underwater in the Yucatan peninsula dates from 12,000 years ago, and includes sequences from Asia and from the Americas. Genetic evidence also suggests that some Native American populations died out.

It seems that with every new DNA discovery, we have to rewrite prehistory or add details. This was the case for a tuft of hair from a Paleo-Eskimo discovered in Greenland from the first group of people to settle there, called the Saqqaq. The hair is about 4,000 years old. Analysis of its DNA revealed characteristics of the man the hair came from, as well as clues to how the Saqqaq got to Greenland.

The Saqqaq was male with coarse, dark hair and skin, brown eyes, shovel-shaped teeth, dry earwax, and, if he had lived longer, would have become bald. He had type A+ blood, he was at increased risk for high blood pressure and diabetes, he had a high tolerance for alcohol, and he may have become addicted to nicotine had it been possible to cultivate it in the frozen wasteland. The most revealing information, however, concerned the Saqqaq Eskimo's origins. His mitochondrial DNA was distinct from that of either Native Americans or modern Eskimos, yet was very similar to that of the Chukchis people of Siberia. Apparently, the ancestors of the Greenland Eskimos crossed the land bridge about 5,400 years ago, separate from the crossings that founded the Native Americans and modern Eskimos. The conclusion: The earliest Eskimos in Greenland came from a different migration than the one that was ancestral to Native Americans and modern Eskimos, and left no present-day descendants. The discovery of the 4,000-year-old non–Native American/non-Eskimo reveals a major limitation of genetic anthropology: We cannot fit in the puzzle pieces that we cannot find, or that no longer echo in modern gene pools.

Ancestry Testing

Today, people can learn some facts about their ancestry by mailing a DNA sample to a company offering genetic genealogy services. Most companies test mitochondrial DNA to trace maternal lineages and Y chromosome DNA sequences to trace paternal lineages, as well as some autosomal sequences. Several markers are checked for each DNA source, and the results compared to growing databases to look for regions of the genome that individuals share. "Deep ancestry" refers to assigning likely origin to a major part of the world— sub-Saharan Africa, for example—or a major population group or subgroup, such as "African American" or "African American with roots in Nigeria or Senegal." As genetic genealogy databases grow with sequenced genomes, ancestry testing will provide more detailed information.

DNA testing can bring the search for our origins into the present, compared to finding fossils or tracing long-ago humanity out of Africa, by revealing whether any two individuals living today share an ancestor. Such tests assign an approximate generation to the "most recent common ancestor" (MRCA). The more markers tested, the more meaningful the results. If two people share all thirty-seven of thirty-seven tested markers, there is a 50 percent chance that their MRCA was no more than two generations ago. The people are so alike because not enough generations have passed for their genomes to have diverged. Sharing twenty-five of twenty-five markers gives a 50 percent chance that the MRCA was not more than three generations ago. If two people share twelve markers, there is a 50% chance that the MRCA was no longer ago than sevengenerations.

For some people, deep ancestry testing may reveal something that they already know from family history, such as Ashkenazi Jews or Caucasian Europeans. In contrast, for African American families who cannot trace their history back before slavery using documents and oral histories, ancestry testing can tell them if their forebears came from any of more than two dozen places along the coast of West Africa and inland.

Ancestry testing companies match new customers' DNA to sequences in their databases, and report back possible distant cousins. A person might then receive e-mail from dozens of fifth cousins—we each have 4,688 of them! Ancestry companies can apply an algorithm to the DNA data and family tree information that includes birth locations to identify shared ancestors.

Ancestry testing has limitations. Testing mtDNA and Y chromosome DNA considers much less than 1 percent of the genome, and traces only some lineages. Another limitation is that a haplogroup may come from more than one geographic region, due to gene flow. In addition, not all human haplogroups have been discovered, so a person may be erroneously placed into one group to which he or she partially matches, because the true haplogroup has not yet been described. *Bioethics: Choices for the Future* on page 318 discusses privacy issues that may arise with ancestry testing.

Key Concepts Questions 16.3

1. Who does "mitochondrial Eve" represent?
2. Describe the main events of how we think people populated the world.
3. Did people get to the Americas?
4. What can we learn from ancestry testing?

Genetic Privacy: A Compromised Genealogy Database

The intent of the 1000 Genomes Project was to collect a large number of human genome sequences to study inherited variation. The database was to be anonymous. Part of the informed consent form reads, ". . . *it will be hard for anyone to find out anything about you personally from any of this research.*" However, online searches shattered the idea of privacy.

Yaniv Erlich, a researcher at the Whitehead Institute, who had worked with databases at banks, and one of his students, Melissa Gymrek, had the idea to try to identify people who'd anonymously donated DNA to the 1000 Genomes Project. They were not interested in identifying the participants, but to see if they could. If so, they could warn government officials that the supposedly anonymous database was not private.

The researchers used short tandem repeats (STRs). Recall from chapter 14 that STRs are the non-protein-encoding short repeated sequences used in forensics investigations. They are also used in genealogy research. The STR patterns that Erlich and Gymrek considered classified Y chromosome haplotypes. Using public genealogy databases, the researchers looked up surnames corresponding to specific Y haplotypes. Further clues came from basic public information such as state of residence and birth year. Often Googling just some of the data led to family websites that instantly confirmed that the researchers had correctly identified an individual from his DNA. Cross-referencing to the DNA sequences of cells that the 1000 Genomes Project had deposited in the Coriell Cell Repositories in New Jersey identified women.

When the researchers had identified 50 people and realized how easy it was, they contacted officials at the National Institutes of Health. The agency took swift measures to hide some of the data, such as year of birth. The researchers and *Science* magazine published the findings to begin a discussion and possibly prevent further exposures of personal genetic information.

The problem of maintaining genetic privacy when participating in research projects that may benefit many will continue, and likely grow. Said Erlich at the time, "This is an important result that points out the potential for breaches of privacy in genome studies." Added Gymrek, "We show that if, for example, your Uncle Dave submitted his DNA to a genetic genealogy database, you could be identified."

Questions for Discussion

1. How can researchers collect enough genome data to better understand human inherited variation, yet protect identities?

2. Ten years from now, storing the information in our genomes may be routine, perhaps even required. What measures can researchers and bioethicists take now to protect genetic privacy?

3. How does genome data differ from other biomedical information, such as cholesterol level or height?

4. Do you think that the popularity of direct-to-consumer ancestry testing contributes to the genetic privacy problem?

5. Why has the Food and Drug Administration controlled some direct-to-consumer genetic tests but not others?

6. Children with rare diseases whose mutations are listed in mutation databases have been identified from information on hometown, age, and other nongenetic characteristics. Suggest a way to better protect their identities.

7. Is complete de-identification in a database possible?

8. Personal genetic information, as well as any other type of information, can be inadvertently shared via social media. What can be done to avoid this problem?

16.4 What Makes Us Human?

We can investigate and identify the traits and abilities that distinguish us from the chimpanzees and bonobos, our closest relatives, because they live today. For the archaic humans that we know about by only a bit of bone, genome sequencing has provided clues to some of their traits, suggesting abilities gained as the archaic humans interbred and evolved into us.

We assess similarity by DNA sequence, numbers of copies of sequences, or sequences missing from the human genome. We share about 98.7 percent sequence similarity, but we also differ in the number of copies of certain DNA sequences. These include insertions and deletions, which are collectively called "indels." Considering indels, our degree of genome similarity

to chimps and bonobos is only about 96.6 percent. The degree of similarity may even be as low as 94 percent if sequences not in the human genome are considered. That is, what *isn't* present defines us as well as what *is* present.

Considering Genomes

Comparisons of the human genome sequence to those of other species are interesting and humbling. For example, our genome is not that different from that of a pufferfish, plus introns and repeated sequences. Our exomes are remarkably similar for animals that live in such different environments.

Overall, the human genome has a more complex organization of the same basic parts that make up the fruit fly and roundworm genomes. For example, the human genome has thirty copies of the gene that encodes fibroblast growth factor, compared to two copies in the fly and worm genomes. This growth factor is important for the development of highly complex organs.

Genome studies indicate that over deep evolutionary time, genes and gene pieces provided vertebrates, including humans, with certain defining characteristics:

- complex neural networks;
- blood clotting pathways;
- acquired immunity;
- refined apoptosis;
- greater control of transcription;
- complex development; and
- more intricate signaling within and among cells.

Comparing the human genome to itself provides clues to evolution, too. The many duplicated genes and chromosome segments in the human genome suggest that it doubled, at least once, since diverging from the genome of a vertebrate ancestor about 500 million years ago. Either the human genome doubled twice, followed by loss of some genes, or one doubling was followed by additional duplication of certain DNA sequences.

The extensive duplication within the human genome is what distinguishes our genome from those of other primates. Some of the doublings are vast. Half of chromosome 20 repeats, rearranged, on chromosome 18. Much of chromosome 2's short arm reappears as almost three-quarters of chromosome 14, and a block on its long arm is echoed on chromosome 12. The gene-packed yet tiny chromosome 22 includes eight huge duplications and several gene families. Repeated DNA sequences in the human genome may provide raw material for future evolution. A second copy of a DNA sequence can mutate, allowing a cell to experiment with a new function while the old one carries on, a little like trying a new car before selling the old one. More often, the twin gene mutates into a silenced pseudogene, leaving a ghost of the gene behind as a similar but untranslated DNA sequence.

A duplication can be located near the original DNA sequence it was copied from, or away from it. A sequence repeated right next to itself is called a tandem duplication, and it usually results from mispairing during DNA replication. A copy of a gene on a different chromosome may arise when messenger RNA is copied (reverse transcribed) into DNA, which then inserts elsewhere among the chromosomes.

Duplication of an entire genome results in polyploidy, discussed in chapter 13. It is common in plants and some insects, but not vertebrates. If a polyploid event was followed by loss of some genes and duplication of others, the result would look much like the modern human genome (**figure 16.17**). The remnants of such an ancient whole-genome duplication would have become further muddled with time, as inversions and translocations altered the ancestral DNA landscape.

Genes That Help to Define Us

Comparison of the chimpanzee and human genomes has revealed "human accelerated regions." These are highly conserved sequences that show signs of positive selection in humans, such as an amino acid change seen in all human

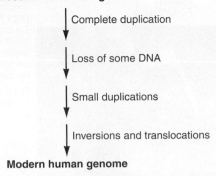

Ancestral vertebrate genome

Complete duplication

Loss of some DNA

Small duplications

Inversions and translocations

Modern human genome

Figure 16.17 Evolution of the human genome. The many duplicated DNA sequences in the human genome suggest a complete duplication followed by other chromosome-level events.

groups but not in the chimp or orangutan versions of the same gene. (Recall from chapter 15 that positive selection maintains a beneficial trait.) Signs of positive selection in the human genome can complement views of our ancestry from fossils.

Relatively few DNA sequence differences distinguish us from our closest known extinct relatives, the Denisovans and Neanderthals: 31,000 single nucleotide changes (SNPs and mutations), and fewer than 125 small losses or gains of DNA bases (indels). Only 96 of the DNA differences in the human genome alter amino acids in proteins (are nonsynonymous) and can therefore be subject to natural selection. Subtracting Denisovan DNA sequences that are conserved in ape and monkey genomes, indicating their antiquity, reduces the list of distinctive human mutations to 23. Eight of them have to do with brain function and development, neural connectivity and synapses, and two of these mutations are implicated in autism. About 3,000 sequence differences are in parts of the human genome that regulate gene expression, so just a few distinctions can nonetheless have large effects.

Traits that may be uniquely human include spoken language, abstract reasoning ability, highly opposable thumbs, and larger frontal lobes of the brain. We can't know whether the Denisovans and Neanderthals were able to speak and the extent of their abilities to think. Following are examples of traits determined by single genes that help to distinguish us from our closest relatives (**figure 16.18**).

Speech

A family living in London whose members have unintelligible speech led to the discovery of a single gene (*FOXP2*, OMIM 605317) that controls speaking ability. The gene is present, but different, in chimps.

Different Developmental Timetables

More primitive primates lack or have very little fetal hemoglobin (see figure 11.2). In more recently evolved and more complex primates, fetal hemoglobin extends the fetal period,

Speech
FOXP2 gene

Abundant sweat glands

Big brain
Single genes, copy number variants, and gene expression patterns contribute to human cognition.

Fetal hemoglobin extends prenatal brain development.

Bare skin
Silenced keratin genes

Walking + running
Ability to put one foot in front of the other; *AHI*1 gene

Figure 16.18 What makes us human? A collection of traits distinguishes us from our closest primate relatives.

allowing time for substantial brain growth and development before birth. With larger brains came greater skills. Other single genes make possible the longer childhood and adolescence of humans compared to chimps.

Bare Skin

Nearly all mammals have thick, abundant hairs that insulate, protect, and are important in social displays and species recognition. Chimpanzees and gorillas express a keratin gene whose counterpart in humans is not expressed.

Our naked skins make sense in terms of evolution in a few ways. Relatively hairless skin enables us to sweat. When our forebears moved onto the plains as the forests shrank, about 2.5 to 3 million years ago, individuals with less hair and more abundant sweat glands could travel farther in search of food, or to avoid becoming food. Natural selection favored them. Today, our skin is peppered with sweat glands that pump out quarts of sweat a day, compared to our furred fellow mammals, who have sparser and less efficient sweat glands. Lack of hair may have enabled our ancestors to shed skin parasites such as lice. Human skin is thin yet strong, thanks to a unique combination of keratin proteins that fill the flattened bricklike skin cells.

Walking

Walking requires the ability to place one foot in front of the other. People who have Joubert syndrome (OMIM 608629) can't do this. Nerve cell fibers cannot cross from their origin on one side of the brain to the other, so moving just one arm or leg is impossible; both move at once. The part of the brain that controls posture, balance, and coordination is compromised. The gene that causes Joubert syndrome, *AHI1*, is identical in all modern human groups examined, but has different alleles in chimps, gorillas, and orangutans. Perhaps in the lineage leading to humans, the gene came to control walking by making it possible to place one foot in front of the other.

Running

Homo erectus was the first hominin to be able to run long distances, thanks to specific anatomical adaptations. The nuchal ligament that connects the skull to the neck became more highly developed in *H. erectus,* keeping the head in place while running. Leg muscles were more highly developed than those of chimps or australopithecines, acting as springs. *H. erectus* originated large buttocks, whose muscles contract during running. All three of these structures are not merely the result of being able to walk, but enabled early *Homo,* and us, to run. This skill would have helped our ancestors to escape predators, find food, and locate new homes.

A Big Brain

Only a few genes may control the difference in brain size between us and chimps. About 2.4 million years ago, a gene called *MYH16* underwent a nonsense mutation, which prevented production of a type of muscle protein called a myosin. The mutation is seen in all modern human populations, but not in other primates. Without this type of myosin, jaws developed less, which allowed expansion of the bony plates of the skull, permitting greater brain growth. Researchers nicknamed the mutation "room for thought." Fossil evidence indicates that the switch from "big jaw, small brain" to "small jaw, big brain" happened when *Homo* gradually replaced *Australopithecus,* about 2 million years ago. The genetic analysis is new, but the idea isn't. Charles Darwin wrote in 1871 that different-sized jaw muscles were at the root of the distinction between apes and humans.

The major reason why humans and chimps look and behave differently but are genetically so similar seems to have more to do with gene expression than genome sequence. For example, a study comparing gene expression in the liver and brain found many more differences between the two species in the brain than in the liver, suggesting that our mental capabilities have extended beyond those of chimps more than have the functions of our livers. The brain, therefore, is a big part of what makes us human.

A larger brain presumably led to greater cognitive skills. In this area, gene copy number may be more important than the nature of the genes. At least 134 genes are present in more copies in the human genome than in genomes of apes. Many of these genes affect brain structure or function. Some of the genes foster long-term memory, and others, when mutant, cause intellectual disability or impair language skills. Single genes implicated in fueling human brain growth control the migration of nerve cells in the front of the fetal brain.

Recent mutations may have propelled our brain power, too. A mutation in a gene that encodes an enzyme called a fatty acid desaturase enables the digestive system to use long-chain fatty acids, which are found in fish and shellfish and are useful in brain development. The mutation occurred in Africa from 180,000 to 80,000 years ago, which was when our ancestors lived near the lakes of Central Africa. Positive selection greatly increased the prevalence of the mutation, but at a cost—the encoded enzyme also causes inflammation and may be responsible for the increased prevalence of hypertension, stroke, type 2 diabetes, and coronary heart disease among African Americans.

Smell

Our chemical senses—smell and taste—have decreased as our reliance on them has decreased. The sense of smell derives from a 1-inch-square patch of tissue high in the nose that consists of 12 million cells that bear odorant receptor (OR) proteins. (In contrast, a bloodhound has 4 billion such cells!) Molecules from a smelly substance bind to combinations of these receptors, which then signal the brain in a way that creates the perception of the associated odor.

The 906 human odorant receptor genes are clustered, and more than half are pseudogenes—their sequences are similar to active "smell" genes, but are riddled with nonsense mutations. Perhaps they are remnants of a distant past, when we depended more upon our chemical senses for survival, and natural selection silenced them. Natural selection has also acted positively to retain OR genes that continue to function. While the pseudogenes contain many diverse SNPs, the functional OR genes are remarkably alike in sequence among individuals, indicating a successful function. In addition, the nucleotide differences that persist among the retained OR genes actually alter the encoded amino acids, suggesting that natural selection favored these sequences.

How will evolution mold humanity in the future? Perhaps the most profound characteristic that distinguishes us from our immediate ancestors is ability to alter the environment, which has led to our domination of the planet. "Anatomically modern humans"—us—exist in varieties of skin color, body size, intellectual ability, and hundreds of other observable and measurable traits. Yet at the same time, people from different populations are meeting and mixing their genomes at unprecedented levels, thanks to transportation and communication over vast distances. The current state of humanity is far different from the days of the australopithecines, who dwelled in geographical areas so dispersed that they existed as an overlapping series of distinct species. Today, phenotypes persist that in millennia past would have disappeared due to negative natural selection, thanks to health care and technology.

What will humans be like 10,000 years from now? Will admixture increase so that we all have genome parts that were once associated with specific population groups? Will new technologies and medical care improve our genomes, or allow gene variants that would have been lethal in the past to persist and be passed to future generations? Will we manipulate our genomes to direct our own future evolution? Understanding our origins poses many questions about our futures.

Key Concepts Questions 16.4

1. How does the human genome differ from the genomes of other animals, in a general sense?

2. What are some genes that provide traits that are uniquely human?

3. How do we influence our own future evolution?

Summary

16.1 Human Origins

1. DNA provides information on evolution because sequence similarities indicate descent from shared ancestors.

2. Humans are hominins, a taxonomic group which includes our ancestors who lived after the split from other African apes about 6 million years ago.

3. The australopithecines preceded and then coexisted with early *Homo,* who lived in caves, had strong family units, and used tools, then lived in societies, and used fire. *Homo sapiens idaltu* lived about 156,000 years ago, and looked like us.

4. *Homo sapiens idaltu,* the Neanderthals, and the Denisovans were archaic humans. Neanderthals and Denisovan DNA sequences persist in the genomes of some modern human populations, indicating ancient inbreeding.

5. The diverse genomes of modern indigenous peoples provide clues to the genomes of our ancestors.

16.2 Methods to Study Molecular Evolution

6. **Molecular evolution** considers differences at the genome, chromosome, protein, and DNA sequence levels with mutation rates to estimate species relatedness.

7. Genes in the same order on chromosomes in different species show **synteny**.

8. For a highly conserved gene DNA sequence is similar or identical in different species, indicating importance and shared ancestry.

9. Evolutionary tree diagrams represent gene sequence information from several species, using molecular clocks based on mutation rates.

10. Molecular clocks based on mtDNA date recent events through the maternal line because this DNA mutates faster than nuclear DNA. Y chromosome genes trace paternal lineage. Markers (SNPs, STRs, and CNVs) in mtDNA, Y chromosome DNA, and autosomal DNA are used to study human origins and expansions. Many linked markers inherited together form **haplogroups**.

11. Changing haplotype patterns indicate genetic **admixture** which, using crossover rates and generation times, can estimate when populations mixed.

16.3 The Peopling of the Planet

12. The rate of mtDNA mutation and current mtDNA diversity can be extrapolated to hypothesize that a theoretical first woman lived, in Africa, about 200,000 years ago. *Homo sapiens* began to leave Africa about 56,000 years ago.

13. A series of migrations and founder effects peopled the planet, with genetic diversity decreasing from that of the ancestral African population, but new mutations occurring.

14. After the last ice age, people crossed the Bering Strait from Siberia, occupying the Americas.

15. Humans and chimps share 98.7 percent of their protein-encoding gene sequences. Indels, introns, and repeats create genome differences between humans and chimps, which also differ in gene expression.

16. The human genome shows many signs of past duplication.

16.4 What Makes Us Human?

17. Humans and chimps share 98.7 percent of their protein-encoding gene sequences. Indels, introns, and repeats create genome differences between humans and chimps, which also differ in gene expression.

18. The human genome shows many signs of past duplication.

19. Humans differ from other apes in adaptations that allowed greater brain growth, less hairy skins and the adaptations that they provide, and the abilities to speak, walk, and run.

20. We affect our own evolution by treating disease and altering the environment.

www.mhhe.com/lewisgenetics11

Answers to all end-of-chapter questions can be found at **www.mhhe.com/lewisgenetics11**. You will also find additional practice quizzes, animations, videos, and vocabulary flashcards to help you master the material in this chapter.

Review Questions

1. Arrange the following primates in the order in which they lived, indicating any that may have overlapped in time.
   a. *Homo erectus*
   b. *Australopithecus afarensis*
   c. Neanderthals
   d. *Ardipithecus*
   e. *Homo sapiens idaltu*
   f. the Denisova hominin

2. Explain how the evolutionary tree diagram in figure 16.1 indicates that we did not descend directly from chimpanzees.

3. How were australopithecines like chimps but also like humans?

4. What evidence led researchers to hypothesize that there was only one species of early *Homo*?

5. Why are Neanderthal DNA sequences in the genomes of modern Europeans but not in modern Africans?

6. What is the evidence that native Africans have the most ancient roots?

7. Give an example of how a single gene difference can have a profound effect on the phenotypes of two species.

8. Give an example of a trait in any of the hominins discussed in the chapter that illustrates positive selection.

9. Explain what comparing genomes from indigenous peoples to other modern peoples can reveal about evolution.

10. What is compared in molecular evolution studies?

11. Give an example of molecular evidence that is consistent with fossil or other evidence.

12. What do "highly conserved" DNA or amino acid sequences indicate about evolution?

13. Explain why a nonsynonymous mutation in an exon is likely to have a greater potential impact on evolution than a synonymous mutation in an exon.

14. What is a limitation of an evolutionary tree diagram constructed using DNA or protein sequence data?

15. Describe the type of information that Y chromosome and mitochondrial DNA sequences provide, and why it is incomplete.

16. How can haplotypes provide information on genetic admixture and ancient migrations?

17. How is the information in haplotype changes conceptually opposite that in evolutionary tree diagrams?

18. What information can ancestry testing provide and what is a limitation of it?

19. List three defining characteristics of vertebrates, as indicated from genome studies, and of humans.

20. How can we affect our own future evolution?

Applied Questions

1. Select an example from this chapter and explain how it illustrates one of the forces of evolutionary change discussed in chapter 15 (natural selection, nonrandom mating, migration, genetic drift, or mutation).

2. What is a limitation of designating species based only on fossil evidence?

3. Hypothesize why many species of *Australopithecus* lived at one time, but only one species of *Homo* lives today.

4. A geneticist aboard a federation starship must deduce how closely related humans, Klingons, Romulans, and Betazoids are. Each organism walks on two legs, lives in complex societies, uses tools and technologies, looks similar, and reproduces in the same manner. Each can interbreed with any of the others. The geneticist finds the following data:

   - Klingons and Romulans each have 44 chromosomes. Humans and Betazoids have 46 chromosomes. Human chromosomes 15 and 17 resemble part of the same large chromosome in Klingons and Romulans.

   - Humans and Klingons have 97 percent of their chromosome bands in common. Humans and Romulans have 98 percent of their chromosome bands in common, and humans and Betazoids show 100 percent correspondence. Humans and Betazoids differ only by an extra segment on chromosome 11, which appears to be a duplication.

   - The cytochrome *c* amino acid sequence is identical in humans and Betazoids, differs by one amino acid between Humans and Romulans, and differs by two amino acids between humans and Klingons.

   - The gene for collagen contains 50 introns in humans, 50 introns in Betazoids, 62 introns in Romulans, and 74 introns in Klingons.

   - Mitochondrial DNA analysis reveals many more individual differences between Klingons and Romulans than between humans and Betazoids.

   a. Suggest a series of chromosomal abnormalities or variants that might explain the karyotypic differences among these four types of organisms.

   b. Which are our closest relatives among the Klingons, Romulans, and Betazoids? What is the evidence for this?

   c. Are Klingons, Romulans, humans, and Betazoids distinct species? What information reveals this?

   d. Which of the evolutionary tree diagrams is consistent with the data?

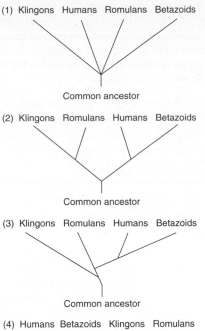

(1) Klingons Humans Romulans Betazoids

Common ancestor

(2) Klingons Romulans Humans Betazoids

Common ancestor

(3) Klingons Romulans Humans Betazoids

Common ancestor

(4) Humans Betazoids Klingons Romulans

Common ancestor

5. A man with white skin checks the box on forms about personal information for "African American," claiming that we are all, if we go back far enough in our family trees, from Africa. Is he correct?

6. In Central Africa, the Mbuti Pygmies are hunter-gatherers who live amid agricultural communities of peoples called the Alur, Hema, and Nande. Researchers compared autosomal STRs, mitochondrial DNA, and Y chromosome DNA haplogroups among these four types of people. The pygmies had the most diverse Y chromosomes, including about a third of the sequence that was the same as those among the agricultural groups, who had greater mtDNA diversity than the pygmies. None of the agricultural males had pygmy Y DNA sequences. Create a narrative of gene flow to explain these findings.

7. Suggest an explanation for the persistence of pale skin in Ethiopia, even though the phenotype makes a person more susceptible to skin damage from the sun.

8. Name a technology that has contributed to human admixture.

Web Activities

1. Go to http://www.peoplesoftheworld.org/ or a similar website. Select an indigenous people, do further research, and describe their habitat and culture, and any distinctive health strengths or problems. To what extent do you think genetics is responsible for the state of their health? Explain your reasoning.

2. Answer the following questions after reading the website for the Genographic Project, or from the chapter.

   a. Explain why analyzing your mtDNA or Y chromosome DNA cannot provide a complete picture of your ancestry.

   b. Explain how a female can trace her paternal lineage if she doesn't have a Y chromosome.

   c. Would you want your ancestry information and identity posted on the Genographic database or an ancestry company's database so that cousins can contact you?

Forensics Focus

1. How can the technology used to describe the 4,000-year-old Greenland Eskimo be applied to forensics techniques used on crime scene evidence?

2. In many African cultures, "family" is not dictated by genetics, but by who cares for whom. Any adult can be "mother" or "father" to any child. Since the early 1990s, many parts of East Africa have been under civil war. Thousands of Africans have asked to be admitted to the United States to join relatives. The "family reunification resettlement program" enables parents, siblings, and children of U.S. citizens to come to the United States. In 2008, addressing rumors that many people were lying that they were related to people in the United States, the State Department began to ask refugees from Kenya to voluntarily provide a DNA sample, to be compared to that of the U.S. citizen claimed to be a relative. When the DNA testing turned up many cases of people claiming to be family who were not blood relatives, the resettlement program was stopped.

   a. Do you think that DNA testing should have been imposed on people seeking asylum in the United States?

   b. Should the people have been compelled to have their DNA tested?

   c. How should cultural definitions, such as that of "family," be handled?

   d. How should the situation be resolved?

Case Studies and Research Results

1. The Australian aborigines are one of the oldest peoples outside of Africa, tracing their ancestry back 55,000 years. When geneticists first approached them in the 1990s about studying their DNA, the people became upset. They regarded ancestry genetic research as a threat to their traditional belief systems, as challenging their identity and claims to land, and feared stigma from genetic test results. They also objected strenuously to one research project referring to Aborigines as a "vanishing people."

   An Aboriginal genome was sequenced in 2011, from a hair sample from a young man from the 1920s donated to a museum. The researchers went into the Aborigine community and got consent. Now the researchers would like to sequence the genomes of additional members of the community whose DNA was sampled years ago, before completion of the human genome project.

   Should the researchers request new informed consent to sequence the genomes in the samples, and if so, what information should they provide? Should they sequence genomes from samples from people who have died or cannot be located?

2. "Neanderthals are not totally extinct; they live on in some of us," said Svante Paabo, the leader of the Neanderthal Genome Project. What does he mean?

3. The 4,000-year-old Saqqaq Paleo Eskimo from Greenland is known only from a tuft of hair. How did researchers learn about other characteristics?

4. A Y chromosome haplotype has mutations for the *SRY* gene and genes called *M96* and *P29*. Modern Africans have three variants of this haplotype. Two variants are only in Africans, but the third variant, E3, is also seen in western Asia and parts of Europe. Researchers examined specific subhaplotypes (variations of the variations) and found that one type, E-M81, accounts for 80 percent of the Y chromosomes sampled in northwest Africa, falling sharply in incidence to the east, and not present in sub-Saharan Africa. That same haplotype is found in a small percentage of the Y chromosomes in Spain and Portugal. Consult a map, and propose a scenario for this gene flow. What further information would be useful in reconstructing migration patterns?

5. Roland has always considered himself African American, but ancestry testing of his Y chromosome indicates a Chinese background. He is very upset, concluding that he is not African American after all. Explain how he has misinterpreted the test results or the test results could be in error.

6. What technique did researchers use to estimate that a large movement of people from west Eurasia into

eastern Africa occurred about 1,000 B.C., and that movement from eastern to southern Africa happened about 500 A.D.?

7. "Glossogenetics" is the study of associating language with patterns of genetic variation. Explain the finding that languages in a particular geographic area track with Y chromosome sequences, but not with mtDNA sequences.

8. For more than 20 million years, lice have lived on the skins of primates. Researchers compared a 1,525-base-pair sequence of mtDNA among modern varieties of lice, and, applying the mutation rate, derived an evolutionary tree. It depicts a split in the louse lineage, with one group of head and body lice living throughout the world, and another group of only head lice living in the Americas.

   a. What events in human evolution roughly correspond to the branch points in the louse evolutionary tree?

   b. What might be the significance of the similarity between the evolutionary trees for lice and humans?

   c. What is the evidence that lice moved from archaic humans to modern humans?

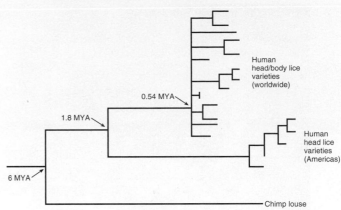

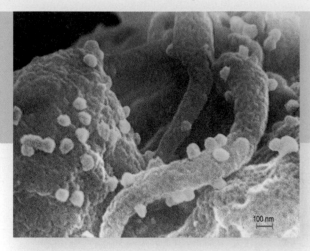

These HIV particles (green) attach to a white blood cell (pink). In people with a certain mutation, cells lack one of two types of receptors that HIV requires to enter.

100 nm

Genetics of Immunity

Learning Outcomes

17.1 The Importance of Cell Surfaces

1. List the components of the immune system.

2. Describe the basis of blood groups.

3. Explain what human leukocyte antigens are and what they indicate about health.

17.2 The Human Immune System

4. Distinguish among physical barriers, innate immunity, and adaptive immunity.

5. Distinguish between the humoral and cellular immune responses.

17.3 Abnormal Immunity

6. Discuss conditions that result when the immune system is underactive, overactive, and misdirected.

17.4 Altering Immunity

7. Describe how medical technologies boost or suppress immunity to prevent or treat disease.

8. Explain the requirements for the body to accept an organ from another person.

17.5 Using Genomics to Fight Infection

9. Discuss how we can use knowledge of the genomes of pathogens.

The **BIG** Picture

The immune system enables us to share the planet with other organisms, while ignoring the members of the microbiome that normally live in and on our bodies. Genes control the immune response. We can alter immunity to enhance health.

Mimicking a Mutation to Protect Against HIV

In 2008, a 40-year-old man received a stem cell transplant to treat leukemia at a hospital in Berlin. He had been HIV positive for at least a decade, and had taken anti-HIV drugs for 4 years. Leukemia was his first HIV-related illness. Known at first as "the Berlin patient," Timothy Brown became the star of a groundbreaking experiment.

Brown's stem cell donor was both a tissue match for Brown and was one of the 0.5 percent of Caucasians who are genetically resistant to HIV infection. As a "*CCR5 delta 32*" homozygote, the donor's cells did not have a type of receptor, called CCR5, that HIV must bind to enter. (Figure 17.11 shows this receptor.) Could a transplant of the man's stem cells into Brown fight his leukemia *and* his HIV infection?

So far, the answer is yes! The transplanted stem cells eventually replaced Brown's blood and bone marrow with HIV-resistant cells, and all signs of the infection vanished. As Brown made headlines, researchers were already thinking about how to mimic the natural mutation that protects against HIV. One approach is "genome editing," illustrated in figure 19.7.

In genome editing, researchers take samples of the cells most disabled with virus (CD4 T cells) from patients, culture the cells in the laboratory, and use special enzymes called "zinc finger nucleases" to remove the *CCR5* gene. Then the manipulated cells are infused back into the patients, who stop taking their HIV medications to see if the doctored cells help. So far in a few dozen patients the strategy appears to be safe, and potentially effective. When patients discontinue their anti-HIV drugs after the cell infusion, the infection is still present but definitely dampened, and gradually the fixed cells appear to take over.

17.1 The Importance of Cell Surfaces

We share the planet with plants, microbes, fungi, and other animals, but we can become ill when some of them, or their parts, enter our bodies. The human immune system protects us against this happening. It is a mobile army of about 2 trillion cells, the biochemicals they release, and the organs where they are produced and stored. Genes control many aspects of immune system function, including susceptibility to infection.

Protection against infection is based on the ability of the immune system to recognize "foreign" or "nonself" cell surfaces that are not part of the body. These include surfaces of certain microorganisms such as bacteria and yeasts; nonliving "infectious agents" such as viruses and prions; and tumor cells and transplanted cells. If stimulated, the immune system launches a highly coordinated attack that includes both general and highly specific responses. Organisms or infectious agents that cause disease are called pathogens. **Clinical Connection 17.1** highlights one commonly encountered type of pathogen—viruses. **Figure 17.1** shows another—bacteria. A bacterium is a cell; a virus is simpler than a cell.

Understanding how genes control immunity makes it possible to enhance or redirect the system's ability to fight disease. Mutations can impair immune function, causing immune deficiencies, autoimmune disorders, allergies, and cancer. Genes affect immunity by conferring susceptibilities or resistances to certain infectious diseases. Like other inherited characteristics, degree of immune protection varies from person to person. One individual may suffer frequent respiratory infections, yet another is rarely ill. A study of the immune systems of people who survived the 1918 flu pandemic revealed that many decades later, they can still rapidly destroy a flu virus. Yet 50 million people died of that flu, many in just days.

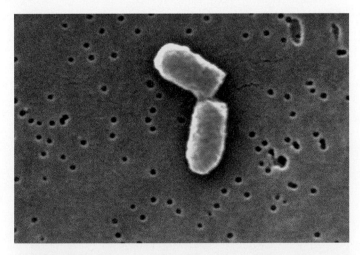

Figure 17.1 A bacterial pathogen. *Escherichia coli* is a normal resident of the human small intestine, but under certain conditions can produce a toxin that causes severe diarrhea ("food poisoning") and can damage the kidneys.

A few types of genes encode proteins that powerfully affect immunity. **Antibodies** are proteins that directly attack foreign **antigens,** which are any molecules that can elicit an immune response. Most antigens are proteins or carbohydrates. Genes also specify the cell surface antigens that mark the body's cells as "self."

Blood Groups

Some of the antigens that dot our cell surfaces determine blood types. Figures 5.3 and 5.4 describe the familiar ABO blood types. We actually have thirty-three major blood types based on protein and carbohydrate antigens on the surfaces of red blood cells, and more may be added as the human genome sequence is annotated. Each blood type includes many subtypes, generating hundreds of ways that the topographies of our red blood cells differ from individual to individual. **Table 17.1** lists a few blood groups.

For blood transfusions, blood is typed and matched from donor to recipient. For more than a century, an approach called serology typed blood according to red blood cell antigens. A newer way to type blood is to identify the *instructions* for the cell-surface antigens—that is, the genes that encode these proteins. This approach, termed genotyping, uses a tiny device that detects 100 distinct DNA "signatures" for blood types. Genotyping is especially useful for people who have a chronic disorder that requires multiple transfusions, such as leukemia or sickle cell disease. They produce so many antibodies against so many types of donor blood that it is often difficult to determine their blood types by serology.

The Major Histocompatibility Complex

Many proteins on our cell surfaces are encoded by genes that are part of a 6-million-base-long DNA sequence on the short arm of chromosome 6 called the major histocompatibility complex (MHC). This region includes about 70 genes, and confers

Table 17.1	Blood Groups
Blood Group (OMIM)	**Description**
MN (111300)	Codominant alleles *M, N,* and *S* determine six genotypes and phenotypes. The antigens bind two glycoproteins.
Lewis (111100)	Allele *Le* encodes fucosyltransferase (FUT3) that adds an antigen to the sugar fucose, which the product of the *H* gene places on red blood cells. *H* gene expression is necessary for the ABO phenotype (see figure 5.3). People with *LeLe* or *Lele* have the Lewis antigen on red blood cells and in saliva. People of genotype *lele* do not.
Secretor (182100)	People with the *Se* allele secrete A, B, and H antigens into body fluids.

Viruses

A viral infection can make us feel terrible, but many of the aches and pains are in fact actions of the human immune system.

A virus is a single or double strand of RNA or DNA wrapped in a protein coat, and in some types, an outer envelope, too. A virus can reproduce only if it enters and uses a host cell's energy resources, protein synthetic machinery, and secretion pathway. It is a very streamlined structure. A virus may have only a few protein-encoding genes, but many copies of the same protein can assemble to form an intricate covering, like the panes of glass in a greenhouse. Ebola viruses, for example, have only seven types of proteins, but these assemble into a structure capable of reducing a human body to little more than a bag of blood and decomposed tissue. In contrast, the smallpox virus has more than 100 different types of proteins, and HIV also has a complex structure (see figures 17.11 and 17.12).

Viruses are with us all the time—not only when we are ill. Part of the DNA sequence of some human chromosomes includes viral DNA sequences that are vestiges of past infections, perhaps passed, silently, from distant ancestors. Many DNA viruses reproduce by inserting their DNA into the host cell's DNA. In contrast, an RNA virus must first copy its RNA into DNA before it can insert into a human chromosome. A viral enzyme called reverse transcriptase copies viral RNA into DNA. Certain RNA viruses are called retroviruses because they transmit genetic information opposite the usual direction—instead of DNA. to RNA to protein, viral RNA is copied into DNA, which may then be copied back into RNA to guide the synthesis of viral proteins. HIV is a retrovirus.

Once viral DNA integrates into the host cell's DNA, it can either remain and replicate along with the host's DNA without causing harm, or it can take over and kill the cell. Activated viral genes direct the host cell to replicate viral DNA and then transcribe and translate it, producing viral proteins. The cell bursts, releasing many new viruses.

Diverse viruses infect all types of organisms. Their genetic material cannot repair itself, so the mutation rate may be high. This is one reason why we cannot develop an effective vaccine against HIV or the common cold, and why new influenza vaccines must be developed each year.

Influenza viruses have RNA as their genetic material and are of three strains: A, B, and C. Influenza A is the most common and comes from birds, sometimes passing through pigs. Each viral strain has subtypes, based on two types of glycoproteins on their surfaces. Specifically, influenza A has many copies of any of sixteen variants of a large surface glycoprotein called a hemagglutinin (HA), and many copies of any of nine variants of another surface glycoprotein called a neuraminidase (NA) (**figure 1a**). The 1918 Spanish flu and the swine flu of 2009 were "H1N1," whereas the "bird flu" of 2004 was "H5N1." Vaccines typically consist of two types of influenza A and one type of the less common influenza B. Influenza is easily transmitted in a sneeze (figure 1b), but about a third of people infected with influenza virus do not develop symptoms.

Questions for Discussion

1. How does the structure of a virus differ from that of a cell?

2. How does an RNA virus insert its genetic material into a human chromosome?

3. Why is it difficult to create a vaccine that protects against HIV?

4. What part of an influenza virus is used to manufacture a vaccine?

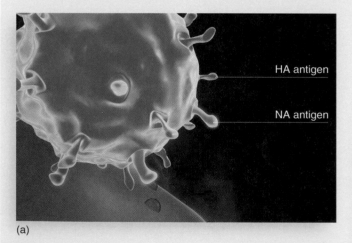

HA antigen

NA antigen

(a)

(b)

Figure 1 **The flu.** **(a)** Influenza viruses are classified by variants of large surface glycoproteins called hemagglutinins (HAs), and neuraminidases (NAs). **(b)** Flu is easily transmissible in a sneeze. This is one of several reasons why vaccination is important.

about 50 percent of the genetic influence on immunity. MHC genes are classified into three groups based on their functions.

MHC class III genes encode proteins that are in plasma (the liquid portion of blood) and provide nonspecific immune functions. MHC classes I and II genes encode the **human leukocyte antigens** (**HLAs**), so-named because they were first studied in leukocytes (white blood cells). The HLA proteins link to sugars, forming branchlike glycoproteins that extend from cell surfaces.

The proteins that the class I and II HLA genes encode differ in the types of immune system cells they alert. Some HLA glycoproteins bind bacterial and viral proteins, displaying them like badges to alert other immune system cells. This action, called antigen processing, is often the first step in an immune response. The cell that displays the foreign antigen is called an **antigen-presenting cell**. **Figure 17.2** shows how a large cell called a macrophage displays bacterial antigens. Certain white blood cells called T cells (or T lymphocytes) are also antigen-presenting cells. Dendritic cells are antigen-presenting cells found in places where the body contacts the environment, such as in the skin and in the linings of the respiratory and digestive tracts. Dendritic cells signal T cells, initiating an immune response.

A person's HLA "type" identifies all of his or her cells as "self," or belonging to the same individual. In addition to common HLA markers are more specific markers that distinguish particular tissue types within an individual. Class I HLA genes include three genes that vary greatly and are found on all cell types, and three other genes that are more restricted in their distribution. Class II includes three major genes whose encoded proteins are found mostly on antigen-presenting cells.

An individual's overall HLA type is based on the six major HLA genes. So variable are these genes that only 2 in every 20,000 unrelated people match for the six major HLA genes by chance. When transplant physicians attempt to match donor tissue to a potential recipient, they determine how alike the two individuals are for these six genes. Usually at least four of the genes must match for a transplant to have a reasonable chance of success. Before DNA profiling, HLA typing was the predominant type of blood test used in forensic and paternity cases to rule out involvement of certain individuals. However, HLA genotyping has become very complex because hundreds of alleles are now known.

A few disorders are very strongly associated with inheriting particular HLA types. This is the case for ankylosing spondylitis, which inflames and deforms vertebrae. A person with either of two subtypes of an HLA called B27 is 100 times as likely to develop the condition as someone who lacks either form of the antigen. HLA-associated risks are not absolute. More than 90 percent of people who suffer from ankylosing spondylitis have the B27 antigen, which occurs in only 5 percent of the general population. However, 10 percent of people who have ankylosing spondylitis do *not* have the B27 antigen, and some people who have the antigen never develop the disease.

Key Concepts Questions 17.1

1. What is the overall function of the immune system?
2. What are the major components of the immune system?
3. How are genes part of an immune response?
4. Distinguish between viruses and bacteria.
5. What are blood types and HLA types?

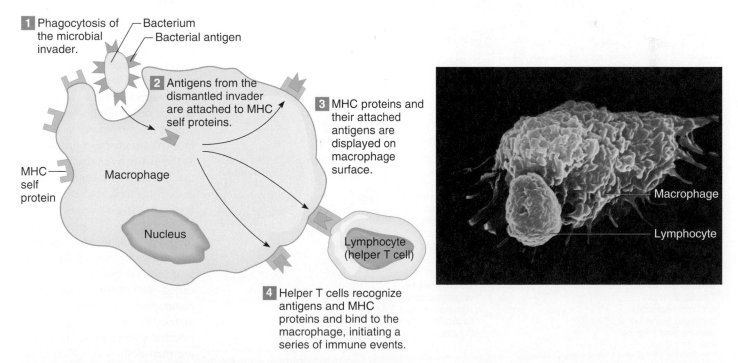

1 Phagocytosis of the microbial invader.

— Bacterium
— Bacterial antigen

2 Antigens from the dismantled invader are attached to MHC self proteins.

3 MHC proteins and their attached antigens are displayed on macrophage surface.

MHC self protein

Macrophage

Nucleus

Lymphocyte (helper T cell)

4 Helper T cells recognize antigens and MHC proteins and bind to the macrophage, initiating a series of immune events.

Macrophage

Lymphocyte

Figure 17.2 **Macrophages are antigen-presenting cells.** A macrophage engulfs a bacterium, then displays foreign antigens on its surface, which are held in place by major histocompatibility complex (MHC) self proteins. This "antigen presentation" sets into motion many immune reactions, including binding to a lymphocyte.

17.2 The Human Immune System

The human immune system is a network of vessels called lymphatics that transport lymph fluid to bean-shaped structures throughout the body called lymph nodes. The spleen and thymus gland are also part of the immune system (**figure 17.3**).

Lymph fluid carries white blood cells called lymphocytes and the wandering, scavenging macrophages that capture and degrade bacteria, viruses, and cellular debris. Figure 2.3 shows a macrophage engulfing bacteria. **B cells** and **T cells** are the two major types of lymphocytes.

The genetic connection to immunity is the proteins required to carry out an immune response. The immune response attacks pathogens, cancer cells, and transplanted cells with two lines of defense—an immediate generalized **innate immune response,** and a more specific, slower **adaptive immune response**. These defenses act after various physical barriers block pathogens. **Figure 17.4** summarizes the basic components of the immune system, discussed in the following sections. ("Immune response" is used synonymously with "immunity.")

Physical Barriers and Innate Immunity

Several familiar structures and fluids keep pathogens from entering the body in the innate immune response: unbroken skin, mucous membranes such as the lining inside the mouth, earwax, and waving cilia that push debris and pathogens up and out of the respiratory tract. Most microbes that reach the stomach perish in a vat of churning acid or are flushed out in diarrhea. These physical barriers are nonspecific. That is, they keep out anything foreign, not just particular pathogens.

If a pathogen breaches these physical barriers, innate immunity provides a rapid, broad defense. The term *innate* refers to the fact that these general defenses are in the body, ready to function should infection threaten without requiring specific stimulation. A central part of the innate immune response is **inflammation,** a process that creates a hostile environment for certain types of pathogens at an injury site. Inflammation sends in cells that engulf and destroy pathogens. Such cells are called phagocytes, and their engulfing action is phagocytosis. Certain types of white blood cells and the large, wandering macrophages are phagocytes. Also at the infection site, plasma accumulates, which dilutes toxins and brings in antimicrobial chemicals. Increased blood flow with inflammation warms the area, turning it swollen and red.

In addition to inflammation, three classes of proteins participate in innate immunity. These are the complement system, collectins, and cytokines. Mutations in the genes that encode these proteins lower resistance to infection.

The **complement** system consists of plasma proteins that assist, or complement, several other defenses. Some complement proteins puncture bacterial plasma membranes, bursting the cells. Other complement proteins dismantle viruses or trigger release of histamine from mast cells, another type of immune system cell that is involved in allergies. Histamine dilates blood vessels, increasing fluid flow to the infected or injured area. Still other complement proteins attract phagocytes to an injury site.

Collectins broadly protect against bacteria, yeasts, and some viruses by detecting slight differences in their surfaces from human cells. Groups of human collectins correspond to the surfaces of different pathogens, such as the distinctive sugars on yeast, the linked sugars and lipids of certain bacteria, and the surface features of some viruses.

Cytokines play roles in both innate and adaptive immunity. As part of the innate immune response, cytokines called **interferons** alert other components of the immune system to the presence of cells infected with viruses. These cells are then destroyed, which limits the spread of the viral infection. **Interleukins** are cytokines that cause fever, temporarily triggering

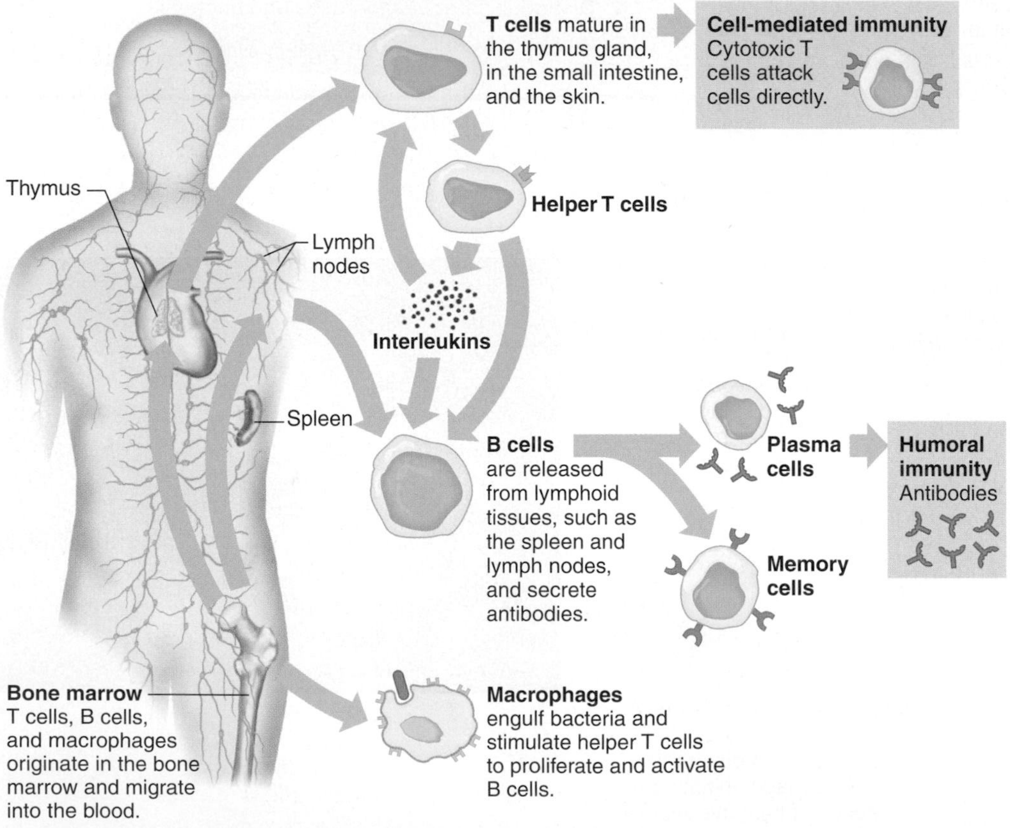

Figure 17.3 The immune system produces diverse cells. T cells, B cells, and macrophages build an overall immune response. All three types of cells originate in the bone marrow and circulate in the blood.

T cells mature in the thymus gland, in the small intestine, and the skin.

Cell-mediated immunity Cytotoxic T cells attack cells directly.

Helper T cells

Interleukins

B cells are released from lymphoid tissues, such as the spleen and lymph nodes, and secrete antibodies.

Plasma cells

Memory cells

Humoral immunity Antibodies

Macrophages engulf bacteria and stimulate helper T cells to proliferate and activate B cells.

Thymus

Lymph nodes

Spleen

Bone marrow T cells, B cells, and macrophages originate in the bone marrow and migrate into the blood.

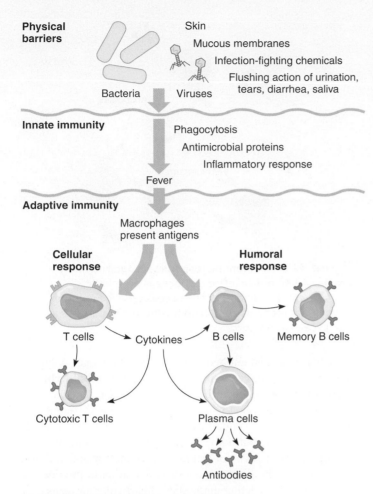

Physical barriers — Skin, Mucous membranes, Infection-fighting chemicals, Flushing action of urination, tears, diarrhea, saliva

Bacteria Viruses

Innate immunity — Phagocytosis, Antimicrobial proteins, Inflammatory response

Fever

Adaptive immunity — Macrophages present antigens

Cellular response / Humoral response

T cells → Cytokines → B cells → Memory B cells

Cytotoxic T cells

Plasma cells → Antibodies

Figure 17.4 **Levels of immune protection.**
Pathogens such as viruses first must breach physical barriers, then encounter nonspecific cells and molecules of the innate immune response. If this is ineffective, the adaptive immune response begins: Antigen-presenting cells stimulate T cells to produce cytokines, which activate B cells to divide and differentiate into plasma cells, which secrete antibodies. Once activated, these cells "remember" the pathogen, allowing faster responses to subsequent encounters.

a higher body temperature that directly kills some infecting bacteria and viruses. Fever also counters microbial growth indirectly, because higher body temperature reduces the iron level in the blood. Because bacteria and fungi require more iron as the body temperature rises, they cannot survive in a fever-ridden body. Phagocytes also attack more vigorously when the temperature rises. Tumor necrosis factor is another type of cytokine that activates other protective biochemicals, destroys certain bacterial toxins, and attacks cancer cells. Many of the aches and pains we experience from an infection are actually due to the innate immune response, not directly to the actions of the pathogens.

Adaptive Immunity

Adaptive immunity must be stimulated into action. It may take days to respond, compared to minutes for innate immunity. Adaptive immunity is highly specific and directed.

B cells and T cells carry out the adaptive immune response, which has two components. In the **humoral immune response,** B cells produce antibodies after activation by T cells. ("Humor" means fluid; antibodies are carried in fluids.) In the **cellular immune response,** T cells produce cytokines and activate other cells. B and T cells differentiate in the bone marrow and migrate to the lymph nodes, spleen, and thymus gland, as well as circulate in the blood and tissue fluid.

The adaptive arm of the immune system has three basic characteristics. It is *diverse,* vanquishing many types of pathogens. It is *specific,* distinguishing the cells and molecules that cause disease from those that are harmless. The immune system also *remembers,* responding faster to a subsequent encounter with a foreign antigen than it did the first time. The first assault initiates a **primary immune response**. The second assault, based on the system's "memory," is a **secondary immune response**. The immune system's memory is why we get some infections, such as chickenpox, only once. However, upper respiratory infections and influenza recur because the causative viruses mutate, presenting a different face to our immune systems each season.

The Humoral Immune Response—B Cells and Antibodies

An antibody response begins when an antigen-presenting macrophage activates a T cell. This cell in turn contacts a B cell that has surface receptors that can bind the type of foreign antigen the macrophage presents. The immune system has so many B cells, each with different combinations of surface antigens, that one or more is nearly always available that corresponds to a particular foreign antigen. Turnover of B cells is high. Each day, millions of B cells perish in the lymph nodes and spleen, while millions more form in the bone marrow, each with a unique combination of surface molecules.

Once an activated T cell finds a B cell match, it releases cytokines that stimulate the B cell to divide. Soon the B cell gives rise to two types of cells (**figure 17.5**). The first, **plasma cells,** are antibody factories, each secreting 1,000 to 2,000 identical antibodies per second into the bloodstream. Plasma cells live only days. These cells provide the primary immune response. Plasma cells derived from different B cells secrete different antibodies. Each type of antibody corresponds to a specific part of the pathogen, like hitting a person in different parts of the body. This multi-pronged attack is called a polyclonal antibody response (**figure 17.6**). The second type of B cell descendant, **memory cells,** are far fewer and usually dormant. They respond to the foreign antigen faster and with more force should it appear again. This is a secondary immune response. Memory B cells are what enabled survivors of the 1918 flu pandemic to resist infection.

An antibody molecule is built of several polypeptides and is therefore encoded by several genes. The simplest type of antibody molecule is four polypeptide chains connected by disulfide (sulfur-sulfur) bonds, forming a shape like the letter Y (**figure 17.7**). A large antibody molecule might consist of three, four, or five such Ys joined.

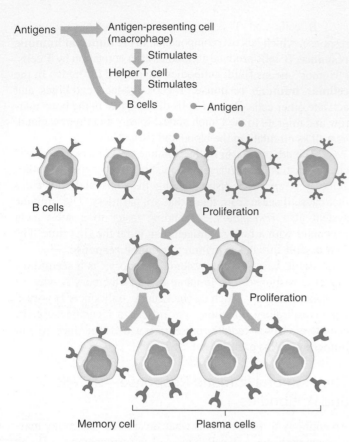

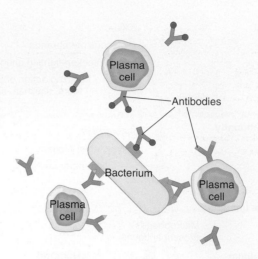

Figure 17.6 An immune response recognizes many targets. A humoral immune response is polyclonal, which means that different plasma cells produce antibody proteins that recognize and bind to different features of a foreign cell's surface.

Figure 17.5 Production of antibodies. In the humoral immune response, B cells proliferate and mature into antibody-secreting plasma cells. Only the B cell that binds the antigen proliferates; its descendants may develop into memory cells or plasma cells. Plasma cells greatly outnumber memory cells.

In a Y-shaped antibody subunit, the two longer polypeptides are called **heavy chains,** and the other two **light chains**. The lower portion of each chain is an amino acid sequence that is very similar in all antibody molecules, even in different species. These areas are called constant regions, and they provide the activity of the antibody. The amino acid sequences of the upper portions of each polypeptide chain, the variable regions, can differ greatly among antibodies. These parts provide the specificities of particular antibodies to particular antigens.

Antibodies can bind certain antigens because of the three-dimensional shapes of the tips of the variable regions.

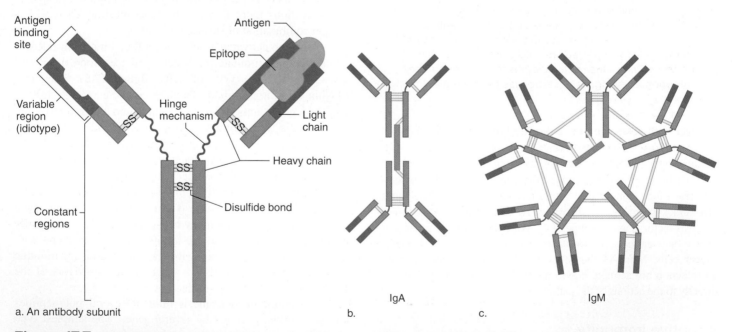

Figure 17.7 Antibody structure. The simplest antibody molecule **(a)** consists of four polypeptide chains, two heavy and two light, joined by pairs of sulfur atoms that form disulfide bonds. Part of each polypeptide chain has a constant sequence of amino acids, and the remainder varies. The tops of the Y-shaped molecules form antigen binding sites. **(b)** IgA consists of two Y-shaped subunits, and IgM **(c)** consists of five subunits.

These specialized ends are **antigen binding sites,** and the parts that actually contact the antigen are called **idiotypes**. The parts of the antigens that idiotypes bind are **epitopes**. An antibody contorts to form a pocket around the antigen.

Antibodies have several functions. Antibody-antigen binding may inactivate a pathogen or neutralize the toxin it produces. Antibodies can clump pathogens, making them more visible to macrophages, which then destroy them. Antibodies also activate complement, extending the innate immune response. In some situations, the antibody response can be harmful.

Antibodies are of five major types, distinguished by where they act and what they do (**table 17.2**). (Antibodies are also called immunoglobulins, abbreviated *Ig*.) Different antibody types predominate in different stages of an infection.

How can a human body manufacture seemingly limitless varieties of antibodies from the information in a limited number of antibody genes? This great diversity is possible because parts of different antibody genes combine. During the early development of B cells, sections of their antibody genes move to other chromosomal locations, creating new genetic instructions for antibodies.

The assembly of antibody molecules is like putting together many different outfits from the contents of a closet containing 200 pairs of pants, a drawer containing fifteen different shirts, and four belts. Specifically, each variable region of a heavy chain and a light chain consists of three sections, called V (for variable), D (for diversity), and J (for joining). The *V, D,* and *J* genes—several of each—for the heavy chains are on chromosome 14, and the corresponding genes for the light chains are on chromosomes 2 and 22. *C* (constant) genes encode the constant regions of each heavy and light chain. A promoter sequence precedes the *V* genes and an enhancer sequence precedes the *C* genes. These control sequences oversee the mixing

and matching of the *V, D,* and J genes. **Figure 17.8** shows how the genetic instructions for the antibody parts are combined in different ways to encode the heavy and light polypeptide chains.

Enzymes cut and paste the pieces of antibody gene parts. The number of combinations of parts to build antibodies is so great that virtually any antigen that a person with a healthy immune system might encounter will elicit an immune response.

The Cellular Immune Response—T Cells and Cytokines

T cells provide the cellular immune response. It is called "cellular" because the T cells themselves travel to where they act, unlike B cells, which secrete antibodies into the bloodstream. T cells descend from stem cells in the bone marrow, then travel to the thymus gland ("T" refers to thymus). As the immature T cells, called thymocytes, migrate toward the interior of the thymus, they display diverse cell surface receptors. Then selection happens. As the wandering thymocytes touch lining cells in the gland that are studded with "self" antigens, thymocytes that do not attack the lining cells begin maturing into T cells, whereas those that harm the lining cells die by apoptosis—in

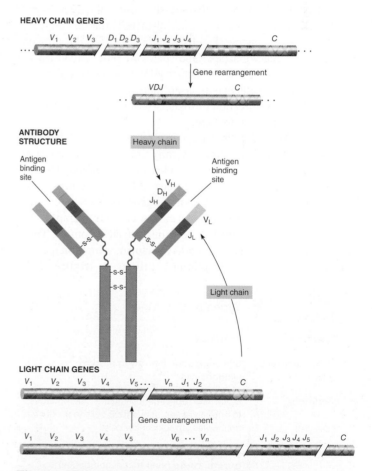

Figure 17.8 Antibody diversity. The human immune system can produce antibodies to millions of possible antigens because each polypeptide is encoded by more than one gene. That is, the many components of antibodies can combine in many ways.

Table 17.2	Types of Antibodies	
Type*	Location	Functions
IgA	Milk, saliva, urine, and tears; respiratory and digestive secretions	Protects against pathogens at points of entry into body
IgD	On B cells in blood	Stimulates B cells to make other types of antibodies, particularly in infants
IgE	In secretions with IgA and in mast cells in tissues	Acts as receptor for antigens that cause mast cells to secrete allergy mediators
IgG	Blood plasma and tissue fluid; passes to fetus	Protects against bacteria, viruses, and toxins, especially in secondary immune response
IgM	Blood plasma	Fights bacteria in primary immune response; includes anti-A and anti-B antibodies of ABO blood groups

*The letters *A, D, E, G,* and *M* refer to the specific conformation of heavy chains characteristic of each class of antibody.

great numbers. Gradually, T cells-to-be that recognize self persist, while those that harm body cells are destroyed.

Several types of T cells are distinguished by the types and patterns of receptors on their surfaces, and by their functions. Helper T cells have many functions: They recognize foreign antigens on macrophages, stimulate B cells to produce antibodies, secrete cytokines, and activate another type of T cell called a cytotoxic T cell (also called a killer T cell). Regulatory T cells help to suppress an immune response when it is no longer required. The cytokines that helper T cells secrete include interleukins, interferons, tumor necrosis factor, and colony-stimulating factors, which stimulate white blood cells in bone marrow to mature (**table 17.3**). Cytokines interact with and signal each other, sometimes in complex cascades.

Distinctive surfaces distinguish subsets of helper T cells. Certain antigens called cluster-of-differentiation antigens, or CD antigens, enable T cells to recognize foreign antigens displayed on macrophages. One such cell type, called a CD4 helper T cell, is an early target of HIV. Considering the critical role helper T cells play in coordinating immunity, it is little wonder that HIV infection ultimately topples the entire system, a point we will return to soon.

Cytotoxic T cells lack CD4 receptors but have CD8 receptors. These cells attack virally infected and cancerous cells by attaching to them and releasing chemicals. They do this by linking two surface peptides to form structures called T cell receptors that bind foreign antigens. When a cytotoxic T cell encounters a nonself cell—a cancer cell, for example—the T cell receptors draw the two cells into physical contact. The T cell then releases a protein called perforin, which pierces the cancer cell's plasma membrane, killing it (**figure 17.9**).

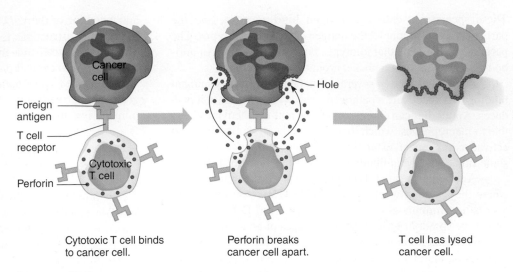

Cytotoxic T cell binds to cancer cell.

Perforin breaks cancer cell apart.

T cell has lysed cancer cell.

Figure 17.9 **Death of a cancer cell.** A cytotoxic T cell binds to a cancer cell and injects perforin, a protein that pierces (lyses) the cancer cell's plasma membrane. The cancer cell dies, leaving debris that macrophages clear away.

Cytotoxic T cell receptors also attract cells that are covered with certain viruses, destroying the cells before the viruses on them can enter, replicate, and spread the infection. Cytotoxic T cells continually monitor body cells, recognizing and eliminating virally infected cells and tumor cells.

Table 17.4 summarizes types of immune system cells.

Table 17.4	Types of Immune System Cells
Cell Type	**Function**
Macrophage	Presents antigens
	Performs phagocytosis
Dendritic cell	Presents antigens
Mast cell	Releases histamine in inflammation
	Releases allergy mediators
B cell	Matures into antibody-producing plasma cell or into memory cell
T cells	
Helper	Recognizes nonself antigens presented on macrophages
	Stimulates B cells to produce antibodies
	Secretes cytokines
	Activates cytotoxic T cells
Cytotoxic	Attacks cancer cells and cells infected with viruses upon recognizing antigens
Neutrophil	Attacks bacteria
Natural killer	Attacks cancer cells and cells infected with viruses without recognizing antigens; activates other white blood cells
Suppressor	Inhibits antibody production

Table 17.3	Types of Cytokines
Cytokine	**Function**
Colony-stimulating factors	Stimulate bone marrow to produce lymphocytes
Interferons	Block viral replication, stimulate macrophages to engulf viruses, stimulate B cells to produce antibodies, attack cancer cells
Interleukins	Control lymphocyte differentiation and growth, cause fever that accompanies bacterial infection
Tumor necrosis factor	Stops tumor growth, releases growth factors, stimulates lymphocyte differentiation, dismantles bacterial toxins

Key Concepts Questions 17.2

1. What are the three major parts of the immune response?
2. How does innate immunity differ from adaptive immunity?
3. Describe antibody structure and function.
4. What do helper T cells do?
5. What do cytotoxic T cells do?

17.3 Abnormal Immunity

The immune system continually adapts to environmental change. Because the immune response is so diverse, its breakdown affects health in many ways. Immune system malfunction may be inherited or acquired, and immunity may be too weak, too strong, or misdirected. Abnormal immune responses may be multifactorial, with several genes contributing to susceptibility to infection, or caused by mutation in a single gene.

Inherited Immune Deficiencies

The more than twenty types of inherited immune deficiencies affect innate and/or adaptive immunity (**table 17.5**). These conditions can arise in several ways.

In chronic granulomatous disease, white blood cells called neutrophils engulf bacteria, but, due to deficiency of an enzyme called an oxidase, they cannot produce the activated oxygen compounds that would kill bacteria. Because the oxidase is made of four polypeptide chains, four genes encode it, and there are four ways to inherit the disease, all X-linked. A very rare autosomal recessive form of the disease is caused by a defect in the part of the host cell that encloses bacteria. Antibiotics and gamma interferon are used to prevent bacterial infections in these patients, and a bone marrow or an umbilical cord stem cell transplant can cure the disease.

Mutations in genes that encode cytokines or T cell receptors impair cellular immunity, which primarily targets viruses and cancer cells. Because T cells activate the B cells that manufacture antibodies, abnormal cellular immunity (T cell function) disrupts humoral immunity (B cell function). Mutations in the genes that encode antibody segments, that control how the segments join, or that direct maturation of B cells mostly impair immunity against bacterial infection. Inherited immune deficiency can also result from defective B cells, which usually increases vulnerability to certain bacterial infections.

Severe combined immune deficiencies (SCIDs) affect both humoral and cellular immunity. About half of SCID cases are X-linked. In a less severe form, the individual lacks B cells but has some T cells. Before antibiotic drugs became available, children with this form of SCID died before age 10 of overwhelming bacterial infection. In a more severe form of X-linked

Table 17.5 Inherited Immune Deficiencies

Disease	OMIM	Inheritance*	Defect
Chronic granulomatous disease	306400	ar, AD, Xlr	Abnormal phagocytes can't kill engulfed bacteria
Immune defect due to absence of thymus	242700	ar	No thymus, no T cells
Neutrophil immuno-deficiency syndrome	608203	ar	Deficiencies of T cells, B cells, and neutrophils
SCID			
Adenosine deaminase deficiency	102700	ar	No T or B cells
Adenosine deaminase deficiency with sensitivity to ionizing radiation	602450	ar	No T, B, or natural killer cells
IL-2 receptor mutation	300400	Xlr	No T, B, or natural killer cells
X-linked lymphoproliferative disease	308240	Xlr	Absence of protein that enables T cells to bind B cells
X1	300400	Xlr	Abnormal interleukin-2

*ar = autosomal recessive

AD = autosomal dominant

Xlr = X-linked recessive

SCID = severe combined immune deficiency

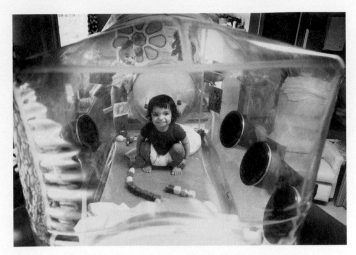

Figure 17.10 David Vetter, the original "bubble boy," had severe combined immune deficiency X1. Because his T cells could not mature, he was virtually defenseless against infection. Gene therapy can now treat this disease.

SCID, lack of B and T cells causes death by 18 months of age, usually of severe and diverse infections.

A young man named David Vetter taught the world about the difficulty of life without immunity years before AIDS arrived. David had an X-linked recessive form of SCID, called SCID-X1, that caused him to be born without a thymus gland. His T cells could not mature and activate B cells, leaving him defenseless in a germ-filled world. Born in Texas in 1971, David spent his short life in a vinyl bubble, awaiting a treatment that never came (**figure 17.10**). As he reached adolescence, David wanted to leave his bubble. He did, and received a bone marrow transplant from his sister. Sadly, her bone marrow contained Epstein-Barr virus. Her healthy immune system could handle it and she had no symptoms, but the virus caused lymphoma, a cancer of the immune system, in David. He died in just a few weeks. Today experimental gene therapy can treat SCID-X1, discussed in chapter 20.

Acquired Immune Deficiency Syndrome

AIDS is not inherited, but acquired by infection with HIV, a virus that gradually shuts down the immune system. The effect of HIV on a human body is especially astounding because the virus is so simple. Its genome is a millionth the size of ours, and its nine genes, consisting of about 9,000 RNA bases, encode only fifteen proteins! But HIV affects more than 200 human proteins as it invades the immune system.

HIV infection begins as the virus enters macrophages, impairing this first line of defense. In these cells, and later in helper T cells, the virus adheres with its surface protein, called gp120, to two coreceptors on the host cell surface, CD4 and CCR5 (**figure 17.11**). (CCR5 is the glycoprotein altered by mutation discussed in the chapter opener.) Another glycoprotein, gp41, anchors gp120 molecules into the viral envelope. When the virus binds both coreceptors, virus and cell surface contort in a way that enables viruses to enter the cell. Once in the cell, reverse transcriptase copies the viral RNA into DNA, which replicates to form a DNA double helix. The viral DNA then enters the nucleus and inserts into a chromosome. As the viral genes are transcribed and translated, the cell fills with viral pieces, which are assembled into complete new viral particles that eventually bud from the cell (**figure 17.12**).

Once helper T cells start to die at a high rate, bacterial infections begin, because B cells aren't activated to produce antibodies. Much later in infection, HIV variants arise that can bind to a receptor called CXCR4 on cytotoxic T cells, killing them. Loss of these cells renders the body very vulnerable to viral infections and cancer.

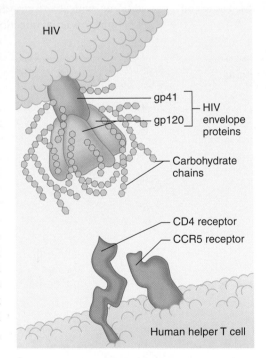

a.

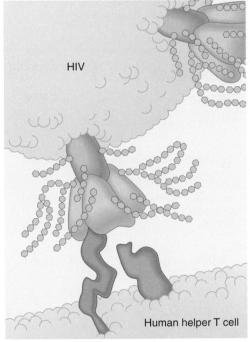

b.

Figure 17.11 HIV binds to a helper T cell. (a) The part of HIV that binds to helper T cells is called gp120 (gp stands for glycoprotein). **(b)** The carbohydrate chains that shield the protein part of gp120 move aside as they approach the cell surface, and the viral molecule can now bind to a CD4 receptor. Binding to the CCR5 receptor is also necessary. Then the viral envelope fuses with the plasma membrane and the virus enters. (The size of HIV is greatly exaggerated.) The man described in the chapter opener received stem cells from a donor whose cells lack CCR5. Years later, his leukemia and viral load are at undetectable levels.

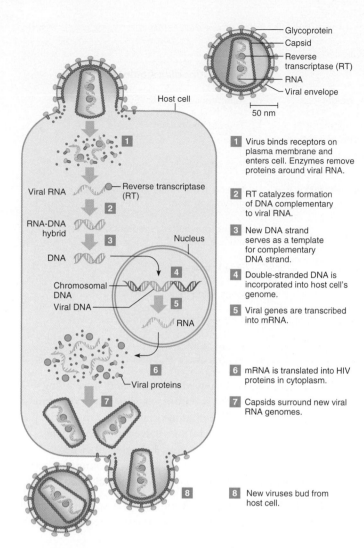

Viral RNA		Reverse transcriptase (RT)
RNA-DNA hybrid		
DNA		Nucleus
Chromosomal DNA		
Viral DNA		RNA
		Viral proteins
Host cell		

Glycoprotein
Capsid
Reverse transcriptase (RT)
RNA
Viral envelope
50 nm

1 Virus binds receptors on plasma membrane and enters cell. Enzymes remove proteins around viral RNA.

2 RT catalyzes formation of DNA complementary to viral RNA.

3 New DNA strand serves as a template for complementary DNA strand.

4 Double-stranded DNA is incorporated into host cell's genome.

5 Viral genes are transcribed into mRNA.

6 mRNA is translated into HIV proteins in cytoplasm.

7 Capsids surround new viral RNA genomes.

8 New viruses bud from host cell.

Figure 17.12 **How HIV infects.** HIV integrates into the host chromosome, then commandeers transcription and translation, ultimately producing more virus particles.

HIV replicates quickly, changes quickly, and can hide. The virus mutates easily because it cannot repair replication errors and errors happen frequently—1 per every 5,000 or so bases—because of the "sloppiness" of reverse transcriptase in copying viral RNA into DNA. The immune system cannot keep up; antibodies against one viral variant are useless against the next. For several years, the bone marrow produces 2 billion new T and B cells a day. A million to a billion new HIV particles bud daily from infected cells.

So genetically diverse is the population of HIV in a human host that, within days of the initial infection, variants arise that resist the drugs used to treat AIDS (see figure 15.12). HIV's changeable nature is why combining drugs with different actions is the most effective way to slow the disease into a chronic, lifelong, but treatable illness, instead of a killer (**table 17.6**). Several classes of drugs have cut the death rate from AIDS dramatically. They work at different points of infection: blocking binding or entry of the virus into T cells, replicating viral genetic material, and processing viral proteins.

Table 17.6	Anti-HIV Drugs	
Drug Type	**Mechanism**	
Reverse transcriptase inhibitor	Blocks copying of viral RNA into DNA	
Protease inhibitor	Blocks shortening of certain viral proteins	
Fusion inhibitor	Blocks ability of HIV to bind a cell	
Entry inhibitor	Blocks ability of HIV to enter a cell	

Clues to developing new drugs to treat HIV infection come from people at high risk who resist infection. Researchers identified variants of four receptors or the molecules that bind to them that block HIV from entering cells by looking at the DNA of people who had unprotected sex with many partners but who never became infected. Some of them were homozygous recessive for the 32-base deletion in the *CCR5* gene described in the chapter opener. The abnormal CCR5 coreceptors are too stunted to reach the infected cell's surface, so HIV has nowhere to dock. Heterozygotes, with one copy of the deletion, can become infected, but they remain healthy for several years longer than people who do not have the deletion. Curiously, the same mutation may have enabled people to survive various plagues in Europe during the Middle Ages. Apparently, more than one pathogen uses the CCR5 entryway into human cells.

Autoimmunity

In **autoimmunity,** the immune system produces antibodies that attack the body's own tissues. These antibodies are called **autoantibodies**. About 5 percent of the population has an autoimmune disorder. The signs and symptoms resulting from autoimmune disorders reflect the cell types under attack (**table 17.7**).

Most autoimmune disorders are not inherited as single-gene diseases. However, the fact that different autoimmune disorders affect members of the same family, and may respond to the same drugs, suggests that these conditions stem from shared susceptibilities. For example, in autoimmune polyendocrinopathy syndrome type I (OMIM 240300), caused by a mutation in a single gene on chromosome 21, autoantibodies attack endocrine glands in a sequence, so that different members of a family may have very different symptoms. However, the genetics of autoimmunity is usually more complex than this. Genome-wide association studies have identified dozens of genes that are each associated with more than one autoimmune disorder. On the other hand, an autoimmune disorder may result from the actions of variants of several genes that each contributes to susceptibility, perhaps in the presence of a specific environmental trigger such as a food. This is the case for the digestive disorder Crohn's disease, which has been associated with thirty-two genome regions.

Some of the more common autoimmune disorders may actually arise in several ways when parts of the immune

Table 17.7   Autoimmune Disorders

Disorder	Symptoms	Autoantibodies Against
Diabetes mellitus (type 1)	Thirst, hunger, weakness, weight loss	Pancreatic beta cells
Graves disease	Restlessness, weight loss, irritability, increased heart rate and blood pressure	Thyroid gland cells
Hemolytic anemia	Fatigue, weakness	Red blood cells
Multiple sclerosis	Weakness, poor coordination, failing vision, disturbed speech	Myelin in the white matter of the central nervous system
Myasthenia gravis	Muscle weakness	Neurotransmitter receptors on skeletal muscle cells
Rheumatic fever	Weakness, shortness of breath	Heart valve cells
Rheumatoid arthritis	Joint pain and deformity	Cells lining joints
Systemic lupus erythematosus	Red facial rash, fever, weakness, joint pain	Connective tissue
Ulcerative colitis	Lower abdominal pain	Colon cells

response are overactive. This is the case for systemic lupus erythematosus, better known as "lupus." A butterfly-shaped rash is characteristic on the cheeks, but the disease also produces autoantibodies that affect the connective tissue of many organs. These are the kidneys, joints, lungs, brain, spinal cord, and the heart and blood vessels. A person may need dialysis when the kidneys are involved, blood pressure medication to counter increasing pressure in the lungs, and drugs to minimize buildup of fatty deposits on artery walls. Lupus can also cause strokes, memory loss, fever, seizures, headache, and psychosis.

Lupus involves several aspects of the immune response, including cell surface characteristics, secretion of interferons, production of autoantibodies, activation of B and T cells, antigen presentation, adhesion of immune system cells to blood vessel linings, inflammation, removal of complexes of immune cells and foreign antigens, and cytokine production. Therefore, it isn't surprising that variants of at least ten different genes can predispose a person to developing this condition. Perhaps inheriting susceptibility in three or four of the genes causes lupus.

How does the immune system turn against itself? Autoimmunity may arise in several ways:

- A virus replicating in a cell incorporates proteins from the cell's surface onto its own. When the immune system "learns" the surface of the virus to destroy it, it also learns to attack human cells that normally bear the protein.
- Some cells that should have died in the thymus somehow escape the massive die-off, persisting to attack "self" tissue later on.
- A nonself antigen coincidentally resembles a self antigen, and the immune system attacks both. In rheumatic fever, for example, antigens on heart valve cells resemble those on *Streptococcus* bacteria; antibodies produced to fight a strep throat also attack the heart valve cells.

- If X inactivation is skewed, a female may have a few cells that express the X chromosome genes of one parent (see figure 6.10). The immune system may respond to these cells as foreign if they have surface antigens that are not also on the majority of cells. Skewed X inactivation may explain why some autoimmune disorders are much more common in females.

Clinical Connection 17.2 highlights a special situation in which two immune systems must coexist—pregnancy.

Allergies

An allergy is an immune system response to a substance, called an allergen, that does not actually present a threat. Many allergens are particles small enough to be carried in the air and enter a person's respiratory tract. The size of the allergen may determine the type of allergy. For example, grass pollen is large and remains in the upper respiratory tract, where it causes hayfever. But allergens from house dust mites, cat dander, and cockroaches are small enough to infiltrate the lungs, triggering asthma. Asthma is a chronic disease in which contractions of the airways, inflammation, and accumulation of mucus block air flow.

Both humoral and cellular immunity take part in an allergic response (**figure 17.13**). Antibodies of class IgE bind to mast cells, sending signals that open the mast cells, which releases allergy mediators such as histamine and heparin. Allergy mediators cause inflammation, with symptoms that may include runny eyes from hay fever, narrowed airways from asthma, rashes, or the overwhelming body-wide allergic reaction called anaphylactic shock. Allergens also activate a class of helper T cells that produce cytokines, whose genes are clustered on chromosome 5q. Regions of chromosomes 12q and 17q have genes that control IgE production.

A Special Immunological Relationship: Mother-to-Be and Fetus

The immune system recognizes "self" cell surfaces and protects the body from foreign, "nonself" cells and molecules. This is very helpful when the nonself triggers are parts of infecting bacteria, but what tempers the immune system of a pregnant woman to accept cells from her fetus? Half of a fetal genome comes from the father, and so fetal cell surfaces likely include some antigens from him that would be "foreign" to the mother-to-be. Similarly, some of her antigens might be foreign to the fetus. Yet pregnant woman and fetus routinely swap cells. Most pregnancies retain fetal cells. The presence of cell populations from more than one individual in one body is called microchimerism ("little mosaic"). We do not understand how the immune system evolved tolerance between pregnant woman and fetus, but following are three examples of the immunological "crosstalk" between the two.

"T Regs"

Samples of lymph nodes from fetuses indicate that up to 1 percent of the cells are maternal. The woman's cells stimulate the fetal immune system to produce "regulatory T cells," called "T Regs," which dampen the fetal immune response. The maternal immune system similarly produces T Regs that inhibit response to fetal cells. In one experiment, fetal lymph node samples did not react against cells from the pregnant woman unless the fetal regulatory T cells were removed. Children retain these cells for several years. It may be possible to stimulate production of T Regs later in life to help a recipient's body accept an organ transplant.

Scleroderma

People who have scleroderma describe the condition as "the body turning to stone." The skin hardens into an armorlike texture (**figure 1**). Scleroderma usually begins in middle age, and affects mostly women. It was long thought to be autoimmune, but discovery of Y chromosomes in skin cells from scleroderma patients who are mothers of sons revealed a very different source of the illness—lingering cells from a fetus. Cells from female fetuses can presumably have the same effect but cannot be distinguished from the mother on the basis of a sex chromosome check.

The degree of genetic difference between a mother and a son may play a role in development of scleroderma. Mothers who have the condition tend to have cell surfaces that are more similar to those of their sons than mothers who do not have scleroderma. Perhaps the similarity of cell surfaces enabled the fetal cells to escape destruction by the mother's immune system.

Rh Incompatibility

"Rh," the rhesus factor discovered in rhesus monkeys, is a blood group (OMIM 111700). A person is Rh+ if red blood cells have a surface molecule called the RhD antigen. Rh type is important

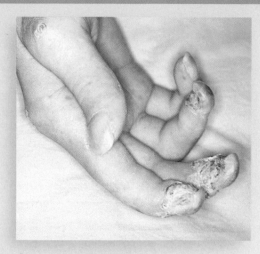

Figure 1 **An autoimmune disorder—maybe.** Scleroderma hardens the skin. Some cases appear to be caused by a long-delayed immune response to cells retained from a fetus decades earlier.

when an Rh+ man and an Rh− woman conceive a child who is Rh+ (**figure 2**). The woman's immune system manufactures antibodies against the few fetal cells that enter her bloodstream. Not enough antibodies form to harm the fetus that sets off the reaction, but the number of antibodies continues to increase. If she carries another Rh+ fetus, the antibodies can attack the fetal blood supply, causing potentially fatal hemolytic disease of the fetus and newborn. It can be treated at birth with a transfusion of Rh− blood.

Fortunately, natural and medical protections make this complication rare today. If a woman's ABO blood type is O and the fetus is A or B, her anti-A or anti-B antibodies attack the fetal cells in her circulation before her immune system produces anti-Rh antibodies. Also, if a pregnant woman alerts her physician to a potential incompatibility, she can be given RhoGAM, which is antibody against the Rh antigen. It shields fetal cells so her system does not manufacture the harmful antibodies. When she becomes pregnant again, fetal DNA in her circulation can be tested to see if it is Rh− or Rh+. If the second fetus is Rh−, she does not need RhoGAM. A first Rh+ fetus developing in an Rh− mother can be affected if her blood has been exposed to Rh+ cells in amniocentesis, a blood transfusion, an ectopic (tubal) pregnancy, a miscarriage, or pregnancy termination.

Questions for Discussion

1. Explain why a maternal immune system might attack a fetus.
2. How do regulatory T cells help a pregnant woman's body accept the fetus?
3. Explain how scleroderma may result from retained fetal cells in the maternal body.
4. What is Rh incompatibility?

(Continued)

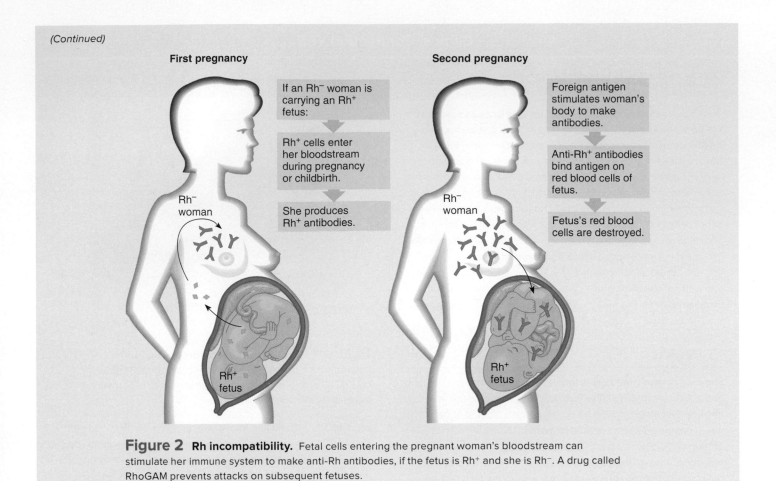

(Continued)

First pregnancy

Rh⁻ woman

Rh⁺ fetus

If an Rh⁻ woman is carrying an Rh⁺ fetus:

Rh⁺ cells enter her bloodstream during pregnancy or childbirth.

She produces Rh⁺ antibodies.

Second pregnancy

Rh⁻ woman

Rh⁺ fetus

Foreign antigen stimulates woman's body to make antibodies.

Anti-Rh⁺ antibodies bind antigen on red blood cells of fetus.

Fetus's red blood cells are destroyed.

Figure 2 **Rh incompatibility.** Fetal cells entering the pregnant woman's bloodstream can stimulate her immune system to make anti-Rh antibodies, if the fetus is Rh⁺ and she is Rh⁻. A drug called RhoGAM prevents attacks on subsequent fetuses.

The misdirected immune response of an allergy may be due to a mutation. Half the normal amount of a skin protein called filaggrin can lead to development of the allergic conditions atopic dermatitis (a type of eczema), asthma, peanut allergy, and hay fever (**figure 17.14**). Filaggrin is a gigantic protein that binds to the keratin proteins that form most of the outermost skin layer, the epidermis. Filaggrin normally breaks down, releasing amino acids that rise to the top of the skin and keep it moisturized. The epidermis forms a barrier that keeps out irritants, pathogens, and allergens.

People with the rare inherited disease ichthyosis vulgaris (OMIM 146700) have two mutations in the filaggrin gene, and experience severe skin flaking. Many more individuals—possibly one in ten of us—has a mutation that causes ichthyosis that is so mild that we just treat it with skin lotion. When researchers realized that people with either form of ichthyosis very often also have the itchy red inflamed skin of atopic dermatitis, they realized there could be a connection: When the epidermis is cracked, allergens enter and reach deeper skin layers, where they encounter dendritic cells, the sentries of the immune system. Once activated, these cells signal inflammation. Atopic dermatitis results. The dendritic cells also activate immune memory, so that when years later the person inhales the same allergens that

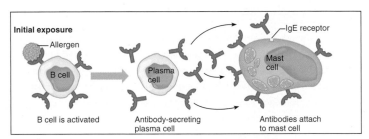

Initial exposure

Allergen

B cell

B cell is activated

Plasma cell

Antibody-secreting plasma cell

IgE receptor

Mast cell

Antibodies attach to mast cell

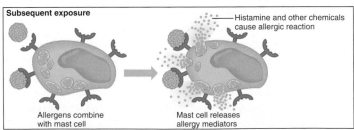

Subsequent exposure

Allergens combine with mast cell

Histamine and other chemicals cause allergic reaction

Mast cell releases allergy mediators

Figure 17.13 **Allergy.** In an allergic reaction, an allergen such as pollen activates B cells, which divide and give rise to antibody-secreting plasma cells. The antibodies attach to mast cells. When the person encounters allergens again, the allergens combine with the antibodies on the mast cells, which then burst, releasing the chemicals that cause itchy eyes and a runny nose.

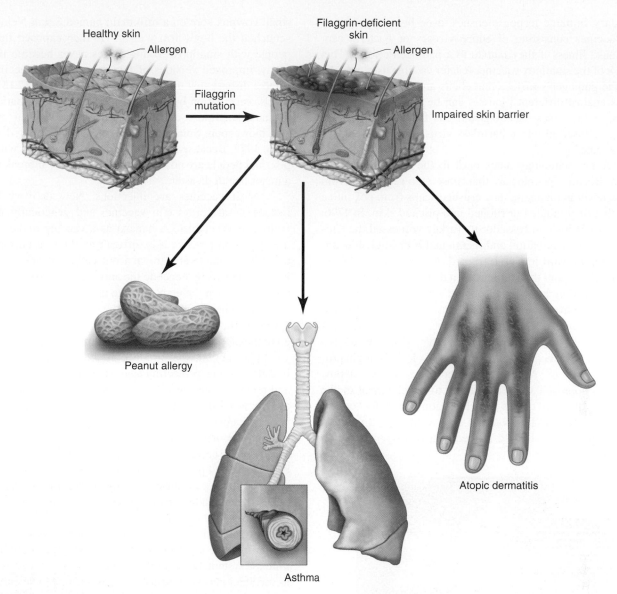

Figure 17.14 **Mutation in the filaggrin gene sets the stage for allergy.** Atopic dermatitis, peanut allergy, asthma, and hay fever may result from lack of filaggrin protein to protect the skin.

once crossed the broken skin, an immune response ensues in the airways, causing a form of asthma due to a skin barrier deficiency. Researchers are testing whether treating breaks in the skin early in life can prevent atopic dermatitis, asthma, and other allergies.

17.4 Altering Immunity

Medical technologies can alter or augment immune system functions. Vaccines trick the immune system into acting early. Antibiotic drugs, which are substances derived from organisms such as fungi and soil bacteria, have been used for decades to assist an immune response. Cytokines and altered antibodies are used as drugs. Transplants require suppression of the immune system so that the body will accept a nonself replacement part.

Vaccines

A **vaccine** is an inactive or partial form of a pathogen that stimulates the immune system to alert B cells to produce antibodies. When the person then encounters the natural pathogen, a secondary immune response ensues, even before symptoms arise. Vaccines consisting of entire viruses or bacteria can, rarely, cause illness if they mutate to a pathogenic form. This was a risk of the smallpox vaccine. A safer vaccine uses only the part of the pathogen's surface that elicits an immune response. Vaccines against different illnesses can be combined into one injection, or the genes encoding antigens from several pathogens can be inserted into a harmless virus and delivered as a "super vaccine."

Vaccine technology dates back to the eleventh century in China. Because people saw that those who recovered from smallpox never got it again, they crushed scabs from pox into a powder that they inhaled or rubbed into pricked skin. In 1796, the wife of a British ambassador to Turkey witnessed the Chinese method of vaccination and mentioned it to English country physician Edward Jenner. Intrigued, Jenner was vaccinated the Chinese way, and then thought of a different approach.

It was widely known that people who milked cows contracted a mild illness called cowpox, but did not get smallpox. The cows became ill from infected horses. Because the virus seemed to jump from one species to another, Jenner wondered whether exposing a healthy person to cowpox lesions might protect against smallpox. A slightly different virus causes cowpox, but Jenner's approach worked, leading to development of the first vaccine (the word comes from the Latin *vacca,* for "cow").

Jenner tried his first vaccine on a volunteer, 8-year-old James Phipps. Jenner dipped a needle in pus oozing from a small cowpox sore on a milkmaid named Sarah Nelmes, then scratched the boy's arm with it. He then exposed the boy to people with smallpox. Young James never became ill. Eventually, improved versions of Jenner's smallpox vaccine eradicated a disease that once killed millions (**figure 17.15**). By the 1970s, vaccination became unnecessary. Several nations have resumed smallpox vaccination in case the virus is ever used as a bioweapon. Smallpox has not naturally infected a human since 1977. Because many doctors are unfamiliar with smallpox, and people are no longer vaccinated, an outbreak could be a major health disaster.

Most vaccines are injections. New delivery methods include nasal sprays (flu vaccine) and genetically modified fruits and vegetables. A banana as a vaccine makes sense in theory, but in practice it is difficult to obtain a uniform product. Edible plants are grown from cells that are given genes from pathogens that encode the antigens that evoke an immune response. When the plant vaccine is eaten, the foreign antigens stimulate phagocytes beneath the small intestinal lining to "present" the antigens to nearby T cells. From here, the antigens go to the bloodstream, where they stimulate B cells to divide to yield plasma cells that produce antibodies that coat the small intestinal lining, protecting against pathogens in food. Potatoes and tomatoes have also been genetically modified to function as vaccines. Edible vaccines are still experimental.

Whatever the form of vaccine, it is important that a substantial proportion of a population be vaccinated to control an infectious disease. This establishes "herd immunity"—that is, if unvaccinated people are rare, then if the pathogen appears, it does not spread, because so many people are protected. If the population includes unvaccinated individuals who come into contact, the disease can spread.

An infectious disease such as flu that is mild or harmless to most people can kill a person who has a compromised immune system. Diseases that had been nearly eradicated thanks to vaccination have returned in areas where people either refuse to have their children vaccinated, or cannot due to war. Pertussis, measles, and polio have returned to certain parts of the world.

Immunotherapy

Immunotherapy amplifies or redirects the immune response. It originated in the nineteenth century to treat disease. Today, a few immunotherapies are in use, with more in clinical trials.

Monoclonal Antibodies Boost Humoral Immunity

When a B cell recognizes a single foreign antigen, it manufactures a single, or monoclonal, type of antibody. A large amount of a single antibody type could target a particular pathogen or cancer cell because of the antibody's great specificity.

In 1975, British researchers Cesar Milstein and George Köhler devised monoclonal antibody (MAb) technology, which mass-produces a single B cell, preserving its specificity and amplifying its antibody type. First, they injected a mouse with a sheep's red blood cells (**figure 17.16**). They then isolated a

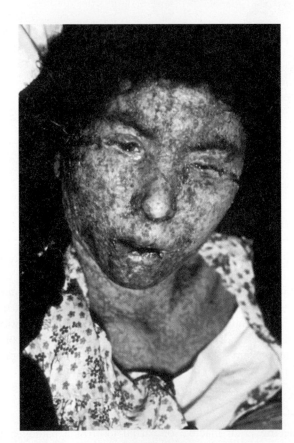

Figure 17.15 **Smallpox.** This woman had such a severe case of smallpox that the lesions joined.

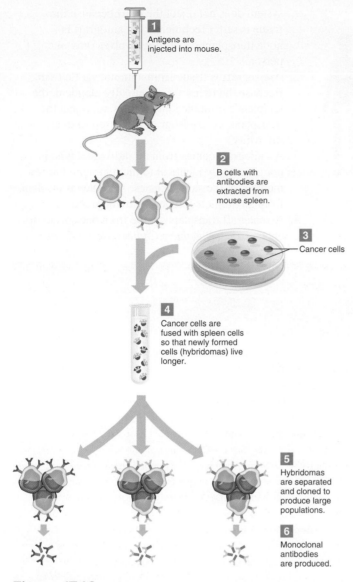

1 Antigens are injected into mouse.

2 B cells with antibodies are extracted from mouse spleen.

3 Cancer cells

4 Cancer cells are fused with spleen cells so that newly formed cells (hybridomas) live longer.

5 Hybridomas are separated and cloned to produce large populations.

6 Monoclonal antibodies are produced.

Figure 17.16 Monoclonal antibody technology.
Monoclonal antibodies are pure preparations of a single antibody type that recognize a single antigen type. They are useful in diagnosing and treating disease because of their specificity.

single B cell from the mouse's spleen and fused it with a cancerous white blood cell from a mouse. The fused cell, called a hybridoma, had a valuable pair of talents. Like the B cell, it produced large numbers of a single antibody type. Like the cancer cell, it divided continuously.

MAbs are used in basic research, veterinary and human health care, agriculture, forestry, and forensics. They can diagnose everything from strep throat to turf grass disease. In a home pregnancy test, a woman places drops of her urine onto a paper strip containing a MAb that binds hCG, the "pregnancy" hormone. The color changes if the MAb binds its target. In cancer diagnosis, if a MAb attached to a fluorescent dye and injected into a patient or applied to a sample of tissue or body fluid binds its target—an antigen found mostly or only on cancer cells—fluorescence indicates disease. MAbs linked

to radioactive isotopes or to drugs deliver treatment to cancer cells. The MAb drug trastuzumab (Herceptin) blocks receptors on certain breast cancer cell surfaces, preventing them from receiving signals to divide. MAbs used in humans are made to more closely resemble natural human antibodies than the original mouse-derived products, which caused allergic reactions. More than 30 MAb-based treatments are available for human conditions.

Cytokines Boost Cellular Immunity

As coordinators of immunity, cytokines are used to treat a variety of conditions. However, it has been difficult to develop these body chemicals into drugs because they act only for short periods. They must be delivered precisely where they are needed, or overdose or side effects can occur.

Interferons (IFs) were the first cytokines tested on a large scale, and today are used to treat cancer, genital warts, multiple sclerosis, and some other conditions. Interleukin-2 (IL-2) is a cytokine that is administered intravenously to treat kidney cancer recurrence. Colony-stimulating factors, which cause immature white blood cells to mature and differentiate, are used to boost white blood cell levels in people with suppressed immune systems, such as individuals with AIDS or those receiving cancer chemotherapy. Treatment with these factors enables a patient to withstand higher doses of a conventional drug.

Because excess of another cytokine, tumor necrosis factor (TNF), underlies some disorders, blocking its activity treats some conditions. One drug consists of part of a receptor for TNF. Taking the drug prevents TNF from binding to cells that line joints, relieving arthritis pain. Excess TNF in rheumatoid arthritis prevents the joint lining cells from secreting lubricating fluid.

Transplants

When a car breaks down, replacing the damaged part often fixes the trouble. The same is sometimes true for the human body. Hearts, kidneys, livers, lungs, corneas, pancreases, skin, and bone marrow are routinely transplanted, sometimes several organs at a time. Although transplant medicine had a shaky start, many problems have been solved. Today, thousands of transplants are performed annually and recipients gain years of life. The challenge to successful transplantation lies in genetics because individual inherited differences in cell surfaces determine whether the body will accept tissue from a particular donor.

Transplant Types

Transplants are classified by the relationship of donor to recipient (**figure 17.17**):

1. An autograft transfers tissue from one part of a person's body to another. A skin graft taken from the thigh to replace burned skin on the chest, or a leg vein that replaces a coronary artery, are autografts. The immune

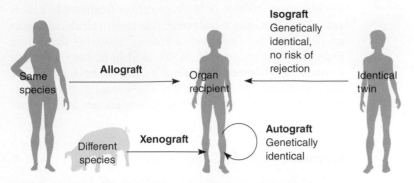

Figure 17.17 Transplant types. An autograft is within an individual. An isograft is between identical twins. An allograft is between members of the same species, and a xenograft is between members of different species.

system does not reject the graft because the tissue is self. (Technically, an autograft is not a transplant because it involves only one person.)

2. An isograft is tissue from a monozygotic twin. Because the twins are genetically identical, the recipient's immune system does not reject the transplant. Ovary isografts are used to treat infertility.

3. An allograft comes from an individual who is not genetically identical to the recipient, but is a member of the same species. A kidney transplant from an unrelated donor is an allograft.

4. A xenograft transplants tissue from one species to another. (See *Bioethics: Choices for the Future*.)

Bioethics: Choices for the Future

Pig Parts

In 1902, a German medical journal reported an astonishing experiment. A physician, Emmerich Ullman, had attached the blood vessels of a patient dying of kidney failure to a pig's kidney set up by her bedside. The patient's immune system rejected the attachment almost immediately.

Nearly a century later, in 1997, a similar experiment took place. Robert Pennington, a 19-year-old suffering from acute liver failure and desperately needing a transplant, survived for six and a half hours with his blood circulating outside of his body through a living liver removed from a 15-week-old, 118-pound pig named Sweetie Pie. The pig liver served as a bridge until a human liver became available. But Sweetie Pie was no ordinary pig. She had been genetically modified and bred so that her cells displayed a human protein that controlled rejection of tissue transplanted from an animal of another species. Because of this bit of added humanity, plus immunosuppressant

drugs, Pennington's body tolerated the pig liver's help for the few crucial hours. Baboons have also been organ donors (**figure 1**).

Successful xenotransplants would help alleviate the organ shortage. A possible danger of xenotransplants is that people may acquire viruses from the donor organs. Viruses can "jump" species, and the outcome in the new host is unpredictable. For example, a virus called PERV—for "porcine endogenous retrovirus"—can infect human cells in culture. However, several dozen patients who received implants of pig tissue did not show evidence of PERV years later. That study, though, looked only at blood. We still do not know what effect pig viruses can have on a human body. Because many viral infections take years to cause symptoms, introducing a new infectious disease in the future could be the trade-off for using xenotransplants to solve the current organ shortage.

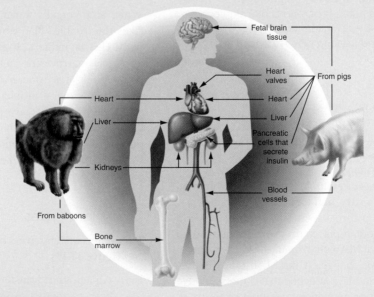

Figure 1 Baboons and pigs can provide tissues and organs for transplant.

(Continued)

Rejection Reactions—Or Acceptance

The immune system recognizes most donor tissue as nonself, and launches a tissue rejection reaction in which T cells, antibodies, and activated complement destroy the foreign tissue. The greater the difference between recipient and donor cell surfaces, the more rapid and severe the rejection reaction. An extreme example is the hyperacute rejection reaction against tissue transplanted from another species. Donor tissue from another type of animal is usually destroyed in minutes as blood vessels blacken and cut off the blood supply.

Physicians use several approaches to limit rejection to help a transplant recipient survive. These include closely matching the HLA types of donor and recipient and stripping donor tissue of antigens. Gene expression microarrays can be used to better match donors to recipients.

Immunosuppressant drugs inhibit production of the antibodies and T cells that attack transplanted tissue. If recipients get bone marrow stem cells from the donors, they need immunosuppressant drugs for only a short time, because along with the organ, the bone marrow stem cells help the recipient's body accept the transplanted tissue. Gene expression profiling can identify transplant recipients unlikely to reject their new organs. Still experimental, this approach can spare some people taking immunosuppressants, which have side effects.

Rejection is not the only problem that can arise from an organ transplant. Graft-versus-host disease can develop when bone marrow transplants are used to correct certain blood deficiencies and cancers. The transplanted bone marrow, which is actually part of the donor's immune system, attacks the recipient—its new body—as foreign. Symptoms include rash, abdominal pain, nausea, vomiting, hair loss, and jaundice.

Sometimes a problem arises if a bone marrow transplant to treat cancer is too closely matched to the recipient. If the cancer returns with the same cell surfaces as it had earlier, the patient's new bone marrow is so similar to the old marrow that it is equally unable to fight the cancer. The best tissue for transplant may be a compromise: different enough to control the cancer, but not so different that rejection occurs.

Key Concepts Questions 17.4

1. How does a vaccine protect against an infectious disease?

2. How are monoclonal antibodies and cytokines used clinically?

3. Describe the types of transplants.

4. What happens in organ rejection?

17.5 Using Genomics to Fight Infection

Immunity against infectious disease arises from interactions of two genomes—ours and the pathogen's. Human genome information is revealing how the immune system halts infectious disease. Information from pathogen genomes reveals how they make us sick.

Researchers can use such genomic information to better understand not only how an infection affects the human body, but how infections spread, causing outbreaks and epidemics. Genomic information can translate into new treatment approaches. The DNA sequence for *Streptococcus pneumoniae,* for example, revealed genetic instructions for a huge protein that enables the bacterium to adhere to human cells. Potential drugs could dismantle this adhesion protein. Sequencing the genomes of pathogens can also help to halt the spread of the diseases they cause, as the following two examples illustrate.

Reverse Vaccinology

Older vaccines consisted of parts of pathogens that were detected using standard microbiological approaches. In a strategy called **reverse vaccinology,** researchers consult genome sequence information to identify genes that encode "hidden" antigens that might serve as the basis for a vaccine. If the proteins that correspond to the genes induce immunity in experimental animals such as mice, the preparation is tested in humans.

The first reverse vaccine targeted meningococcus B, which causes more than half of all cases of meningococcal meningitis, a bacterial infectious disease that inflames the membranes around the brain and/or spinal cord. Meningococcal meningitis can be fatal, or cause deafness or brain damage. Using part of the bacterium as a vaccine is difficult, because some molecules on the surface resemble human proteins so much that a vaccine might direct an immune response against the body. Proteins unique to the bacterium are too varied to easily use as the basis for a vaccine.

To develop a vaccine against meningococcus B, researchers used bioinformatics (computer analysis of DNA sequences) to identify hundreds of bacterial antigens, and tested the antigens in mice to see if they would prevent the infection. A few antigens that were highly specific to the bacterium and protected mice became the basis of the vaccine that is now used for humans. Vaccines against severe acute respiratory syndrome (SARS) and several varieties of influenza were also developed using information from pathogen genomes. Reverse vaccines do not consist of DNA, but rather are based on finding protein antigens that elicit a human immune response using pathogen genome information.

Genome Sequencing to Track Outbreaks

DNA is a changeable molecule, and so over time, even short times, mutation alters genome sequences. Epidemiologists can follow the rapidly changing genome sequences in pathogens taken from body fluids or on objects to identify the source of an outbreak and trace the spread of infectious disease. This approach is an example of a strategy called metagenomics, which sequences random pieces of DNA, or increasingly entire genomes, that are present in

a particular environment, such as a drop of lake water—or a blood or stool sample. Following are two examples of how a genomic approach helped restore public health.

Hospital-Acquired Pneumonia

In June 2011, infection by the drug-resistant bacterium *Klebsiella pneumoniae* killed six patients at the National Institutes of Health's (NIH) Clinical Center in Bethesda, Maryland, and caused life-threatening pneumonia in five others (**figure 17.18**). *K. pneumoniae* is normally found on the skin and in the mouth and intestines, but certain strains, when inhaled, can severely damage the lungs. Researchers from the nearby National Human Genome Research Institute teamed pathogen genome sequencing with classic epidemiological sleuthing to quickly identify the source of the outbreak at the clinical center, halting the spread and saving lives.

K. pneumoniae lung infections kill half of the people they infect, are resistant to many antibiotic drugs, and are so alike genetically that standard microbiological techniques to identify pathogenic variants miss subtle genetic changes. Genome sequencing, however, checks every nucleotide. When researchers

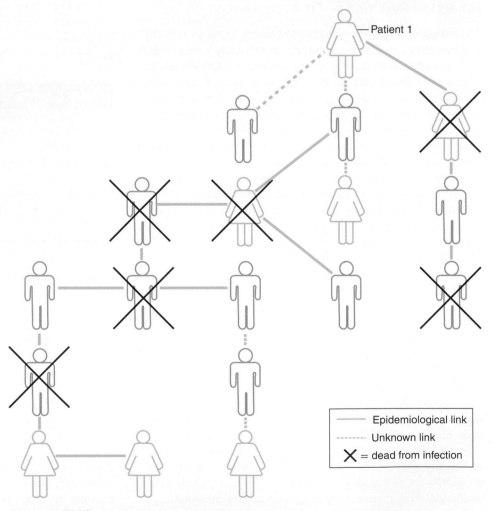

Figure 17.18 On the trail of a pathogen. Researchers compared SNPs at 41 sites in the genome of *Klebsiella pneumoniae* to assemble the chain of infection to trace an outbreak at the National Institutes of Health Clinical Center.

sequenced bacterial genomes from the first ("index") case in the outbreak, who is a 43-year-old woman who recovered, they discovered distinguishing single nucleotide polymorphisms (SNPs) at 41 sites in the 6-million-base genome of *K. pneumoniae*. The bacteria were evolving—mutating—so rapidly that secretions sampled from different parts of the patient's body yielded *K. pneumoniae* with slightly different SNP patterns.

Using an algorithm to compare bacterial sequences, the researchers found that some of the sick people shared SNP patterns with bacteria from the lungs and groin of the index patient, and some infected individuals had SNP patterns like those from the first patient's throat. One patient got the infection from a contaminated ventilator, which provided a clue to transmission of the infection in the hospital.

To see who might be infected but not yet show symptoms, the researchers next sequenced the genomes of all 1,115 patients in the hospital at the time. This part of the investigation showed that five people carried the bacteria without developing symptoms, and they had spread the infection to six others who hadn't developed symptoms but would. Identifying all of these people, and discovering objects that they had touched, enabled the investigators to control the infection.

Using pathogen genome sequencing revealed two new facts about the infection: *K. pneumoniae* survives on inanimate objects and can be transmitted by carriers who never become ill. Comparing pathogen genomes may help to prevent some of the 1.7 million hospital-acquired infections that happen in the United States each year, killing nearly 100,000 people.

Toxic *Escherichia coli*

Pathogen genome sequencing helped epidemiologists control a more widespread outbreak than that of pneumonia at the NIH Clinical Center. It involved *Escherichia coli,* a bacterium that is a normal part of the human intestinal microbiome (see figure 17.1). A strain of the bacterium, called "STEC" for Shiga-toxigenic *Escherichia coli,* produces a toxin that causes severe diarrhea and, in some individuals, a form of kidney failure that can be lethal.

In June 2011, an outbreak in Germany sickened nearly 4,000 people and resulted in 52 fatalities. The specific strain of *E. coli* had never been seen in an epidemic. Identifying the pathogen was a life-and-death race against time, and standard microbiological culturing techniques were too slow. So researchers conducted a metagenomics analysis on stool samples from people who suffered from diarrhea during the outbreak. Each sample contained DNA from thousands of microorganisms, as well as a great deal of DNA from the patient.

Of the 45 samples that the researchers analyzed, 40 included *E. coli* and 27 of those 40 carried shiga toxin genes. The investigators then used traditional epidemiological methods to determine that the affected individuals had all eaten raw bean sprouts in a particular town in Germany, which were ultimately traced to a particular farm where presumably runoff from cattle feces had contaminated the sprouts. Five of the patients' stool samples had genomes from other types of bacteria, including *Salmonella, Campylobacter,* and *Clostridium* species. Unfortunately the infection does not respond to antibiotics, and the only treatment is to keep the patient hydrated.

Since the 2011 cases, epidemiologists have used genomic sleuthing to contain other outbreaks. Because it is now possible to sequence a microbial genome in just hours, it will soon be possible to much more accurately diagnose an infectious disease in a doctor's office. As researchers continue to identify variants of human immune system genes that protect against infection as well as gene variants in the pathogens that provoke the immune response, we will be better able to prevent, contain, and treat infectious diseases.

Key Concepts Questions 17.5

1. How does reverse vaccinology use pathogen genome information?

2. How are pathogen genome sequences used to identify the source of an outbreak of an infectious disease and to trace its spread?

Summary

17.1 The Importance of Cell Surfaces

1. The cells and biochemicals of the immune system distinguish self from nonself, protecting the body against infections and cancer.

2. Genes encode immune system proteins, and may confer susceptibilities to certain infectious diseases.

3. An **antigen** is a molecule that elicits an immune response. Patterns of cell surface proteins and glycoproteins determine blood types. **Human leukocyte antigen** genes encode cell surface antigens that bind foreign antigens that **antigen-presenting cells** display to the immune system.

17.2 The Human Immune System

4. If a pathogen breaches physical barriers, the **innate immune response** produces the redness and swelling of **inflammation,** plus **complement, collectins,** and **cytokines** such as **interferons** and **interleukins**. The response is broad and general.

5. The **adaptive immune response** is slower, specific, and has memory.

6. The **humoral immune response** begins when macrophages display foreign antigens near HLAs. This activates **T cells,** which activate **B cells,** which give rise to **plasma cells** that secrete specific **antibodies.** Some

B cells give rise to **memory cells.** The **primary immune response** is the first reaction to encountering a nonself antigen, and the **secondary immune response** is a reaction to subsequent encounters.

7. An antibody is Y-shaped and has four polypeptide chains, two **heavy chains** and two **light chains.** Each antibody molecule has regions of constant amino acid sequence and regions of variable sequence.

8. The tips of the Y of each antibody subunit form **antigen binding sites,** which include the more specific **idiotypes** that bind foreign antigens at their **epitopes.**

9. Antibodies bind antigens to form immune complexes large enough for other immune system components to detect and destroy. Antibody genes are rearranged during early B cell development, providing instructions to produce a great variety of antibodies.

10. T cells carry out the **cellular immune response.** Their precursors are selected in the thymus to recognize self. Helper T cells secrete cytokines that activate other T cells and B cells. A helper T cell's CD4 antigen binds macrophages that present foreign antigens. Cytotoxic T cells release biochemicals that kill bacteria and destroy cells covered with viruses.

17.3 Abnormal Immunity

11. Mutations in antibody or cytokine genes, or in genes encoding T cell receptors, cause inherited immune deficiencies. Severe combined immune deficiencies affect both branches of the immune system.

12. HIV binds to the coreceptors CD4 and CCR5 on macrophages and helper T cells, and, later in infection, triggers apoptosis of cytotoxic T cells. As HIV replicates, it mutates, evading immune attack. Falling CD4 helper T cell numbers allow opportunistic infections and cancers to flourish. People who cannot produce a complete CCR5 protein resist HIV infection.

13. In **autoimmunity,** the body manufactures **autoantibodies** against its own cells.

14. In people who are susceptible to allergies, allergens stimulate IgE antibodies to bind to mast cells, which causes the cells to release allergy mediators. Certain helper T cells release selected cytokines. Allergies may be a holdover of past immune function.

17.4 Altering Immunity

15. A **vaccine** presents a disabled pathogen, or part of one, to elicit a primary immune response.

16. Immunotherapy enhances or redirects immune function. Monoclonal antibodies are useful in diagnosing and treating some diseases because of their abundance and specificity. Cytokines are used to treat various conditions.

17. Transplant types include autografts (within oneself), isografts (between identical twins), allografts (within a species), and xenografts (between species). A tissue rejection reaction occurs if donor tissue is too unlike recipient tissue.

17.5 Using Genomics to Fight Infection

18. Knowing the genome sequence of a pathogen provides information beyond traditional microbiological classification. Sequencing DNA in the environment is called metagenomics.

19. **Reverse vaccinology** uses pathogen genome sequence information to identify antigens that can form the basis of an effective vaccine.

20. Comparing SNP patterns among samples of bacteria taken from patients in an outbreak or epidemic can reveal the spread of the infection, including inanimate objects and carriers. It can also identify bacterial strains that produce toxins.

www.mhhe.com/lewisgenetics11

Answers to all end-of-chapter questions can be found at **www.mhhe.com/lewisgenetics11**. You will also find additional practice quizzes, animations, videos, and vocabulary flashcards to help you master the material in this chapter.

Review Questions

1. Match the cell type to the type of biochemical it produces.

   1. mast cell
   2. T cell
   3. B cell
   4. macrophage
   5. all cells with nuclei
   6. antigen-presenting cell

   a. antibodies
   b. HLA class II genes
   c. interleukin
   d. histamine
   e. interferon
   f. heparin
   g. tumor necrosis factor
   h. HLA class I genes

2. What does "nonself" mean? Give an example of a nonself cell in your own body.

3. Distinguish between viruses and bacteria.

4. What is the physical basis of a blood type?

5. Distinguish between using serology or genotyping to type blood.

6. Explain why an HLA-disease association is not a diagnosis.

7. Explain how mucus, tears, cilia, and ear wax are part of the immune response.

8. Distinguish between
   a. a T cell and a B cell.
   b. innate and adaptive immunity.
   c. a primary and secondary immune response.
   d. a cellular and humoral immune response.
   e. an autoimmune condition and an allergy.
   f. an inherited and acquired immune deficiency.

9. Which components of the human immune response explain why we experience the same symptoms of an upper respiratory infection (a "cold") when many different types of viruses can cause these conditions?

10. State the function of each of the following immune system biochemicals:
    a. complement proteins
    b. collectins
    c. antibodies
    d. cytokines
    e. filaggrin

11. What does HIV do to the human immune system?

12. Cite three reasons why developing a vaccine against HIV infection has been challenging.

13. What would be the consequences of lacking
    a. helper T cells?
    b. cytotoxic T cells?
    c. B cells?
    d. macrophages?

14. Explain how the immune system can respond to millions of different nonself antigens using only a few hundred antibody genes.

15. How are SCID and AIDS similar and different?

16. What part do antibodies play in allergic reactions and in autoimmune disorders?

17. What do a plasma cell and a memory cell descended from the same B cell have in common? How do they differ?

18. Why is a deficiency of T cells more dangerous than a deficiency of B cells?

19. Cite two explanations for why autoimmune disorders are more common in females.

20. How do each of the following illnesses disturb immunity?
    a. graft-versus-host disease
    b. SCID
    c. scleroderma
    d. AIDS
    e. atopic dermatitis

21. Why is a polyclonal antibody response valuable in the body, but a monoclonal antibody valuable as a diagnostic tool?

22. State how each of the following alters immune system functions:
    a. a vaccine
    b. an antibiotic drug
    c. a cytokine-based drug
    d. an antihistamine drug
    e. a transplant

23. Explain how a "reverse" vaccine is similar to but also different from a traditional vaccine.

24. How can knowing the genome sequence of a pathogen be useful in an outbreak of infectious disease?

Applied Questions

1. "Winter vomiting disease," a form of gastroenteritis sometimes called "stomach flu," is caused by a virus called norovirus. It makes a person miserable for a day or two. Why do some people get the illness every year?

2. Rasmussen's encephalitis causes 100 or more seizures a day. Affected children have antibodies that attack brain cell receptors that normally bind neurotransmitters. Is this condition most likely an inherited immune deficiency, an autoimmune disorder, or an allergy? State a reason for your answer.

3. In the TV program *House*, a talented physician and his staff confront difficult-to-diagnose medical cases. They often have to hypothesize whether symptoms are due to an infection, allergy, poison, autoimmunity, or genetic disease. Discuss how these alternatives might be distinguished.

4. In people with a certain HLA genotype, a protein in their joints resembles an antigen on the bacterium that causes Lyme disease. This infection is transmitted in a tick bite and causes flulike symptoms and joint pain (arthritis). When these individuals become infected, their immune systems attack the bacteria and their joints. Explain why antibiotics treat the early phase of the disease, but not the arthritis.

5. A person exposed for the first time to Coxsackie virus develops a painful sore throat. How is the immune system alerted to the exposure to the virus? When the person encounters the virus again, why don't symptoms develop?

6. A young woman who has aplastic anemia will soon die as her lymphocyte levels drop sharply. What type of cytokine might help her?

7. Tawanda is a 16-year-old with cystic fibrosis. She receives a lung transplant from a woman who has just died in a car accident. Tawanda and the donor share four of the six HLA markers commonly used to match donor to recipient. What action can the transplant team take to minimize the risk that Tawanda's immune system rejects the transplanted lung?

Web Activities

1. Many websites describe products (food supplements) that supposedly "boost" immune system function. Locate such a website and identify claims that are unclear, deceptive, vague, or incorrect. Alternatively, identify a claim that *is* consistent with the description of immune system function in this chapter.

2. Explore www.clinicaltrials.gov. Select and explain clinical trials for an immune deficiency, an autoimmune condition, and an allergy.

Case Studies and Research Results

1. State whether each of the following situations describes an autograft, an isograft, an allograft, or a xenograft.

   a. A man donates part of his liver to his daughter, who has a liver damaged by cystic fibrosis.

   b. A woman with infertility receives an ovary transplant from her identical twin.

   c. A man receives a heart valve from a pig.

   d. A woman who has had a breast removed has a new breast built using her fatty thigh tissue.

2. Mark and Louise are planning to have their first child, but they are concerned because they think that they have an Rh incompatibility. He is Rh^- and she is Rh^+. Will there be a problem? Why or why not?

3. Twenty-four children and teens at a summer camp for people with cystic fibrosis contract severe lung infections from multidrug-resistant bacterium *Mycobacterium abscessus*. Describe a technology that infectious disease experts called in from the Centers for Disease Control and Prevention might use, based on genetics, that can reveal exactly how the infection spread at the camp.

4. In a Dutch family, a father, son, and daughter frequently develop fungal infections (candidiasis) of the mouth, throat, and skin of the feet. All three family members each has at least one autoimmune condition. The mother is unaffected. Investigation of the immune responses of the father and children found that they do not make sufficient helper T cells, nor do they produce enough interferon. They also make autoantibodies against two types of interleukins.

   a. What is the mode of inheritance of the underlying immune system dysfunction that affects the father and two children?

   b. Explain how the immune systems of the father and children are abnormal.

   c. How can autoimmunity indirectly increase the risk of infection?

5. Researchers have successfully treated mice for type 1 diabetes with transplants of beta cells from the pancreases of pigs. What type of transplant is this?

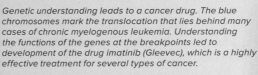

Genetic understanding leads to a cancer drug. The blue chromosomes mark the translocation that lies behind many cases of chronic myelogenous leukemia. Understanding the functions of the genes at the breakpoints led to development of the drug imatinib (Gleevec), which is a highly effective treatment for several types of cancer.

Cancer Genetics and Genomics

Learning Outcomes

18.1 Cancer Is an Abnormal Growth That Invades and Spreads

1. What is a cancerous tumor?

2. Explain how loss of cell cycle control causes cancer.

3. Explain how most cancers are not inherited, but are genetic.

18.2 Cancer at the Cellular Level

4. Describe cancer cells.

5. How can cancer arise from stem cells?

18.3 Cancer Genes and Genomes

6. Distinguish between driver and passenger mutations in cancer.

7. Discuss how mutations in several genes drive cancer.

8. Describe what can happen to chromosomes in cancer cells.

9. Explain how mutations in oncogenes and tumor suppressor genes cause or increase susceptibility to cancer.

10. How do environmental factors contribute to cancer?

18.4 The Challenges of Diagnosing and Treating Cancer

11. How can identifying cancer mutations and gene expression patterns sharpen diagnoses and target treatments?

The **BIG** Picture

A complex chain reaction of changes at the gene, chromosome, and genome levels causes and propels the diverse collection of diseases that we call cancer.

A Genetic Journey to a Blockbuster Cancer Drug

When 23-year-old *Glamour* magazine editor Erin Zammett Ruddy went for a routine physical in November 2001, she expected reassurance that her healthy lifestyle had indeed been keeping her healthy. What she got, a few days later, was a shock. Instead of having 4,000 to 10,000 white blood cells per milliliter of blood, she had more than 10 times that number—and many of the cells were cancerous.

"I had just returned from a nice, long lunch to find a message from my doctor. Could I call back? Something had come up in my blood work," recalled Erin. "I was diagnosed with chronic myelogenous leukemia. CML is cancer, and until very recently, it proved fatal in the vast majority of cases."

Erin's diagnosis came just a few months after a landmark report of a new drug—and, ironically, an article in *Glamour* about three CML survivors. A successful cancer drug typically helps about 20 percent of the patients who take it, often just extending life a few months. But cancer in the blood had vanished in 53 of 54 initial patients, usually quickly. So Erin contacted the lead researcher, Brian Druker, and joined the group. Her cancer was reversed—with just a pill a day, and no side effects.

The drug, imatinib (Gleevec), is now the standard treatment for CML and several other cancers. Today Erin is a successful freelance writer, mother of three, and healthy. The story of the drug's development illustrates how understanding the genetic events that start and propel a cancer can guide development of an effective weapon.

18.1 Cancer Is an Abnormal Growth That Invades and Spreads

One in three people develop cancer. A person may learn that he or she has cancer after noticing symptoms and reporting them to a health care provider, who then orders diagnostic tests. For example, a change in bowel habits, lower abdominal discomfort, blood-tinged stools, and fatigue are warning signs of colon cancer. Or, a cancer diagnosis may follow a routine screening test, such as finding blood in a stool sample or a growth on a colonoscopy scan.

However a cancer is detected, chances are it has been present for years, perhaps decades. A cancer takes time to grow because it is the culmination of a series of genetic and genomic changes—mutations—that enable certain cells to divide more frequently than normal, forming a new growth that enlarges, eventually crowding healthy tissue.

The underlying derangement of the cell cycle that causes and sustains a cancer may be set into motion by inherited cancer susceptibility genes, or, more commonly, from an environmental exposure such as to ultraviolet radiation in sunlight or to toxins in cigarette smoke. The fact that smokers who develop lung cancer have ten times as many mutations in their cancer cells than do nonsmokers who develop lung cancer indicates the powerful role of the environment in causing cancer.

In the past, we have classified the more than 100 types of cancer according to the cell type or tissue in which the tumor originates. In the future, classifying cancers by the specific mutations that occur in their cells to initiate the cancer and drive it may play a larger role in selecting treatments.

Researchers have sequenced the genomes of thousands of tumors to better understand the genetic changes that cause and accompany a cancer. Sequences of mutations in somatic cells, shattering of chromosomes, and changes in gene expression underlie the progression of cancer as it spreads. Mutations may affect the expression of other genes, but so may epigenetic influences, such as DNA methylation and chromatin remodeling (see chapter 11). One researcher calls the accumulating DNA changes that lie behind cancer "genomic scars."

This chapter explores cancer at the cellular, genetic, and genomic levels. **Figure 18.1** summarizes the major concepts. The depiction can be read from either direction, but the central portion—the cellular level—is the way of looking at cancer that may ultimately prove the most useful. All of the genes in which mutations increase susceptibility to, or cause, cancer affect three basic cellular pathways: cell fate, cell survival, and genome maintenance. Cell fate refers to

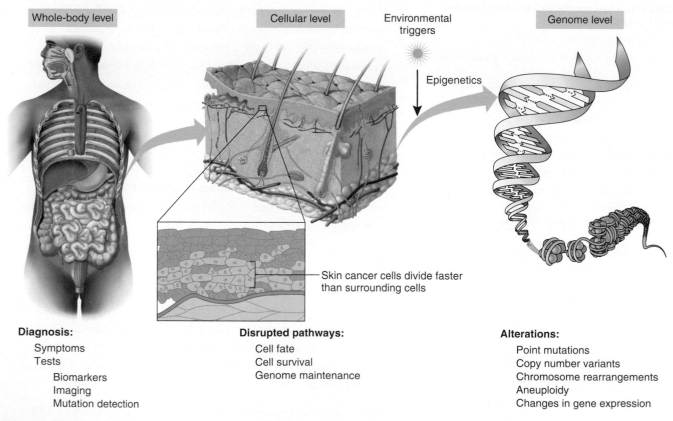

Whole-body level

Cellular level

Environmental triggers

Epigenetics

Genome level

Skin cancer cells divide faster than surrounding cells

Diagnosis:
Symptoms
Tests
    Biomarkers
    Imaging
    Mutation detection

Disrupted pathways:
Cell fate
Cell survival
Genome maintenance

Alterations:
Point mutations
Copy number variants
Chromosome rearrangements
Aneuploidy
Changes in gene expression

Figure 18.1 **Levels of cancer.** It takes years for a cancer to produce symptoms, as more and more cells lose their specializations and divide more frequently than the cells from which they descend. Mutations—from single-base changes to large-scale chromosomal upheavals—drive the disease, accompanied by changes in gene expression, some of which are responses to environmentally induced epigenetic effects.

differentiation (specialization). Cell survival refers to oxygen availability and preventing apoptosis. Genome maintenance refers to the abilities to survive in the presence of reactive oxygen species and toxins, to repair DNA, to maintain chromosome integrity and structure, and to correctly splice mRNA molecules.

Cancer is a complication of being a many-celled organism. Our specialized cells must follow a schedule of mitosis—the cell cycle—so that organs and other body parts grow appropriately during childhood, stay a particular size and shape throughout adulthood, and repair damage by replacing tissue. If a cell in solid tissue escapes normal controls on its division rate, it forms a growth called a tumor (**figure 18.2**). In the blood, a "liquid tumor" divides more frequently than others, taking over the population of blood cells.

A tumor is benign if it grows in place but does not spread into, or "invade," surrounding tissue. A tumor is cancerous, or malignant, if it infiltrates nearby tissue. Pieces of a malignant tumor can enter the bloodstream or lymphatic vessels and travel to other areas, where the cancer cells "seed" the formation of new tumors. The process of spreading is termed **metastasis,** which means "not standing still."

It is metastasis that makes a cancer deadly, because the new growth may be in an inaccessible part of the body, or genetically distinct enough from the original, or primary, tumor that drugs that were effective early in the illness no longer work. Metastases are difficult to detect. If a few sites of metastases appear on a medical scan, there may actually be dozens of the growths in the body.

An early hint at the genetic nature of cancer was the observation that most substances known to be carcinogens (causing cancer) are also mutagens (damaging DNA). Researchers first discovered genes that could cause cancer in humans in 1976, as versions of genes from certain viruses that had been associated with tumors in birds, dating from the mid-1800s. In the 1980s and 1990s, research to identify cancer-causing genes began with rare families that had many young members who had the same type of cancer and specific unusual chromosomes. Finding genes in the affected chromosome regions whose protein products could alter cell cycle control lead to the discovery of more than 100 **oncogenes.** An oncogene is a gene that causes cancer when inappropriately activated.

Family studies also identified more than 30 **tumor suppressor genes,** which cause cancer when they are deleted or inactivated. The normal function of a tumor suppressor gene is to keep the cell cycle running at the appropriate rate for a particular cell type under particular conditions.

Most mutations that cause cancer are in oncogenes or tumor suppressor genes. The effects of mutations in oncogenes are typically dominant, and those of tumor suppressor genes recessive. A third category of cancer genes includes mismatch mutations in DNA repair genes (see section 12.7) that allow other mutations to persist. When these other

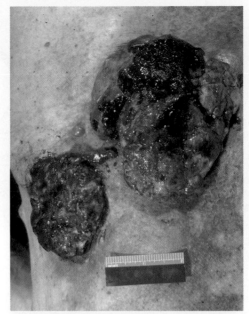

a.

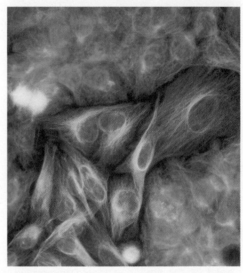

b.

Figure 18.2 Cancer cells stand out. (a) A melanoma is a cancer of the pigment-producing cells (melanocytes) in the skin. This cancer may have any or all of five characteristics, abbreviated abcde: it is asymmetric, has borders that are irregular, color variations, a diameter of more than 5 millimeters, and elevation. **(b)** These melanoma cells stain orange. The different staining characteristics of cancer cells are due to differences in gene expression patterns between the normal and cancerous states.

mutations activate oncogenes or inactivate tumor suppressor genes, cancer can result. Most DNA repair disorders are inherited in a single-gene fashion, and are quite rare. They typically cause diverse and widespread tumors, often beginning at a young age. We return to the genes behind cancer in section 18.3.

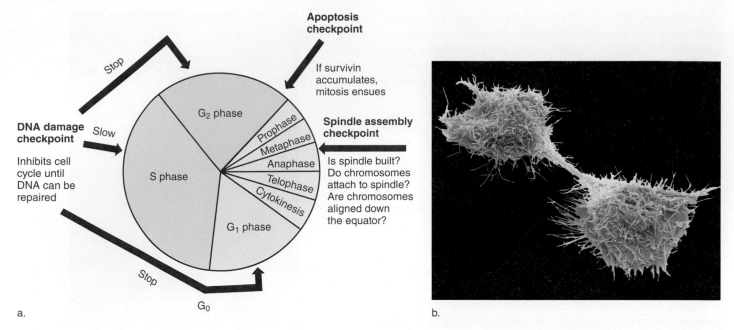

Figure 18.3 Cell cycle checkpoints. **(a)** Checkpoints ensure that mitotic events occur in the correct sequence. Many types of cancer result from faulty checkpoints. **(b)** These fibrosarcoma cancer cells descend from connective tissue cells (fibroblasts) in bone. The photo captures them in telophase of mitosis.

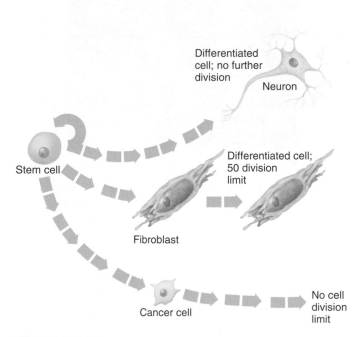

Figure 18.4 Cancer sends a cell down a pathway of unrestricted cell division. Cells may be terminally differentiated and no longer divide, such as a neuron, or differentiated yet still capable of limited cell division, such as a fibroblast (connective tissue cell). Cancer cells either lose specializations or never specialize; they divide unceasingly. (Arrows represent some cell divisions; not all daughter cells are shown.)

Loss of Cell Cycle Control

The most fundamental characteristic of cancer is the underlying disruption of the cell cycle. **Figure 18.3** repeats the cell cycle diagram from chapter 2. Cancer begins when a cell divides more frequently, or more times, than the noncancerous cell it descended from (**figure 18.4**). Mitosis in a cancer cell is like a runaway train, racing along without signals and control points.

The timing, rate, and number of mitoses a cell undergoes depend on protein growth factors and signaling molecules from outside the cell, and on transcription factors from within. Because these biochemicals are under genetic control, so is the cell cycle. Cancer cells probably arise often, because mitoses are so frequent that an occasional cell escapes control. However, the immune system destroys most cancer cells after recognizing tumor-specific antigens on their surfaces.

The discovery of the checkpoints that control the cell cycle revealed how cancer can begin. A mutation in a gene that normally halts or slows the cell cycle can lift the constraint, leading to inappropriate mitosis. Failure to pause long enough to repair DNA can allow a mutation in an oncogene or tumor suppressor gene to persist.

Loss of control over telomere length may also contribute to cancer by affecting the cell cycle. Recall that telomeres, or chromosome tips, protect chromosomes from breaking (see figure 2.18). Human telomeres consist of the DNA sequence TTAGGG repeated thousands of times. The repeats are normally lost from the telomere ends as a cell matures, from 15 to 40 nucleotides per cell division. The more specialized a cell, the shorter its telomeres. The chromosomes in skin, nerve, and muscle cells, for example, have short telomeres. Chromosomes in a sperm cell or oocyte, however, have long telomeres. This makes sense—as the precursors of a new organism, gametes must retain the capacity to divide many times.

Gametes keep their telomeres long using an enzyme, telomerase, that consists of RNA and protein. Part of the RNA—AAUCCC—is a template for the 6-DNA-base repeat TTAGGG that builds telomeres. Telomerase moves down the DNA like a zipper, adding six "teeth" (bases) at a time. Mutation in the gene that encodes telomerase, called *TERT*, causes some cancers.

In normal, specialized cells, telomerase is turned off and telomeres shrink, signaling a halt to cell division when they reach a certain size. In cancer cells, telomerase is turned back on. Telomeres extend, and this releases the normal brake on rapid cell division. As daughter cells of the original abnormal cell continue to divide uncontrollably, a tumor forms, grows, and may spread. Usually the longer the telomeres in cancer cells, the more advanced the disease. However, turning on telomerase production in a cell is not sufficient in itself to cause cancer. Many other things must go wrong for cancer to begin.

Cancer cells can divide continuously if given sufficient nutrients and space. Cervical cancer cells of a woman named Henrietta Lacks, who died in 1951, vividly illustrate the hardiness of these cells. Her cells persist today as standard cultures in many research laboratories. These "HeLa" cells divide so vigorously that when they contaminate cultures of other cells they soon take over.

Cells vary greatly in their capacity to divide. Cancer cells divide more frequently or more times than the cells from which they arise. Yet even the fastest-dividing cancer cells, which complete mitosis every 18 to 24 hours, do not divide as often as some cells in a normal human embryo do. A tumor grows more slowly at first because fewer cells divide. By the time the tumor is the size of a pea—when it is usually detectable—billions of cells are actively dividing. A cancerous tumor eventually grows faster than surrounding tissue because a greater proportion of its cells is dividing.

Inherited Versus Sporadic Cancer

Cancer is *genetic,* because it is caused by changes in DNA, but not usually *inherited.* Only about 10 percent of cases result from inheriting a cancer susceptibility allele from a parent. The inherited allele is a **germline mutation,** meaning that it is present in every cell of the individual, including the gametes. Cancer develops when a second mutation occurs in the other allele in a somatic cell in the affected body part (**figure 18.5**).

The majority of cancers are sporadic, and caused by **somatic mutations** that affect only nonsex cells. A sporadic

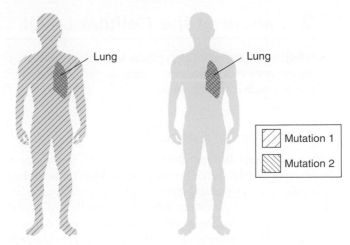

a. Germline (inherited) cancer   b. Sporadic cancer

Figure 18.5 Germline versus sporadic cancer.
(a) In germline cancer, every cell has one gene variant that increases cancer susceptibility, and a somatic mutation occurs in cells of the affected tissue. This type of predisposition to cancer is inherited as a single-gene trait, and is due to an initial germline mutation. **(b)** A sporadic cancer forms when a dominant mutation occurs in a somatic cell or two recessive mutations occur in the same gene on homologous chromosomes in a somatic cell. An environmental factor can cause the somatic mutations of cancer.

cancer may result from a single dominant mutation or from two recessive mutations in copies of the same gene. The cell loses control of its cell cycle and accelerated division of its daughter cells forms the tumor. Eventually, tumor cells may contain dozens of mutations that are not in neighboring, healthy cells.

Germline cancers are rare, but they have high penetrance and tend to strike earlier in life than sporadic cancers. Germline mutations may explain why some heavy smokers develop lung cancer, but many do not; the unlucky ones may have inherited a susceptibility allele. Years of exposure to the carcinogens in smoke eventually cause a mutation in a tumor suppressor gene or oncogene of a lung cell, giving it a proliferative advantage. Without the susceptibility gene, two such somatic mutations are necessary to trigger the cancer. This, too, can be the result of an environmental insult, but it takes longer for two events to occur than one.

Key Concepts Questions 18.1

1. Distinguish between a benign and a cancerous tumor.
2. What is the relationship between genes and cancer?
3. What are the three cellular processes that cancer disrupts?
4. Distinguish between oncogenes and tumor suppressor genes.
5. Distinguish between inherited and sporadic cancers.

18.2 Cancer at the Cellular Level

Cancer begins at the genetic and cellular levels. If not halted by the immune system or treatment, it spreads through tissues to take over organs and organ systems.

Characteristics of Cancer Cells

A cancer cell looks different from a normal cell. Some cancer cells are rounder than the cells they descend from because they do not adhere to surrounding normal cells as strongly as other cells do. Because the plasma membrane is more fluid, different substances cross it. A cancer cell's surface may sport different antigens than are on other cells or different numbers of antigens that are also on normal cells. The "prostate specific antigen" (PSA) blood test that indicates increased risk of prostate cancer, for example, detects elevated levels of this protein that may come from cancer cell surfaces.

When a cancer cell divides, both daughter cells are cancerous, because they inherit the altered cell cycle control. Therefore, cancer is said to be heritable because it is passed from parent cell to daughter cell. A cancer is also transplantable. If a cancer cell is injected into a healthy animal of the same species, it will proliferate there.

A cancer cell is **dedifferentiated,** which means that it is less specialized than the normal cell types near it that it might have descended from. A skin cancer cell, for example, is rounder and softer than the flattened, scaly, healthy skin cells above it in the epidermis, and is more like a stem cell in both appearance and division rate.

Cancer cell growth is unusual. Normal cells in a container divide to form a single layer; cancer cells pile up on one another. In an organism, this pileup would produce a tumor. Cancer cells that grow all over one another are said to lack contact inhibition—they do not stop dividing when they crowd other cells.

Cancer cells have surface structures that enable them to squeeze into any space, a property called **invasiveness.** They anchor themselves to tissue boundaries, called basement membranes, where they secrete enzymes that cut paths through healthy tissue. Unlike a benign tumor, an invasive malignant tumor grows irregularly, sending tentacles in all directions. The cell can move. Mutations affect the cytoskeleton (see figure 2.11), breaking down actin microfilaments and releasing actin molecules that migrate to the cell surface, moving the cell from where it is anchored in surrounding tissue.

Figure 18.6 reviews the stages of a cancer's formation (carcinogenesis). All of these changes that craft a cancer cell from a healthy cell, and the proliferation of cancer cells and eventual invasion and metastasis, take time. Pancreatic cancer, for example, begins 10 to 15 years before the first abdominal pain, then is deadly within 2 years. Lung cancer due to smoking begins with irritation of the lining tissue in respiratory tubes and may not produce symptoms for two decades. Cancer cells on the move eventually reach the bloodstream or lymphatic vessels, which take them to other body parts. This is metastasis.

Once a tumor has grown to the size of a pinhead, interior cancer cells respond to the oxygen-poor environment by secreting a protein, called vascular endothelial growth factor (VEGF). It stimulates nearby capillaries (the tiniest blood vessels) to sprout new branches that extend toward the tumor, bringing in oxygen and nutrients and removing wastes. This growth of new capillary extensions is called **angiogenesis,** and it is critical to a cancer's growth and spread. Capillaries may snake into and out of the tumor. Cancer cells wrap around the blood vessels and creep out upon this scaffolding, invading nearby tissue. In addition to attracting their own blood supply, cancer cells may also secrete hormones that encourage their own growth. This is a new ability because the cells they descend from do not produce these hormones.

When cancer cells move to a new body part, the DNA of secondary tumor cells often mutates, and chromosomes may break or rearrange. Many cancer cells are aneuploid (with missing or extra chromosomes). The metastasized cancer thus becomes a new genetic entity that may resist treatments that were effective against most cells of the original tumor. **Table 18.1** summarizes the characteristics of cancer cells.

Origins of Cancer Cells

Factors that influence whether or not cancer develops include how specialized the initial cell is and the location of that cell in the tissue. Cancer can begin at a cellular level in at least four ways:

- activation of stem cells that produce cancer cells;
- dedifferentiation;
- increase in the proportion of a tissue that consists of stem or progenitor cells; and
- faulty tissue repair.

Dedifferentiation is not an all-or-none phenomenon. Most cancer cells are more specialized than stem cells, but considerably less specialized than the differentiated cells near them in a tissue. From which does the cancer cell arise, the stem cell or the specialized cell? A cancer cell may descend from a stem cell that yields slightly differentiated daughter cells that retain the capacity to self-renew, or a cancer cell may arise from a specialized cell that loses some of its features and can divide. Certain stem cells, called **cancer stem cells,** veer from normal development and produce both cancer cells and abnormal specialized cells. Cancer stem cells are found in cancers of the brain, blood, and epithelium (particularly in the breast, colon, and prostate).

Figure 18.7 illustrates how cancer stem cells may cause brain tumors. In (*a*), as cancer stem cells give rise to progenitors and then differentiated cells (neurons, astrocytes, and oligodendrocytes), a cell surface molecule called CD133 is normally lost (designated CD133$^-$) at the late progenitor stage. In contrast, in (*b*), cancer cells retain the molecule (designated CD133$^+$). Some progenitor cells that descend from a cancer

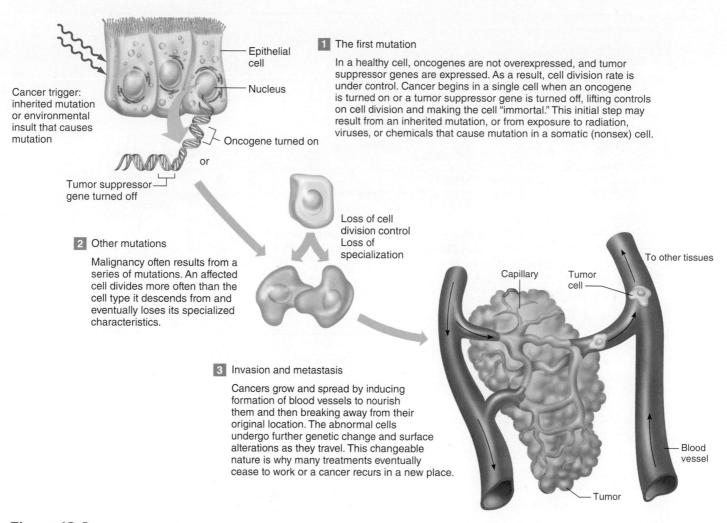

1 The first mutation

In a healthy cell, oncogenes are not overexpressed, and tumor suppressor genes are expressed. As a result, cell division rate is under control. Cancer begins in a single cell when an oncogene is turned on or a tumor suppressor gene is turned off, lifting controls on cell division and making the cell "immortal." This initial step may result from an inherited mutation, or from exposure to radiation, viruses, or chemicals that cause mutation in a somatic (nonsex) cell.

Epithelial cell

Cancer trigger: inherited mutation or environmental insult that causes mutation

Nucleus

Oncogene turned on

or

Tumor suppressor gene turned off

Loss of cell division control
Loss of specialization

2 Other mutations

Malignancy often results from a series of mutations. An affected cell divides more often than the cell type it descends from and eventually loses its specialized characteristics.

Capillary   Tumor cell   To other tissues

3 Invasion and metastasis

Cancers grow and spread by inducing formation of blood vessels to nourish them and then breaking away from their original location. The abnormal cells undergo further genetic change and surface alterations as they travel. This changeable nature is why many treatments eventually cease to work or a cancer recurs in a new place.

Blood vessel

Tumor

Figure 18.6 Steps in the development of cancer.

Table 18.1	Characteristics of Cancer Cells
Oilier, less adherent	
Loss of cell cycle control	
Heritable	
Transplantable	
Dedifferentiated	
Lack contact inhibition	
Induce local blood vessel formation (angiogenesis)	
Invasive	
Increased mutation rate	
Can spread (metastasize)	

stem cell can relentlessly divide, and they ultimately accumulate, forming a brain tumor.

Cancer may also begin when cells lose some of their distinguishing characteristics as mutations occur when they divide. Or, cells on the road to cancer may begin to express "stemness" genes that override signals to remain specialized (**figure 18.8**).

Another possible origin of cancer may be a loss of balance at the tissue level in favor of cells that can divide continually or frequently—like a population growing faster if more of its members are of reproductive age. Consider a tissue that is 5 percent stem cells, 10 percent progenitors, and 85 percent differentiated cells. If a mutation, over time, shifts the balance in a way that creates more stem and progenitor cells, the extra cells pile up, and a tumor forms (**figure 18.9**).

Uncontrolled tissue repair may cause cancer (**figure 18.10**). If too many cells divide to fill in the space left by injured tissue, and those cells keep dividing, an abnormal growth may result.

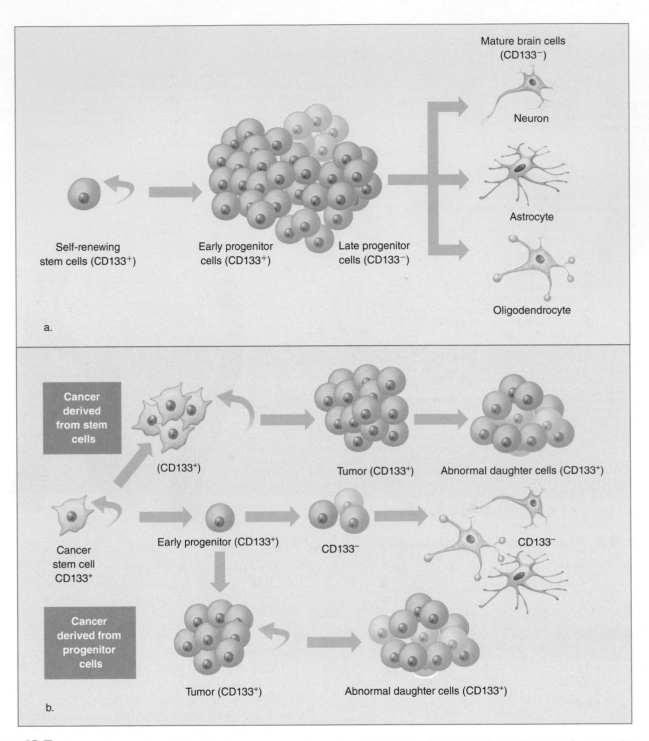

Figure 18.7 Cancer stem cells. **(a)** In the developing healthy brain, stem cells self-renew and give rise to early progenitor cells, which divide to yield late progenitor cells. These late progenitor cells lose the CD133 cell surface marker, and they divide to give rise to daughter cells that specialize as neurons or two types of supportive cells, astrocytes or oligodendrocytes. **(b)** A cancer stem cell can divide to self-renew and give rise to a cancer cell, which in turn can also spawn abnormal daughter cells (top part of part b). Or, an early progenitor cell can give rise to normal differentiated cells (middle part of part b). Or, cancer-causing mutations occur in the cancer stem cell–derived early progenitor cell. In this case, the early progenitors form the tumor, which may spawn some abnormal daughter cells (bottom part of part b). Note that stem cells, cancer stem cells, early progenitor cells, and abnormal daughter cells all have the CD133+ marker, but the differentiated cells do not.

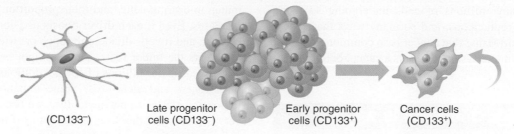

Figure 18.8 **Dedifferentiation reverses specialization.** Mutations in a differentiated cell could reactivate latent "stemness" genes, giving the cell greater capacity to divide while causing it to lose some of its specializations.

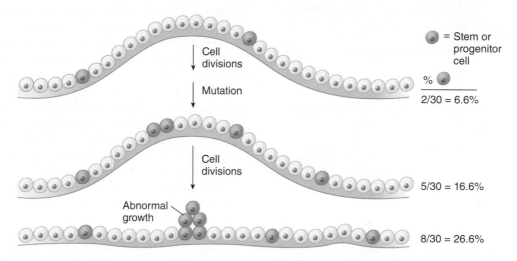

Figure 18.9 **Shifting the balance in a tissue toward cells that divide.** If a mutation renders a differentiated cell able to divide to yield other cells that frequently divide, then over time these cells may take over, forming an abnormal growth.

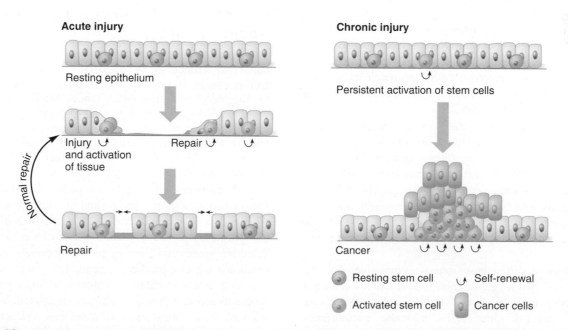

Figure 18.10 **Too much repair may trigger tumor formation.** If epithelium is occasionally damaged, resting stem cells can become activated and divide to fill in the tissue. If injury is chronic, the persistent activation of stem cells to renew the tissue can veer out of control, fueling an abnormal growth.

With so many millions of cells undergoing so many error-prone DNA replications, and so many ways that cancer can arise, it isn't surprising that cancer is so common. Yet most of the time, the immune system destroys a cancer before it progresses very far.

18.3 Cancer Genes and Genomes

Researchers have sequenced the genomes of cells from thousands of tumors. As a result, a more global view is emerging of how mutations cause cancer and keep changing the course of the disease.

Driver and Passenger Mutations

A driver of a vehicle takes it to the destination; a passenger goes along for the ride. In cancer genetics, a **driver mutation** provides the selective growth advantage to a cell that defines the cancerous state. A **passenger mutation** occurs in a cancer cell, but does not cause or propel the cancer's growth or spread. Drivers can be oncogenes or tumor suppressor genes, and may be generated from abnormal chromosomes.

Only about 130 genes may have driver mutations, but genes that can yield passenger mutations, which occur in cancerous as well as noncancerous cells, are much more common. More than 99 percent of the mutations in cancer cells are passengers, just along for the ride.

Tumors vary greatly in the numbers of each type of mutation. A cancer generally has from two to eight driver mutations. The number of passenger mutations increases with age. For a 40-year-old and an 80-year-old with the same type of cancer (as determined histologically), the older person's tumor cells will have many more passenger mutations than the younger person's tumor cells. This makes sense. The passage of time brings replication errors and environmental exposures.

The effect of driver mutations is cumulative. The initial mutation is called a "gatekeeper," and it enables a normal epithelial (lining cell) to divide slightly faster than others. A clone of faster-dividing cells gradually accumulates. (A clone is group of cells that descend from one cell.) Then a second mutation boosts the division rate in the already mutation-bearing cells, and their proportion within the tissue increases. Even if each driver boosts division rate by only 0.4 percent, and if cells divide only once or twice a week, in several years the tumor that a person might be able to feel will already consist of billions of cells. Additional mutations drive invasiveness and metastasis. Some mutations that enable a tumor to metastasize may actually have been present from the very beginning of the disease, almost as if they are waiting to spring into action.

Researchers use several techniques that compare tumors to deduce the mutational steps that drive the disease. One approach is to count and compare mutations in tumor cells from people at different stages of the same type of cancer. The older the tumor, the more genetic changes have accumulated. A mutation present in all stages among several individuals' tumors acts early in the disease process, whereas a mutation seen only in the tumor cells of sicker people acts later.

Another way to identify driver mutations that intervene after the gatekeeper is to study the cancer cells of patients who respond to a drug and then relapse. For example, an experimental drug that helped several patients with metastatic melanoma stopped working after 7 months for some patients, whose skin tumors came back. Their cancer cells had mutations in three genes other than the gatekeeper. These new mutations altered cancer cell surfaces and metabolism in ways that enabled the cells to ignore the drug and keep dividing and spreading. The changeability of a cancer is why a "cocktail" of drugs that act on different cellular pathways (such as cell adhesion or signal transduction) may be the best treatment approach, as it is for HIV infection.

Colon Cancer Illustrates the Stepwise Nature of Cancer

The accumulation of mutations that drive a cancer is perhaps best studied in cancer of the colon (large intestine). Colon cancer is usually not frequent enough in families to suggest Mendelian inheritance, but when individuals with many polyps (benign stalklike growths) called adenomas are included, a single-gene pattern indeed emerges (**figure 18.11**). This condition, called familial adenomatous polyposis (FAP; OMIM 175100), affects 1 in 5,000 people in the United States.

Healthy colon lining cells typically live 3 days. In FAP, the cells do not die on schedule and instead pile up into polyps. FAP begins in early childhood with hundreds of tiny polyps that progress over many years to colon cancer, driven by activated oncogenes and silenced tumor suppressor genes. First, a deletion removes a gatekeeper gene, *APC*. Normally APC protein binds another protein, β-catenin, adding a phosphate that blocks its action. When the *APC* gene is deleted, β-catenin isn't silenced, and instead enters the nucleus and activates genes that promote mitosis. A tumor forms and becomes malignant when other mutations, such as in the driver genes *TGF*-β and

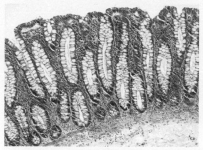

a. Normal colon epithelium

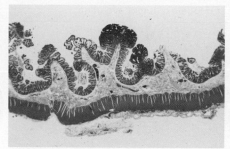

Colon adenoma (polyp)

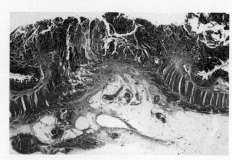

Colon carcinoma (cancer)

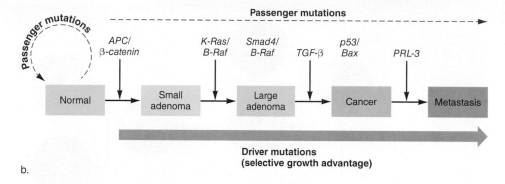

b.

Figure 18.11 Several mutations drive FAP colon cancer. (a) Tumors develop over many years, from normal epithelium, through adenomas, and finally to epithelial cancer, called carcinoma. **(b)** Colon lining cells divide more frequently when the *APC* gene on chromosome 5q is deleted or undergoes a point mutation. Adenomas form. Activation of oncogenes, such as *K-Ras* and *B-Raf,* fuel growth. Mutations in *p53* and other genes push the adenoma cells to become cancerous. Finally, mutations in *PRL-3* trigger metastasis.

p53, push the abnormal cells into further derangement. *TGF*-β normally inhibits mitosis and *p53* normally sends cells to a fate of apoptosis. *PRL-3* is a gene that acts late in the process, enabling the cancer to spread.

The evolution of a cancer is similar to the evolution of species in that changes accumulate over time. Just as linear changes may represent only part of an evolutionary tree diagram, linear progressions of mutations may be a too-simple view of cancer. Researchers now think that in many tumors, cell lineages branch when they acquire new mutations, which may fuel or accompany metastasis (**figure 18.12**).

The following sections look at specific oncogenes and tumor suppressor genes. Chromosome abnormalities can cause these mutations. The chromosomes in cancer cells may be abnormal in number and/or structure. They may bear translocations, inversions, or have pieces missing or extra. A translocation that joins parts of nonhomologous chromosomes can hike expression of a gene enough to turn it into an oncogene (**figure 18.13**). A duplication can increase the number of copies of a particular oncogene from two—one on each of a pair of homologs—to up to 100. A deletion may remove a tumor suppressor gene. A one-time event called chromothripsis shatters several chromosomes and may kill the cell—or trigger cancer.

A Closer Look at Oncogenes

Genes that normally trigger cell division are called **proto-oncogenes.** They are active where and when high rates of cell division are necessary, such as in a wound or in an embryo. When proto-oncogenes are transcribed and translated too rapidly or frequently, or perhaps at the wrong time in development or place in the body, they function as oncogenes ("onco" means cancer). Usually oncogene activation is associated with a point mutation or a chromosomal translocation or inversion that places the gene next to another that is more highly expressed (transcribed). Oncogene activation causes a gain-of-function. In contrast, a tumor suppressor gene mutation is usually a deletion that causes a loss-of-function (see figure 4.8).

Proto-oncogenes can also become oncogenes by being physically next to highly transcribed genes. Three examples of genes that can activate proto-oncogenes are a viral gene, a gene encoding a hormone, and parts of antibody genes.

A virus infecting a cell may insert DNA next to a proto-oncogene. When the viral DNA is rapidly transcribed, the adjacent proto-oncogene (now an oncogene) is also rapidly transcribed. Increased production of the oncogene's encoded protein then switches on genes that promote mitosis,

Models of tumor evolution

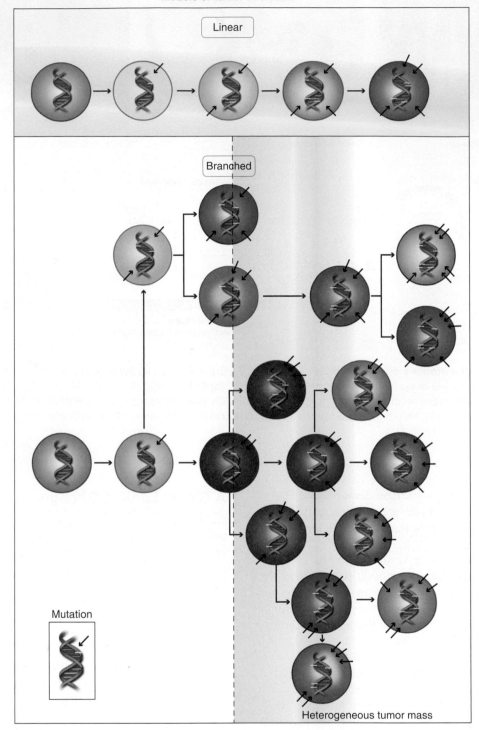

Figure 18.12 The evolution of a cancer. Genetic changes in cancer may occur in a linear and/or branching pattern. A heterogeneous tumor mass evolves as new mutations arise, while old ones remain. Note that the number of mutations increases from left to right

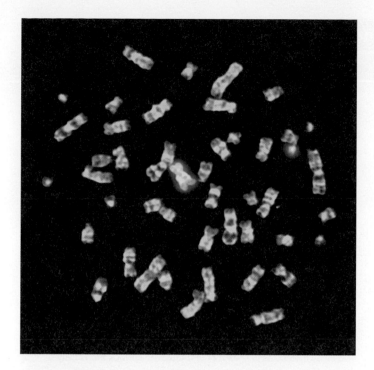

Figure 18.13 Cancer chromosomes. This cancer cell has only one abnormality visible at the chromosomal level (see the chapter opener). Many cancer cells, late in the disease, are riddled with broken and rearranged chromosomes.

triggering the cascade of changes that leads to cancer. Viral damage to a human genome may be what one researcher calls "catastrophic," activating and amplifying oncogenes as well as inverting and translocating chromosomes. Viruses cause cervical cancer, Kaposi sarcoma, and acute T cell leukemia.

A proto-oncogene may be activated when it is moved next to a gene that is normally very actively transcribed. This happens when an inversion on chromosome 11 places a proto-oncogene next to a DNA sequence that controls transcription of the parathyroid hormone gene. When the gland synthesizes the hormone, the oncogene is expressed, too. Cells in the gland divide, forming a tumor.

Antibody genes are among the most highly transcribed, so it isn't surprising that a translocation or inversion that places a proto-oncogene next to an antibody gene causes cancer. Cervical cancer and anal cancer following HPV infection may begin when proto-oncogenes are mistakenly activated with antibody genes. Similarly, in Burkitt lymphoma, a cancer common in Africa, a large tumor develops from lymph glands near the jaw. People with Burkitt lymphoma are infected with the Epstein-Barr virus, which stimulates specific chromosome movements in maturing B cells to assemble antibodies against the virus. A translocation places a proto-oncogene on chromosome 8 next to an antibody gene on chromosome 14. The oncogene is overexpressed, and the cell division rate increases. Tumor cells of Burkitt lymphoma patients have the translocation (**figure 18.14**).

A proto-oncogene may not only move next to another gene, but also be transcribed and translated with it as if they are one gene. The double gene product, called a **fusion protein,** activates or lifts control of cell division. For example, in acute promyelocytic leukemia, a translocation between chromosomes 15 and 17 brings together a gene coding for the retinoic acid cell surface receptor and an oncogene called *myl*. The fusion protein functions as a transcription factor, which, when overexpressed, causes cancer. The nature of this fusion protein explains why some patients who receive retinoid (vitamin A–based) drugs recover. Their immature, dedifferentiated cancer cells, apparently stuck in an early stage of development where they divide frequently, suddenly differentiate, mature, and die. Perhaps the cancer-causing fusion protein prevents affected white blood cells from getting enough retinoids to specialize, locking them in an embryonic-like, rapidly dividing state. Supplying extra retinoids allows the cells to resume their normal developmental pathway. **Clinical Connection 18.1** describes the

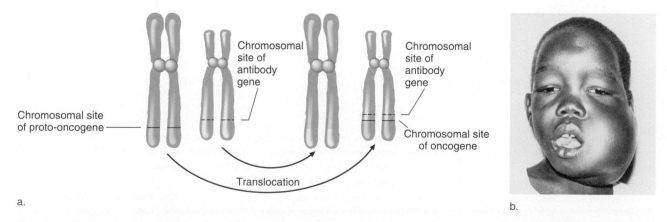

a.

b.

Figure 18.14 A translocation that causes cancer. (a) In Burkitt lymphoma, a proto-oncogene on chromosome 8 moves to chromosome 14, next to a highly expressed antibody gene. Overexpression of the translocated proto-oncogene, now an oncogene, triggers the molecular and cellular changes of cancer. **(b)** Burkitt lymphoma typically affects the jaw.

The Story of Gleevec

The tale of Gleevec began on August 13, 1958, when two men entered hospitals in Philadelphia and reported weeks of fatigue. Each had very high white blood cell counts and were diagnosed with chronic myelogenous leukemia (CML). Too many immature white blood cells were crowding the healthy cells. The men's blood samples eventually fell into the hands of pathologist Peter Nowell and cytogeneticist David Hungerford. They had developed ways to stimulate white blood cells to divide in culture, and they probed the chromosomes of both leukemic and normal-appearing white blood cells in the two tired men and five other patients with CML.

Nowell and Hungerford discovered a small, unusual chromosome that was only in the leukemic cells. This was the first chromosome abnormality to be linked to cancer. Later, it would be dubbed "the Philadelphia chromosome" (Ph[1]). The link between the cancer and the chromosome anomaly held up in other patients. The photo in the chapter opener shows the Philadelphia chromosome.

With refinements in chromosome banding, important details emerged. By 1972, new stains that distinguished AT-rich from GC-rich chromosome regions revealed that Ph[1] is the result of a translocation. By 1984, researchers had homed in on the two genes juxtaposed in the translocation between chromosomes 9 and 22. One gene from chromosome 9 is called the Abelson oncogene (*abl*), and the other gene, from chromosome 22, is called the breakpoint cluster region (*bcr*). Two different fusion genes form. The *bcr-abl* fusion gene is part of the Philadelphia chromosome, and it causes CML. (The other fusion gene does not affect health.)

The discovery that led directly to development of the drug Gleevec was that the encoded fusion protein, called the BCR-ABL oncoprotein, is a form of the enzyme tyrosine kinase, which is the normal product of the *abl* gene. A kinase is an enzyme that transfers a phosphate (PO_4) to another molecule, and is a key part of signal transduction pathways. The cancer-causing form of tyrosine kinase is active for too long, which sends signals into the cell, stimulating it to divide, too many times.

Through the 1980s, drug developers tested more than 400 small molecules in search of one that would block the activity of the errant tyrosine kinase, without derailing other important enzymes. They found what would become Gleevec in 1992. **Figure 1** shows how the drug nestles into the pocket on the tyrosine kinase that must bind ATP to stimulate cell division. With ATP binding blocked, cancer cells

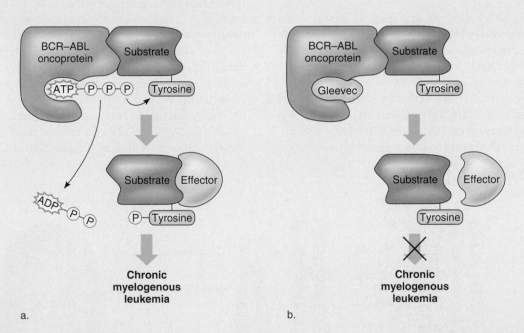

a.

b.

Figure 1 **How Gleevec treats chronic myelogenous leukemia.** In CML, a translocation forms the fusion oncoprotein BCR-ABL, which functions as a tyrosine kinase. A tyrosine (an amino acid) of a substrate molecule picks up a phosphate from the ATP nestled in the oncoprotein, making the substrate able to bind to another protein, called an effector, that triggers runaway cell division **(a).** Gleevec replaces the ATP **(b).** Without phosphorylation of the tyrosine on the substrate, cell division stops.

Source: Adapted from "Drug therapy: Imantinib mesylate—A new oral targeted therapy" by Savage & Antman: New England Journal of Medicine 346: 683–693. Copyright © 2002 Massachusetts Medical Society. All rights reserved. Reprinted by permission.

(Continued)

do not receive the message to divide, and they cease doing so. After passing safety tests, the drug worked so dramatically that it set a new speed record for drug approval—10 weeks.

Patients showed improvement in several ways:

- Hematological remission: The percentage of leukemia cells in the blood fell.
- Cytogenetic remission: The percentage of cells with the Ph[1] chromosome fell.
- Molecular remission: The level of mRNA representing the *bcr-abl* fusion gene fell.

Molecular remission is the goal of CML treatment, but in actuality, fusion gene mRNA rarely reaches undetectable levels. As a result, patients can become resistant to Gleevec if a mutation makes a remaining cell resistant to the drug. Relapse occurs in 3 to 16 percent of patients, depending on how sick they were when diagnosed. Resistance is a result of natural selection. Those few cancer cells able to divide in the presence of the drug eventually take over. Again, genetic research came to the rescue. By discovering how resistant cells evade the drug, researchers tweaked Gleevec, making it bind more strongly, and developed new drugs that fit the slightly altered active site in resistant cancer cells. In some patients leukemia returns if drug treatment stops, although women have been able to stop safely during pregnancy.

Questions for Discussion

1. What type of chromosome abnormality results in the formation of the Philadelphia chromosome?
2. Is CML caused by an activated oncogene or a deleted tumor suppressor gene?
3. How does the abnormal enzyme behind CML cause the cancer?
4. Explain how Gleevec works.
5. Why isn't Gleevec a permanent cure for CML?
6. In the United States the wholesale price of Gleevec for one patient is $76,000 per year. A group of hundreds of physicians and cancer researchers published a paper that calls this price "exorbitant and unjustifiable," pointing out that production costs have plummeted since the drug was introduced in 2001. What factors do you think should be considered in determining the cost of a drug such as Gleevec?

fusion protein that led to development of Gleevec, the drug discussed in the chapter opener.

Another way that an oncogene can cause cancer is by excessive response to a growth factor. In about 25 percent of women with breast cancer, affected cells have 1 to 2 million copies of a cell surface protein called HER2 that is the product of an oncogene. The normal number of these proteins is only 20,000 to 100,000.

The HER2 proteins are receptors for epidermal growth factor. In breast cells, the receptors traverse the plasma membrane, extending outside the cell into the extracellular matrix and also dipping into the cytoplasm. They function as a tyrosine kinase, as is the case for the leukemia described in Clinical Connection 18.1. When the growth factor binds to the tyrosine (an amino acid) of the receptor, the tyrosine picks up a phosphate group, which signals the cell to activate transcription of genes that stimulate cell division. In HER2 breast cancer, too many tyrosine kinase receptors send too many signals to divide.

HER2 breast cancer usually strikes early in adulthood and spreads quickly. However, a monoclonal antibody-based drug called trastuzumab (Herceptin) binds to the receptors, blocking the signal to divide (see figure 17.16). Interestingly, Herceptin works when the extra receptors arise from multiple copies of the gene, rather than from extra transcription of a single *HER2* gene.

A Closer Look at Tumor Suppressor Genes

Some cancers result from loss or silencing of a tumor suppressor gene that normally inhibits expression of genes involved in all of the activities that turn a cell cancerous. Cancer can result when a tumor suppressor gene is deleted or if the promoter region binds too many methyl (CH_3) groups, which blocks transcription.

Wilms' tumor is an example of a cancer that develops from loss of tumor suppression. A gene that normally halts mitosis in the rapidly developing kidney tubules in the fetus is absent. As a result, an affected child's kidney retains pockets of cells dividing as frequently as if they were still in the fetus, forming a tumor. Following are descriptions of specific tumor suppressor genes.

Retinoblastoma (RB)

Retinoblastoma (RB; OMIM 180200) is a rare childhood eye tumor (**figure 18.15**). A Glimpse of History on page 366 reveals that this cancer has been recognized for a long time.

About 1 in 20,000 infants develops RB, and half of them inherit susceptibility to the disorder. They have one germline mutant allele for the *RB1* gene in each of their cells, and then cancer develops in a somatic cell where the

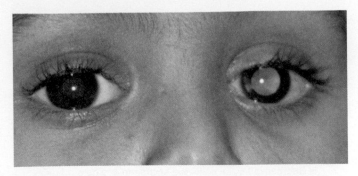

Figure 18.15 Retinoblastoma type 1 is due to mutation in a tumor suppressor gene. The first sign of retinoblastoma is typically an unusual pale area that appears in an eye in a photograph where a tumor reflects light differently than unaffected structures. The fact that only one eye is affected in this child suggests two somatic mutations in the one eye, rather than inherited susceptibility in all cells followed by somatic mutations in both, which might manifest as tumors in both eyes.

second copy of the *RB1* gene mutates. Therefore, inherited retinoblastoma requires two point mutations or deletions, one germline and one somatic. In some sporadic (noninherited) cases, two somatic mutations occur in the *RB1* gene, one on each copy of chromosome 13. Either way, the cancer usually starts in a cone cell of the retina, which provides color vision. Study of retinoblastoma inspired the "two-hit" hypothesis of cancer causation—that two mutations (germline and somatic or two somatic) are required to cause a cancer related to tumor suppressor deletion or malfunction. (A second form of retinoblastoma is caused by mutation in an oncogene, *MYCN*.)

The discovery that many children with RB have deletions in the same region of the long arm of chromosome 13 led researchers to the *RB1* gene and its protein product, which linked the cancer to control of the cell cycle. The RB1 protein normally binds transcription factors so that they cannot activate genes that carry out mitosis. It normally halts the cell cycle at G_1. When the *RB1* gene is mutant or missing, the hold on the transcription factor is released, and cell division ensues.

For many years, the only treatment for retinoblastoma was removal of the affected eye. Today, children with an affected parent or sibling, who have a 50 percent chance of having inherited the mutant *RB1* gene, can be monitored from birth so that noninvasive treatment (chemotherapy) can begin early. Full recovery is common.

Mutations in the *RB1* gene cause other cancers. Some children successfully treated for retinoblastoma develop bone cancer as teens or bladder cancer as adults. Mutant *RB1* genes have been found in the cells of patients with breast, lung, or prostate cancers, or acute myeloid leukemia, who never had the eye tumors. Expression of the same genetic defect in different tissues may cause these cancers.

p53 Normally Prevents Many Cancers

A single gene that causes a variety of cancers when mutant is *p53*. Recall from chapter 12 that the p53 protein transcription factor "decides" whether a cell repairs DNA replication errors or dies by apoptosis. If a cell loses a *p53* gene, or if the gene mutates and malfunctions, a cell with damaged DNA is permitted to divide, and cancer may be the result.

More than half of human cancers arise from a point mutation or deletion in the *p53* gene. This may be because p53 protein is a genetic mediator between environmental insults and development of cancer (**figure 18.16**). A type of skin cancer, for example, is caused by a *p53* mutation in skin cells damaged by an excessive inflammatory response that can result from repeated sunburns. That is, *p53* may be the link between sun exposure and skin cancer.

In most *p53*-related cancers, mutations occur only in somatic cells. However, in the germline condition Li-Fraumeni syndrome (OMIM 151623), family members who inherit a mutation in the *p53* gene have a very high risk of developing cancer—50 percent do so by age 30, and 90 percent by age 70. A somatic *p53* mutation in the affected tissue results in cancer because a germline mutation in the gene is already present.

Stomach Cancer Lifts Cellular Adhesion

E-cadherin normally acts as a cellular adhesion protein found in tissue linings, but it is also a tumor suppressor because when its gene is deleted, cancer results. This was the case for the Bradfield family. Golda Bradfield died of stomach cancer in 1960. By the time some of her grown children developed the cancer too, the grandchildren began to realize that their family had a terrible legacy. Genetic testing revealed familial diffuse gastric cancer (OMIM 192090), caused by an "exon skipping" mutation in the E-cadherin gene (see figure 12.10). This is a missense mutation (single-base change) that deletes an entire exon from the mRNA transcribed from the gene.

Golda's grandchildren had genetic tests. Eleven of them had inherited the mutant gene, but scans of their stomachs did not show any tumors. Still, they all had their stomachs

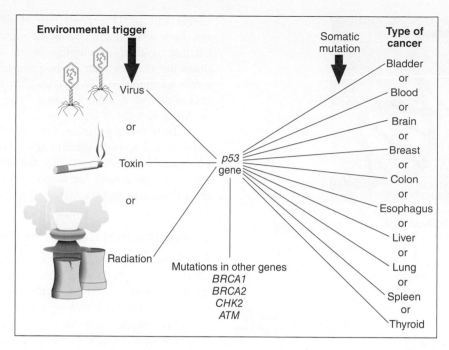

Figure 18.16 *p53* **cancers reflect environmental insults.** The environment triggers mutations or changes in gene expression that lead to cancer. The *p53* gene is a mediator—"the guardian of the genome." The protein products of many genes interact with *p53*.

removed. Most of them already had hundreds of tumors, too tiny to have been seen on medical scans. The cousins without stomachs are doing well. Like people who have their stomachs surgically shrunk to lose weight, the cousins avoid hard-to-digest foods and eat a little at a time, throughout the day. The inconvenience, they say, is a fair trade for eliminating the fear of their grandmother's, parents', and aunts' and uncles' fates—stomach cancer.

BRCA1/BRCA2 Mutations Disrupt Repair

Breast cancer that runs in families may be due to inheriting a germline mutation and then having a somatic mutation occur in a breast cell (a familial form), or two somatic mutations affecting the same breast cell (a sporadic form), as figure 18.5 depicts. However, breast cancer is so common that a family with many affected members may actually have multiple sporadic cases, rather than an inherited form of the disease.

Only about 5 percent of breast cancers are familial, caused by mutations in any of at least 20 genes that increase susceptibility. Most of the genes associated with familial breast cancer encode proteins that interact in ways that enable DNA to survive damage. If DNA cannot be repaired, mutations that directly cause cancer can accumulate and persist.

The two major breast cancer susceptibility genes are *BRCA1* and *BRCA2*. Together they account for 15 to 20 percent of familial cases. Inheriting a mutation in *BRCA1,* which stands for "breast cancer predisposition gene 1," greatly increases the lifetime risk of inheriting breast and ovarian cancer. This risk,

however, varies in different population groups because of the modifying effects of other genes. In the most common *BRCA1* mutation, deletion of two adjacent DNA bases alters the reading frame, shortening the protein. Hundreds of mutations in the gene are known, and most of them are very rare. A *BRCA* mutation is inherited in an autosomal dominant manner, with incomplete penetrance because it increases susceptibility, rather than directly causing cancer.

BRCA1 encodes a protein that interacts with many other proteins that counter DNA damage in several ways. One very important form of protection is the mending of areas of the genome where both DNA strands are broken at the same site. These double-stranded breaks are particularly dangerous because they cut the chromosomes, making rearrangements such as deletions and translocation possible. **Figure 18.17** depicts the central role of *BRCA1* in protecting DNA.

BRCA1 mutations have different incidences in different populations. Only 1 in 833 people in the general U.S. population has a mutant *BRCA1* allele. That figure is more than 1 in 50 among Ashkenazi Jewish people, due to population bottlenecks and nonrandom mating. The *BRCA1* gene was initially discovered in Ashkenazi families in which several members developed the cancer at very young ages. In this population, a woman who inherits a *BRCA1* mutation has up to an 87 percent risk of developing breast cancer over her lifetime, and a 50 percent risk of developing ovarian cancer. For Ashkenazi families without a strong clinical history of breast cancer, the risks are 65 percent for breast cancer and 39 percent for ovarian cancer.

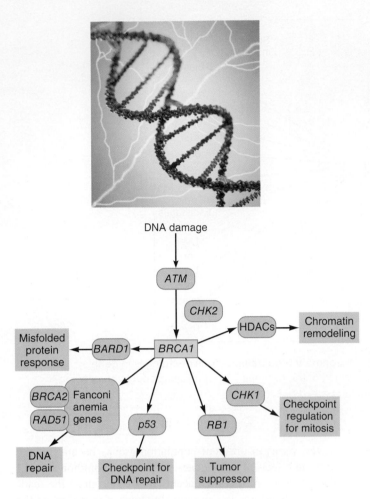

DNA damage

Figure 18.17 The *BRCA1* gene controls many cellular defense mechanisms. *BRCA1* functions as a genetic crossroads in handling DNA damage. Inheriting a *BRCA1* mutation increases susceptibility to several cancers (breast, ovary, cervix, uterine tube, uterus, peritoneum, pancreas, and prostate gland). Mutations in any of the genes with which it interacts increases cancer susceptibility too. (Italicized abbreviations indicate genes. HDACs are histone deacetylases.)

Some women who learn that they have inherited an allele predisposing to breast or ovarian cancer have their breasts and/or ovaries removed. This action makes more sense for an Ashkenazi woman facing an 87 percent lifetime risk of breast or ovarian cancer than it does for a woman in a population group with a much lower risk. The general population risk of actually developing a *BRCA1* cancer if one inherits a mutation is only about 10 percent, based on empirical (observational) evidence. Women with such mutations born after 1940 have a higher risk than those born earlier, suggesting that the environment also plays a role in whether inheriting a mutation causes cancer.

BRCA2 breast cancer is also more common among the Ashkenazim. Ashkenazi women who inherit a mutation in *BRCA2* face a 45 percent lifetime risk of developing breast

cancer and an 11 percent risk of developing ovarian cancer. Men who inherit a *BRCA2* mutation have a 6 percent lifetime risk of developing breast cancer, which is 100 times the risk for men in the general population. Inheriting a *BRCA2* mutation also increases the risk of developing cancers of the colon, kidney, prostate, pancreas, gallbladder, skin, or stomach.

Tests for any cancer susceptibility genes can reveal a "variant of uncertain significance." This means that the gene sequence is not wild type, but the identified variant has not been associated with increased cancer susceptibility or any other disease phenotype.

Environmental Causes of Cancer

Environmental factors contribute to cancer by mutating or altering the expression of genes that control the cell cycle, apoptosis, and DNA repair. Inheriting a susceptibility gene places a person farther along a particular road to cancer, but cancer can happen in somatic cells in anyone. It is more practical, for now, to identify environmental cancer triggers and develop ways to control them or limit our exposure to them, than to alter genes.

Looking at cancer at a population level reveals the interactions of genes and the environment. For example, researchers examined samples of non-Hodgkin's lymphoma tumors from 172 farmers, 65 of whom had a specific chromosomal translocation. The 65 farmers were much more likely to have been exposed for long times to toxic insecticides, herbicides, fungicides, and fumigants, compared to the farmers with lymphoma who did not have the translocation.

Determining precisely how an environmental factor such as diet affects cancer risk can be complicated. Consider the cruciferous vegetables, such as broccoli and brussels sprouts, which are associated with decreased risk of developing colon cancer. These vegetables release compounds called glucosinolates, which in turn activate "xenobiotic metabolizing enzymes" that detoxify carcinogenic products of cooked meat called heterocyclic aromatic amines. With a vegetable-poor, meaty diet, these amines accumulate. They cross the lining of the digestive tract and circulate to the liver, where enzymes metabolize them into compounds that cause driver mutations for colon cancer (**figure 18.18**).

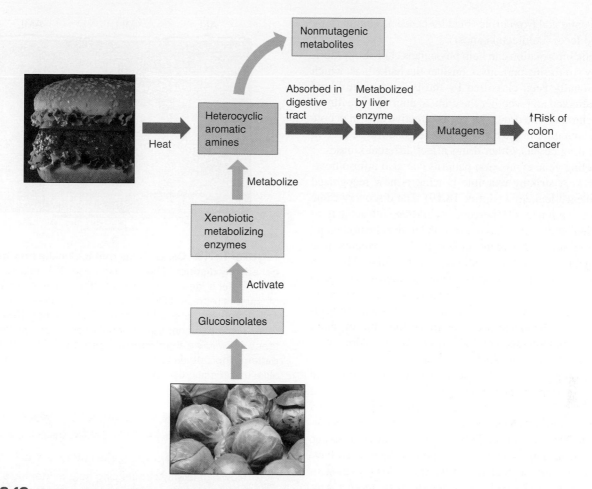

Figure 18.18 **One way that cruciferous vegetables lower cancer risk.** Compounds called heterocyclic aromatic amines form in cooking meat, are absorbed into the digestive tract, and are metabolized by a liver enzyme into mutagens, which may cause colon cancer. Broccoli and brussels sprouts produce glucosinolates, which activate xenobiotic metabolizing enzymes that block part of the pathway that leads to production of the mutagens.

18.4 The Challenges of Diagnosing and Treating Cancer

Eradicating cancer is difficult, if not impossible, because of the diseases' great heterogeneity. Although a handful of driver mutations set a cancer into motion by affecting a small set of cellular processes, this can happen in so many ways that the cancers of no two patients are exactly alike, nor are multiple tumors within the same body genetically alike, or even the cells within a single tumor.

The traditional ways of treating cancer are broad. The oldest approach, surgery, is still effective if it removes a primary tumor before it has invaded healthy tissue and spread through the bloodstream. Chemotherapy and radiation hit all rapidly dividing cells, causing adverse effects by harming healthy cells. Newer approaches that are targeted based on genetic information, such as the tyrosine kinase inhibitor Gleevec (see Clinical Connection 18.1), are highly effective if matched to a sensitive tumor type. For all of these approaches, however, it takes just a few escaped cancer cells to sow the seeds of a future tumor.

The ability to sequence cancer exomes and genomes is providing a new specificity to our approach to both diagnosing and treating these diseases. If several members of a family have cancer of the same body part—the way cancers are currently classified—yet they all test negative for known cancer genes, then sequencing their exomes can reveal candidate genes that are mutant in the affected family members. If researchers can validate that mutations in the candidate gene affect other families, then a new gene can be added to the list of cancer driver genes, and a new drug target emerges.

For people who have been diagnosed with a particular type of cancer, a powerful new way to monitor the disease is to collect and examine DNA from a tumor that ends up in the bloodstream. This "cell-free tumor DNA (ctDNA)" can reveal response to a treatment, for example. If a patient tries a new drug and then the proportion of tumor DNA sequences in the blood plasma increases, the tumor has become resistant to that drug. This blood test, called a "liquid biopsy," is much less invasive than exploratory surgery to monitor a cancer's spread—which is often impossible. Using tumor DNA as a biomarker is more specific than a protein biomarker because a protein biomarker may also be present on healthy cells. Tumor DNA and cells can also be collected from urine (bladder cancer), sputum (lung cancer), and feces (colorectal cancer).

Genetic information can help physicians choose appropriate drugs by stratifying patients. Consider the leukemias, which have traditionally been classified by the type of white blood cell that is affected and whether the clinical course of the illness is acute or chronic. The drug Gleevec is effective only in people who have a mutation in the tyrosine kinase gene whose protein product the drug targets. Other drugs affect different kinases.

Detecting gene expression patterns can also inform treatment choices. A striking example is what is now recognized as mixed lineage leukemia (**figure 18.19**). The discovery came from the observation that 10 percent of children with acute lymphoblastic leukemia (ALL) do not respond to the chemotherapy that helps the other 90 percent, although all the patients have the same symptoms of fever, fatigue, and bruising. Although the leukemic cells all look alike under a microscope, gene expression analysis using DNA microarrays (chips) revealed that in the 10 percent of patients who do not respond to standard ALL drugs, the cancerous cells make too little of about 1,000 proteins and too much of 200 others. These children, who actually have MLL and not ALL, respond to different drugs.

Continuing analysis of cancer cell genomes combined with the annotation of the human genome (discovering what genes do) will tell us more about exactly what happens when a cell becomes malignant. But conquering cancer may be an elusive goal. The DNA of cancer cells mutates in ways that enable cells to pump out any drug sent into them. Cancer cells have redundant pathways, so that if a drug shuts down angiogenesis or invasiveness, the cell completes the task another way. A more realistic goal than eradicating cancer may be to kill enough cancer cells, and sufficiently slow their spread, so that it takes the remainder of a human lifetime for tumors to grow back enough to harm health. In this way, cancer can become a chronic, manageable condition. As long as our cells divide, we are at risk of developing cancer.

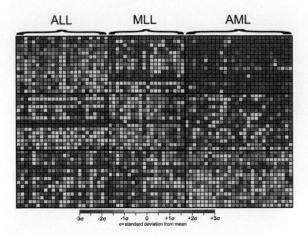

Figure 18.19 **Cancer cells that look alike may be genetically distinct.** These leukemias—ALL, MLL, and AML—differ in gene expression patterns. The columns of squares represent DNA from tumor samples, and the rows compare the activities of particular genes. Red tones indicate higher-than-normal expression and blue tones show lower-than-normal expression. The different patterns indicate distinct cancers, although the cells look alike under a microscope.

Key Concepts Questions 18.4

1. What is a major challenge of curing cancer?

2. How can exome sequencing help to identify a previously unknown cancer gene in a family?

3. How can a liquid biopsy track a cancer?

4. Explain how testing for mutations and for gene expression patterns can help a physician to select a treatment for a patient.

Summary

18.1 Cancer Is an Abnormal Growth That Invades and Spreads

1. Cancer is a genetically driven loss of cell cycle control, creating a population of highly proliferative cells that invades surrounding tissue. Spread of a cancer is called **metastasis**.

2. Activated **oncogenes** and inactivated **tumor suppressor genes** cause cancer.

3. Mutations in genes that encode or control transcription factors, cell cycle checkpoint proteins, growth factors, repair proteins, or telomerase may disrupt the cell cycle, causing cancer.

4. Sporadic cancers result from two **somatic mutations** in the two copies of a gene. They are more common than inherited cancers that are caused by **germline mutations** that confer susceptibility plus somatic mutations in affected tissue.

18.2 Cancer at the Cellular Level

5. A tumor cell divides more frequently or more times than cells surrounding it, has altered surface properties, loses the specializations of the cell type it arose from, and produces daughter cells like itself.

6. A malignant tumor infiltrates tissues and can metastasize by attaching to basement membranes and secreting enzymes that penetrate tissues and open a route to the bloodstream. **Angiogenesis** establishes a blood supply. A cancer cell can travel, establishing secondary tumors.

7. Cell specialization and position within a tissue affect whether cancer begins and persists.

8. **Cancer stem cells** can divide to yield cancer cells and abnormally differentiated cells.

9. A cell that dedifferentiates and/or expresses "stemness" genes can begin a cancer.

10. A mutation that enables a cell to divide continually can alter the percentages of cells in a tissue that can divide, resulting in an abnormal growth.

11. Chronic repair of tissue damage can provoke stem cells into producing an abnormal growth.

18.3 Cancer Genes and Genomes

12. A **driver mutation** provides a selective growth advantage to a cell. A cancerous tumor has two to six driver mutations; about 130 driver genes are known. A **passenger mutation** occurs in a cancer cell but does not cause or contribute to the disease.

13. To decipher the series of mutations that drive a cancer, researchers examine the mutations in cells from patients at various stages of the same type of cancer. Mutations present at all stages are the first to occur.

14. Researchers worked out the multigene, sequential nature of cancer first for FAP colon cancer.

15. Chromosomes in cancer cells may be very abnormal. Chromothripsis is a single, highly destructive event that shatters chromosomes.

16. Cancer is often the result of activation of **proto-oncogenes** to oncogenes, and inactivation of tumor suppressor genes. Mutations in DNA repair genes cause cancer by increasing the mutation rate.

17. Proto-oncogenes normally promote controlled cell growth, but are overexpressed because of a point mutation, placement next to a highly expressed gene, or transcription and translation with another gene, producing a **fusion protein.** Oncogenes may also be overexpressed growth factor receptors.

18. A tumor suppressor is a gene that normally enables a cell to respond to factors that limit its division.

19. Environmental factors contribute to cancer by mutating genes and altering gene expression.

18.4 The Challenges of Diagnosing and Treating Cancer

20. Traditional cancer treatments are surgery, radiation, and chemotherapy. Newer approaches based on identifying mutations and gene expression patterns are used to subtype cancers and better target treatments.

21. Exome sequencing is used to identify new cancer genes. Analyzing cell-free tumor DNA can track cancer progression and drug resistance.

www.mhhe.com/lewisgenetics11

Answers to all end-of-chapter questions can be found at **www.mhhe.com/lewisgenetics11.** You will also find additional practice quizzes, animations, videos, and vocabulary flashcards to help you master the material in this chapter.

Review Questions

1. Distinguish between a benign and a malignant tumor.

2. How does a sporadic cancer differ from a Mendelian disease in terms of predicting risk of occurrence in the relatives of a patient?

3. What are the three basic cellular pathways that cancer disrupts?

4. What is metastasis, and why is it dangerous?

5. Explain the connection between cancer and control of the cell cycle.

6. List four characteristics of cancer cells.

7. Describe four ways that cancer can originate at the cell or tissue level.

8. Distinguish between a driver mutation and a passenger mutation.

9. Describe two abnormalities of chromosomes in cancer.

10. Explain how comparing mutations in cells from the same cancer type at different stages can reveal the sequence of genetic changes that drive the cancer.

11. How can classifying cancer genomically differ from classifying cancer histologically or by initial affected body part?

12. Distinguish between a proto-oncogene and an oncogene.

13. Describe two events that can activate an oncogene.

14. Explain why an oncogene is associated with a gain-of-function and a mutation in a tumor suppressor gene is associated with a loss-of-function.

15. How are retinoblastoma type 1, a *p53*-related cancer, inherited stomach cancer, and *BRCA1* breast cancer similar?

16. What is the role of poor DNA repair in causing cancer?

17. What did investigation of FAP colon cancer reveal about the nature of cancer in general?

18. Distinguish among mutations, altered gene expression, and epigenetic changes in cancer.

19. Explain why not all cancers affecting the same cell type respond the same way to a particular drug.

20. Describe an alternative to performing surgery to determine whether a metastatic tumor is responding to a new drug.

Applied Questions

1. Breast cancer can develop from inheriting a germline mutation and then undergoing a second mutation in a breast cell; or from two mutations in a breast cell, one in each copy of a tumor suppressor gene. Cite another type of cancer, discussed in the chapter, that can arise in these two ways.

2. How do the mechanisms of the drugs Gleevec and Herceptin differ?

3. Why is a "cocktail" of several drugs likely to be more effective at slowing the course of a cancer than using a single, very powerful drug?

4. von Hippel-Lindau syndrome (OMIM 193300) is an inherited cancer syndrome. The responsible mutation lifts control over the transcription of certain genes, which, when overexpressed, cause tumors to form in the kidneys, adrenal glands, and blood vessels. Is the von Hippel-Lindau gene an oncogene or a tumor suppressor? Cite a reason for your answer.

5. The *BRCA2* gene causes some cases of Wilms' tumor and some cases of breast cancer. Explain how the same tumor suppressor mutation can cause different cancers.

6. Ads for the cervical cancer vaccine present the fact that a virus can cause cancer as startling news, when in fact this has been known for decades. Explain how a virus might cause cancer.

7. A tumor is removed from a mouse and broken up into cells. Each cell is injected into a different mouse. Although all the mice used in the experiment are genetically identical and raised in the same environment, the animals develop cancers with different rates of metastasis. Some mice die quickly, some linger, and others recover. What do these results indicate about the characteristics of the original tumor cells?

8. Colon, breast, ovarian, and stomach cancers can be prevented by removing the affected organ. Why is this approach not possible for leukemia?

9. A 53-year-old man has his first colonoscopy. He has no family history of colon cancer. The procedure detects 18 polyps and two "suspicious" growths that do not look like polyps—they seem to blend into the surrounding tissue, rather than projecting like tiny stalks. The gastroenterologist removes the suspicious tissue, and a pathologist determines that the growths are cancerous. The man is very upset, blaming his consumption of a huge, undercooked steak on several occasions over the past month. Is he correct or incorrect that he caused the cancer with his eating habits?

10. Before my mother died of breast cancer in 2000, she suffered through several rounds of chemotherapy, which her physician prescribed because those were the drugs he used most often. When the disease spread to her bones and then her liver, it was clear that the drugs weren't working, but by then the disease had metastasized too widely for further treatment. Describe how a patient with metastatic breast cancer today might have a different treatment experience.

11. The rapid-aging disorder Hutchinson-Gilford progeria syndrome (see the opener to chapter 3) results from an abnormal protein that causes the nuclear membrane to protrude inward, where it contacts the tips of chromosomes, perhaps shortening them. From this information, would you expect cancer to be part of the syndrome? Why or why not?

Web Activities

1. The Cancer Genome Atlas Research Network (cancergenome.nih.gov) is sequencing thousands of genomes of a selection of cancer types.

   a. What are the criteria for a cancer to have its genome sequenced?

   b. How do you think the program should prioritize cancers for study?

   c. What can we learn from sequencing many samples from the same histological cancer type?

   d. What might be the value to a patient of knowing the genome sequence of his or her cancer?

   e. If cancer genome sequencing becomes a clinical tool, why might it be useful to repeat the sequencing every few years, rather than basing treatment decisions for many years based on an initial sequence?

2. Go to the Cancer Quest website (www.cancerquest.org). Select "Cancer Biology" from the drop-down menu and click on Cancer Genes. Select an oncogene or tumor suppressor gene and describe how, when mutant, the gene causes cancer.

3. Consult the websites for the pharmaceutical companies that market Herceptin, Gleevec, bevacizumab (Avastin), or any other cancer drug and explain how the drug works.

Case Studies and Research Results

1. Most cases of retinoblastoma are due to a germline mutation in the cancer susceptibility tumor suppressor gene *RB1*. Researchers discovered extra copies of an oncogene, called *MYCN,* in children who had a tumor in one eye, but no affected relatives, and who did not have mutations in *RB1. MYCN* sometimes becomes overexpressed in later stages of RB1, but in the newly recognized type of disease, called MYCN RB, the oncogene is the gatekeeper mutation. The mutation is somatic. That is, the *MYCN* mutation originated in an eye cell of the affected child.

   a. Which form of retinoblastoma, the one due to the oncogene or the tumor suppressor, is likely to recur in a sibling of the affected child? Explain your answer.

   b. For one form of RB, ophthalmologists typically remove the affected eye but they treat the other form less aggressively, with chemotherapy. For which type of RB does surgical removal of the affected eye make the most sense if the goal is to maintain some vision, and why?

2. In a large family, fifteen people over four generations develop the skin cancer melanoma. None of the affected relatives have mutations in known melanoma genes. Researchers sequenced the exomes of several affected family members and discovered a mutation in the *TERT* gene. Explain how the cancer likely arises.

3. Describing tumors by their mutations is challenging some classifications based on traditional histology (what cells look like and how they take up stains). The Clinical Lung Cancer Genome Project compared mutations in 1,255 lung tumors, and found that the same histologically defined lung cancers can have different mutations in different genes, and that some tumors classified as different by histology in fact harbor the same

mutations. What further information is needed in order to apply these observations to clinical practice?

4. When I had thyroid cancer in 1993, I had to wait on the operating table while slices of my two tumors were sent to the pathology lab. The growths looked so unusual that the surgeon could not tell whether they were of a type that tended to spread or a less dangerous type. So the pathologist examined the tissue microscopically to tell that one tumor was papillary, and the other follicular—both not dangerous. What is a newer technology that could have distinguished the tumor types at the time of biopsy weeks earlier?

5. DeShawn takes the drug Gleevec to treat his leukemia, and it has worked so well that he thinks he is cured. He stops taking the drug, and 4 months later his leukemia returns. This time, the cancer cells do not have the BCR-ABL mRNA characteristic of the disease. Explain what has happened.

6. The genomes of three patients with acute myeloid leukemia are sequenced and mutations in the following genes noted:

patient 1	*IDH1* and *NPM1*
patient 2	just *IDH1*
patient 3	*IDH1, NPM1,* and *IDH2*
patient 4	*IDH1, NPM1, IDH2,* and *FLT3*

   Explain how these patients can have the same diagnosis, yet mutations in different genes.

7. Elsie finds a small lump in her breast and goes to her physician, who takes a medical and family history. She mentions that her father died of brain cancer, a cousin had leukemia, and her older sister was just diagnosed with a tumor of connective tissue. The doctor assures her that the family cancer history doesn't raise the risk that her breast lump is cancerous, because the other cancers were not in the breast. Is the doctor correct? Why or why not?

Transgenic pigs given a bacterial digestive enzyme excrete genetically modified, less-polluting manure. The author stands on a pile of the nonmodified material at the University of Georgia.

Genetic Technologies: Patenting, Modifying, and Monitoring DNA

Learning Outcomes

19.1 Patenting DNA

1. State the criteria for a patentable invention.

2. Discuss the history of patenting organisms and DNA.

19.2 Modifying DNA

3. Distinguish between recombinant DNA and transgenic organisms.

4. Describe applications of recombinant DNA technology.

19.3 Monitoring Gene Function

5. Explain how a DNA microarray is used to monitor gene expression.

19.4 Gene Silencing and Genome Editing

6. Describe ways to decrease expression of a specific gene.

7. Define genome editing.

The **BIG** Picture

Ancient biotechnologies gave us bakeries and breweries, foods and medicines. Modern biotechnologies manipulate DNA to give us new ways to study, monitor, and treat disease, and alter foods and the environment.

Improving Pig Manure

Pig manure presents a serious environmental problem. The animals do not make an enzyme to extract the mineral nutrient phosphorus from a compound called phytate in cereal grains, so they are given dietary phosphorus supplements. As a result, their manure is full of phosphorus. The element washes into natural waters, contributing to fish kills, oxygen depletion in aquatic ecosystems, algal blooms, and even the greenhouse effect. But biotechnology may have solved the "pig poop" problem.

In the past, pig raisers have tried various approaches to keep their animals healthy and the environment clean, such as feeding animal by-products from which the pigs can extract more phosphorus and giving supplements of the enzyme phytase, which liberates phosphorus from phytate. But consuming animal by-products can introduce prion diseases, and giving phytase before each meal is costly. A "phytase transgenic pig," however, is genetically modified to secrete bacterial phytase in its saliva, which enables the animal to excrete low-phosphorus manure. It is called an Enviropig.

A transgenic organism has a gene in each of its cells from an organism of another species. The Enviropig has a phytase gene from the bacterium *E. coli,* as well as a promoter DNA sequence from a mouse that controls secretion of phytase from the salivary glands. Enviropig's manure has 75 percent less phosphorus than unaltered pig excrement.

19.1 Patenting DNA

DNA is the language of life, the instruction manual for keeping an organism alive. Yet we also use DNA. Manipulating DNA is part of **biotechnology,** which is the use or alteration of cells or biological molecules for specific applications, including products and processes. Biotechnology is an ancient art as well as a modern science, and is familiar as well as futuristic. Using yeast to ferment fruit into wine is a biotechnology, as is extracting and using biochemicals from organisms.

The terminology for biotechnology can be confusing. The popular terms "genetic engineering" and "genetic modification" refer broadly to any biotechnology that manipulates DNA. It includes altering the DNA of an organism to suppress or enhance the activities of its own genes, and combining the genetic material of different species. Organisms that harbor DNA from other species are termed **transgenic** and their DNA is called **recombinant DNA**. The Enviropig described in the chapter opener is transgenic.

Creating transgenic organisms is possible because all life uses the same genetic code—that is, the same DNA triplets encode the same amino acids (**figure 19.1**). Mixing DNA from different species may seem unnatural, but in fact DNA moves and mixes between species in nature—bacteria do it, and it is why we have viral DNA sequences in our chromosomes. But human-directed genetic modification usually gives organisms traits they would not have naturally, such as fish that can tolerate very cold water, tomatoes that grow in salt water, and bacteria that synthesize human insulin.

What Is Patentable?

Creating transgenic organisms raises legal questions, because the design of novel combinations of DNA may be considered intellectual property, and therefore patentable. To qualify for patent protection, a transgenic organism, as any other invention, must be new, useful, and not obvious to an expert in the field. A corn plant that manufactures a protein naturally found in green beans but not

in corn, thereby making the corn more nutritious, is an example of a patentable transgenic organism. A DNA sequence might be patentable if it is part of a medical device used to diagnose an inherited or infectious disease. DNA-based tests, for example, can identify specific mutations that cause cystic fibrosis. Another test can distinguish among 13 bacterial pathogens and even identify strains that are resistant to specific antibiotic drugs. DNA is also patentable as a research tool, as are algorithms used to extract information from DNA sequences and databases built of DNA sequences. The Technology Timeline highlights some of the events and controversies surrounding patenting of genetic material.

Patent law has had to evolve to keep up with modern biotechnology. In the 1980s, when sequencing a gene was painstakingly slow, only a few genes were patented. In the mid-1990s, with faster sequencing technology and shortcuts to finding the protein-encoding parts of the genome, the U.S. National Institutes of Health and biotech companies began seeking patent protection

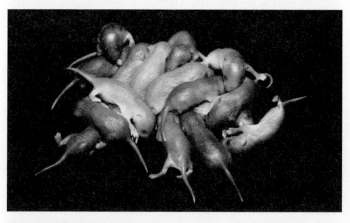

Figure 19.1 The universality of the genetic code makes biotechnology possible. The greenish mice contain the gene encoding a jellyfish's green fluorescent protein (GFP). Researchers use GFP to mark genes of interest. The GFP mice glow less greenly as they mature and more hair covers the skin. The non-green mice are not genetically modified.

Technology Timeline

PATENTING LIFE AND GENES

1790	U.S. patent act enacted. A patented invention must be new, useful, and not obvious.
1873	Louis Pasteur is awarded first patent on a life form, for yeast used in industrial processes.
1930	New plant variants can be patented.
1980	First patent awarded on a genetically modified organism, a bacterium given four DNA rings that enable it to metabolize components of crude oil.
1988	First patent awarded for a transgenic organism, a mouse that manufactures human protein in its milk. Harvard University granted patent for "OncoMouse" transgenic for human cancer.
1992	Biotechnology company awarded patent for all forms of transgenic cotton. Groups concerned that this will limit the rights of subsistence farmers contest the patent several times.
1996–1999	Companies patent partial gene sequences and certain disease-causing genes for developing specific medical tests.
2000	With gene and genome discoveries pouring into the Patent and Trademark Office, requirements tightened for showing utility of a DNA sequence.
2003	Attempts to enforce patents on non-protein-encoding parts of the human genome anger researchers who support open access to the information.
2007	Patent requirements must embrace new, more complex definition of a gene.
2009	Patents on breast cancer genes challenged.
2010	Direct-to-consumer genetic testing companies struggle to license DNA patents for multigene and SNP association tests. Patents on breast cancer genes invalidated.
2011	U.S. government considers changes to gene patent laws.
2013	U.S. Supreme Court declares genes unpatentable.

for thousands of short DNA sequences, even if their functions weren't known. Because of the flood of applications, the U.S. Patent and Trademark Office tightened requirements for usefulness. Today, with entire genomes being sequenced much faster than it once took to decipher a single gene, a DNA sequence alone does not warrant patent protection. It must be useful as a tool for research or as a novel or improved product, such as a diagnostic test or a drug. In the United States, more than one in five human genes is patented in some way, yet only a few gene patents have been challenged.

DNA patenting became very controversial in 2009, when several groups, including the American Civil Liberties Union and the Public Patent Foundation, challenged patents on two breast cancer genes (BRCA1 and BRCA2) held by biotechnology company Myriad Genetics and the University of Utah. The company did not license the full gene sequences to other companies, so patients were forced to take Myriad's test for increased familial breast cancer risk, costing more than $3,000 to sequence the entire genes. The patents discouraged research and prevented patients from getting second opinions. In 2010, a federal judge in the United States ruled seven patents on the genes "improperly granted" because they are based on a "law of nature." In 2011, the court invalidated the patents on the two genes, but a federal appeals court overruled that action, claiming that an isolated gene is not the same as a gene in a cell, which is part of a chromosome.

In 2013 the U.S. Supreme Court ruled that "a naturally occurring DNA segment is a product of nature and not patent eligible merely because it has been isolated." The court did allow patenting of complementary DNA (cDNA), which is synthesized in a laboratory using an enzyme (reverse transcriptase) that makes a DNA molecule that is complementary in sequence to a specific mRNA. The cDNA represents only the exons of a gene because the introns are spliced out during transcription into mRNA. A cDNA is considered not to be a product of nature because its exact sequence is not in the genome of an organism.

Key Concepts Questions 19.1

1. How do modern applications of biotechnology differ from ancient applications?

2. What are the requirements for a patented invention?

3. Why did the U.S. Supreme Court rule that a gene's DNA sequence is not patentable?

19.2 Modifying DNA

Recombinant DNA technology adds genes from one type of organism to the genome of another. It was the first gene modification biotechnology, and was initially done in bacteria to produce peptides and proteins useful as drugs. When bacteria bearing recombinant DNA divide, they yield many copies of the "foreign" DNA, and under proper conditions they produce many copies of the protein that the foreign DNA specifies. Recombinant DNA technology is also known as gene cloning. "Cloning" in this context refers to making many copies of a specific DNA sequence.

Recombinant DNA

In February 1975, molecular biologists convened at Asilomar, on California's Monterey Peninsula, to discuss the safety and implications of a new type of experiment: combining genes of two species. Would experiments that deliver a cancer-causing virus be safe? The researchers discussed restricting the types of organisms and viruses used in recombinant DNA research and brainstormed ways to prevent escape of a resulting organism from the laboratory. The guidelines drawn up at Asilomar outlined measures of "physical containment," such as using specialized hoods and airflow systems, and "biological containment," such as weakening organisms so that they could not survive outside the laboratory.

Recombinant DNA technology turned out to be safer than expected, and it spread to industry faster and in more diverse ways than anyone had imagined. However, recombinant DNA-based products have been slow to reach the marketplace because of the high cost of the research and the long time it takes to develop a new drug. Today, several dozen such drugs are available, and more are in the pipeline. Recombinant DNA research initially focused on providing direct gene products such as peptides and proteins. These included insulin, growth hormone, and clotting factors. However, the technology can target carbohydrates and lipids by affecting the genes that encode enzymes required to synthesize them.

Constructing and Selecting Recombinant DNA Molecules

Manufacturing recombinant DNA molecules requires **restriction enzymes** that cut donor and recipient DNA at the same sequence; DNA to carry the donor DNA (called **cloning vectors);** and recipient cells (bacteria or other cultured single cells). After inserting donor DNA into vectors, cells are selected to receive the gene of interest, which they use to manufacture the desired protein.

Restriction enzymes are naturally found in bacteria, where they cut DNA of infecting viruses, protecting the bacteria. Methyl (CH_3) groups shield the bacterium's own DNA from its restriction enzymes. Bacteria have hundreds of types of restriction enzymes. Some of them cut DNA at particular sequences of four, five, or six bases that are symmetrical in a specific way—the recognized sequence reads the same, from the 5′ to 3′ direction, on both strands of the DNA. For example, the restriction enzyme EcoR1, whose actions are depicted in **figure 19.2**, cuts at the sequence GAATTC. The complementary sequence on the other strand is CTTAAG, which, read backwards, is GAATTC. (You can try this with other sequences to see that it rarely works this way.) In the English language, this type of symmetry is called a palindrome, referring to a sequence of letters that reads the same in both directions, such as "Madam, I'm Adam." Unlike the language comparison, however, palindromic sequences in DNA are on complementary strands.

The cutting action of some restriction enzymes on double-stranded DNA creates single-stranded extensions. They are called "sticky ends" because they are complementary to each other, forming hydrogen bonds as their bases pair. Restriction enzymes

Figure 19.2

Restriction enzyme recognition sequence

G A A T T C
C T T A A G

G A A T T C
C T T A A G

a.

1 Restriction enzymes cut DNA at specific sequences.

Sticky end

G
C T T A A

A A T T C
G

G
C T T A A

A A T T C
G

b.

Sticky end

Donor DNA

A A T T C
G

G
C T T A A

2

DNAs from two sources cut with the same restriction enzyme have "sticky" ends. Two pieces of DNA hydrogen bond, and ligase seals the sugar-phosphate backbone.

G A A T T C
C T T A A G

G A A T T C
C T T A A G

Host DNA   Donor DNA   Host DNA

c.

3 Recombinant DNA molecule

Figure 19.2 **Recombining DNA uses restriction enzymes to insert a foreign DNA sequence.** These enzymes can be used as molecular scissors because they cut DNA from any source at the same base sequence.

work as molecular scissors in creating recombinant DNA molecules because they cut at the same sequence in any DNA source. That is, the same sticky ends result from the same restriction enzyme, whether the DNA is from a mockingbird or a maple.

Another natural "tool" used in recombinant DNA technology is a cloning vector. This structure carries DNA from the cells of one species into the cells of another. A vector can be any piece of DNA into which other DNA can insert. A commonly used vector is a **plasmid,** which is a small circle of double-stranded DNA that exists naturally in some bacteria, yeasts, plant cells, and cells of other types of organisms. Viruses that infect bacteria, called bacteriophages, are another type of vector, manipulated to transport DNA but not cause disease. Disabled retroviruses are used as vectors too, as are DNA sequences from bacteria and yeast called artificial chromosomes.

Choice of cloning vector must consider size of the gene, which must be short enough to insert into the vector. Gene size is typically measured in kilobases (kb), which are thousands of bases. Cloning vectors can hold up to about 2 million DNA bases.

To create a recombinant DNA molecule, a restriction enzyme cuts DNA from a donor cell at sequences that bracket the gene of interest (**figure 19.3**). The enzyme leaves single-stranded ends on the cut DNA, each bearing a characteristic base sequence. Next, a plasmid is isolated and cut with the same restriction enzyme used to cut the donor DNA. Because the same restriction enzyme cuts both the donor DNA and the plasmid DNA, the same complementary single-stranded base sequences extend from the cut ends of each. When the cut plasmid and the donor DNA are mixed, the single-stranded sticky ends of some plasmids base pair with the sticky ends of the donor DNA. The result is

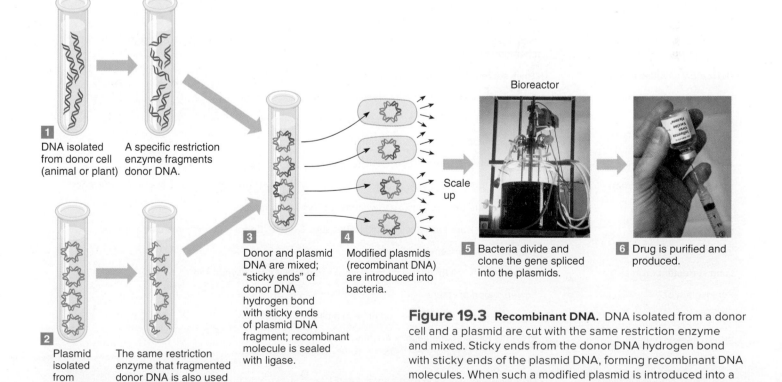

1 DNA isolated from donor cell (animal or plant)   A specific restriction enzyme fragments donor DNA.

2 Plasmid isolated from bacterium   The same restriction enzyme that fragmented donor DNA is also used to open plasmid DNA.

3 Donor and plasmid DNA are mixed; "sticky ends" of donor DNA hydrogen bond with sticky ends of plasmid DNA fragment; recombinant molecule is sealed with ligase.

4 Modified plasmids (recombinant DNA) are introduced into bacteria.

Bioreactor

Scale up

5 Bacteria divide and clone the gene spliced into the plasmids.

6 Drug is purified and produced.

Figure 19.3 **Recombinant DNA.** DNA isolated from a donor cell and a plasmid are cut with the same restriction enzyme and mixed. Sticky ends from the donor DNA hydrogen bond with sticky ends of the plasmid DNA, forming recombinant DNA molecules. When such a modified plasmid is introduced into a bacterium, it is mass produced as the bacterium divides.

a recombinant DNA molecule, such as a plasmid carrying the human insulin gene. The plasmid and its human gene can now be transferred into a cell, such as a bacterium or a white blood cell.

Selecting Recombinant DNA Molecules

Much of the effort in recombinant DNA technology is in identifying and separating cells that contain the gene of interest. This entails distinguishing cells bearing the gene from cells that lack plasmids or that have taken up "empty" plasmids that do not contain the gene. Researchers have clever ways of separating the useful cells. One separation strategy uses plasmids that have an antibiotic resistance gene as well as a gene that encodes an enzyme that catalyzes a reaction that produces a blue color. When the antibiotic is applied, only cells that have plasmids survive. If a human gene inserts and interrupts the gene for the enzyme, the bacterial colony that grows is not blue, and is therefore easily distinguished from the blue bacterial cells that have not taken up the human gene.

When cells containing the recombinant plasmid divide, so does the plasmid. Within hours, the original cell gives rise to many cells harboring the recombinant plasmid. The enzymes, ribosomes, energy molecules, and factors necessary for protein synthesis transcribe and translate the plasmid DNA and its foreign gene, producing the desired protein. Then the protein is separated, collected, purified, and packaged to create a product, such as a new drug.

Products from Recombinant DNA Technology

In basic research, recombinant DNA technology provides a way to isolate genes from complex organisms and observe their functions on the molecular level. Recombinant DNA has many practical uses, too. The first was to mass-produce protein-based drugs.

Drugs manufactured using recombinant DNA technology are pure, and are the human version of the protein. Before recombinant DNA technology was invented, human growth hormone came from cadavers, follicle-stimulating hormone came from the urine of postmenopausal women, and clotting factors were pooled from hundreds or thousands of donors. These sources introduced great risk of infection, especially after HIV and hepatitis C became more widespread.

The first drug manufactured using recombinant DNA technology was insulin, which is produced in bacterial cells *(E. coli)*. Before 1982, people with type 1 diabetes mellitus obtained the insulin that they injected daily from pancreases removed from cattle in slaughterhouses. Cattle insulin is so similar to the human peptide, differing in only two of its fifty-one amino acids, that most people with diabetes could use it. However, about one in twenty patients is allergic to cow insulin because of the slight chemical difference. Until recombinant DNA technology was developed, the allergic patients had to use expensive combinations of insulin from other animals or human cadavers. **Table 19.1** lists some drugs produced using recombinant DNA technology.

Table 19.1	Drugs Produced Using Recombinant DNA Technology
Drug	**Use**
Atrial natriuretic peptide	Dilates blood vessels, promotes urination
Colony-stimulating factors	Help restore bone marrow after marrow transplant; restore blood cells following cancer chemotherapy
Deoxyribonuclease (DNase)	Thins secretions in lungs of people with cystic fibrosis
Epidermal growth factor	Accelerates healing of wounds and burns; treats gastric ulcers
Erythropoietin (EPO)	Stimulates production of red blood cells in cancer patients
Factor VIII	Promotes blood clotting in treatment of hemophilia
Glucocerebrosidase	Corrects enzyme deficiency in Gaucher disease
Human growth hormone	Promotes growth of muscle and bone in people with very short stature due to hormone deficiency
Insulin	Allows cells to take up glucose in treatment of type 1 diabetes
Interferons	Treat genital warts, hairy cell leukemia, hepatitis C and B, Kaposi sarcoma, multiple sclerosis
Interleukin-2	Treats kidney cancer recurrence
Lung surfactant protein	Helps lung alveoli to inflate in infants with respiratory distress syndrome
Renin inhibitor	Lowers blood pressure
Somatostatin	Decreases growth in muscle and bone in pituitary giants
Superoxide dismutase	Prevents further damage to heart muscle after heart attack
Thrombin	Stops postsurgical bleeding
Tissue plasminogen activator	Dissolves blood clots in treatment of heart attack, stroke, and pulmonary embolism

Insulin is a simple peptide and is therefore straightforward to mass-produce in bacteria. Some drugs, however, require that sugars be attached, or must fold in specific, intricate ways to function. These molecules must be produced in eukaryotic cells that readily carry out these modifications. Yeast cells, Chinese hamster ovary cells, insect cells, and even carrot and tobacco cells have been used to produce human proteins.

Drugs developed using recombinant DNA technology must compete with conventional products. Deciding whether a recombinant drug is preferable to an existing similar drug is often a matter of economics. For example, interferon β-1b treats a type of multiple sclerosis, but this recombinant drug costs close to $70,000 per year. British researchers calculated that more people would be served if funds were spent on improved supportive care for many rather than on this costly treatment for a few.

Tissue plasminogen activator (tPA), a recombinant clot-busting drug, also has cheaper alternatives. If injected within 4 hours of a heart attack, tPA dramatically limits damage to the heart muscle by restoring blood flow. It costs $2,000 a shot. An older drug, streptokinase, is extracted from unaltered bacteria and is nearly as effective, at $300 per injection. Patients who have already received streptokinase and could have an allergic reaction if they were to use it again can benefit from tPA. *Bioethics: Choices for the Future* considers another drug derived from recombinant DNA technology: erythropoietin (EPO).

Safer vaccines are created using recombinant DNA technology. Recall that a vaccine is a live, killed, or part of a

---

EPO: Built-in Blood Cell Booster or Performance-Enhancing Drug?

"Cycle, run, and swim longer and faster than anyone else!" proclaims a website selling a product that supposedly boosts levels of erythropoietin. "EPO" is a glycoprotein hormone that the kidneys produce in response to low levels of oxygen in the blood. The hormone travels to the bone marrow and binds receptors on cells that give rise to red blood cell progenitors. Soon, more red blood cells enter the circulation, carrying more oxygen to the tissues (**figure 1**). Enhanced stamina results.

The value of EPO as a drug became evident after the invention of hemodialysis to treat kidney failure in 1961. Dialysis removes EPO from the blood, causing severe anemia. But boosting EPO levels proved difficult because levels in human plasma are too low to pool from donors. Instead, in the 1970s, the U.S. government obtained EPO from South American farmers with hookworm infections and Japanese aplastic anemia patients, who secrete abundant EPO into urine. But when the AIDS epidemic came, biochemicals from human body fluids were no longer safe to use. Recombinant DNA technology solved the EPO problem. It is sold under various names to treat anemia in dialysis and AIDS patients and is given with cancer chemotherapy to avoid the need for transfusions.

Figure 1 At least two genes control EPO secretion. Certain variants of these genes increase the number of red blood cells, increasing endurance but also raising risk of heart attack and stroke.

EPO's ability to increase the oxygen-carrying capacity of blood under low oxygen conditions is why athletes train at high altitudes to increase endurance. Since the early 1990s, athletes have abused EPO to reproduce this effect, at great risk. EPO thickens the blood, raising the risk of a blockage that can cause a heart attack or stroke, especially when intense, grueling exercise removes water from the bloodstream. Excess EPO caused sudden death during sleep for at least eighteen cyclists. Olympic athletes now take urine tests that detect recombinant EPO, which has a slightly different configuration of sugars than the form an athlete's kidneys naturally produce. Famed cyclists Floyd Landis and Lance Armstrong admitted to EPO abuse, after years of denial.

People with familial erythrocytosis get the effects of extra EPO naturally. Type 1 (OMIM 133100) is autosomal dominant, and is caused by mutation in the EPO receptor. Affected individuals have large and abundant red blood cells, but low blood serum levels of EPO. A member of a family from Finland with this condition won several Olympic medals for skiing thanks to his inborn ability. An autosomal recessive form of the condition, type 2 (OMIM 263400), increases the level of EPO in the bloodstream. Both forms of erythrocytosis usually have no symptoms, but increase the risk of circulation blocked by the sluggish, oxygen-laden blood.

Questions for Discussion

1. Was it ethical in the 1970s to obtain EPO from sick, poor people in South America and Japan to treat people in the United States?

2. Should using a substance made naturally in the body be considered performance enhancement?

3. Should tests be developed to identify athletes whose genes, anatomy, or physiology give them a competitive advantage? Why or why not?

4. When developing drugs that use recombinant DNA technology, should researchers consider how the product could be abused?

pathogen that stimulates an immune response that protects a person who encounters the actual pathogen. Traditional influenza vaccine, for example, is cultured in eggs, to which some people are allergic. The process is also very time-consuming. An alternative flu vaccine became available in 2013 that consists of the genes that encode the hemagglutinin proteins from two influenza A strains and one influenza B strain (see Clinical Connection 17.1), and is effective against H1N1 and H3N2 infection. The influenza virus genes are introduced aboard a virus called a baculovirus that readily infects the cells of certain insects—the fall armyworm in this case. The new "eggless vaccine" not only avoids the use of eggs, but also does not use antibiotics, preservatives, or live flu viruses, and can be manufactured faster than conventional flu vaccine.

A new vaccine that protects against malaria is based on altering a bacterium (*Pantoea agglomerans*) that normally inhabits the mosquito gut. Mosquitoes transmit *Plasmodium falciparum*, the parasite that causes malaria. The bacteria are given genes from *E. coli* bacteria that enable them to produce proteins that tear apart the insect's intestines. Using recombinant bacteria is easier than genetically modifying mosquitoes to prevent malaria.

Transgenic Organisms

Eukaryotic cells growing in culture are generally better at producing human proteins than are prokaryotic cells such as bacteria. An even more efficient way to express some recombinant genes is in a body fluid of a transgenic animal, such as the saliva of the Enviropig. The fact that the cells secreting the human protein are part of an animal more closely mimics the environment in the human body. The genetic change must be introduced into a fertilized ovum so that it is present in every cell of the transgenic organism.

Transgenic sheep, cows, and goats have all expressed human genes in their milk, including genes that encode clotting factors, clot busters, and the connective tissue protein collagen. Production of human antibodies in rabbit and cow milk illustrates the potential value of transgenic animals. These antibodies are used to treat cancers. Recall from figure 17.8 that antibodies are assembled from the products of several genes. Researchers attach the appropriate human antibody genes to promoters for milk proteins. (Promoters are the short sequences at the starts of genes that control transcription rates.) These promoters normally oversee production of abundant milk proteins. The mammary gland cells of transgenic animals can assemble antibody parts to secrete the final molecules—just as if they were being produced in a plasma cell in the human immune system.

Several techniques are used to insert DNA into animal cells to create transgenic animals. Chemicals and brief jolts of electricity open transient holes in plasma membranes that admit "naked" DNA, or a gunlike device is used to shoot tiny metal particles coated with DNA inside cells. DNA may also cross the plasma membrane in tiny fatty bubbles called liposomes.

Getting foreign DNA into a fertilized ovum is the first step in creating a transgenic organism. It is quite a technical challenge, especially for nonanimal cells, which have cell walls in addition to plasma membranes. The recombinant DNA must enter the nucleus, replicate with the cell's own DNA, and be transmitted when the cell divides. Finally, for an animal, an organism must be regenerated from the fertilized ovum, which means gestation in a surrogate mother. If the trait is dominant, the transgenic organism must express it in the appropriate tissues at the right time in development to be useful. If the trait is recessive, crosses between heterozygotes may be necessary to yield homozygotes that express the trait. Then the organisms must pass the characteristic on to the next generation.

Animal Models

Herds of transgenic farm animals supplying drugs in their milk have not become important sources of pharmaceuticals—they are too difficult to maintain. Transgenic animals are more useful as models of human disease (**figure 19.4**). Inserting the mutant human beta globin gene that causes sickle cell disease into mice, for example, results in a mouse model of the disorder. Drug candidates can be tested on these animal models and abandoned if they cause significant side effects before testing in humans.

Transgenic animal models, however, have limitations. Researchers cannot control where a transgene inserts in a genome, and how many copies do so. The level of gene expression necessary for a phenotype to emerge may also differ in the model and humans. This was the case for a mouse model of familial Alzheimer disease (OMIM 104760). The transgene has the exact same DNA sequence that disrupts amyloid precursor protein in a Swedish family with the condition, but apparently did nothing to the mice—until researchers increased transcription rate tenfold. Only then did the telltale plaques and tangles, and neuron cell death, appear in the mouse brains.

Animal models might not mimic the human condition exactly because of differences in their rates of development. For example, for some inherited diseases that do not cause symptoms until adulthood in humans, mice simply do not live long enough to evaluate the phenotype. A transgenic monkey, for example, is a more accurate model of Huntington disease than a mouse, because as a primate the monkey is much more similar to a human in life span, metabolism, reproduction, behavior, and cognition.

Figure 19.4 Animal models mimic human disease.
Transgenic mice that have a mutation that causes Huntington disease in humans are tested for coordination on a rotating drum with a grooved surface.

Genetically Modified Foods

Traditional agriculture is the controlled breeding of plants and animals to select individuals with certain combinations of inherited traits that are useful to us, such as seedless fruits and lean meat. It is a form of genetic modification based on phenotype, such as taste or appearance, and is therefore both subjective and imprecise, affecting many genes. In contrast, DNA-based techniques manipulate one or a few specific genes at a time. Organisms altered to have genes from other species or to over- or underexpress their own genes are termed "genetically modified" organisms or GMOs. An organism given genes from another species is transgenic.

Golden rice is a well known GM crop that uses genes from corn and bacteria to produce twenty-three times as much beta carotene (a vitamin A precursor) as unaltered rice. It contains no allergens or toxins. Developed by the not-for-profit International Rice Research Institute in the Philippines, the rice was meant to improve human nutrition in the many nations where rice is a dietary staple. In Africa and Asia, 2 million people die each year because lack of vitamin A impairs their immunity to infectious diseases.

Some nations outlaw GM foods, but people in the United States have been eating them for years, apparently safely. A government report found that 70% of processed foods in supermarkets contain at least one GM ingredient. However, some people object to GM foods, for a variety of reasons. Officials in France and Austria have called such crops "not natural," "corrupt," and "heretical." An enraged consumer declared on a TV news program, "I will not eat food that contains DNA!" **Figure 19.5** is an artist's depiction of fears of GMOs.

Other reasons for objecting to GM foods are more practical. Labeling a food that contains a nutrient not normally in it—like a protein from peanuts in corn—could prevent allergic reactions. Many GM crops are designed to resist certain herbicides, encouraging reliance on use of that herbicide. Bioethical concerns are raised when the same company that creates the GM crop also manufactures and markets the herbicide to which it is resistant, forcing farmers to buy the products.

An ecological-based objection to GM plants is that field tests may not adequately predict the effects on ecosystems. Buffer zones of non-GM plants are routinely planted around fields of GM varieties, but wind pollination can take GM plants far. A sampling of 400 canola plants growing along roads in North Dakota found that 86 percent of them had a gene indicating that they descended from GM plants.

Another science-based objection to GM crops is that overreliance on them may lead to genetic uniformity, which is just as dangerous as traditional monoculture. Genetic sameness creates vulnerability in a population should environmental conditions change, and no variants are there to survive natural selection. Some GM organisms, such as fish that grow to twice normal size or can survive at temperature extremes, can disrupt natural ecosystems. **Table 19.2** lists some GM organisms.

Bioremediation

Recombinant DNA technology and transgenic organisms provide processes as well as products. In **bioremediation,** bacteria or plants with the ability to detoxify certain pollutants are released or grown in a particular area. Natural selection has sculpted such organisms, perhaps as adaptations that render them unpalatable to predators. Bioremediation uses genes that enable an organism to metabolize a substance that, to another species, is a toxin. The technology uses unaltered organisms, and also transfers "detox" genes to other species so that the protein products can more easily penetrate a polluted area.

Nature offers many organisms with interesting tastes. A type of tree that grows in a tropical rainforest on an island near Australia, for example, accumulates so much nickel from soil that slashing its bark releases a bright green latex ooze. Genes from this tree can be used to clean up nickel-contaminated soil.

Figure 19.5 An artist's view of biotechnology. Artist Alexis Rockman vividly captures fears of biotechnology, including a pig used to incubate spare parts for sick humans, a muscle-boosted boxy cow, a featherless chicken with extra wings, a mini-warthog, and a mouse with a human ear growing out of its back (this last one really happened).

Table 19.2	Some Genetically Modified Organisms
Altered Trait	**Organism**
Less sugar	Grapes
Resist viral infection	Cassava, papaya, plum
Resist mad cow disease	Cattle
Tolerate an herbicide	Sugar beets, corn, soybeans
More iron and vitamin A	Bananas, rice
Altered fatty acid composition	Canola
Faster growth	Salmon

Bioremediation can tap the metabolisms of transgenic microorganisms, sending them into plants whose roots then distribute the detox proteins in the soil. For example, transgenic yellow poplar trees can thrive in mercury-tainted soil if they have a bacterial gene that encodes an enzyme that converts a highly toxic form of mercury in soil to a less toxic gas. The tree's leaves then release the gas.

Cleaning up munitions dumps from wars is another use of bioremediation, such as deploying bacteria that normally break down trinitrotoluene, or TNT, the major ingredient in dynamite and land mines. The enzyme that provides this ability is linked to the *GFP* gene (see figure 19.1). Bacteria spread in a contaminated area glow near land mines, revealing the locations more clearly than a metal detector could. Once the land mines are removed, the bacteria die as their food vanishes.

Key Concepts Questions 19.2

1. Explain the steps of recombinant DNA technology.
2. Discuss applications of recombinant DNA technology.
3. What are some examples of transgenic plants and animals?

19.3 Monitoring Gene Function

We usually cannot do very much about the gene variants that we inherit. Gene expression, in contrast, is where we can make a difference by controlling our environment. Monitoring gene expression requires detecting the mRNAs in particular cells under particular conditions. To do this, devices called gene expression DNA microarrays, or gene chips, detect and display the mRNAs in a cell. The creativity of the technique lies in choosing the types of cells to interrogate.

Evaluating a spinal cord injury illustrates the basic steps in creating a DNA microarray to assess gene expression. Researchers knew that in the hours after such a devastating injury, immune system cells and inflammatory biochemicals flood the affected area, but it took **gene expression profiling** to reveal just how fast healing begins.

A **microarray** is a piece of glass or plastic that is about 1.5 centimeters square—smaller than a postage stamp. Many small pieces of DNA (oligonucleotides) of known sequence are attached to one surface, in a grid pattern. The researcher records the position of each DNA piece in the grid. In many applications, a sample from an abnormal situation (such as disease, injury, or environmental exposure) is compared to a normal control. **Figure 19.6** compares cerebrospinal fluid (CSF; the liquid that bathes the spinal cord) from an injured person (sample A) to fluid from a healthy person (sample B). Messenger RNAs are extracted from the samples and complementary DNAs (cDNAs) are made. Researchers make cDNAs from mRNA using an enzyme from a retrovirus, reverse transcriptase. A cDNA includes codons for a mature mRNA, but not sequences for promoters and introns, so it represents the exons of a gene.

The cDNAs from the injury sample are labeled with a red fluorescent dye, and the cDNAs from the control sample are labeled with a green fluorescent dye. These labeled DNAs are then applied to the microarray, which displays thousands of genes likely to be involved in a spinal cord injury, or even the entire human exome. Considering so many DNA sequences allows for surprises, avoiding the assumption that we know what to look for.

DNA that binds to complementary sequences on the grid fluoresce in place. A laser scanner then detects and converts the results to a colored image. Each position on the microarray can bind DNA pieces from both samples (injured and healthy), either, or neither. The scanner also detects fluorescence intensities, which provides information on how strongly the gene is expressed (how much mRNA is in the sample).

A computer algorithm interprets the pattern of gene expression. For the spinal cord example, the visual data mean the following:

- Red indicates a gene expressed in CSF only when the spinal cord is injured (and presumably leaking inflammatory molecules).
- Green indicates a gene expressed in CSF only when the spinal cord is intact.
- Yellow indicates positions where both red- and green-bound dyes fluoresce, representing genes that are expressed whether or not the spinal cord has been injured.
- Black, or a lack of fluorescence, corresponds to DNA sequences that are not expressed in CSF, because they do not show up in either sample.

The color and intensity pattern of the microarray provides a glimpse of gene expression following spinal cord injury. The technique is even more powerful when it is repeated at different times after injury. When researchers did exactly that on injured rats, they discovered genes expressed just after the injury whose participation they never suspected. Their microarrays, summarized in **table 19.3**, revealed waves of expression of genes involved in healing. Analysis on the first day indicated activation of the same suite of genes whose protein products heal injury to the deep layer of skin—a total surprise that suggested new points for drugs to intervene.

Table 19.3	Gene Expression Profiling Chronicles Repair After Spinal Cord Injury
Time After Injury (rats)	**Type of Increased Gene Expression**
Day 1	Protective genes to preserve remaining tissue
Day 3	Growth, repair, cell division
Day 10	Repair of connective tissues
	Angiogenesis
Days 30–90	Blood vessels mature
	New type of connective tissue associated with healing

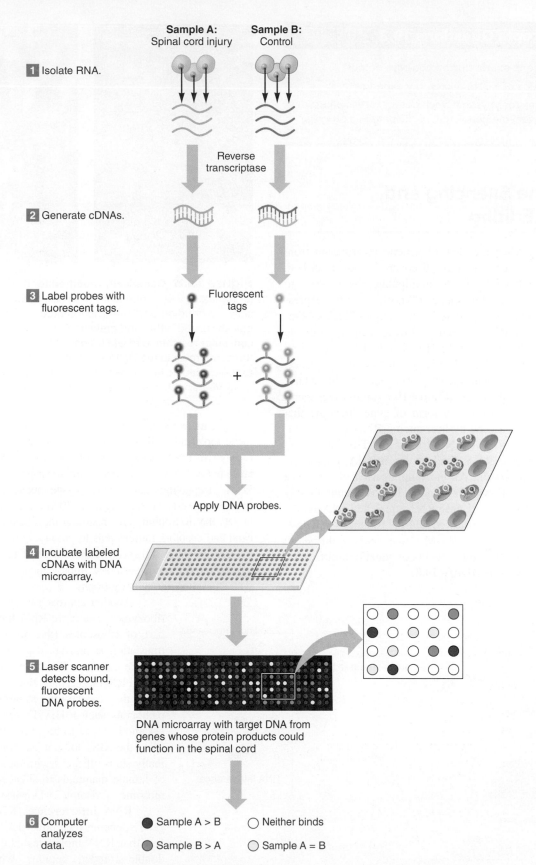

Sample A:
Spinal cord injury

Sample B:
Control

1 Isolate RNA.

Reverse
transcriptase

2 Generate cDNAs.

3 Label probes with
fluorescent tags.

Fluorescent
tags

+

Apply DNA probes.

4 Incubate labeled
cDNAs with DNA
microarray.

5 Laser scanner
detects bound,
fluorescent
DNA probes.

DNA microarray with target DNA from
genes whose protein products could
function in the spinal cord

6 Computer
analyzes
data.

● Sample A > B   ○ Neither binds

● Sample B > A   ○ Sample A = B

Figure 19.6 **A DNA microarray experiment reveals gene expression in response to spinal cord injury.** In this example, the red label represents DNA from a patient with a spinal cord injury, and the green label represents control DNA from a healthy person. DNA targets on the microarray that bind red but not green can reveal new points of intervention for drugs.

Figure 19.8 **Genetically modified foods.**
(a) Tomatoes were genetically modified to ripen more slowly, extending shelf life. The tomatoes softened too quickly, tasted bitter, and grew unpopular with rising consumer resistance to GM foods. (b) More than 85% of the corn grown in the United States is genetically modified to resist specific herbicides. We have been eating it for many years, with no apparent ill effects.

19.4 Gene Silencing and Genome Editing

Since the 1960s, when the flow of genetic information from DNA to RNA to protein was discovered and described, researchers have been trying to manipulate the process to diminish ("knock down") or silence ("knock out") the expression of specific genes. **Gene silencing** techniques block synthesis of, or degrade, mRNA. **Genome editing** techniques create double-stranded breaks in the DNA double helix, enabling insertion of a desired DNA sequence or removal of a sequence. Different approaches to gene silencing and genome editing have had varying degrees of success. **Figure 19.7** summarizes them.

Antisense technology is a form of gene silencing that blocks expression of a gene by introducing RNA that is complementary to the gene's mRNA transcript. The introduced RNA, called antisense RNA, binds to the mRNA, preventing its translation into protein. An early application of antisense technology was the FlavrSavr tomato. It was meant to stay fresh longer because the antisense RNA squelched activity of a ripening enzyme. Ripening slowed, but the tomatoes still softened. Nearly 2 million cans of GM tomato paste sold before production halted in 1997. Even results of genetic modification aren't entirely predictable (**figure 19.8**).

A more recent variation of antisense technology uses synthetic molecules called morpholinos that consist of sequences of twenty-five DNA bases attached to organic groups that are similar to but not exactly the same as the sugar-phosphate backbone of DNA. Morpholinos can block splice-site mutations that would otherwise delete entire exons. Figure 12.10 illustrates a morpholino that targets the dystrophin gene, restoring the function of the skipped exon and enabling muscle cells to produce some dystrophin. The morpholino-based drug is being tested to see if it can restore enough dystrophin to provide sustained improvement in mobility in boys with Duchenne muscular dystrophy.

Another approach to gene silencing uses ribozymes. These are RNA molecules that are part of ribosomes (the organelles on which translation occurs), and they have catalytic activity, like enzymes. Ribozymes fit the shapes of certain RNA molecules. Because ribozymes cut RNA, they can be used to destroy RNAs from pathogens, such as HIV. Figure 13.14 illustrates yet another way to turn off gene expression—using the XIST long, noncoding RNA that normally shuts off one X chromosome in the cells of female mammals to silence the extra chromosome of trisomy 21 Down syndrome.

RNA interference (**RNAi**) is another gene silencing technique that is based on the fact that RNA molecules can fold into short, double-stranded regions where the base sequence is complementary. **Figure 19.9** shows how this bonding produces a hairpin shape. Short, double-stranded RNAs introduced into cells can have great effects.

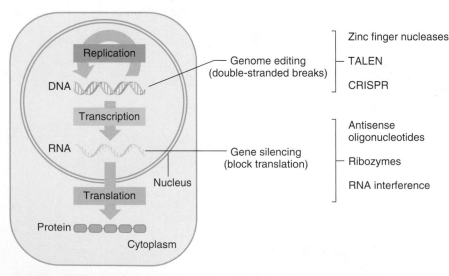

Figure 19.7 **Gene silencing and genome editing.** Gene silencing techniques prevent mRNA from being translated into protein. Genome editing techniques enable researchers to remove double-stranded segments of DNA, so that they can induce deletions or swap in genes.

GACAUUCGCAGCUAUAGCUGCGAAUAGU →

Figure 19.9 Hairpins. RNA hydrogen bonds with itself, forming hairpin loops. RNAi uses similar molecules to "knock down" expression of specific genes by binding their mRNAs.

The Nobel Prize in Physiology or Medicine was awarded in 2006 to Andrew Fire and Craig Mello for explaining how RNAi works. They discovered that short, double-stranded RNAs sent into cells separate into single strands, one of which binds its complement in mRNA, preventing it from being translated. The small RNAs that carry out RNA interference are called small interfering RNAs (siRNAs).

Several proteins and protein complexes orchestrate RNAi, and are also part of microRNA function (**figure 19.10** and see section 11.2). First, an enzyme called Dicer cuts long, double-stranded RNAs into pieces twenty-one to twenty-four nucleotides long. These pieces contact a group of proteins that form an RNA-induced silencing complex, or RISC. One strand of the double-stranded short RNA, called the guide strand, adheres. Now, as part of RISC, the guide strand finds its complementary RNA and binds. Then another part of RISC, a protein called argonaute, degrades the targeted RNA,

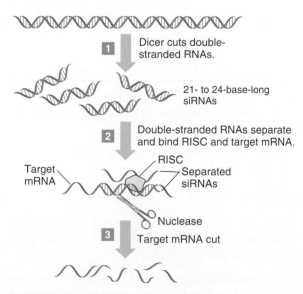

1 Dicer cuts double-stranded RNAs.

21- to 24-base-long siRNAs

2 Double-stranded RNAs separate and bind RISC and target mRNA.

Target mRNA

RISC

Separated siRNAs

Nuclease

3 Target mRNA cut

Figure 19.10 RNA interference. Dicer cuts double-stranded parts of RNA molecules, which then associate with RNA-induced silencing complexes (RISCs). The RNA is open, revealing single strands that locate and bind specific mRNAs. Nucleases then break down the targeted mRNAs, preventing their translation into protein.

preventing its translation into protein. SiRNAs act in the nucleus too, where they alter methylation and the binding of histones to certain genes.

Shortly after RNA interference was discovered in 1998, biotechnology and pharmaceutical companies began developing the approach to silence genes. Potential applications ranged from creating vaccines by knocking down expression of key genes in viruses that cause disease, such as AIDS, polio, and hepatitis C; to treating cancer by silencing oncogenes; to creating a better-tasting decaf coffee by silencing an enzyme required for caffeine synthesis in coffee plants. However, problems arose in clinical trials of RNAi-based drugs. In the human body, the synthetic "small interfering RNAs" that carry out RNAi can inflame the liver, rather than reach their intended targets. For now, RNAi may be better suited as a research tool to see what happens when gene expression is turned off in animal models of human disease and in human cells growing in culture.

More recently developed than RNAi and other gene silencing technologies are genome editing approaches. These harness DNA-cutting enzymes that generate double-stranded breaks, so that genes can be swapped in or out. They are research tools for now.

Zinc finger nuclease technology uses protein motifs (parts of proteins that have characteristic shapes) called zinc fingers that consist of a beta-pleated sheet linked to an alpha helix by a zinc atom (see figure 10.18). Different zinc fingers bind different three-base DNA sequences. If zinc fingers bind, a nuclease (called FokI) cuts the DNA, enabling the zinc fingers to be used like scissors. In a similar technique, called TALEN, or transcription activator-like effector nuclease technology, a restriction enzyme from a bacterium (*Xanthomonas*) that is a plant pathogen cuts DNA on both strands.

CRISPR technology uses an enzyme, called CAS9, which is guided by a short RNA to complementary DNA that it then cuts on both strands. CRISPR stands for "clustered regularly interspaced short palindromic repeats," and refers to short DNA repeats in certain bacteria that enable the bacteria to excise viral DNA. CRISPR uses RNA rather than proteins, as zinc finger technology and TALEN do, which is cheaper and easier to use. CRISPR can cut out more than one gene at a time, enabling researchers to investigate gene interactions. Both gene silencing and genome editing techniques are limited by "off-target effects"—doing something other than what we want them to do.

The next chapter explores how genetic technologies are used to diagnose and treat disease.

Key Concepts Questions 19.4

1. How do antisense technologies, ribozymes, and RNA interference silence gene expression?

2. How are zinc finger nuclease and CRISPR technologies used to edit genomes?

Summary

19.1 Patenting DNA

1. **Biotechnology** alters cells or biochemicals to provide a product. It extracts natural products, alters an organism's DNA, and combines DNA from different species.

2. A **transgenic organism** has DNA from a different species. **Recombinant DNA** comes from more than one type of organism. Both are possible because of the universality of the genetic code.

3. Patented DNA must be useful, novel, and nonobvious. A gene sequence by itself can no longer receive patent protection in the United States.

19.2 Modifying DNA

4. Recombinant DNA technology mass-produces proteins in bacteria or other single cells. Begun hesitantly in 1975, the technology has matured into a valuable method to produce proteins.

5. To construct a recombinant DNA molecule, **restriction enzymes** cut the gene of interest and a **cloning vector** such as a **plasmid** at a short palindromic sequence, creating complementary "sticky ends." The DNAs are mixed and vectors that pick up foreign DNA selected.

6. Genes conferring antibiotic resistance and color changes in growth media are used to select cells containing recombinant DNA. Useful proteins are isolated and purified.

7. A multicellular transgenic organism has an introduced gene in every cell. Heterozygotes for a transgene are bred to yield homozygotes. Some transgenic animals are used to model human disease. Transgenic plants are genetically modified to have traits from other species.

19.3 Monitoring Gene Function

8. **DNA microarrays** hold DNA pieces to which fluorescently labeled cDNA probes from samples are applied. They are used in **gene expression profiling.**

19.4 Gene Silencing and Genome Editing

9. **Gene silencing** uses antisense molecules, ribozymes, and **RNA interference (RNAi)** to block translation of mRNAs.

10. **Genome editing** uses enzymes to create double-stranded breaks in DNA, enabling researchers to add or delete genes. **Zinc finger nuclease technology** and **CRISPR technology** are used to edit genomes.

www.mhhe.com/lewisgenetics11

Answers to all end-of-chapter questions can be found at **www.mhhe.com/lewisgenetics11**. You will also find additional practice quizzes, animations, videos, and vocabulary flashcards to help you master the material in this chapter.

Review Questions

1. Cite three examples of a DNA sequence that meets requirements for patentability.

2. Why is a DNA sequence not patentable, but a DNA sequence that is used to diagnose a disease in a medical test patentable?

3. Describe the functions of each of the following tools used in a biotechnology:

   a. restriction enzymes
   b. cloning vectors
   c. DNA microarrays
   d. short nucleic acid sequences

4. How are cells containing recombinant DNA selected?

5. List the components of an experiment to produce recombinant human insulin in *E. coli* cells.

6. Why would recombinant DNA technology and creation of transgenic organisms be restricted or impossible if the genetic code were not universal?

7. What is an advantage of a drug produced using recombinant DNA technology compared to one extracted from natural sources?

8. How are transgenic animals better models of human disease than animal models whose DNA is unaltered? What are limitations of transgenic animal models?

9. How is genetic modification of a crop usually more precise and predictable than using conventional breeding to create a new plant variety?

10. Explain the advantages of using a DNA microarray that covers all of the protein-encoding genes in the human genome (the "exome"), rather than selected genes whose protein products are known to take part in the disease process being investigated.

11. Describe how a technology to silence a gene or edit a genome can be used to treat a disease.

Applied Questions

1. Phosphorus in pig excrement pollutes aquatic ecosystems, causing fish kills and algal blooms, and contributes to the greenhouse effect. *E. coli* produces an enzyme that breaks down phosphorus. Describe the steps to create a transgenic pig that secretes the bacterial enzyme, and therefore excretes less polluting feces.

2. Do you agree with the Supreme Court that a gene is not patentable but a cDNA is? Cite a reason for your answer.

3. Which (if any) objection to GMOs do you agree with, and why?

4. Genetic modification endows organisms with novel abilities. From the following three lists (choose one item from each list), devise an experiment to produce a particular protein, and suggest its use.

Organism	Biological Fluid	Protein Product
Pig	Milk	Human beta globin chains
Cow	Semen	Human collagen
Goat	Milk	Human EPO
Chicken	Egg white	Human tPA
Aspen tree	Sap	Human interferon
Silkworm	Blood plasma	Jellyfish GFP
Rabbit	Honey	Human clotting factor
Mouse	Saliva	Human alpha-1-antitrypsin

5. Collagen is a connective tissue protein that is used in skincare products, shampoo, desserts, and in artificial skin. For many years, it was obtained from the hooves and hides of cows collected from slaughterhouses. Human collagen can be manufactured in transgenic mice. Describe the advantages of the mouse source of collagen.

6. People did not object to the production of human insulin in bacterial cells used to treat diabetes, yet some people object to mixing DNA from different animal and plant species in agricultural biotechnology. Why do you think that the same general technique is perceived as beneficial in one situation, yet a threat in another?

7. A human oncogene called *ras* is inserted into mice, creating transgenic animals that develop a variety of tumors. Why are mouse cells able to transcribe and translate human genes?

8. In a DNA microarray experiment, researchers attach certain DNA pieces to the grid. For example, to study an injury, genes known to be involved in the inflammatory response might be attached. How might this approach be limited?

9. In the summer of 2013, 400 protesters destroyed fields of "golden rice" growing in the Philippines, upset that the rice would harm people's health, destroy the environment, and make money for herbicide producers. Do research on golden rice, or another GM crop or animal, and discuss these three issues: safety, environmental concerns, and business or economic repercussions.

10. Dairy cows are given bovine somatotropin (a growth hormone) that is manufactured in *E. coli*, which increases their milk yield. The milk has the same nutrient composition as milk from cows not given the hormone, but may contain traces of the growth hormone. The hormone is broken down by the human digestive tract. The only adverse effect of the hormone is mastitis (inflammation) of the cows' udders. The FDA approved the product in 1993 in the United States, but Canada and the European Union ban it because of the mastitis in cows. Is bovine somatotropin a product of recombinant DNA technology, antisense technology, or a transgenic procedure?

11. Lance Armstrong admitted in early 2013 to having taken recombinant EPO during all seven of his Tour de France wins. What effect did taking this hormone have on his body? Did the fact that the EPO was made using recombinant DNA technology make a difference in its effect on him?

Web Activities

1. Go to Nobelprize.org, nobelprizes.com, or another website and describe the work that led to discovery of RNA interference (2006) and use of green fluorescent protein from jellyfish to mark gene expression (2008).

2. Use the Web to identify three drugs made using recombinant DNA technology, and list the illnesses they are used to treat.

3. Go to the GMO Compass website (http://www.gmo-compass.org/eng/home/). Click on GMO Food Database and describe a genetic modification to a plant that you enjoy eating. Would the fact that a food is GM bother you?

4. Go to the Jackson Laboratory website (http://www.jax .org/), search among the types of transgenic mice, and describe the genetic change in one of them that is used in human disease research.

5. Recombinant DNA technology is used to manufacture human growth hormone (hGH), which is used to treat some forms of dwarfism. However, "anti-aging" clinics and websites sell what they claim is hGH to healthy consumers, although studies indicate that the only benefit is a slight increase in muscle mass. Possible side effects are serious, and include diabetes, breast development in men, joint pain, fluid retention, and shortened life. Legislation is pending to classify hGH as a controlled substance, limiting its distribution.

   a. Consult a website selling hGH and list the claims and warnings. Which do you think are accurate?
   b. Do you think laws should restrict access to hGH by people who do not have the medical conditions for which it is prescribed? Cite a reason for your answer.

Case Studies and Research Results

1. Nancy is a transgenic sheep who produces human alpha-1-antitrypsin (AAT) in her milk. This protein, normally found in blood serum, enables the microscopic air sacs in the lungs to inflate. Without it, inherited emphysema results. Donated blood cannot yield enough AAT to help the thousands of people who need it. Describe the steps taken to enable Nancy to secrete human AAT in her milk.

2. To investigate causes of acne, researchers used DNA microarrays that cover the entire human genome. Samples came from facial skin of people with flawless complexions and from people with severe acne. In the simplified portion of a DNA microarray shown, one sample is labeled green and comes from healthy skin; a second sample is labeled red and represents skin with acne. Sites on the microarray where both probes bind fluoresce yellow. The genes are indicated by letter and number.

   a. Which genes are expressed in skin whether or not a person has acne?

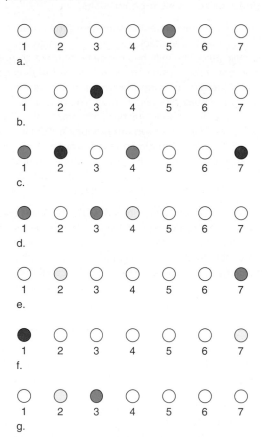

   b. Which genes are expressed only when acne develops?
   c. List three DNA pieces that correspond to genes that are not expressed in skin.
   d. How would you use microarrays to trace changes in gene expression as acne begins and worsens?
   e. Design a microarray experiment to explore gene expression in response to sunburn.

3. When four boys in a clinical trial to test a morpholino-based drug to treat Duchenne muscular dystrophy (see figure 12.10) performed slightly better on a walking test than they had before taking the drug, the news media widely reported the "breakthrough." However, the study was conducted over a period of only a few months, and the Food and Drug Administration would not let the clinical trial proceed until the boys' mobility was assessed for at least 2 years. The reason was that new studies of the disease's "natural history"—when symptoms appear—suggested that the improvement could have just been part of the normal fluctuation of symptoms. In addition, measuring levels of dystrophin did not predict improvement in symptoms. The delay was not widely reported. How can the news media balance the urge to report news fast, particularly very good or bad news, with the need to wait until findings are put into perspective?

4. Most viral infections of the lower respiratory tract (the lungs and airways) in babies are caused by rhinovirus (the common cold virus), influenza virus, or respiratory syncytial virus (RSV). Little is known about RSV, and treatment is nonspecific—just supportive. A study showed that white blood cells from babies infected with RSV express 2,000 genes at significantly different levels than white blood cells from healthy infants.

   a. What technology discussed in the chapter could be used to assess the differences among the three types of infections?
   b. What further information would reveal more about RSV infection?
   c. How would you develop a diagnostic test to distinguish RSV from rhinovirus and flu?
   d. Design an experiment to enable physicians to determine which babies with RSV will develop symptoms severe enough to require supplemental oxygen and hospitalization.

5. Why was Dolly the sheep, who was cloned from a cell taken from a six-year-old ewe, not patentable, but a sheep bearing a human gene is?

This book has followed Max Randell, center, since he was 3 years old, after he had received gene therapy for Canavan disease. His mother Ilyce is on the right, and on the left is younger brother Alex, who will one day be a neuroscientist. Max is currently in high school.

Genetic Testing and Treatment

Learning Outcomes

20.1 Genetic Counseling

1. Describe the services that a genetic counselor provides.

20.2 Genetic Testing

2. Describe types of genetic tests that are done at different stages of human prenatal development and life.

3. Discuss the benefits and limitations of direct-to-consumer genetic testing.

4. Explain how pharmacogenetics and pharmacogenomics personalize drug treatments.

20.3 Treating Genetic Disease

5. Describe three approaches to correcting inborn errors of metabolism.

6. Explain how an existing drug can be "repurposed" to treat a genetic disease.

7. Discuss how gene therapy has become safer and applicable to more diverse diseases.

The BIG Picture

DNA-based tests are no longer entering health care in a trickle but in a torrent as researchers identify the functions and variants of more genes, and develop faster ways to sequence DNA. Genetic testing to predict and diagnose disease and genetic approaches to treat disease are becoming a standard part of health care.

Fighting Canavan Disease

Max Randell was not expected to survive his first 2 years. Today he is a teen, thanks to gene therapy.

In early 1998, Ilyce and Mike Randell, of Buffalo Grove, Illinois, were worried about their 5-month-old son. He could not hold up his head, roll over, or reach for objects, and he was not very responsive. When doctors diagnosed Canavan disease (OMIM 271900), in which brain neurons lose their fatty sheaths and the brain slowly degenerates, they suggested that Max move to a nursing home. Instead, Max became the youngest person at the time to receive gene therapy for a degenerative brain disease.

Max had his first gene therapy, to test safety, in 1998 at Yale University. Soon after, he became able to use a walker, and his vision greatly improved. He had a second gene therapy in 2001. Each time, billions of viral particles, each carrying a corrected copy of the gene that was mutant in each of Max's cells, were delivered in catheters introduced through six holes drilled into his skull. Newer versions of the therapy use a different virus that can be delivered by intravenous injection.

Max's brother Alex does not have Canavan disease, which both parents carry. Today Max's movements are very limited, but he is very alert and aware. He answers questions slowly, with a blink for "yes" and widening eyes for "no." He earns very high grades in school. The Randell brothers adore each other, and Alex plans to become a neuroscientist, to help people like Max.

20.1 Genetic Counseling

Over the past few decades, the field of human genetics has evolved from an academic life science, to a medical specialty, to a source of personal information that ordinary people can access. This chapter presents the types of genetic services that a health care consumer might encounter: genetic counseling, genetic testing, and protein and gene-based therapies.

Our genomes are windows into who we are, filled with clues to health, ancestry, how we differ from each other and how we are also very similar. Genetic tests provide views of our genomes at several levels. They may detect variants in a single gene, abnormal chromosomes, variability across wide swaths of the genome, the entire protein-encoding part of the genome (the exome), or the whole genome. The U.S. National Institutes of Health maintains information about more than 10,000 DNA-based tests in the Genetic Testing Registry (https://www.ncbi.nlm.nih.gov/gtr/). Health care consumers, physicians, and researchers use the registry.

A **genetic counselor** is a health care professional with a master's degree who helps patients and their families navigate the confusing path of genetic testing (**figure 20.1**). Such tests identify genotypes that cause, contribute to, or raise the risk of developing a specific disease. Genetic counseling addresses medical, psychological, sociological, and ethical issues, and a genetic counselor has medical, scientific, and communication skills. A counselor can interpret a DNA test, explain uncertainties, and suggest ways to cope with anxiety, fear, or guilt associated with taking genetic tests—or not taking them.

Genetic counseling began in pediatrics and prenatal care (see *A Glimpse of History*). Today it embraces diseases of adults too, branching into such specialties as cancer, cardiovascular disease, neurology, hematology, and ophthalmology. A genetic counselor can explain the difference between susceptibility from a gene variant that contributes a small amount to risk and a single-gene diagnostic test with a risk based on Mendelian inheritance. Education and public policy may be part of the job. Genetic counselors in New York,

A GLIMPSE OF HISTORY

In 1947, geneticist Sheldon Reed coined the term "genetic counseling" for the advice he gave to physician colleagues on how to explain heredity to patients with single-gene diseases. In 1971, the first class of specially trained genetic counselors graduated from Sarah Lawrence College, in Bronxville, New York. Today, thirty-two programs in the United States offer a master's degree in genetic counseling, and many other countries have programs.

Reasons to seek genetic counseling:

Family history of abnormal chromosomes

Elevated risk of single-gene disorder

Family history of multifactorial disorder

Family history of cancer

Genetic counseling sessions:

Family history

Pedigree construction

Information provided on specific disorders, modes of inheritance, tests to identify at-risk family members

Testing arranged, discussion of results

Links to support groups, appropriate services

Follow-up contact

Figure 20.1 The genetic counseling process.

for example, hold "DNA days" to educate state legislators. Genetic counselors are integral parts of research teams conducting exome and genome sequencing projects, searching the scientific literature and disease databases to assign functions to genes.

Many genetic counselors work directly with patients as parts of health care teams, typically at medical centers. A consultation may entail a single visit to explore a test result, such as finding that a pregnant woman is a carrier for spinal muscular atrophy, or a several-month-long relationship as the counselor guides a decision to take (or not take) a test for an adult-onset disorder, such as Huntington disease or susceptibility to *BRCA* breast cancer.

The knowledge that a genetic counselor imparts is similar to what you have read in this book, but personalized and applied to a specific disorder. A counselor might explain Mendel's laws, but substitute a family's condition for pea color. Or, a counselor might explain how an inherited susceptibility can combine with a controllable environmental factor, such as smoking or poor diet, to affect health.

A genetic counseling session begins with a discussion of the family's health history. Using an online tool or pencil and paper, the counselor constructs a pedigree, then deduces and explains the risks of disease for particular family members (figure 20.1). The counselor may initially present possibilities and defer discussion of specific risks and options until test results are available. The counselor also explains which second-degree relatives—aunts, uncles, nieces, nephews, and cousins—might benefit from being informed about a test result. The genetic counselor provides detailed information on the condition and refers the family to support groups. If a couple wants to have a biological child who does not have the illness, a discussion of assisted reproductive technologies (see chapter 21) might be helpful.

A large part of the genetic counselor's job is to determine when specific biochemical, gene, or chromosome tests are appropriate, and to arrange for people to take the tests. The counselor then interprets test results and helps the patient or family choose among medical options, in consultation with other medical specialists. Until the recent availability of genetic tests for more conditions, people most often sought genetic counseling for either prenatal diagnosis or a disease in the family.

Prenatal genetic counseling typically presents population (empiric) and family-based risks, explains tests, and discusses whether the benefits of testing outweigh the risks. The couple, or woman, decides whether noninvasive prenatal testing of cell-free fetal DNA, amniocentesis, chorionic villus sampling, maternal serum screening, ultrasound, or no testing is best for them. Part of a prenatal genetic counseling session is to explain that tests that rule out some conditions do not guarantee a healthy baby. If a test reveals that a fetus has a serious medical condition, the counselor discusses possible outcomes, treatment plans, and the option of ending the pregnancy.

Genetic counseling when an inherited disease is in a family is more tailored to the individual situation than general prenatal counseling. For recessive disorders, the affected individual is usually a child. Illness in the first affected child is often a surprise, and especially if this is a first child, recognition of a problem may take months and a diagnosis, years. However, family exome analysis (see section 4.5), combined with rigorous mutation database searching, can now diagnose in minutes what once took years.

Communicating the risk subsequent children face may be difficult. Many people think that if one child has an autosomal recessive condition, then the next three will be healthy. Actually, each child has a 1 in 4 chance of inheriting the illness. Counseling for subsequent pregnancies requires great sensitivity. Some people will not terminate a pregnancy when the fetus has a condition that already affects their living child, yet some people will see that as the best option. Genetic counselors must respect these feelings, and tailor the discussion accordingly, while still presenting all the facts and choices.

Genetic counseling for adult-onset disorders does not have the problem of potential parents making important decisions for existing or future children, but presents the conflicting feelings of people choosing whether or not to find out if a disease is likely in their future. Often, they have seen loved ones suffer with the illness. This is the case for Huntington disease (see the opener to chapter 4). Predictive tests are introducing a new type of patient, the "genetically unwell" or those in a "premanifest" state—people with mutant genes but no symptoms (yet). Such a disease-associated genotype indicates elevated risk, but is not a medical diagnosis, which is based on symptoms and results of other types of tests.

When genetic counseling began in the 1970s, it was "nondirective," meaning that the practitioner presented options but did not offer an opinion or suggest a course of action. That approach is changing as the field moves from analyzing hard-to-treat, rare single-gene disorders to considering inherited susceptibilities to more common illnesses that are more treatable, and for which lifestyle changes might realistically alter the outcome. A more recent description of a genetic counseling session is "shared deliberation and decision making between the counselor and the client."

Genetic counselors regularly communicate with physicians and other health care professionals. They are important parts of teams at molecular diagnostic testing laboratories, where they guide physicians in ordering and interpreting tests. Before a doctor orders a test, the counselor helps to interpret risks from the patient's pedigree, discusses the pros and cons of the appropriate test, and raises ethical issues that might arise when other family members are considered. While the test is under way, the genetic counselor ensures that time constraints are respected, such as an advancing pregnancy, and updates the physician. Once test results are in, the counselor may request a repeat if they are inconsistent with the patient's symptoms; interpret the results; suggest additional tests; and alert the physician if the patient might be a candidate for participating in a clinical trial.

The United States has about 3,000 genetic counselors, and most of them practice in urban areas. Finding a genetic counselor with a specific expertise can be difficult. For example, only 400 genetic counselors in the United States are specially trained in cancer genetic counseling. Due to the shortage of counselors and demand for their services, other professionals, such as physicians, nurses, social workers, and PhD geneticists, may provide counseling. Dietitians, physical therapists, psychologists, and speech-language pathologists also discuss genetics with their patients, although they may not be specifically trained to do so. Genetic testing companies may connect patients with genetic counselors for phone or e-mail consultations. However, some states require that genetic counselors be licensed, requiring specific training.

As genetic testing becomes more commonplace, the need for genetic counselors and other genetics-savvy professionals to help individuals and families best use the new information will increase. Medical schools offer courses in which students interpret their own genetic tests and even exome or genome sequences. Physicians attend 2-day workshops to have their genomes sequenced and interpreted, and are given iPads loaded with their results. The goal of this continuing medical education is to learn how to incorporate genetic and genomic testing into clinical practice.

> ### Key Concepts Questions 20.1
>
> 1. What services and types of information does a genetic counselor provide?
> 2. To what issues must a genetic counselor be sensitive?
> 3. Where do genetic counselors work?
> 4. How is the medical field preparing for increased use of genetic and genomic testing?

20.2 Genetic Testing

Genetic tests are administered at all stages of human existence, and for a variety of reasons (**table 20.1**). Identifying mutations can help in diagnosis and choosing treatments. Unlike the results of a cholesterol check or an X ray, however, results of a genetic test can have effects beyond the individual, to family members who share genotypes that affect health.

As the pace of exome and genome sequencing accelerates, and such testing becomes more widely available, it may soon be more economical to obtain all the information and parse it for results relevant to an individual, than to do single-gene tests. The following sections consider genetic tests according to a developmental timeframe, with references to topics already covered.

Genetic Testing From Fertilized Ovum to the Elderly

Some prenatal genetic tests have been in use for decades; others are new or in development.

Preconception and Prenatal Testing

When a direct-to-consumer genetic testing company was awarded a patent in 2013 for "gamete donor selection based on genetic calculations," many people interpreted the idea as a method to create designer babies. The invention is actually a computer program that predicts the results of meiosis in gametes of an individual. According to the patent language, the invention analyzes possible gamete genotypes of "the recipient" (presumably a woman) and "a plurality of donors" (presumably men), considering penetrance and epistasis. Recall from chapter 5 that penetrance refers to the frequency that a genotype will manifest as a particular phenotype, and epistasis refers to gene-gene interactions.

The algorithm reports the likelihood that any two people of opposite sex can produce a child with a particular combination of genotypes, interpreted as possible traits. For example, a couple might ask to have their DNA analyzed to predict the likelihood of their having a child with red hair, green eyes, a small nose, and low risk of *BRCA* breast cancer, Alzheimer disease, and hereditary hemochromatosis. The computer program also detects if potential parents are close blood relatives. The patent refers to a "hypothetical offspring" rather than a "designer baby" because testing an oocyte or sperm for certain inherited traits would destroy it in the process.

Although we still can't order up a particular baby, it has been possible for many years to collect sperm and separate them into fractions that are enriched for X-bearing or Y-bearing cells, to attempt to conceive a girl or boy, respectively. Using a technique called flow cytometry, sperm are labeled with fluorescent markers. X-bearing sperm cells glow more intensely than Y-bearing sperm because the X chromosome is so much larger. Flow cytometry assigns each type of sperm a positive or negative charge, and uses the distinction to separate and collect them. The approach is not as specific as selecting a single sperm, but can increase the probability of having, for example, a daughter in a family that has an X-linked condition, rather than a son who would face a 50:50 chance of inheriting the illness.

During the first few days following conception, when the embryo consists of only a few cells, sampling one of them can reveal mutations, and then the rest of the embryo placed in the uterus to continue development. This technique, called preimplantation genetic diagnosis, is discussed further in chapter 21. Genetic testing techniques used later in pregnancy are discussed in chapter 13. They are chorionic villus sampling, amniocentesis, maternal serum markers, and noninvasive prenatal testing (which tests cell-free fetal DNA).

A new application of testing chromosomes, called rescue karyotyping, is done on tissue samples that were stored when a woman's uterus was scraped following a miscarriage. In the past, such material was not routinely subject to any form of genetic testing if the woman had had fewer than three pregnancy losses. However, the technology to detect microdeletions and microduplications (also called copy number variants) improved for use in diagnosing children with developmental and other disabilities (discussed later in this section), giving researchers the idea to seek chromosomal clues in evidence

Table 20.1 Genetic Testing

Test	Description	Chapter
PRENATAL		
Sperm selection	Enrich for X- or Y-bearing sperm, then intrauterine insemination or *in vitro* fertilization	6, 21
Preimplantation genetic diagnosis	Test DNA and chromosomes of 1 cell of 8-celled embryo; implant remaining 7-celled healthy embryo in uterus to continue development.	21
Chorionic villus sampling	Test DNA and chromosomes of chorionic villus cell	13
Rescue karyotyping	Detect small deletions and duplications in archived cells from past unexplained pregnancy losses	13
Noninvasive prenatal diagnosis	Test cell-free fetal DNA	13
Maternal serum markers	Measure levels of biomarkers in pregnant woman's blood	13
Amniocentesis	Test DNA and chromosomes of amniocytes	13
NEWBORNS		
Screening	Screen metabolite and DNA in heelstick blood sample for 50-plus actionable conditions	20
Genome sequencing	Sequence genomes of large samples of newborns to assess clinical value	20
CHILDREN		
Chromosome microarray analysis	Detect small deletions and duplications associated with certain phenotypes	8
Exome sequencing	Diagnose unrecognized syndromes or atypical cases; family comparisons reveal *de novo* or inherited mutations in children	1, 4, 8
ADULTS		
Dor Yeshorim	Carrier tests for Jewish genetic diseases; preconception	15
Sickle cell disease	Test athletes; carriers at risk for symptoms	11, 12
Comprehensive carrier testing	Up to 500 tests for heterozygotes for single-gene diseases; preconception	20
Population carrier screen	Tests for heterozygotes for diseases more prevalent in certain population groups	15
Ancestry testing	Y chromosome and mitochondrial sequences point to geographic origin; DNA markers identify distant cousins	16
Forensics testing	Copy numbers of short tandem repeats (STRs) in crime scene or disaster evidence	14
Military	Identify remains; risk for depression, PTSD; rapid infection diagnosis	1
Susceptibility	*BRCA* cancers, Alzheimer disease	5, 18
Pharmacogenetics	Drug efficacy, adverse effects, and dose	20
Genome-wide association studies	Identify genes contributing small degrees to a phenotype	7
Paternity	Half of a child's genome from father's genome	14
Centenarians	Identify gene variants that extremely old people share	3
POSTHUMOUS		
Disease diagnosis	Identify in remains mutation discovered after death	2
History	Identify remains	5
Human origins	Compare genomes of modern and archaic humans	16

from past pregnancies. Such information can indicate increased risk for future pregnancies—or, more often, alleviate concern by revealing an aneuploid (extra or missing chromosome), which is not likely to repeat.

Genetic Testing of Newborns

The goal of newborn screening is to identify infants who are at very high risk of developing certain inherited diseases that are "actionable"—that is, parents and health care professionals can provide treatments and services to improve the quality of life for the child. Testing a drop of a newborn's blood for metabolites (biochemicals that indicate an inborn error of metabolism) has been routine for decades (**figure 20.2**). For most conditions it uses an analytical chemistry technique called mass spectrometry, but DNA testing has been added over the years. The number of conditions screened for in this way has steadily grown, from the first test for phenylketonuria (PKU) to more than fifty conditions today.

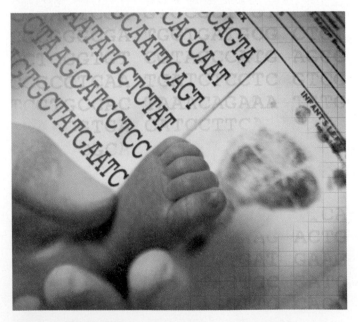

Figure 20.2 **Newborn screening.** In addition to taking a newborn baby's footprint, a few drops of blood are sampled with a heel prick for metabolic and genetic testing.

A newborn genetic screening test that indicates increased risk for a specific condition is followed up with diagnostic tests that look for evidence of the disease. Early accurate diagnosis of a genetic disease can lead to treatment, enrollment in a clinical trial to test a new therapy, and provides information on risks to future pregnancies. Genetics adds precision to a diagnosis. The downside of newborn screening is that follow-up tests may *not* show an abnormality, and the family may then experience a situation called "patients in waiting," when the genotype associated with an illness is present, but the phenotype isn't—yet. Studies have found that this incomplete knowledge may interfere with parent-child bonding and cause great anxiety for years.

The challenges of newborn screening for genetic disease will be amplified with the much more complete information that genome sequencing provides. Five-year pilot projects began in 2013 to assess whether exome and genome sequencing improve health care of newborns. The programs are also investigating the ethical, social, and legal implications of gathering so much genetic information.

Genetic Testing of Children

Tests on children range from single-gene tests that are done because symptoms match those of a known disorder, to chromosome tests, to exome and genome sequencing.

A **chromosomal microarray (CMA) test** detects very small deletions and duplications that are associated with autism, developmental delay, intellectual disability, behavioral problems, and other phenotypes. Finding a deletion or duplication might be comforting because it is a possible explanation for the child's problems, but it wouldn't necessarily change treatment.

Since 2011, several projects have been using whole exome sequencing to diagnose children whose symptoms do not match known syndromes, such as Nicholas Volker's intestinal disease (see Clinical Connection 1.1) and Bea Reinhoff's unusual features (see section 4.5). Thousands of children have now received diagnoses. The knowledge helps not only them and their health care providers, but each new diagnosis becomes a new test that can help other children.

Exome studies are revealing that some children could not be diagnosed because they actually have *two* genetic diseases. For example, a 9-year-old boy was very weak and had episodes when he would stop breathing. He also had droopy eyelids, difficulty feeding, increased respiratory secretions, and at 8 months developed an enlarged heart. His sister had died at 20 months when she stopped breathing when she had a fever. The boy's exome sequence revealed that he had congenital myasthenic syndrome (OMIM 601592), in which a fever causes too-rapid breakdown of a neurotransmitter. A drug prevented the boy from dying during a fever, as his sister had. But his exome sequence also revealed mutations in a second gene, *ABCC9,* that enlarged his heart (OMIM 608569).

Genetic testing of children for diseases that will not cause symptoms for many years is very controversial. For Huntington disease (see the opener to chapter 4), it is generally agreed not to do a presymptomatic test in anyone under age 18, because it can be so upsetting. (Juvenile HD, discussed in the chapter 4 opener, is an exception.) The legality of direct-to-consumer DNA testing for health traits is under question, but from 2008 until 2013, even though company websites stated that children should not be tested, parents could and did send in children's DNA under the parents' names. Some companies marketed to parents to test their children for traits such as athletic ability (see Applied Question 3). *Bioethics: Choices for the Future* in chapter 1 describes one of the first efforts to incorporate genetic testing into a university curriculum.

Genetic Testing of Adults

Like children, adults take single-gene tests as part of diagnostic workups based on symptoms or other test results. They may also take genetic tests to detect increased risk of developing a particular cancer, such as a *BRCA* test. Many adults take direct-to-consumer DNA-based ancestry tests, and DNA testing is commonly done in forensics. In the military, genetic testing is used to identify remains; to detect a disease that might put a soldier at risk, such as sickle cell disease or a susceptibility to develop depression; and to diagnose a communicable disease on the battlefield.

Many adults begin to think about genetic testing when they are considering having children, and wonder what traits and illnesses they might pass on. In the past, carrier tests focused on specific population groups in which a disease is more common, for economic reasons. In the 1970s, carrier testing for sickle cell disease targeted African Americans, while testing for Tay-Sachs disease recruited Ashkenazi Jews (see table 15.2).

Faster DNA sequencing has lowered the cost of carrier tests to the point that it may be more economical to test everyone for many diseases with one blood sample per person. Researchers have developed "preconception comprehensive carrier screening" tests that detect heterozygotes for nearly 500 single-gene diseases that affect children. The selected diseases are severe and their mutations highly penetrant, so that knowing a genotype reliably predicts a phenotype.

Another group of adults who are having their exomes and genomes sequenced are medical students and physicians. The goal is to learn how to provide such information to their patients. *Bioethics: Choices for the Future* considers

---

Bioethics: Choices for the Future

Incidental Findings: Does Sequencing Provide Too Much Information?

In medical practice, an "incidentaloma" occurs when a diagnostic work-up for one condition discovers another – such as an X-ray to rule out pneumonia revealing lung cancer. A genetic "incidental" (or "unsolicited") finding arises when sequencing a person's exome to discover a mutation that accounts for one set of symptoms identifies a mutation that indicates a second condition that may or may not already have produced symptoms.

Incidental genetic findings were first identified in children having their exomes sequenced to explain unfamiliar combinations of developmental delay, intellectual disability or other neurological symptoms, and/or birth defects. An early case illustrated how life-saving unexpected information can be. A 2-year-old had severe feeding problems, seizures, failure to thrive, developmental delay, and intellectual disability. Doctors had ruled out infection, Angelman syndrome, Rett syndrome, and mitochondrial disease. Family exome sequencing (see section 4.5) revealed an autosomal dominant mutation that originated in the boy, in a gene, *SYNGAP1* (OMIM 603384), which affects synapse formation. That explained the developmental delay, intellectual disability, and seizures. But the exome sequence also revealed a mutation in the connective tissue protein fibrillin, which causes Marfan syndrome, for which the boy had no symptoms (see figure 5.7). An ultrasound of his heart indeed revealed an enlarged aortic root. A burst aorta can be the first, and deadly, symptom of the syndrome. Drug treatment may have saved the boy's life.

As more people have their exomes sequenced, incidental findings are inevitable, because we all have mutations. How should a clinician determine which results to test for or report to a patient? That is, should all of the information in an exome be deciphered?

In 2013, the American College of Medical Genetics and Genomics recommended reporting 57 conditions that are prevalent, caused by mutation of a single gene, and "actionable." The list had 25 syndromes that included cancers, 8 connective tissue disorders, and malignant hyperthermia, in which exposure to a certain anesthetic can be lethal. The organization advised that patients be offered the opportunity to "opt-out" of knowing incidental findings when the clinician delivers results. After much feedback, however, the recommendations were changed so that the discussion of what a patient wants to know comes *before* the sequencing. If a patient does not want to know about a particular condition, it isn't tested for among the exome results.

Questions For Discussion

1. What factors should a physician take into account when discussing with a patient which findings to report?

2. Should everyone have their exome sequenced and the information entered into a database, even if they don't want to know the results, to speed gene discovery and development of new drugs and diagnostics?

3. Many people have taken genetic tests for mutations in *APOE4* that greatly increase the risk of developing Alzheimer disease. A more recently discovered gene has variants that protect against the disease. How can clinicians prevent harm from the incompleteness of genetic information?

4. How can an overextended health care system that can barely handle people who are ill right now provide for people who have taken genetic tests that indicate they may become sick in the future?

a problem that can arise with exome and genome sequencing: detecting and reporting unexpected results, called incidental (or unsolicited) findings, about conditions that were not the reason for the sequencing.

Genetic testing isn't only for the young. We have a lot to learn from the genomes of those who have survived past age 100 with good health. Genetic testing after death is important too. The opener to chapter 2 describes a girl with Rett syndrome diagnosed years after her death from the DNA in a saved baby tooth, and chapter 16 discusses DNA sequences from historical figures and even from our prehistorical ancestors, the Neanderthals and Denisovans. It is only when our distant ancestors lived too long ago for their DNA to have persisted that we can learn about them only from their fossilized bones.

A Closer Look at Three Types of Genetic Tests

Health care professionals recommend genetic tests, but tests have been offered direct-to-consumer (DTC) from companies. Although the legality of some tests may not be settled, past examples of DTC marketing of genetic tests may be important in the future, when exome and genome sequencing may become commonplace. With that caveat in mind, following are three types of genetic tests that have raised intriguing issues.

DTC Testing for Health Conditions

Companies have marketed DTC DNA-based tests for traits, susceptibilities, and genetic diseases. The tests range from the obvious (eye color) to the dubious (athletic ability) to the serious (cancers).

In the United States, the Clinical Laboratory Improvement Amendments, or CLIA, control genetic testing of body materials for the prevention, diagnosis, or monitoring response to treatment of a disease or health impairment. The CLIA regulations, instituted in 1988, added "specialty areas" in 1992 to cover very complex tests, such as those involving immunology or toxicology. These did not include genetic testing, which at the time was very limited. Since then, the genetics community has repeatedly asked that genetic tests be included as a specialty area, to no avail. State regulations can override CLIA, but only if they are equally or more stringent. Therefore, the regulation of genetic tests remains somewhat unclear. Availability of exome and genome sequencing will complicate matters.

Tests that offer genetic information, but are not intended to be used to diagnose a disease, do not come under the CLIA regulations. The distinction between information and diagnosis is often based on careful wording. One company, for example, ". . . scans your DNA for genetic risk markers associated with both common and uncommon health conditions." "Association" means a relationship between one piece of information and another—it is not a correlation or a cause, nor a diagnosis. Some companies have had CLIA certification for some of their tests but not others, but even that regulation is not always clear. For example, CLIA certifies that laboratories are safe and that a particular test measures what it claims to measure. However, the regulations do not require that knowing the measurement, such as the concentration of a molecule in blood, leads to actions that improve health.

Even DTC tests for very well-studied mutations can lead to complications and confusion. This is the case for hereditary hemochromatosis (HH, OMIM 235200). In this autosomal recessive "iron overload" disease, cells in the small intestine absorb too much iron from food. Early signs and symptoms of HH include chronic fatigue, increased susceptibility to infection, hair loss, infertility, muscle pain, and feeling cold. Over many years, the excess iron is deposited throughout the body, damaging vital organs. HH is incompletely penetrant— that is, many people who have the disease-associated genotype do not have symptoms, especially women who lose the extra iron in the monthly menstrual flow.

Diagnosis of symptomatic HH is important, because lowering the body's iron levels is easy—have blood removed periodically. However, diagnosis is based on an increase in the level of the iron-carrying protein ferritin in the blood and confirmed with a liver biopsy, rather than on a genetic test because of the incomplete penetrance. Yet a website states: "Here is an easy way to get direct testing without a prescription and results sent directly to the patient only." A customer might not know that inheriting the genotype for HH is not the same as having the disease.

An advantage of DTC genetic testing companies is that they are amassing tremendous data stores, which can be mined to make discoveries much faster than is possible with traditional research methods. Most company websites also provide information about genetics. However, some companies prey on consumer unfamiliarity with genetics. This was the case for certain "nutrigenetics" companies.

Nutrigenetics Testing

"Nutrigenetics" DTC websites have offered genetic tests along with general questionnaires about diet, exercise, and lifestyle habits. The companies return supposedly personalized profiles with dietary suggestions, often with an offer of a pricey package of exactly the supplements that an individual purportedly needs to prevent realization of his or her genetic fate.

After the media spread the word of these services, the U.S. Government Accountability Office tested the tests. An investigator took two DNA samples—one from a 9-month-old girl and the other from a 48-year-old man—and created fourteen lifestyle/dietary profiles for these "fictitious consumers"— twelve for the female, two for the male. The samples were sent to four nutrigenetics companies, none of which asked for a health history. Here is an example of the information sent to the companies:

- The DNA from the man was submitted as being from a 32-year-old male, 150 pounds, 5'9", who smokes, rarely exercises, drinks coffee, and takes vitamin supplements.
- The DNA from the baby girl was submitted as being from a 33-year-old woman, 185 pounds, 5'5", who

smokes, drinks a lot of coffee, doesn't exercise, and eats a lot of dairy, grains, and fats.

- The same baby girl DNA was also submitted as that of a 59-year-old man, 140 pounds, 5′7″, who exercises, never smoked, takes vitamins, hates coffee, and eats a lot of protein and fried foods.

The elevated risks found for the three individuals were exactly the same: osteoporosis, hypertension, type 2 diabetes, and heart disease. One company offered the appropriate multivitamin supplements for $1,200, which the investigation found to be worth about $35. Recommendations stated the obvious, such as advising a smoker to quit. The advice tracked with the fictional lifestyle/diet information, and *not* genetics. Concluded the study: "Although these recommendations may be beneficial to consumers in that they constitute common sense health and dietary guidance, DNA analysis is not needed to generate this advice." Some of the suggestions could even be dangerous, such as vitamin excesses in people with certain medical conditions.

Matching Patient to Drug

People react differently to the same dose of the same drug because we differ in the rates at which our bodies respond to and metabolize drugs. Genetic tests can highlight these differences. A **pharmacogenetic** test detects a variant of a single gene that affects drug metabolism, and a **pharmacogenomic** test detects variants of multiple genes or gene expression patterns that affect drug metabolism. Pharmaceutical and biotechnology companies use these tests in developing drugs, and physicians are increasingly using these tests in prescribing drugs.

Genetic testing to guide drug selection offers several advantages:

- identifying patients likely to suffer an adverse reaction to a drug;
- selecting the drug most likely to be effective;
- monitoring response to drug treatment; and
- predicting the course of the illness (prognosis).

An early use of pharmacogenetics was in breast cancer; women with the HER2 subtype respond to the drug trastuzumab (Herceptin). A pharmacogenomic example is the use of a DNA microarray depicting the expression of eighteen genes to predict whether a person is likely to respond to certain drugs used to treat hepatitis C, which have severe side effects. A pharmocogenetic/genomic approach might have averted disaster in 2004, when widespread use of a type of arthritis drug called a COX-2 inhibitor caused heart damage in some patients. Several drugs were discontinued or their use restricted, robbing many people with arthritis of their benefits. Tests that detect specific variants of genes that encode proteins called cytochromes (P450, 2D6, and 2C19) are now used to predict who will develop adverse effects from these drugs.

One of the first drugs to be described using pharmacogenetics was the blood thinner warfarin (also known as Coumadin). This drug has a very small range of concentration in which it keeps blood at a healthy consistency, but people can vary up to tenfold in the dose required. Too little drug allows dangerous clotting; too much causes dangerous bleeding. In the past, physicians would give an initial standard dose, based on a patient's age, gender, health status, weight, and ethnicity, then monitor the patient for a few weeks to check for too much clotting or bleeding, tweaking the dose until it was about right. This general approach led to hospitalization for abnormal bleeding in 43,000 of the 2 million people prescribed the drug each year.

A "pharmacogenetic algorithm" is now used to prescribe warfarin. It considers two genes: two variants of *CYP2C9* and one variant of *VKORC1* that are associated with increased sensitivity to the drug. People with these gene variants require lower doses of warfarin. The new, genetic way of testing for warfarin response is especially helpful for the 50 percent of patients who fall at the extremes of the range of drug concentration that is effective. However, considering clinical information remains an extremely important part of determining the dose of warfarin in particular, and other drugs in general. This is because variants of genes other than the ones that are tested for can affect how an individual human body metabolizes a drug.

Key Concepts Questions 20.2

1. On what biological process is "gamete donor selection based on genetic calculations" based?

2. What genetic tests are used on embryos and fetuses?

3. In what circumstances are genetic tests done on children?

4. Why might adults take genetic tests?

5. How can direct-to-consumer genetic tests confuse consumers?

6. Explain how pharmacogenetics and pharmacogenomics help physicians select the best drugs for patients.

20.3 Treating Genetic Disease

Tests for genetic diseases greatly outnumber treatments, because treatments are very challenging to develop. They must correct the abnormality in the appropriate cells and tissues to prevent or minimize symptoms, while not harming other parts of the body. Treatments for single-gene diseases have evolved through several stages, in parallel to development of new technologies:

- removing an affected body part;
- replacing an affected body part or biochemical with material from a donor;
- delivering pure, human proteins derived from recombinant DNA technology;
- refolding correctly a misfolded protein; and
- gene therapy, to supply wild type alleles.

Drugs

Preventing a disease phenotype may be as straightforward as adding digestive enzymes to applesauce for a child with cystic fibrosis, or giving a clotting factor to a boy with hemophilia. Inborn errors of metabolism are particularly treatable when the biochemical pathways are well understood and enzymes can be replaced.

Lysosomal storage diseases are a well-studied subclass of inborn errors of metabolism. A deficient or abnormal enzyme leads to buildup of the substrate (the molecule that the enzyme acts on) and deficit of the breakdown product of the substrate. Recall from figure 2.7 that a lysosome is an organelle that dismantles debris. It houses 43 types of enzymes, and each breaks down a specific molecule. Case Studies and Research Results #5 describes one condition.

Treatment of type 1 Gaucher disease (OMIM 230800), a lysosomal storage disease, illustrates three general approaches to counteracting an inborn error of metabolism that affects an enzyme, summarized in **table 20.2** and **figure 20.3**. In type 1 Gaucher disease, the enzyme glucocerebrosidase is deficient or absent. As the substrate builds up because there is little or no enzyme to break it down, lysosomes swell, ultimately bursting cells. Symptoms include an enlarged liver and spleen, bone pain, and deficiencies of blood cells. Too few red blood cells cause the fatigue of anemia; too few platelets cause easy bruising and bleeding; and too few white blood cells increase the risk of infection. The disease is very variable in age of onset, severity of symptoms, and rate of progression.

Early treatments for Gaucher disease corrected affected body parts: removing the spleen, replacing joints, transfusing blood, or transplanting bone marrow. In 1991, *enzyme replacement therapy* became available, which supplies recombinant glucocerebrosidase. This treatment is effective but costs about $550,000 a year, and takes several hours to infuse twice a month.

In 2003 came a different approach: *substrate reduction therapy*. This is a drug taken by mouth that decreases the amounts of the substrate, the molecule on which the deficient enzyme acts. A third approach is called *pharmacological chaperone therapy*, in which an oral drug binds a patient's misfolded enzyme, stabilizing it sufficiently to allow some function. Clinical Connection 4.2 discusses a drug that correctly refolds the abnormal CFTR protein that causes many cases of cystic fibrosis.

Repurposed drugs can treat genetic disease. Certain antibiotics suppress nonsense mutations, preventing proteins from being prematurely shortened. The chapter 3 opener describes a failed cancer drug that helps children who have an accelerated aging disorder. A drug that treats erectile dysfunction by increasing blood flow to the penis improves leg muscle function in boys who have Becker muscular dystrophy. Repurposing a drug is much more economical than developing a new one.

Gene Therapy

Gene therapy delivers working copies of genes to specific cell types or body parts, typically aboard modified viruses. More than 2,000 clinical trials of gene therapies have been conducted worldwide since 1990. The first approval, in Europe, was in 2012 for a rare enzyme deficiency that causes very painful pancreatitis. **Tables 20.3** and **20.4** list some general concerns and requirements related to gene therapy.

Researchers had expected that the sequencing of the human genome would accelerate the pace of gene therapy development. Instead, new information about the complexity of gene interactions, and a few gene therapies that harmed patients, led to a

Table 20.2	Lysosomal Storage Disease Treatments	
Treatment	**Mechanism**	
Enzyme replacement therapy	Recombinant human enzyme infused to compensate for deficient or absent enzyme	
Substrate reduction therapy	Oral drug that reduces level of substrate so enzyme can function more effectively	
Pharmacological chaperone therapy	Oral drug that binds to patient's misfolded protein, restoring function	

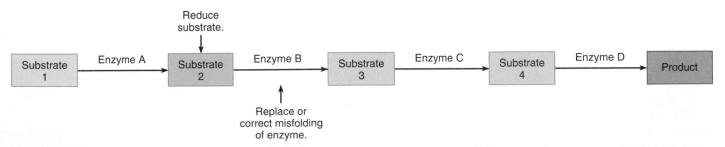

Figure 20.3 **Counteracting a metabolic abnormality.** Treatments of lysosomal storage diseases are based on understanding metabolic pathways, in which a series of enzyme-catalyzed reactions leads to formation of a product. If any enzyme is deficient or its activity blocked, the substrate builds up and the product is deficient. Enzyme replacement therapy delivers an absent or deficient enzyme. Substrate reduction therapy decreases the amount of substrate and pharmacological chaperone therapy corrects misfolded proteins. Points of intervention vary with the specific condition.

Table 20.3	Gene Therapy Concerns

Scientific	**Bioethical**
1. Which cells should be treated, and how?	1. Does the participant in a gene therapy trial truly understand the risks?
2. What proportion of the targeted cell population must be corrected to alleviate or halt progression of symptoms?	2. If a gene therapy is effective, how will recipients be selected, assuming it is expensive at first?
3. Is overexpression of the therapeutic gene dangerous?	3. Should rare or more common disorders be the focus of gene therapy research and clinical trials?
4. Is it dangerous if the altered gene enters cells other than the intended ones?	4. What effect should deaths among volunteers have on research efforts?
5. How long will the affected cells function?	5. Should clinical trials be halted if the delivered gene enters the germline?
6. Will the immune system attack introduced or altered cells?	
7. Is the targeted DNA sequence in more than one gene?	

Table 20.4	Requirements for Approval of Clinical Trials for Gene Therapy

1. Knowledge of the defect, how it causes symptoms, and the course of the illness (natural history)

2. An animal model and/or cultured human cells

3. No alternate therapies, or patients for whom existing therapies are not possible or have not worked

4. Experiments as safe as possible

Gene therapy strategies also vary in invasiveness (**figure 20.4**). Cells can be altered outside the body and then infused into the bloodstream. This is called *ex vivo* ("outside the body") **gene therapy**. In the more invasive *in vivo* ("in the living body") **gene therapy**, the gene and its vector are introduced directly into the body. A catheter might be used to deliver the gene to the liver or the brain, for example. There, the vector must enter the appropriate cells and the human DNA be transcribed into mRNA and translated into protein. Then, the protein must do its job.

reevaluation of the idea that we can augment or replace a gene's function with predictable effects. Since 2008, the field has been reborn with a few successes.

Providing functional genes to treat an inherited disorder may provide a longer-lasting effect than treating symptoms or supplying a protein. The first gene therapy efforts were for well-studied inherited disorders, even though they are very rare. Today, about 8 percent of gene therapy clinical trials are for single-gene diseases. About 65 percent target cancers, and about 8 percent treat cardiovascular disease.

Types and Targets of Gene Therapy

Gene therapy approaches vary in the way that healing genes are delivered and where they are sent. **Germline gene therapy** alters the DNA of a gamete or fertilized ovum, so that all cells of the individual have the change. The transgenic organisms discussed in chapter 19 had germline gene therapy. The correction is heritable, passing to offspring. Germline gene therapy is not being done in humans, although it is a popular plot in fiction.

Somatic gene therapy corrects only the cells that an illness affects. It is nonheritable: A recipient does *not* pass the genetic correction to offspring. Clearing lungs congested from cystic fibrosis with a nasal spray containing functional *CFTR* genes is an example of somatic gene therapy.

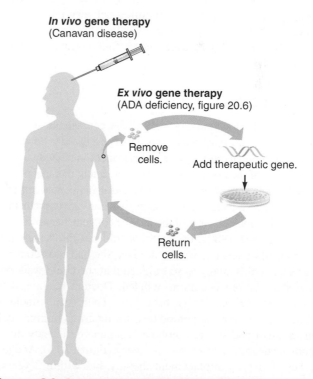

In vivo gene therapy
(Canavan disease)

Ex vivo gene therapy
(ADA deficiency, figure 20.6)

Remove cells.

Add therapeutic gene.

Return cells.

Figure 20.4 Gene therapy invasiveness. Therapeutic genes are delivered to cells removed from the body that are then returned (*ex vivo* gene therapy) or delivered directly to an interior body part (*in vivo* gene therapy).

Researchers obtain therapeutic genes using the polymerase chain reaction, recombinant DNA technologies, and other techniques that cut DNA. In the future, researchers—and, someday, clinicians—may deliver synthetic genes. The origin should not make a difference—DNA is DNA and the genetic code universal.

The next step in gene therapy, gene transfer, usually delivers the healing DNA with other DNA that is mobile, called a vector. Most gene therapy vectors are disarmed viruses. Researchers remove the viral genes that cause symptoms or alert the immune system and add the corrective gene. Different viral vectors are useful for different treatments. A certain virus may transfer its cargo with great efficiency to a specific cell type, but carry only a short DNA sequence. Another virus might carry a large piece of DNA but enter many cell types, causing "off-target" side effects. Even if a viral vector goes where intended, it must enter enough cells to alleviate symptoms or slow or halt progression of the disease. Finally, a viral vector must not integrate into a gene that harms the patient, such as an oncogene or tumor suppressor gene, which could cause cancer.

Some gene therapies use viruses that normally infect the targeted cells. For example, a herpes simplex virus delivers the gene encoding a pain-relieving peptide to nerve endings in skin. Researchers team parts of viruses to target a certain cell type. Adeno-associated virus (AAV), for example, infects many cell types, but adding a promoter from a parvovirus gene restricts it to red blood cell progenitors in bone marrow. Different subtypes of AAV home to different body parts. AAV9 can cross the blood-brain barrier and is given by intravenous infusion, replacing earlier versions that required delivery into the brain through catheters. Fatty structures called liposomes are also used as vectors. **Figure 20.5** shows four targets of somatic gene therapies.

Initial Success

Any new medical technology begins with courageous volunteers who know that they may risk their health. Gene therapy, however, is unlike conventional drug therapy in that it alters an individual's genotype in part of the body. Because the potentially therapeutic gene is usually delivered with other DNA, and it may enter cell types other than those affected in the disease, reactions are unpredictable. Following is a look at some of the pioneers of gene therapy.

In the late 1980s, the DeSilvas did not think their little girl, Ashanthi ("Ashi"), would survive. She had near-continual coughs and colds, and was so fatigued that she could walk only a few steps before becoming winded. Doctors diagnosed the obvious: asthma, an allergy, bronchitis. Then Ashi's uncle, an immunologist, suggested blood tests for inherited immune deficiencies. Ashi had severe combined immune deficiency due to adenosine deaminase (ADA) deficiency (**figure 20.6**). At age 2, she began enzyme replacement therapy, but within 2 years it stopped working. She would likely die of infection.

Then her physician heard about a clinical trial of gene therapy that would give her white blood cells functional ADA genes. Ashi was chosen, and on September 14, 1990, at 12:52

Endothelium. The tile-like endothelium that forms capillaries can be genetically altered to secrete proteins into the circulation.

Muscle. Immature muscle cells (myoblasts) given healthy dystrophin genes may treat muscular dystrophy.

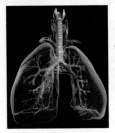

Liver. To treat certain inborn errors of metabolism, only 5 percent of the liver's 10 trillion cells need to be genetically altered.

Lungs. Gene therapy can reach damaged lungs through an aerosol spray. Enough cells would have to be reached to treat hereditary emphysema (alpha-1-antitrypsin deficiency) or cystic fibrosis.

Figure 20.5 Some sites of gene therapy.

P.M, she sat up in bed at the National Institutes of Health in Bethesda, Maryland, and began receiving her own corrected white blood cells intravenously. An ongoing multinational trial is following 42 children given an improved form of the gene therapy, and 31 of them are completely well. The new approach alters stem cells instead of the more specialized T cells, and adds a drug to the regimen that creates more space in the bone marrow for corrected cells to accumulate.

Gene Therapy Setbacks

In September 1999, 18-year-old Jesse Gelsinger died 4 days after receiving gene therapy, from an overwhelming immune response to the virus used to introduce the therapeutic gene. He had ornithine transcarbamylase deficiency (OTC) (OMIM 311250). In this X-linked recessive disorder, a person cannot make a liver enzyme required to adequately break down dietary proteins (**figure 20.7**). Instead of being excreted in urine, the nitrogen released from the amino acids combines with hydrogen to form ammonia (NH_3), which rapidly accumulates in the bloodstream and travels to the brain, with devastating effects. It usually causes irreversible coma within 72 hours of birth. Half of affected babies die within a month, and another quarter

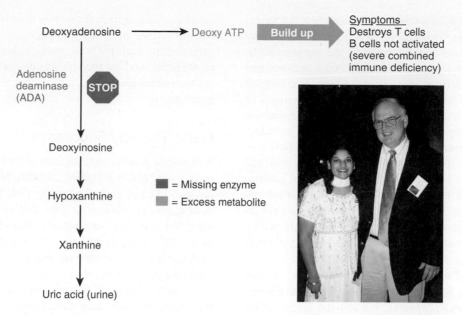

Figure 20.6 ADA deficiency. Absence of the enzyme adenosine deaminase (ADA) causes deoxy ATP to build up, which destroys T cells, which then cannot stimulate B cells to secrete antibodies. The result is severe combined immune deficiency (SCID). In 1990, Ashi DeSilva became the first person to have gene therapy. Today she is healthy. This photo, with Dr. Michael Blaese, was taken when she was 17.

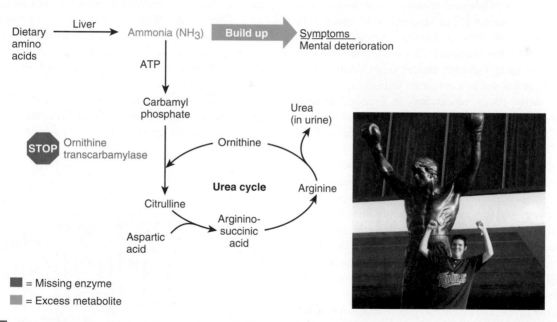

Figure 20.7 OTC deficiency. Lack of the enzyme ornithine transcarbamylase prevents nitrogen stripped from dietary proteins from leaving the body in urine. Instead, the nitrogen combines with hydrogen to form ammonia, which damages the brain, causing coma. Jesse Gelsinger had a mild case because he was a mosaic, with the mutation in only some of his cells. He had gene therapy and died 4 days later of an overwhelming immune response. The gene-bearing viruses introduced into his liver targeted the intended liver cells, but also entered immune system cells.

by age 5. The survivors can control their symptoms by following a very restrictive low-protein diet and taking drugs that bind ammonia.

Jesse wasn't diagnosed until he was 2, when he went into a coma and entered Children's Hospital of Philadelphia. He had a mild case because he was a mosaic—some of his cells could produce the enzyme. At age 10, Jesse's diet lapsed and he was

hospitalized again. When he went into a coma in December 1998 after missing a few days of his medications, he considered volunteering for a gene therapy trial the doctor had mentioned. When Jesse turned 18 the next June, he underwent tests at the University of Pennsylvania and was admitted into the clinical trial. He was jubilant. He knew he might not directly benefit, but he had wanted to help babies who die of the condition.

The gene therapy was an adenovirus carrying a functional human *OTC* gene. This virus had already been used in many gene therapy experiments and did not have the genes to replicate and cause respiratory symptoms. Seventeen patients had been treated, without serious side effects, when Jesse entered the hospital to receive more than a trillion altered viruses in an artery leading into his liver. It was Monday, September 13. That night, Jesse developed a high fever. By Tuesday morning, the whites of his eyes were yellow, indicating that his liver was struggling to dismantle the hemoglobin released from burst red blood cells. A flood of hemoglobin meant a flood of protein, elevating the ammonia level in his liver to ten times normal levels by mid-afternoon. Jesse became disoriented, then comatose. His lungs and then other vital organs began to fail, and by Friday he was brain dead. His dedicated and devastated medical team stood by as his father turned off life support.

The autopsy and analysis of tests done before, during, and after the procedure showed that the adenovirus had entered not only the hepatocytes as expected, but also the macrophages that function as sentries for the immune system. In response, interleukins flooded Jesse's body, and inflammation raged. Although afterward parents of children with OTC deficiency implored government officials to continue to fund the research, the death of Jesse Gelsinger led to suspension of many gene therapy trials. The death drew particular attention to safety because, unlike most other volunteers, Jesse had not been very ill. However, some gene therapy clinical trials continued in other countries, and that led to the second setback.

Since the early 1990s, researchers in France had been working on gene therapy for X-linked severe combined immune deficiency (SCID-X1). This is the disease that claimed the life of David Vetter, the "bubble boy" (see figure 17.10). In SCID-X1, T cells lack certain cytokine receptors, which prevents the immune system from recognizing infection. Most children die in infancy. *Ex vivo* gene therapy removes a boy's T cell progenitor cells from bone marrow, gives them wild type alleles, and infuses the corrected cells back into the body. The researchers used a retrovirus, which only enters dividing cells. If the healing viruses could infect the T cell progenitor cells, the cells would differentiate into mature T cells capable of alerting the immune system.

In 1999, the researchers gave the gene therapy for SCID-X1 to two babies. Both boys initially did well, their skin infections and persistent diarrhea clearing up. Even their tiny thymus glands hummed to life and grew, and their T cells showed the correction. Encouraged, the researchers treated more boys, and published a report in 2002 in a prominent medical journal. Soon after the article appeared, blood work on one of the first two boys showed a sharp increase in the numbers of a subtype of T cell. In a short time, this abnormality progressed to leukemia. Later, researchers discovered that retroviruses had inserted into chromosome 11, activating a proto-oncogene. This effect was not seen in mice.

The researchers hoped the leukemia was a fluke, but by the end of the year, another boy had it. By then, ten boys in France and ten in England had received the gene therapy for SCID-X1. Seventeen of them regained immunity, but five developed leukemia and one boy died of it. Leukemia induced by retroviruses inserting into oncogenes has also happened in gene therapy trials for Wiskott-Aldrich syndrome (see the chapter 6 opener) and for chronic granulomatous disease. Researchers have since altered the viruses to not disrupt cancer-causing genes and to "self-inactivate" after they deliver their cargo.

Gene Therapy Successes

A dramatic gene therapy success is for Leber congenital amaurosis type 2 (LCA2; OMIM 204100). More than 200 individuals who lived in dark shadows or total blindness can now see. A little girl, just days after gene therapy, could navigate a curb using her eyes, not her cane. Another child could see his food in a dimly lit restaurant near the hospital where days earlier he'd had his eye operated on. For years he had had to feel and smell the items on his plate to identify them.

At the backs of the eyes in a person with LCA2, cells that make up a thin layer called the retinal pigment epithelium (RPE) cannot make an enzyme necessary to convert vitamin A to a form that the rods and cones, the cells that transmit light energy to the brain, can use (**figure 20.8**). The visual world fades until sight is completely gone by early adulthood.

One of the first success stories for gene therapy for LCA2 was that of Corey Haas, who was 8 years old when he was treated, in his first eye at Children's Hospital of Philadelphia. Corey's parents first noticed that their baby son didn't make eye contact, yet stared at brightly lit bulbs. He was a clumsy toddler. Several doctors said the boy was just very nearsighted, but then an educator visiting the Haas home mentioned another child with similar visual problems

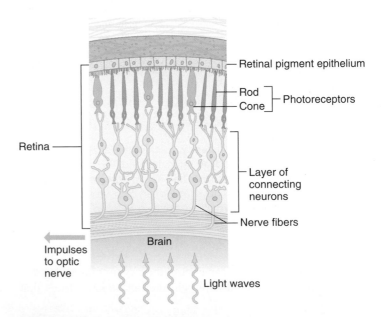

Figure 20.8 LCA2. In Leber congenital amaurosis type 2, a missing enzyme prevents cells of the retinal pigment epithelium from activating vitamin A so that it can nourish the rods and cones, which send visual signals to the brain.

who was seeing a doctor in Boston. That physician ultimately diagnosed Corey. Luckily, she had been on a team that had conducted successful gene therapy on a breed of sheepdog that had the very same disease. The doctor also knew that her colleagues in Philadelphia were looking for someone about Corey's age to be part of a clinical trial of gene therapy that had already restored vision to sheepdogs and several young adults (**figure 20.9**).

Three days after his birthday, in September 2008, Corey had the gene therapy on his worse eye, his left; his second would be done later, if the treatment worked. The surgeon cut a tiny pocket in the eye and injected several billion copies of adeno-associated virus (AAV), each one carrying a wild type version of the *RPE65* gene that was mutant in the boy. AAV is safer than AV, but can deliver genes only to nondividing cells, where the gene persists outside the chromosome as part of a ring of DNA. Just 4 days later, on a brilliant afternoon, the Haas family visited the Philadelphia zoo. When Corey looked up at the giant balloon hovering over the entranceway, he screamed! It was the first time he had been able to see bright sunlight.

Gene therapy has also been effective for adrenoleukodystrophy (ALD). This is the disease of peroxisomes that affected Lorenzo Odone, whose parents devised a dietary oil to help him, described in chapter 2. Peroxisomes are tiny sacs inside cells that house enzymes. They have porthole-like openings formed by a protein called ABCD1. A mutation in the *ABCD1* gene on the X chromosome prevents the portholes from admitting an enzyme needed to process fats called very long chain fatty acids, which are used to make myelin. The affected brain cells, called microglia, descend from progenitor cells in the bone marrow, which makes gene therapy for ALD much less invasive than for Canavan disease, described in the chapter opener.

b.

a.

Figure 20.9 Sheepdogs and people get LCA2

(a) Kristina Narfström, the veterinary researcher who discovered the disease in dogs, poses with Pluto, who had gene therapy when he was 4 years old. (b) Corey Haas lived in an ever-narrowing visual world until gene therapy restored his ability to see. He is age 10 in this photo, taken when he was between gene therapies. Today the two eyes are treated within days of each other.

The gene therapy for ALD fixes bone marrow cells outside the body, and introduces them into the bloodstream, where they go to the brain. Here they give rise to corrected microglia. The viral vector is HIV, stripped of the genes that cause AIDS. HIV is a type of retrovirus called a lentivirus, and is an excellent vector because it does not insert into proto-oncogenes, is very efficient, can carry large genes, and enters many types of cells. For several treated boys, blood levels of the crippling fats fell so greatly and brain neurons gained enough myelin so that they could attend school—with only about 15 percent of their microglia corrected! Teaming the gene therapy for ALD with newborn screening may completely prevent symptoms of this otherwise devastating disease.

The early gene therapies and many of the current ones address single-gene (Mendelian) disorders. The approach is also useful for cancers (which are due to somatic mutations) and even diseases that aren't inherited at all. Following are two exciting new examples.

Xerostomia ("dry mouth") develops in 40,000 people in the United States each year as a side effect of radiation treatment for head and neck cancer. The damaged salivary glands and resulting lack of saliva can lead to tooth decay, mouth sores, and gum disease, making eating quite difficult and painful. Salivary gland gene therapy may also alleviate the dry mouth of Sjögren's syndrome, an autoimmune condition that affects 4 million people in the United States.

Gene therapy for dry mouth revives remaining salivary gland cells by delivering a gene called aquaporin-1, aboard a viral vector (AV or AAV). Aquaporin-1 proteins form water channels that release fluid from salivary gland cells, into the mouth. The gene therapy is easily applied by slipping a tube into a natural opening near an upper molar that leads to one of the paired parotid salivary glands. The researcher who developed this gene therapy, a dentist and biochemist, tried it on himself, reporting that it feels like a poke followed by a sensation like a snake slithering along the back of the teeth. Even though the gene therapy uses adenovirus, which caused the death of Jesse Gelsinger, the viral vector appears safe in the confines of the salivary gland. If gene therapy vectors escape, the person would simply spit them out. Patients in clinical trials for dry mouth gene therapy are making more saliva and feeling much better.

A gene therapy approach called chimeric antigen receptor (CAR) technology combines two parts of the immune system to alter a leukemia patient's T cells in a way that enables them to destroy the cancer cells. A chimera is a mythical creature consisting of part lion, snake, and goat. A CAR is a protein in the plasma membrane of a T cell consisting of a T cell receptor that dips into the cell (see figure 17.9) and part of an antibody that juts from the cell (see figure 17.7). The T cell receptor handles signals while the antibody directs the T cell to a specified target, such as cell surface molecules (antigens) that are more abundant on cancer cells.

A billion or so CAR-bearing cells binding cancer cells act like flares revealing a military target, signaling other components of the immune response to destroy the cancer. This strategy is gene therapy because the chimeric antigen doesn't exist in nature—the T cells receive DNA sequences encoding the T cell receptor part and the antibody part aboard HIV, used as the vector.

Several dramatic cases illustrate effectiveness of the CAR technique. Emma Whitehead was 6 years old and near death when she had the gene therapy in the spring of 2012. She improved so fast that the following fall she was able to return to school. The CAR strategy is being tested on several types of cancers.

Summary

20.1 Genetic Counseling

1. **Genetic counselors** provide information on inheritance patterns for specific illnesses, disease risks and symptoms, and available tests and treatments.
2. Prenatal counseling and counseling a family coping with a particular disease pose different challenges.
3. Genetic counselors interpret genetic tests and assist other health care professionals in incorporating genetic information into their practices.

20.2 Genetic Testing

4. Genetic tests are performed before birth, on newborns, on children, and in adults. A **chromosomal microarray (CMA) test** detects small deletions and duplications.
5. The Clinical Laboratory Improvement Amendments regulate some genetic tests.
6. DTC tests may provide confusing, inaccurate, or incomplete information.
7. **Pharmacogenetic** and **pharmacogenomic** tests provide information on how individuals metabolize certain drugs.

20.3 Treating Genetic Disease

8. Enzyme replacement therapy, substrate reduction therapy, and pharmacological chaperone therapy treat biochemical imbalances.
9. Drugs may be repurposed to treat genetic diseases.
10. Gene therapy delivers genes and encourages production of a needed substance at appropriate times and in appropriate tissues, in therapeutic (not toxic) amounts.
11. **Germline gene therapy** affects gametes or fertilized ova, affects all cells of an individual, and is transmitted to future generations. It is not performed in humans. **Somatic gene therapy** affects somatic cells and is not passed to offspring.
12. *Ex vivo* **gene therapy** is applied to cells outside the body that are then reimplanted or reinfused into the patient. *In vivo* **gene therapy** delivers gene-carrying vectors directly into the body.
13. Several types of vectors are used to deliver therapeutic genes. Viruses are most commonly used. Some gene therapies target stem or progenitor cells, because they can divide and move.
14. After initial success in 1990, many gene therapy clinical trials halted after a death in 1999. Safer vectors and better understanding of disease have made gene therapies more successful.

www.mhhe.com/lewisgenetics11

Answers to all end-of-chapter questions can be found at **www.mhhe.com/lewisgenetics11**. You will also find additional practice quizzes, animations, videos, and vocabulary flashcards to help you master the material in this chapter.

Review Questions

1. What is unique about the services that a genetic counselor provides compared to those of a nurse or physician?
2. How are the consequences of a genetic test different from that of a cholesterol check?
3. Compare and contrast the types of information from preconception comprehensive carrier screening, prenatal diagnosis, and newborn screening in terms of the types of actions that can be taken in response to receiving the results.
4. Describe a situation in which a DNA test performed on human remains provides interesting or helpful information.
5. Using information from this or other chapters, or the Internet, cite DNA-based tests given to a fetus, a newborn, a young adult, and a middle-aged person.
6. Explain how a pharmacogenetic test can improve quality of life for a person with cancer.

7. List an advantage and a limitation of a direct-to-consumer genetic test.

8. Explain why the removal of blood in people with hereditary hemochromatosis is not gene therapy.

9. Distinguish among enzyme replacement therapy, substrate reduction therapy, and pharmacological chaperone therapy.

10. What technology described in chapter 19 would be used to carry out germline gene therapy, if this were permitted in humans?

11. Explain how somatic or germline gene therapy can affect evolution.

12. What factors would a researcher consider in selecting a viral vector for gene therapy?

13. Identify two complications that slowed the development of gene therapy.

14. Explain how one of the gene therapies described in the chapter works.

Applied Questions

1. Discuss the challenges that a genetic counselor faces in explaining to parents-to-be a prenatal diagnosis of trisomy 21 in a fetus with no family history, compared to a prenatal diagnosis of translocation Down syndrome in a family with a reproductive history of pregnancy loss and birth defects. (Chapter 13 discusses Down syndrome.)

2. Suggest how newborn screening might be changed to minimize the numbers of "patients-in-waiting."

3. A company tests for variants of the gene *ACTN3* (alpha-actinin 3), which encodes a protein that binds actin, a cytoskeletal protein. One genotype is more common among elite sprint athletes, and another among endurance athletes. Some parents are testing their young children for these gene variants and using the results to decide whether the child should pursue a sport that

entails sprinting or endurance. Would you have your child tested? Cite a reason for your answer.

4. Explain why high penetrance is an important criterion for including a disease in newborn screening.

5. "Personalized medicine" is a popular if vague phrase. Describe one such approach mentioned in the chapter.

6. Should an online dating service include analysis based on the patent for "gamete donor selection based on genetic calculations"? Cite a reason for your answer.

7. Describe two genetic testing situations that can cause great anxiety.

8. Discuss the role of economics in population-level genetic testing.

Web Activities

1. Go to clinicaltrials.gov and search under "gene therapy." Describe one. Include the mode of inheritance, age of onset, symptom severity, variability, existing treatments (if any), and how the gene therapy works.

2. Enter terms related to human reproduction (such as "sperm" or "gamete") into the search field at the U.S. Patent and Trademark Office (www.uspto.gov), and describe a reproductive technology.

3. Go to the Pharmacogenomics Knowledge Base (http://www.pharmgkb.org/) and click on "Important PGx gene." Select a gene and discuss its importance in health care.

4. Go to www.genedx.com. Describe a genetic test that this company offers.

Case Studies and Research Results

1. If you were the genetic counselor for the following patients, how would you answer their questions or address their concerns? (See other chapters for specific information.)

   a. A couple in their early forties is expecting their first child. Amniocentesis indicates that the fetus is XXX. When they learn of the abnormality, the couple asks to terminate the pregnancy, fearing severe birth defects caused by the extra chromosome.

   b. Two people of normal height have a child with achondroplastic dwarfism, an autosomal dominant trait. Will future children have the condition (see figure 5.1)?

   c. A couple has results from tests they have taken from a direct-to-consumer company. Tests based on genome-wide association studies indicate that they each have inherited susceptibility to asthma as well as gene variants that in some populations are associated with autism. Both also have several gene variants that are found in lung cancers. Each has a few recessive mutant alleles, but not in the same genes. On the basis of these results, they do not think that they are "genetically healthy" enough to have children.

2. Jill and Scott S. had thought 6-month-old Dana was developing just fine until Scott's sister, a pediatrician, noticed that the baby's abdomen was swollen and hard. Knowing that the underlying enlarged liver and spleen could indicate an inborn error of metabolism, Scott's sister suggested the child undergo several tests.

   Dana had inherited sphingomyelin lipidosis, also known as Niemann-Pick disease type A (OMIM 257200). Both parents were carriers, but Jill had tested negative when she took a Jewish genetic disease panel during her pregnancy because her mutation was very rare and not part of the test panel. Dana was successfully treated with a transplant of umbilical cord blood cells from a donor. She caught up developmentally and became more alert. Monocytes, a type of white blood cell, from the cord blood traveled to her brain and manufactured the deficient enzyme. Dietary therapy does not work for this condition because the enzyme cannot cross from the blood to the brain. Monocytes, however, can enter the brain.

   a. Explain the effect of Dana's treatment on her phenotype and genotype.

   b. Why did the transplant have to come from donated cord blood, and not from Dana's own, which had been stored?

   c. If you were the genetic counselor, what advice would you give this couple if they conceive again?

3. A clinical trial of gene therapy for the clotting disorder hemophilia B introduced the gene for factor IX in AAV into the livers of several men. Researchers halted the trial when they detected the altered gene in semen of treated men. Why did they take this precaution?

4. Tanisha and Jamal were concerned that their son Perry did not make eye contact or respond as his older siblings had, but the pediatrician assured them that the boy was fine, just easily distracted. In school, though, Perry could not focus, and had angry outbursts. Tanisha, a nurse, learned about chromosomal microarray (CMA) technology in a genetics class. She had Perry tested, which revealed a small duplication. The report stated that the finding was a "variant of uncertain significance." The duplication was abnormal, but it hadn't been associated with a syndrome. Tanisha wanted to include the test results in Perry's school records, in case a situation should arise in which a medical explanation might be helpful. But Jamal feared that the genetic test result might stigmatize the boy. Do you agree with either parent? Cite a reason for your answer.

5. Eliza O'Neill's first symptoms of Sanfilippo syndrome type A (OMIM 252900) were slight developmental delay, hyperactivity, and not interacting with her peers at preschool. A pediatrician who noticed that Eliza's forehead jutted out a little ordered an MRI scan, which revealed fluid at the back of her brain and flattened vertebrae in her neck, features of the syndrome. A friend of the family who is a geneticist thought these characteristics fit the pattern of Sanfilippo and suggested a urine test, which showed the telltale buildup of heparan sulfate, caused by deficiency of a lysosomal enzyme (see figure 2.7).

   Eliza was diagnosed at age four, on July 17th, 2013. Recalled her father Glenn, "In one terrifying instant, we were told that we would have to watch Eliza fade away before our eyes." As the heparan sulfate builds up, the child loses the ability to speak and move. Seizures begin, and life typically ends in early adolescence.

   Describe how one of the technologies described in the chapter might be able to help Eliza. Learn more about her at www.facebook.com/ElizaOStory.

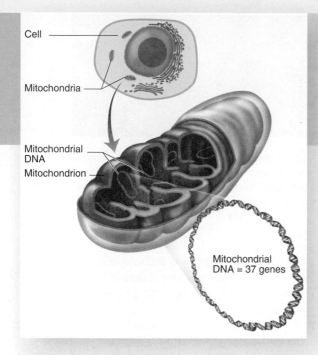

Cell

Mitochondria

Mitochondrial DNA

Mitochondrion

Mitochondrial DNA = 37 genes

Mitochondrial replacement creates a fertilized ovum that has genetic material from three individuals—a male genome, a female genome, and 37 mitochondrial genes from another individual.

Reproductive Technologies

Learning Outcomes

21.1   Savior Siblings and More

1. Explain how a child can be conceived to provide tissue for an older sibling.

2. Define *assisted reproductive technology.*

21.2   Infertility and Subfertility

3. Distinguish infertility from subfertility.

4. Describe causes of infertility in the male.

5. Describe causes of infertility in the female.

6. List infertility tests.

21.3   Assisted Reproductive Technologies

7. Describe assisted reproductive technologies that donate sperm, uterus, or oocyte.

8. List the steps of *in vitro* fertilization.

9. Explain how preimplantation genetic diagnosis avoids the birth of a child with a particular genetic disease.

10. Explain how testing a polar body can reveal information about a genotype of a fertilized ovum.

21.4   Extra Embryos

11. Discuss uses for extra embryos resulting from assisted reproductive technologies.

The **BIG** Picture

Assisted reproductive technologies provide intriguing and sometimes complex variations on the process of conceiving a child and carrying it to term.

Replacing Mitochondria

The tiny mitochondrial genome of 16,569 DNA base pairs can cause severe illness when any of its 37 genes, which encode a few proteins, rRNAs, and tRNAs, is mutant. Mothers transmit these diseases because sperm do not deliver their mitochondria when they fertilize an oocyte. However, the inheritance pattern is complex, because cells have many mitochondria, mitochondria have several copies of their minuscule genomes, and heteroplasmy (different genetic variants in a cell) is common, as section 5.2 discusses. Mitochondrial disorders cause fatigue and exercise intolerance, and can also affect the liver, kidney, endocrine system, brain, and lungs. Some conditions cause deafness and blindness.

Mitochondrial replacement can prevent inheriting mutations carried in these organelles. Researchers do it two ways: (1) transfer the male and female pronuclei from an *in vitro* fertilized ovum into an oocyte donated from a healthy woman that has had its nucleus removed, so that the pronuclei share a cytoplasm with donor mitochondria; and (2) transfer the spindle apparatus, to which mitochondria cling, from a donor cell to a fertilized ovum. Whichever way it happens, the resulting first cell has two haploid genomes plus the small amount of mitochondrial DNA from the donor, a third individual.

Mitochondrial replacement has worked in monkeys but in other animals has led to developmental problems, including accelerated aging, poor growth, lowered fertility, and early death. Some researchers caution that the rarity of mitochondrial diseases, compared to the high mutation rate of mitochondrial DNA, suggests that embryos that inherit the disorders are naturally weeded out. Another complicating factor may be epigenetic effects from a cytoplasm on a nucleus from a different individual.

When the Food and Drug Administration held a meeting to discuss the science behind mitochondrial manipulation, a public comment session dealt with the bioethics of creating a "three-parent" embryo. Another compelling objection came from a young woman who has suffered with a mitochondrial disease for more than a decade. She concluded, *"Oh gosh! It would be so great if I was listening to all this research and it was about therapies for those of us with mitochondrial disease, helping those of us whose lives are severely affected. But this might be the gateway to that."*

Figure 21.1 **Savior siblings.** Adam Nash was conceived and selected to save his sister Molly's life. He is also a much-loved sibling and son. Several other families have since conceived a child to help another.

21.1 Savior Siblings and More

A couple in search of an oocyte donor advertises in a college newspaper seeking an attractive, bright young woman from an athletic family. A cancer patient has ovarian tissue removed and frozen before undergoing treatment. Two years later, she has a strip of the frozen tissue thawed and oocytes separated and fertilized in a laboratory dish with her partner's sperm. Several cleavage embryos develop and two are implanted in her uterus. She becomes a mother—of twins. A man paralyzed from the waist down has sperm removed and injected into his partner's oocyte. He, too, becomes a parent when he thought he never would.

Lisa and Jack Nash wanted to have a child for a different reason. Their daughter Molly, born on July 4, 1994, had Fanconi anemia (OMIM 227650). This autosomal recessive condition would destroy her bone marrow and her immunity. An umbilical cord stem cell transplant from a sibling could likely cure her, but Molly had no siblings. Nor did her parents wish to have another child who would have a one in four chance of inheriting the disorder, as Mendel's first law dictates. Technology offered a solution.

In late 1999, researchers at the Reproductive Genetics Institute at Illinois Medical Center mixed Jack's sperm with Lisa's oocytes in a laboratory dish. After allowing fifteen of the fertilized ova to develop to the 8-cell stage, researchers separated and applied DNA probes to one cell from each embryo. A cell that had wild type Fanconi anemia alleles and that matched Molly's human leukocyte antigen (HLA) type was identified and its seven-celled remainder implanted into Lisa's uterus. Adam was born in late summer. A month later, physicians infused his umbilical cord stem cells into Molly, saving her life (**figure 21.1**). The Nashes were initially sharply criticized for intentionally conceiving a "savior sibling." As others

followed their example, conceiving and selecting a child to provide cells for a sibling became more accepted.

Increased knowledge of how the genomes of two individuals come together and interact has spawned several novel ways to have children. **Assisted reproductive technologies (ARTs)** replace the source of a male or female gamete, aid fertilization, or provide a uterus. These procedures were developed to treat infertility, but are increasingly including genetic testing too. In the United States, the government does not regulate ARTs, but the American Society for Reproductive Medicine provides voluntary guidelines. The United Kingdom has pioneered ARTs and its Human Fertilisation and Embryology Authority has served as a model for government regulation. A great advantage of the British regulation of reproductive health services and technologies is that databases include success rates of the different procedures. Another advantage is that access to reproductive technology is not limited to those who can afford it.

The landscape of assisted reproductive technologies is constantly changing, fed by imagination as well as by new discoveries. In early 2007, for example, bioethicists in the Netherlands published a controversial proposal: Select two savior sibling embryos. Permit one embryo to continue developing to be born, and use the other embryo to derive and store embryonic stem cells that could one day provide healing cells if the older child, who is a clone, becomes sick. Later that same year, the invention of induced pluripotent stem cells (reprogramming; see figure 2.22) provided a similar source of cells from the patient, not requiring use of a spare embryo at all.

Key Concepts Questions 21.1

1. What are assisted reproductive technologies?
2. What is a savior sibling?

21.2 Infertility and Subfertility

Infertility is the inability to conceive a child after a year of frequent intercourse without the use of contraceptives. Some specialists use the term *subfertility* to distinguish those individuals and couples who can conceive unaided, but for whom this may take longer than average. On a more personal level, infertility is a seemingly endless monthly cycle of raised hopes and crushing despair. In addition to declining fertility, as a woman ages, the incidence of pregnancy-related problems rises, including chromosomal anomalies, fetal deaths, premature births, and low-birth-weight babies. Older fathers are at increased risk of having children who have autism or schizophrenia. Sperm motility declines with age.

One in six couples has difficulty conceiving or giving birth to children. Physicians who specialize in infertility treatment can identify a physical cause in 90 percent of cases. Of these cases, 30 percent of the time the problem is primarily in the male, and 60 percent of the time it is primarily in the female. When a physical problem is not obvious, the cause is usually a mutation or chromosomal aberration that impairs fertility in the male. The statistics are somewhat unclear, because in 20 percent of the 90 percent, both partners have a medical condition that could contribute to infertility or subfertility. A common combination is a woman with an irregular menstrual cycle and a man with a low sperm count.

Male Infertility

Infertility in the male is easier to detect but sometimes harder to treat than female infertility. Four in 100 men in the general population are infertile, and half of them do not make any sperm, a condition called azoospermia. Some men have difficulty fathering a child because they produce fewer than the average 20 to 200 million sperm cells per milliliter of ejaculate. This condition, called oligospermia, has several causes. If a low sperm count is due to a hormonal imbalance, administering the appropriate hormones may boost sperm output. Sometimes a man's immune system produces IgA antibodies that cover the sperm and prevent them from binding to oocytes. Male infertility can also be due to a varicose vein in the scrotum. This enlarged vein emits heat near developing sperm, which prevents them from maturing. Surgery can remove a scrotal varicose vein.

Most cases of male infertility are genetic. About a third of infertile men have small deletions of the Y chromosome that remove the only copies of key genes whose products control spermatogenesis. Other genetic causes of male infertility include mutations in genes that encode androgen receptors or protein fertility hormones, or that regulate sperm development or motility. **Clinical Connection 21.1** describes a type of autosomal recessive male infertility that is unusual in that it is not part of a syndrome.

For many men with low sperm counts, if they have at least 60 million sperm cells per ejaculate, fertilization is likely eventually. To speed conception, a man with a low sperm count can donate several semen samples over a period of weeks at a fertility clinic. The samples are kept in cold storage, then pooled. Some of the seminal fluid is withdrawn to leave a sperm cell concentrate, which is then placed in the woman's body. It isn't very romantic, but it is highly effective at achieving pregnancy. Men who actually want a very low sperm count—those who have just had a vasectomy for birth control—can use an at-home test kit to monitor their sperm counts. Fewer than 12 to 16 million sperm cells per milliliter of seminal fluid makes pregnancy unlikely.

Sperm quality is more important than quantity. Sperm cells that are unable to move or are shaped abnormally cannot reach an oocyte. Inability to move may be due to a hormone imbalance, and abnormal shapes may reflect impaired apoptosis (programmed cell death) that normally removes such sperm. The genetic package of an immobile or abnormally shaped sperm cell can be injected into an oocyte and sometimes this leads to fertilization. However, even sperm that look and move normally may be unable to fertilize an oocyte.

Female Infertility

Abnormalities in any part of the female reproductive system can cause infertility (**figure 21.2**). Many women with subfertility or infertility have irregular menstrual cycles, making it difficult to pinpoint when conception is most likely. In an average menstrual cycle of 28 days, ovulation usually occurs around the 14th day after menstruation begins. At this time a woman is most likely to conceive.

For a woman with regular menstrual cycles who is under 30 years old and not using birth control, pregnancy typically happens within 3 or 4 months. A woman with irregular menstrual periods can tell when she is most fertile by using an ovulation predictor test, which detects a peak in the level of luteinizing hormone that precedes ovulation by a few hours. Another way to detect the onset of ovulation is to record body temperature each morning using a digital thermometer with subdivisions of hundredths of a degree Fahrenheit, which can indicate the 0.4 to 0.6 rise in temperature when ovulation starts. Several apps track a woman's menstrual cycle, enabling her to predict the time of ovulation. Sperm can survive in a woman's body for up to 5 days, but the oocyte is only viable for 24 to 48 hours after ovulation.

The hormonal imbalance that usually underlies irregular ovulation has various causes. These include a tumor in the ovary or in the pituitary gland in the brain that controls the reproductive system, an underactive thyroid gland, or use of steroid-based drugs such as cortisone. If a nonpregnant woman produces too much prolactin, the hormone that promotes milk production and suppresses ovulation in new mothers, she will not ovulate.

Fertility drugs can stimulate ovulation, but they can also cause women to "superovulate," producing and releasing more than one oocyte each month. A commonly used drug, clomiphene, raises the chance of having twins from 1 to 2 percent to 4 to 6 percent. If a woman's ovaries are completely inactive or absent (due to a birth defect or surgery), she can become

The Case of the Round-Headed Sperm

In fewer than a tenth of a percent of men who are infertile, sperm cells lack the tip, called the acrosome, which contains the enzymes that break through the layers surrounding an oocyte. This condition is called "globozoospermia" (**figure 1**). An Ashkenazi Jewish family led researchers to a gene that, when mutant, causes an autosomal recessive form of male infertility due to sperm with round rather than oval heads

The family went to a center for reproductive medicine in Brussels, the Netherlands. Of the six sons, three were infertile (**figure 2**). Four daughters were fertile. The affected sons' sperm were misshapen, and the mode of inheritance recessive, because the parents were fertile.

Researchers suspected consanguinity—a shared ancestor would increase the risk of inheriting a very rare autosomal recessive condition if the mutation is in the family. But the family denied knowing a relative who had married a relative. Reasoning that perhaps DNA could reveal consanguinity that the family did not know about, researchers scanned the genomes of all six sons for regions of homozygosity that would indicate they had relatives marrying relatives not too far back in the family tree (see chapter 7).

A region of homozygosity in this case was defined as 25 consecutive SNPs that were homozygous. The genomes of all six sons were riddled with these regions, suggesting that at some point, cousin married cousin or an aunt/uncle wed a nephew/niece. One region of homozygosity was seen in all three infertile brothers, but was heterozygous in two of the three fertile brothers. The remaining brother was homozygous wild type for the region.

Next, the researchers scrutinized the part of the long arm of chromosome 3 where the region of homozygosity lay. It houses fifty genes, only one of which is expressed in the testes. This gene is called "spermatogenesis-associated protein 16," or *SPATA16* (OMIM 102530). It has eleven exons, and the mutation in the Ashkenazi family is a single base change, from G to A, at the 848th position in the gene, near the end of exon 4. The mutation affects the splicing out of introns as the gene is transcribed.

The wild type protein product of the *SPATA16* gene is transported from the Golgi apparatus into vesicles that take it to the acrosome as it telescopes out of the front end of a sperm cell. By attaching the gene for the jellyfish's green fluorescent protein (see figure 19.1) to the wild type *SPATA16* gene in cells growing in culture, researchers visualized the protein being transported to the forming acrosome in immature sperm.

Questions for Discussion

1. What is the evidence that the infertility in the family is inherited as an autosomal recessive trait?

2. What is the evidence for inheritance of the mutation from shared ancestors?

3. Discuss how researchers discovered the gene that is mutant in the family.

4. Which chapters in the book discuss the concepts behind the findings described in this Clinical Connection?

5. The day is fast approaching when exome and genome sequencing will become available. What would you do, or not do, if genome information reveals that someone you love is not related to you in the way that you thought?

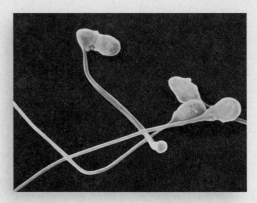

Figure 1 **A sperm cell's streamlined form facilitates movement.** A misshapen sperm cannot fertilize an oocyte.

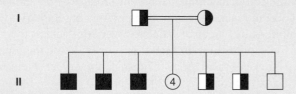

Figure 2 **Inheriting infertility.** In a family with autosomal recessive globozoospermia, three of six sons are infertile.

pregnant only if she uses a donor oocyte. Some cases of female infertility are due to "reduced ovarian reserve"—too few oocytes. This is typically discovered when the ovaries do not respond to fertility drugs. Signs of reduced ovarian reserve are an ovary with too few follicles (observed on an ultrasound scan) or elevated levels of follicle-stimulating hormone on the third day of the menstrual cycle.

The uterine tubes are a common site of female infertility because fertilization usually occurs in open tubes. Blockage can prevent sperm from reaching the oocyte, or entrap a fertilized ovum, keeping it from descending into the uterus. If an embryo begins developing in a blocked tube and is not removed and continues to enlarge, the tube can burst and the woman can die. Such a "tubal pregnancy" is called an ectopic pregnancy.

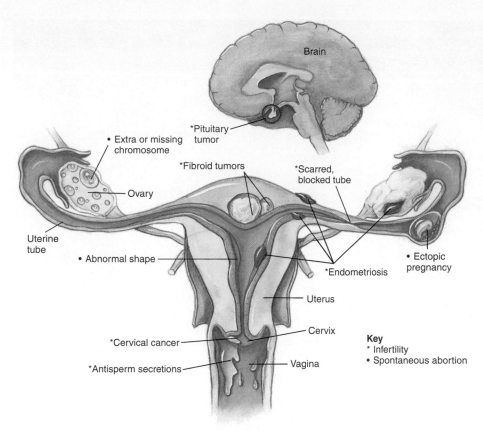

Figure 21.2 Sites of reproductive problems in the female.

that have an abnormal chromosome number, which often causes spontaneous abortion because defects are too severe for development to proceed for long. The cause is usually misaligned spindle fibers when the second meiotic division begins, causing aneuploidy (extra or missing chromosomes). Perhaps the longer exposure of older oocytes to harmful chemicals, viruses, and radiation contributes to the risk of meiotic errors. Losing very early embryos may appear to be infertility because the bleeding accompanying the aborted embryo resembles a heavy menstrual flow.

Infertility Tests

A number of medical tests can identify causes of infertility. The man is checked first, because it is easier, less costly, and less painful to obtain sperm than oocytes.

Sperm are checked for number (sperm count), motility, and morphology (shape). An ejaculate containing up to 40 percent unusual forms is still considered normal, but many more than this can impair fertility. A urologist performs sperm tests. A genetic counselor can evaluate Y chromosome deletions associated with lack of sperm. If a male cause of infertility is not apparent, a gynecologist checks the woman to see that reproductive organs are present and functioning.

Some cases of subfertility or infertility have no clear explanation. Psychological factors may be at play, or it may be that inability to conceive results from consistently poor timing. Sometimes a subfertile couple adopts a child, only to conceive one of their own shortly thereafter; many times, infertility remains a lifelong mystery.

Uterine tubes can also be blocked due to a birth defect or, more likely, from an infection such as pelvic inflammatory disease. A woman may not know she has blocked uterine tubes until she has difficulty conceiving and medical tests uncover the problem. Surgery can open blocked uterine tubes.

Excess tissue growing in the uterine lining may make it inhospitable to an embryo. This tissue can include benign tumors called fibroids or areas of thickened lining from a condition called endometriosis. The tissue can grow outside of the uterus too, in the abdominal cavity. In response to the hormonal cues to menstruate, the excess lining bleeds, causing cramps. Endometriosis can hamper conception, but curiously, if a woman with endometriosis conceives, the cramps and bleeding usually disappear after the birth.

Secretions in the vagina and cervix may be hostile to sperm. Cervical mucus that is thick or sticky due to infection can entrap sperm, keeping them from moving far enough to encounter an oocyte. Vaginal secretions may be so acidic or alkaline that they weaken or kill sperm. Douching daily with an acidic solution such as acetic acid (vinegar) or an alkaline solution such as bicarbonate, can alter the pH of the vagina so that in some cases it is more receptive to sperm cells. Too little mucus can prevent conception too; this is treated with low daily doses of oral estrogen. Sometimes mucus in a woman's body has antibodies that attack sperm. Infertility may also result if the oocyte does not release sperm-attracting biochemicals.

One reason the incidence of female infertility increases with age is that older women are more likely to produce oocytes

---

Key Concepts Questions 21.2

1. What are causes of male infertility?
2. What are causes of female infertility?
3. Describe medical tests used to identify the causes of infertility.

21.3 Assisted Reproductive Technologies

Many people with fertility problems who do not choose to adopt children use alternative ways to conceive. Several of the ARTs were developed in nonhuman animals (see the Technology Timeline on page 412). In the United States, slightly more

LANDMARKS IN REPRODUCTIVE TECHNOLOGY

	In Nonhuman Animals	In Humans
1782	Intrauterine insemination (IUI) in dogs	
1790		Pregnancy reported from intrauterine insemination
1890s	Birth from embryo transplantation in rabbits	IUI by donor
1949	Cryoprotectant successfully freezes animal sperm	
1951	First calf born after embryo transplantation	
1952	Live calf born after insemination with frozen sperm	
1953		First reported pregnancy after insemination with frozen sperm
1959	Live rabbit offspring produced from *in vitro* ("test tube") fertilization (IVF)	
1972	Live offspring from frozen mouse embryos	
1976	Intracytoplasmic sperm injection (ICSI) in hamsters	First reported commercial surrogate motherhood arrangement in the United States
1978	Transplantation of ovaries from one cow to another	Baby born after *in vitro* fertilization (IVF) in United Kingdom
1980		Baby born after IVF in Australia
1981	Calf born after IVF	Baby born after IVF in United States
1982	Sexing of embryos in rabbits	
	Cattle embryos split to produce genetically identical twins	
1983		Embryo transfer after uterine lavage
1984		Baby born in Australia from frozen and thawed embryo
1985		Baby born after gamete intrafallopian transfer (GIFT)
		First reported gestational-only surrogacy arrangement in the United States
1986		Baby born in the United States from frozen and thawed embryo
1989		First preimplantation genetic diagnosis (PGD)
1992		First pregnancies from intracytoplasmic sperm injection
1994		62-year-old woman gives birth from fertilized donated oocyte
1995	Sheep cloned from embryo cell nuclei	Babies born following ICSI
1996	Sheep cloned from adult cell nucleus	
1998	Mice cloned from adult cell nuclei	Baby born 7 years after his twin
1999	Cattle cloned from adult cell nuclei	
2000	Pigs cloned from adult cell nuclei	
2001		Sibling born following PGD to treat sister for genetic disease
		Human preimplantation embryo cloned, survives to 6 cells
2003		3,000-plus preimplantation genetic diagnoses performed to date

LANDMARKS IN REPRODUCTIVE TECHNOLOGY

	In Nonhuman Animals	In Humans
2004	Woman pays $50,000 to have her cat cloned	First birth from a woman who had ovarian tissue preserved and implanted on an ovary after cancer treatment
2005	Dog cloned	
2011		First children born free of single-gene disease following sequential polar body analysis
2013		First woman conceives from stored ovarian tissue.

than 1 percent of the approximately 4 million births a year are from ARTs, and worldwide ART accounts for about 250,000 births a year.

This section describes types of ARTs. The different procedures can be performed on material from the parents-to-be ("nondonor") or from donors, and may be "fresh" (collected just prior to the procedure) or "frozen" (preserved in liquid nitrogen). Except for intrauterine insemination, the ARTs cost thousands of dollars and are not typically covered by health insurance in the United States.

Donated Sperm—Intrauterine Insemination

The oldest assisted reproductive technology is **intrauterine insemination (IUI),** in which a doctor places donated sperm into a woman's cervix or uterus. (It used to be called artificial insemination.) The success rate is 5 to 15% per attempt. The sperm are first washed free of seminal fluid, which can inflame female tissues. A woman might seek IUI if her partner is infertile or has a mutation that the couple wishes to avoid passing to their child. Women also undergo IUI to be a single parent without having sex, or a lesbian couple may use it to have a child.

The first documented IUI in humans was done in 1790. For many years, physicians donated sperm, and this became a way for male medical students to earn a few extra dollars. By 1953, sperm could be frozen and stored and IUI became much more commonplace. Today, donated sperm are frozen and stored in sperm banks, which provide the cells to obstetricians who perform the procedure. IUI costs on average $865 per cycle, according to the American Society of Reproductive Medicine, with higher charges from some facilities for sperm from donors who have professional degrees because those men are paid more for their donations. Additional fees are charged for a more complete medical history of the donor, for photos of the man at different ages, and for participation in a "consent program" in which the donor's identity is revealed when his offspring turns 18 years old. If ovulation is induced to increase the chances of success of IUI, additional costs may exceed $3,000.

A couple who chooses IUI can select sperm from a catalog that lists the personal characteristics of donors, such as blood type, hair and eye color, skin color, build, educational level, and interests. Some traits have nothing to do with genetics. If a couple desires a child of one sex—such as a daughter to avoid passing on an X-linked disorder—sperm can be separated into fractions enriched for X-bearing or Y-bearing sperm.

Problems can arise in IUI if a donor learns that he has an inherited disease. For example, a man developed cerebellar ataxia (OMIM 608029), a movement disorder, years after he donated sperm. Eighteen children conceived using his sperm face a 1 in 2 risk of having inherited the mutant gene. Over-enthusiastic sperm donors can lead to problems. One man, listed in the Fairfax Cryobank as "Donor 401," earned $40,000 donating sperm while in law school. He was quite attractive and popular, and forty-five children were conceived with his sperm. When a few of the families he started appeared on a talk show, several other families tuning in were shaken to see so many children who resembled their own. Cases came to light of males fathering more than 150 offspring, prompting sperm banks to limit sales of a particular male's sperm cells. The website http://www.donorsiblingregistry.com has enabled more than 10,000 half-siblings who share sperm donor fathers to meet.

A male's role in reproductive technologies is simpler than a woman's. A man can be a genetic parent, contributing half of his genetic self in his sperm, but a woman can be both a genetic parent (donating an oocyte) and a gestational parent (donating the uterus).

A Donated Uterus—Surrogate Motherhood

If a man produces healthy sperm but his partner's uterus cannot maintain a pregnancy, a surrogate mother may help by being inseminated with the man's sperm. When the child is born, the surrogate mother gives the baby to the couple. In this variation of the technology, the surrogate is both the genetic and the gestational mother. Attorneys usually arrange surrogate relationships. The surrogate mother signs a statement signifying her intent to give up the baby. In some U.S. states, and in some nations, she is paid for her 9-month job, but in the United Kingdom compensation is illegal. This is to prevent wealthy couples from taking advantage of women who become surrogates for the money.

A problem with surrogate motherhood is that a woman may not be able to predict her responses to pregnancy and childbirth in a lawyer's office months before she must hand over the baby. When a surrogate mother changes her mind, the results are wrenching for all. A prominent early case involved Mary Beth Whitehead, who carried the child of a married man for a fee and then changed her mind about giving up the baby. The courts eventually awarded custody to the father and his wife. The woman who raises the baby may feel badly too, especially when people say she is not the "real" mother.

Another type of surrogate mother lends only her uterus, receiving a fertilized ovum conceived from a man and a woman who has healthy ovaries but lacks a functional uterus. This variation is an "embryo transfer to a host uterus," and the pregnant woman is a "gestational-only surrogate mother." She turns the child over to the biological parents. About 1,600 babies are born in the United States to gestational surrogates each year.

In Vitro Fertilization

In *in vitro* fertilization (IVF), which means "fertilization in glass," sperm and oocyte join in a laboratory dish. Soon after, the embryo that forms is placed in a uterus. If all goes well, it implants into the uterine lining and continues development until a baby is born.

Louise Joy Brown, the first "test-tube baby," was born in 1978, amid great attention and sharp criticism. A prominent bioethicist said that IVF challenged "the idea of humanness and of our human life and the meaning of our embodiment and our relation to ancestors and descendants." Yet Louise is, despite her unusual beginnings, an ordinary young woman. More than 5 million children have been born following IVF.

A woman might undergo IVF if her ovaries and uterus work but her uterine tubes are blocked. Using a laparoscope, which is a lit surgical instrument inserted into the body through a small incision, a physician removes several of the largest oocytes from an ovary and transfers them to a culture dish. If left in the body, only one oocyte would exit the ovary, but in culture, many oocytes can mature sufficiently to be fertilized *in vitro*. Chemicals, sperm, and other cell types similar to those in the female reproductive tract are added to the culture. An acidic solution may be applied to the zona pellucida, which is the layer around the egg, to thin it to ease the sperm's penetration.

Sperm that cannot readily enter the oocyte may be sucked up into a tiny syringe and microinjected into the female cell. This technique, called **intracytoplasmic sperm injection (ICSI)**, is more effective than IVF alone and has become standard at some facilities (**figure 21.3**). ICSI is very helpful for men who have low sperm counts or many abnormal sperm. It makes fatherhood possible for men who cannot ejaculate, such as those who have suffered spinal cord injuries. ICSI has been performed on thousands of men with about a 30 percent success rate.

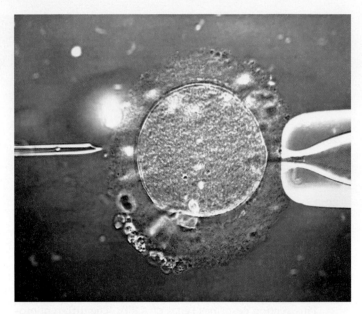

Figure 21.3 ICSI. Intracytoplasmic sperm injection (ICSI) enables some infertile men, men with spinal cord injuries, or men with certain illnesses to become fathers. A single sperm cell is injected into the cytoplasm of an oocyte.

Two to five days after sperm wash over the oocytes in the dish, or are injected into them, a blastocyst is transferred to the uterus. If the hormone human chorionic gonadotropin appears in the woman's blood a few days later, and its level rises, she is pregnant.

IVF costs from on average $8,158 per cycle. Medications can add $3,000 to $5,000 to the cost. ICSI adds another $1,544 on average. Children born following IVF have a slight increase in the rate of birth defects (about 8 percent) compared to children conceived naturally (about 3 percent). This difference may be due to the medical problems that caused the parents to seek IVF, the tendency of IVF to interfere with the parting of chromosome pairs during meiosis, closer scrutiny of IVF pregnancies, and/or effects on imprinting from the time spent in culture. Children born after IVF on average have higher birth weights. An increase in birth defects among multiple births since 1984 is possibly due to an increase in the use of ARTs.

In the past, several embryos were implanted to increase the success rate of IVF, but this led to many multiple births, which are riskier than single births. In some cases, physicians had to remove embryos to make room for others to survive. To avoid the multiples problem, and because IVF has become more successful as techniques have improved, guidelines now suggest transferring only one embryo.

Embryos resulting from IVF that are not soon implanted in the woman can be frozen in liquid nitrogen ("cryopreserved" or "vitrified") for later use. Cryoprotectant chemicals are used to prevent salts from building up or ice crystals from damaging delicate cell parts. Freezing takes a few hours; thawing about a half hour. The longest an embryo has been frozen, stored, and then successfully revived is 13 years; the "oldest" pregnancy using a frozen embryo occurred 9 years after the freezing.

So many people have had IVF since Louise Joy Brown was born that researchers have developed algorithms to predict the chances that the procedure will be successful and lead to a birth for a particular couple. Overall the chances of a live birth following IVF are about 25 percent, but this prediction varies greatly, depending on certain risk factors that lower the likelihood of success. These include:

- maternal age—success is 30 to 40 percent for women (oocyte donors) under age 34, but only 5 to 10 percent for women over 40;
- increased time being infertile;
- number of previous failed IVF attempts;
- number of previous IVF attempts;
- use of a woman's own oocytes rather than a donor's; and
- infertility with a known cause.

A website (http://www.ivfpredict.com) assesses these risks. In one example, a couple had been infertile for 11 years. They attempted IVF four times that resulted in two failures and two spontaneous abortions. They had used the woman's eggs and ICSI because too many sperm were abnormal. The chance of success per IVF attempt is about 8 percent, but if they use a donor oocyte, the likelihood of success doubles.

Gamete and Zygote Intrafallopian Transfer

IVF may fail because of the artificial environment for fertilization. A procedure called **GIFT,** which stands for **gamete intrafallopian transfer,** improves the setting. (Uterine tubes are also called fallopian tubes.) Fertilization is assisted in GIFT, but it occurs in the woman's body rather than in glassware.

In GIFT, several of a woman's largest oocytes are removed. The man submits a sperm sample, and the most active cells are separated from it. The collected oocytes and sperm are deposited together in the woman's uterine tube, at a site past any obstruction that might otherwise block fertilization. GIFT is about 22 percent successful.

A variation of GIFT is **ZIFT,** which stands for **zygote intrafallopian transfer**. In this procedure, an IVF ovum is introduced into the woman's uterine tube. Allowing the fertilized ovum to make its own way to the uterus increases the chance that it will implant. ZIFT is also 22 percent successful.

GIFT and ZIFT are done less frequently than IVF. These procedures may not work for women who have scarred uterine tubes. The average cost of GIFT or ZIFT is $15,000 to $20,000.

Bioethics: Choices for the Future on page 416 considers the unusual situation of collecting gametes from a person shortly after the person has died.

Oocyte Banking and Donation

Oocytes can be stored, as sperm are, but the procedure may create problems. Because an oocyte is the largest type of cell, it contains a large volume of water. Freezing can form ice crystals that damage cell parts. Candidates for preserving oocytes include women who wish to have children later in life and women who will contact toxins or teratogens in the workplace or in chemotherapy.

Oocytes are frozen in liquid nitrogen at $-30°C$ to $-40°C$, when they are at metaphase of the second meiotic division. At this time, the chromosomes are aligned along the spindle, which is sensitive to temperature extremes. If the spindle comes apart as the cell freezes, the oocyte may lose a chromosome, which would devastate development. Another problem with freezing oocytes is retention of a polar body, leading to a diploid oocyte. Only 100 babies have been born using frozen oocytes despite two decades of attempts. The probability of achieving pregnancy using a frozen oocyte with current technology is only about 3 percent, and the technique is still investigational. However, websites offering "egg freezing" claim high rates of success, which can mean anything from fertilization to a birth.

To avoid the difficulty of freezing oocytes, strips of ovarian tissue can be frozen, stored, thawed, and reimplanted at various sites, such as under the skin of the forearm or abdomen or in the pelvic cavity near the ovaries. The tissue ovulates and the oocytes are collected and fertilized *in vitro*. The first child resulting from fertilization of an oocyte from reimplanted ovarian tissue was born in 2004. The mother, age 25, had been diagnosed with advanced Hodgkin's lymphoma. The harsh chemotherapy and radiation cured her cancer, but destroyed her ovaries. Beforehand, five strips of tissue from her left ovary were frozen. Later, several pieces of ovarian tissue were thawed and implanted in a pocket that surgeons crafted on one of her shriveled ovaries, near the entrance to a uterine tube. Menstrual cycles resumed, and shortly thereafter, the woman became pregnant with her daughter, who is healthy. Freezing ovarian tissue may become routine for cancer patients of childbearing age.

Women who have no oocytes or wish to avoid passing on a mutation can obtain oocytes from donors, who are typically younger women. Some women become oocyte donors when they undergo IVF and have "extra." The potential father's sperm and donor's oocytes are placed in the recipient's uterus or uterine tube, or fertilization occurs in the laboratory and a blastocyst is transferred to the woman's uterus. A program in the United Kingdom funds IVF for women who cannot afford the procedure if they donate their "extras." The higher success rate of using oocytes from younger women confirms that it is the oocyte that age affects, and not the uterine lining.

Embryo adoption is a variation on oocyte donation. A woman with malfunctioning ovaries but a healthy uterus carries an embryo that results when her partner's sperm is used in intrauterine insemination of a woman who produces healthy oocytes. If the woman conceives, the embryo is gently flushed out of her uterus a week later and inserted through the cervix and into the uterus of the woman with malfunctioning ovaries. The child is genetically that of the man and the woman who carries it for the first week, but is born from the woman who cannot produce healthy oocytes. "Embryo adoption" also describes use of IVF "leftovers."

Removing and Using Gametes After Death

A gamete is a packet containing one copy of a person's genome. If after death gametes are collected and combined with an opposite gamete type, the deceased person can become a new parent. This "postmortem gamete retrieval" has happened for years, but nearly always for men.

One of the first cases of postmortem sperm removal affected Bruce and Gaby V. In their early thirties, they had delayed becoming parents, confident that their good health would make pregnancy possible later. Then Bruce suddenly died of an allergic reaction to a drug. Gaby knew how much Bruce had wanted to be a father, so she asked the medical examiner to collect Bruce's sperm. The sample was sent to the California Cryobank, where it lay deeply frozen for more than a year. In the summer of 1978, the sperm were defrosted and used to fertilize one of Gaby's oocytes. On March 17, Bruce and Gaby's daughter was born. It was the first case of postmortem sperm retrieval in which the father did not actively participate in the decision. Since 1990, U.S. servicemen who feared infertility from exposure to chemical or biological weapons have taken advantage of sperm bank discounts to the military, preserving their sperm before deploying.

Postmortem sperm retrieval raises legal and ethical issues based on timing. In one case, a woman conceived twins using her husband's preserved sperm 16 months after he died of leukemia; he had stated his wishes for her to do so. The Social Security Administration refused to provide survivor benefits to their daughters, claiming that the husband was not a father, but a sperm donor. The Massachusetts Superior Court reversed this decision. In New Jersey, a mother claimed Social Security benefits for twins conceived after her husband's death. An appeals court upheld the denial of benefits, claiming that the children had to have been dependents at the time of their father's death.

The first case of postmortem oocyte retrieval was reported in 2010. A 36-year-old woman stood up on a plane following many hours of sleeping in one position, and her heart stopped. By the time a doctor on board restarted it, the woman's brain had been robbed of oxygen for several precious minutes. The plane made an emergency landing and she was taken to a hospital, where she was placed on a respirator. Scans showed blood clots in her lungs that had caused the collapse. By the fourth day, the woman's brain was dangerously swelling, and by the ninth day, her brain

activity was nearly nil, although she could still open her eyes and move spontaneously. Her husband, parents, and in-laws asked that the tubes keeping her alive be withdrawn. Then, several hours after this was done, they changed their minds, and asked that the breathing tube be reinserted so that the woman's oocytes could be retrieved. They had no idea how difficult this would be.

The physicians, knowing the complexities of the medical situation, wanted to know one other person's opinion—the patient's. So they consulted with the young woman's gynecologist, who had no record or recollection of the patient stating she wished to have children. The young woman was not completely brain dead, so the decisions would not be the same as for donating an organ after death. If her oocytes were to be used to give her husband a child, assisted reproductive technologies would obviously be necessary—*in vitro* fertilization and a surrogate mother. Even before that could happen, though, the woman would have to undergo 2 weeks of hormone treatments to ovulate, during which time she would have to lie flat, which could cause her death. For these practical reasons, and the fact that the woman had never stated that she wished to be a parent, the family elected to turn off life support, and she quickly died.

Like other assisted reproductive technologies, postmortem gamete retrieval is not regulated at the federal level in the United States. Bioethicists have identified situations to avoid:

- someone other than a spouse wishing to use the gamete;
- a too-hasty decision based on grief; and
- use of the gamete for monetary gain.

Questions for Discussion

1. How does the case of the 36-year-old woman whose oocytes were retrieved following her brain death differ from that of a pregnant woman in a coma who is kept alive for several weeks so that her baby can be born?

2. The people described in this essay did not have other children. How might the situation differ for a couple who already have children?

3. Do you think that Social Security or another benefit system should cover fetuses, embryos, or gametes?

4. How might postmortem gamete retrieval be abused?

In another technology, cytoplasmic donation, older women have their oocytes injected with cytoplasm from the oocytes of younger women to "rejuvenate" the cells. Although resulting children conceived through IVF appear to be healthy, they are being monitored for a potential problem—heteroplasmy, or two sources of mitochondria in one cell (see figure 5.11). Researchers do not yet know the health

consequences, if any, of having mitochondria from the donor cytoplasm plus mitochondria from the recipient's oocyte. These conceptions also have an elevated incidence of XO syndrome, which often causes spontaneous abortion.

Because oocytes are harder to obtain than sperm, oocyte donation technology has lagged behind that of sperm banks, but is catching up. One IVF facility that has run a donor oocyte program since 1988 has a brochure that describes 120 oocyte donors of various ethnic backgrounds, like a catalog of sperm donors. The oocyte donors are young and have undergone extensive medical and genetic tests. Recipients may be up to 55 years of age.

Preimplantation Genetic Diagnosis

Prenatal diagnostic tests such as amniocentesis and chorionic villus sampling can be used in pregnancies achieved with assisted reproductive technologies. A test called **preimplantation genetic diagnosis** (**PGD**) detects genetic and chromosomal abnormalities *before* pregnancy starts. The couple selects a very early "preimplantation" embryo that tests show has not inherited a specific detectable genetic condition. "Preimplantation" refers to the fact that the embryo is tested at a stage prior to when it would implant in the uterus. PGD was used to select Adam Nash, whose umbilical cord stem cells cured his sister's Fanconi anemia (see figure 21.1). PGD has about a 29 percent success rate. It adds on average $3,550 to the cost of IVF.

PGD is possible because one cell, or blastomere, can be removed for testing from an 8-celled embryo, and the remaining seven cells can complete development normally in a uterus. Before the embryo is implanted into the woman the single cell is karyotyped, or its DNA amplified and probed for genes that the parents carry. It may soon be more economical to sequence the exome or genome in place of these more specific tests. Embryos that pass these tests are selected

to complete development or are stored. At first, researchers implanted the remaining seven cells, but letting the selected embryo continue developing in the dish until day 5, when it is 80 to 120 cells, is more successful. Obtaining the cell to be tested is called "blastomere biopsy" (**figure 21.4**). Accuracy in detecting a mutation or abnormal chromosome is about 97 percent. Errors generally happen when a somatic mutation affects the sampled blastomere but not the rest of the embryo. Amplification of the selected blastomere DNA may cause such a somatic mutation.

PGD is not new. The first children who had PGD were born in 1989. In these first cases, probes for Y chromosome–specific DNA sequences were used to select females, who could not inherit the X-linked conditions their mothers carried. The alternative to PGD would have been to face the 25 percent chance of conceiving an affected male.

In March 1992, the first child was born who underwent PGD to avoid a specific inherited disease. Chloe O'Brien was checked as an 8-celled preimplantation embryo to see if she had escaped the cystic fibrosis that affected her brother. Since then, PGD has helped to select thousands of children free of several dozen types of inherited illnesses. It has been used for the better-known single-gene disorders as well as for rare ones.

Like most ARTs, use of PGD has expanded as it has become more accurate and more familiar. Today it has taken on a quality-control role in addition to being a tool to detect and prevent rare diseases. PGD is increasingly being used to screen early embryos derived from IVF for normal chromosome number before implanting them into women. This selection process should increase the chances of successful live births, but in the first large trial, PGD actually lowered the birth rate—perhaps the intervention harms the embryos. In the Netherlands, researchers examined *all* of the cells of several preimplantation

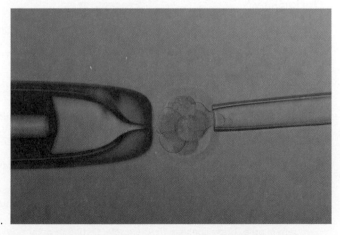

Preimplantation embryo

One cell removed for genetic analysis

a.

Seven cells complete development

DNA probes

If genetically healthy, preimplantation embryo is implanted into woman and develops into a baby.

If genetic disease is inherited, preimplantation embryo is *not* implanted into woman.

b.

c.

Figure 21.4 **Preimplantation genetic diagnosis (PGD) probes disease-causing genes or chromosome aberrations in an 8-celled preimplantation embryo.** (a) A single cell is separated and tested to see if it contains a disease-causing genotype or atypical chromosome. (b) If it doesn't, the remaining seven cells divide a few more times and are transferred to the oocyte donor to complete development. (c) This preimplantation embryo is held still by suction applied on the left. On the right, a pipette draws up a single blastomere. *In vitro* fertilization took place 45 hours previously.

human embryos and found that some cells can have normal chromosomes and others not. Therefore, the assumption underlying PGD—that a sampled cell represents all the cells of an embryo—may need to be reexamined.

PGD can introduce a bioethical "slippery slope" when it is used other than to ensure that a child is free of a certain disease. Some people may regard gender selection using PGD as a misuse of the technology. A couple with five sons might, for example, use PGD to select a daughter. But this use of technology might just be a high-tech version of age-old human nature, according to one physician who performs PGD. "From the dawn of time, people have tried to control the sex of offspring, whether that means making love with one partner wearing army boots, or using a fluorescence-activated cell sorter to separate X- and Y-bearing sperm. PGD represents a quantum leap in that ability—all you have to do is read the X and Y chromosome paints," he says.

While PGD used solely for family planning is certainly more civilized than placing baby girls outside the gates of ancient cities to perish, the American Society for Reproductive Medicine endorses the use of PGD for sex selection only to avoid passing on an X-linked disease, which was the first application of the technology. Yet even PGD to avoid disease can be controversial. In the United Kingdom, where the government regulates reproductive technology, inherited cancer susceptibility is an approved indication for PGD. These cancers do not begin until adulthood, the susceptibility is incompletely penetrant (not everyone who inherits the disease-associated genotype will actually develop cancer), and the cancer may be treatable.

Table 21.1 summarizes the assisted reproductive technologies.

Sequential polar body analysis

A technique called **sequential polar body analysis** may substitute for PGD and provides genetic information even earlier in development. The approach is based on the fact that meiosis completes in the female only as a secondary oocyte is fertilized (see figure 3.10).

If a woman is heterozygous for a mutation, then an oocyte would inherit the mutation and its associated polar bodies would inherit the wild type allele, or vice versa. This is in accordance with Mendel's first law, gene segregation. The timetable of female meiosis is important, too. The first polar body forms as the developing oocyte leaves the ovary. That polar body is not accessible, and it would not show the effects of crossing over. A second polar body, however, which forms at fertilization, can be tested (**figure 21.5**). Researchers

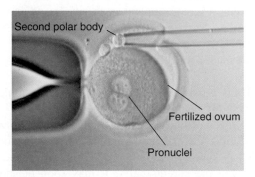

Figure 21.5 Deducing a fertilized ovum's genotype by testing a polar body. Because of Mendel's law of segregation, if a polar body associated with a fertilized ovum has a mutation that the woman carries, wild type for that gene is inferred for the secondary oocyte.

Table 21.1	Some Assisted Reproductive Technologies
Technology	**Procedures**
GIFT	Deposits collected oocytes and sperm in uterine tube.
IVF	Mixes sperm and oocytes in a dish. Chemicals simulate intrauterine environment to encourage fertilization.
IUI	Places or injects washed sperm into the cervix or uterus.
ICSI	Injects immature or rare sperm into oocyte, before IVF.
Oocyte freezing	Oocytes retrieved and frozen in liquid nitrogen.
Ovulation induction	Drugs control timing of ovulation in order to perform a particular procedure.
PGD	Searches for specific mutant allele in sampled cell of 8-celled embryo. Its absence indicates remaining 7-celled embryo can be nurtured and implanted in woman, and child will be free of genetic condition.
Sequential polar body analysis	Genetic testing of polar body attached to just-fertilized ovum enables inference that fertilized ovum is free of a family's mutation.
Surrogate mother	Woman carries a pregnancy for another.
ZIFT	Places IVF ovum in uterine tube.

can sequence linked markers or even entire genomes of polar bodies to determine whether crossing over has occurred, to be certain that the fertilized ovum has not inherited the mutation.

Because sequential polar body analysis is still experimental, researchers follow it up with PGD to test their predictions and ensure that IVF embryos transferred to the woman's uterus to develop are free of the family's mutation. The idea of probing polar bodies dates from the 1980s, but the technique first led to results in 2011. So far mutations behind more than 150 single-gene diseases have been detected at this initial stage of prenatal development—the very first cell.

Key Concepts Questions 21.3

1. What is intrauterine insemination?
2. Distinguish between a genetic and gestational surrogate mother and a gestational surrogate mother.
3. List the steps of IVF.
4. Describe GIFT and ZIFT.
5. What is embryo adoption?
6. Explain how PGD enables the avoidance of having a child with a chromosomal abnormality or a specific single-gene disease.
7. How can genetic testing of a polar body reveal information about a fertilized ovum?

21.4 Extra Embryos

Sometimes assisted reproductive technologies leave "extra" oocytes, fertilized ova, or very early embryos. **Table 21.2** lists the possible fates of this biological material.

In the United States, nearly half a million embryos derived from IVF sit in freezers; some have been there for years. Most couples who donate embryos to others do so anonymously, with no intention of learning how their genetic offspring are raised. Scott and Glenda Lyons chose a different path when they learned that their attempt at IVF had yielded too many embryos.

Table 21.2	Fates of Frozen Embryos
1.	Store indefinitely.
2.	Store and destroy after a set time.
3.	Donate for embryonic stem cell derivation and research.
4.	Thaw later for use by biological parents.
5.	Thaw later for use by other parents.
6.	Discard.

In 2001, two of Glenda's eighteen embryos were transferred to her uterus, and developed into twins Samantha and Mitchell. Through a website where couples chat about fertility issues, Scott and Glenda met and selected Bruce and Susan Lindeman to receive fourteen remaining embryos. This second couple had tried IVF three times, with no luck. The Lyons's frozen embryos were shipped cross-country to a clinic where two were implanted in Susan's uterus. In July 2003, Chase and Jack Lindeman were born—genetic siblings of Samantha and Mitchell Lyons. But there were still embryos left. The Lyonses allowed the Lindemans to send twelve embryos to a third couple, who used two to have twin daughters in August 2004. They are biological siblings of Samantha and Mitchell Lyons and Chase and Jack Lindeman (**figure 21.6**).

Another alternative to disposing of fertilized ova and embryos is to donate them for use in research. The results of experiments sometimes challenge long-held ideas, indicating that we still have much to learn about early human prenatal development. This was the case for a study from Royal Victoria

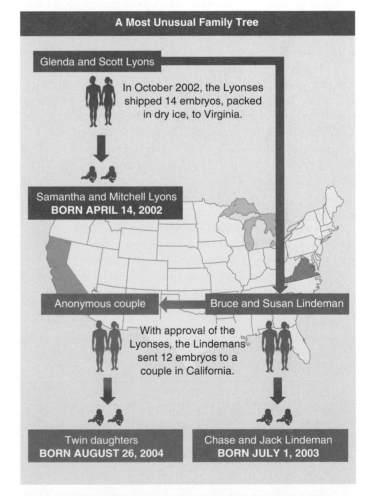

Figure 21.6 Using extra embryos. Six children resulted from Glenda and Scott Lyons's embryos. The Lyonses had a boy and a girl, then donated embryos to the Lindemans, who had twin boys. Finally, a couple in California used the Lyons's embryos to have twin daughters.

Hospital in Montreal. Researchers examined the chromosomes of sperm from a man with XXY syndrome. Many of the sperm would be expected to have an extra X chromosome, due to nondisjunction (see figure 13.12), which could lead to a preponderance of XXX and XXY offspring. Surprisingly, only 3.9 percent of the man's *sperm* had extra chromosomes, but five out of ten of his spare *embryos* had an abnormal X, Y, or chromosome 18. That is, even though most of the man's sperm were normal, his embryos weren't. The source of reproductive problems in XXY syndrome, therefore, might not be in the sperm, but in early embryos—a finding that was previously unknown and not expected, and was only learned because of observing early human embryos.

In another study, Australian researchers followed the fates of single blastomeres that had too many or too few chromosomes. They wanted to see whether the abnormal cells preferentially ended up in the inner cell mass, which develops into the embryo, or the trophectoderm, which becomes extra-embryonic membranes. The study showed that cells with extra or missing chromosomes become part of the inner cell mass much more frequently than expected by chance. This finding indicates that the ability of a blastomere sampled for PGD to predict health may depend on whether it is fated to be part of the inner cell mass.

Using fertilized ova or embryos designated for discard in research is controversial. Without regulations on privately funded research, ethically questionable experiments can happen. For example, researchers reported at a conference that they had mixed human cells from male embryos with cells from female embryos, to see if the normal male cells could "save" the female cells with a mutation. Sex was chosen as a marker because the Y chromosome is easy to detect, but the idea of human embryos with mixed sex parts caused a public outcry.

ARTs introduce ownership and parentage issues (**table 21.3**). Another controversy is that human genome information is providing more traits to track and perhaps control in coming generations. When we can routinely scan the human genome in gametes, fertilized ova, or early embryos, who will decide which traits are worth living with, and which aren't?

ARTs operate on molecules and cells, but affect individuals and families. Ultimately, by introducing artificial selection, these interventions may affect the gene pool.

Table 21.3	Assisted Reproductive Disasters

1. A physician used his own sperm to perform intrauterine insemination on 15 patients, telling them that he had used sperm from anonymous donors.

2. A plane crash killed the wealthy parents of two early embryos stored at −320°F (−195°C) in a hospital. Adult children of the couple were asked to share their estate with two 8-celled siblings-to-be.

3. Several couples planning to marry discovered that they were half-siblings. Their mothers had been inseminated with sperm from the same donor.

4. Two Rhode Island couples sued a fertility clinic for misplacing several embryos.

5. Several couples sued a fertility clinic for implanting their oocytes or embryos in other women without donor consent. One woman requested partial custody of the resulting children if her oocytes were taken, and full custody if her embryos were used, even though the children were of school age and she had never met them.

6. A man sued his ex-wife for possession of their frozen fertilized ova. He won, and donated them for research. She had wanted to be pregnant.

7. The night before *in vitro* fertilized embryos were to be implanted in a 40-year-old woman's uterus after she and her husband had spent 4 years trying to conceive, the man changed his mind, and wanted the embryos destroyed.

Key Concepts Questions 21.4

1. What can be done with extra fertilized ova and early embryos?
2. What can we learn from using early embryos in research?

Summary

21.1 Savior Siblings and More

1. **Assisted reproductive technologies (ARTs)** replace what is missing in reproduction, using laboratory procedures and people other than the infertile couple.

21.2 Infertility and Subfertility

2. **Infertility** is the inability to conceive a child after a year of unprotected intercourse. Subfertile individuals or couples manufacture gametes, but take longer than usual to conceive.

3. Causes of infertility in the male include low sperm count, a malfunctioning immune system, a varicose vein in the scrotum, structural sperm defects, drug exposure, vasectomy reversal, and abnormal hormone levels. Mutation may impair fertility.

4. Causes of infertility in the female include absent or irregular ovulation, blocked uterine tubes, an inhospitable or malshaped uterus, antisperm secretions, or lack of sperm-attracting biochemicals. Early pregnancy loss due to abnormal chromosome number is more common in older women and may appear to be infertility.

21.3 Assisted Reproductive Technologies

5. In **intrauterine insemination (IUI)**, donor sperm are placed into a woman's reproductive tract in a clinical setting.

6. A gestational and genetic surrogate mother provides her oocyte. Then intrauterine insemination is performed with sperm from a man whose partner cannot conceive or carry a fetus. The surrogate provides her uterus for 9 months. A gestational-only surrogate mother receives a fertilized ovum that resulted from *in vitro* **fertilization (IVF)** of a secondary oocyte by a sperm that came from the couple who asked her to carry the pregnancy.

7. In IVF, oocytes and sperm join in a dish, fertilized ova divide a few times, and embryos are placed in the woman's body, circumventing blocked tubes or malfunctioning sperm. **Intracytoplasmic sperm injection (ISCI)** introduces immature or nonmotile sperm into oocytes.

8. Embryos can be frozen and thawed and then complete development when placed in a woman's uterus.

9. **Gamete intrafallopian transfer (GIFT)** introduces oocytes and sperm into a uterine tube past a blockage; fertilization occurs in the woman's body. **Zygote intrafallopian transfer (ZIFT)** places an early embryo in a uterine tube.

10. Oocytes can be frozen and stored. In embryo adoption, a woman undergoes IUI. A week later, the embryo is washed out of her uterus and put into the reproductive tract of the woman whose partner donated the sperm.

11. Seven-celled embryos can develop normally if a blastomere is removed at the 8-cell stage and cleared for abnormal chromosomes or genes. This is **preimplantation genetic diagnosis (PGD)**.

12. **Sequential polar body analysis** infers absence of a mutation in a fertilized ovum by checking a second polar body.

21.4 Extra Embryos

13. Extra fertilized ova and early embryos generated in IVF are used, donated to couples, stored, donated for research, or discarded. They enable researchers to study aspects of early human development that they could not investigate in other ways.

www.mhhe.com/lewisgenetics11

Answers to all end-of-chapter questions can be found at **www.mhhe.com/lewisgenetics11**. You will also find additional practice quizzes, animations, videos, and vocabulary flashcards to help you master the material in this chapter.

Review Questions

1. Which assisted reproductive technologies might help the following couples? (More than one answer may fit some situations.)

   a. A woman is born without a uterus, but manufactures healthy oocytes.
   b. A man has cancer treatments that damage his sperm.
   c. A genetic test reveals that a woman will develop Huntington disease. She wants a child, but does not want to pass on the disease.
   d. Two women wish to have and raise a child together.
   e. A man and woman are carriers of sickle cell disease. They do not want to have an affected child or terminate a pregnancy to avoid the birth of an affected child.
   f. A woman's uterine tubes are scarred and blocked.
   g. A young woman must have radiation to treat ovarian cancer, but wishes to have a child.

2. Why are men typically tested for infertility before women?

3. What are some of the causes of infertility among older women?

4. Cite a situation in which both man and woman contribute to subfertility.

5. How does ZIFT differ from GIFT and IVF?

6. Explain how preimplantation genetic diagnosis is similar to and different from the prenatal diagnosis techniques described in chapter 13.

7. How do each of the following ARTs deviate from the normal biological process?

   a. IVF
   b. GIFT
   c. embryo adoption
   d. gestational-only surrogacy
   e. intrauterine insemination
   f. cytoplasmic donation
   g. mitochondrial replacement

8. Explain how PGD works, and list two events in early prenatal development that might explain cases where the PGD result is inaccurate.

9. Why is it much easier to freeze and revive early embryos than oocytes?

10. Describe a scenario in which each of the following technologies is abused:

    a. surrogate mother
    b. gamete donation
    c. IVF
    d. PGD

11. Explain how sequential polar body analysis is based on Mendel's first law.

Applied Questions

1. Of the three individuals who contribute genetic material to an embryo that has had mitochondrial replacement, which provides the most DNA?

2. Neil Patrick Harris is an actor who had twins with his partner, David Burtka. They used a surrogate mother, who carried two embryos that had been fertilized *in vitro*, one with Neil's sperm and the other with David's sperm. In terms of genetics, how closely are the babies, a boy and a girl, related to each other? (The fathers do not know who fathered which child.)

3. Some ARTs were invented to help people who could not have children for medical reasons, or to avoid conceiving a child with a genetic disease in the family. With time, as the technologies became more familiar, in the United States people with economic means began to use them for other reasons, such as a celebrity who does not wish to lose her shape during pregnancy. Remembering that the U.S. government does not regulate ARTs, do you think that any measures should be instituted to select candidates for ARTs? How can these technologies be made more affordable?

4. How might exome and genome sequencing be incorporated into ARTs?

5. Singer Sir Elton John and his partner David Furnish are parents. For their first child, Sir Elton wanted his sperm to be used because he thinks that his songwriting talent is inherited. Which ART could they have used to have son Zachary?

6. A man reads his medical chart and discovers that the results of his sperm analysis indicate that 22 percent of his sperm are shaped abnormally. He wonders why the physician said he had normal fertility if so many sperm are abnormally shaped. Has the doctor made an error?

7. An embryo bank in Texas offers IVF leftover embryos, which would otherwise remain in the deep freeze or be discarded, to people wanting to have children, for $2,500 each. The bank circumvents bioethical concerns by claiming that it sells a service, not an embryo. People in favor of the bank claim that purchasing an embryo is not different from paying for sperm or eggs, or an adopted child. Those who object to the bank claim that it makes an embryo a commodity.

   a. Do you think that an embryo bank is a good idea, or is it unethical?

   b. Whose rights are involved in the operation of the embryo bank?

   c. Who should be liable if a child that develops from the embryo has an inherited disease?

   d. Is the bank elitist because the cost is so high?

8. A newspaper columnist wrote that frozen human embryos are "microscopic Americans." Former president George W. Bush called them "unique and genetically complete, like every other human being." A stem cell researcher referred to embryonic stem cells as "like any other cell in an adult, no different from the skin cells you rub off with a towel after a shower."

   a. What is your opinion of the status of an 8-celled human embryo?

   b. Does the status of a cell from an 8-celled human embryo in culture depend upon whether that cell is tested and discarded, or part of the embryo that is implanted into a woman?

   c. Do you think that there is any harm in an influential person, such as a journalist, politician, or researcher, stating his or her opinion of the status of an embryo?

   d. How might the following individuals respond to or feel about these definitions?

      i. A woman who ended a pregnancy, for whatever reason

      ii. A couple who have had a spontaneous abortion

      iii. A person with a disease that one day may be treated with stem cells

      iv. A couple who have tried to conceive for a decade

9. At the same time that 62- and 63-year-old women gave birth, actors Tony Randall and Anthony Quinn became fathers at ages 77 and 78—and didn't receive nearly as much criticism as the women. Do you think this is an unfair double standard, or a criticism based on biology?

10. Many people spend thousands of dollars pursuing pregnancy. What is an alternative solution to their quest for parenthood?

11. An Oregon man anonymously donated sperm that were used to conceive a child. The man later claimed, and won, rights to visit his child. Is this situation for the man more analogous to a genetic and gestational surrogate mother, or an oocyte donor who wishes to see the child she helped to bring into existence?

12. Big Tom is a bull with valuable genetic traits. His sperm are used to conceive 1,000 calves. Mist, a dairy cow with exceptional milk output, has many oocytes removed, fertilized *in vitro*, and implanted into surrogate mothers. With their help, Mist becomes the genetic mother of 100 calves—many more than she could give birth to naturally. Which two ARTs are based on these examples from agriculture?

13. State who the genetic parents are and who the gestational mother is in each of the following cases:

   a. A man was exposed to unknown burning chemicals and received several vaccines during the first Gulf war, and abused drugs for several years before and after that. Now he wants to become a father, but he is concerned that past exposures to toxins have damaged his sperm. His wife undergoes intrauterine insemination with sperm from the husband's brother, who has led a calmer and healthier life.

b. A 26-year-old woman has her uterus removed because of cancer. However, her ovaries are intact and her oocytes are healthy. She has oocytes removed and fertilized *in vitro* with her husband's sperm. Two resulting embryos are implanted into the uterus of the woman's best friend.

c. Max and Tina had a child by IVF and froze three extra embryos. Two are thawed years later and implanted into the uterus of Tina's sister, Karen. Karen's uterus is healthy, but she has ovarian cysts that often prevent her from ovulating.

d. Forty-year-old Christopher wanted children, but did not want a partner. He donated sperm, which were used for intrauterine insemination of a mother of one. The woman carried the resulting fetus to term for a fee, and gave birth to a daughter.

e. Two men want to raise a child together. They go to a fertility clinic, have their sperm collected, mixed, and used to inseminate a friend. Nine months later she gives the baby to them.

14. Delaying childbirth until after age 35 is associated with certain physical risks, yet an older woman is often more mature and financially secure. Many women delay childbirth so that they can establish careers.

Suggest societal changes, perhaps using an ART, that would allow women to more easily have children and less interrupted careers.

15. An IVF attempt yields twelve more embryos than the couple who conceived them can use. What could they do with the extras?

16. What do you think children born of an ART should be told about their origins and when should this happen?

17. An IVF program in India offers preimplantation genetic diagnosis to help couples who already have a daughter to conceive a son. The reasoning is that because having a male heir is of such great importance in this society, offering PGD can enable couples to avoid aborting second and subsequent female pregnancies. Do you agree or disagree that PGD should be used for sex selection in this sociological context?

18. ICSI can help men who have small deletions in their Y chromosomes that stop sperm from maturing to become parents, but this passes on their infertility. Suggest an ART that they could use to prevent male infertility.

19. Novelist Jodi Picoult wrote about a savior sibling in *My Sister's Keeper*. She referred to preimplantation genetic diagnosis as "genetic engineering." Is this correct? Why or why not?

Web Activities

1. Invent a situation for a couple trying to have a baby with IVF and use http://www.ivfpredict.com to estimate their chances of success.

2. The U.S. government bans use of federal funds to create human embryos for research purposes, but does not regulate the human reproductive technology field at all. Consult websites to learn about regulations of stem cell research and ARTs in other nations.

3. Go to the Centers for Disease Control and Prevention website. Click on ART Trends, and use the information to answer the following questions.

a. Since 1996, to what extent has the use of ARTs in the United States increased?

b. Which is more successfully implanted into an infertile woman's uterus, a fresh or frozen donor oocyte?

c. Which is more successfully implanted into an infertile woman's uterus, a donated oocyte or one of her own?

d. What are two factors that could complicate data collection on ART success rates?

4. A company called Extend Fertility provides oocyte freezing services, telling women to "set your own biological clock." The home page states, "Today's women

lead rich and busy lives—obtaining advanced degrees, pursuing successful careers, and taking better care of ourselves. As a result of this progress, many of us choose to have children later than our mothers did."

Look at this or another egg-freezing website. Discuss how it might be viewed by the following individuals:

a. A 73-year-old father of a healthy baby

b. A 26-year-old woman, married with no children but who wants them, facing 6 months of chemotherapy

c. An orphaned 10-year-old in Thailand

d. A healthy 28-year-old woman in the United States who wants to earn degrees in medicine, law, and business before becoming pregnant

e. A young mother in Mexico who is giving her son up for adoption because she cannot afford to raise him

5. Read the posts on surrogatemother.com and describe a match between a potential surrogate and a couple or individual wishing to use her services.

6. Describe the types of data that an ovulation app records.

7. Describe a success story at the Donor Sibling Registry (https://www.donorsiblingregistry.com).

Case Studies and Research Results

1. Doola is 28 years old and is trying to decide if she and her husband are ready for parenthood when she learns that her 48-year-old mother has Alzheimer disease. The mother's physician tells Doola that because of the early onset, the Alzheimer disease could be inherited through a susceptibility gene. Doola is tested and indeed has the same dominant allele. She wants to have a child right away, so that she can enjoy many years as a mother. Her husband David feels that it wouldn't be fair to have a child knowing that Alzheimer disease likely lies in Doola's future.

    a. Who do you agree with, and why?

    b. David is also concerned that Doola could pass on her Alzheimer gene variant to a child. Which technology might help them avoid this?

    c. Is Doola correct in assuming that she is destined to develop Alzheimer disease?

2. Natallie Evans had to have her ovaries removed at a young age because they were precancerous, so she and her partner had IVF and froze their embryos for use at a later time. Under British law, both partners must consent for the continued storage of frozen embryos. Evans and her partner split, and he revoked his consent. She sued for the right to use the embryos. She told the court, "I am pleased to have the opportunity to ask the court to save my embryos and let me use them to have the child I so desperately want."

    What information should the court consider in deciding this case? Whose rights do you think should be paramount?

3. One-year-old twins Violet and Kieran graced the cover of *The New York Times Magazine,* above the headline "Meet the Twiblings." Their gestation required one sperm donor, one oocyte donor, and two uterus donors (surrogate mothers). Melanie and Michael, the twins' parents, had attempted IVF six times without success, so they tried a more complicated combination of ARTs. They found two surrogate mothers because a twin pregnancy is riskier than two singleton pregnancies. Because Melanie's "advanced maternal age" of 41 might explain why IVF hadn't worked, the couple also used an oocyte donor. Michael's sperm was mixed with the donated oocytes in a lab dish, very early embryos were implanted in the two surrogates, and 9 months later, Violet and Kieran were born.

    a. How are the children twins, but also not twins?

    b. Which two adults in the case are the genetic parents?

    c. If Melanie and Michael conceive, what percentage of the genome would Violet and Kieran each share with the new family member?

4. Another article in *The New York Times* introduced healthy toddler twins and their younger sibling who had been born without inheriting the prion disease (see table 10.6) that was devastating the brain of their maternal grandfather, and would one day do so to their mother. A case report in a neurology journal described how researchers used sequential polar body analysis and IVF to ensure that the children did not inherit the disease. Describe how determining a genotype of a polar body can provide information about a fertilized ovum.

CHAPTER

22

What can we learn from our genome sequences?

Genomics

Learning Outcomes

22.1 From Genetics to Genomics

1. Explain how linkage studies led to the idea to sequence the human genome.

2. Distinguish between the two approaches used to sequence the human genome.

3. Explain why genome sequencing requires multiple copies of a genome.

22.2 Analysis of the Human Genome

4. What types of information are included in genome annotation?

5. What are limitations of exome and genome sequencing?

22.3 Personal Genome Sequencing

6. What types of information did the first sequenced genomes provide?

7. Discuss practical aspects of implementing exome and genome sequencing in health care.

Just over a decade ago the first human genomes were sequenced. Today the cost has plummeted to the point that personal genome sequencing is not only possible, but may soon become widely used to help diagnose disease, inform treatment choices, and reveal our origins.

Will you have your genome sequenced? Getting the most useful information as possible from our genome sequences will depend upon many people permitting their information to be collected and analyzed, even if it remains anonymous. Genomics may impact the future for many of us.

100,000 Genomes and Counting

Genome researchers like numbers. They spent the 1990s anticipating the sequencing of the 3.2 billion bits of information that comprise a human genome. Even before that was done, the focus had already turned to identifying the points where human genomes vary, which led to the 3.5 million variant bases of the HapMap project. Meanwhile, as technology improved, the feasibility of routinely sequencing human genomes grew.

The first people to have their genomes sequenced were celebrities, curious journalists or scientists, or intrigued millionaires. Craig Venter, who led one of the original sequencing teams, was first to have his genome sequence published, in late 2007. Six months later came the genome sequence of James Watson. Ten prominent people kicked off the Personal Genome Project ("PGP-10") led by Harvard University geneticist George Church. The goal of PGP-10: developing a new approach to preventing and treating disease by integrating genome information with health, lifestyle, ancestry, and microbiome information.

The "1000 Genomes Project" examined 2,500 genomes from 27 populations to identify 99 percent of human genetic diversity. It tracked genetic variants that affect more than 1 percent of any population, including 15 million SNPs, a million copy number variants, and 20,000 chromosome alterations. The effort has enabled researchers to identify genetic variants that affect health. Genome projects will blossom over the coming years.

22.1 From Genetics to Genomics

Genetics is a young science, genomics younger still. As one field has evolved into another, milestones have come at oddly regular intervals. A century after Gregor Mendel announced and published his findings, the genetic code was deciphered; a century after his laws were rediscovered, the human genome was sequenced.

Geneticist H. Winkler coined the term *genome* in 1920. A hybrid of "gene" and "chromosome," genome then denoted a complete set of chromosomes and genes. The modern definition refers to all the DNA in a haploid set of chromosomes. The term *genomics,* credited to T. H. Roderick in 1986, indicates the study of genomes. Thoughts of sequencing genomes echoed through much of the twentieth century, as researchers described the units of inheritance from several different perspectives.

Beginnings in Linkage Studies

Sequencing the human genome unofficially began in the 1980s with deciphering signposts along the chromosomes and developing shortcuts to handle so much information. Many of the initial steps and tools grew from existing technology. Linkage maps and studies of rare families that had chromosome abnormalities and specific syndromes enabled researchers to assign some genes to their chromosomes. It was an ambitious start. Then automated DNA sequencing took genetic analysis to a new level—information.

The evolution of increasingly detailed genetic maps is similar to zooming in on a geographical satellite map (**figure 22.1**). A cytogenetic map (of a chromosome) is like a map of California within a map of the United States, highlighting only the largest cities. A linkage map is like a map that depicts the smaller cities and large towns, and a physical map is similar to a geographical map indicating all towns. Finally, a sequence map is the equivalent of a Google map showing all of the specific buildings in an area.

Before the human genome was sequenced, researchers matched single genes to specific diseases using an approach called positional cloning. The technique began with examining a particular phenotype corresponding to a Mendelian disorder in large families. The phenotype was easily matched to a chromosome segment if all the affected individuals shared a chromosome abnormality that relatives without the phenotype did not have. But abnormal chromosomes are rare. Another way that positional cloning located medically important genes on chromosomes used linkage maps that showed parts of a chromosome shared by only the individuals in a family who had the same

syndrome. Then researchers isolated pieces of the implicated chromosome, determined short DNA sequences corresponding to the region of interest, and overlapped the pieces, to gradually identify the gene behind a particular phenotype. Positional cloning was an indirect method that, without knowing the nature of a gene's protein product, nonetheless narrowed down its chromosomal locus.

Throughout the 1980s and 1990s, positional cloning experiments discovered the genes behind Duchenne muscular dystrophy, cystic fibrosis, Huntington disease, and many others. So slow was the process that it took a decade to go from discovery of a genetic marker for Huntington disease to discovery of the gene.

Sequencing the Human Genome

The idea to sequence the human genome occurred to many researchers at about the same time, but with different goals (see the Technology Timeline on page 427). It was first brought up at a meeting held by the Department of Energy (DOE) in

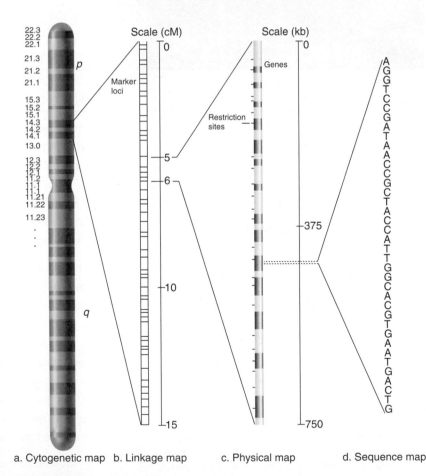

a. Cytogenetic map   b. Linkage map   c. Physical map   d. Sequence map

Figure 22.1 Different levels of genetic maps are like zooming up the magnification on a geographical map. (a) A cytogenetic map, based on associations between chromosome aberrations and syndromes, can distinguish DNA sequences at least 5,000 kilobases (kb) apart. (b) A linkage map derived from recombination data distinguishes genes hundreds of kb apart. (c) A physical map constructed from overlapped DNA pieces distinguishes genes tens of kb apart. (d) A sequence map depicts the order of nucleotide bases. (cM = centiMorgans)

EVOLUTION OF GENOME PROJECTS AND RELATED TECHNOLOGIES

1977	Two methods of DNA sequencing invented. The Sanger method persists for detailed sequencing of individual genes.
1985–1988	Idea to sequence human genome suggested at several scientific meetings.
1988	Congress authorizes the Department of Energy and the National Institutes of Health to fund the human genome project.
1989	Researchers at Stanford and Duke Universities invent DNA microarrays.
1990	Human genome project officially begins.
1991	Expressed sequence tag (EST) technology identifies protein-encoding DNA sequences.
1992	First DNA microarrays available.
1993	Need to automate DNA sequencing recognized.
1994	U.S. and French researchers publish preliminary map of 6,000 genetic markers, one every 1 million bases along the chromosomes.
1995	Emphasis shifts from gene mapping to sequencing. First genome sequenced: *Hemophilus influenzae*.
1996	Resolution to make all data public and updated daily at GenBank website. First eukaryote genome sequenced—yeast.
1998	Public Consortium releases preliminary map of pieces covering 98 percent of human genome. Millions of sequences in GenBank. Directions for DNA microarrays posted on Internet. First multicellular organism's genome sequenced: roundworm.
1999	Rate of filing of new sequences in GenBank triples. Public Consortium and two private companies race to complete sequencing. First human chromosome sequenced (22).
2000	Completion of first draft human genome sequence announced. Microarray technology flourishes. First plant and fungal genomes sequenced.
2001	Two versions of draft human genome sequence published.
2003	Finished human genome sequence announced to coincide with fiftieth anniversary of discovery of DNA structure. Human exome sequence available on DNA microarrays. HapMap project identifies linkage patterns.
2004	Final version of human genome sequence published.
2005	Annotation of human genome sequence continues. Number of species with sequenced genomes soars. High-throughput, next-generation DNA sequencing introduced.
2007	Detailed analysis of 1 percent of the human genome (ENCODE project) reveals that most of it is transcribed. First individual human genome sequenced. Human Microbiome Project begins.
2008	First genome synthesized: *Mycoplasma genitalium*.
2010	Several human genomes sequenced. First synthetic genome supports a bacterial cell. By now, 4,000 bacterial and viral and 250 eukaryotic species' genomes have been sequenced.
2012	1000 Genomes Project completes cataloging of human genetic diversity.
2014	Tens of thousands of human genomes and hundreds of thousands of human exomes sequenced.
?	Personal human genome sequencing is a routine part of health care.

1984 to discuss the long-term population genetic effects of exposure to low-level radiation. In 1985, researchers meeting at the University of California, Santa Cruz, called for an institute to sequence the human genome, because sequencing of viral genomes had shown that it could be done. The next year, virologist Renato Dulbecco proposed that the key to understanding the origin of cancer lay in knowing the human genome sequence. Later that year, scientists packed a room at the Cold Spring Harbor Laboratory on New York's Long Island to discuss the feasibility of a project to sequence the human genome. At first those against the project outnumbered those

for it 5 to 1. The major fear was the shifting of goals of life science research from inquiry-based experimentation to amassing huge amounts of data—ironic considering the importance of bioinformatics today.

A furious debate ensued. Detractors claimed that the project would be more gruntwork than a creative intellectual endeavor, comparing it to conquering Mt. Everest just because it is there. Practical benefits would be far in the future. Some researchers feared that such a "big science" project would divert government funds from basic research and AIDS. Finally, the National Academy of Sciences convened a committee representing both sides to debate the feasibility, risks, and benefits of the project. The naysayers were swayed to the other side. In 1988, Congress authorized the National Institutes of Health (NIH) and the DOE to fund the $3 billion, 15-year human genome project, which began in 1990 with James Watson at the helm. The project set aside 3 percent of its budget for the Ethical, Legal and Social Implications (ELSI) Research Program. It has helped ensure that genetic information is not used to discriminate. Eventually, an international consortium as well as a private company, Celera Genomics, sequenced the human genome. They worked separately, and one effort finished a few months before the other, but the accomplishment is referred to here as "the human genome project."

A series of technological improvements sped the genome project. In 1991, a shortcut called expressed sequence tag (EST) technology enabled researchers to quickly pick out genes most likely to be implicated in disease. This was a foreshadowing for future efforts to focus on the exome, which is the part of the genome that encodes protein and is thought to be responsible for 85 percent of single-gene diseases. ESTs are cDNAs made from the mRNAs in a cell type that is abnormal in a particular illness, such as an airway lining cell in cystic fibrosis. ESTs therefore represent gene expression. Also in 1991, researchers began using DNA microarrays to display short DNA molecules. Microarray technology became important in DNA sequencing (tiling arrays) as well as in assessing gene expression (expression arrays).

Computer algorithms assembled many short pieces of DNA with overlapping end sequences into longer sequences (**figure 22.2**). When the project began, researchers cut several genomes' worth of DNA into overlapping pieces of about 40,000 bases (40 kilobases), then randomly cut the pieces into small fragments. The greater the number of overlaps, the more complete the final assembled sequence. The sites of overlap had to be unique sequences, found in only one place in the genome. Overlaps of repeated sequences found in several places in the genome could lead to more than one derived overall sequence—a little like searching a document for the word "that" versus searching for an unusual word, such as "dandelion." Searching for "dandelion" is more likely to lead to a specific part of a document, whereas "that" may occur in several places—just like repeats in a genome.

The use of unique sequences is why the human genome project did not uncover copy number variants. For example, the sequence CTACTACTA would appear only as CTA. Researchers did not at first appreciate the fact that repeats are a different form of information and source of variation than DNA base sequences. A balance was necessary between using DNA pieces large enough to be unique, but not so large that the sequencing would take a very long time.

Two general approaches were used to build the long DNA sequences to initially derive the sequence of the human genome (**figure 22.3**). The "clone-by-clone" technique the U.S. government-funded group used aligned DNA pieces one chromosome at a time. The "whole-genome shotgun" approach Celera Genomics used shattered the entire genome, then used an algorithm to identify and align overlaps in a continuous sequence. Whole-genome shotgun sequencing can be compared to cutting the binding off a large book, throwing it into the air and freeing every page, and reassembling the dispersed pages in order. A "clone-by-clone" dismantling of the book would divide it into bound chapters. Whole-genome shotgunning is faster, but it misses some sections (particularly repeats) that the clone-by-clone method detects.

Technical advances continued. In 1995, DNA sequencing was automated, and software was developed that could rapidly locate the unique sequence overlaps among many small pieces of DNA and assemble them, eliminating the need to gather large guidepost pieces. In 1999 the race to sequence the human genome became intensely competitive. The battling factions finally called a truce. On June 26, 2000, Craig Venter from Celera Genomics and Francis Collins, representing the International Consortium (and now director of the NIH), flanked President Clinton in the

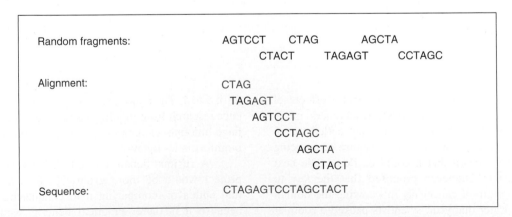

Figure 22.2 Deriving a DNA sequence. Automated DNA sequencers first determine the sequences of short pieces of DNA, or sometimes of just the ends of short pieces. Then algorithms search for overlaps. By overlapping the pieces, the software derives the overall DNA sequence.

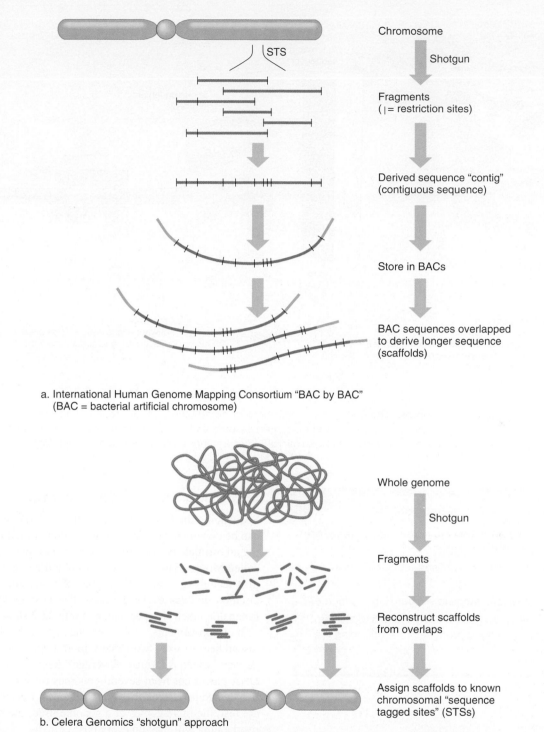

Chromosome

Shotgun

Fragments
(| = restriction sites)

Derived sequence "contig"
(contiguous sequence)

Store in BACs

BAC sequences overlapped
to derive longer sequence
(scaffolds)

a. International Human Genome Mapping Consortium "BAC by BAC"
(BAC = bacterial artificial chromosome)

Whole genome

Shotgun

Fragments

Reconstruct scaffolds
from overlaps

Assign scaffolds to known
chromosomal "sequence
tagged sites" (STSs)

b. Celera Genomics "shotgun" approach

Figure 22.3 **Two routes to sequencing the human genome. (a)** The International Consortium began with known chromosomal sites and overlapped large pieces, called contigs, that were reconstructed from many small, overlapping pieces. "STS" stands for "sequence tagged site," which refers to specific known parts of chromosomes. A BAC (bacterial artificial chromosome) is a cloning vector that uses bacterial DNA. **(b)** Celera Genomics shotgunned several copies of a genome into small pieces, overlapped them to form scaffolds, and then assigned scaffolds to known chromosomal sites. They used some Consortium data.

White House rose garden to unveil the "first draft" of the human genome sequence. The milestone capped a decade-long project involving thousands of researchers, culminating a century of discovery. The historic June 26 date came about because it was the only opening on the White House calendar! In other words, the work was monumental; its announcement, somewhat staged. **Figure 22.4** is a conceptual overview of genome sequencing.

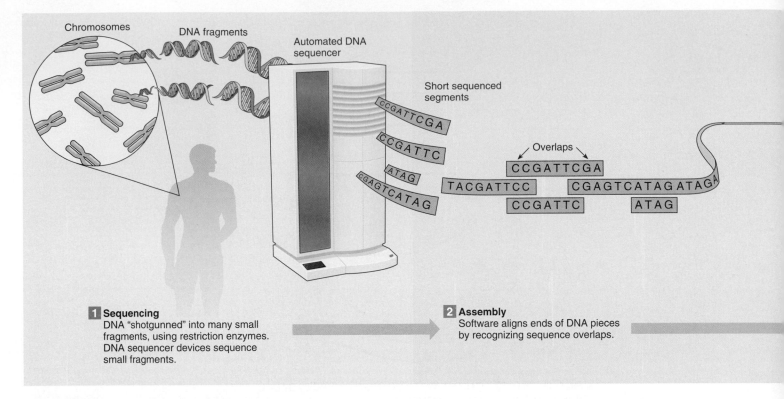

Figure 22.4 **Sequencing genomes.** The first-generation, whole-genome "shotgun" approach to genome sequencing overlapped DNA pieces cut from several copies of a genome, then assembled the overall sequence. The "next-generation" genome sequencing techniques used today are based on microfluidics and nanomaterials and are much faster.

Text in figure:
- Chromosomes
- DNA fragments
- Automated DNA sequencer
- Short sequenced segments
- CCGATTCGA
- CCGATTC
- ATAG
- CGAGTCATAG
- Overlaps
- CCGATTCGA
- TACGATTCC
- CGAGTCATAGATAGA
- CCGATTC
- ATAG
- **1** **Sequencing** DNA "shotgunned" into many small fragments, using restriction enzymes. DNA sequencer devices sequence small fragments.
- **2** **Assembly** Software aligns ends of DNA pieces by recognizing sequence overlaps.

Key Concepts Questions 22.1

1. Describe the research that led to the idea to sequence the human genome.

2. What were the initial goals of sequencing the human genome?

3. What are the general steps in sequencing a genome?

4. Distinguish between the clone-by-clone and whole-genome shotgunning approaches to sequencing the human genome.

22.2 Analysis of the Human Genome

Before the first drafts of the human genome were done, researchers were already working on the next stages by identifying sites of variation and continuing the discovery of functions of individual genes. **Table 22.1** describes several genome-related projects. A Glimpse of (Pre)history on page 432 introduces comparative genomics, which attempts to reconstruct past biological events from patterns of shared DNA sequences among modern species.

Improving Speed and Coverage

Sequencing the first human genomes took 6 to 8 years; today it can be done in a day. The cost has fallen dramatically too, beating an estimate commonly used in the computer hardware field called Moore's law. The law states that the rate of improvement in a technology (such as number of transistors on integrated circuits) doubles every 2 years. The cost of sequencing has fallen at a much sharper rate, as **table 22.2** shows.

Improvements in sequencing technology enabled researchers to work with many more copies of an individual's genome, which is termed "coverage." Recall from chapter 9 that DNA pieces cut from several genomes must be sequenced and overlapped to derive the overall sequence. Because some pieces are lost, the more copies of a genome used, the more likely the overlapping will pick up every base.

At least twenty-eight genome copies are necessary to ensure that most sequences are represented in the final sequence—it is a little like backing up a computer's contents on an external hard drive to be certain no information is lost. A genome with 40-fold coverage, for example, means each site in the genome is read on average forty times. High coverage is needed to detect a rare sequence. For example, the first exome experiments to diagnose a genetic disease, the case of Nicholas Volker told in Clinical Application 1.1, had 34-fold coverage.

Sequencing genomes provides much more information than sequencing exomes, which are just the exon (coding) parts of protein-encoding genes. Knowing the sequences surrounding

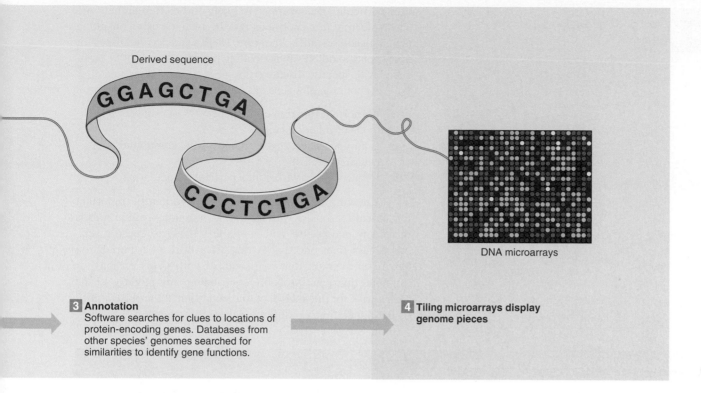

Derived sequence

GGAGCTGA

CCCTCTGA

DNA microarrays

3 **Annotation**
Software searches for clues to locations of protein-encoding genes. Databases from other species' genomes searched for similarities to identify gene functions.

4 **Tiling microarrays display genome pieces**

Table 22.1	Selected Projects to Analyze Human Genomes
HapMap (hapmap.ncbi.nlm.nih.gov)	From 2003–2005, assembled SNP maps to establish haplotypes for use in gene identification
ENCODE (Encyclopedia of DNA Elements) (www.genome.gov/10005107)	Began in 2003 with "parts list" of representative 1% of genome. Went on to analyze functional coding as well as noncoding sequences in entire genome
Personal Genomes Project (www.personalgenomes.org)	Began in 2005, at Harvard University, to collect genome sequence and health data from volunteers willing to post their information (see table 22.3)
Human Variome Project (www.humanvariomeproject.org)	Began in 2006 to collect, organize, and share data on genetic variation, especially affecting health, through disease-specific variation databases
1000 Genomes Project (http://www.1000genomes.org/)	Began in 2007 to sequence 2,500 human genomes seeking variants present in >1% of people in several populations, to catalog human genetic variation
ClinSeq (www.genome.gov/25521305)	Began in 2009, comparing genomes of 1,500 participants to their health histories, as pilot study of adding genome information to clinical care
Clinical Genome Resource (www.genome.gov)	Begun in 2013 to develop a framework to evaluate clinically relevant gene variants

the exons can detect "structural variants" such as inversions and balanced translocations (see figures 13.16 and 13.20), which move or flip DNA but do not alter the base sequence. Exome sequencing arrays must be designed to distinguish between pairs of SNPs on the same homologous chromosomes (in "*cis*" configuration) or on different homologs (in "*trans*" configuration). The *cis* configuration indicates linkage, which is important in predicting transmission of a genotype to offspring.

The Goal: Annotation

Just as finding a book written in a foreign language is meaningless unless translated, knowing the sequence of a human genome is not useful unless we know what the information means. "Annotation" in linguistics means "a note of explanation or comment added to a text or diagram." In genomics, annotations are descriptions of genes, and what the significance

Table 22.2	Cost of Sequencing a Human Genome
Year	**Cost**
1990	$1 billion
2001	$100 million
2003	$30 million
2005	$10 million
2008	$1 million
2009	$100,000
2010	$50,000
2011	$10,000
2012	$8,000
2014	<$4,000

of a particular gene variant is likely to be. Researchers are accomplishing this enormous task by meticulously consulting the published scientific literature, DNA sequence databases for every gene, and SNPs in noncoding regions that might be associated with specific disease risks.

Annotation of a gene variant might include:

- the normal function of the gene;
- mode of inheritance;
- genotype (heterozygote, homozygote, compound heterozygote); and
- frequency of a variant in a particular population.

Knowing the frequency of a gene variant is important for logical reasons. A variant that is common—which means that many people live with it—is less likely to cause a serious illness than one that is rare. If a third of a population has a gene variant that is associated with hypertension, for example, that gene probably contributes only slightly to the overall risk for this multifactorial condition. Otherwise a third of the population would have severe hypertension. However,

A GLIMPSE OF (PRE)HISTORY: COMPARATIVE GENOMICS

A goal of the human genome project was to sequence genomes of other modern species to learn more about the diversity and evolution of life on Earth. Thousands of species have had their genomes sequenced. The first were viruses and bacteria, because they have small genomes. Next came the genomes of animals important to and similar to us, such as mice, rats, chimps, cats, and dogs—more than 135 vertebrate species. Most informative, however, have been the genomes of species that represent evolutionary crossroads. These are organisms that introduced a new trait or were the last to have an old one.

Comparing genomes of modern species enables researchers to infer evolutionary relationships from DNA sequences that are conserved (shared) and presumably selected through time. **Figure 1** shows one way of displaying short sequence similarities, called a pictogram. DNA sequences from different species are aligned, and the bases at different points indicated. A large letter A, C, T, or G indicates, for example, that all species examined have the same base at that site. A polymorphic site, in contrast, has different bases for different species.

Comparative genomics uses conserved sequences to identify biologically important genome regions, assuming that persistence means evolutionary success. But there are exceptions—the human genome has some conserved sequences with no apparent function. Either we haven't discovered the functions, or genomes include "raw material" for future functions. About 6 percent of the human genome sequence is highly conserved. The following examples illustrate the types of information inferred from conserved DNA sequences.

The Minimum Gene Set Required for Life

The smallest microorganism known to be able to reproduce is *Mycoplasma genitalium*. It infects cabbage, citrus fruit, corn, broccoli, honeybees, and spiders, and

causes respiratory illness in chickens, pigs, cows, and humans. Researchers call its tiny genome the "near-minimal set of genes for independent life." Of 480 protein-encoding genes, 265 to 350 are essential. Considering how *Mycoplasma* uses its genes reveals the fundamental challenges of being alive. *M. genitalium* was the first organism to have its genome synthesized. Researchers created a "synthetic genome" consisting of the 582,970 bases, in four pieces, of the tiny bacterium's

a. Not highly conserved

b. Highly conserved

Figure 1 **A pictogram indicates conservation of the DNA sequence.** These pictograms are for short sequences in corresponding regions of the human, mouse, rat, and chicken genomes. A large letter means that all four species have the same base at that site. If four letters appear in one column, then the species differ. Pictogram **(a)** is not highly conserved; **(b)** is.

(Continued)

genome and delivered it to *E. coli* or yeast cells that had had their genomes removed.

Fundamental Distinctions Among the Three Domains of Life

Methanococcus jannaschii is a microorganism that lives at the bottom of 2,600-meter-tall "white smoker" chimneys in the Pacific Ocean, at high temperature and pressure and without oxygen. Like bacteria, the archaea lack nuclei but they replicate DNA and synthesize proteins in ways similar to multicellular organisms, and so they are considered a third form of life. The genome sequence confirmed this designation.

The Simplest Organism with a Nucleus

The yeast *Saccharomyces cerevisiae* is single-celled with only about 6,000 genes, but a third of them have counterparts among mammals, including at least seventy genes implicated in human diseases. Understanding what a gene does in yeast can provide clues to how it affects human health. For example, counterparts of mutations in cell cycle control genes in yeast cause cancer in humans.

The Basic Blueprints of an Animal

The genome of the tiny, transparent, 959-celled nematode worm *Caenorhabditis elegans* is packed with information on what it takes to be an animal. The worm's signal transduction pathways, cytoskeleton, immune system, apoptotic pathways, and brain proteins are very similar to our own. Curiously, the fruit fly (*Drosophila melanogaster*) genome has 13,601 genes, fewer than the 18,425 in the much simpler worm. Of 289 disease-causing genes in humans, 177 have counterparts in *Drosophila*. The fly is a model for testing new treatments.

From Birds to Mammals

The sequencing of the chicken genome marked a number of milestones. It was the first agricultural animal, the first bird, and, as such, the first direct descendant of dinosaurs. The genome of the red jungle fowl *Gallus gallus*, the ancestor of the domestic chicken, is remarkably like our own, minus many repeats, but its genome organization is intriguing. Like other birds, fishes, and reptiles, but not mammals, the chicken genome is distributed among very large macrochromosomes and tiny microchromosomes. Repeats may have been responsible for the larger sizes of mammalian chromosomes.

From Chimps to Humans

Most comparisons of the human genome to those of other species seek similarities. Comparisons of our genome to that of the chimpanzee, however, seek genetic *differences,* which may reflect what makes humans unique (see section 16.4). The human and chimp genomes differ by 1.2 percent, equaling about 40 million DNA base substitutions. Within those differences, as well as copy number distinctions, may lie the answers to compelling medical questions. Why do only humans get malaria and Alzheimer disease? Why is HIV infection deadlier than the chimp version, SIV?

the reverse is not true—a rare variant can be harmless, and just unusual. Many annotations take these types of data into consideration to provide an estimate of how likely the gene variant or genotype is to cause disease—such as "likely pathogenic," "elevates risk by 4 percent," or a "variant of uncertain significance." **Table 22.3** offers some examples of these interpretations.

To be most valuable, an annotated genome sequence is compared to as much health and family history information that a person can provide. The old-fashioned pedigree is still an invaluable tool and the starting point for many investigations, for this is the information that most people already know—who has what in the family. Genome annotations are also including microbiome data (see section 2.5), because the genes of the organisms that live in and on us can affect our own gene expression.

Limitations of Exome and Genome Sequencing

Whole exome and genome sequencing provide a wealth of information unimaginable a generation ago, but neither provides a complete picture of health. Some limitations arise only from exome sequencing, but some apply to both exome and genome sequencing. As fantastic as these technologies are, they come with caveats, which tap into topics from past chapters.

Exome sequences include only the 1 to 2 percent of the genome that encodes protein. Family exome sequencing has been extremely successful in identifying diseases in children whose mutations arose *de novo,* as section 4.5 describes. Genome analysis software can add other types of sequences, such as microRNAs (which control expression of suites of genes, see section 11.2), long noncoding RNAs, and other regions that a researcher specifies.

Both exome and genome sequencing will not detect the few genes that reside in mitochondria, rather than in the nucleus (see figure 5.10). DNA repeats will not show up on exome or genome sequences beyond one copy without other types of tests, because having multiples of a sequence does not alter the sequence. Such repeats include copy number variants and triple repeat mutations (see the chapter 4 opener and Clinical Connection 12.1). Deletions will not show up either. In addition, exons (protein-encoding gene parts) buried within highly repeated introns may not be detected.

Exome and genome sequencing are also blind to uniparental disomy, the rare situation in which a child inherits two alleles of a gene from one parent (see figure 13.25) rather than one from each. The two copies are identical in DNA sequence, and so "count" only once in the overall sequence.

Table 22.3   A Gallery of Genomes

Following are descriptions of the most important findings in the genomes of five people who have participated in the Personal Genome Project.

Individual	Gender	Findings
hu04FD18	XY	Hypertrophic cardiomyopathy (enlarged heart), dominant mutation of "uncertain pathogenicity"
		Peroxisome disease similar to adrenoleukodystrophy, asymptomatic carrier
		Age-related macular degeneration (ARMD; visual loss), mutation "likely pathogenic"
		Hypertension, homozygous recessive for common mutation increases risk 8%
		Increased plasma triglycerides due to *ApoA5* mutation, not associated with increased risk of coronary artery disease
hu704A85	XX	Biotinidase deficiency, asymptomatic carrier
		Adenosine deaminase deficiency, asymptomatic carrier
		Stiff arteries, lifetime risk of heart attack increased 0.5–3% above population risk of 15%
		Nephropathy (kidney disease) mutation, risk increased <0.1%
		Hypertension, increased risk associated with variant in noncoding DNA
hu3073E3	XY	Baldness (confirmed with use of mirror)
		Ichthyosis vulgaris. Heterozygous for filaggrin mutation has 30–50% risk of atopic dermatitis (eczema), which individual already has
		Polydactyly mutation, but known in only one family
		Age-related macular degeneration mutation, "likely pathogenic"
hu132B5C	XX	Esophageal cancer, dominant mutation increases risk 4-fold
		Crohn disease, mutation in *NOD2* gene "likely pathogenic"
		Down syndrome risk to offspring 0.4% due to mutation in *MTRR* gene (age is more important risk factor)
		Hypertension, 4% increase in risk due to SNPs in noncoding region of angiotensin II gene
Hu04DF3C	XY	Galactosemia deficiency, asymptomatic carrier
		Intellectual disability, dominant variant in *CDH* gene, "not statistically significant" increase in risk
		Lumbar disc disease increased from 4% in general population to 11%; individual already has it

Perhaps the most important limitation of both exome and genome sequencing is at the level beyond the single gene. Once the genome is fully annotated, additional computational power may be required to determine and describe all possible combinations of gene variants, incorporating control sequences such as those that encode microRNAs.

How do genes regulate one another? One gene affecting the expression of another can explain why siblings with the same single-gene disease suffer to a different extent. For example, a child with severe spinal muscular atrophy, in which an abnormal protein shortens axons of motor neurons, may have a brother who also inherits SMA but has a milder case thanks to inheriting a variant of a second gene that extends axons. Computational tools will be needed to sort out networks of interacting genes, sometimes called "connectomes." What was once regarded as simple epistatic interactions—gene affecting gene—may in reality be the tip of an iceberg of complex networks of genes whose expression, which changes in tissue and time, influence each other.

Finally, epigenetic changes induced by environmental factors provide a layer of information that must be applied to DNA information. These are the influences that are perhaps the most important, because we can act on many of them.

22.3 Personal Genome Sequencing

Nearly three decades ago, when "the human genome project" was just an idea, probably the most important word, in hindsight, was "the." Today, with increasing focus on how we differ genetically from each other, the age of personal genome sequencing is here. Genome sequencing can provide a canvas on which other types of information can be painted, to give a fuller picture of how our bodies function and malfunction. Although genome sequencing is important for investigating our ancestry and our diversity today, it will perhaps be most practically valuable in health care, which will have to adapt to a new type of medical information—and a deluge of it.

Practical Medical Matters

The field of human genetics for many years was strictly an academic discipline, a biological science. Until the acceleration of gene discoveries in the 1990s and the sequencing of the human genome in the early 2000s, human genetics as a medical specialty was a very small field, with a few knowledgeable physicians and families with very rare diseases making up much of the small patient population. With the introduction of direct-to-consumer genetic testing in 2008, the possibility of testing genes not only for disease-causing mutations, but for variants that indicate only risk, was suddenly available to anyone—without requiring medical expertise, until the Food and Drug Administration intervened in 2013. Even with regulatory restrictions, many people have taken these tests.

Genetic and genomic testing as part of health care must meet certain practical criteria. The most important criterion for a regulated DNA test or treatment is clinical utility. Does benefit outweigh risk? Is it as effective as an existing, approved test or treatment? Will it help people who cannot use the existing test or treatment? Efficacy must be demonstrated, not just assumed. For example, molecular evidence may indicate that, hypothetically, people with a particular genotype might respond better to a particular drug than people with different genotypes. A clinical trial must evaluate the drug in both groups of people to demonstrate that prescribing-by-genotype—pharmacogenetics—is actually helpful on the whole-person level. The challenge of validating genetic testing arose with newborn screening (see section 20.2), when clinicians realized the importance of following up initial identification of infants at high risk of developing a genetic disease with definitive diagnostic tests.

A DNA test result alone is not sufficient to diagnose a disease, but may support a clinical diagnosis based on symptoms and the results of other types of tests. For example, a person might be a heterozygote for familial hypercholesterolemia, but is not diagnosed with the condition unless the serum cholesterol level becomes elevated or a cardiovascular condition develops. However, knowing that a mutation is present can coax a person to seek further testing. This is the case for the family described in chapter 18 with stomach cancer. Relatives who inherited the mutation had scans that revealed tiny tumors—they had cancer already but didn't know it.

The uncertainty in genetic and genomic testing that makes further diagnostic testing necessary arises from the complications of Mendel's laws discussed in chapter 5. A DNA test alone is not sufficient for diagnosis because of incomplete penetrance (genotype does not always foretell phenotype), variable expressivity (different severities in different individuals), epistasis (gene-gene interactions), genetic heterogeneity (mutation in more than one gene causing a phenotype), and environmental influences (epigenetics).

How will electronic medical records handle the nuances of genomic data? Will records include entire sequences of the DNA bases A, T, C, and G, or just the diagnostic report that includes gene variants known to be pathogenic? How will the records embrace future discoveries that impact the stored data or diagnoses? How will diagnostic codes work? While these matters are under discussion, the medical profession has had to catch up both to the profusion of new genetic tests and genomic technologies, and with the fact that many patients are very knowledgeable about DNA—in general and sometimes their own. As recently as 2013, less than 0.1 percent of graduating medical students were choosing medical genetics as a specialty. It is more likely that genomics will become incorporated into specialties, as it already is in oncology and pediatrics. More and more medical schools are having students analyze their own genomes, and physicians are attending continuing medical education programs to learn genomics. Clinical geneticists, genetic counselors, and molecular pathologists are the specialists who are leading the way in the new genomic medicine.

Types of Information in Human Genomes

The human genome sequenced by the public consortium was actually composites of different individuals. The first two genomes from specific individuals to be sequenced, of genome research pioneers Craig Venter and James Watson, yielded few medical surprises. Instead, they showed that we had greatly underestimated genetic variation by focusing only on the DNA sequence. The numbers of copies of short sequences—copy number variants, or CNVs—contribute significantly to genetic variation.

"Back in 2001, we thought we differed from chimps by 1.27% of our genomes. Now we know that we differ from each other by as much as 1 to 3%. If we count all the differences, we are about 5 to 6% from the chimp. In the way we put sequences in public databases, we lost the insertions and deletions," said Venter to an American Society of Human Genetics meeting after he'd had a year to think about what his personal genome sequencing had revealed. He already knew much of it from his family history and personal experience.

Venter has gene variants associated with increased risk of Alzheimer disease and cardiovascular disease. He has alleles for dry earwax, blue eyes, lactose intolerance, a preference for activities in the evening, and tendencies toward antisocial behavior, novelty seeking, and substance abuse. He metabolizes caffeine fast, which he also knew. "I can have two double lattes and wash it down with a Red Bull and not be affected by it," he said.

James Watson, according to his genome sequence, carries a dozen rare recessive disorders that would affect glycogen storage, vision, and DNA repair if homozygous, and he is at elevated risk for twenty other disorders. He elected not to learn his status for the *ApoE4* gene variant that increases risk of Alzheimer disease because his grandmother had the condition, but people inferred the result from the surrounding published parts of his genome (a deduction called imputation that compares the surrounding sequence to that of other people's genomes). He also has a "variant of uncertain significance" for the *BRCA1* cancer susceptibility gene and is a slow metabolizer of beta blockers and antipsychotic medications, indicating that he could overdose on normal weight-based dosages.

Science journals deemed Watson's results "of thin clinical value" and yielding "few biological insights." Watson and Venter differ in inherited drug responses, supporting the value of pharmacogenetics/genomics (discussed in chapter 20). Said Venter, "You probably wouldn't suspect this based on our appearance—we are both bald, white scientists."

The third person to have his genome sequenced was called, simply, "YH." He is Han Chinese, an East Asian population that accounts for 30 percent of modern humanity. He has no inherited diseases in his family, but his genome includes 116 gene variants that cause recessive disorders, as well as many risk alleles. He shares with Craig Venter a tendency to tobacco addiction and high-risk alleles for Alzheimer disease.

An overall comparison of the first three genome sequences of individuals provides a peek at our variation. Each man has about 1.2 million SNPs, but a unique collection. Each has only 0.20 to 0.23 percent of SNPs that are nonsynonymous, meaning that they alter an encoded amino acid, and the men share only 37 percent of these more meaningful SNPs. The math indicates, therefore, that about 0.07 percent of our SNPs may affect our phenotypes.

After the first three human genome sequences were published, others followed from specific ethnic groups. Then followed people who could afford the initially high cost: journalists who were paid to write about the experience, and celebrities. Reasons varied, as they still do. The late Steve Jobs (founder of Apple) and late journalist Christopher Hitchens had their cancer genomes sequenced to guide drug choices.

Scholar Henry Louis Gates Jr. had his genome sequenced to trace his African roots. An actress and a rock star reportedly did it to better understand mental illness in their families.

One geneticist had his genome sequenced to serve as a control for a project to sequence the genomes of all the citizens of Qatar. He discovered that he has gene variants for baldness (which he knew from looking in the mirror), a recessive disease that affects children, Viking ancestors, and most important, a blood clotting disorder that explained why he bleeds profusely when injured. Despite learning interesting information, the researcher voiced fears: his family learning things they didn't want to know, and even someone using his DNA sequence to frame him for a crime.

Another geneticist who had his genome sequenced early on and wrote about it, in a prominent medical journal, is Stephen Quake. He co-invented a next-generation DNA sequencing device. He thereby joined the ranks of scientists who experimented on themselves, such as Jonas Salk, who took his own polio vaccine. The central question: *What could Stephen Quake's genome sequence tell him that would be useful?*

Quake's quest began where most genetic studies begin: a family history. The 19-year-old son of his first cousin had died of a heart condition, and other relatives had heart and blood vessel troubles. Yet Stephen Quake is healthy—he exercises, eats well, doesn't smoke, and his conventional disease risk factors are ones he cannot control, such as male sex and age. He also had standard blood tests and tests based on his family history of cardiovascular disease: An electrocardiogram to assess heart function, an echocardiogram to assess heart structure, and a battery of exercise exams such as stress tests.

What did Dr. Quake learn about himself from his genome sequence? He indeed has gene variants—some common, some novel—that raise his risk of developing heart and blood vessel disease. He is a carrier of hereditary hemochromatosis (see section 20.2), epidermolysis bullosa (a condition in which the skin layers separate), and biotinidase deficiency (see Clinical Connection 2.1). More important than confirming what he already knew from family history and identifying recessive disorders that will not affect him was the drug information. Taking a statin drug would keep his cholesterol levels down and not cause the side effect of muscle damage. Based on his blood work a statin didn't seem necessary, but his family history and genome sequence convinced his personal physician otherwise, and so the statin drug may save his life. He also learned that should his inherited propensity for developing blood clots come true, the drug clopidogrel (Plavix) would not work, and he would need to take low doses of the blood thinner warfarin (Coumadin). Should he develop type 2 diabetes, as his genome sequence suggests, the drug metformin (Glucophage) would likely not work.

Do You Want Your Genome Sequenced?

The number of human genomes sequenced will soon exceed one million, and the number of exomes sequenced is more than that. As the cost of sequencing continues to drop, and as health care practitioners are having their own genomes sequenced and learning how to interpret them, chances are that many of us will face the choice of knowing our own genome sequences (**figure 22.5**).

Figure 22.5 Will you have your genome sequenced?

Here are some final thoughts to consider in deciding whether or not, or when, to have your own genome sequenced, and some more general questions:

- Who should decide which gene variants are reported to a patient—the patient, or a health care provider?
- What criteria should a health care provider consider in deciding which genome findings to report to a patient?
- Should health care providers inform a patient of a gene variant that adds only slightly to the risk of developing a particular multifactorial disorder, such as hypertension?
- Should only actionable genome findings be reported to patients?
- How can a health care consumer stay up to date on research results that might be relevant?
- Does a patient have an ethical obligation to share genome results with relatives who may also be at elevated risk of developing an illness?

- Should a parent have a child's genome sequenced?
- Should all newborns have their genomes sequenced?

Other concerns about genome sequencing are societal. Who will pay for genome sequencing? How can the 3,000 genetic counselors and 1,200 clinical geneticists in the United States handle routine genome sequencing and interpretation? Will societal pressure drive people to have their genomes sequenced, much as nearly everyone now has e-mail? Could governments make genome sequencing compulsory, and could the practice introduce new ways to discriminate against people? We have come a very long way from comparing genome sequencing to conquering Mt. Everest.

What will the coming flood of genetic and genomic information ultimately mean? Will it tell us where we came from more than family lore and documents? Will physicians consult strings of A, C, T, and G to determine how best to treat their patients, or will signs, symptoms, family history, and a patient's observations turn out to be more valuable types of information? Only time will tell.

I hope that this book has offered you glimpses of the future and stimulated you to think about the choices that genetic and genomic technologies will present. For weekly coverage of human genetics, read my blog DNA Science (http://blogs.plos.org/dnascience/), at Public Library of Science.

Key Concepts Questions 22.3

1. What can personal information from genome sequences add to health care, and what complications might it introduce?
2. What types of information have people learned from having their genomes sequenced?
3. What societal issues does genome sequencing raise?

Summary

22.1 From Genetics to Genomics

1. Genetic maps have increased in detail and resolution, from cytogenetic and linkage maps to physical and sequence maps.
2. Positional cloning discovered individual genes by beginning with a phenotype and gradually identifying a causative gene, localizing it to part of a chromosome.
3. The human genome project began in 1990 under the direction of the DOE and NIH. Technological advances sped the sequencing.
4. Several copies of a genome are cut and the pieces sequenced, overlapped, and aligned to derive the continuous sequence. For the human genome, the International Consortium used a chromosome-by-chromosome approach and Celera Genomics used whole-genome shotgunning.

22.2 Analysis of the Human Genome

5. After sequencing genomes became possible, attention turned to refining the process, cataloging human variation, and discovering gene functions.

6. The cost of sequencing decreased and the speed of sequencing increased, as researchers annotated genes with information on gene function, mode of inheritance and frequencies of variants, and genotypes.
7. Exome and genome sequencing alone will not detect copy number variants, mitochondrial DNA, uniparental disomy, or gene-gene and gene-environment interactions.
8. Genome information used in medical tests and treatments must be validated for clinical utility (be novel or at least as useful as an existing test or treatment).

22.3 Personal Genome Sequencing

9. An individual human genome sequence can confirm what is known from family history, detect disease-associated recessive alleles, detect gene variants that contribute to risk of having or developing a trait or illness, and predict drug responses.
10. Widespread availability of genome sequencing will raise personal as well as societal questions.

www.mhhe.com/lewisgenetics11

Answers to all end-of-chapter questions can be found at
www.mhhe.com/lewisgenetics11. You will also find additional
practice quizzes, animations, videos, and vocabulary
flashcards to help you master the material in this chapter.

Review Questions

1. Distinguish between the exome and the genome.

2. How did family linkage patterns and chromosomal aberrations lay the groundwork for sequencing the human genome?

3. Describe how the four levels of genetic maps differ, and what new types of maps depict.

4. Explain how positional cloning was used to identify disease-causing genes.

5. Why was there initial disagreement over whether the human genome should be sequenced?

6. Why did the human genome project not detect copy number variants?

7. Explain why discovery of the human genome sequence was just a starting point.

8. Why is it important to use many copies of a genome when deriving the sequence?

9. Explain why a gene variant that is present at a high frequency in a population is less likely to be harmful than a very rare variant in that population.

10. Why is it helpful to include microbiome information in a genome annotation document?

11. Give an example of evolutionary information deduced from comparative genomics.

12. Explain why an inversion might escape detection in exome sequencing but not in genome sequencing.

13. List three limitations of exome and/or genome sequencing.

14. List three criteria for genome information to have clinical utility.

15. Give examples of types of information that people can learn from knowing their genome sequences.

Applied Questions

1. How should genetic or genomic information be incorporated into clinical diagnosis? Discuss the significance of the differences between Mendelian and multifactorial traits, and the effect of penetrance and epistasis.

2. Compare and contrast DNA testing to an X ray or other standard medical diagnostic test.

3. Some researchers are suggesting that to cut costs and make genome technology available to more people, testing should only look for genes most likely to be mutant in a particular person. What is an advantage and a disadvantage of this strategy?

4. Opinions range widely on the value of having one's genome sequenced. Said James Watson, "Every time someone goes into a children's hospital with a serious disease, it would be immoral not to sequence him." Yet

another geneticist said "Genome sequencing will yield an excess of information that is useless, uninterpretable, and possibly damaging to the patient." Still another geneticist said "We should have our genomes sequenced as early as possible." What is your opinion?

5. An article in the *Journal of the American Medical Association* quoted two statements that medical students hear:

   "Don't order a test unless the results will affect your plan of action."

   "The diagnosis is usually in the patient's history."

   How might the availability of genome sequencing change these statements?

6. Discuss a medical situation, perhaps in your own family, that might benefit from exome or genome sequencing.

Web Activities

1. Read about the Electronic Medical Record and Genomics Network (http://www.genome.gov/27540473), and describe how it is evolving from a research tool to a clinical guideline.

2. Explore the website for the Personal Genome Project (http://www.personalgenomes.org/) and describe a person's genome, similar to the entries in table 22.3.

3. Go to www.genome.org and describe a "Genome Advance of the Month" and why it is important.

4. Go to the Ethical, Legal and Social Implications (ELSI) Research Program website (http://www.genome.gov/10001618). Discuss one societal concern arising from genomics, and how it might affect you.

5. Go to www.medomics.com. What is a diagnostic genome?

Case Studies and Research Results

1. After the tsunami that devastated Japan in 2011, many organisms never before seen washed up on shore, thrown from the deep sea and some quite bizarre in appearance. Researchers collected specimens and sequenced DNA to try to classify the animals. Consider the following 8-base sequence that is similar among the species:

Fish 1 / Fish 2 / Fish 3 / Fish 4 — Position 1 2 3 4 5 6 7 8

   a. Write the DNA sequences for the two most closely related fishes.
   b. Which position(s) in the sequence are highly conserved?
   c. Which position(s) in the sequence are the least conserved?
   d. Which site is probably not essential, and how do you know this?
   e. A coelacanth has C T A C T G G T for this section of the genome. Which of the mystery fishes is the coelacanth's closest relative?

2. Some researchers have said that healthy lifestyle habits (not smoking, following a healthy diet, exercising regularly, regular wellness exams, and diagnosing and treating common diseases early) would save more lives than sequencing everyone's genome and predicting diseases from the information. Suggest an experiment to evaluate these two approaches to staying healthy.

3. Alexis and Noah Beery are fraternal twins who were diagnosed at age 2 with dopa-responsive dystonia (OMIM 128230). Due to deficiency of the neurotransmitter dopamine, the twins had uncontrollable movements and tremors that would worsen throughout the day. For a time, the twins responded to L-dopa, a drug used to treat Parkinson disease, but by age 14, new symptoms arose. Alexis's throat would spasm, obstructing her breathing, and her brother developed severe hand tremors and attention deficit. The twins' father, a researcher at a biotechnology company, arranged whole-genome sequencing for the family—himself and his wife, the twins, and an older brother.

   Genome sequencing revealed that the twins are deficient in three neurotransmitters—serotonin and norepinephrine in addition to dopamine. Each twin is a compound heterozygote for the *SPR* gene, which encodes an enzyme that each type of neurotransmitter requires for its synthesis in the body. Happily, a drug that is a precursor to serotonin was well-studied: 5-HTP. It along with L-dopa enabled the twins to lead normal lives. In addition, the mutation that they inherited from their mother explains the fibromyalgia that she and some of her relatives have.

   a. Does 5-HTP correct the phenotype or the genotype?
   b. What genotypes are possible for the older brother?
   c. If the Beerys have another child, what is the risk that he or she inherits the condition that affects the twins?
   d. Is the mother's mutation that causes her fibromyalgia recessive or dominant?
   e. What term describes the relationship between the twins' hormonal problem and their mother's fibromyalgia (see chapter 12)?

Glossary

Pronunciations are provided for technical terms.

KEY ə = eh
- – = long vowel sound
- ˘ = short vowel sound
- ʺ = heavy accent
- ʹ = light accent
- ^ = aw

A

acrocentric (ăk′rō-sĕn′trĭk) A chromosome in which the centromere is near one end.

adaptive immunity (ĭ-myōō′nĭ-tē) A slow, specific immune response following exposure to a foreign antigen.

adenine (ăd′en-ēn′) One of two purine nitrogenous bases in DNA and RNA.

admixture New combinations of gene variants that arise when individuals from two previously distinct populations have children together.

affected sibling pair study A gene identification approach that looks for gene variants that siblings with a particular condition or trait share but that siblings who do not share the trait do not share.

allelic (ə-lēl ik) **disorders** Different diseases caused by mutations in the same gene.

allele (ə-lēl′) An alternate (variant) form of a gene.

alternate splicing Building different proteins by combining exons of a gene in different ways.

amino (ə-mē′nō) **acid** A small organic molecule that is a protein building block.

amniocentesis (ăm′nē-ō-sĕn-tē′sĭs) A test that examines fetal chromosomes and biochemicals in amniotic fluid.

anaphase (ănə-fāz′) Stage of mitosis when the centromeres of replicated chromosomes part.

aneuploid (ăn′yū-ploid′) A cell with one or more extra or missing chromosomes.

angiogenesis (anʺje-o-jen′ə-sis) Extension of blood vessels.

antibody (ăn′tē-bŏd′ē) A multisubunit protein, produced by B cells, that binds a specific foreign antigen, alerting the immune system or destroying the antigen.

anticodon (ăntē-kō′dŏn) A three-base sequence on one loop of a transfer RNA molecule that is complementary to an mRNA codon, and joins the appropriate amino acid and its mRNA.

antigen (ăn′tē-jən) A molecule that elicits an immune response.

antigen (ăn′tē-jən) **binding sites** Specialized ends of antibody chains.

antigen (ăn′tē-jən) **-presenting cell** A cell displaying a foreign antigen.

antiparallel The head-to-tail position of the entwined chains of the DNA double helix.

apoptosis (āpō-tō′sis) A form of cell death that is a normal part of growth and development.

assisted reproductive technologies (**ARTs**) Procedures that replace a gamete or the uterus to help people with fertility problems have children.

autoantibodies (ô′tō-ăn′tē-bŏdēz) Antibodies that attack the body's own cells.

autoimmunity (ô′tō-ĭ-myōō′nĭ-tē) An immune attack against one's own body.

autophagy (au-toph′ -a-gy) A process in which a cell dismantles its own debris.

autosomal (ôtə-sōməl) **dominant** The mode of inheritance in which one autosomal allele causes a phenotype. Such a trait can affect males and females and does not skip generations.

autosomal (ôtə-sōməl) **recessive** The mode of inheritance in which two autosomal alleles are required to cause a phenotype. Such a trait can affect males and females and can skip generations.

autosome (ôtə-sōm) A chromosome that does not have a gene that determines sex.

B

balanced polymorphism (pŏl′ē-môr′fizəm) Maintenance of a harmful recessive allele in a population because the heterozygote has a reproductive advantage.

base excision repair Removal of up to five contiguous DNA nucleotides to correct oxidative damage.

B cell A type of lymphocyte that secretes antibody proteins in response to nonself antigens displayed on other immune system cells.

bioethics A field that addresses personal issues that arise in applying medical technology and genetic information.

bioremediation Use of plants or microorganisms to detoxify environmental pollutants.

biotechnology The alteration of cells or biochemicals with a specific application.

blastocyst (blăs′tō-sĭst′) A hollow ball of cells descended from a fertilized ovum.

blastomere (blăstō′-mēr′) A cell of a blastocyst.

C

cancer (kăn′sər) A group of disorders resulting from loss of cell cycle control.

cancer stem cells Stem cells that divide and yield cancer cells and abnormal specialized cells.

carbohydrate (karʺbo-hī′-drāt) An organic compound that consists of carbon, hydrogen, and oxygen in a 1:2:1 ratio. Includes sugars and starches.

carcinogen (kar-sĭnə-jən) A substance that causes cancer.

cell (sel) The fundamental unit of life.

cell (sel) **cycle** A cycle of events describing a cell's preparation for division and division itself.

cell-free fetal DNA Small pieces of fetal DNA in a woman's bloodstream used for testing for genetic disease.

cellular adhesion A precise series of interactions among the proteins that connect cells.

cellular immune (ĭ-myōōn′) **response** T cells release cytokines to stimulate and coordinate an immune response.

centriole (sĕn′-trē-ohl) A structure in cells that organizes microtubules into the mitotic spindle.

centromere (sĕn′-trō-mîr) The largest constriction in a chromosome, located at a specific site in each chromosome type.

centrosome (sĕn′-trō-sōm) A structure built of centrioles and proteins that organizes microtubules into a spindle during cell division.

chaperone protein A protein that binds a polypeptide and guides folding.

chorionic villus (kôrē-ŏn′ĭk vĭl-us) **sampling** (**CVS**) A prenatal diagnostic technique that analyzes chromosomes in chorionic villus cells, which, like the fetus, descend from the fertilized ovum.

chromatid (krō′ mə-tĭd) A single, very long DNA molecule and its associated proteins, forming a longitudinal half of a replicated chromosome.

chromatin (krō′ mə-tĭn) DNA and its associated proteins.

chromatin (krō′ mə-tĭn) **remodeling** Adding or removing chemical groups to or from histones, which can alter gene expression.

chromosome (krō′ mə-sōm′) A highly wound continuous molecule of DNA and the proteins associated with it.

chromosomal microarray analysis A technique that detects small copy number variants.

cleavage (klēvĭj) A series of rapid mitotic cell divisions after fertilization.

clines (klĭnz) Allele frequencies that change from one geographical area to another.

cloning vector A piece of DNA used to transfer DNA from a cell of one organism into the cell of another.

coding strand The side of the DNA double helix for a particular gene from which RNA is not transcribed.

codominant A heterozygote in which both alleles are fully expressed.

codon (kō′dŏn) A continuous triplet of mRNA that specifies a particular amino acid.

coefficient of relatedness The proportion of genes that two people related in a certain way share.

collectins (ko-lek′tinz) Immune system molecules that detect viruses, bacteria, and yeasts.

comparative genomics (jə-nō′miks) Identifying conserved DNA sequences among genomes of different species.

comparative genomic hybridization (CGH) A technique using fluorescent labels to detect copy number variants.

complement Plasma proteins that have a variety of immune functions.

complementary base pairs The pairs of DNA bases that hydrogen bond; adenine bonds to thymine and guanine to cytosine.

complementary DNA (cDNA) A DNA molecule that is the complement of an mRNA, copied using reverse transcriptase.

compound heterozygote An individual with two different mutations in the same gene.

concordance (kən-kôr′dens) A measure indicating the degree to which a trait is inherited; percentage of twin pairs in which both members express a trait.

conditional mutation (myōō-tā′shən) A genotype that is expressed only under certain environmental conditions.

conformation The three-dimensional shape of a molecule.

consanguinity (kŏnsăn-gwin′i-tē) Blood relatives having children together.

copy number variant A DNA sequence present in different numbers of copies in different individuals; repeats.

CRISPR A genome editing technology that uses a bacterial enzyme and RNA to make double-stranded breaks at selected sites in a genome.

critical period The time during prenatal development when a structure is sensitive to damage from a mutation or an environmental intervention.

crossing over An event during prophase I when homologs exchange parts.

cytogenetics (sītō-jə-nĕt′iks) Matching phenotypes to detectable chromosomal abnormalities.

cytokine (sītō-kīn′) A biochemical that a T cell secretes that controls immune function.

cytokinesis (sī-tō-kin-ē′-sis) Division of the cytoplasm and its contents.

cytoplasm (sī′-tō-plăzm) Cellular contents other than organelles.

cytosine (sī′-tō-sēn) One of the two pyrimidine nitrogenous bases in DNA and RNA.

cytoskeleton (sī-tō-skĕl′i-tn) A framework of protein tubules and rods that supports the cell and gives it a distinctive form.

D

dedifferentiated A cell less specialized than the cell it descends from. A characteristic of a cancer cell.

deletion mutation (myōō-ta′-shən) A missing sequence of DNA or part of a chromosome.

deoxyribonucleic (dē-ŏksē-rībō-nōō-klā′ik) **acid (DNA)** The genetic material; the biochemical that forms genes.

deoxyribose (dē-ŏksē-rī′bōs) 5-carbon sugar in a DNA nucleotide.

differentiation Cell specialization, reflecting differential gene expression.

dihybrid cross Breeding individuals heterozygous for two traits.

diploid (dĭp′loid) A cell containing two sets of chromosomes.

dizygotic (dīzī-gŏt′ik) **(DZ) twins** Twins that originate as two fertilized ova; fraternal twins.

DNA *See* **deoxyribonucleic acid**.

DNA microarray A set of target genes embedded in a glass chip, to which labeled cDNAs from a clinical sample bind and fluoresce. Microarrays show patterns of gene expression and gene variants present.

DNA polymerase (pə-lim′ər-ās) **(DNAP)** An enzyme that adds new bases to replicating DNA and corrects mismatched base pairs.

DNA probe A labeled short sequence of DNA that binds its complement in a biological sample.

DNA profiling A biotechnology that detects differences in the number of copies of certain DNA repeats among individuals. Used to rule out or establish identity.

DNA replication Construction of a new DNA double helix using the information in parental strands as a template.

dominant A gene variant expressed when present in one copy.

driver mutation A mutation that provides the selective growth advantage of a cancer cell.

duplication An extra copy of a DNA sequence, usually caused by misaligned pairing in meiosis.

E

ectoderm (ĕktō-dûrm) The outermost primary germ layer of the primordial embryo.

embryo (ĕm′brē - ō′) In humans, prenatal development until the end of the eighth week, when all basic structures are present.

embryonic (ĕmbrē-ŏn′ik) **stem (ES) cell** A cell, derived in laboratory culture from inner cell mass cells of very early embryos, that can self-renew and some of its daughters differentiate as any cell type.

empiric risk Probability that a trait will recur based upon its incidence in a population.

endoderm (ĕn′-dō-dûrm) The innermost primary germ layer of the primordial embryo.

endoplasmic reticulum (ĕndō-plăzmik rə-tik′u-ləm) **(ER)** An organelle consisting of a labyrinth of membranous tubules on which proteins, lipids, and sugars are synthesized.

enzyme (ĕnzim) A type of protein that speeds the rate of a specific biochemical reaction.

epigenetic (ĕpē-jə-nĕt′ik) A layer of information placed on a gene that modifies expression, not changing the DNA sequence, such as methylation.

epistasis (ĕpē-stā-sis) A gene masking the expression of another.

epitope (ep′i-tōp) Part of an antigen that an antibody binds.

equational division The second meiotic division, producing four cells from two.

euchromatin (yōō-krō′mə-tin′) Parts of chromosomes that do not stain and that contain active genes.

eugenics (yōō-jĕn′iks) The control of individual reproductive choices to achieve a societal goal.

eukaryotic cell (yōō-kar′ē-ŏt′ik sel) A complex cell containing organelles, including a nucleus.

euploid (yōō′-ploid) A somatic cell with the normal number of chromosomes for the species.

ex vivo **gene therapy** Replacing mutant genes in cells growing in the laboratory and then introducing the cells into a patient.

excision repair Enzyme-catalyzed removal of pyrimidine dimers in DNA.

exome (x-ōm) The 1 to 2 percent of the genome that encodes protein.

exon (x-on) A part of a gene that encodes amino acids.

exon skipping A mutation at a splice site that removes an exon (which specifies contiguous amino acids) from the encoded protein.

expanding repeat A short DNA sequence that is present in a certain range of copy numbers in wild type individuals but, when expanded, causes a disease phenotype.

expressivity Degree of severity of a phenotype.

F

family exome analysis A strategy that identifies gene variants in all protein-encoding genes in a person whose symptoms have eluded diagnosis and in family members, then identifying the gene that causes the mutant phenotype.

fetus (fē′təs) The prenatal human after the eighth week of development, when structures grow and specialize.

founder effect A type of genetic drift in which a few individuals found a new settlement, perpetuating a subset of alleles from the original population.

frameshift mutation (myōō-tā′shən) A mutation that alters a gene's reading frame.

fusion protein A protein that forms from translation of transcripts from two genes.

G

G$_0$ phase An offshoot of the cell cycle in which the cell remains specialized but does not replicate its DNA or divide.

G$_1$ phase The stage of the cell cycle following mitosis in which the cell resumes synthesis of proteins, lipids, and carbohydrates.

G$_2$ phase The stage of the cell cycle following S phase but before mitosis, when certain proteins are synthesized.

gamete (găm′ēt) A sex cell.

gamete intrafallopian (găm′ēt intrə-fə-lōpē-ən) **transfer (GIFT)** An infertility treatment in which sperm and oocytes are placed in a woman's uterine tube.

gastrula (găstrə-lə) A three-layered embryo.

gene (jēn) A sequence of DNA that instructs a cell to produce a particular protein.

gene expression Transcription of a gene's DNA into RNA.

gene expression profiling Use of DNA microarrays to detect the types and amounts of cDNAs reverse transcribed from the mRNAs in a particular cell source.

gene pool All the genes in a population.

gene silencing Techniques that block transcription or degrade mRNA, preventing translation into protein.

gene therapy Introducing a functioning gene to compensate for the effects of a mutation.

genetic (jə-nĕt′ĭk) **code** The correspondence between specific mRNA triplets and the amino acids they specify.

genetic counselor A medical specialist who calculates risk of recurrence of inherited disorders in families, applying the laws of inheritance to pedigrees and interpreting genetic test results.

genetic determinism Attributing a trait to a gene or genes.

genetic drift Changes in allele frequencies in a population due to random sampling from a larger population.

genetic heterogeneity A phenotype that can be caused by variants of any of several genes.

genetic load The collection of deleterious recessive alleles in a population.

genetic marker A DNA sequence near a gene of interest that is co-inherited unless separated by a crossover.

genetics The study of inherited variation.

genome (jē′nōm) The complete set of genetic instructions in the cells of a particular type of organism.

genome editing Creating double-stranded breaks in the DNA double helix, enabling insertion or removal of a specific DNA sequence.

genome-wide association study (GWAS) A study in which millions of variants (single nucleotide polymorphisms or copy number variants) that form haplotypes are compared between people with a condition and unaffected individuals to identify parts of the genome that might contribute to a phenotype.

genomic (jē nō′m ĭk) **imprinting** Differing of the phenotype depending upon which parent transmits a particular allele.

genomics (jē nōm ĭks) The study of the functions and interactions of many genes or other DNA sequences, or comparing genomes.

genotype (jē n′ə- tĭp) The allele combinations in an individual that cause traits or disorders.

genotypic (jēn′ə- tĭp′ĭk) **ratio** The ratio of genotype classes expected in the progeny of a cross.

germline gene therapy Genetic alterations of gametes or fertilized ova, which perpetuate the change throughout the organism and transmit it to future generations.

germline mutation A mutation that is in every cell in an individual because it was present in the fertilized ovum.

Golgi (gōl′jē) **apparatus** An organelle, consisting of flattened, membranous sacs, that packages secretion components.

gonadal mosaicism Having two or more genetically distinct cell populations in an ovary or testis.

gonads (gō′-nadz) Paired structures in the reproductive system where sperm or oocytes are manufactured.

growth factor A protein that stimulates mitosis.

guanine (gwa′ nēn) One of the two purine nitrogenous bases in DNA and RNA.

H

haploid (hăp′ loid) A cell with one set of chromosomes.

haplogroup (hăp′ lō-groōp) A specific set of markers on the Y or mitochondrial chromosome that are inherited together and therefore can be used to trace ancestry.

haplotype (hăp′ lō tĭp) A series of known DNA sequences or single nucleotide polymorphisms linked on a chromosome.

Hardy-Weinberg equilibrium An idealized state in which allele frequencies in a population do not change from generation to generation.

heavy chains The two longer polypeptide chains of an antibody subunit.

hemizygous (hĕm′ ē-zī′ gəs) The sex that has half as many X-linked genes as the other; a human male is hemizygous.

heredity Transmission of inherited traits from generation to generation.

heritability An estimate of the proportion of phenotypic variation in a group due to genes.

heterochromatin (hĕtə-rō-krō′mətĭn) Dark-staining chromosome parts that have few protein-encoding genes.

heterogametic (hĕt′ə-rō-gə-mē′tĭk) The sex with two different sex chromosomes; a human male.

heteroplasmy (hĕt′ə-rō-plăz-mē) Mitochondria in the same cell having different alleles of a particular gene.

heterozygous (hĕtə-rō-zī′gəs) Having two different alleles of a gene.

histone (hĭs′tōn) A type of protein around which DNA entwines in a regular pattern.

hominins (hŏm′ə-nĭnz) Animals ancestral to humans only.

homogametic (hō′mō-gə-mē′tĭk) The sex with identical types of sex chromosomes; the human female.

homologous (hō-mŏl′ə-gəs) **pairs** Chromosomes with the same gene sequence.

homozygosity (hōmō-zī-gəs′-ĭ-tē) **mapping** An approach to gene discovery that correlates stretches of homozygous DNA base sequence in the genomes of related individuals to certain traits or disorders.

homozygous (hōmō-zī′ gəs) Having two identical alleles of a gene.

hormone (hor′ mōn) A biochemical produced in a gland and carried in the blood to a target organ, where it exerts an effect.

human leukocyte antigen (lōōkə-sīt′ ăn′tĭ-jən) **(HLA) complex** Genes closely linked on the short arm of chromosome 6 that encode cell surface proteins important in immune system function.

humoral (yōō′ mər-əl) **immune response** Process in which B cells secrete antibodies into the bloodstream.

I

idiotype (ĭd′ēo-tĭp) Part of an antibody molecule that binds an antigen.

incidence The number of new cases of a disease during a certain time in a population.

incomplete dominance A heterozygote intermediate in phenotype between either homozygote.

independent assortment The random arrangement of homologous chromosome pairs, in terms of maternal or paternal origin, down the center of a cell in metaphase I. Inheritance of a gene on one chromosome does not influence inheritance of a gene on a different chromosome. (Mendel's second law)

induced pluripotent stem (iPS) cells Somatic cells that are reprogrammed toward an alternative developmental fate by altering their gene expression.

infertility The inability to conceive a child after a year of unprotected intercourse.

inflammation Part of the innate immune response that causes an infected or injured area to swell with fluid, turn red, and attract phagocytes.

innate immunity (ĭ-myōō′nĭ-tē) Components of immune response that are present at birth and do not require exposure to an environmental stimulus.

inner cell mass A clump of cells on the inside of the blastocyst that will continue developing into an embryo. Source of embryonic stem cells.

insertional translocation A rare type of translocation in which a part of one chromosome is part of a nonhomologous chromosome.

insertion mutation (myōō-tā′-shən) A mutation that adds DNA bases.

interferon (in″tər-fēr′on) A type of cytokine.

interleukin (in″tər-loo′kin) A type of cytokine.

intermediate filament A type of cytoskeletal component made of different proteins in different cell types.

interphase (ĭn′tər-fāz′) Stage when a cell is not dividing.

intracytoplasmic (ĭn′trə-sītō-plăzmĭk) **sperm injection (ICSI)** An infertility treatment that injects a sperm cell nucleus into an oocyte, to overcome poor or absent sperm motility.

intrauterine (ĭn′trə-yōō′tər-ĭn) **insemination (IUI)** An infertility treatment that places donor sperm in the cervix or uterus.

intron (ĭn trŏn) Part of a gene that is transcribed but is excised from the mRNA before translation into protein.

in vitro (ĭn vē′trō) **fertilization (IVF)** Placing oocytes and sperm in a laboratory dish with appropriate biochemicals so that fertilization occurs, then, after a few cell divisions, transferring the embryos to a woman's uterus.

in vivo **gene therapy** Introduction of vectors carrying therapeutic human genes directly into the body part where they will act.

isochromosome A chromosome that has two copies of one arm but none of the other, as a result of cell division along the wrong plane.

K

karyotype (kă′rē-ō-tīp) A size-order chromosome chart.

L

law of independent assortment *See* **independent assortment**.

law of segregation *See* **segregation**.

lethal allele (ə-lēl′) An allele that causes death before reproductive maturity or halts prenatal development.

ligase (lī′gās) An enzyme that catalyzes the formation of covalent bonds in the sugar-phosphate backbone of a nucleic acid.

light chains The two shorter polypeptide chains of an antibody subunit.

linkage Genes on the same chromosome.

linkage disequilibrium Extremely tight linkage between DNA sequences.

linkage maps Maps that show gene order on chromosomes, determined from crossover frequencies between pairs of genes.

lipid (lĭp′ĭd) A type of organic molecule that has more carbon and hydrogen atoms than oxygen atoms. Includes fats and oils.

long noncoding RNA An RNA molecule in the nucleus that associates with chromatin and likely helps control gene expression, especially in the brain.

lysosome (lī′sō-sōm) A saclike organelle containing enzymes that degrade debris.

M

macroevolution (măk′rō-ĕv′ə-lōōshən) Genetic change sufficient to form a new species.

manifesting heterozygote (hĕt′ə-rō-zīgōt) A female carrier of an X-linked recessive gene who expresses the phenotype because the normal allele is inactivated in some tissues.

meiosis (mī-ō′sĭs) Cell division that halves the number of chromosomes to form haploid gametes.

memory cells B or T cell descendants that carry out a secondary immune response.

mesoderm (mĕz-ō-dûrm) The middle primary germ layer.

messenger RNA (mRNA) A molecule of RNA complementary in sequence to the template strand of a gene that specifies a protein product.

metacentric (mĕtə-sĕn′trĭk) A chromosome with the centromere approximately in the center.

metagenomics Sequencing all of the genomes present in a sample of a particular environment.

metaphase (mĕtə-fāz) The stage of mitosis when chromosomes align along the center of the cell.

metastasis (mĕtə-stā′-sĭs) Spread of cancer from its site of origin to other parts of the body.

microbiome (mahy-kroh-bahy′-ohm) All of the organisms that live in and on another organism.

microevolution Change of allele frequency in a population.

microfilament A solid rod of actin protein that forms part of the cytoskeleton.

microRNA A 21- or 22-base-long RNA that binds to certain mRNAs, blocking their translation into protein.

microtubule (mī̆krō-tōōbyōōl) A hollow structure built of tubulin protein that forms part of the cytoskeleton.

mismatch repair Proofreading of DNA for misalignment of short, repeated segments.

missense (mĭs′sĕns) **mutation** A single base change mutation that alters an amino acid.

mitochondrion (mītō-kŏn′drē-ən) An organelle consisting of a double membrane that houses enzymes that catalyze reactions that extract energy from nutrients.

mitosis (mī-tōsĭs) Division of somatic (nonsex) cells.

mode of inheritance The pattern in which a gene variant passes from generation to generation. It may be dominant or recessive, autosomal or X- or Y-linked.

molecular evolution Changes in protein and DNA sequences over time used to estimate how recently species diverged from a shared ancestor.

monohybrid (mŏn′ō-hībrĭd) **cross** A cross of two individuals who are heterozygous for a single trait.

monosomy (mŏn′ō-sō′mē) A human cell with 45 (one missing) chromosomes.

monozygotic (mŏnō-zī-gŏt′ĭk) (**MZ**) **twins** Twins that originate as a single fertilized ovum; identical twins.

morula (môr′ yə-lə) The very early prenatal stage that resembles a mulberry.

multifactorial A trait or illness determined by several genes and the environment.

mutagen (myōō′tə-jən) A substance that changes, adds, or deletes a DNA base.

mutant (myōōt′nt) An allele that differs from the normal or most common allele in a population that alters the phenotype.

mutation (myōō-tā′shən) A change in a protein-encoding gene that affects the phenotype and is rare in a population.

N

natural selection Differential survival and reproduction of individuals with particular phenotypes in particular environments, which may alter allele frequencies in subsequent generations.

neural (nōōr′əl) **tube** A structure in the embryo that develops into the brain and spinal cord.

neurexin A protein in the presynaptic membrane of a neuron that uses the neurotransmitter glutamate and is involved in autism.

neuroglia Several types of cells in the nervous system that support neurons.

neuron (nōrin′) A nerve cell.

neuroligin A protein in the postsynaptic membrane of a neuron that uses the neurotransmitter glutamate and is involved in autism.

neurotransmitter A molecule that transmits messages in the nervous system.

nitrogenous (nī-trŏj′ə-nəs) **base** A nitrogen-containing base that is part of a nucleotide.

nondisjunction (nŏndĭs-jŭngk′shən) The unequal partitioning of chromosomes into gametes during meiosis.

noninvasive prenatal testing A general term that includes ultrasound imaging and testing cell-free fetal DNA in maternal plasma.

nonsense mutation (myōōtā′shən) A point mutation that changes an amino-acid-coding codon into a stop codon, prematurely terminating synthesis of the encoded protein.

nonsense suppression Action of a drug that enables protein synthesis to ignore a nonsense mutation (stop codon not at the end of a gene).

nonsynonymous codon (kō′don) A codon that encodes a different amino acid from another codon.

nucleic (nōō-klē′ĭk) **acid** DNA or RNA.

nucleolus (nōō-klē′ə-ləs) A structure in the nucleus where ribosomes are assembled from ribosomal RNA and protein.

nucleosome (nōō′-klē-ō-sōm) A unit of chromatin structure.

nucleotide (nōō-klēō-tĭd) The building block of a nucleic acid, consisting of a phosphate group, a nitrogenous base, and a 5-carbon sugar.

nucleotide (nōō-klēō-tĭd) **excision repair** Replacement of up to 30 nucleotides to correct DNA damage of several types.

nucleus (nōō-klēəs) A large, membrane-bounded region of a eukaryotic cell that houses DNA.

O

oncogene (ŏn′kə-jēn) A gene that normally controls the cell cycle, but causes cancer when overexpressed.

oocyte (ō′ə-sīt) The female gamete (sex cell).

oogenesis (ōə-jĕn′ĭ-sĭs) Oocyte development.

oogonium (o-o-gō′-ni-um) A cell in the ovary that gives rise to an oocyte, in meiosis.

open reading frame A sequence of DNA that does not include a stop codon.

organelle (ŏr′gə-nĕl′) A specialized structure in a eukaryotic cell that carries out a specific function.

ovaries (o′və-rēz) The female gonads.

P

paracentric (para sĕn′-trĭk) **inversion** An inverted chromosome that does not include the centromere.

passenger mutation A mutation in a cancer cell that does not provide selective growth advantage.

pedigree A chart of symbols connected by lines that depict the genetic relationships and transmission of inherited traits in related individuals.

penetrance Percentage of individuals with a genotype who have an associated phenotype.

pericentric (pər-ē sĕn-trik) **inversion** An inverted chromosome that includes the centromere.

peroxisome (pə-rŏk′sǐ-sōm) An organelle consisting of a double membrane that houses enzymes with various functions.

pharmacogenetics (farm a kō jə-nĕt-ĭks) Testing for single gene variants that affect drug metabolism.

pharmacogenomics (farm a kō jə-nōm-ĭks) Testing for variants of many genes or gene expression profiles that affect drug metabolism.

phenocopy (fē′ nō-kŏp′ē) An environmentally caused trait that occurs in a familial pattern, mimicking inheritance.

phenotype (fē′ nō-tīp) The expression of a gene in traits or symptoms.

plasma cell A cell descended from a B cell that produces abundant antibodies of a single type.

plasma membrane (plăz′mə mĕm′brān) The selective barrier around a cell, consisting of proteins, glycolipids, glycoproteins, and lipid rafts on or in a phospholipid bilayer.

plasmid (plăz′ mǐd) A small circle of double-stranded DNA found in some bacteria. Used as a vector in recombinant DNA technology.

pleiotropic (plēə-trōpǐk) A single-gene disorder with several symptoms. Different symptom subsets may occur in different individuals.

point mutation (myōō-tā′ shən) A single base change in DNA.

polar body A product of female meiosis that contains little cytoplasm and does not continue to develop into an oocyte.

polygenic (pŏlē-jēn′ ĭk) Traits determined by more than one gene.

polymerase (pŏlə′-mə-r′ās) **chain reaction (PCR)** A nucleic acid amplification technique in which a DNA sequence is replicated in a test tube to rapidly produce many copies.

polymorphism (pŏlē-môr′ fǐz əm) A DNA base or sequence at a certain chromosomal locus that varies in a small percentage of individuals in a population.

polyploid (pŏl′ē-ploid) A cell with one or more extra sets of chromosomes.

population A group of interbreeding individuals.

population bottleneck Decrease in allele diversity resulting from an event that kills many members of a population, followed by restoration of population numbers.

population genetics (jə-nĕt′ĭks) The study of allele frequencies in different groups of individuals.

population study Comparison of disease incidence in different groups of people.

preimplantation genetic (jə-nĕt′ĭk) **diagnosis (PGD)** Removing a cell from an 8-celled embryo and testing it for a mutation to deduce the genotype of the embryo.

prevalence The number of cases of a disease in a population at a particular time.

primary germ layers The three layers of an embryo.

primary immune response Immune system's response to initial encounter with a nonself antigen.

primary (1°) structure The amino acid sequence of a protein.

progenitor cell A cell whose descendants can follow any of several developmental pathways, but not all.

prokaryotic cell (prō-kārē - ŏt′ik sĕl) A cell that does not have a nucleus or other organelles. One of the three domains of life. Bacteria.

promoter A control sequence near the start of a gene.

pronuclei (prō-nōō′klēī) DNA packets in the fertilized ovum.

prophase (prō′fāz) The first stage of mitosis or meiosis, when chromatin condenses.

prospective study A study that follows two or more groups.

proteasome (prō-tēə-sōm) A multiprotein structure in a cell shaped like a barrel through which misfolded proteins pass and are refolded or dismantled.

protein A type of macromolecule that is the direct product of genetic information; a chain of amino acids.

proteome (prō′tē-ōm) The set of proteins a cell produces.

proteomics (prōtē - ō′ mǐks) Study of the proteins produced in a particular cell type under particular conditions.

proto-oncogene (prōtō-ŏn′kə-jēn) A gene that normally controls the cell cycle, but when overexpressed causes cancer.

pseudogene (sōō′ dō jēn) A gene that does not encode protein, but whose sequence very closely resembles that of a coding gene.

Punnett square A diagram used to follow parental gene contributions to offspring.

purine (pyōō r′ēn) A nucleic acid base with a two-ring structure; adenine and guanine are purines.

pyrimidine (pǐ-rǐm′ǐ-dēn) A nucleic acid base with a single-ring structure; cytosine, thymine, and uracil are pyrimidines.

Q

quantitative trait loci Genes that determine polygenic traits.

quaternary (4°) structure A protein that has more than one polypeptide subunit.

R

reading frame The point in a DNA sequence from which contiguous triplets encode amino acids of a protein. A DNA sequence has three reading frames.

recessive An allele whose expression is masked by another allele.

reciprocal translocation A chromosome aberration in which two nonhomologous chromosomes exchange parts, conserving genetic balance but rearranging genes.

recombinant (rē-kŏm′bə-nənt) A series of alleles on a chromosome that differs from the series of either parent.

recombinant (rē-kŏm′bə-nənt) **DNA** Transferring genes between species.

reduction division The first meiotic division, which halves the chromosome number.

replication fork Locally opened portion of a replicating DNA double helix.

restriction enzyme An enzyme, typically from bacteria, that cuts DNA at a specific sequence, and used to create recombinant DNA molecules.

reverse vaccinology Creating a vaccine using pathogen genome sequence information to identify antigens that stimulate a human immune response.

ribonucleic (rǐ bō-nōō-klē′ĭk) **acid (RNA)** A nucleic acid whose bases are A, C, U, and G.

ribose (rǐ′bōs) A 5-carbon sugar in RNA.

ribosomal (rǐ′bōs-ō′məl) **RNA (rRNA)** RNA that, with proteins, comprises ribosomes.

ribosome (rǐ′bō sōm) An organelle consisting of RNA and protein that is a scaffold and catalyst for protein synthesis.

risk factor A characteristic or experience associated with increased likelihood of developing a particular medical condition.

RNA interference (RNAi) Introduction of a small interfering RNA molecule that binds to and prevents translation of a specific mRNA.

RNA polymerase (pŏl′ə-mə-rās) **(RNAP)** An enzyme that adds RNA nucleotides to a growing RNA chain.

Robertsonian (Răb-ərt-sō′-nē-ən) **translocation** A chromosome aberration in which two short arms of nonhomologous chromosomes break and the long arms fuse, forming one unusual, large chromosome.

run of homozygosity Regions of the genome in which contiguous SNPs (single nucleotide polymorphisms) are homozygous, indicating a shared ancestor with another person with the same pattern.

S

S phase The stage of interphase when DNA replicates.

secondary immune response Immune system activation in response to a second or subsequent encounter with a pathogen.

secondary (2°) structure Folds in a polypeptide caused by attractions between amino acids close together in the primary structure.

segregation The distribution of alleles of a gene into separate gametes during meiosis. (Mendel's first law)

self-renewal Defining property of a stem cell; the ability to yield a daughter cell like itself.

semiconservative replication DNA synthesis along each half of the double helix.

sequential polar body analysis Testing the DNA of a second polar body to infer the genotype of its associated fertilized ovum.

sex chromosome (krō′mə-sōōm) A chromosome containing genes that specify sex.

sex-influenced trait Phenotype caused when an allele is recessive in one sex but dominant in the other.

sex-limited trait A trait that affects a structure or function present in only one sex.

sex ratio Number of males divided by number of females multiplied by 1,000 for people of a certain age in a population.

short tandem repeats (STRs) Repeats of 2 to 10 DNA bases that are compared in DNA profiling.

signal transduction A series of biochemical reactions and interactions that pass information from outside a cell to inside, triggering a response.

single nucleotide polymorphism (nōōklēō-tĭd pŏlē-môr′ fĭz′əm) (**SNP**) Single base sites that differ among individuals.

sirtuin An enzyme that regulates energy use that may help to maintain health in the elderly.

somatic cell (sō-măt′ĭk sĕl) A nonsex cell, with 23 pairs of chromosomes in humans.

somatic (sō-măt′ĭk) **gene therapy** Genetic alteration of a specific cell type, not transmitted to future generations.

somatic mutation (sō-măt′ĭk myōō-tā′shən) A genetic change in a nonsex cell.

spermatogenesis (spər-măt′ə-jĕn′ĭ-sis) Sperm cell differentiation.

spermatogonium (sper″mah-to-gō′ ne-um) An undifferentiated cell in a seminiferous tubule that can give rise to a sperm cell in meiosis.

spermatozoon (spər-măt′ə-zō′ŏn) (sperm) A mature male reproductive cell (meiotic product).

spindle A structure composed of microtubules that pulls sets of chromosomes apart in a dividing cell.

splice-site mutation A single base mutation at a site in a gene that controls intron removal, resulting in extra or absent amino acids in the protein product.

spontaneous mutation (myōō-tā′sheən) A genetic change that results from mispairing when the replication machinery encounters a base in its rare tautomeric form.

***SRY* gene** The sex-determining region of the Y. If the *SRY* gene is activated, the gonad develops into a testis; if not, an ovary forms under direction of other genes.

stem cells Cells that give rise to other stem cells, as well as to cells that differentiate.

submetacentric (sŭb mĕt-ə-sĕn′trĭk) A chromosome in which the centromere establishes a long arm and a short arm.

sugar-phosphate backbone The "rails" of a DNA double helix, consisting of alternating deoxyribose and phosphate groups, oriented opposite one another.

synapse The space between two neurons that a neurotransmitter must cross to transmit a message.

synonymous codons (kō d ŏnz) DNA triplets that specify the same amino acid.

synteny (sĭn′tə-nē) Correspondence of genes on the same chromosome in several species.

T

tandem duplication A duplicated DNA sequence next to the original sequence.

T cell A type of lymphocyte that produces cytokines and coordinates the immune response.

telomerase (tə-lŏm′ə-rās) An enzyme, including a sequence of RNA, that adds DNA to chromosome tips.

telomere (tĕl′ə-mîr) A chromosome tip.

telophase (tĕlə-fāz) The stage of mitosis or meiosis when daughter cells separate.

template strand The DNA strand carrying the information to be transcribed.

teratogen (tə-răt′ə-jən) A substance that causes a birth defect.

tertiary (3°) structure Folds in a polypeptide caused by interactions between amino acids and water. This draws together amino acids that are far apart in the primary structure.

testes (tes′tēz) The male gonads.

thymine (thī′mēn) One of the two pyrimidine bases in DNA.

transcription Manufacturing RNA from DNA.

transcription factor A protein that activates the transcription of certain genes.

transfer RNA (tRNA) A type of RNA that connects mRNA to amino acids during protein synthesis.

transgenic organism (trăns-jĕn′ĭk) An individual with a genetic modification, typically introduction of a gene from another species, in every cell.

transition A point mutation altering a purine to a purine or a pyrimidine to a pyrimidine.

translation Assembly of an amino acid chain according to the sequence of base triplets in a molecule of mRNA.

translocation Exchange between nonhomologous chromosomes.

translocation carrier An individual with exchanged chromosomes but no signs or symptoms. The person has the usual amount of genetic material, but it is rearranged.

transposon (trăns-pōzŏn) A gene or DNA segment that moves to another chromosome.

transversion A point mutation altering a purine to a pyrimidine or vice versa.

trisomy (trī sō′mē) A human cell with 47 chromosomes (one extra).

tumor suppressor gene (tōōmər səprĕs′ər jĕn) A recessive gene whose normal function is to limit the number of divisions a cell undergoes.

U

uniparental disomy (yû-ni-pə′rent-əl dī sō mē) (**UDP**) Inheriting two copies of the same gene from one parent.

uracil (yōōr′ə-sĭl) One of the four types of bases in RNA; a pyrimidine.

V

vaccine (vak-sē′n) An inactive or partial form of a pathogen that stimulates antibody production.

vesicles (ves-ə-kulz) Bubble-like membrane-bounded organelles that participate in secretion.

virus (vī rəs) An infectious particle built of nucleic acid in a protein coat.

W

wild type The most common phenotype in a population for a particular gene.

X

X inactivation The inactivation of one X chromosome in each cell of a female mammal, occurring early in embryonic development.

X-linked Genes on an X chromosome.

X-Y homologs (hŏm′ə-lôgz) Y-linked genes that are similar to genes on the X chromosome.

Y

Y-linked Genes on a Y chromosome.

Z

zinc finger nuclease technology A genome editing technique that uses protein motifs called zinc fingers to bind specific DNA triplets, enabling a nuclease to cut the DNA.

zygote (zī′gŏt) A prenatal human from the fertilized ovum stage until formation of the primordial embryo, at about two weeks.

zygote intrafallopian transfer (zī′ gŏt ĭn′trə-fə-lō′ pē-ən) (**ZIFT**) An assisted reproductive technology in which an ovum fertilized *in vitro* is placed in a woman's uterine tube.

Credits

Line Art and Text

Chapter 4
TA 4.2: Cystic Fibrosis Foundation. Patient Registry 2006 Annual Report.

Chapter 6
TA 6.1: Courtesy Professor Jennifer A. Marshall-Graves, Australian National University.
TA 6.3: Courtesy David Page, Massachusetts Institute of Technology, Howard Hughes Medical Institute Investigator.

Chapter 7
Figure 7.2: Data from Gordon Mendenhall, Thomas Mertens, and Jon Hendrix, "Fingerprint Ridge Count" in *The American Biology Teacher*, vol. 51, no. 4, April 1989, pp. 204–206.
Figure 7.14: Graph from www.unitedhealthfoundation.org. Reprinted by permission.

Chapter 8
Table. 8.1: Psyciatric Genomics, Inc., Gaithersburg, MD. The information was collated from the Surgeon General's 1999 Report on Mental Health.
Figure 8.8: © Robert Gilliam

Chapter 12
Figure 12.10: Adapted from Exon Skipping, © 2013 Prosensa.

Chapter 13
Figure 13.6: Source of data from *Color Atlas of Genetics* by Eberhard Passage, p. 401. Copyright © 2001. Thieme Medical Publishers, Inc.

Chapter 14
Figure 14.10: Copyright 2008 National Academy of Sciences. All rights reserved. http://www.koshland-science-museum.org.

Chapter 16
Figure 16.14: Adapted from Hua Liu, Frank Prugnolle, Andrea Manica and Francois Balloux, "A Geographically Explicit Genetic Model of Worldwide Human-Settlement History," *The American Journal of Human Genetics*, Vol. 79: 2, 2006.

Chapter 18
Figure 18.8: Adapted from "Neurobiology: At the root of brain cancer," Michael F. Clarke, *Nature* 432:281–282. Macmillan Publishers Ltd, November 18, 2004.
Figure 18.14a: Adapted from "Tissue repair and stem cell renewal in carcinogenesis," Philip A. Beachy, Sunil S. Karhadkar, and David M. Berman, *Nature* 432:324–331. Macmillan Publishers Ltd, 2004.

Clinical Connection 18.1 Figure 1: Adapted from "Drug therapy: Imantinib mesylate—A new oral targeted therapy" by Savage & Antman: *New England Journal of Medicine* 346:683–693. Copyright © 2002 Massachusetts Medical Society.

Photo Credits

Image Researcher: Toni Michaels/PhotoFind, L.L.C.
Author photo: Courtesy, Wendy Josephs

Front Matter
Page v(top left): © Tom Grill/Corbis RF; p. v(top right): © Tom Grill/Corbis RF; p. v(center left): Courtesy, Jane Mervar; p. v(center right): © CDC/C. Goldsmith, P. Feorino, E.L. Palmer, W.R. McManus; p. v(bottom left): © Dr. Gopal Murti/ Science Source; p. v(bottom right): Courtesy Barry Palevitz; p. vi(1): © Tom Grill/Corbis RF; p. vi(2): © JJ pixs/Alamy; p. vi(3): Photo Courtesy of The Progeria Research Foundation; p. vi(4): Courtesy, Jane Mervar; p. vii(5): Courtesy, Gavin R. Stevens Foundation; p. vii(7): © Corbis RF; p. vii(8): © Sami Sarkis/Photographer's Choice/Getty Images RF; p. vii(9): © Dr. Gopal Murti/Science Source; p. viii(11): Courtesy, Bomber Command Museum of Canada; p. viii(12): © Dr. P. Marazzi/ Science Source; p. viii(13): © National Library of Medicine; p. viii(14): © Tom Grill/Corbis RF; p. viii(15): © Digital Vision/Getty RF; p. viii(16): © David Gray/Reuters/Landov; p. ix(17): © CDC/C. Goldsmith, P. Feorino, E.L. Palmer, W.R. McManus; p. ix(18): © Addenbrookes Hospital/ Science Source; p. ix(19): Courtesy Barry Palevitz; p. ix(20): Courtesy of the Randell family. Photo by Greg Gomez; p. ix(22): © Don Carstens/ Artville RF; p. xiv(top): Courtesy, Jane Mervar; p. xiv(bottom): © Centers for Disease Control (CDC); p. xv(top): All rights reserved, Clinic for Special Children; p. xv(center): Courtesy, Lori Sames. Photo by Dr. Wendy Josephs; p. xv(bottom): Courtesy of the Author; p. xix: © Clive Chilvers/Alamy.

Chapter 1
Opener: © Tom Grill/Corbis RF; 1.1: © BananaStock/PunchStock RF; 1.2: © CNRI/ Science Source; 1.5a: © Lester V. Bergman/ Corbis; 1.5b: © Steve Mason/Getty RF; 1.7: © Leslie Close/AP Images; 1.8: © Clive Chilvers/ Alamy; p. 10: © Greg Kuchik/Getty Images RF; p. 12: © Jane Ades, NHGRI.

Chapter 2
Opener: © JJ pixs/Alamy; 2.1a: © Biophoto Associates/Science Source; 2.3: © SPL/Science Source; p. 19: © SPL/Science Source; 2.4(top): © David M. Phillips/The Population Council/ Science Source; 2.4(left): © K.R. Porter/Science Source; 2.4(right): © EM Research Services,

Newcastle University RF; 2.7: © Prof. P. Motta & T. Naguro/SPL/Science Source; 2.8: © Bill Longcore/Science Source; 2.11: © Dr. Gopal Murti/Science Source; p. 28: Courtesy, Lori Sames. Photo by Dr. Wendy Josephs; 2.13b: © Mediscan/Medical-on-Line/Alamy; 2.15b: © SPL/Science Source; 2.16(all): © Ed Reschke; 2.18: © Health Protection Agency/Science Source; 2.19(top): © David McCarthy/Science Source; 2.19(center): © Peter Skinner/Science Source.

Chapter 3
Opener: Photo Courtesy of The Progeria Research Foundation; p. 49(top): Illustration by Nicolaas Hartsoeker from Essai de Dioptrique, 1695; 3.10: © Prof. P.M. Motta/Univ. "La Sapienza", Rome/Science Source; p. 52: Courtesy of the Author; 3.14(1): © Petit Format/ Nestle/Science Source; 3.14(2): © P.M. Motta & J. Van Blerkom/SPL/Science Source; 3.14(3): © Petit Format/Nestle/Science Source; p. 55: Photo courtesy of Malte Spielmann; 3.17: Courtesy, Brittany and Abby Hensel; 3.18a: © Geoff Tompkinson/Science Source; 3.18a: © Petit Format/Nestle/Science Source; 3.18b: © Petit Format/SPL/Science Source; 3.19: © Nestle/Petit Format/Science Source; 3.22: © Wealan Pollard/ age fotostock RF.

Chapter 4
Opener: Courtesy, Jane Mervar; p. 70: © Pixtal/ age Fotostock RF; p. 76: Courtesy, Cystic Fibrosis Foundation; 4.15: © Paul Burns/Getty Images RF; 4.18: Courtesy, Lisa Hane.

Chapter 5
Opener: Courtesy, Gavin R. Stevens Foundation; 5.1a: © Kevin Winter/Getty Images; 5.2: © Alfred Pasieka/SPL/Science Source; 5.7: © Columbia University, ho/AP Images; p. 96: © Marc Phares/ Science Source.

Chapter 6
Figure 6.1: © Biophoto Associates/Science Source; p. 113: © Dr. Walter Just; 6.6: Courtesy, Dr. Mark A. Crowe; 6.8b: Courtesy, Richard Alan Lewis M.D., M.S., Baylor College of Medicine; 6.9a: Photo courtesy of Pragna I. Patel, Ph.D/ Baylor College of Medicine; 6.11: © Animal Attraction/OS50/Getty RF; 6.12: Designed by Mark Sherman. Provided by Arthur Riggs and Craig Cooney; 6.14: © The McGraw-Hill Companies, Inc.; 6.15b: © Carla D. Kipper; 6.15c: © Dan James/Caters News Agency.

Chapter 7
Opener: © Corbis RF; 7.1: © Brand X Pictures RF; p. 132: © The McGraw-Hill Companies, Inc./ Al Telser, photographer; 7.4b: © SWNS.com; 7.3: © The McGraw-Hill Companies, Inc./Photo by David Hyde and Wayne Falda; 7.5: Courtesy,

Smile Train, Mark Atkinson, photographer; 7.8: © Bob Kreisel /Alamy; 7.13: © Darryl Leniuk/age fotostock.

Chapter 8

Opener: © Sami Sarkis/Photographer's Choice/ Getty Images RF; 8.2: © Stanford University Center for Narcolepsy; 8.5: © Ryan McVay/Getty Images RF.

Chapter 9

Opener: © Dr. Gopal Murti/Science Source; 9.4a-b: © Science Source; 9.5: © A. Barrington Brown/Science Source; 9.9b: © Mark Boulton/Science Source; 9.18(top): © James Cavallini/Science Source; 9.18(bottom): © Custom Medical Stock Photo/Alamy RF.

Chapter 10

Figure 10.20: © The Nobel Foundation, 1976.

Chapter 11

Opener: Courtesy, Bomber Command Museum of Canada; 11.4a: © Petit Format/Science Source; 11.4b: © PhotoDisc/Getty RF; 11.11(inset): © BSIP SA/Alamy; 11.12a: © E.H. Gill/Custom Medical Stock Photo.

Chapter 12

Opener: © Dr. P. Marazzi/Science Source; 12.1: © Mike Wintroath/AP Images; 12.3: © Science Photo Library RF/Getty Images RF; 12.4: © Biophoto Associates/Science Source; 12.5a-c: Courtesy Shiro Ikegawa, RIKEN; p. 223: Courtesy, Rebekah Lieberman; p. 226 (All): From R. Simensen, R. Curtis Rogers, "Fragile X Syndrome," American Family Physician, 39:186 May 1989. @ American Academy of Family Physicians; 12.12: © St Bartholomew's Hospital/Science Source; 12.13a: © Burke Triolo Productions/Getty Images RF; 12.13b: © CNRI/SPL/Custom Medical Stock Photo; 12.17: © Wellcome Image Library/Custom Medical Stock Photo.

Chapter 13

Opener: © National Library of Medicine; 13.3: © Biophoto Associates/Science Source. Colorization by Mary Martin; 13.5c: © Dr. Najeeb Layyous/ Science Source; 13.7a: Drawing by Walther

Flemming, 1882; 13.7b: © CNRI/Science Source; 13.8: © James King-Holmes/Science Source; 13.11: CNRI/Science Source; 13.13: © Angela Hampton/Bubbles Photolibrary/Alamy; 13.14: © Stockbyte/Veer RF; 13.15: © Wellcome Image Library/Custom Medical Stock Photo; 13.17: Courtesy, Kathy Naylor; 13.20b: Courtesy Lawrence Livermore National Laboratory; 13.24: Courtesy, Ring Chromosome 20 Foundation.

Chapter 14

Opener: © Tom Grill/Corbis RF; 14.1: © Comstock/Punchstock RF; 14.8: © Trinity Mirror/ Mirrorpix/Alamy; 14.9: Image courtesy of Esther N. Signer; 14.12: © Lisa Zador/Getty Images RF.

Chapter 15

Opener: © Digital Vision/Getty RF; 15.2: © Purestock/SuperStock RF; p. 284: All rights reserved, Clinic for Special Children, 2013; 15.9: © Goodshoot/Alamy RF; 15.10: © Brand X Pictures/ Getty Images RF; 15.11: © Science Source; 15.12: © CDC/C. Goldsmith, P. Feorino, E. L. Palmer, W. R. McManus; 15.14a: © CDC/ Steven Glenn, Laboratory & Consultation Division; 15.14b: © CDC/Sickle Cell Foundation of Georgia: Jackie George, Beverly Sinclair/photo by Janice Haney Carr; p. 293: Centers for Disease Control (CDC); 15.17: © The Daily Progess, Dan Lopez/AP Images; p. 297: Courtesy, Marie Deatherage.

Chapter 16

Opener: © David Gray/Reuters/Landov; 16.2: © Jon Foster/National Geographic Image Collection/Alamy; 16.3: © Volker Steger/Science Source; 16.4: © Danita Delimont/Alamy; 16.6: © Knut Finstermeier, Max Planck Institute for Evolutionary Anthropology; 16.7: © Max Planck Institute for Evolutionary Anthropology; 16.9: © Andrea Solero/AFP/Getty Images; 16.10: © Peter Johnson/Corbis; 16.11a: © Joe Blossom Alamy; 16.11b: © Trajano Paiva/Alamy; 16.11c: © Daniela Duncan/Getty Images; 16.18: © Ingram Publishing RF.

Chapter 17

Opener: © CDC/C. Goldsmith, P. Feorino, E.L. Palmer, W.R. McManus; 17.1: © CDC/Janice Haney Carr; p. 328(a): © MedicalRF.com;

p. 328(b): © Custom Medical Stock Photo/Alamy RF; 17.2: © CNRI/Science Source; 17.10: © Bettmann/Corbis; p. 339: © Courtesy, Dr. Maureen Mayes; 17.15: © CDC/ Carl Flint; Armed Forces Institute of Pathology.

Chapter 18

Opener: © Addenbrookes Hospital/Science Source; 18.2a: © The McGraw-Hill Companies, Inc.; 18.2b: © Nancy Kedersha/Science Source; 18.3b: © Steve Gschmeissner/SPL/Getty Images RF; 18.11(left): © The McGraw-Hill Companies, Inc./Al Telser, photographer; 18.11(center, right): © Biophoto Associates/Science Source; 18.13: © Addenbrookes Hospital/Science Source; 18.14b: © Centers for Disease Control; 18.15: © Custom Medical Stock Photo; 18.17: © Comstock Images/ Jupiter Images; 18.18(top): © PhotoDisc/Getty RF; 18.18(bottom): © Jack Star/PhotoLink/Getty Images RF.

Chapter 19

Opener: Courtesy Barry Palevitz; 19.1: © Eye of Science/Science Source; 19.3(5): © Maximilian Stock Ltd./SPL/Science Source; 19.3(6): © CDC/Jim Gathany; p. 379: © Image Source/ Getty RF; 19.4(1-3): Courtesy MRC Harwell; 19.5: © Alexis Rockman, 2000; 19.8(left): © Corbis RF; 19.8(right): © Dorling Kindersley/ Getty Images RF.

Chapter 20

Opener: Courtesy of the Randell family. Photo by Greg Gomez; 20.1: © mediacolor's/Alamy; 20.2: National Human Genome Research Institute; 20.5: © MedicalRF.com/Getty Images RF; 20.6: © Anna Powers; 20.7: Photo compliments of Mickie Gelsinger; 20.9a-b: Courtesy, Wendy Josephs.

Chapter 21

Figure 21.1: © Keri Pickett/Pickett Pictures; p. 410: © Tony Brain/SPL/Science Source; 21.3: © Science Photo Library/Getty Images RF; 21.4c: © Pascal Goetgheluck/Science Source; 21.5: Courtesy, Dr. Anver Kuliev.

Chapter 22

Opener: © Don Carstens/Artville RF.

Index

Note: Page numbers followed by *f* and *t* indicate figures and tables, respectively. Clinical Connection is indicated by *cc*.

A

ABCC9 gene, 394
ABCD1 gene, 403
ABO blood types, 107, 327
    codominance and, 91–92, 92*f*
    genotypes for, 92, 92*f*
    migration and, 282
Aborigines, Australian, 324
Accutane. *See* Isotretinoin
Acetylase(s), 203, 204*f*
Acetyl groups, binding to histones, 203, 203*f*–204*f*
Achondroplasia, 90, 90*f*, 218–219
Achromatopsia, 283
    type 2, in Plain people, 285*t*
Acquired immune deficiency syndrome, 336–337, 336*f*–337*f*.
    See also HIV
Acridines, mutagenicity of, 220
Acrosome, 48, 49*f*
Actin, 27, 27*f*
Actinin 3, 130
Acute exertional rhabdomyolysis, 87
Acute promyelocytic leukemia, 363
Adaptive immunity, 330–331, 331*f*
Addiction. *See also* Drug addiction
    genetic markers for, 161
Adenine (A), 3, 4*f*, 166, 168, 168*f*, 182*f*
Adeno-associated virus, as vector for gene therapy, 400, 403
Adenoma(s), colonic, 360–361, 361*f*
Adenosine deaminase (ADA) deficiency, 335*t*, 400, 401*f*
Adenosine triphosphate (ATP), 23
Adenovirus, as vector for gene therapy, 402–403
Admixture, genetic, 313–314, 314*f*
A DNA, 166
Adoption study(ies), 137–138
Adrenoleukodystrophy, 23
    gene therapy for, 403
Adult-onset inherited disorders, 62
Affected sibling pair study, 140, 141*t*, 147
African sleeping sickness, 300
Aging, 62–64
    accelerated, disorders resembling, 62–63, 63*t*
Agriculture, and lactose tolerance, 279
AHI1 gene, 320
AIDS (acquired immune deficiency syndrome), 336–337,
    336*f*–337*f*
Aitken, John, 114
Albinism, 73, 74*f*
    epistasis in, 92
    in Hopi Indians, 280
    pedigree analysis for, 81, 82*f*
Alcohol, teratogenicity of, 60–61
Alcoholism, 153
Aldrich, Robert, 110
Alkylating agents, mutagenicity of, 219–220
Allantois, 53–54
Allele(s)
    definition of, 4
    dominant, 71
        definition of, 5
    lethal, 90, 97*t*
    multiple, 91, 97*t*
    presence of, versus expression of, 5
    recessive, 71
        definition of, 5
Allele frequency(ies), 264
    constant, 265–267
    ethnic groups and, 273
    factors affecting, 280, 294*f*, 295*t*

    genetic drift and, 282–286, 283*f*, 294*f*, 295*t*
    migration and, 281–282, 281*f*–282*f*, 294*f*, 295*t*
    mutations and, 286–287, 287*f*, 294*f*, 295*t*
    natural selection and, 287–295, 287*f*, 294*f*, 295*t*
    nonrandom mating and, 280–281, 280*f*, 294*f*, 295*t*
Allelic disorders, 217–218, 217*t*
Allergen(s), 338, 340*f*
Allergy(ies), 338–341, 340*f*–341*f*
Allison, Anthony, 292
Allograft(s), 344, 344*f*
Alopecia areata, pedigree analysis for, 81, 82*f*
Alpha-fetoprotein, 57
    maternal serum, 243*t*, 244
Alpha globin, gene for, mutations of, 219, 220*f*
Alpha helix, 192, 192*f*, 195
Alpha radiation, 221
5-Alpha reductase deficiency, 114, 115*f*
Alpha synuclein, 194
Alpha thalassemia, 219, 292
Alport syndrome, 215*t*
ALS. *See* Amyotrophic lateral sclerosis (ALS)
Alternate splicing, 186, 205, 205*f*
Alu repeat(s), 208
Alzheimer, Alois, 62
Alzheimer disease, 8, 62, 96*cc*, 194, 194*t*
    early-onset, 80–81, 203
    familial, 236, 300
        mouse model of, 380
    genes associated with, 96*cc*, 97*t*
    trisomy 21 and, 247
Ambras syndrome, 237–238, 237*f*
Amelogenesis imperfecta, 128
Ames test, 220
Amines, trace, 153
Amino acid(s), 3, 168, 180
    binding to tRNA, 183, 183*f*–184*f*
    codons for, 187*t*, 189
    metabolism, inborn errors of, 19
Amino groups, 180
Amish, Old Order, 282
    genetic disorders in, 284*cc*–285*cc*, 285*t*
Amniocentesis, 54, 240, 241*f*, 242–243, 392, 393*t*
Amnion, 56, 56*f*
Amniotic cavity, 52
Amniotic fluid, 54, 54*f*, 59
Amniotic sac, 53–54
Amphetamines, 153
Amyloid plaque(s), 96*cc*
Amyloid precursor protein, 96*cc*, 97*t*, 203
Amyotrophic lateral sclerosis (ALS), 36, 234–235
    animal model of, 213, 213*f*
Anandamide, 153
Anaphase, 30–31
Anaphase I, 46, 46*f*
Anaphase II, 46, 47*f*
Ancestry informative markers, 313
Ancestry testing, 317, 393*t*, 395
Andrews, Tommie Lee, 270*cc*
Androgen insensitivity syndrome, 114, 115*f*
Aneuploidy, 245–247, 245*t*, 246*f*, 258*t*, 411
    autosomal, 245, 247–250
    in cancer, 356
    sex chromosome, 245, 250, 252*t*
        female, 250–252, 252*t*
        male, 252–253, 252*t*
Angelman syndrome, 125–126, 126*f*, 128, 259
Angiogenesis, 356
Angiotensin-converting enzyme (ACE), gene for, variants,
    and athletic ability, 146–147
Animal models, 213, 213*f*, 380, 380*f*
Ankylosing spondylitis, 329
Annotation, 4

    definition of, 431
    in genomics, 431–433
Anorexia nervosa, 147
Anosmia, 129
Antennapedia, 55
Antibiotic resistance, 290–291, 291*f*
Antibody(ies), 327, 330*f*, 331–333, 332*f*. *See also* Monoclonal
    antibody(ies)
    functions of, 333
    human, production in transgenic animals, 380
    production of, 331, 331*f*–332*f*
    structure of, 331–332, 332*f*, 333, 333*f*
    subunits, 331–332, 332*f*
    types of, 333, 333*t*
    variable regions, 332–333, 332*f*
Anticipation, 225, 225*f*
Anticodon(s), 183, 183*f*–184*f*
Antidepressant(s), mechanism of action of, 154, 155*f*
Antigen(s), 327
Antigen binding sites, 332–333, 332*f*
Antigen-presenting cell(s), 329, 329*f*
Antigen processing, 329
Antimony, 93
Antiparallelism, 168, 169*f*
Antisense technology, 384, 384*f*
Aortic aneurysm, 215*t*, 216
APC gene, 360–361, 361*f*
APOC-3 gene, 64*t*
APOE gene, 8
APOE4 gene, 96*cc*, 97*t*
ApoL1 gene, 300
Apolipoprotein(s) (Apo), 132*cc*
    ApoC3, 64*t*
    ApoE, 132*cc*
    ApoE4, gene for, 96*cc*, 97*t*
Apoptosis, 28–29, 29*f*, 32–33, 33*f*
Apoptosis checkpoint, 31, 32*f*, 354*f*
Appetite, genes/proteins affecting, 142–143, 143*t*
Aquaporin-1, 403
Archaea, 16, 433
Ardi, 303, 304*f*
Ardipithecus kadabba, 303, 304*f*
Ardipithecus ramidus, 303, 304*f*–305*f*
Argininosuccinic aciduria, 180–181, 180*f*
Argonaute, 385
Armstrong, Lance, 379, 387
Arsenic, 93
Artificial selection, 288–289, 288*f*, 295
Ashkenazi Jewish populations, genetic disorders among,
    285–286, 286*t*
Assisted reproductive technologies (ART), 408, 411–419,
    418*t*, 423
    problems with, 420, 420*t*
Association(s), 141
Asthma, 338, 340–341, 341*f*
Ataxia-oculomotor apraxia syndrome, 107
Ataxia-telangiectasia (AT), 63*t*, 232
Atherosclerosis, 210
Athletic ability, 130, 146–147
    genetic testing for, 405
Atom(s), 6*f*
Atopic dermatitis, 340–341, 341*f*
ATP7A gene, mutations of, 217*t*
Atrial natriuretic peptide (ANP), recombinant, 378*t*
Attention-deficit/hyperactivity disorder, 227
Auriculocondylar syndrome, 67
Australopithecus, 303–305, 304*f*
Australopithecus afarensis, 304, 304*f*–305*f*
Australopithecus anamensis, 304, 304*f*
Australopithecus garhi, 304*f*, 305
Autism spectrum disorder, 157–159, 227
    genes associated with, 161

genetics of, 157
heritability of, 157
Autografts, 343–344, 344f
Autoimmune disorders, 337–338, 338t
Autoimmune polyendocrinopathy, type I, 337
Autoimmunity, 337–338
Autophagy, 23
Autosomal dominant inheritance, 69, 74–75, 74f, 74t, 81, 82f
Autosomal recessive inheritance, 69, 74–75, 77t, 81, 82f
and autism, 157–158
Autosome(s), 4
aneuploidy, 245, 247–250
Avery, Oswald, 165, 167t
Azoospermia, 409

B

Baboons, transplants from, 344
Bacteria, 16. *See also* Microbiome
antibiotic resistance in, 290–291, 291f
pathogenic, 327, 327f
use in bioremediation, 381–382
Bacterial infection(s), in cystic fibrosis, 76cc
Bacterial transformation, 164–165, 164f–165f
Bacteriophage(s), as cloning vectors, 377
Balanced polymorphism, 291–292, 292f, 292t, 293f
Baldness, pattern, 122
Banks, Shirley, 212–213
Bardet-Biedly syndrome, 27
Barr body(ies), 123, 248, 249f
Base(s), 3, 4f, 165–166
in DNA, 182, 182f
nitrogenous, 168, 168f
in RNA, 182, 182f
Base excision repair, 230, 231f
Base pairs, 168, 168f, 169, 170f–171f. *See also*
Complementary base pairs
Bateson, William, 70, 101–102
B cells (B lymphocytes), 330, 330f, 331, 331f, 334t
BCR-ABL, 364cc
B DNA, 166, 166f
Beaudet, Alfred, 259
Becker muscular dystrophy, 222–223
drug therapy for, 398
Beckwith-Wiedemann syndrome, 125
Bedouin community, genetic testing in, 88
Beery, Alexis and Noah, 439
Behavior(s), 149
genes and, 149
Behavioral disorders, 149–150
genetic studies of, clinical benefits, 158–159
prevalence of, in U.S., 150t
Bell curve, for distribution of phenotypes of polygenic traits, 133
Beta amyloid, 62, 194
Beta cells, pancreatic, 201–202, 201f
Beta globin gene, mutation of, 214, 214f
Beta-pleated sheet, 192, 192f, 195
Beta radiation, 221
Beta thalassemia, 214, 217
B3GALT6 gene, mutations of, 217
Binding proteins, in DNA replication, 172, 173f
Binet, Alfred, 151
Bioethics, 3
of banking stem cells, 37
of cloning, 52
of gamete collection/use after death, 416
of genetic information use, 275
of genetic testing, 11–12
of genome data collection, 318
and incidental findings of genetic tests, 395
of infidelity testing, 175
of population screening for neural tube defects, 297
of prenatal diagnosis of trisomy 21, 250
of recombinant erythropoietin, 379
of xenografts, 344–345
Biological organization, levels of, 5, 6f
Bioremediation, 381–382
Biotechnology, 375, 381f
and patent law, 375–376
Biotinidase deficiency, 19, 436

Bipolar disorder, 155, 157, 162
Birth defects, 59–61
Bitter taste receptor gene, 64t, 299
Blastocyst, 52, 53f, 54t
Blastomere(s), 51
Blastomere biopsy, 417, 417f
Bleb(s), 32, 33f
Blighted ovum, 49
Blindness, mutations causing, 95, 95f
Blood clotting, 132cc
Blood groups, 327, 327t
Bloodsworth, Kirk, 269
Blood type(s)
Bombay phenotype and, 92
codominance and, 91–92, 92f
Duffy, 103
genotypes for, 92, 92f
Bloom syndrome, 179
in Ashkenazi Jewish populations, 286t
Blue diaper syndrome, 73cc
Blue people of Troublesome Creek, 228
Blue person disease, 228–229
Body mass index (BMI), 142, 142f
heritability for, 142
Body weight, 147
environmental factors affecting, 143–144
genes affecting, 142–143, 143t
genome-wide association study of, 143
Bombay phenotype, 92, 107
Bone marrow transplantation, 345
Bouchard, Thomas, 138
Bradfield, Golda, 366
Brain, 149
cerebral organoids and, 58
development of, 319–320
critical period for, 60
study using organoids, 58
in drug addiction, 153, 153f
embryology of, 57
primate, developmental timetables of, 319–320
size of, 320–321
BRCA1 gene, 129, 283, 367–368, 368f
patenting of, 376
BRCA2 gene, 367–368, 372
patenting of, 376
Breast cancer
in Ashkenazi Jewish populations, 286t
BRCA1, 367–368
BRCA2, 367–368
familial, 367
HER2, 365
microRNAs in, 210
sporadic, 367
Brittle bone disease. *See* Osteogenesis imperfecta
Brown, Louise Joy, 414
Brown, Timothy, 326
Buck, Carrie, 295, 296f
Buck, Vivian, 295, 296f
Bulbourethral gland(s), 43, 43f
Bunker, Chang and Eng, 57
Burbank, Luther, 295
Burkholderia cepacia, 76cc
Burkitt lymphoma, 363, 363f
Burning man syndrome, 26
Burtka, David, 422
Bush, George W., 422
Bystander effect, 221

C

Caenorhabditis elegans, 433
Calcium homeostasis modulator 1, gene for, 97t
Canavan disease, 301, 389, 389f
in Ashkenazi Jewish populations, 286t
Cancer, 25, 33, 352–355
at cellular level, 352–353, 352f, 356–360, 357f–359f
classification of, 352
diagnosis of, 369–370, 370f
environment and, 352, 366, 367f, 368
evolution of, 361, 362f
genes and, 352

gene therapy for, 403–404
at genome level, 352f
germline (inherited), 355, 355f
levels of, 352, 352f
loss of cell cycle control in, 354–355, 354f
metastases, 353, 356, 357f, 372
microRNAs and, 205
RNA interference and, 205
sporadic, 355, 355f
treatment of, 369–370
at whole-body level, 352f
Cancer cell(s)
characteristics of, 356, 357t
cytotoxic T cells and, 334, 334f
dedifferentiation, 356–357, 359f
invasiveness, 356, 357f
origins of, 356–360, 357f–359f
Cancer Genome Atlas Research Network, 372
Cancer Quest, 372
Cancer stem cells, 356–357, 358f
Capacitation, 51
Capillary electrophoresis, 271, 272f
Carbohydrates, 17
metabolism of, inborn errors of, 19
Carbon(s), of sugars in nucleic acids, numbering, 168, 169f
Carcinogen(s), 220, 353
Carcinogenesis, 356, 357f
Cardiovascular disease, 62, 132cc–133cc, 146
risk factors for, 132cc
Carrier(s), 69
with disease symptoms, 77
genetic testing for, 393t, 395
Case-control study, 140, 141t
Caspase(s), 32, 33f
Cat(s)
calico, 123, 123f
cloned, 52
domestication of, 288
multitoed, 300
tortoiseshell, 123
Cataplexy, 150
Catechol-*O*-methyltransferase, genotype, and acadeic success, 162
CCR5 mutation, 299–300, 326
protective effect against HIV, 213, 281, 326, 337
CCR5 receptor
in HIV infection, 336–337, 336f
in plague resistance, 337
CDK5RAP2 gene, 58
CD4 receptor, in HIV infection, 336–337, 336f
Celera Genomics, 428–429
Cell(s), 2, 6f, 16
animal, 20f
chemical constituents of, 17–18
components of, 16–27
differentiation of, 5, 33, 33f–34f
energy production in, 23–24, 23f
genome in, 5
number of, in body, 5
preparation, for chromosome analysis, 241–242
types of, 5
Cell–cell communication, 24–25
Cell cycle, 29–32, 29f
checkpoints, 31, 32f, 354, 354f
control of, 31–32
Cell death, 28–29. *See also* Apoptosis; Necrosis
Cell division, 28–32. *See also* Meiosis; Mitosis
Cell fate, cancer and, 352–353, 352f
Cell-free tumor DNA, 369–370
Cell lineage(s), 33–35, 34f
Cell-mediated immunity (CMI), 330f
Cell survival, cancer and, 352–353, 352f
Cellular adhesion, 25
Cellular immune response, 331, 331f, 333–334
Centenarians, 63–64
genetic testing of, 393t, 396
Centimorgan(s), 102
Central dogma, 181
Central nervous system (CNS),
embryology of, 57
Centriole(s), 20f, 30, 30f–31f

Centromere(s), 30, 30*f*, 31, 31*f*, 171*f*, 207*t*, 208, 238–239, 238*f*–239*f*
    position of, 239, 240*f*
Centromere protein A (CENP-A), 238
Centrosome(s), 30
Cerebellar ataxia, 413
Cervix, uterine, 44, 44*f*
CETP gene, 64*t*
CFTR. *See* Cystic fibrosis transductance regulator
Chain termination, 176
Chands syndrome, 86
Channelopathies, 24, 26*cc*
Chantix. *See* Varenicline (Chantix)
Chaperone proteins, 190, 191*f*, 193
Charcot-Marie-Tooth disease, 224
Chargaff, Erwin, 166, 167*t*
Chase, Martha, 165, 166*f*, 167*t*
Check Orphan, 87
Cheetah, loss of genetic diversity in, 285, 286*f*
Chelators, 214
Chemicals, mutagenic, 219–221
Chernobyl nuclear disaster, 220
Chicken, genome of, 433
Child(ren), genetic testing of, 393*t*, 394–395
Child abuse, posttraumatic stress disorder after, 148–149
Chimeric antigen receptor (CAR) technology, 403–404
Chimeric RNA transcripts, 205
Chimpanzee(s)
    gene expression in, comparison with humans, 320
    genome of, and human genome, comparison of, 6, 319, 433
    individual diversity of, 7
China, one-child policy, 116
Chloride channel(s)
    abnormalities, 26
    in cystic fibrosis, 26
Cholera, 294
Cholesterol, serum levels, 132*cc*, 133
Cholesteryl ester transfer protein gene, 64*t*
Chondrodysplasia, 215*t*
Chorion, 54*f*, 56, 56*f*
Chorionic villi, 53–54
Chorionic villus sampling, 54, 240–241, 241*f*, 242–243, 392, 393*t*
Chromatid(s), 30, 30*f*, 31, 31*f*, 45, 170, 171*f*
Chromatin, 170, 171*f*
Chromatin remodeling, 203–204, 203*f*–204*f*
Chromosomal imbalance(s), maternal age effect and, 50–51, 240, 241*f*, 248
Chromosomal microarray (CMA) test, 393*t*, 394
Chromosome(s), 3*f*, 4, 18, 70, 171*f*, 238–240. *See also* Sex chromosome(s)
    abnormalities, 244–259, 245*t*
        and cancer, 361, 363*f*
        organizations for families with, 245*t*
        shorthand notation for, 243*t*
    acentric, 257, 257*f*, 258*t*
    acrocentric, 239, 240*f*
    banding, 242–243
    comparison of, 311–312, 312*t*
    dicentric, 257, 257*f*, 258*t*
    direct visualization of, 240–243
    duplication of, 29
    extra, indirect detection of, 240, 243–244
    fetal, testing of, 240
    homologous, 71, 71*f*
    indirect detection of, 240, 243–244
    long arm *(q)*, 238, 238*f*, 239, 243*f*
    metacentric, 239, 240*f*
    preparation, for analysis, 241–242
    replication, 30, 30*f*, 45
    short arm *(p)*, 238, 238*f*, 239, 243*f*
    shorthand notation for, 243*t*
    staining of, 242
    structure of, abnormalities, 253–258, 254*f*, 258*t*
    submetacentric, 239, 240*f*
    telocentric, 239, 240*f*
    tests, 240
    unreplicated, 30*f*
    visualization, 241–242, 242*f*
Chromosome 15. *See also* Angelman syndrome; Autism spectrum disorder; Prader-Willi syndrome
    isodicentric, 262

Chromosome 21, 238. *See also* Trisomy 21 Down syndrome
Chromosome 22, 238
Chromosome number, 29, 44, 46
    atypical, 244–253, 245*t*, 258*t*
Chromosome pairs, 29, 44
Chromothripsis, 361
Chronic granulomatous disease, 335, 335*t*
Chronic myelogenous leukemia, 351, 364*cc*–365*cc*
Church, George, 425
Cigarette smoke, teratogenicity of, 60
Cilia, 27, 27*f*
    motile, 27, 27*f*
    primary, 27, 27*f*
Ciliopathy, 27
Circadian rhythms, disruption, genes associated with, 161
cis configuration, 102, 102*f*, 431
Cleavage, 51–52, 53*f*, 54*f*
Cleft lip, recurrence, empiric risk of, 135, 135*t*, 136*f*
Cline(s), 282, 282*f*
Clinical Connection
    cardiovascular health and disease, 132
    colorblindness, 119
    DNA profiling, 270
    exome sequencing, 10
    faulty ion channels, 26
    fragile X syndrome, 226–227
    genetic disorders in Plain populations, 284–285
    genetics of Alzheimer disease, 96
    Gleevec, 364–365
    homeotic mutations, 55
    immune diseases, 339
    inborn errors of metabolism, 19
    infertility, 410
    maternal-fetal relationship, 339–340
    Rh incompatibility, 339–340
    single-gene inheritance, 73
    treatment of cystic fibrosis, 76
    viruses, 328
Clinical Genome Resource, 431*t*
Clinical Laboratory Improvement Amendments, 396
Clinical Lung Cancer Genome Project, 373
Clinic for Special Children, 284*cc*–285*cc*
ClinSeq, 431*t*
Clinton, Bill, 428–429
Clitoris, 44, 44*f*
Clock genes, 151
Clomiphene, 409
Cloning, 36, 52
    gene, 376
    positional, 426
Cloning vector(s), 376–377, 377*f*
Clopidogrel (Plavix), 436
CML. *See* Chronic myelogenous leukemia
CNTNAP2 gene, 284*cc*
CNV. *See* Copy number variants
Cobalamin C deficiency, 109
Cocaine, 153
Cockayne syndrome, 63*t*, 232
Coding strand, of DNA, 181, 182*f*, 185
CODIS, 270–271, 275
Codominance, 91–92, 92*f*, 97*t*
Codon(s), 183, 187, 187*t*
    for amino acids, 187*t*, 189
    nonsynonymous, 189
    synonymous, 189
Coefficient of relatedness, 136, 137*t*, 142*t*
Cohort study, 140, 141*t*
Colchicine, 241
Collagen
    conformation of, 215, 216*f*
    disorders of, 215–216, 215*t*
    distribution in body, 215
    functions of, 215
    synthesis of, 215
Collectin(s), 330
Collins, Francis, 428–429
Colon cancer, 360–361, 361*f*
    inherited, 232
Colony-stimulating factor(s), 334*t*
    immunotherapy with, 343
    recombinant, 378*t*

Colorblindness, 118, 119*cc*
    blue, 119*cc*
    green, 119*cc*
    pedigree analysis for, 80
    red, 119*cc*
Color vision, 119*cc*
Columbus, Christopher, 9–10
Comparative genomic hybridization, 254
Comparative genomics, 430, 432–433
Complement, 330, 333
Complementary base pairs, 169–170, 170*f*
Complementary DNA (cDNA), 382
    patenting of, 376
Complete dominance, 91
Compound heterozygote, 91
Concordance, of traits, in twins, 138, 138*t*, 142*t*
Conditional probability, 81–83, 83*f*, 90*f*
Cone(s), 95, 95*f*, 119*cc*
Conformation(s)
    protein, 192
    of RNA, 182–183
Congenital adrenal hyperplasia, 114, 115*f*
Congenital myasthenic syndrome, 394
Congenital rubella syndrome, 61
Connective tissue, 6*t*, 17*f*
Connectome(s), 202, 434
Consanguinity, 75, 281, 410*cc*
    and 5-alpha reductase deficiency, 114
Contact inhibition, 32
Control experiment(s), 164
Cook, Robin, *Chromosome Six*, 345
Cooley, Thomas, 214
Copy number variants, 137, 139, 139*f*, 139*t*, 157, 225–227, 253–254, 286–287, 435–436
Corn, genetically modified, 384, 384*f*
Corona radiata, 51, 51*f*
Correlation(s), 141
Cortical dysplasia-focal epilepsy syndrome, 284*cc*
Coumadin. *See* Warfarin
Cousin(s)
    first, marriage of, pedigree analysis for, 82*f*
    marriage between, 281
COX-2 inhibitors, adverse effects of, genetic testing to predict, 397
Creutzfeldt-Jakob disease (CJD), 194*t*
    variant, 194
Crick, Francis, 164, 167, 167*f*, 167*t*, 170, 181, 187, 198
Cri-du-chat syndrome, 253–254, 254*f*
Crigler-Najjar 1 syndrome, in Plain people, 285*t*
Criminality, heritability of, 146
CRISPR technology, 384*f*, 385
Cristae, 23, 23*f*
Critical period, 60*f*
Crohn's disease, 337
Crop(s), genetically modified, 381, 381*t*, 384, 384*f*, 387
Crossing over, 45–46, 46*f*, 47, 102, 102*f*
    distance between genes and, 102–103, 103*f*
Cruciferous vegetables, and cancer prevention, 368, 369*f*
Cryptorchidism, in dogs, 129
ctDNA. *See* Cell-free tumor DNA
Cyanosis, 228, 228*f*
Cyclin(s), 32
CYP2A6 gene, and nicotine metabolism, 162
Cystic fibrosis, 7, 26, 26*f*, 69, 74–75, 76*cc*, 216*t*, 217, 222
    carrier frequency, 267–268, 267*t*, 268*f*
    and diarrheal disease, 292*t*, 294
    double delta genotype, 91
    epidemiology of, 76*cc*
    gene for, discovery of, 426
    gene therapy for, 399
    genotype, and clinical course, 91
    pedigree analysis for, haplotypes in, 105, 105*f*
    treatment of, 76*cc*
    uniparental disomy in, 259, 259*f*
Cystic fibrosis transductance regulator, 26, 26*f*, 294
    gene for *(CFTR)*, 217
        mutations of, 76*cc*
    misfolding of, 193
Cytochrome *c*, evolution of, 312, 312*t*
Cytogenetic map(s), 426, 426*f*

Cytogenetics, 238
  cell preparation for, 241–242
  chromosome preparation for, 241–242
Cytokine(s), 330–331, 331*f*, 333–334, 338
  immunotherapy with, 343
  types of, 334, 334*t*
Cytokinesis, 28–29, 29*f*–30*f*, 31
Cytoplasm, 18, 20*f*
Cytoplasmic donation, 416
Cytosine (C), 3, 4*f*, 166, 168, 168*f*, 182*f*
Cytoskeleton, 25–27, 27*f*
Cytosol, 18
Cytotoxic T cells, 334, 334*t*

D

Dalton, John, 119*cc*
Damage tolerance, 231, 231*f*
Darwin, Charles, 136, 151
Daughter cells, 29, 31, 47*f*
Davenport, Charles, 295
Deacetylase(s), 203, 204*f*
Deatherage-Newsome, Blaine, 297, 297*f*
Death receptor(s), 32, 33*f*
Dedifferentiation, of cancer cell, 356–357, 359*f*
Deinococcus radiodurans, 230
Deletion mutation(s), 222*t*, 224, 245*t*, 253–254, 254*f*, 258*t*
Dendritic cells, 329, 334*t*
Denisovans, 307*f*, 308–309, 308*f*, 319
Density shift experiments, 172
Dentin, 206, 206*f*
Dentinogenesis imperfecta, 206, 206*f*
Dentin phosphoprotein (DPP), 206, 206*f*
Dentin sialophosphoprotein (DSP), 206, 206*f*
Deoxyribonuclease (DNase), recombinant, 378*t*
Deoxyribose, 165, 168, 168*f*, 182, 182*f*
  carbons of, numbering, 168, 169*f*
Dependence, drug, 153
Depression, 154
Dermatoglyphics, 133
Descendants of Niall, 299
DeSilva, Ashi, 400, 401*f*
Deuteranopia, 119*cc*
Diabetes mellitus, 62, 202
  type 1, 338*t*
  type 2, 131
Diarrheal disease, cystic fibrosis and, 292*t*, 294
Dicer, 385, 385*f*
Dichromats, 119*cc*
DiGeorge syndrome, 55, 261–262
Dihybrid(s), 78
Dihybrid cross, 78, 79*f*
  linked genes and, 101, 101*f*
  unlinked genes and, 101, 101*f*
Dihydrotestosterone, 114, 115*f*
Dikika infant, 305
Diploid cells, 16, 44, 45*f*, 48, 48*f*, 49, 50*f*
Disaster victims, DNA profiling of, 274, 274*t*
Disease(s). *See also specific disease*
  environment and, 7–8
  factors contributing to, 5
  genetic, treatment of, 397–404
  human, animal models of, 380, 380*f*
  monofactorial (Mendelian), 72
  multifactorial, 72
  single-gene, 8, 8*t*, 43, 72
    paternal age and, 50–51
    treatment of, 397–404
Diseasome, 8, 9*f*, 202
DMD gene, mutations of, 217*t*
DNA (deoxyribonucleic acid), 1–4, 3*f*, 18, 163*f*.
  See also Mitochondrial DNA
  in bacterial transformation, 165, 165*f*
  cell-free fetal, analysis, 244, 244*f*
  centromere-associated, 238
  chromosomal, 238
  coding strand, 181, 182*f*, 185
  coiling of, 170
  conserved sequences, 311
    pictograms for, 432, 432*f*
  double helix of, 3, 4*f*

directionality in, 168–169, 169*f*
  discovery of, 167, 167*f*
  formation of, 168–169
  fetal, testing, 244, 244*f*
  Hershey-Chase blender experiments with, 165, 166*f*
  lagging strand, 173*f*
  leading strand, 173*f*
  length of, 170
  linker regions, 170, 171*f*
  mitochondrial, mutations of, 230
  modification of, 376–382. *See also* Recombinant DNA;
    Transgenic organism(s)
  palindromic sequences, 376, 377*f*
  patenting, 375–376
  patenting of, 375–376
  replication, 3, 4*f*, 170–176, 181*f*
    accuracy of, 230
    semiconservative, 172, 172*f*
    steps of, 172–173, 172*f*–173*f*
  and RNA, comparison of, 165, 182, 182*f*, 182*t*
  sequences
    comparison of, 6
    derivation of, 428, 428*f*
  sequencing, 176–177, 176*f*
    massively parallel, 177
    next-generation, 176–177
    Sanger, 176
  in sperm, 48
  sticky ends, 376–377, 377*f*
  structure of, 167, 167*f*, 168–170, 182, 182*f*
  template strand, 181, 182*f*, 185
  viral, in human genome, 207, 208*f*
  X-ray diffraction pattern of, 166–167, 166*f*
DNA amplification, 173, 175
DNA damage checkpoint, 31, 32*f*, 354*f*
DNA dragnet, 275
DNA fingerprinting, 269
DNA microarrays, 382–384, 383*f*, 388
DNA polymerase (DNAP), 172, 173*f*, 230
  proofreading function of, 172–173
DNA probes, 312
DNA profiling, 7, 263–264, 268–276, 270*cc*
  of disaster victims, 274, 274*t*
  in identification of victims, 273–274
  of natural disaster victims, 274, 274*t*
  population statistics and, 270*cc*, 271–273, 273*f*
  in reuniting Holocaust survivors, 274
  of World Trade Center victims, 274
DNA repair, 229–232
  disorders of, 231–232, 232*f*, 353
  types of, 230–231
DNA Science, 437
DNA transfer, 275–276
Dog(s)
  artificial selection of, 288–289, 288*f*
  domestication of, 288
  narcolepsy in, 150–151, 150*f*
Dolly the sheep, 36, 388
Domain(s), 16
Dominance, 75–77. *See also* Trait(s), autosomal dominant
  complete, 91
  incomplete, 91–92, 91*f*, 97*t*
Donohue, R. P., 103
donorsiblingregistry.com, 413, 423
Dopamine, nicotine addiction and, 153–154, 154*f*
Dopa-responsive dystonia, 439
Dor Yeshorim, 281, 393*t*
Down, John Langdon Haydon, 247
Down syndrome, 152, 157, 242. *See also* Trisomy 21 Down
  syndrome
  diagnosis of, 239
  translocation, 246, 254–256
Driver mutations, 360
Drosophila melanogaster (fruit fly), 433
  homeotic mutations of, 55
  linkage in, 102
Drug(s)
  of abuse, 153–154
  for genetic diseases, 398
  morpholino-based, 384, 388
  prescribing, personal genotypes and, 133

production of, using recombinant DNA technology,
    378–380, 378*t*
  repurposed, 398
  RNAi-based, 385
  selection, genetic testing for, 397
Drug addiction, 152–154
  brain in, 153, 153*f*
  definition of, 153
  proteins involved in, 153
Druker, Brian, 351
Dry mouth, gene therapy for, 403
Duchenne muscular dystrophy, 8, 16, 16*f*, 216*t*, 384, 388
  gene for, discovery of, 426
  treatment of, 222–223, 224*f*
Ductus deferens, 43, 43*f*
Duffy blood type, 103
Dugdale, Richard, 301
Dulbecco, Renato, 427
Duplication(s), 222*t*, 245*t*, 253–254, 254*f*, 258*t*, 319, 361
Dutch Famine Study, 199
Dutch Hunger Winter, 199, 199*f*
Dystrophic epidermolysis bullosa, 215*t*
Dystrophin
  mutations affecting, 222–223
  production
    drug therapy promoting, 223, 224*f*
    morpholino-based drug and, 384, 388

E

Earle, Pliny, 80
Ebu Gogo, 302–303, 302*f*
E-cadherin, 366–367
EcoR1, 376, 377*f*
Ectoderm, 52–53, 54*f*
Ectopic pregnancy, 410
Edward syndrome. *See* Trisomy 18
Egg. *See* Oocyte(s); Ovum
Egypt, Ptolemy Dynasty, pedigree for, 82*f*
Ehlers-Danlos syndrome, 215*t*, 216*f*, 217, 217*f*, 234
  progeroid type, 217, 217*f*
Ejaculation, 43
Electronic Medical Record and Genomics Network, 439
Elementen, 69–70
Elliptocytosis, 104
Ellis-van Creveld syndrome, in Plain people, 285*t*
Ellobius lutescens, 113, 113*f*
Embryo(s), 59*f*. *See also* Cleavage; Implantation
  cryopreservation of, 414
  definition of, 51
  development of, 57
  extra, 419–420, 419*f*
  formation of, 52–53
  frozen, fates of, 419*t*
  human, status of, 422
  primordial, 53, 54*f*
  supportive structures for, development of, 53–56
Embryo adoption, 415
Embryonic disc, 52, 54*t*
Embryonic induction, 57
Embryonic stem (ES) cells, 35–36, 36*t*
Empiric risk, 135–136, 142*t*
ENCODE project, 208, 431*t*
Endoderm, 53, 54*f*
Endogamy, 281
Endometriosis, 411
Endoplasmic reticulum (ER), 18–22, 24
  functions of, 20, 22, 24*t*
  structure of, 18, 24*t*
Endorphins, 153
Endosome, 23
Energy, production of, mitochondrial, 23–24, 23*f*
Enhanceosome, 203, 204*f*
Enkephalins, 153
Environment
  and cancer, 352, 366, 367*f*, 368
  and cloning, 52
  and disease, 7–8
  effects on gene action, 1–2, 5, 5*f*
  and gene expression, 199
  and genes, 131

Environment—*Cont.*
   and intellectual development, 5
   and longevity, 63
   and schizophrenia, 156
Environmental epigenetics hypothesis, 145
Enviropig, 374
Enzyme(s), 17–19, 164. *See also specific enzyme*
   lysosomal, 22*f*, 23
   peroxisomal, 23
Enzyme replacement therapy, 398, 398*f*, 398*t*
EPAS1 gene, 288, 288*f*
Epidermal growth factor (EGF), 32
   recombinant, 378*t*
Epidermolysis bullosa, 436
Epidermolytic hyperkeratosis, 235
Epididymis, 43, 43*f*
Epigenetic change, 123–124, 200, 434–435
   and posttraumatic stress disorder, 149
   and schizophrenia, 156–157
Epigenetics, 53
Epigenome, 53
Epistasis, 92, 95–96, 97*t*, 141, 392, 434–435
   and heritability, 137
Epithelial tissue (epithelium), 6*t*, 17*f*
Epitopes, 332*f*, 333
Equational division, 45, 45*f*
Equator, of cell, 30
Erlich, Yanev, 318
Erythropoietin (EPO), recombinant, 378*t*, 379, 387
Escherichia coli, 327*f*
   Shigatoxigenic (STEC), 347
   toxic, outbreaks, tracking, 347
Eskimos, 317, 324
Estradiol, maternal serum, 243*t*
Ethical, Legal and Social Implications (ELSI) Research
   Program, 428, 439
Euchromatin, 238, 238*f*
Eugenics, 295–298, 296*f*, 296*t*, 301
Eukarya, 16
Eukaryotes, 16
Eukaryotic cell, 16, 18*f*
Eumelanin, 73
Euploidy, 245, 246*f*
Eve, mitochondrial, 314, 315*f*
Evolution, 264, 303. *See also* Molecular evolution
   of cancer, 361, 362*f*
   of cytochrome *c*, 312, 312*t*
   definition of, 6
   of human genome, 319, 319*f*
   of lactose tolerance, 279
   of lice, 325
   of species, 6
Evolutionary tree(s), 312
   for lice, 325
Excision repair, 230, 230*f*–231*f*
Exercise, and gene expression, 210–211
Exome, definition of, 4, 181
Exome sequencing, 8–9, 10*cc*, 83–84, 83*f*, 89, 152, 157, 158*f*,
   227, 430–431. *See also* Genetic testing
   for diagnosis of childhood disease, 393*t*, 394
   limitations of, 433–434
Exon(s), 186, 186*f*, 205, 205*f*
Exon skipping, 222–223
Expanding repeats, 224–225, 225*f*
Expressed sequence tag (EST), 428
Expressivity, 93, 95–96, 97*t*, 435
Eye color, 73–74, 74*f*

F

Fabry disease, 123, 128–129
Face blindness, 107–108
Facial features, genome-wide association study of, 140
Factor VIII, recombinant, 378*t*
Factor XI deficiency, 222
FalavrSavr tomato, 384, 384*f*
Fallopian tube(s). *See* Uterine tube(s)
Familial adenomatous polyposis (FAP), 360–361, 361*f*
Familial advanced sleep phase syndrome, 151, 151*f*
Familial amyotrophic lateral sclerosis, 194*t*
Familial diffuse gastric cancer, 366–367

Familial dysautonomia, 222–223
   in Ashkenazi Jewish populations, 286*t*
Familial erythrocytosis
   type 1, 379
   type 2, 379
Familial hypercholesterolemia, 19, 132*cc*, 216*t*, 221*f*, 224
   genotypes, 91, 91*f*
   incomplete dominance in, 91, 91*f*
Family, as level of genetic organization, 5
Family exome analysis, 83–84, 83*f*
Fanconi anemia, 408
   type C, in Ashkenazi Jewish populations, 286*t*
Farmers, early, 309, 309*f*
Fatal familial insomnia, 151, 194*t*
Fatty acid desaturase, 321
Fava beans, 229
FBN1 gene, mutations of, 217*t*
Fecal transplantation, 38
Ferritin, structure of, 192
Fertility drugs, 409–410
Fertilization, 44, 49–51, 51*f*, 53*f*, 54*t*
Fetal alcohol syndrome (FAS), 60–61, 61*f*, 152
Fetal period, 51
Fetus, 58, 59*f*
   definition of, 51
   growth of, 58–59
   supportive structures for, development of, 53–56
Fever, 331
F₁ (first filial) generation, 69
F₂ (second filial) generation, 69
FGFR3 gene, mutations of, 217*t*
FGMO Compass, 387
Fibrillin, 93–95
Filaggrin, 340, 341*f*
Fingerprint patterns, 133–134, 133*f*
Fire, Andrew, 385
Fisher, Ronald Aylmer, 295
FKBP5 gene, 148
Flagellum, of sperm, 48
Flemming, Walter, 242*f*
Flores, little people of, 302–303, 302*f*
Flow cytometry, 392
Fluorescence in situ hybridization (FISH), 242, 242*f*, 256,
   256*f*, 312
Flu vaccine, production of, recombinant DNA
   technology for, 380
FMR1 gene, 226
Folic acid, and prevention of neural tube defects, 57
Food(s), genetically modified, 381, 381*t*, 384,
   384*f*, 387
Ford, Cara, 258
Ford, Stewart, 258
Forensic science, 7. *See also* DNA profiling
   DNA testing in, 393*t*, 395
Forkhead box O3 gene, 64*t*
Founder effect, 282–283, 283*t*, 295*t*, 314–315
   in Plain populations, 284*cc*–285*cc*
FOXO3 gene, 64*t*
FOXP2 gene, 319
Fragile X-associated tremor/ataxia syndrome, 226–227
Fragile X syndrome, 152, 157, 226*cc*–227*cc*
Frameshift mutations, 188*f*
Framingham Heart Study, 140
Franklin, Lonnie David, Jr., 275
Franklin, Rosalind, 166, 166*f*, 167, 167*t*
Friedman, Jeffrey, 142–143
Frontotemporal dementia, 194*t*, 197
Frozen Ark project, 179
Fruit fly(ies). *See Drosophila melanogaster* (fruit fly)
Fugate, Martin, 228
Furnish, David, 422
Furrow, 30, 30*f*
Fusion gene, 224
Fusion protein, 363–365

G

Gain-of-function mutation(s), 77, 77*f*, 213
   dominant toxic, 225
Galactokinase deficiency, 281–282, 282*f*, 283
Gallus gallus, genome of, 433

Galton, Francis, 151, 295
Gamete(s), 43–44
   collection/use after death, 416
   maturation of, 47–51
   mutations in, parental age and, 50–51
   random combination of, 71, 71*f*
Gamete intrafallopian transfer (GIFT), 415, 418*t*
Gamma radiation, 221
GAN. *See* Giant axonal neuropathy
Garrod, Archibald, 164
Gastric cancer, familial diffuse, 366–367
Gastrula, 53, 54*t*
Gates, Henry Louis, Jr., 436
Gaucher disease, 224
   in Ashkenazi Jewish populations, 286*t*
   type 1, treatment of, 398, 398*f*, 398*t*
GBA gene, mutations of, 217*t*
Gelsinger, Jesse, 400–402, 401*f*
Gene(s), 3*f*, 70, 168. *See also specific gene*
   and behavior, 149
   clock, 151
   cloning, 376
   comparison of, 311
      interspecies, 311
   definition of, 2, 163, 208
   disease-causing, counterparts in nonhumans, 433
   encoding more than one protein, 205–206, 206*f*
   and environment, 131
   fusion, 224
   homeotic, 53, 55*cc*
   identification of, in genome-wide association study,
      140, 140*f*
   in introns, 205–206, 206*f*
   linked, inheritance of, 100–101, 101*f*
   and longevity, 63–64, 64*t*
   meaning of, 163
   mitochondrial, 98–100, 98*f*, 100*f*
      mutations of, 98, 219
      products of, 98
      proteins encoded by, 98
   modifier, 92
   patenting of, 375–376
   pleiotropic, 93–94, 93*f*–94*f*
   protein-encoding, 4
   pseudoautosomal, 112
   size
      and cloning vector selection, 377
      measurement of, 377
   in subtelomere, 239
   variant, annotation of, 431–433
   X-linked, 117
   Y-linked, 117
Gene chips, 382–384
Gene doping, 209–210
genedx.com, 405
Gene expression, 5, 90–98
   definition of, 8
   and disease, 8, 9*f*
   embryonic, 53
   environment and, 199
   human, in transgenic animals, 380
   in humans vs chimps, 320
   at molecular level, 200
   monitoring, 382–384
   at organ level, 200–202, 201*f*
   patterns, in genome-wide association studies,
      139, 139*t*
   profiling, 210
      after spinal cord injury, 382, 382*t*, 383*f*
   regulation of, 184, 203–205
   timing of, 200
Gene flow, 264, 280
Gene pool, definition of, 6, 264
General intelligence *(g),* 151–152
Gene silencing, 384–385, 384*f*
   by methyl groups, 124, 124*f*
Gene therapy, 23, 41, 398–404
   for Canavan disease, 389
   clinical trials of, 405
      requirements for, 399, 399*t*
   concerns about, 399, 399*t*

ex vivo, 399, 399f
germline, 399
historical perspective on, 400–402
in vivo, 399, 399f
setbacks in, 400–402
somatic, 399–400, 400f
targets for, 400, 400f
successes, 402–404
targets of, 399–400
types of, 399–400
for Wiskott-Aldrich syndrome, 110
Genetic admixture, 313–314, 314f
Genetic Alliance, 108
Genetically modified foods, 381, 381t, 384, 384f, 387
Genetically modified organisms (GMOs), 381, 381f, 381t
Genetic code, 186–189, 187t, 229
degeneracy of, 189
nonoverlapping nature of, 188
triplet nature of, 187, 188f
universality, 188, 375
Genetic counseling, 390–392, 390f, 405
about adult-onset disorders, 391
about inherited disease, 391
historical perspective on, 390
prenatal, 391
Genetic counselor(s), 390–392, 390f
Genetic determinism, 130, 147
definition of, 5
Genetic diversity, meiosis and, 44, 46
Genetic drift, 264, 282–286, 283f
Genetic engineering, 375
Genetic equilibrium, 265
Genetic heterogeneity, 89, 94–96, 97t, 141, 435
Genetic information, maximizing, in genes, 205–206
Genetic load, 286
Genetic map(s), 426, 426f
Genetic markers, 105
in genome-wide association studies, 139–142, 139t
Genetic modification, 375
Genetic privacy, 274–276, 318
Genetics
applications of, 7–9
definition of, 2
levels of, 3–7, 3f
Genetic Science Learning Center (Eccles Institute of Human Genetics), 262
Genetic testing, 11, 390, 392–397, 393t, 405, 435
of adults, 393t, 395–396
of centenarians, 393t, 396
of children, 393t, 394–395
direct-to-consumer, 396, 435
incidental findings of, ethical considerations with, 395
neonatal, 393t, 394
posthumous, 393t, 396
preconception, 392–394, 393t, 395
prenatal, 392–394, 393t
and privacy, 11–12
Genetic Testing Registry, 390
Genographic Project, 324
Genome(s). See also Human genome
annotations, 431–433, 434t
comparison of, 6–7, 311
definition of, 2, 426
Denisovan, 308–309
global perspective on, 9–11
Khoisan, 310
maintenance, cancer and, 352–353, 352f
Neanderthal, 307–309
sequencing, 1–2. See also Genetic testing; Personal genome sequencing
approaches for, 428, 430f–431f
in newborns, 393t, 394
sequencing of, limitations of, 433–434
synthetic, 432–433
Genome editing, 248, 326, 384–385, 384f
genome.org, 439
Genome-wide associations, 105
Genome-wide association study (GWAS), 139–142, 139t, 142t, 145, 393t
approach for, 139–140, 140f
of autoimmune disorders, 337

bias in, 141
limitations of, 141
Genomic imprinting, 124–126, 125f
abnormalities, 125–126, 126f, 259
cloning and, 52
Genomic medicine, 435
Genomics, 200, 435
applications of, 7–9
definition of, 2, 426
in fighting infection, 345–347
Genomic testing, 435
Genotype, 90
definition of, 5, 71
lethal, 90
Genotype frequency(ies), 264
migration and, 281–282, 281f–282f
Genotype-Tissue Expression (GTEx), 210
Genotyping, of blood, 327
George III (King of England), 93, 93f
German measles, fetal effects of, 61
Germ cells, 16
Germ layer(s), primary, 53, 54f, 57
derivatives of, 53, 54f
Germline cells, 44
Gerstmann-Sträussler-Scheinker disease, 194t
Ghrelin, 143, 143t
GHR gene, 64, 64t
Giant axonal neuropathy, 27–28, 41
Gibberellin, 70
Gigaxonin, 28
Globin, variants, 228–229, 228t
Globin chain switching, 200–201, 200f
Globozoospermia, 410cc
Glossogenetics, 325
Glucocerebrosidase, recombinant, 378t
Glucophage. See Metformin (Glucophage)
Glucose-6-phosphate dehydrogenase (G6PD) deficiency, 11, 229, 229f, 292t
Glucosinolates, 368, 369f
Glutaric acidemia, type 1, 284cc
Glutaric aciduria, type 1, 277
Glycolipid(s), formation of, 22
Glycome, 202
Glycoprotein(s), formation of, 22
Golden rice, 387
Golgi apparatus, 20f–21f, 22, 22f, 24
functions of, 22–23, 24t
structure of, 22, 24t
Gonad(s), 43
indifferent, 111, 111f
Gonadal mosaicism, 218
Gopher Kids Study, 13
gotchaDNA.com, 175
G₀ phase, 29–30, 29f
G₁ phase, 29–30, 29f
G₂ phase, 29–30, 29f
Graft-versus-host disease, 345
Graves disease, 338t
Green fluorescent protein (GFP)
in bacteria used in bioremediation, 382
in mice, 375f
Griffith, Frederick, 164–165, 167t
Grim Sleeper, 275
Growth factor(s), 32
Growth hormone (GH), recombinant human, 378t, 387
Growth hormone receptor (GHR), 64, 64t
Guanine (G), 3, 4f, 166, 168, 168f, 182f
Guthrie, Arlo, 163
Guthrie, Woody, 163
Gut microbiome, and body weight, 144
Gymrek, Melissa, 318

H

Haas, Corey, 402–403, 403f
Hair color, 5f
Hairpin loops, 384, 385f
Hammer, Michael, 274
Hannah's Hope Foundation, 28
Haplogroups, 313
in Native Americans, 316

Haploid cells, 16, 44, 45f, 46, 47f, 48, 48f, 49, 50f
Haplotype(s), 104–105, 105f
HapMap project, 425, 431t
Hardy, Godfrey Harold, 265
Hardy-Weinberg equilibrium, 265–267, 265f–266f, 267t, 280–281, 294f. See also DNA profiling
application of, 267–268, 267t, 268f
and DNA profiling, 272
Harris, Neal Patrick, 422
Hartsoeker, Niklass, 49
Hay fever, 340
Heart
embryology of, 57
enlarged, ABCC9 gene mutation and, 394
Heart attack(s), familial hypercholesterolemia and, 19
Heart health, 132cc–133cc
Heavy chains, 332, 332f, 333, 333f
Height, 137
bell curve for, 134, 134f
variations in, 134, 134f
Heinz bodies, 229, 229f
HeLa cells, 355
Helicase, 172, 173f
annealing, 173
Helix-turn-helix, 184
Helper T cells, 334, 334t, 338
in HIV infection, 336–337, 336f
Heme, biosynthesis, disorders of, 93, 94f
Hemings, Sally, 7
Hemizygous males, 117
Hemochromatosis, hereditary, 108, 436
diagnosis of, 396
genetic testing for, 396
Hemoglobin, 215
adult, 200f, 201
in embryo, 200f, 201
fetal, 200f, 201, 319–320
mutations of, 227–229
structure of, 192, 200, 200f
Hemoglobin C, 228, 228t, 292
Hemoglobin M, 228
Hemoglobin S, 228, 228t
Hemolytic anemia, 229, 338t
Hemophilia, pedigree analysis for, 80
Hemophilia A, 120, 129, 216t, 224
carrier frequency, 268, 268f
in females, 123
Hemophilia B, 218–219
gene therapy for, 406
inheritance of, 118–120, 118f
Hensel, Abigail and Brittany, 57, 57f
Hepatitis B, 61
Herceptin. See Trastuzumab (Herceptin)
HERC2 gene, 74, 74f
Hereditary hemochromatosis, 108, 436
diagnosis of, 396
genetic testing for, 396
Hereditary nonpolyposis colorectal cancer (HNPCC), 232
Heredity, definition of, 2
Heritability, 136–137, 136f, 142t, 145–146
of human traits, 136, 136t
Hermaphroditism, 114
Herpes simplex virus (HSV)
fetal/neonatal effects of, 61
as vector for gene therapy, 400
HER2 protein, 365
Herrick, James, 215
Hershey, Alfred, 165, 166f, 167t
Heterochromatin, 238, 238f
Heteroplasmy, 99–100, 100f, 407, 416
Heterozygote(s), 71, 268, 269f
compound, 91
manifesting, 123
Heterozygous advantage, 291, 292t
Heterozygous individual(s), 71
HGPS. See Hutchinson-Gilford progeria syndrome
High-density lipoproteins (HDL), 132cc
and longevity, 63
Histones, 170, 171f, 203
acetylated, and transcription, 203, 203f–204f
methylated, and transcription, 203

Hitchens, Christopher, 436
Hitler, Adolph, 163
HIV, 326*f*
  fetal effects of, 61
  genetic diversity of, 290, 290*f*, 337
  infection, 336–337, 336*f*–337*f*
    drug therapy for, 337, 337*t*
    stages of, 290, 290*f*
  mutation with protective effect against, 213, 281, 326
  natural selection and, 290, 290*f*
  replication, 336–337, 337*f*
  as vector for gene therapy, 403–404
HLA. *See* Human leukocyte antigen(s) (HLA)
Hodge, Nancy, 270*cc*
Holmes, Oliver Wendell, Jr., 296*f*
Holocaust survivors, DNA profiling of, 274
Homeobox, 55
Homeotics, 53, 55*cc*
Hominini, 303
Homo spp., early, 305–306, 306*f*
Homo erectus, 304*f*, 305–306, 306*f*, 320
Homo floresiensis, 302–303, 302*f*
Homo habilis, 304*f*, 305
Homolog(s), 44
Homologous pairs, 44
Homo sapiens, 303
Homo sapiens idaltu, 306–307, 307*f*, 314
Homosexuality, 115
Homozygosity mapping, 140, 141*t*
Homozygote(s), 71, 268, 269*f*
Homozygous individual(s), 71
Homunculus, 49
Hope for Trisomy 13 + 18, 245*t*
Hopi Indians, albinism in, 280
Hormone(s), 32
Housekeeping genes, 200
Human accelerated regions, 319
Human chorionic gonadotropin (hCG), 52
  beta, 243*t*
Human Chromosome Launchpad, 262
Human endogenous retroviruses (HERVs), 207
Human genome, 3*f*
  analysis of, 430–435
    projects for, 431*t*
  and chimpanzee, comparison of, 6, 319, 433
  comparison to genomes of other animals, 318–319
  duplication within, 319
  evolution of, 319, 319*f*
  non-protein-encoding parts of, 206–207, 207*t*
  protein-encoding genes of, 206
  sequencing of, 426–430. *See also* Personal genome
      sequencing
    cost of, 430, 432*t*
    coverage of, 430–431
    speed of, 430–431
Human Genome Landmarks, 262
Human genome project, 428–429, 429*f*
Human immunodeficiency virus (HIV). *See* HIV
Human leukocyte antigen(s) (HLA), 329, 349
Human Microbiome Project, 37
Human origins, 303–311
Humans
  archaic, 306–309, 307*f*
  characteristics of, 318–321
  definition of, 318–321
  migrations (peopling of the world), 314–317, 315*f*–316*f*
  modern, 309–310
  unique traits of, 319
Human Variome Project, 431*t*
Humoral immunity, 331–333, 331*f*–332*f*
Humor immunity, 330*f*
Hungerford, David, 364*cc*
Hunter syndrome, 123, 128
Huntingtin, 193
Huntington disease, 8, 69, 90, 126, 193, 194*t*, 216*t*, 235
  animal models of, 380, 380*f*
  gain-of-function mutation in, 77, 77*f*
  gene for, discovery of, 426
  genetic testing for, 394
  inheritance of, 74, 74*f*
  juvenile, 68, 68*f*, 87

penetrance, 93, 95
  variable expressivity, 95
Hutchinson-Gilford progeria syndrome, 42–43, 63, 63*t*, 217,
    235, 372
Hybrid(s), 70–71
Hydatidiform mole, 125
Hydrogen bonds, 167, 169, 170*f*
Hydrophilic surface(s), 24, 25*f*
Hydrophobic surface(s), 24, 25*f*
21-Hydroxylase deficiency, 114
Hypertrichosis, 237–238, 237*f*
  congenital generalized, 120, 121*f*
Hypocretin, 150–151
Hypomania, 155
Hypomyelination with atrophy of the basal ganglia and
    cerebellum, 197
Hypoxanthine-guanine phosphoribosyltransferase (HGPRT),
    123–124
  deficiency, 19
Hypoxia inducible factor 2, 288, 288*f*

I

Ichthyosis, 117–118, 117*f*
Ichthyosis vulgaris, 340
ICM. *See* Inner cell mass
Identity, establishing, 7
Ideogram, 243, 243*f*
Idiotypes, 332*f*, 333
Imatinib (Gleevec), 32, 351, 351*f*, 364*cc*–365*cc*
Immune deficiencies, inherited, 335–336, 335*t*
Immune response, 330
  primary, 331
  secondary, 331
Immune system, 327, 330–335, 330*f*
  abnormalities, 335–341
  altering, 341–345
  memory, 331
Immunity, abnormal, 335–341
Immunoglobulin(s)
  IgA, 332*f*, 333*t*
  IgD, 333*t*
  IgE, 333*t*, 338
  IgG, 333*t*
  IgM, 332*f*, 333*t*
Immunotherapy, 342–343
Implantation, 52, 53*f*
Imprinted Gene Catalogue, 128
Imputation, 436
Inborn errors of metabolism, 19*cc*, 77, 164, 180–181, 180*f*
  treatment of, 398
Incidence, 135
Incidental, genetic, 395
Incidentaloma, 395
Incomplete dominance, 97*t*
Incontinentia pigmenti, 120, 120*f*, 123
Independent assortment, 46, 47*f*
  and linkage, comparison of, 103, 103*f*
Indigenous peoples, 310, 310*f*, 324
Individual, as level of genetic organization, 5
Induced pluripotent stem (iPS) cells, 35–36, 36*t*, 58, 408
  reprogramming of, 36, 210
Infectious disease(s)
  emerging, 301
  outbreaks, tracking, 346–347, 346*f*
  prevention of, genomics and, 345–347
  treatment of, genomics and, 345–347
Infertility, 409–411
  definition of, 409
  female, 409–411, 411*f*
  male, 224, 409
  tests for, 411
Infidelity DNA testing, 175
Inflammation, 330
Influenza vaccine, 342, 346
  production of, recombinant DNA technology for, 380
Influenza virus(es), 328*cc*
Ingram, V. M., 215
Inheritance
  autosomal dominant, 69, 74–75, 74*f*, 74*t*, 81, 82*f*
  autosomal recessive, 69, 74–75, 77*t*, 81, 82*f*

Mendel's laws of, 69–71, 71*f*, 72*t*, 74–75. *See also* Law
    of independent assortment; Law of segregation
  mitochondrial, 98–100, 98*f*, 100*f*
  modes of, 69, 74–75
  of more than one gene, 78–79, 80*f*
  multifactorial, 136–137
  multigene, 78–79, 80*f*
  single-gene, 68–70, 72–74
    factors affecting, 96, 97*t*
  of single-gene traits, solving problems in, 75
Inhibin A, 243*t*
Initiation complex, 189, 190*f*–191*f*
Innate immunity, 330–331, 331*f*
Inner cell mass, 52, 54*t*
  and stem cell production, 35–36
Innocence Project, 263–264, 277
Insertional translocation(s), 256
Insertion mutation(s), 221*f*, 222*t*, 224
Insomnia, 151, 162
Insulin, 202
  posttranslational modification, 190
  recombinant human, 378, 378*t*
Insulin-like growth factor(s), IGF-2, and schizophrenia, 199
Intellectual disability, 152
  aneuploidy and, 245
  genes associated with, 161
  studies of, exome sequencing in, 161
Intelligence, 151–152
  heritability of, 152
Intelligence quotient. *See* IQ
Interactome(s), 202
Interferon(s) (IFN), 330, 334*t*
  immunotherapy with, 343
  recombinant, 378*t*, 379
Interleukin(s) (IL), 330–331, 334*t*
  IL-2
    abnormal, 335*t*
    immunotherapy with, 343
    receptor mutation, 335*t*
    recombinant, 378*t*
Intermediate filament(s), 25, 27, 27*f*, 28
International 22q11.2 Deletion Syndrome
    Foundation, 245
Interphase, 29–30, 29*f*–30*f*
  of meiosis, 45
Intersex, 114
Intracytoplasmic sperm injection (ICSI), 414–415, 414*f*
Intrauterine growth retardation (IUGR), 62
Intrauterine insemination (IUI), 418*t*
  with donated sperm, 413
Intron(s), 186, 186*f*, 205
  genes in, 205–206, 206*f*
Inversion(s), 245*t*, 256–257, 257*f*, 258*t*
  and cancer, 361, 363
  paracentric, 257, 257*f*
  pericentric, 257, 257*f*
In vitro fertilization (IVF), 414–415, 418*t*
  success, prediction of, 415
Ion channel(s), 26*cc*
  abnormal, 26*cc*
IQ, 151
  and success, 151, 152*f*
Irons, Ernest, 215
Ishiguro, Kazuo, *Never Let Me Go*, 345
Isochromosome(s), 245*t*, 257, 258*f*
Isoforms, protein, 205
Isograft(s), 344, 344*f*
Isotretinoin, teratogenicity of, 61
Ivacaftor (Kalydeco), 76*cc*

J

Jackson Laboratory, 387
Jacobs syndrome, 243*t*
Jacon, François, 184
Jefferson, Thomas, 7, 277
Jeffreys, Alec, 269, 269*f*, 274–275
Jenner, Edward, 342
jhdkids.com, 87
Jobs, Steve, 436
John, Elton, 422

Joubert syndrome, 320
Jukes study, 301
Jumping Frenchmen of Maine, 73cc

K

Kallmann syndrome, 121, 197
Kalydeco. See Ivacaftor (Kalydeco)
Karyotype(s), 239–240, 239f, 241–242
    abnormalities, 244
    definition of, 4
Keratin, 320
Khan, Genghis, 280–281
Khoisan, 310, 310f
Kilobase (kb), 168, 168f
Kinase(s), 32
King, Taylor, 41
Klebsiella pneumoniae, outbreaks, tracking, 346–347, 346f
Klinefelter syndrome, 242, 243t, 252
Knoblich, Juergen, 58
Köhler, George, 342–343
Kuru, 194, 194f

L

Lacks, Henrietta, 355
Lactase deficiency, 19, 279, 287
Lactase persistence, 279, 287–288
Lactocyte(s), 22
Lactose intolerance, 19, 279, 287
Lactose tolerance, evolution of, 279
Lamin A, 63
lamin A gene, 235
    mutations of, 217
Lancaster, Madeline, 58
Landis, Floyd, 379
Laron syndrome, 64
Law of independent assortment, 77–78, 78f, 272
Law of segregation, 70–71, 71f, 72t
    and uniparental disomy, 258–259
Leader sequence, 188
Leber congenital amaurosis, 89, 89f, 95
    type 2, 402, 402f
        gene therapy for, 402–403, 403f
Lee, Pearl, 214
Leight syndrome, 99–100
Lens crystallins, 94
Lentivirus, 403
Leptin, 142–143, 143t
Leptin receptor(s), 143, 143t
Lesch-Nyhan syndrome, 19
    carriers, test for, 123–124
Lethal allele(s), 90, 97t
Lethal equivalents, 287
Leucine zippers, 184
Leukemia(s). See also Chronic myelogenous leukemia
    gene therapy for, 403–404
    mixed lineage, 203–204, 370, 370f
Levene, Phoebus, 165, 167t
Lewis blood group, 327t
Lewy bodies, 198
Lewy body dementia, 194t
Lice, evolution of, 325
Liebenberg syndrome, 55, 55f
    in Plain people, 285t
Lieberman, Lynn, 223
Lieberman, Rebekah, 223
Li-Fraumeni syndrome, 366
Ligand, 25
Ligase, 172–173, 173f
Light chains, 332, 332f, 333, 333f
Limb formation, critical period for, 60
Lindeman, Bruce and Susan, 419, 419f
Linkage, 100–105
    crossing over and, 102, 102f
    definition of, 101
    and independent assortment, comparison of, 103, 103f
    LOD score and, 105
Linkage disequilibrium (LD), 104, 266
    in dog breeds, 289
Linkage maps, 102–103, 426, 426f

determination, using logic, 104
Lipid(s), 17, 21f
    metabolism, 132cc
        inborn errors of, 19
    secretion of, 22
    synthesis of, 21f
Lipidome, 202
Lipid rafts, 24
Lipoprotein(s), 132cc
Lipoprotein lipase (LPL), 132cc
Living to 100 Life Expectancy Calculator, 66
Lloyd, Dale II, 87
LOD score, 105
Loeys-Dietz syndrome, 83–84
Longevity, 63–64, 64t
Long noncoding RNAs, 207t, 208
Long QT syndrome, 108–109
Long-QT syndrome, 26
Lorenzo's Oil (film), 23
Loss-of-function mutation(s), 77, 77f, 213
Louis XVI (King of France), son of, identification of remains of, 100
Louis XVII, remains of, identification of, 100
Low-density lipoprotein (LDL), 132cc
Low-density lipoprotein (LDL) receptor, mutations affecting, 221f
Lowe syndrome, 129
LSD, 153
Lucy, 304–305, 305f
Lung cancer
    genome-wide association study of, 141
    risk, genes affecting, 146
    risk factors for, 131, 131f
    smoking and, 154, 355
Lupus. See Systemic lupus erythematosus
Lymph, 330, 330f
Lymphatics, 330, 330f
Lymph node(s), 330, 330f
Lymphocyte(s), 330, 330f, 331, 331f
Lymphoma(s), 55
Lymphoproliferative disease, X-linked, 335t
Lynch syndrome, 232
Lyons, Glenda and Scott, 419, 419f
Lysosomal storage disease, 23, 123
    treatment of, 398, 398f, 398t
Lysosome, 20f–22f, 23
    functions of, 22f, 23, 24t
    structure of, 23, 24t

M

MAbs. See Monoclonal antibody(ies)
MacLeod, Colin, 165, 167t
Macroevolution, 265
Macromolecule(s), 6f, 17–18
Macrophage(s), 329, 329f, 334t
Mad cow disease, 194
MAGEL2 gene, 227
Major depressive disorder, 154
Major histocompatibility complex (MHC), 327–329, 329f
Malaria, 7
    prevention, in United States, historical perspective on, 293
    prevention of, recombinant DNA technology for, 380
    resistance to, hemoglobin mutations and, 228
    sickle cell disease and, 291–293, 292f–293f
Malignant hyperthermia, 197
    in Plain people, 285t
Malnutrition
    fetal effects of, 61
    microbiome and, 41
    prenatal
        and adult-onset disorders, 62
        and adult phenotype, 199
Mania, 155
Manifesting heterozygote, 123
Maple syrup urine disease, 19, 277
    in Plain populations, 284cc–285cc
Map units, 102
Marfan syndrome, 83–84, 93–94, 94f, 95, 216, 216t, 395
Marie Antoinette, son of, identification of remains of, 100

Marijuana, 153
Marshall-Graves, Jennider A., 113–114
Mast cell(s), 334t, 338
Mastromarino, Michael, 40
Maternal age
    and chromosomal abnormalities in fetus, 50–51, 240, 241f, 248
    and infertility, 411
    and pregnancy problems, 409
    and trisomy 21, 240, 241f, 248
Maternal age effect, 50–51, 240, 241f, 248
Maternal serum markers, 243–244, 243t, 392, 393t
Matthaei, Heinrich, 189
Maturation, 62–64
McCarty, Maclyn, 165, 167t
McClintock, Barbara, 208
McGovern, James B., 108
MDP syndrome, 235
MECP2 gene, 15
medomics. com, 439
Megabase (mb), 168, 168f
Meiosis, 29, 44–47, 45f–47f, 71, 71f
    errors in, 245–246, 246f
    and genetic diversity, 44, 46
    and mitosis, comparison of, 45t
    and mutations, 50–51, 213
    in oogenesis, 49, 49f–50f
    in spermatogenesis, 48, 48f
Meiosis I, 45–46, 45f–46f, 48, 48f, 49–50, 50f
    nondisjunction at, 246, 246f
Meiosis II, 45–46, 45f, 47f, 48, 48f, 49–50, 50f
    nondisjunction at, 246, 246f
Melanin, 73, 134
Melanocyte(s), 73, 134
Melanoma, 353f
Melanosome(s), 73, 134
MELAS, 98–99
Mello, Craig, 385
Membrane(s), biological, 24–25
    structure of, 24
Memory B cells, 331, 331f–332f
Mendel, Gregor, 69–70, 70f
Mendelian Inheritance in Man, 73cc
Mendel's laws of inheritance, 69–71, 71f, 72t, 74–75. See also Law of independent assortment; Law of segregation
Mennonites, 282
    genetic disorders in, 284cc–285cc, 285t
Menstrual cycle, 44
    irregularities, and infertility, 409
Mervar, Erica, 68, 224
Mervar, Jacey, 68, 87, 224
Mervar, Jane, 68, 224
Mervar, Karl, 68, 68f
Mervar, Karli, 68, 68f
Meselson, Matthew, 172
Mesoderm, 53, 54f
Messenger RNA (mRNA), 3, 18, 20, 21f, 183, 183t, 185
    cap, 185, 186f
    and microRNAs, 204–205
    poly A tail, 186, 186f
    processing of, 185–186, 186f
    in translation, 189, 190f–191f
Metagenomics, 9–11, 38, 346
Metaphase, 31f
Metaphase I, 46, 46f–47f
Metaphase II, 46, 47f
Metastasis, 353, 356, 357f, 372
Metformin (Glucophage), 436
Methanococcus jannaschii, 433
Methicillin-resistant Staphylococcus aureus (MRSA), 291
Methyl groups
    binding
        and epigenetic changes, 200
        in genome-wide association studies, 139, 139t
    binding to histones, 203, 203f
    and embryonic gene expression, 53
    gene silencing by, 124, 124f
Methylome, 203
Methyltransferase(s)
    and bipolar disorder, 157
    and schizophrenia, 156–157

Microarray(s), 428. *See also* Chromosomal microarray (CMA) test
  definition of, 382
  DNA, 382–384, 383*f*
Microbiome
  gut, 37–38
  human, 37–38
  and malnutrition, 41
Microcephaly, 58
Microchimerism, 339*cc*
Microdeletions, 254
Microevolution, 264–265
Microfilament(s), 20*f*, 25, 27, 27*f*
MicroRNAs, 203–205, 204*f*, 433–434
  in breast cancer, 210
Microsatellites, 231–232, 269*t*. *See also* Short tandem repeats (STRs)
Microtubule(s), 20*f*, 25, 27, 27*f*
  centriole, 30
Middlesex (novel), 114
Miescher, Friedrich, 164, 167*t*
Migraine, 131
Migration, 264, 281–282, 281*f*–282*f*
  ancient human, 314–317, 315*f*–316*f*, 324–325
Mild cognitive impairment, 96*cc*
Military, genetic testing used in, 393*t*, 395
Milk, secretion of, 18, 21*f*, 22
Milstein, Cesar, 342–343
Mineral(s), metabolism, inborn errors of, 19
Minisatellites, 220, 269*t*. *See also* Variable number of tandem repeats (VNTRs)
Mismatch repair, 230–231, 231*f*
Mitochondria. *See* Mitochondrion (pl., mitochondria)
Mitochondrial disorders, 98–99
  heteroplasmic, 99–100, 100*f*
Mitochondrial DNA, 24, 98, 99*f*, 99*t*, 312–313, 313*f*
  cloning and, 52
  forensic applications of, 100, 271
  historical applications of, 100
  of Native Americans, 316
Mitochondrial DNA, forensic applications of, 274
Mitochondrial Eve, 314, 315*f*
Mitochondrial inheritance, 98–100, 98*f*, 100*f*
Mitochondrial myopathy(ies), 98
Mitochondrial myopathy encephalopathy lactic acidosis syndrome. *See* MELAS
Mitochondrial replacement, 99, 407–408, 407*f*
Mitochondrion (pl., mitochondria), 20*f*–21*f*, 23–24, 23*f*, 98
  fragment, 22*f*
  functions of, 23, 24*t*, 98
  proteins of, 193
  of sperm, 48, 49*f*, 98
  structure of, 23, 24*t*
Mitogenomes, 313
Mitosis, 28–33, 29*f*–31*f*, 354, 354*f*
  in cancer cells, 354, 354*f*
  control of, 31–32
  errors in, 247
  and meiosis, comparison of, 45*t*
  and mutations, 213
  in spermatogenesis, 47, 48*f*
MKRN3 gene, 124
MN blood group, 327*t*
Modifier gene(s), 92
Molecular clocks, 312
Molecular evolution, study of, 311–314
Molecule(s), 6*f*
MoM value, 244
Monoamine oxidase A, 161–162
Monoclonal antibody(ies), 342–343, 343*f*
Monod, Jacques, 184
Monohybrid cross, 70, 70*f*
Monosomy(ies), 245–246, 245*t*, 246*f*
Mood disorders, 154–155
Moore's law, 430
Morgan, Thomas Hunt, 102
Morpholinos, 384, 388
Morton, Holmes, 284*cc*
Morula, 51–52, 54*t*
Mosaicism, 214, 247
  chromosomal, 241
  gonadal, 218

Turner syndrome, 251–252
  X inactivation and, 122, 122*f*
Mossy foot. *See* Podoconiosis
Most recent common ancestor, 317
MRCA. *See* Most recent common ancestor
mRNA. *See* Messenger RNA (mRNA)
MRSA. *See* Methicillin-resistant *Staphylococcus aureus* (MRSA)
mtDNA. *See* Mitochondrial DNA
Müllerian ducts, 111, 111*f*
Mullis, Kary, 175
Multigene family(ies), 239
Multiple endocrine neoplasia, 51
Multiple sclerosis, 338*t*
  heritability of, 145
Multiple unrelated affected subjects, analysis, in autism, 157, 158*f*
Multipotent cell(s), 34
Mundlos, Stefan, 55
Muscular dystrophy, 72, 217, 234
Muscular tissue (muscle), 6*t*, 17*f*
Mutagen(s)
  accidental exposure to, 220
  as carcinogens, 353
  definition of, 218
  intentional use to cause mutations, 219–220
  natural exposure to, 220–221
Mutant(s), definition of, 213
Mutation(s), 213–214, 265
  and allele frequencies, 286–287, 287*f*, 294*f*, 295*t*
  and cancer, 352, 356–357, 357*f*, 359*f*, 360
  causes of, 218–221
  chromosomal, 238
  cloning and, 52
  conditional, 229
  definition of, 4, 71, 213
  de novo, 218–219, 254, 256
  disease-causing, 216, 216*t*
  effects of, 4, 213
  expanding, 222*t*
  frameshift, 222*t*, 224
  gain-of-function, 77, 77*f*, 213, 225
  germline, 213
    and cancer, 355, 355*f*
  homeotic, 53, 55*cc*
  induced, 219–221
  interspecies similarities in, 311, 311*f*
  loss-of-function, 77, 77*f*, 213
  missense, 221–222, 221*f*, 222*t*, 224
  nonsense, 221*f*, 222, 222*t*, 224
  parental age and, 50–51
  point, 227
  position of, 227–229
  protection against, 229
  silent, 221, 229
  somatic, 213
    and cancer, 355, 355*f*
  splice-site, 222–223
  spontaneous, 218–219, 218*f*
    rates of, 218–219
  types of, 213, 221–227
Mutational hot spots, 219
Myasthenia gravis, 338*t*
MYCN gene, 366, 373
Mycobacterium abscessus, 76*cc*, 350
Mycoplasma genitalium, 432–433
Myelin, 149, 149*f*
MYH16 gene, 320
Myoglobin, structure of, 192
Myotonic dystrophy, 225, 225*f*
  type 1, 225
  type 2, 225

N

Nail-patella syndrome, inheritance of, 104, 104*f*
Narcolepsy, 150–151
  in dogs, 150–151, 150*f*
Narfström, Kristina, 403*f*
Nash, Adam, 408, 408*f*, 417
Nash, Lisa and Jack, 408
Nash, Molly, 408, 408*f*
Nathans, Jeremy, 119*cc*

National Center for Biotechnology Information, 108
National DNA Index (NDIS), 271
National Down Syndrome Society, 245*t*
National Organization for Rare Diseases, 87
Native Americans, origins of, 315–317, 316*f*
Natural killer cells, 334*t*
Natural selection, 265, 281, 286
  and allele frequencies, 287–295, 287*f*, 294*f*, 295*t*
  and antibiotic resistance, 290–291, 291*f*
  and HIV, 290, 290*f*
  negative, 287–288
  positive, 287–288, 288*f*
  tuberculosis and, 289–290
Naura, Western Samoa, obesity in, 143
Naylor, Ashley, 253, 254*f*
Neanderthals, 307–308, 307*f*–308*f*, 319, 324
  Altai, 309
Necrosis, 32
Neel, James, 144
Nelmes, Sarah, 342
Nervous tissue, 6*t*, 17*f*
Neural tube, 57
Neural tube defects, 57, 244, 301
  empiric risk of, 135
  screening for, ethical considerations in, 297
Neurexin(s), in autism, 158, 159*f*
Neurofibrillary tangles, 62, 96*cc*, 194
Neurofibromatosis (NF), type 1, 206, 216*t*, 218
Neurofibromin, 205–206
Neuroglia, 149
Neuroligins, in autism, 158, 159*f*
Neuron(s), 149, 149*f*
Neuronal ceroid lipofuscinosis, 41
Neurotransmission, 149, 149*f*
Neurotransmitter(s), 149, 149*f*
Neutrophil(s), 334*t*
Newborn(s), genetic testing of, 393*t*, 394
Newborn screening, 393*t*, 394, 394*f*, 405
Next-generation sequencing, 176–177
Nicholas II (Tsar of Russia), heteroplasmy in family of, 99
Nicotine
  addiction, 153–154, 154*f*
  metabolism, *CYP2A6* genotype and, 162
Nicotinic receptor, 153–154, 154*f*
Niemann-Pick disease, type A, 406
  in Ashkenazi Jewish populations, 286*t*
Nirenberg, Marshall, 189
Nobelprize.org, 387
nobelprizes.com, 387
Noncoding RNA (ncRNA), 206–208, 207*t*
Nondisjunction, 246–247, 246*f*
  and sex chromosome aneuploidy, 250, 252*t*
Nonrandom mating, 264, 280–281, 280*f*
Noonan syndrome, 126
Norovirus, 349
Notochord, 57
Nowell, Peter, 364*cc*
Nuclear envelope, 18, 20*f*–21*f*, 30*f*
Nuclear lamina, 18
Nuclear pores, 18, 20*f*–21*f*
Nucleic acid(s), 17–18, 164–165. *See also* DNA (deoxyribonucleic acid); RNA (ribonucleic acid)
  components of, 165
Nuclein, 164
Nucleolus, 18, 20*f*, 30*f*, 239
Nucleoplasm, 18
Nucleosome(s), 170, 171*f*
Nucleotide(s), 168, 168*f*
Nucleotide excision repair, 230, 231*f*
Nucleus, cell, 16, 18, 20*f*–21*f*, 30*f*
  functions of, 24*t*
  structure of, 24*t*
Nutrigenetics, 396–397
Nutrition, prenatal, and adult phenotype, 199

O

Obesity, 142–143, 145
  heritability of, 143
O'Brien, Chloe, 417
OCA2 gene, 73–74, 74*f*

Octaploidy, 245
Odone, Lorenzo, 403
Odorant receptors, 321
Office of Rare Diseases Research, 87, 108
Ohno, Susumo, 113
Okazaki fragments, 172, 173*f*
Oligospermia, 409
Oncogene(s), 353, 361–365
Online Mendelian Inheritance in Man (OMIM), 4, 161
Oocyte(s), 43–44, 51
    age-related changes in, 411, 415
    development of, 49, 50*f*
    donation of, 415–417
    formation of, 49–50
    freezing, 415, 418*t*
    number of, 49–50
    postmortem retrieval, 416
    primary, 49, 50*f*
    secondary, 49, 49*f*–50*f*, 51–52
Oogenesis, 49–50, 50*f*
Oogonium, 49, 50*f*
Open chromatin, 203, 211
Open reading frame, 188
Operon(s), 184
Opium, 153
Opsin(s), 119*cc*
Orexin, 151
Organ(s), 5, 6*f*, 16
Organelle(s), 6*f*, 16, 18–24, 20*f*, 24*t*
    functions of, 18, 24*t*
    structure of, 24*t*
Organism(s), 6*f*
    single-celled, 16
Organogenesis, 57
Organoids, 58
    cerebral, 58
Organ system(s), 5, 6*f*, 16
Orgasm, 43–44
Ornithine transcarbamylase deficiency, 400–402, 401*f*
Orrorin tugensis, 303, 304*f*
Osteoarthritis, 62, 215*t*
Osteogenesis imperfecta, 62, 95, 216
    type I, 212–213, 212*f*, 215*t*
Osteoporosis, 5
Ötzi the Ice Man, 309–310, 309*f*
Ovalocytosis, 300
Ovarian tissue, reimplantation of, 415
Ovary(ies), 44, 44*f*, 50*f*
Ovulation, 49–50, 50*f*
    detection of, 409
    induction of, 409, 418*t*
    irregular, 409
Ovum, 44, 49, 50*f*
    fertilized, 54*t. See also* Fertilization

P

Pääbo, Svante, 308, 324
Page, David, 113–114
Painter, Theophilus, 242
Palindrome(s), 112, 114, 376, 377*f*
    and mutations, 219, 219*f*
Pancreas
    cell lineages in, 201–202, 201*f*
    endocrine, 201, 201*f*
    exocrine, 201, 201*f*
Paracentric inversion(s), 257, 257*f*
Parental progeny, 102, 102*f*
Parent-of-origin effects, 124–126
Parkinson disease, 7, 40, 146, 194, 194*t*
    early-onset, 197
    fetal implants in, 198
Paroxysmal extreme pain disorder, 26
Passenger mutations, 360
Patau syndrome. *See* Trisomy 13
Patent(s), 375–376
Paternal age, and risk in offspring, 409
Paternal age effect, 50–51, 126
Paternity testing, 393*t*
Pathogen(s), 327
Pauling, Linus, 167, 215

PCSK9 gene, 145
Peanut allergy, 340, 341*f*
Pea plants
    linked genes in, discovery of, 101–102, 102*f*
    Mendel's studies of, 69–71, 69*f*–70*f*, 77–78, 79*f*
Pedigree(s), 5, 433
    analysis, 79–83
    for autosomal dominant traits, 81, 82*f*
    for autosomal recessive traits, 81, 82*f*
    definition of, 79–80
    genealogical, 80, 82*f*
    historical perspective on, 80–81
    inconclusive, 81, 82*f*
    symbols/lines used in, 80, 81*f*
Peling skin syndrome, 86
Penetrance, 93, 95–96, 97*t*, 392
    incomplete, 95–96, 125, 435
Penicillamine, 19
Penis, 43, 43*f*
Pennington, Robert, 344
Peptide(s), 181
Peptide bonds, 181, 189, 191*f*
Peptide fingerprints, 215
Perforin(s), 334, 334*f*
Pericentric inversion(s), 257, 257*f*
period 1 gene, 151
Periodic fever, familial, in Plain people, 285*t*
Peroxisome, 20*f*, 23, 403
    fragment, 22*f*
    functions of, 23, 24*t*
    structure of, 23, 24*t*
Personal Genome Project, 425, 431*t*, 439
Personal genome sequencing, 435–437
    concerns about, 437–438
PERV. *See* Porcine endogenous retrovirus
p53 gene, 360–361, 361*f*, 366, 367*f*
P₁ (parental) generation, 69
Phagocyte(s), 32, 33*f*, 331
Pharmacogenetics, 8, 393*t*, 397, 435–436
Pharmacogenomics, 436
Pharmacogenomics Knowledge Base, 405
Pharmacogenomic test(s), 397
Pharmacological chaperone therapy, 398, 398*t*
Phenocopy(ies), 95, 97*t*, 141
Phenotype, 168
    definition of, 5, 71
    gene expression and, 90–98
    mutant, 71
    variable expressivity, 93
    wild type, 71
    X inactivation and, 123–124, 123*f*
Phenotype frequency(ies), 264
Phenotypic ratio(s), 71, 72*t*
    linked genes and, 101, 101*f*
    single-gene, factors affecting, 90–98, 97*t*
    unlinked genes and, 101, 101*f*
Phenylketonuria, 194*t*, 286–287, 292*t*
    frequency in various populations, 264, 264*t*
    newborn screening for, 394
Phenylthiocarbamide, ability to taste, 277
Pheomelanin, 73
Philadelphia chromosome, 364*cc*
Phipps, James, 342
Phocomelia, 60, 95
Phosphate group(s), 168, 168*f*
    binding to histones, 203, 203*f*
Phosphodiester bonds, 168, 168*f*
Phospholipid bilayer, 24, 25*f*
Photolyases, 230
Photopigments, 119*cc*
Photoreactivation, 230, 231*f*
Photoreceptor cells, 95, 95*f*, 119*cc*
Physical barriers, in immune protection, 330–331, 331*f*
Physical map(s), 426, 426*f*
Picoult, Jodi, *My Sister's Keeper,* 423
Pictogram, 432, 432*f*
Pierson syndrome, in Plain people, 285*t*
Pig(s)
    phytase transgenic, 374
    transplants from, 344
Pig manure, genetically modified, 374, 374*f*

PIK3R1 gene, 197
Pima Indians, obesity among, 143–144
Pingelapese blindness, 283
Pitchfork, Colin, 269
PITX1 gene, 55
Placenta, 53–54
Plain Populations, genetic disorders in, 284*cc*–285*cc*, 285*t*
Plant(s)
    genetically modified, 381, 381*t*
    use in bioremediation, 381–382
Plasma cells, 331, 331*f*–332*f*
Plasma membrane, 18, 20*f*–21*f*, 24–25, 25*f*
Plasmid(s), 377–378, 377*f*
plastin 3 gene, 92
Plavix. *See* Clopidogrel (Plavix)
Pleiotropy, 93–94, 93*f*–94*f*, 95–96, 97*t*
    and autism, 158
Pletcher, Finley, 89
Pletcher, Jennifer, 89
Pluripotent cell(s), 34. *See also* Induced pluripotent stem
    (iPS) cells
Pluto (dog), 403*f*
Pneumonia, hospital-acquired, outbreaks, tracking,
    346–347, 346*f*
Podoconiosis, 7–8, 8*f*
Point mutations, 221
Polar body(ies), 49, 49*f*–50*f. See also* Sequential polar body
    analysis
POLD1 gene, 235
Polycystic kidney disease, 27, 62
Polydactyly, 5*f*
    incomplete penetrance, 93
    pedigree analysis for, 81, 82*f*
    variable expressivity, 93
Polyglutamine diseases, 225
Polymerase, 172
Polymerase chain reaction (PCR), 173–175, 174*f*, 174*t*
Polymorphism(s), 213–214
    balanced, 291–292, 292*f*, 292*t*, 293*f*
    definition of, 213
Polypeptide(s), 181
    synthesis of, 189–190, 191*f*
Polyploidy, 44, 245, 245*f*, 245*t*, 258*t*
Population(s), definition of, 5–6, 264, 264*f*
Population bottleneck(s), 282–286, 295*t*, 299
    in cheetahs, 285, 286*f*
Population genetics, 6, 264–265
    definition of, 264
Porcine endogenous retrovirus, 344
Porphyria(s), 93, 94*f*, 108–109
Porphyria variegata, 93, 93*f*–94*f*, 283
Porphyrin, 93, 94*f*
Positional cloning, 426
positiveexposure.org, 235
Postmortem interval (PMI), 210
Posttranslational modifications, 190
Posttraumatic stress disorder (PTSD)
    causes of, 148
    characteristics of, 148
    child abuse and, 148–149
Potassium channel(s), abnormalities, 26
p53 protein, 231
Prader-Willi syndrome, 125–126, 126*f*, 129, 227, 259
Precocious puberty, central, 124
Preeclampsia, 122
Pregnancy-associated plasma protein (PAPP-A), 243*t*
Preimplantation genetic diagnosis, 392, 393*t*, 417–418, 417*f*, 418*t*
Pre-mRNA, 186, 186*f*
Prenatal development, 51–59, 54*t*
Prenatal diagnosis, noninvasive, 244, 392, 393*t*
Presenilin 1, 236
    gene for, 97*t*
Presenilin 2, 300
    gene for, 97*t*
Prevalence, 135
Primaquine, 229
Primase, 172, 173*f*
Primate(s), taxonomy of, 303, 304*f*
Primer(s), for PCR, 173–175, 174*f*
Primitive streak, 57
Prion(s), 193–195, 195*f*, 292*t*

Prion diseases, 193–194, 194t, 234
Privacy
    ancestry testing and, 318
    DNA profiling and, 274–276
    genetic testing and, 11–12
PRL-3 gene, 361, 361f
Probiotics, 38
Procollagen, 215
Product rule, 78–79, 80f, 272
Progenitor cell(s), 33–35, 33f–34f
Progeria, 42–43, 42f, 62–63, 170
Progerin, 63
Prokaryotes, 16
Prokaryotic cell, 16, 18f
Promoter(s), 184, 184f–185f, 203, 380
Pronucleus (pl., pronuclei), 51
Prophase, of mitosis, 30, 30f
Prophase I, 45, 46f
Prophase II, 46, 47f
Prosopagnosia, 107–108
Prostate cancer, 202
Prostate gland, 43, 43f
Prostate-specific antigen (PSA), 205, 356
Protanopia, 119cc
Proteasome, 193, 193f, 197
Protein(s), 2, 4, 17–18
    centromere-associated, 238
    comparison of, 312
    conformations, 192
    cytoskeletal, 25
    diversity of, 181t
    folding of, 22, 192, 192f, 193
    functions of, 168, 181t
    fusion, 363–365
    isoforms, 205
    membrane, 24, 25f
    metabolism, inborn errors of, 19
    misfolding of, 193
        disorders associated with, 193, 194t
    in phospholipid bilayer, 24, 25f
    posttranslational modification, 190
    primary structure of, 192, 192f
    processing of, 192–195
    quaternary structure of, 192, 192f
    secondary structure of, 192, 192f
    structure of, 192, 192f
    synthesis of, 3, 4f, 18, 20, 21f, 30, 181, 181f, 184, 186,
        189–190, 190f–191f
    tertiary structure of, 192, 192f
Proteome(s), age-related changes in, 202, 202f
Proteomics, 202
Proto-oncogenes, 361–363
PSA-linked molecule, 205
PSEN1 gene, mutations of, 217t
Pseudoautosomal region(s), PAR1 and PAR2, 112, 112f
Pseudogenes, 207–208, 207t, 208, 224
Pseudohermaphroditism, 114–115
Pseudomonas, 76cc
5p⁻ syndrome, 253–254, 254f
PTC gene, 277
Pulmonary fibrosis, idiopathic, 145–146
Punnett, R. C., 101–102
Punnett square, 71, 72f
    in tracking more than one trait, 79, 80f
Purines, 168, 168f
Pygmies, 323
Pyloric stenosis, empiric risk of, 135–136
Pyrimidine dimers, 230–231
Pyrimidines, 168, 168f
Pythagoras, 229

Q

Quake, Stephen, 436
Quantitative traitloci (QTLs), 133
Quinn, Anthony, 422

R

Race(s)
    definition of, 134

skin color and, 134–135
Radiation
    acute exposure to, 220
    ionizing, 221
    mutagenicity of, 219–221
    natural environmental sources of, 220
RAG1 gene, 284cc
Rainbows Down Under, 245t
Randall, Tony, 422
Randell, Alex, 389, 389f
Randell, Ilyce, 389, 389f
Randell, Max, 389, 389f
Randell, Mike, 389
Ransome, Joseph, 119cc
Rape, and eugenics, 297–298
Rasmussen's encephalitis, 349
RB1 gene, 365–366, 373
Reading frame(s), 187–188, 188f
    open, 224
Receptor(s), 25
Recessive analysis, 157, 158f
Recessive congenital methemoglobinemia, 228
Recessiveness, 75–77. *See also* Trait(s), autosomal recessive
Reciprocal translocation, 256, 256f
Recombinant DNA, 375–380, 387
    drugs produced using, 378–380, 378t
    manufacture of, 376–378, 377f
    products of, 378–380, 378t
    safety of, 376
    selection of, 378
Recombinant progeny, 102–103
Recombination mapping, 104, 105f
Reduced ovarian reserve, 410
Reduction division, 45, 45f
Reed, Sheldon, 390
Regulatory T cells, 339cc
Relationship, degrees of, 136, 137f, 137t
Release factor(s), 190, 191f
Renin inhibitor, recombinant, 378t
Repeats, 207t, 208, 319, 428. *See also* Expanding repeats;
    Short tandem repeats (STRs); Variable number of tandem
    repeats (VNTRs)
    in DNA profiling, 269–271, 269t
    mutations in, 219, 219f
Replication bubbles, 173, 173f
Replication fork, 172
Reproductive system, 43–44
    female, 44, 44f
    male, 43, 43f
Rescue karyotyping, 392–394, 393t
Respiratory syncytial virus (RSV), 388
Restriction enzyme(s), 376–377, 377f
Resveratrol, 62
Retinal (protein), 119cc
Retinitis pigmentosa, 129, 145, 234
    X-linked, 234
Retinoblastoma, 365–366, 366f, 373
RetNet, 108
RET oncogene, mutations of, 217t
Retrovirus, 207, 328cc. *See also* Porcine endogenous
    retrovirus
Rett syndrome, 15–16, 120, 124, 157, 184, 208
Reverse PCR, 174f
Reverse transcriptase, 328cc, 336–337, 337f, 382
Reverse vaccinology, 346
R gene, of peas, 70
R groups, 180, 192f
Rhabdomyolysis, acute exertional, 87
Rh blood type, 104
Rheumatic fever, 338, 338t
Rheumatoid arthritis, 338t, 343
Rh incompatibility, 339cc–340cc, 350
Rhodopsin, 119cc
Ribose, 165, 182, 182f
Ribosomal RNA (rRNA), 183, 183f, 207, 207t
    in translation, 189, 190f
Ribosome(s), 16, 18, 20, 20f, 22, 183, 183f, 186
    functions of, 24t
    large subunit of, 183, 183f, 189, 190f–191f
    P site, 189, 191f
    A site, 189, 191f

small subunit of, 183, 183f, 189, 190f
    structure of, 24t
Ribozyme(s), 183, 189
    in gene silencing, 384, 384f
Rienhoff, Bea, 83–84, 83f, 88, 152
Rienhoff, Hugh, 83–84, 83f, 88
Ring chromosome, 245t, 257–258, 258f, 258t
Ring Chromosome 20 Foundation, 258
RNA (ribonucleic acid), 3, 4f, 18, 165
    conformation of, 182–183
    and DNA, comparison of, 165, 182, 182f, 182t
    functions of, 182
    processing of, 185–186
    structure of, 181–183, 182f
    types of, 181–183
RNA-induced silencing complex (RISC), 385, 385f
RNA interference (RNAi), 205
    in gene silencing, 384–385, 384f–385f
RNA polymerase, 181, 184–185, 184f
RNA primer, 172
RNA tie club, 189
Robertsonian translocation, 255, 255f
Rockman, Alexis, 381f
Rod(s), 95, 95f, 119cc
Roderick, T. H., 426
Rothmund-Thomson syndrome, 63, 63t
Rough endoplasmic reticulum (RER), 18, 20, 20f–21f, 22
Rubella. *See* German measles
Ruddy, Erin Zammett, 351
Running, 320
Runs of homozygosity, 285
Russell, Karen, "Sleep Donation," 162
Russell-Silverman syndrome, 128

S

Saccharomyces cerevisiae, 433
Sahelanthropus tchadensis, 303, 304f
Salk, Jonas, 436
Salla disease, in Plain people, 285t
Sames, Hannah, 28, 41
Sames, Lori, 28
Sander, Frederick, 176
Saqqaq, 317, 324
Sargasso Sea, metagenomics of, 9–10
Satellite(s), 239
Savior siblings, 408, 408f
Scaffold proteins, 170, 171f
Schizophrenia, 153, 155–157, 227
    empiric risks for, 156, 156t
    environmental factors and, 156
    epigenetic changes and, 156–157
    genetic associations, 156
    prenatal starvation and, 199
    risk factors for, 156
    signs and symptoms, 155, 156f
Scleroderma, 339cc
Sclerosteosis, 86
SCN1A gene, 40
Scrapie, 193–194, 195f
Scrotum, 43, 43f
Secretion, 18–23, 21f
Secretor blood group, 327t
Secretor gene, 107
Segmental progeroid syndromes, 62–63, 63t
Selective serotonin reuptake inhibitors (SSRIs), mechanism
    of action of, 154, 155f
Self-renewal, of stem cells, 16, 33, 34f
Semenya, Caster, 129
Seminal fluid, 43
Seminal vesicle(s), 43, 43f
Seminiferous tubules, 43, 43f
Sequence map(s), 426, 426f
Sequential polar body analysis, 418–419, 418f, 418t, 424
Serotonin transporter gene, *SLC6A4,* 161–162
Sethi, Seema, 210
Severe acute respiratory syndrome (SARS), vaccine
    against, 346
Severe combined immunodeficiency (SCID), 284cc,
    335–336, 335t
    X-linked (SCID-XI), 335–336, 335t, 402

Sex
    heterogametic, 111–112
    homogametic, 111–112
Sex cell(s), 43
Sex chromosome(s), 4, 111–112, 111*f*
    aneuploidy. *See* Aneuploidy, sex chromosome
    traits inherited on, 117–121
Sex determination, 59, 111, 111*f*, 117, 117*f*
Sex-influenced traits, 122
Sex-limited traits, 122
Sex ratio, 128
    definition of, 116
    primary, 116
    secondary, 116
    tertiary, 116
Sex selection, preimplantation testing for, 423
Sexual development, 59, 111, 111*f*
    male, mutations affecting, 114, 115*f*
Sexual identity, 115, 116*t*
Sheep, transgenic, 388
Sherpa people, 288, 288*f*
Shoah project, 274
SHORT syndrome, 197
Short tandem repeats (STRs), 175, 269*t*, 270–271,
    271*f*–272*f*, 318
    for animals, 277
Sickle cell disease, 11, 72, 214, 214*f*, 215, 217, 222, 228, 292*t*
    carriers, with disease symptoms, 77, 87
    conditional probability illustrated for, 81–83, 83*f*
    genetic testing for, 393*t*, 395
    and malaria, 291–293, 292*f*–293*f*
Siemens, Hermann, 138
Signal sequence, 193
Signal transduction, 25
Single nucleotide polymorphisms (SNP), 139, 139*f*, 139*t*, 147,
    213, 227, 436
    in genome-wide association studies, 139–140, 140*f*
Sirtuins, 62
Sister chromatid(s), 30*f*, 238, 238*f*
Site-directed mutagenesis, 220
Skin, human, 320
Skin color
    heritability of, 136
    variations in, 134, 134*f*
Sleep, 150–151
    depression and, 154
Sleep paralysis, 150
Sleep-wake cycle, 151
    disrupted, 151, 151*f*
Small interfering RNAs (siRNAs), 385
Small nuclear ribonucleoproteins (snRNPs), 186
Smallpox, 342, 342*f*
Smell, sense of, 321
Smith-Lemli-Opitz syndrome, 292*t*
Smoking, 60, 153–154, 355
Smooth endoplasmic reticulum (SER), 20*f*–21*f*, 22
Sodium channel(s)
    abnormalities, 26
    in cystic fibrosis, 26
Solomon Islanders, with blond hair, 141–142, 141*f*
Somatic cell nuclear transfer, 36, 52
Somatic cells, 16
Somatostatin, recombinant, 378*t*
Somatotropin, bovine, 387
Sox9 gene, 238
SPATA16 gene, 410*cc*
Speciation, 265
Species, 303
Speech, 319
Sperm, 43–44, 51
    misshapen, 410*cc*
    postmortem removal, 416
    quality, factors affecting, 409
    structure of, 48, 49*f*
Spermatid(s), 48, 48*f*
Spermatocyte(s)
    primary, 48, 48*f*
    secondary, 48, 48*f*
Spermatogenesis, 47–48, 48*f*
Spermatogonium (pl., spermatogonia), 47–48, 48*f*
Spermatozoa, 48

Sperm count, low, 409
Sperm donors, 413
Sperm selection, 392, 393*t*
S phase, 29–30, 29*f*, 172, 181, 238
Spina bifida, 301
Spinal cord, embryology of, 57
Spinal cord injury, gene expression profiling after, 382, 382*t*, 383*f*
Spinal muscular atrophy 1, 434
    epistasis in, 92
Spindle
    meiotic, 45
    mitotic, 30, 30*f*, 31
Spindle assembly checkpoint, 31, 32*f*, 354*f*
Spindle fiber(s), 46, 46*f*
Spondyloepimetaphyseal dysplasia with joint laxity, type 1,
    217, 217*f*
Spontaneous abortion
    chromosomal abnormalities and, 245–247
    of Down syndrome fetus, 255–256
    lethal allele combinations and, 90
    of XO fetuses, 251
SRY gene, 59, 111, 111*f*, 112, 112*f*, 113–114, 117*f*
Stahl, Franklin, 172
Staniford, Tom, 235
Start codon(s), 188, 188*f*, 189
Starvation, prenatal, and schizophrenia, 199
Statins, 132*cc*, 436
Stem cell(s), 5, 33–36, 33*f*–34*f*, 54*f*. *See also* Cancer stem cells
    adult, 35–36, 36*t*
    applications of, 36
    banking of, 37
    definition of, 16
    in disease research, 36
    in drug development, 36
    in health care, 36–37
    oocyte-producing, 50
    sources of, 35–36
    in tissue/organ creation, 36
    treatments based on, 35, 35*f*
    from umbilical cord, 54
Stem cell therapy, for Wiskott-Aldrich syndrome, 110
Stem cell tourism, 36
Stevens, Gavin, 89, 89*f*
Stevens, Jennifer and Troy, 89
Stickler syndrome, 215*t*
Sticky ends, 376–377, 377*f*
Stomach cancer, 366–367
Stomach flu, 349
Stop codon(s), 188–190, 191*f*
    in nonsense mutations, 222
Strauss, Kevin, 284*cc*
Streptococcus pneumoniae, 345
    type R, 164–165, 164*f*–165*f*
    type S, 164–165, 164*f*
STRs. *See* Short tandem repeats (STRs)
Sturtevant, Alfred, 102
Stuttering, 142
Subfertility, definition of, 409
Substrate reduction therapy, 398, 398*f*, 398*t*
Subtelomere(s), 239, 239*f*
Sugar(s), synthesis of, 21*f*
Sugar-phosphate backbone, 168, 168*f*
Superoxide dismutase, recombinant, 378*t*
Support Organization for Trisomy 18, 13 and Related
    Disorders (SOFT), 245*t*
Suppressor cells, 334*t*
Surfactant, 54–56
    recombinant, 378*t*
Surrogate motherhood, 413–414, 418*t*, 423–424
Survivins, 31, 32*f*
Susceptibility, genetic testing for, 393*t*, 395
Sutton, Josiah, 263
Sweetie Pie (pig), 344
Synapse(s), 149, 149*f*
Synapsis, 45
Syndactyly, 29*f*
Syndrome X, 300
SYNGAP1 gene, 395
Synteny, 312
Synuclein, 40
Systemic lupus erythematosus, 338, 338*t*

T

TALEN, 384*f*, 385
Tandem duplication, 224
Taq1 polymerase, in PCR, 175
TATA binding protein, 184, 184*f*
TATA box, 184, 184*f*, 203, 204*f*
Tau protein, 62, 96*cc*, 194
Tautomers, 218
TAW2R16 gene, 64*t*
Tay-Sachs disease, 23, 75, 90–91, 235, 281
    in Ashkenazi Jewish populations, 286*t*
    carrier frequency, 268
    genetic testing for, 395
T cells (T lymphocytes), 329, 329*f*, 330, 330*f*, 331, 331*f*,
    333–334, 334*t*
    types of, 334
Telomerase, 32, 355
Telomere(s), 31–32, 32*f*, 173, 207*t*, 208, 238–239, 238*f*–239*f*,
    354–355
    cloning and, 52
Telophase, 31, 31*f*, 354, 354*f*
Telophase I, 46, 46*f*
Telophase II, 46, 47*f*
Template strand, of DNA, 181, 182*f*, 185
Teratogens, 60–61
    occupational exposure to, 61
Teratoma, 125, 126*f*
Terminator sequence(s), 184, 185*f*
TERT gene, 355, 373
Test cross, 71, 72*f*
Testes, 43, 43*f*
Testosterone, 114, 115*f*
Tetrachromats, 119*cc*
Tetrahydrocannabinol (THC), 153
Tetraploidy, 245
Thalassemia, 214
Thalidomide, teratogenicity of, 60, 95
1000 Genomes Project, 318, 425, 431*t*
Thrifty gene hypothesis, 144
Thrombin, recombinant, 378*t*
Thymine (T), 3, 4*f*, 166, 168, 168*f*, 182, 182*f*, 185
Thymine dimers, 230, 230*f*
Thymocytes, 333
Tibetan plateau, Sherpa people of, 288, 288*f*
Tissue(s), 6*f*
    definition of, 5
    types of, 5, 6*t*, 16, 17*f*
Tissue plasminogen activator, recombinant, 378*t*, 379
Titin, 197
Tolerance, drug, 153
Torsion dystonia, 277
Totipotent cell(s), 34
Tourette syndrome, 73*cc*
Tracking Rare Incidence Syndromes (TRIS), 245*t*
Trait(s)
    autosomal dominant, 74–75, 74*f*, 74*t*
    autosomal recessive, 74–75, 77*t*
    behavioral, 2, 131
    complex, 131
    dominant, 70, 73*cc*
    inherited, 2, 2*f*
    Mendelian (single-gene), 4–5, 5*f*, 43, 69–70, 72,
        73*cc*, 131, 133
        inheritance, solving problems in, 75
    monogenic, 131
    multifactorial, 4–5, 5*f*, 130–131
        continuous variation, 133
        evaluation, terminology used in, 142, 142*t*
        investigation, 135–142
        polygenic, 131
    polygenic, 131, 133–135
        continuous variation, 133
        investigation, 135–142
        multifactorial, 131
        recurrence risk, prediction of, 135–142
    recessive, 70, 73*cc*
    sex-influenced, 122
    sex-limited, 122
    uniquely human, 319
    X-linked, 103, 117

Trait(s)—*Cont.*
  dominant, 120, 120*f*, 120*t*, 121*f*
    recessive, 117–120, 117*t*
  Y-linked, 117
trans configuration, 102, 102*f*, 431
Transcript(s), mRNA, 183
Transcription, 3, 4*f*, 181, 181*f*, 185*f*
  elongation stage, 184, 185*f*
  initiation, 184, 185*f*
  steps in, 184–185
  termination, 184–185, 185*f*
Transcription factor(s), 112–114, 183–184, 203
  binding domains, 184
  functions of, 184, 184*f*
  mutations in, 184
Transfer RNA (tRNA), 18, 183, 183*f*, 183*t*, 184*f*, 189, 207, 207*t*
  in translation, 189–190, 190*f*–191*f*
Transforming growth factor (TGF) receptor, TGFβR, 95
Transgender, 115
Transgenic organism(s), 374–375, 375*f*, 380–382, 388
Transition, 221
Translation, 3, 4*f*, 181, 181*f*, 185–186, 187*f*, 189–190, 190*f*–191*f*
  elongation stage, 189–190, 190*f*–191*f*
  initiation, 189, 190*f*–191*f*
  termination, 190, 191*f*
Translocation(s), 239–240, 245*t*, 254–256, 258*t*
  and cancer, 361–363, 363*f*
  insertional, 256
  reciprocal, 256, 256*f*
  Robertsonian, 255, 255*f*
Translocation carrier, 255, 255*f*
Transplants, 343–345, 350
  rejection, 345, 349
  types of, 343–345
Transposon(s), 207*t*, 208, 224
Transversion, 221
Transvestism, 115
Trastuzumab (Herceptin), 343, 365, 397
T Regs, 339*cc*
Tricellulin, 236
Trichloroethylene, 146
Trichothiodystrophy, 63*t*, 231–232
Trichromats, 119*cc*
Triggering receptor expressed on myeloid cells, gene for, 97*t*
Trio analysis, 157, 158*f*
Triplet repeat disorders, 224–225, 226*cc*, 235
Triploidy, 245–246
Triplo-X, 252
Trisomy(ies), 240, 245–246, 245*t*, 246*f*
  mosaicism for, 247
Trisomy 13, 244, 247, 247*t*, 249–250
Trisomy 16, 246
Trisomy 18, 244, 247, 247*t*, 248–249, 251, 251*f*
Trisomy 21, 247, 247*t*
Trisomy 21 Down syndrome, 51, 244, 246–248, 248*f*, 384
  and Alzheimer disease, 247
  genome editing in, 248, 249*f*
  maternal age and, 240, 241*f*, 248
  prenatal diagnosis of, and prevalence of syndrome, 250
Trisomy 18 Foundation, 245*t*
Trophoblast, 52, 54*f*, 54*t*
Trps1 gene, 237–238
TRPV4 gene, mutations of, 217*t*
TUBB4A gene, 197
Tuberculosis
  in HIV-infected (AIDS) patients, 289
  and natural selection, 289–290
  treatment of, 289–290, 289*f*
Tubulin, 25, 27, 27*f*
Tumor(s)
  benign, 353
  malignant, 353. *See also* Cancer
Tumor necrosis factor, 331, 334*t*
  drugs targeted at, 343
Tumor suppressor gene(s), 353, 365–368
Turner, Henry, 251
Turner syndrome. *See also* XO syndrome
  isochromosome-associated, 257

Tutankhamun, 7
Twin(s)
  conjoined ("Siamese"), 56–57
  dizygotic (DZ; fraternal), 56, 56*f*, 138
  monozygotic (MZ; identical), 56–57, 56*f*, 138
Twin study(ies), 115, 138–139, 146
  of homosexuality, 115
  of sleep, 150
Typhoid fever, 294
TYRP1 gene, 142

U

Ubiquitin, 193, 193*f*, 197
UCP 2, 3, 4 genes, 64*t*
Ulcerative colitis, 338*t*
Ullman, Emmerich, 344
Ullrich syndrome, 251. *See also* XO syndrome
Ultrasound, fetal, 241*f*, 244
Umbilical cord, 53–54
Uncoupling proteins, 64*t*
Unfolded protein response, 193
Uniparental disomy, 258–259, 259*f*, 433
UNIQUE–the Rare Chromosome Disorder Support Group, 245*t*
United Mitochondrial Disease Foundation, 108
UPD. *See* Uniparental disomy
Uracil (U), 182, 182*f*
Urethra, male, 43, 43*f*
Uterine tube(s), 44, 44*f*
  blocked, 410–411
Uterus, 44, 44*f*
UV radiation, mutagenicity, 230

V

Vaccine(s), 341–342
  edible, 342
  production of, recombinant DNA technology for, 379–380
  reverse, 346
Vagina, 44, 44*f*
Valproic acid, teratogenicity of, 60
Vanishing twin phenomenon, 57
van Leeuwenhoek, Anton, 49
Varenicline (Chantix), 161
Variable number of tandem repeats (VNTRs), 269–270, 269*t*, 271*f*
Varicose vein, scrotal, 409
Vector(s)
  cloning, 376–377, 377*f*
  for gene therapy, 400
Venter, Craig, 425, 428–429
  genome of, 435–436
Vesicle(s)
  functions of, 22, 24*t*
  secretory, 22–23
  structure of, 22, 24*t*
Vetter, David, 336, 336*f*, 402
Victoria (Queen of England), and hemophilia, 80, 118–120, 118*f*
Virus(es), 328*cc*. *See also specific virus*
  and cancer, 361–363
  DNA, 207, 328*cc*
  RNA, 207, 328*cc*
  teratogenicity of, 61
  as vectors for gene therapy, 400
Vision, genes affecting, 95, 95*f*
Vitamin(s), metabolism, inborn errors of, 19
Vitamin A, teratogenicity of, 61
Vitamin C, prenatal effects of, 61
Vitiligo, 282
Vlax Roma, 282, 282*f*
VNTRs. *See* Variable number of tandem repeats (VNTRs)
Volker, Nicholas, 9–10, 83–84, 152, 217, 430
von Hippel-Lindau syndrome, 372

W

Waardenburg syndrome, 311
Walking, 320

War, and eugenics, 297–298
Warfarin, 436
  prescription, pharmacogenetic algorithm for, 397
WAS gene, 110
Watson, James, 164, 167, 167*f*, 167*t*, 169–170, 181, 425, 428, 438
  genome of, 435–436
Watson, Max, 109
Weinberg, Wilhelm, 265
Werewolves, 237–238, 237*f*
Werner syndrome, 63, 63*t*
Whitehead, Emma, 404
Whitehead, Mary Beth, 414
Whole exome sequencing, 227
Wild type, 71
Wilkins, Maurice, 166–167, 167*t*
Wilms' tumor, 365
Wilson disease, 19, 19*f*
Winkler, H., 426
Winslow, Pax, 109
Winslow, Shiloh, 108–109
Winter vomiting disease, 349
Wiskott, Alfred, 110, 110*f*
Wiskott-Aldrich syndrome, 110, 110*f*
Wnt4 gene, 111
Wobble position, 189
Wolffian ducts, 111, 111*f*
World Trade Center victims, DNA profiling of, 274
Wrongful birth, 301

X

45,X, 246
X chromosome, 4, 59, 111, 111*f*, 112, 239, 239*f*
  inactivation, cloning and, 52
Xenograft(s), 344–345, 344*f*
Xeroderma pigmentosum, 232, 232*f*, 236
Xerostomia, gene therapy for, 403
XIAP gene, 10
X inactivation, 122–124, 122*f*–123*f*
  and phenotype, 123–124, 123*f*
  skewed, and autoimmune disorders, 338
XIST gene, 122
XIST long noncoding RNA, genome editing with, in trisomy 21, 248, 249*f*, 384
X-linked inheritance, 110*f*
  dominant, 120, 120*f*, 120*t*, 121*f*
  recessive, 117–120, 117*t*
  solving problems in, 121
XO syndrome, 242, 243*t*, 250–252
XX male syndrome, 112
XXY syndrome, 242, 243*t*, 252
  reproductive problems in, 420
The XXYY Project, 245*t*
XXYY syndrome, 252–253
XY female syndrome, 112
XYY syndrome, 253

Y

Y chromosome, 4, 59, 111, 111*f*, 112, 112*f*, 239
  anatomy of, 112, 112*f*
  DNA sequence of, 112, 112*f*, 114, 114*f*
  DNA sequences, study of, 313, 313*f*
  male-specific region (MSY), 112, 112*f*
  mammals without, 113, 113*f*
  Niall, 299
Yolk sac, 53–54, 54*f*

Z

Zinc finger nuclease technology, 384*f*, 385
Zinc fingers, 184, 385
ZIT, 418*t*
Zona pellucida, 51, 51*f*
Zygote, 53*f*
  definition of, 51
Zygote intrafallopian transfer (ZIFT), 415